Cardiac Remodeling

Fundamental and Clinical Cardiology

Editor-in-Chief
Samuel Z. Goldhaber, M.D.
Harvard Medical School
and Brigham and Women's Hospital
Boston, Massachusetts

1. Drug Treatment of Hyperlipidemia, *edited by Basil M. Rifkind*
2. Cardiotonic Drugs: A Clinical Review, Second Edition, Revised and Expanded, *edited by Carl V. Leier*
3. Complications of Coronary Angioplasty, *edited by Alexander J. R. Black, H. Vernon Anderson, and Stephen G. Ellis*
4. Unstable Angina, *edited by John D. Rutherford*
5. Beta-Blockers and Cardiac Arrhythmias, *edited by Prakash C. Deedwania*
6. Exercise and the Heart in Health and Disease, *edited by Roy J. Shephard and Henry S. Miller, Jr.*
7. Cardiopulmonary Physiology in Critical Care, *edited by Steven M. Scharf*
8. Atherosclerotic Cardiovascular Disease, Hemostasis, and Endothelial Function, *edited by Robert Boyer Francis, Jr.*
9. Coronary Heart Disease Prevention, *edited by Frank G. Yanowitz*
10. Thrombolysis and Adjunctive Therapy for Acute Myocardial Infarction, *edited by Eric R. Bates*
11. Stunned Myocardium: Properties, Mechanisms, and Clinical Manifestations, *edited by Robert A. Kloner and Karin Przyklenk*
12. Prevention of Venous Thromboembolism, *edited by Samuel Z. Goldhaber*
13. Silent Myocardial Ischemia and Infarction: Third Edition, *Peter F. Cohn*
14. Congestive Cardiac Failure: Pathophysiology and Treatment, *edited by David B. Barnett, Hubert Pouleur and Gary S. Francis*
15. Heart Failure: Basic Science and Clinical Aspects, *edited by Judith K. Gwathmey, G. Maurice Briggs, and Paul D. Allen*
16. Coronary Thrombolysis in Perspective: Principles Underlying Conjunctive and Adjunctive Therapy, *edited by Burton E. Sobel and Desire Collen*
17. Cardiovascular Disease in the Elderly Patient, *edited by Donald D. Tresch and Wilbert S. Aronow*
18. Systemic Cardiac Embolism, *edited by Michael D. Ezekowitz*
19. Low-Molecular-Weight Heparins in Prophylaxis and Therapy of Thromboembolic Diseases, *edited by Henri Bounameaux*
20. Valvular Heart Diseases, *edited by Muayed Al Zaibag and Carlos M. G. Duran*
21. Implantable Cardioverter-Defibrillators: A Comprehensive Textbook, *edited by N. A. Mark Estes, Antonis S. Manolis, and Paul J. Wang*
22. Individualized Therapy of Hypertension, *edited by Norman M. Kaplan and C. Venkata S. Ram*
23. Atlas of Coronary Balloon Angioplasty, *Bernhard Meier and Vivek K. Mehan*

24. Lowering Cholesterol in High-Risk Individuals and Populations, *edited by Basil M. Rifkind*

25. Interventional Cardiology: New Techniques and Strategies for Diagnosis and Treatment, *edited by Christopher J. White and Stephen Ramee*

26. Molecular Genetics and Gene Therapy of Cardiovascular Diseases, *edited by Stephen C. Mockrin*

27. The Pericardium: A Comprehensive Textbook, *David H. Spodick*

28. Coronary Restenosis: From Genetics to Therapeutics, *edited by Giora Z. Feuerstein*

29. The Endothelium in Clinical Practice: Source and Target of Novel Therapies, *edited by Gabor M. Rubanyi and Victor J. Dzau*

30. Molecular Biology of Cardiovascular Disease, *edited by Andrew R. Marks and Mark B. Taubman*

31. Practical Critical Care in Cardiology, *edited by Zab Mohsenifar and P. K. Shah*

32. Intravascular Ultrasound Imaging in Coronary Artery Disease, *edited by Robert J. Siegel*

33. Saphenous Vein Bypass Graft Disease, *edited by Eric R. Bates and David R. Holmes, Jr.*

34. Exercise and the Heart in Health and Disease: Second Edition, Revised and Expanded, *edited by Roy J. Shephard and Henry S. Miller, Jr*

35. Cardiovascular Drug Development: Protocol Design and Methodology, *edited by Jeffrey S. Borer and John C. Somberg*

36. Cardiovascular Disease in the Elderly Patient: Second Edition, Revised and Expanded, *edited by Donald D. Tresch and Wilbert S. Aronow*

37. Clinical Neurocardiology, Louis R. Caplan, *J. Willis Hurst, and Mark Chimowitz*

38. Cardiac Rehabilitation: A Guide to Practice in the 21st Century, *edited by Nanette K. Wenger, L. Kent Smith, Erika Sivarajan Froelicher, and Patricia McCall Comoss*

39. Heparin-Induced Thrombocytopenia, *edited by Theodore E. Warkentin and Andreas Greinacher*

40. Silent Myocardial Ischemia and Infarction: Fourth Edition, *Peter F. Cohn*

41. Foundations of Cardiac Arrhythmias: Basic Concepts and Clinical Approaches, *edited by Peter M. Spooner and Michael R. Rosen*

42. Interpreting Electrocardiograms: Using Basic Principles and Vector Concepts, *J. Willis Hurst*

43. Heparin-Induced Thrombocytopenia: Second Edition, *edited by Theodore E. Warkentin and Andreas Greinacher*

44. Thrombosis and Thromboembolism, *edited by Samuel Z. Goldhaber and Paul M. Ridker*

45. Cardiovascular Plaque Rupture, *edited by David L. Brown*

46. New Therapeutic Agents in Thrombosis and Thrombolysis: Second Edition, Revised and Expanded, *edited by Arthur A. Sasahara and Joseph Loscalzo*

47. Heparin-Induced Thrombocytopenia: Third Edition, *edited by Theodore E. Warkentin and Andreas Greinacher*

48. Cardiovascular Disease in the Elderly, Third Edition, *edited by Wilbert Aronow and Jerome Fleg*

49. Atrial Fibrillation, *edited by Peter Kowey and Gerald Naccarelli*

50. Heart Failure: A Comprehensive Guide to Diagnosis and Treatment, *edited by G. William Dec*

51. Phamacoinvasive Therapy in Acute Myocardial Infarction, *edited by Harold L. Dauerman and Burton E. Sobel*

52. Clinical, Interventional, and Investigational Thrombocardiology, *edited by Richard Becker and Robert A. Harrington*

53. Pediatric Heart Failure, *Robert Shaddy and Gil Wernovsky*

54. Cardiac Remodeling: Mechanisms and Treatment, *edited by Barry Greenberg*

Cardiac Remodeling
Mechanisms and Treatment

edited by

Barry Greenberg
University of California
San Diego, California, U.S.A.

CRC Press
Taylor & Francis Group
Boca Raton London New York

CRC Press is an imprint of the
Taylor & Francis Group, an **informa** business
A TAYLOR & FRANCIS BOOK

First published 2006 Taylor & Francis Group

Published 2019 by CRC Press
Taylor & Francis Group
6000 Broken Sound Parkway NW, Suite 300
Boca Raton, FL 33487-2742

© 2006 by Taylor & Francis Group, LLC
CRC Press is an imprint of Taylor & Francis Group, an Informa business

First issued in paperback 2019

No claim to original U.S. Government works

ISBN-13: 978-0-367-45402-9 (pbk)
ISBN-13: 978-0-8247-2387-3 (hbk)

Visit the Taylor & Francis Web site at
http://www.taylorandfrancis.com

and the CRC Press Web site at
http://www.crcpress.com

Library of Congress Cataloging-in-Publication Data

Catalog record is available from the Library of Congress

Cardiac remodeling is, of course, a prototypic response and the changes that are described in this book reflect transformations that occur throughout nature. It seems fitting, then, to dedicate this text to my family: Jennifer, Lauren, and Miranda, who have changed me and remodeled my heart in ways that I could never have really anticipated or thought possible.

Series Introduction

Taylor & Francis Group has focused on the development of various series of beautifully produced books in different branches of medicine. These series have facilitated the integration of rapidly advancing information for both the clinical specialist and the researcher.

My goal as editor-in-chief of the Fundamental and Clinical Cardiology Series is to assemble the talents of world-renowned authorities to discuss virtually every area of cardiovascular medicine. In the current monograph, Barry Greenberg has edited a much-needed and timely book. Future contributions to this series will include books on molecular biology, interventional cardiology, and clinical management of such problems as coronary artery disease and ventricular arrhythmias.

Samuel Z. Goldhaber

Preface

Over the last several decades there has been a striking increase in the prevalence of heart failure in the United States and other industrialized nations. One of the more astonishing aspects of this pandemic is that it has occurred during a time when the incidence of other major cardiovascular diseases is declining. Although there are several explanations for this paradox (including, most notably, the aging of the population), the one that is perhaps most vexing is that advances in the treatment of coronary disease, myocardial infarction, and valvular heart disease have improved the immediate survival of patients, who are then at increased risk for the future development of heart failure. The major reason for the progression to heart failure in these patients is that the initial injury to the left ventricle initiates a complex series of structural changes throughout the myocardium that ultimately leads to global deterioration in cardiac function. From a genetic and cellular perspective these changes tend to be prototypic in nature, occurring regardless of the nature of the initiating injury. Ultimately, they result in an increase in left ventricle cavity size and muscle mass, deposition of fibrous tissue throughout the myocardium, and changes in chamber configuration. These structural changes are linked to a progressive deterioration in left ventricle systolic and diastolic function. Cardiac remodeling is the "umbrella" term that has been most commonly used to describe this complex and multi-faceted process. It stands to reason, then, that defining the causes and mechanisms responsible for cardiac remodeling is essential for understanding the pathogenesis of heart failure. The aim of this book is to provide a comprehensive overview of this process. The various chapters include information that should be helpful to clinicians and researchers alike as they provide insight into the causes, mechanisms, and current (as well as prospective future) treatments of cardiac remodeling.

Over time different paradigms have been used to describe the syndrome of heart failure. During the early part of the latter century and lasting into the 1960s heart failure was perceived to be a cardio-renal syndrome. According to this paradigm heart failure was caused by abnormalities in the pumping function of the heart that resulted in salt and water retention by the kidney. This paradigm was modified over the next decade when it was recognized that the load (or stress) placed on the walls of the heart was an important determinant of function, and that it also produced important changes in cardiac structure. Indeed, the genesis of the concept of cardiac remodeling can be attributed to observations made during this period. The role of the peripheral circulation in modulating wall stress and cardiac structure and function then gave rise to the cardio-circulatory model of heart failure.

As researchers began to study the factors that controlled vascular tone it soon became apparent that neurohormonal systems played an important role in this process. With further study of the renin–angiotensin–aldosterone and sympathetic nervous systems, evidence accumulated that their effects reached well beyond those on the peripheral circulation alone. Angiotensin II, aldosterone, and norepinephrine were recognized as having widespread systemic and cardiac effects that contributed to the pathogenesis of heart failure. One of the most important of these was to promote the growth of cardiac cells, which as it turns out is a fundamental component of cardiac remodeling. These insights gave rise to the neurohormonal paradigm of heart failure, and they have provided a rationale for the testing of neurohormonal antagonists in patients with this condition. Several of these agents have now been found not only to favorably alter the remodeling process but also to favorably alter the clinical course of patients with heart failure. As the neurohormonal paradigm became ascendant, the direct and powerful relationship between cardiac remodeling and the development of heart failure was recognized.

Although there will probably be further insights into the pathogenesis and treatment of heart failure that are based on the neurohormonal paradigm, it is likely that future mechanistic and therapeutic "breakthroughs" will come from yet another paradigm. This is the evolving genetic paradigm of heart failure, and it will rely heavily on the insights gained from the study of gene structure and function. Much of the basic work in this area will incorporate techniques of molecular biology to approach problems related to pathophysiology. This paradigm has already provided new insights into the mechanisms responsible for cardiac remodeling, and it has resulted in a variety of useful experimental models for testing emerging therapeutic approaches.

Since the overall aim of the book is to provide a comprehensive description of cardiac remodeling, the reader will note that the authors of the various chapters in this book will alternate between the four paradigms of heart failure that are described above. The reason for this is that, while all are helpful, none, by themselves, are entirely sufficient to account for the complex series of changes that take place as the heart remodels.

The initial section of the text is focused on the pathophysiology of cardiac remodeling and the mechanisms involved in the deleterious changes in cardiac structure that ultimately result in heart failure. This section is introduced by a chapter contributed by Dr. Jay Cohn, one of the first clinicians/scientists to recognize the importance of remodeling in the pathogenesis of heart failure. It is followed by chapters describing the molecular, cellular, tissue, and organ changes that occur as the heart remodels. In addition, there are detailed descriptions of the model systems that have been developed to further study the remodeling process and the effects of various interventions aimed at halting or reversing it. This segues nicely into a series of chapters dealing with the causes of remodeling and, in particular, the critical role played by neurohormonal activation in the pathogenesis of this process. The clinical aspects of remodeling, including information about natural history, detection, and treatment options, then follow. One of the most exciting aspects of clinical cardiology to emerge over the last several decades is the recognition that therapeutic interventions can inhibit and sometimes even reverse the remodeling process. Chapters outlining the role of medical and surgical therapies and the emerging role of devices in treating remodeling and heart failure are included, and there is a final chapter describing the future prospects for the treatment of cardiac remodeling.

The importance of understanding and treating myocardial remodeling has never been greater, considering the societal impact of heart failure throughout the industrialized world. Given the intimate connection between progressive cardiac remodeling and deterioration in cardiac function, it is essential that we more effectively recognize and treat this process. This goal takes on added urgency since it is anticipated that there will be a doubling in the heart failure population over the next three to four decades.

Barry Greenberg, MD

Acknowledgments

For some years now, I have been fascinated by the structural changes that can and do occur in the heart and their great importance in determining the clinical course of a wide variety of cardiac ailments. My interest in this area was first stimulated by the work of Drs. William Grossman, William Gaasch, and, particularly, the seminal studies that were carried out by Drs. Janice and Marc Pfeffer. Drs. Mel Cheitlin, Kanu Chatterjee, William Parmley, and Shahbuddin (Sabu) Rahimtoola helped point me in the right direction early in my eareer. Along with Drs. J. David Bristow and Barry Massie, I had the opportunity to study the effects of the vasodilator hydralazine in volume overloaded patients with aortic insufficiency and observed firsthand that the remodeling process could be reversed—a result that alerted me to the great plasticity of cardiac tissue. I continue to learn more about the heart and how it remodels from my colleagues at the University of California, San Diego, including (in no particular order), Drs. Kenneth Chien, Kirk Peterson, Wolfgang Dillmann, Kirk Knowlton, Andrew McCullough, Francisco Villarreal, and Jeffrey Omens. I also gratefully acknowledge the important role that my patients have had over the years in teaching me about how the heart remodels and what this really means.

Contents

Series Introduction v
Preface vii
Acknowledgments xi
Contributors xxi

Part I. The Process of Remodeling

1. **Overview: Cardiac Remodeling and Its Relationship to the Development of Heart Failure** *1*
   Jay N. Cohn and Inder S. Anand
   Structural Changes in Remodeling 2
   Etiologic Considerations 3
   Cardiac Remodeling and Cardiac Function 4
   Adaptive vs. Maladaptive Remodeling 5
   Remodeling and Prognosis 5
   Therapy to Reverse or Slow Remodeling 5
   References 6

2. **Ventricular Remodeling in Pressure vs. Volume Overload** *9*
   Blase A. Carabello
   Introduction 9
   Hypertrophy vs. Remodeling 9
   Hypertrophy and Demand 10
   Mechanism of Pressure vs. Volume Overload Hypertrophy
     Development 17
   Conclusions 21
   References 22

3. **Changes in the Interstitial Matrix During Myocardial Remodeling** *27*
   Anne M. Deschamps and Francis G. Spinale
   Introduction 27

Basement Membrane 27
Collagen Network 28
Cardiac Disease States and Myocardial Remodeling 28
MI and Remodeling 30
Matrix MMPs 30
Summary: A Need for Further Clinical and
   Basic Research 38
Abbreviations 39
References 39

4. Cardiac Myocyte Structural Remodeling *45*
A. Martin Gerdes and Maurice S. Holder
Introduction 45
Technical Considerations in Assessment of Myocyte Shape
   Alterations 46
Alterations in Myocyte Shape in Hypertrophy
   and Failure 48
Molecular Mechanisms of Myocyte Lengthening and Transverse
   Growth 52
Myocyte Slippage 52
Summary 54
References 54

5. Myofilament Remodeling in the Failing Human Heart *57*
Peter VanBuren and Martin M. LeWinter
Introduction 57
The Thick Filament 57
The Thin Filament 61
Summary 67
References 67

6. Alterations in Cardiac Myocyte Cytoskeleton *73*
Henk L. Granzier
Introduction 73
Microtubules 73
Intermediate Filaments 75
Actin Filaments 76
Titin 77
Summary 84
References 84

**7. Importance of Myocyte Loss and Regeneration During the Cardiac
   Remodeling Process** *91*
Shaila Garg and Jagat Narula
Heart Muscle Cell Loss in Myocardial Remodeling 91
Myocyte Regeneration and Remodeling 97

Conclusions 99
References 100

8. **Remodeling from Compensated Hypertrophy
   to Heart Failure** *103*
   Stefan Hein and Jutta Schaper
   General Morphological Considerations 104
   Studies in Patients 108
   General Clinical Considerations 115
   References 118

9. **Myocardial Remodeling: Physiological and Pathological** *121*
   Jeffrey S. Borer and Karl H. Schuleri
   Introduction 121
   Physiological Remodeling 124
   Pathological Remodeling 128
   Conclusion 134
   References 134

10. **Experimental Animal Models of Cardiac Remodeling** *139*
    Elaine J. Tanhehco and Hani N. Sabbah
    Introduction 139
    Chronic Rapid Pacing 141
    Myocardial Infarction 143
    Coronary Microembolization 145
    Other Models of Remodeling 147
    Conclusion 150
    References 151

11. **The Use of Genetically Modified Mice in the Investigation of Cardiac
    Remodeling** *159*
    *Daniel J. Lips, Rutger J. Hassink, Aart Brutel de la Riviere, and
    Pieter A. Doevendans*
    Introduction 159
    Cardiac Remodeling and Myocardial Ischemia 160
    Myocardial Ischemia and Apoptosis 166
    Myocardial Hypertrophy 170
    Hemodynamics 175
    Conclusion 177
    References 179

Part II. Causes of Remodeling

12. **The Renin–Angiotensin–Aldosterone System in Cardiac Remodeling**   *189*
    Barry Greenberg
    Introduction 189

The Classical Systemic and the Local Tissue-Based
Renin–Angiotensin–Aldosterone Systems 190
Cardiac Remodeling and the Effects of Angiotensin II in the
Heart 191
Effects of Angiotensin II on Cardiac Fibroblasts 192
Effects of Angiotensin II on Cardiac Myocytes 193
Cross-Talk Between the RAAS and Other Systems 195
Additional Components of the Renin–Angiotensin–Aldosterone
System 196
Effects of Inhibiting the RAAS on Cardiac Remodeling 202
Conclusions 203
References 204

13. **Adrenergic Receptor Signaling and Cardiac Remodeling** *215*
Gerald W. Dorn II and Lynne E. Wagoner
Cardiac Remodeling as an Adrenergic Receptor-Mediated
Response 215
Apoptosis as a Mechanism for Cardiac Hypertrophy
Decompensation 218
PKC Signaling in Cardiac Hypertrophy 221
β-Adrenergic Signaling in Heart Failure 223
Modulation of Receptor Signaling in the Remodeling Heart 230
Conclusions 232
References 233

14. **The Role of Inflammatory Mediators in Cardiac Remodeling** *241*
Gabor Szalai, Yasushi Sakata, Jana Burchfield,
Shintaro Nemoto, and Douglas L. Mann
Introduction 241
Overview of the Biology of Proinflammatory Cytokines 241
Effect of Inflammatory Mediators on Cardiac Remodeling 242
Interactions Between the Renin-Angiotensin System and
Proinflammatory Cytokines in Adverse Cardiac
Remodeling 251
Conclusion 254
References 255

15. **Oxidative Stress in Heart Failure: Impact on Cardiac Function and
Ventricular Remodeling** *259*
Luciano C. Amado, Anastasios P. Saliaris, and Joshua M. Hare
Introduction 259
Biochemical Mechanisms of Oxidative Stress 259
Evidence for Increased Oxidative Stress in HF 261
Sources of Oxidative Stress in HF 261
Consequences of Oxidative Stress on Cardiac
Remodeling 268

Natural Antioxidant Defense Mechanisms 271
Antioxidant Enzyme Systems 271
Antioxidant Therapies for HF 272
Antioxidant Therapies in Development 273
Conclusions 275
References 275

Part III. Clinical Aspects and Treatment of Cardiac Remodeling

16. **Natural History of Cardiac Remodeling** *285*
    Gary S. Francis and W. H. Wilson Tang
    Introduction 285
    The Index Event 285
    Myocardial Fibrosis and Diastolic Dysfunction 288
    Other Examples of LV Remodeling—Valvular Heart
        Disease and Other Causes 289
    The Role of LV Geometry 289
    Post-Infarction Remodeling 290
    Adaptation 293
    Prognosis 294
    Summary 295
    References 296

17. **Assessment of Ventricular Remodeling in Heart Failure** *303*
    Inder S. Anand
    Introduction 303
    Echocardiography 306
    Radionuclide Ventriculography 308
    Magnetic Resonance Imaging 312
    Conclusions 318
    References 319

18. **Treatment of Remodeling: Inhibition of the
    Renin–Angiotensin–Aldosterone System** *325*
    Richard D. Patten and Marvin A. Konstam
    Molecular Pathways of the RAS 325
    Experimental Models and Inhibition of the RAAS 327
    Clinical Studies 333
    Summary 342
    References 343

19. **Effects of Adrenergic Blockade on Cardiac Remodeling** *349*
    Henry Ooi and Wilson S. Colucci
    Introduction 349
    Pharmacology of β-Blockers 350
    β-Adrenergic Receptor Blockade in Heart Failure 352

Studies of the Effects of β-Blockade on Remodeling in Heart
    Failure 355
Time Course of Improvement 364
Who Benefits the Most from β-Blockade? 364
Comparative Effects of β-Adrenergic Receptor Blockade and Ace
    Inhibition on Remodeling 366
β-Adrenergic Receptor Blockade Following MI 367
Which β-Adrenergic Receptor Blocker to Use? 370
α₁-Adrenergic Receptor Blockade 374
Central Sympathetic Nervous System Inhibition 374
Conclusion 375
References 376

**20. The Effects of Cardiac Revascularization on the
    Remodeling Process** *389*
Victor A. Ferrari, Craig H. Scott, and Martin St. John Sutton
Introduction 389
Ventricular Remodeling 390
Postinfarction Remodeling 390
Clinical Studies with Noninvasive Imaging 394
Revascularization and Infarct Artery Patency 395
Diagnosis/Prognosis 399
Therapy 402
Conclusion 407
References 407

**21. Reverse Remodeling After Heart Valve
    Replacement and Repair** *417*
Lynne Hung and Shahbudin H. Rahimtoola
Introduction 417
Reverse Ventricular Remodeling 417
Aortic Stenosis 422
Aortic Regurgitation 425
Mitral Stenosis 429
Mitral Regurgitation 431
Conclusions 435
Appendix 435
References 436

22. Cardiac Remodeling After LVAD Placement *441*
Kenneth B. Margulies
Introduction 441
LVAD-Associated Regression of Pathological Structure 443
LVAD-Associated Improvement in Abnormal
    Function 445
LVAD-Associated Changes in Molecular Phenotype 450

From Reverse Remodeling to Recovery with LVADs 452
Conclusion 455
References 455

23. **Cardiac Resynchronization Therapy in Heart Failure: A Powerful
    Tool for Reverse Ventricular Remodeling?** *461*
    Chu-Pak Lau
    Introduction 461
    Electromechanical Changes 461
    Acute Effects of CRT 466
    LV Reverse Remodeling 467
    Clinical Trials 471
    Implantation 473
    Interface of Device and HF Management 474
    Conclusion 477
    References 477

24. **Other Surgical or Device-Based Approaches to Treating Cardiac
    Remodeling** *483*
    V. Dor, M. Di Donato, M. Sabatier, and F. Civaia
    The Left Ventricle After Myocardial Infarction 483
    Surgery of the Ischemic Wall 487
    Results 492
    Summary 499
    Other Surgical Techniques in Process 499
    Mechanical Assistance of the Failing Heart 501
    Conclusion 501
    References 502

25. **Gene Transfer and Left Ventricular Remodeling** *505*
    H. Kirk Hammond and David M. Roth
    Introduction 505
    Overview of Cardiac Gene Transfer 505
    Gene Transfer and Left Ventricular Remodeling 510
    Summary and Future Directions 525
    References 525

26. **Stem Cell/Myoblast Transplantation for Myocardial Regeneration**   *531*
    Donald Orlic and Marc S. Penn
    Introduction 531
    Embryonic Origin of Stem Cells 532
    Biology of Adult Stem Cells 532
    Multiple Stem Cell Populations Exist in Bone Marrow 533
    Cytokines Mobilize Stem Cells into Peripheral Blood 534
    Bone Marrow Stem Cell Regeneration of Tissues Following Ischemic
      Injury 535

Myocardial Infarction 536
Myocardial Regeneration 537
Instrumentation to Image Transplanted Cells in Myocardium 537
Bone Marrow Stem Cells for the Regeneration of Infarcted
    Myocardium 538
Myocardial Infarction: Regeneration by Adult Differentiated
    Cells 542
Autologous Transplantation of Differentiated Cells into Ischemic
    Myocardium 543
Autologous Cell Transplantation as a Platform for Gene
    Transfer 543
Clinical Trials in Cell Therapy 544
Conclusion 546
Glossary 546
References 547

27. Future Directions in Cardiac Remodeling *553*
Barry Greenberg
Neurohormonal Blockade 553
Preventing Loss and Replacing Cardiac Myocytes 555
Targeting the Extracellular Matrix 555
Device Therapy 556
Cardiac Surgery 557
Concluding Thoughts 557

Index *559*

Contributors

Luciano C. Amado   Department of Medicine, Johns Hopkins Medical Institutions, Baltimore, Maryland, U.S.A.

Inder S. Anand   Division of Cardiology, University of Minnesota Medical School, and VA Medical Center, Minneapolis, Minnesota, U.S.A.

Jeffrey S. Borer   Division of Cardiovascular Pathophysiology and The Howard Gilman Institute for Valvular Heart Diseases, Weill Medical College of Cornell University, New York, New York, U.S.A.

Aart Brutel de la Riviere   Department of Cardio-Thoracic Surgery, Heart Lung Center Utrecht, Utrecht, The Netherlands

Jana Burchfield   Winters Center for Heart Failure Research, The Cardiology Section, Department of Medicine, Houston Veterans Affairs Medical Center and Baylor College of Medicine, Houston, Texas, U.S.A.

Blase A. Carabello   Department of Medicine, Baylor College of Medicine, Michael E. DeBakey Houston VA Medical Center, Houston, Texas, U.S.A.

F. Civaia   Cardiac Imaging, Centre Cardio Thoracique, Monaco

Jay N. Cohn   Cardiovascular Division, Department of Medicine, University of Minnesota Medical School, Minneapolis, Minnesota, U.S.A.

Wilson S. Colucci   Boston University School of Medicine, Boston Medical Center, Boston, Massachusetts, U.S.A.

Anne M. Deschamps   Medical University of South Carolina, Charleston, South Carolina, U.S.A.

M. Di Donato   Hemodynamic Department, Centre Cardio Thoracique, Monaco

Pieter A. Doevendans   Interuniversity Cardiology Institute The Netherlands, Utrecht, The Netherlands

V. Dor   Thoracic and Cardio-Vascular Surgery, Centre Cardio Thoracique, Monaco

Gerald W. Dorn II Department of Internal Medicine, University of Cincinnati, Cincinnati, Ohio, U.S.A.

Victor A. Ferrari Noninvasive Imaging Laboratory, Cardiac Imaging Program, Cardiovascular Medicine Division, University of Pennsylvania Medical Center, Philadelphia, Pennsylvania, U.S.A.

Gary S. Francis Department of Cardiovascular Medicine, Cleveland Clinic Foundation, Cleveland, Ohio, U.S.A.

Shaila Garg Departments of Internal Medicine and Cardiology, Drexel University College of Medicine, Philadelphia, Pennsylvania, U.S.A.

A. Martin Gerdes Cardiovascular Research Institute, University of South Dakota and Sioux Valley Hospital and Health Systems, Sioux Falls, South Dakota, U.S.A.

Henk L. Granzier Washington State University, Pullman, Washington, U.S.A.

Barry Greenberg Advanced Heart Failure Treatment Program, University of California, San Diego, California, U.S.A.

Joshua M. Hare Department of Medicine, Johns Hopkins Medical Institutions, Baltimore, Maryland, U.S.A.

Rutger J. Hassink Department of Cardio-Thoracic Surgery, Heart Lung Center Utrecht, Utrecht, The Netherlands

Stefan Hein Department of Cardiac Surgery, Kerckhoff-Clinic, Bad Nauheim, Hessen, Germany

Maurice S. Holder Cardiovascular Research Institute, University of South Dakota and Sioux Valley Hospital and Health Systems, Sioux Falls, South Dakota, U.S.A.

Lynne Hung Griffith Center, Division of Cardiovascular Medicine, Department of Medicine, LAC-USC Medical Center, Keck School of Medicine, University of Southern California, Los Angeles, California, U.S.A.

H. Kirk Hammond Department of Medicine, University of California, San Diego, California, U.S.A.

Marvin A. Konstam New England Medical Center, Boston, Massachusetts, U.S.A.

Chu-Pak Lau Cardiology Division, University of Hong Kong, Queen Mary Hospital, Hong Kong

Martin M. LeWinter Department of Medicine and Molecular Physiology and Biophysics, University of Vermont, Burlington, Vermont, U.S.A.

Daniel J. Lips Department of Cardiology, Heart Lung Center Utrecht, Utrecht, The Netherlands

Douglas L. Mann   Winters Center for Heart Failure Research, The Cardiology Section, Department of Medicine, Houston Veterans Affairs Medical Center and Baylor College of Medicine, Houston, Texas, U.S.A.

Kenneth B. Margulies   Cardiovascular Institute, University of Pennsylvania, Philadelphia, Pennsylvania, U.S.A.

Jagat Narula   Departments of Internal Medicine and Cardiology, Drexel University College of Medicine, Philadelphia, Pennsylvania, U.S.A.

Shintaro Nemoto   Winters Center for Heart Failure Research, The Cardiology Section, Department of Medicine, Houston Veterans Affairs Medical Center and Baylor College of Medicine, Houston, Texas, U.S.A.

Henry Ooi   Boston University School of Medicine, Boston Medical Center, Boston, Massachusetts, U.S.A.

Donald Orlic   Cardiovascular Branch, National Heart, Lung, and Blood Institute, National Institutes of Health, Bethesda, Maryland, U.S.A.

Richard D. Patten   New England Medical Center, Boston, Massachusetts, U.S.A.

Marc S. Penn   Departments of Cardiovascular Medicine and Cell Biology, Cleveland Clinic Foundation, Cleveland, Ohio, U.S.A.

Shahbudin H. Rahimtoola   Griffith Center, Division of Cardiovascular Medicine, Department of Medicine, LAC-USC Medical Center, Keck School of Medicine, University of Southern California, Los Angeles, California, U.S.A.

David M. Roth   Department of Anesthesia, University of California, San Diego, California, U.S.A.

M. Sabatier   Hemodynamic Department, Centre Cardio Thoracique, Monaco

Hani N. Sabbah   Henry Ford Health System, Detroit, Michigan, U.S.A.

Yasushi Sakata   Winters Center for Heart Failure Research, The Cardiology Section, Department of Medicine, Houston Veterans Affairs Medical Center and Baylor College of Medicine, Houston, Texas, U.S.A.

Anastasios P. Saliaris   Department of Medicine, Johns Hopkins Medical Institutions, Baltimore, Maryland, U.S.A.

Jutta Schaper   Department of Experimental Cardiology, Max-Planck-Institute, Bad Nauheim, Hessen, Germany

Karl H. Schuleri   Division of Cardiovascular Pathophysiology and The Howard Gilman Institute for Valvular Heart Diseases, Weill Medical College of Cornell University, New York, New York, U.S.A.

Craig H. Scott   Noninvasive Imaging Laboratory, Cardiac Imaging Program, Cardiovascular Medicine Division, University of Pennsylvania Medical Center, Philadelphia, Pennsylvania, U.S.A.

Francis G. Spinale   Medical University of South Carolina, Charleston, South Carolina, U.S.A.

Martin St. John Sutton   Noninvasive Imaging Laboratory, Cardiac Imaging Program, Cardiovascular Medicine Division, University of Pennsylvania Medical Center, Philadelphia, Pennsylvania, U.S.A.

Gabor Szalai   Winters Center for Heart Failure Research, The Cardiology Section, Department of Medicine, Houston Veterans Affairs Medical Center and Baylor College of Medicine, Houston, Texas, U.S.A.

W. H. Wilson Tang   Department of Cardiovascular Medicine, Cleveland Clinic Foundation, Cleveland, Ohio, U.S.A.

Elaine J. Tanhehco   Henry Ford Health System, Detroit, Michigan, U.S.A.

Peter VanBuren   Department of Medicine and Molecular Physiology and Biophysics, University of Vermont, Burlington, Vermont, U.S.A.

Lynne E. Wagoner   Department of Internal Medicine, University of Cincinnati, Cincinnati, Ohio, U.S.A.

1

Overview: Cardiac Remodeling and Its Relationship to the Development of Heart Failure

Jay N. Cohn
Cardiovascular Division, Department of Medicine, University of Minnesota Medical School, Minneapolis, Minnesota, U.S.A.

Inder S. Anand
Division of Cardiology, University of Minnesota Medical School, and VA Medical Center, Minneapolis, Minnesota, U.S.A.

Heart failure has traditionally been considered a clinical syndrome in which the symptoms define the disorder. Indeed, the clinical syndrome is often designated as "congestive heart failure" because symptoms are usually accompanied by fluid retention leading to peripheral edema and pulmonary congestion. This clinical syndrome drove its management during most of the 20th century as exercise intolerance, fatigue, and congestion were the hallmarks of the disease that required treatment. The end of the 20th century and the beginning of the twenty-first century have seen a dramatic change in our understanding and management of the disease. Although symptoms may still be the major reason patients seek medical help, we now recognize that the disease process long precedes the development of symptoms, additionally, the fact that its progression may be independent of symptom worsening or symptom relief has complicated the clinical understanding of the disease. Structural changes in the left ventricle are key to this new insight into the syndrome (1).

Although clinicians in the past focused on the function of the left ventricle and its favorable response to vasodilator and inotropic drug therapy, it is now recognized that structural changes in the left ventricle may represent the disease process and symptoms may or may not coexist (2). This revision in concept has transformed heart failure from a clinical syndrome into a syndrome of ventricular dysfunction. Recent guidelines have recognized this transition by identifying an asymptomatic phase of left ventricular dysfunction as a stage of heart failure (3). The emphasis to the practicing community must be that the ventricular dysfunction needs to be treated to prevent its progression and that symptom relief represents only one item on the agenda facing a physician managing the syndrome.

STRUCTURAL CHANGES IN REMODELING

The structural change in the left ventricle that characterizes heart failure involves myocytes and interstitium, and perhaps the vasculature. The myocyte hypertrophy of pressure overload is characterized by an increase in the transverse diameter of individual myocytes, an increase in the cross-section area, a consequent increase in wall thickness, an increase in mass with a normal chamber dimension, and a normal ejection fraction (Fig. 1) (4,5). Heart failure may ultimately develop in such a ventricle as a result of delayed relaxation resulting in the syndrome of heart failure with preserved systolic function (6). This concentric hypertrophy is a form of ventricular remodeling that, early in its natural history, is not associated with any of the progressive features identified in the dilated ventricle with a reduced ejection fraction. It is in the latter condition, in which individual myocytes lengthen, fibers elongate, wall stress rises, myocyte shortening, and circumferential fiber shortening are reduced, symptoms of heart failure may develop, and life expectancy is shortened (Fig. 1). The role of the interstitium and collagen degradation and synthesis in the progression of this remodeling process is still not fully understood, but it is intuitive that a remodeling of the collagen network must accompany remodeling of bundles of myocytes to accommodate the enlarging and structurally altered fibers. This collagen network remodeling must involve degradation of existing collagen (matrix metallo-proteinase activity) and synthesis of new collagen (7).

As myocytes and fibers lengthen, the left ventricular chamber enlarges and a physiologically appropriate stroke volume can be delivered with considerably less shortening (1). Thus, a low ejection fraction is an obligatory accompaniment of a dilated ventricle in the absence of gross valvular regurgitation. In the past, a reduction of ejection fraction was thought to signify systolic dysfunction; in the current

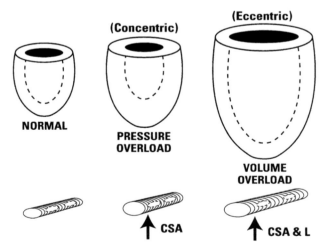

Figure 1 Schematic representation of myocyte change in left ventricular concentric and eccentric hypertension. In pressure overload hypertrophy, the myocyte cross-sectional area increases and the ventricular wall becomes thicker during the compensatory phase. In volume overload hypertrophy, ventricular volume and wall thickness increase proportionally; this is associated with a corresponding proportional increase in both myocyte length and cross-sectional area. *Source*: From Ref. 5.

era it may be more appropriate to identify it as a manifestation of left ventricular remodeling or dilation.

The vasculature also may play a key role in the remodeling process. Enlargement of myocytes and fibers without vascular growth may result in inadequate perfusion that might contribute to impairment of myocardial function (8). Such inadequacy of perfusion may predominate in the subendocardial layer of the myocardium that is exposed to intracavitary pressure. Increased left ventricular diastolic pressure may impede subendocardial perfusion and thus contribute to ischemia. Furthermore, adequacy of microvascular perfusion may be dependent on intact endothelial function that provides nitric oxide secretion to maintain a balance between arterial caliber and flow requirements. Remodeled ventricles may be accompanied by endothelial dysfunction that can further compromise myocardial perfusion (9).

The relationship between left ventricular remodeling and symptoms of heart failure remain elusive. Individuals with a low ejection fraction and a dilated left ventricle may be asymptomatic with respectable exercise tolerance or may, in contrast, present with NYHA Class IV heart failure (10). Some of this variation in clinical presentation may result from differences in left ventricular shortening, particularly during exercise. Why some individuals with a dilated left ventricle maintain adequate contraction and cardiac output and others have reduced wall motion and low output remains a mystery. Genetic, metabolic, and mechanical factors may be involved. Attempts to identify the metabolic and genetic determinants of the transition between left ventricular hypertrophy and failure have been the focus of efforts in a number of laboratories (11). An alternative mechanism, however, attributes the clinical syndrome not to a change in myocardial mechanics but rather to alterations in peripheral blood flow and organ function. The kidney is paramount in this process as a result of renal sodium retention resulting in circulatory congestion and the onset of symptoms that usually respond best to diuretic therapy (12). Such sodium retention and symptom development then alert the physician to the presence of a structurally remodeled left ventricle for which therapy needs to be introduced.

ETIOLOGIC CONSIDERATIONS

The classic process of ventricular remodeling is elicited by a myocardial infarction that results in a necrotic segment of myocardium and a surrounding border zone of impaired perfusion (13). The remote areas of well-perfused myocardium may undergo a structural process of growth and remodeling that may in part be mediated by physical forces related to the nonfunctional segment of the cardiac chamber (14). In addition, however, hormonal activation and gene reprogramming may be involved in these structural changes (15).

Coronary artery disease does not appear to be the sole mechanism for inducing remodeling. Pressure and volume overload, regardless of physiologic mechanism, are adequate to initiate a similar progressive process. The left ventricular remodeling that accompanies nonischemic cardiomyopathy may be induced or perpetuated by changes in myocardial shortening or by dilation and chamber distortion. Regardless of the initiating stimulus, however, the remodeling process may progress in patients with dilated cardiomyopathy so that the end-stage ventricle may exhibit the same structural and functional abnormalities as the end-stage regionally ischemic ventricle (16).

Hormonal stimulation has been identified as a key contributor to the progressive left ventricular structural remodeling process that accounts for symptoms and mortality in heart failure. The remodeling process is predominantly a growth-mediated response and it is now known that hormonal activation can directly stimulate such growth of both myocytes and the collagen matrix (15,17). As these hormonal systems are so mutually interactive and are often stimulated in concert (18), it has been difficult to identify a single hormonal system that may be primarily responsible for the remodeling process. The sympathetic nervous system (19), the renin–angiotensin–aldosterone system (20), and endothelin (21) are just a few of the systems that have been demonstrated to be activated in heart failure, and can be shown in experimental models to induce the structural changes that are observed in the clinical syndrome. The putative role of these neurohormonal systems has led to the testing of a variety of pharmacologic inhibitors of these systems for a favorable effect on the clinical syndrome. Indeed, angiotensin-converting enzyme inhibitors (22), angiotensin-receptor blockers (23), β-adrenoreceptor inhibitors (24–26), and aldosterone inhibitors (27,28) have all been shown to exert a favorable effect on the course of the disease and probably on the course of structural remodeling of the left ventricle.

The distinction between load-generated stimulation and hormonal-mediated stimulation of growth, remains controversial. The hormonal systems that have been shown to contribute to structural remodeling are predominantly vasoconstrictor hormones that may also raise blood pressure and increase workload. Those neurohormonal inhibitors that favorably affect the course of left ventricular remodeling are predominantly vasodilator drugs that would be expected to reduce the load on the left ventricle. Whether the load reduction can account for the pharmacologic efficacy is uncertain, but since the blood pressure fall is usually very small it is likely that the load effects are less important than the direct effects of hormone inhibition.

CARDIAC REMODELING AND CARDIAC FUNCTION

The clinical symptoms of heart failure are directly related to impaired function of the left ventricle. The traditional view is that the fatigue of exercise intolerance is related to inadequate cardiac output response to exercise, and the dyspnea is related to an abnormal rise in left atrial and pulmonary capillary pressure. Although this simple mechanistic view cannot be defended, it is nonetheless clear that the impaired function of the ventricle, at rest and during stress is the fundamental cause of the symptoms, whether mediated directly by left ventricular hemodynamics or indirectly by mechanical changes on ventilation, renal sodium retention, or neurohormonal activation.

The mechanisms of symptom development in previously asymptomatic patients with remodeled ventricles and the exacerbation of symptoms in previously stable symptomatic patients, are poorly understood. Ventricular dilation and reduced ejection fraction are not remarkably dissimilar in asymptomatic and symptomatic patients. Some attribute symptom development to sodium retention, related either to sodium ingestion or subtle worsening of sodium excretion, whereas others attribute it to subtle changes in left ventricular diastolic function causing a rise in left ventricular filling pressure. Others suggest that subtle worsening of myocyte shortening may worsen systolic performance, even though a further reduction of ejection fraction at rest cannot be documented.

Regardless of mechanism, pharmacotherapy aimed at these pathophysiologic abnormalities is effective in reversing symptoms. Diuretics used to reduce filling pressure are the hallmark of this therapy, but vasodilator drugs and inotropic drugs to improve left ventricular emptying and further reduce filling pressure are often dramatically effective.

It is clear that relief of symptoms from the above pharmacotherapy is not necessarily effective in reversing or slowing the structural remodeling process that, likely, is progressive. There is a clear dissociation between symptom relief by improving left ventricular function and slowing long-term progression of the disease, which involves inhibition of left ventricular remodeling.

ADAPTIVE VS. MALADAPTIVE REMODELING

Growth and remodeling of the myocardium are the appropriate structural responses to pressure and volume overload that burdens the left ventricle. A thickened myocardial wall is necessary to normalize wall stress induced by pressure overload and thus, allow fiber shortening to be maintained despite a higher workload (29). Similarly, ventricular dilation would be an appropriate compensatory response of a ventricle faced with a volume overload from aortic or mitral valve regurgitation. These conditions require a larger-than-normal stroke volume to maintain an adequate forward stroke volume. Therefore, these remodeling processes in the left ventricle may be viewed as compensatory or adaptive. Indeed, they may be critical to supporting clinical functional stability in these situations, but these structural changes of hypertrophy and dilation may still exert long-term adverse effects on survival.

In conditions such as myocardial infarction, regional myocardial ischemia, and nonischemic forms of myocarditis and cardiomyopathy, the observed structural changes in the myocardium may be largely maladaptive from the beginning. The structural changes may not be necessary to maintain an adequate stroke volume. Indeed, early intervention to slow or prevent the structural remodeling does not appear to have an adverse effect on clinical symptoms or disease progression (30–32).

REMODELING AND PROGNOSIS

The relationship between left ventricular remodeling [reduced ejection fraction (EF), increased transverse diameter of the ventricle in diastole (LVIDd)] and prognosis has been established in a number of controlled clinical trials as well as in observational studies. The greater the remodeling is the shorter the life expectancy and more likely the need for hospitalization for heart failure. In the valsartan-heart failure trial (Val-HeFT) the lower the EF and the higher the LVIDd at baseline the higher was the mortality and morbidity (33). The two measurements were in fact additive predictors of poor outcome, perhaps because imprecision of echocardiography renders either of the measurements somewhat less reliable guides to remodeling than a combination of the two.

In recent years, natriuretic peptide levels in blood have become available for independent assessment of left ventricular remodeling or function. B-type natriuretic peptide (BNP) has been assessed in recent controlled trials. The level of BNP has served as a strong correlate with left ventricular remodeling and has provided additive prognostic information, presumably as an additional marker for the severity of left ventricular disease (34,35).

THERAPY TO REVERSE OR SLOW REMODELING

The benefits of the therapy for heart failure were in the past assessed by hemodynamic measurements and short-term symptom relief. In recent years, therapeutic efforts have focused on reduction in mortality. The major factor in reducing mortality especially that from cardiac pump failure, is slowing or reversing the remodeling process. This may be viewed as a treatment to slow down the progression of the disease.

Although it is intuitive that heart failure worsens progressively and that left ventricular structural and functional abnormalities progress, the only controlled data come from the original V-HeFT. Patients receiving only digoxin and placebo, studied before the era of acceptance of vasodilator/ACE inhibitor therapy, were monitored sequentially over five years. Left ventricular EF was measured sequentially by a radionuclide technique. In the subgroup randomized to placebo, therefore, we had the opportunity to monitor the progression of heart failure. More than 50% of this cohort died during the five-year period, but even in the survivors there was a progressive decline in EF (36). Thus, remodeling is a progressive, life-threatening process that continues in the face of digoxin and diuretic therapy.

In V-HeFT, the combination of isosorbide dinitrate and hydralazine (now been studied in a combination tablet called BiDil) as well as the ACE inhibitor enalapril produced a significant and sustained improvement in EF and this improvement was powerfully related to improved survival from these drugs (36). Subsequently, a slowing of remodeling has been demonstrated by beta-blockers and angiotensin-receptor blockers, which also improve outcome. Regression or slowing of remodeling appears to be a powerful surrogate for longer survival and reduced morbidity in heart failure (37). In fact, slowing of remodeling may be the biological effect of the interventions, and reduction of morbidity and mortality, a surrogate for the favorable mechanistic effect.

In recent years, devices have also been introduced to slow or halt remodeling. The most direct are interventions to compress or constrain the left ventricle so that it cannot dilate (38). Clinical trials of this strategy to directly attack remodeling are currently underway. Furthermore, cardiac resynchronization therapy with bi-ventricular pacing also may result in a slowing of remodeling (39). Additional data on outcome with such device therapy is awaited.

In conclusion, cellular hypertrophy and structural remodeling of the left ventricle is a fundamental process that characterizes the progression of cardiac dysfunction in patients with heart failure. The link between remodeling and poor prognosis, and between regression of remodeling and a favorable effect of therapy on outcome have raised the possibility that remodeling can serve as a guide to progression of heart failure and that drug or device therapy aimed directly at the structural remodeling may be an appropriate strategy.

REFERENCES

1. Cohn JN. Structural basis for heart failure Ventricular remodeling and its pharmacological inhibition. Circulation 1995; 91(10):2504–2507.
2. Cohn JN. Critical review of heart failure: the role of left ventricular remodeling in the therapeutic response. Clin Cardiol 1995; 18(9 suppl 4):IV4– IV12.
3. Hunt SA, Baker DW, Chin MH, et al. ACC/AHA guidelines for the evaluation and management chro of nic heart failure in the adult. Executive summary, a report of the American College of Cardiology/American Heart Association Task Force on Practice Guidelines (Committee to Revise the 1995 Guidelines for the Evaluation and Management of Heart Failure): developed in collaboration with the International Society for Heart and Lung Transplantation; endorsed by the Heart Failure Society of America. Circulation 2001; 104(24):2996–3007.

4. Grossman W, Jones D, McLaurin LP. Wall stress and patterns of hypertrophy in the human left ventricle. J Clin Invest 1975; 56(1):56–64.

5. Gerdes AM. The use of isolated myocytes to evaluate myocardial remodeling. Trends Cardiovasc Med 1992; 2:152–155.

6. Zile MR. Heart failure with preserved ejection fraction: is this diastolic heart failure? J Am Coll Cardiol 2003; 41(9):1519–1522.

7. Spinale FG, Coker ML, Thomas CV, Walker JD, Mukherjee R, Hebbar L. Time-dependent changes in matrix metalloproteinase activity and expression during the progression of congestive heart failure: relation to ventricular and myocyte function. Circ Res 1998; 82(4):482–495.

8. Anversa P, Loud AV, Levicky V, Guideri G. Left ventricular failure induced by myo-cardial infarction. II. Tissue morphometry. Am J Physiol 1985; 248(6 Pt 2):H883–H889.

9. Karam R, Healy BP, Wicker P. Coronary reserve is depressed in postmyocardial infarc-tion reactive cardiac hypertrophy. Circulation 1990; 81(1):238–246.

10. Franciosa JA, Park M, Levine TB. Lack of corelation between exercise capacity and indexes of left ventricular performances in heart failure. Am J Cardiol 1981; 47:33–39.

11. Katz AM. Cardiomyopathy of overload. A major determinant of prognosis in congestive heart failure. N Engl J Med 1990; 322(2):100–110.

12. Harris P. Evolution and the cardiac patient. Cardiovasc Res 1983; 17(6):313–319, 373–378, 437–445.

13. Hutchins GM, Bulkley BH. Infarct expansion versus extension: two different complica-tions of acute myocardial infarction. Am J Cardiol 1978; 41(7):1127–1132.

14. Weisman HF, Healy B. Myocardial infarct expansion, infarct extension, and reinfarc-tion: pathophysiologic concepts. Prog Cardiovasc Dis 1987; 30(2):73–110.

15. Swynghedauw B. Molecular mechanisms of myocardial remodeling. Physiol Rev 1999; 79(1):215–262.

16. Florea VG, Mareyev VY, Samko AN, Orlova IA, Coats AJ, Belenkov YN. Left ventri-cular remodelling: common process in patients with different primary myocardial disorders. Int J Cardiol 1999; 68(3):281–287.

17. Weber KT, Sun Y, Tyagi SC, Cleutjens JP. Collagen network of the myocardium: func-tion, structural remodeling and regulatory mechanisms. J Mol Cell Cardiol 1994; 26(3):279–292.

18. Anand IS, Ferrari R, Kalra GS, Wahi PL, Poole-Wilson PA, Harris PC. Edema of cardiac origin. Studies of body water and sodium, renal function, hemodynamic indexes, and plasma hormones in untreated congestive cardiac failure. Circulation 1989; 80(2): 299–305.

19. Cohn JN, Levine TB, Olivari MT, et al. Plasma norepinephrine as a guide to prognosis in patients with chronic congestive heart failure. N Engl J Med 1984; 311(13):819–823.

20. Swedberg K, Eneroth P, Kjekshus J, Wilhelmsen L. Hormones regulating cardiovascular function in patients with severe congestive heart failure and their relation to mortality. CONSENSUS Trial Study Group. Circulation 1990; 82(5):1730–1736.

21. Mulder P, Richard V, Derumeaux G, et al. Role of endogenous endothelin in chronic heart failure: effect of long-term treatment with an endothelin antagonist on survival, hemodynamics, and cardiac remodeling. Circulation 1997; 96(6):1976–1982.

22. Garg R, Yusuf S. Overview of randomized trials of angiotensin-converting enzyme inhi-bitors on mortality and morbidity in patients with heart failure. Collaborative Group on ACE Inhibitor Trials. JAMA 1995; 273(18):1450–1456.

23. Cohn JN, Tognoni G. A randomized trial of the angiotensin-receptor blocker valsartan in chronic heart failure. N Engl J Med 2001; 345(23):1667–1675.

24. The CIBIS II Investigators. The Cardiac Insufficiency Bisoprolol Study II (CIBIS-II): a randomised trial. Lancet 1999; 353(9146):9–13.

25. The MERIT-HF Investigators. Effect of metoprolol CR/XL in chronic heart failure: Metoprolol CR/XL Randomised Intervention Trial in Congestive Heart Failure (MERIT-HF). Lancet 1999; 353:2001–2007.

26. Packer M, Coats AJ, Fowler MB, et al. Effect of carvedilol on survival in severe chronic heart failure. N Engl J Med 2001; 344(22):1651–1658.

27. Pitt B, Zannad F, Remme WJ, et al. The effect of spironolactone on morbidity and mortality in patients with severe heart failure. Randomized Aldactone Evaluation Study Investigators. N Engl J Med 1999; 341(10):709–717.

28. Pitt B, Remme WJ, Zannad F, et al. Eplerenone, a selective aldosterone blocker, in patients with left ventricular dysfunction after myocardial infarction. N Engl J Med 2003; 348(14):1309–1321.

29. Gunther S, Grossman W. Determinants of ventricular function in pressure-overload hypertrophy in man. Circulation 1979; 59(4):679–688.

30. Sharpe N, Smith H, Murphy J, Greaves S, Hart H, Gamble G. Early prevention of left ventricular dysfunction after myocardial infarction with angiotensin-converting-enzyme inhibition. Lancet 1991; 337(8746):872–876.

31. The ACE Inhibitor Myocardial Infarction Collaborative Group. Indications for ACE inhibitors in the early treatment of acute myocardial infarction: systematic overview of individual data from 100,000 patients in randomized trials. Circulation 1998; 97(22): 2202–2212.

32. Doughty RN, Whalley GA, Walsh H, Gamble G, Sharpe N. Effects of carvedilol on left ventricular remodeling in patients following acute myocardial infarction: the CAPRI-CORN echo substudy. Circulation 2001; 104 (suppl)(17):II-517. Abstract.

33. Wong M, Staszewsky L, Latini R, et al. Severity of left ventricular remodeling defines outcomes and response to therapy in heart failure: Val-HeFT echocardiography data. J Am Coll Cardiol 2003; Submitted for publication.

34. Anand IS, Latini R, Wong W, et al. Relationship between changes in ejection fraction, BNP and norepinephrine over time and the effect of valsartan in Val-HeFT. Eur Heart J 2003.

35. Anand IS, Fisher LD, Chiang YT, et al. Changes in brain natriuretic peptide and norepinephrine over time and mortality and morbidity in Val-HeF. Circulation 2003; 107: 1276–1281.

36. Cintron G, Johnson G, Francis G, Cobb F, Cohn JN. Prognostic significance of serial changes in left ventricular ejection fraction in patients with congestive heart failure. Circulation 1993; 87(6 suppl):VI17–VI23.

37. Anand IS, Florea VG, Fisher L. Surrogate end points in heart failure. J Am Coll Cardiol 2002; 39(9):1414–1421.

38. Saavedra WF, Tunin RS, Paolocci N, et al. Reverse remodeling and enhanced adrenergic reserve from passive external support in experimental dilated heart failure. J Am Coll Cardiol 2002; 39(12):2069–2076.

39. St. John Sutton MG, Plappert T, Abraham WT, et al. Effect of cardiac resynchronization therapy on left ventricular size and function in chronic heart failure. Circulation 2003; 107(15):1985–1990.

2

Ventricular Remodeling in Pressure vs. Volume Overload

Blase A. Carabello
Department of Medicine, Baylor College of Medicine, Michael E. DeBakey Houston VA Medical Center, Houston, Texas, U.S.A.

INTRODUCTION

The heart's task is simple enough. It is responsible for providing the body's tissues with adequate nutrients via the bloodstream. While the task remains the same throughout life, the cardiac output required to satisfy this mission and the work required to generate it vary with both physiologic and pathophysiologic conditions that develop in the course of existence. While the biologic pathways available for the heart to adapt to these changing conditions are rich in their diversity, the integrated responses of these systems yield only a few specific coping mechanisms. These include the activation of the Frank–Starling mechanism, neurohumoral activation, and hypertrophy (and remodeling). Because sarcomeres normally operate at about 90% of their maximum length (1), the Frank–Starling mechanism is limited in scope and primarily acts to adjust moment-to-moment variations in stroke volume. Neurohumoral control also modulates changes in cardiac output during exercise. Both modalities become activated in myocardial failure where they are initially compensatory. However, it is now clear that persistent activation of most neurohumoral systems is deleterious and results in further myocardial deterioration (2–4). Therefore, it is not surprising that blockade of these systems in heart failure has led to improved life span and life quality (5,6). Regulation of cardiac mass is the third mechanism for adjusting to changes in demand for cardiac output and forms the focus of this chapter.

HYPERTROPHY VS. REMODELING

Once the terms "hypertrophy" and "remodeling" had distinct definitions. Hypertrophy referred to an increase in mass and weight irrespective of ventricular geometry. Remodeling implied a change in geometry without a change in weight. Thus, a ventricle that is enlarged in size but suffered wall thinning would have the same cardiac weight (no hypertrophy) but would have undergone substantial remodeling. However, today the term "remodeling" refers to a change in geometry whether or not there has been a change in weight. In fact, in almost all cases of remodeling, whether eccentric or concentric, there is usually an increase in muscle and overall cardiac weight.

HYPERTROPHY AND DEMAND

Physiologic Hypertrophy

Although evidence for myocyte mitosis exists (7), it is generally held that cardiomyocytes become terminally differentiated shortly after birth. Thus, additional increments in cardiac mass accrue primarily from hypertrophy, a process whereby additional sarcomeres are added to existing myocytes, rather than from hyperplasia. When sarcomeres are added in series, the myocyte increases in length, culminating in increased chamber volume (eccentric hypertrophy). If additional sarcomeres are added in parallel, myocytes and the chamber wall they comprise increase in thickness (concentric hypertrophy). Throughout life, a ratio of left ventricular weight in grams to body weight in kilograms of about 4:1 is maintained in most mammalian species including man. Obviously then, as body weight increases normally from about 3.5 kg at birth to 80 kg in adulthood, there is an equivalent 20- to 25-fold increase in cardiac mass, almost all of which occurs through the hypertrophic process. This huge increase in cardiac mass during normal body growth is obviously normal and not associated with any deleterious consequences. Indeed the process is necessary for life. This point is necessary to emphasize that hypertrophy by itself is not necessarily pathologic. In normal growth, both eccentric and concentric hypertrophy must occur to maintain normal ventricular function. A good approximation of afterload on the ventricle is the wall stress as estimated by the Laplace equation: stress = pressure × radius/2 thickness. With its growth, eccentric hypertrophy increases the stroke volume of the heart enabling it to increase pumping capacity (stroke volume) to match the increase in nutrient demand of the body. However, as eccentric hypertrophy develops, the increase in the radius in the Laplace equation would have the consequence of increasing afterload on the ventricle, in turn impairing shortening of the myocytes, reducing ventricular ejection. Fortunately, as eccentric hypertrophy develops, it is offset by a concomitant increase in wall thickness provided by concurrent concentric hypertrophy. Concentric hypertrophy provides an increase in the denominator of the Laplace equation, in turn normalizing the increase in stress that would occur if only eccentric hypertrophy developed.

After adulthood is reached, additional demands on the heart may engender additional increases in ventricular mass. For instance, there is a substantial difference between the cardiac demands of a sedentary worker and those of a trained athlete. Athletic training may lead to more than a 50% increase in left ventricular mass (8). Isotonic activity such as long distance running typically causes eccentric hypertrophy while isometric exercise such as weight training produces concentric hypertrophy (9). In both cases, ventricular function, both systolic and diastolic, is normal (8,9). Thus, increases in left ventricular mass, even far above that achieved at maximal body growth can be entirely free of any discernable abnormality, confirming that an increase in cardiac mass by itself need not be pathologic.

Compensatory Hypertrophy

As noted above, physiologic hypertrophy develops in response to the demands of growth and exercise. The term "compensatory hypertrophy" refers to an increase in mass that occurs in response to abnormal loads placed upon the heart. Although debated for at least the last 30 years, the increase in mass is considered by many to be beneficial because it enables the heart to maintain normal cardiac output despite the increase in workload (10,11). The pressure demand of aortic stenosis or of hypertension causes concentric hypertrophy to occur while the volume overload of mitral regurgitation is compensated by eccentric hypertrophy. Likewise, following myocardial infarction, the remaining myocardium experiences a volume overload leading to eccentric hypertrophy. As with physiologic hypertrophy, compensatory hypertrophy may be consistent with normal muscle function (12,13). In pressure overload, the increase in thickness value in the Laplace equation allows the ventricle to maintain normal wall stress. In volume overload, the increase in ventricular volume allows the ventricle to pump additional volume to compensate for that lost to regurgitation demanded by increased pathophysiologic needs (i.e., anemia) (14).

Debate about the compensatory nature of hypertrophy stems from two sources. First, the presence of hypertrophy in cardiac diseases greatly increases the mortality rate (15–17). Second, there is an enormous disparity of muscle function among a plethora of hypertrophy models. In some models of even extreme hypertrophy, muscle function is normal (12), whereas in other models of modest hypertrophy, muscle function is abnormal (18,19). In some models of overload, hypertrophy did not have to be present for normal ventricular function, at least for overloads of short duration (20,21); in others, the absence of hypertrophy lead to rapid deterioration (22,23). The differences are accounted for by the differences in the kind of load imposed, by the ventricle that was overloaded, and by the species studied. We do know that in pressure overload in man and large mammals that inadequate hypertrophy increases wall stress and causes ventricular dysfunction and that normal function can co-exist with hypertrophy (24,25).

Pathologic Hypertrophy

If the pressure or volume overload is severe and prolonged enough, eventually muscle dysfunction develops and thus, the hypertrophied myocardium assumes pathologic characteristics, exhibiting systolic or diastolic dysfunction or both. Thus, there is a transition from compensatory to pathologic hypertrophy. The causes of this transition to failure have been the subject of intense investigation. However, no clear or simplistic explanation has been forthcoming. Hypotheses for the transition relate to the magnitude of hypertrophy that develops, to calcium handling, to myofibrillar loss, to neurohumoral activation, and to ischemia.

Magnitude of Hypertrophy

Grossman (26) and others have hypothesized that in hemodynamic overload, the thickness, radius, and pressure terms of the Laplace equation set up a feed-back loop in which, increases in the numerator terms are offset by changes in the denominator so that the wall stress (afterload) is maintained at a normal level (Fig. 1) (10). Thus, an increase in the radius or pressure terms in the numerator would cause just enough of an increase in the thickness term in the denominator to keep stress normal. Indeed in some patients this relationship does exist (10). Wall stress is normalized which helps maintain both normal ejection performance and contractility and thus, the resultant hypertrophy would still be considered to be compensatory. However, in other patients, the hypertrophy that develops is inadequate to normalize stress, stress

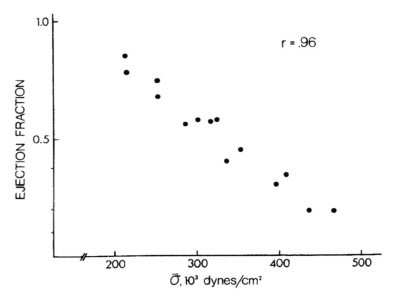

Figure 1 The relationship between ejection fraction and wall stress (σ) in patients with aortic stenosis is demonstrated. In some patients enough hypertrophy has developed to normalize stress, and the ejection performance of those patients is maintained in the normal range. In other patients, hypertrophy is inadequate to normalize stress; stress increases and ejection fraction falls. *Source*: From Ref. 10.

increases and ejection fraction falls (10,24) (Fig. 1). Such inadequate hypertrophy has been associated with a densification of microtubules that comprise the myocardium's cytoskeleton (27). Increased microtubules in turn act as an internal stent that inhibits myocyte shortening. Thus, in this situation of afterload excess, microtubular changes may be responsible for the transition to failure.

In other patients, especially older women, the amount of concentric hypertrophy that develops is more than is needed to compensate load, wall stress is decreased and ejection performance is supra-normal (28). While enhanced systolic function is not harmful, the excessive increase in wall thickness impairs left ventricular filling leading to diastolic dysfunction and thus is pathologic.

Why is there such a variation in the amount of hypertrophy that develops in response to a pressure overload? These differences could be due to differences in the severity of the overload or in the rate and duration of its development. Delineating the various contributions to these differences would be very difficult to test in man where duration and progression of the overload are usually unknown. However, in a recent study, Koide et al. (12) performed identical aortic banding in similar adult dogs. The gradient across the band was increased gradually but identically in all animals over time. Some dogs developed adequate hypertrophy and normal ejection performance and normal contractility at both the chamber and myocyte levels; other dogs had inadequate hypertrophy and reduced ejection performance, reduced contractility, and even overt heart failure recapitulating the variation in hypertrophy seen in man (Figs. 2 and 3). Of interest is the fact that the animals with inadequate hypertrophy had increased wall stress even before banding compared to the dogs with adequate hypertrophy. Thus, the inadequate hypertrophy group was already responding to its ambient load with a muted response. These results suggest an

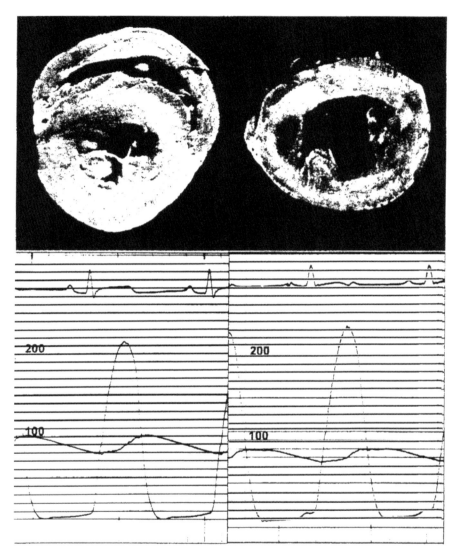

Figure 2 The response to identical aortic banding in two dogs is demonstrated. Despite identical trans-band gradient and left ventricular pressure in both animals, there was a dramatic difference in the hypertrophic response. *Source*: From Ref. 12.

inherently different set point among the dogs for regulating how the mechanical signal for hypertrophy (wall stress) is transduced into the amount of hypertrophy that develops. Thus, hypertrophy magnitude is probably an inherited trait and involves the genetic expression of the various pathways that lead to hypertrophy. Further in this model (that reflects the hypertrophic variation seen in man), extensive hypertrophy was necessary for maintenance of both myocyte and ventricular contractility.

The magnitude of hypertrophy that develops in volume overload also affects the level of compensation that occurs. Many pathologic conditions require increased cardiac output, imposing a volume overload on the heart. Most of these diseases result in the additional volume being delivered into the aorta where they widen pulse

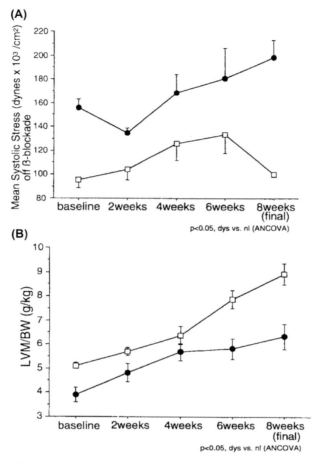

Figure 3 Wall stress (**A**) and left ventricular mass (**B**) are shown for dogs that maintained normal ventricular function (*open squares*) and those that developed ventricular dysfunction (*closed circles*). Despite higher stress, the dysfunction group never developed enough ventricular mass to normalize stress and had persistently less mass than the normal function group. These results amplify the findings from Figure 2 and indicate a difference in the transduction of signal to hypertrophy between the two groups of dogs. *Source*: From Ref. 12.

pressure. Pulse pressure varies directly with stroke volume and with aortic stiffness. Thus, the increased stroke volume demanded by anemia, heart block, aortic insufficiency, etc. results in widened stroke volume, in turn causing systolic hypertension. Therefore, many so-called volume overloads are in fact combined pressure and volume overloads (29). Not surprisingly the heart responds by generating both eccentric and concentric hypertrophy. For example, wall thickness in the typical patient with aortic regurgitation where systolic blood pressure averages 150 mm Hg is about 11 mm (30). On the other hand mitral regurgitation (in the absence of hypertension) represents "pure" volume overload because in this condition, the excess volume is pumped into the left atrium and systolic pressure is either normal or subnormal. In mitral regurgitation, systolic pressure averages about 110 mm Hg and average wall thickness is <0.8 mm. Mitral regurgitation has traditionally been viewed as an unloading lesion in which, the second pathway for left ventricular ejection into

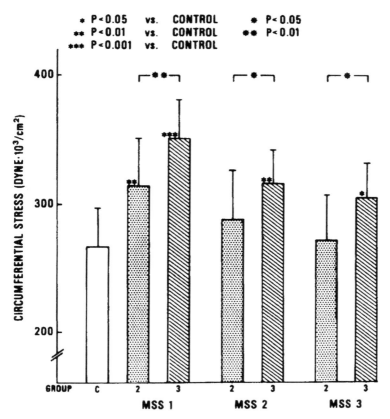

Figure 4 Different methods for calculating mean systolic wall stress (MSS1, MSS2, MSS3) for normal (control) subjects (C), for patients with mitral regurgitation with normal ventricular function (2), and patients with dysfunction (3) are shown. Although mitral regurgitation is often viewed as an unloading lesion, in group 3 patients, wall stress in actually abnormally increased. *Source*: From Ref. 31.

the left atrium decreases left ventricular afterload. However, over time, the increase in the radius term of the Laplace equation in mitral regurgitation may become so large that systolic wall stress is actually abnormally high (31) (Fig. 4). In acute mitral regurgitation, systolic wall stress is sub-normal because the pressure term is reduced and the heart is still small in volume. As mitral regurgitation becomes chronic, the left ventricle enlarges and the radius term returns to normal. With further passage of time and subsequent further increases in ventricular radius, wall stress becomes greater than normal.

Thus, on one hand the lesion of mitral regurgitation tends to unload the left ventricle by affording a second pathway for ejection that lowers systolic blood pressure. On the other hand, the remodeling process actually leads to an increase in afterload because of the failure of left ventricular thickness to normalize wall stress as the increase in Laplacian radius outpaces the reduction in the pressure term of the equation. The geometry of the left ventricle in mitral regurgitation is in fact unique, having the highest radius to thickness ratio and the lowest mass-to-volume ratio of all valvular heart diseases (30).

The conundrum of hypertrophy in pure volume overload is: bad for systole, good for diastole. Most cardiac diseases cause both systolic and diastolic dysfunction and often, diastolic dysfunction precedes the development of systolic dysfunction. Mitral regurgitation is one of the few diseases in which diastolic function is actually supranormal (32,33). Wall thinning in mitral regurgitation allows for enhanced compliance, permitting the increased filling volume of the left ventricle to be accommodated at fairly normal filling pressure, in turn helping to protect the patient from pulmonary edema. However, the lack of concentric remodeling affects systole adversely as noted above.

The Laplace equation predicts that an increase in either the pressure or the radius terms in the numerator will increase wall stress and afterload. Either should incite compensatory hypertrophy but taking the above data collectively, it appears that an increase in the pressure helps to generate concentric remodeling while an increase in the radius does not.

Calcium Handling

Abnormalities in calcium handling have been found in both pressure and volume overload. The force–frequency relationship is blunted in both pressure overload and volume overload. In normal subjects, contractile force increases with heart rate; in hypertrophy, especially when associated with failure, this relationship is blunted.

Peak force occurs at a lower heart rate and the force itself is sub-normal (34,35). The force–frequency relationship is dependent upon excitation–coupling, and upon the shuttling of calcium from the sarcoplasmic reticulum and the cytoplasm. In turn, calcium handling is dependent upon several proteins, which are reduced in heart failure and hypertrophy including SERCA 2, phospholambin, and the Na^+–Ca^{2+} exchanger (36,37). While these abnormalities are clearly important it is not clear which, if any, presages the transition to failure.

Loss and Alterations in Contractile Proteins

In both pressure and volume overload there is evidence of loss of contractile proteins that helps explain some of the muscle dysfunction observed (38,39). Reduced contractile proteins may reflect apoptosis in some cases. In others, the loss is reversible indicating that other processes are involved (39).

Ischemia

It is clear that ischemia develops in some patients with concentric hypertrophy. About 35% of patients with aortic stenosis present with classic angina and overall about 50% of patients complain of angina sometime in the course of the disease even though most have no flow limiting lesions in their epicardial coronary arteries. The cause of this ischemia has been elusive. The presence of angina has not been related well to the magnitude of the hypertrophy (40). Limited coronary flow reserve must play a part (41) but limited flow reserve occurs both in patients with and those without angina. Reduced diastolic perfusion time has correlated best with angina occurrence (42). When ischemia does occur, it occurs with activity. As such ischemia almost surely causes exercise induced ventricular dysfunction (43). But, because coronary flow in such patients is normal at rest, it is harder to postulate ischemia as a cause for resting ventricular dysfunction. It is possible that exercise-induced ischemia leads to stunning, which then could cause resting dysfunction but this hypothesis remains unproven.

MECHANISM OF PRESSURE VS. VOLUME OVERLOAD HYPERTROPHY DEVELOPMENT

The myocardial proteins are in constant flux. The contractile proteins turnover is approximately every 5 to 10 days. Left ventricular mass is maintained by equal rates of protein synthesis and degradation. An increase in mass must accrue either from an increase in synthesis rate or from a decrease in degradation rate. While it seems intuitive that the processes of hypertrophy would entail increased protein synthesis, this concept is not necessarily so.

Hypertrophic Signaling and Protein Synthesis

The steps leading up to a change in myocardial protein turnover encompass at least a dozen signaling pathways and are well beyond the scope of this chapter. However, the central theme of this chapter is that the hemodynamic consequences (increased wall stress) of pressure and volume overload represent a mechanical signal that is in some way transduced into the biochemical events that lead to hypertrophy (44). This signaling may take place at focal adhesion complexes on the cell surface (45). These areas are physically in touch with the extracellular matrix that receives the mechanical stresses placed upon the ventricle and are also rich in integrins, receptors that penetrate the cell surface, capable of triggering a variety of pathways that lead to hypertrophy. Recent studies indicate that mechanical deformation placed upon the cell in diastole (mimicking the stress and strain of volume overload) activate different signaling cascades than does deformation in systole which mimics the effects of pressure overload (46) (Fig. 5). In these studies, Yamamoto et al. (46) used a myocyte preparation in which strain could be introduced at different parts of the cell contraction cycle. Four signaling pathways involved in hypertrophy were examined: the MAP kinase p38 pathway, the extra-cellular signaling related kinase pathway (ERK), the JN kinase pathway (JNK), and the MEK 1/2 pathways. Systolic and diastolic strain perturbation phosphorylated MAPK 38 and JNK pathways equally. However, systolic strain had a much larger effect on ERK and MEK than did diastolic strain. It is not surprising then that these differently activated signaling pathways might activate different hypertrophic processes.

Pressure Overload Hypertrophy

As noted above, the contractile proteins turn over fairly rapidly, thus measurable amounts of new myosin are produced in a time span as short as six hours. By infusing a radioactively tagged amino acid such as leucine into an experimental animal, its rate of incorporation into newly manufactured myosin can be measured. Using this technique, we compared the rate of myosin heavy chain synthesis in control dogs, in dogs with acute pressure overload and in dogs with acute volume overload. As shown in Figure 6, protein synthesis rate rose by 35% within six hours of acute pressure overload while no increase in protein synthesis rate could be detected in the acute volume overload of severe mitral regurgitation (47). In subsequent experiments, we created a step change increase in the pressure gradient across a trans-aortic band in dogs (48). Protein synthesis rate increased acutely and remained elevated for 10 days following the increase in pressure overload, (Fig. 7) after which the hypertrophic response was completed and myosin heavy chain synthesis rates returned to normal. Increased protein synthesis was accomplished both by enhanced translational efficiency

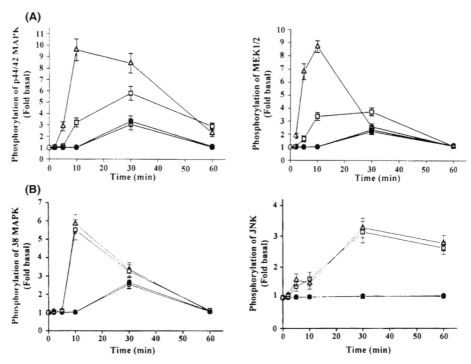

Figure 5 (**A**) Differential expression of MAP kinase pathways according to diastolic versus systolic loading is demonstrated. Systolic loading activated the p44/42 (ERK1/2) more than did diastolic loading (*left panel*). Similar differential activation was seen for MEK1/2 (*right panel*). *Source*: From Ref. 46. (**B**) A schema similar to (**A**) is demonstrated for 38MAPK (*left panel*) and JNK (*right panel*). Diastolic strain and systolic strain activated each pathway similarly. *Key*: -○-, strain; -●-, pacing; -△-, strain imposed during systolic phase; -□-, strain imposed during diastolic phase. *Source*: From Ref. 46.

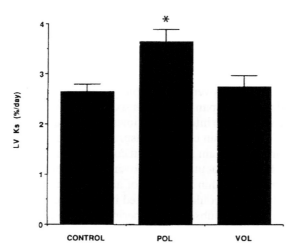

Figure 6 The rate of left ventricular myosin heavy chain synthesis (LV k_s) is demonstrated for normal dogs (CONTROL), for acute severe pressure overload (POL), and for acute severe volume overload (VOL). A rapid increase in protein synthesis is demonstrated for acute pressure overload but not for acute volume overload. *Source*: From Ref. 47.

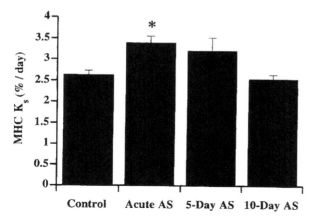

Figure 7 Myosin heavy chain synthesis rate (MHC K_s) in dogs is shown following a stepped increase in aortic constriction. It rose abruptly but fell gradually as the hypertrophic process became complete. *Source*: From Ref. 48.

(Fig. 8; increased message per ribosome) and by translational capacity (Fig. 9, increased ribosomal number), without evidence of increased transcription (MHC message did not increase). Ribosomal RNA did increase indicating increased capacity for protein synthesis as new ribosomes were produced.

Volume Overload Hypertrophy

In the data presented above, there was an acute increase in myosin heavy chain synthesis rate following a pressure overload. However, there was no increase in synthesis rate (Fig. 7) in equally severe volume overload where mitral regurgitation yielded a regurgitant fraction of 0.60.

In subsequent studies, we did find an increase in left ventricular mass of about 30% (49). Examining MHC synthesis rate at two weeks, one month, and three months following creation of severe mitral regurgitation, there never was a detectable increase

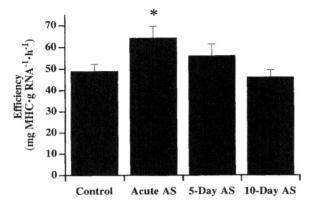

Figure 8 Protein synthesis efficiency expressed as myosin heavy chain (MHC) message per ribosome (RNA) is shown. The increased synthesis rate was in part accomplished by enhanced translation efficiency. *Source*: From Ref. 48.

(A)

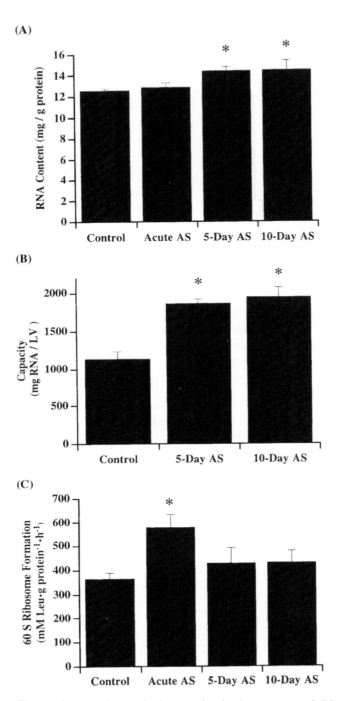

Figure 9 Protein synthesis capacity is shown as total RNA, RNA/LV as measures of ribosomal number (**A** and **B**), and ribosomal synthesis rate (**C**). Capacity for protein synthesis increased following imposition of the pressure overload. *Source*: From Ref. 48.

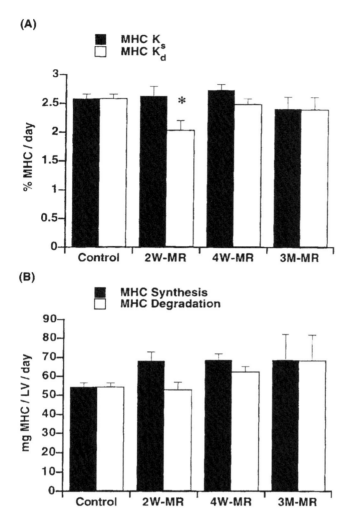

Figure 10 Myosin heavy chain (MHC) synthesis (k_s) and degradation (k_d) rates for dogs with severe mitral regurgitation are expressed as %/day (**A**) and as mgs/day (**B**). K_s never increased. Because LV mass did increase, we calculated that k_d must have decreased. *Source*: From Ref. 49.

in this constant (Fig. 10). Likewise there was no detectable increase in ribosomal RNA or in MHC messenger RNA. As the accumulation of left ventricular mass must imply an imbalance between synthesis and degradation rates, by deduction hypertrophy developed in this volume overload model, not by an increase in MHC synthesis, but rather, by a decrease in degradation rate. Borer et al. (50) have noted similar findings in their rabbit model of mitral regurgitation.

CONCLUSIONS

The heart is composed of terminally differentiated myocytes that are mostly unable to divide. Thus, the major mechanism by which the heart accomplishes normal growth is through hypertrophy. It is postulated that hemodynamic stresses and strains (among other agents) initiate signaling both for the development of concentric and eccentric hypertrophy. In the process of normal growth, the resultant hypertrophy is both a normal and necessary accomplice. Hypertrophy also accompanies abnormal loading conditions for which the hypertrophy is a major form of compensation by which the increased cardiac mass permits increased pumping capacity of the heart. At some point, however, hypertrophy in response to abnormal load takes on pathologic characteristics leading to diastolic and/or systolic myocardial dysfunction and heart failure.

Hypertrophy can arise either by an increase in protein synthesis rate or by a decrease in degradation rate. It appears that pressure and volume overloads activate different signaling cascades resulting in different mechanisms of hypertrophy. In pressure overload, signaling primarily causes an increase in synthesis rate; the magnitude of the response is in part related to the magnitude of the stimulus but also appears modified by genetic differences in the signaling pathways. However, in volume overload, hypertrophy results primarily from a decrease in degradation rate. Thus, it appears that if Laplacian stress is the signal for hypertrophy, the pressure and radius terms in the stress equation numerator effect signaling differently. Increased pressure triggers concentric hypertrophy presumably by increasing protein synthesis rate while increased radius primarily triggers eccentric hypertrophy presumably by decreasing degradation rate. In this later situation, the thickness term is not affected, eventually resulting in increased systolic and diastolic stress with adverse consequences for systole while perhaps benefiting diastolic properties. Once these processes are more clearly defined, they may offer the potential for pharmacologic targeting.

REFERENCES

1. Sonnenblick EH, Ross J Jr, Covell JW, Spotnitz HM, Spiro D. The ultrastructure of the heart in systole and diastole: Changes in sarcomere length. Circ Res 1967; 21: 423–431.
2. Yamazaki T, Komuro I, Kudoh S, Zou Y, Shiojima I, Mizuno T, Takano H, Hiroi Y, Ueki K, Tobe K. Angiotensin II play mediates mechanical stress-induced cardiac hypertrophy. Circ Res 1995; 77:258–265.
3. Sadoshima J, Izumo S. Molecular characterization of angiotensin II-induced hypertrophy of cardiac myocytes and hyperplasia of cardiac fibroblasts: a critical role of the AT1 receptor subtype. Circ Res 1993; 73:413–423.
4. Tsutsui H, Spinale FG, Nagatsu M, Schmid PG, Ishihara K, DeFreyte G, Cooper G IV, Carabello BA. Effects of chronic β-adrenergic blockade on the left ventricular and cardiocyte abnormalities of chronic canine mitral regurgitation. J Clin Invest 1994; 93: 2639–2648.
5. CONSENSUS Trial Study Group. Effects of enalapril on mortality in severe congestive heart failure: Results of the Cooperative North Scandinavian Enalapril Survival Study (CONSENSUS). N Engl J Med 1987; 316:1429–1435.
6. CIBIS-II investigators and committees. The cardiac insufficiency bisoprolol study II (CIBIC-11): a randomized trial. Lancet 1999; 353:9–13.
7. Nadal-Ginard B, Kajstura J, Leri A, Anversa P. Myocyte death, growth, and regeneration in cardiac hypertrophy and failure. Circ Res. 2003 Feb 7; 92(2):139–150.
8. Hoogsteen J, Hoogeveen A, Schaffers H, Wijn PF, van der Wall EE. Left atrial and ventricular dimensions in highly trained cyclists. Int J Cardiovasc Imaging. 2003 Jun; 19(3):211–217.

9. D'Andrea A, Limongelli G, Caso P, Sarubbi B, Della Pietra A, Brancaccio P, Cice G, Scherillo M, Limongelli F, Calabro R. Association between left ventricular structure and cardiac performance during effort in two morphological forms of athlete's heart. Int J Cardiol. 2002; 86(2–3):177–184.

10. Gunther S, Grossman W. Determinants of ventricular function in pressure-overload hypertrophy in man. Circulation 1979; 59:679–688.

11. Huber D, Brimm J, Kock R, Grimm J, Koch R, Krayenbushi HP. Determinants of ejection performance in aortic stenosis. Circulation 1981; 64:126–134.

12. Koide M, Nagatsu M, Zile MR, Hamawaki M, Swindle MM, Keech G, DeFreyte G, Tagawa H, Cooper G 4th, Carabello BA. Premorbid determinants of left ventricular dysfunction in a novel mode of gradually induce pressure overload in the adult canine. Circulation. 1997; 95(6):1349–1351.

13. Donner R, Carabello BA, Black I, Spann JF. Left ventricular wall stress in compensated aortic stenosis in children. Am J Cardiol 1983; 51(6):946–951.

14. Carabello BA. Mitral regurgitation, Part 2: Proper timing of mitral valve replacement. Mod Concepts Cardiovasc Disease. 1988 Nov; Volume 57, Number 11.

15. Kannel WB, Gordon T, Offutt D. Left ventricular hypertrophy by electrocardiogram. Prevalence, incidence, and mortality in the Framingham study. Ann Intern Med 1969; 71:89–105.

16. Levy D, Garrison RJ, Savage DD, Kannel WB, Castelli WP. Prognostic implications of echocardiographically determined left ventricular mass in the Framingham heart study. N Engl J Med 1990; 322:1561–1566.

17. Vakili BA, Okin PM, Devereux RB. Prognostic implications of left ventricular hypertrophy. Am Heart J 2001; 141:334–341.

18. Carabello, BA, Nakano K, Corin W, Biedermann R, Spann JF, Jr. Left ventricular function in experimental volume overload hypertrophy. Am J Physiol 1989; 256 (Heart Circ Physiol 25):H974–H981.

19. Cooper G IV, Tomanek RJ, Ehrhardt JC, Marcus ML. Chronic progressive pressure overload of the cat right ventricle. Circ Res 1981; 48:488–497.

20. Hill JA, Karimi M, Kutschke W, Davison RL, Zimmerman K, Wang Z, Kerber BE, Weiss RM. Cardiac hypertrophy is not a required compensatory response to short-term pressure overload. Circulation 2000; 101:2863–2869.

21. Esposito G, Rapacciuolo A, Naga Prasad SV, Takaoka H, Thomas SA, Koch WJ, Rockman HA. Genetic alterations that inhibit in vivo pressure-overload hypertrophy prevent cardiac dysfunction despite increased wall stress. Circulation 2002; 105:85–92.

22. Bardoff C, Ruetten H, Mueller S, Stahmer M, Gehring D, Jung F, Ihling C, Zeiher AM, Dimmeler S. Fas receptor signaling inhibits glycogen synthase kinase 3 beta and induces cardiac hypertrophy following pressure overload. J Clin Invest 2002; 109:373–381.

23. Rogers JH, Tamirisa P, Kovacs A, Weinheimer C, Courtois M, Blumer KJ, Keily DP, Muslin AJ. RGS4 causes increased mortality and reduced cardiac hypertrophy in response to overload. J Clin Invest 1999; 104:567–576.

24. Carabello BA, Green LH, Grossman W, Cohn LH, Koster JK, Collins JJ Jr. Hemodynamic determinants of prognosis of aortic valve replacement in critical aortic stenosis and advanced congestive heart failure. Circulation 1980; 62:42–48.

25. Wisenbaugh T, Allen P, Cooper G IV, Holzgrefe H. Contractile function, myosin ATPase activity and isozymes in the hypertrophied pig left ventricle after a chronic progressive pressure overload. Circ Res 1983; 53:332–341.

26. Grossman W, Jones D, McLaurin LP. Wall stress and patterns of hypertrophy in the human left ventricle. J Clin Invest 1975:56–64.

27. Tsutsui H, Ishihara K, Cooper G IV. Cytoskeletal role in the contractile dysfunction of hypertrophied myocardium. Science. 1993 Apr 30; 260(5108):682–687.

28. Carroll JD, Carroll EP, Feldman T, Ward DM, Lang RM, McGaughey D, Karp RB. Sex-associated differences in left ventricular function in aortic stenosis of the elderly. Circulation 1992; 86:1099–1107.

29. Wiensbaugh T, Spann JF, Carabello BA. Differences in myocardial performance and load between patients with similar amounts of chronic aortic versus chronic mitral regurgitation. J Am Coll Cardiol 1984; 3:916–923.

30. Carabello BA. The relationship of left ventricular geometry and hypertrophy to left ventricular function in valvular heart disease. J Heart Valve Dis 1995; 4(suppl 2):S132–S138; discussion S138–S139.

31. Corin WJ, Monrad ES, Murakami T, Nonogi H, Hess OM, Krayenbushi HP. The relationship of afterload to ejection performance in chronic mitral regurgitation. Circulation 1987; 76:59–67.

32. Zile MR, Tomita M, Nakano K, Mirsky I, Usher B, Lindroth J, Carabello BA. Effects of left ventricular volume overload produced by mitral regurgitation on diastolic function. Am J Physiol 1991; 261(5 Pt 2):H1471–H1480.

33. Corin WJ, Murakami T, Monrad ES, Hess OM, Krayenbuehl HP. Left ventricular passive diastolic properties in chronic mitral regurgitation. Circulation 1991; 83(3): 797–807.

34. Schotthauer K, Schottman J, Ber DM, Bers DM, Maier LS, Schult U, Minami K, Just H, Hasenfuss G, Pleske B. Frequency-dependent changes in the contribution of SR Ca to CA transients in failing human myocardium assessed with ryanodine. J Mol Cell Cardiol 1998; 30:1285–1294.

35. Mulieri LA, Leavitt BJ, Martin BJ, Haeberle JR, Alpert NR. Myocardial force-frequency defect in mitral regurgitation heart failure is reversed by forskolin. Circulation 1993; 88(6):2700–2704.

36. He H, Giordano FJ, Hilal-Dandan R, Choi DJ, Rockman HA, McDonough PM, Bluhm WF, Meyer M, Sayen MR, Swanson E, Dillman WH. Overexpression of the rat sarcoplasmic reticulum Ca ATPase gene in the heart of transgenic mice accelerates calcium transients and cardiac relaxation. J Clin Invest 1997; 100:380–389.

37. Hasenfuss G, Schillinger W, Lehnart SE, Preuss M, Pleske B, Maier LS, Prestle J, Minami K, Just H. Relationship between Na^+-Ca^{2+} exchanger protein levels and diastolic function of failing human myocardium. Circulation 1999; 99:641–648.

38. Kostin S, Pool L, Elsasser A, Hein S, Drexler HC, Arnon E, Hayakawa Y, Zimmermann R, Bauer E, Klovekorn WP, Schaper J. Myocytes die by multiple mechanisms in filing human hearts. Circ Res 2003; 92(7):715–724. Epub 2003 Mar 20.

39. Spinale FG, Ishihra K, Zile M, DeFryte G, Crawford FA, Carabello BA. Structural basis for changes in left ventricular function and geometry because of chronic mitral regurgitation and after correction of volume overload. J Thorac Cardiovasc Surg 1993; 106(6): 1147–1157.

40. Julius BK, Spillman M, Vassali G, Villari G, Eberli FR, Hess OM. Angina pectoris in patients with aortic stenosis and normal coronary arteries: mechanisms and pathophysiological concepts. Circulation 1997; 95:892–898.

41. Marcus ML, Doty DB, Hiratzka LF, Wright CB, Eastham CL. Decreased coronary reserve: a mechanism for angina pectoris in patients with aortic stenosis and normal coronary arteries. N Engl J Med 1982; 307:1362–1366.

42. Rajappan K, Rimoldi OE, Dutka DP, Ariff B, Pennell DJ, Sheridan DJ, Camioi PG. Mechanisms of coronary microcirculatory dysfunction in patients with aortic stenosis and angiographically normal coronary arteries. Circulation 2002; 105:470–476.

43. Nakano K, Corin WJ, Spann JF Jr, Biederman RW, Denslow S, Carabello BA. Abnormal subendocardial blood flow in pressure overload hypertrophy is associated with pacing-induced subendocardial dysfunction. Circ Res 1989; 65:555–1564.

44. Sugden PH. Signalling in myocardial hypertrophy. Circ Res 1999; 84:633–646.

45. Borg TK, Burgess ML. Holding it all together: organization and functions of the extracellular matrix of the heart. Heart Failure 1993; 8:230–238.

46. Yamamoto K, Dang Q, Maeda Y, Huang H, Kelley RA, Lee RT. Regulation of cardiomyocyte mechanotransduction by the cardiac cycle. Circulation 2001; 103:1459–1464.

47. Imamura T, McDermott PJ, Kent RL, Nagatsu M, Cooper G 4th, Carabello BA. Acute changes in myosin heavy chain synthesis rate in pressure versus volume overload. Circ Res 1994; 75(3):418–425.

48. Nagatomo Y, Carabello BA, Hamawaki M, Nemoto S, Matsuo T, McDermott PJ. Translational mechanisms accelerate the rate of protein synthesis during canine pressure-overload hypertrophy. Am J Physiol 1999; 277(6 Pt 2):H2176–H2184.

49. Matsuo T, Carabello BA, Nagatomo Y, Koide M, Hamawaki M, Zile MR, McDermott PJ. Mechanisms of cardiac hypertrophy in canine volume overload. Am J Physiol 1998; 275(1 Pt 2):H65–H74.

50. Borer JS, Carter JN, Jacobson MH, Herrold EM, Magid NM. Myofibrillar protein synthesis rates in mitral regurgitation. Circulation 1997; 96(suppl 1):I–469.

3

Changes in the Interstitial Matrix During Myocardial Remodeling

Anne M. Deschamps and Francis G. Spinale
Medical University of South Carolina, Charleston, South Carolina, U.S.A.

INTRODUCTION

The myocardial extracellular matrix (ECM) was previously thought to serve solely as a means to align cells within the tissue. Emerging evidence, however, suggests that the ECM plays a complex and divergent role in influencing cell behavior. It is now recognized that the ECM is a major influence of cell migration, proliferation, adhesion, and cell-to-cell signaling (1,2). Both biological and physical stimuli, which are operative through the ECM, directly affect cardiac myocyte structure and function. The development of left ventricular (LV) dilation has been shown to be associated with discontinuity and disruption of the supporting fibrillar collagen network, abnormalities in the degree of collagen crosslinking, and defects in basement membrane structure and function (3–8). This chapter introduces the players of the ECM and describes some of the cardiac disease states associated with LV remodeling, placing a particular emphasis on postmyocardial infarction (MI) remodeling.

BASEMENT MEMBRANE

The basement membrane forms a specialized type of ECM that separates the stroma from the cell membrane. It is a dense, sheet-like structure in which numerous proteins play a role. Collagen type IV is the most abundant protein found within the basement membrane where it forms a covalently stabilized polygonal framework (9). A second polymer network of laminin self-assembles and is bridged to the collagen type IV network by nidogen, a dumbbell-shaped sulfated glycoprotein (9). Minor components of the basement membrane include agrin, osteopontin, fibulin, type XV collagen, and type XVII collagen (10). A 50% reduction in adhesion of basement membrane proteins was observed in myocytes from cardiomyopathic tissue when compared with control tissue, suggesting that changes within the basement membrane can contribute to LV remodeling during the course of cardiomyopathic disease (4).

COLLAGEN NETWORK

Collagen is a major structural protein of the ECM. In the myocardium, fibroblasts or myofibroblasts typically secrete collagen polypeptide chains known as α chains. Three α chains are then wound tightly around each other to form a tensile, triple helical super structure. Collagen type I and type III constitute the majority of the fibrillar collagen content of the myocardium, whereas collagen type IV is an important network-forming collagen found in the basement membrane.

The collagen network within the LV is highly organized. It is composed of three distinct units: (1) endomysial collagen, which surrounds and connects individual myocytes, (2) perimysial collagen, which surrounds and connects bundles of cardiomyocytes, and (3) epimysial collagen, which forms an outer layer (1,11,12) (Fig. 1).

It seems plausible, then, that alterations in the content and quality of the collagen network can lead to changes in LV structure and function. Increases in collagen, as seen with pressure overload hypertrophy (POH) lead to an increased wall stiffness, a less compliant LV, and an evidence of abnormalities in diastolic function, whereas degradation of existing collagen leads to a dilated LV phenotype and more compliant, but abnormally functioning ventricle.

CARDIAC DISEASE STATES AND MYOCARDIAL REMODELING

Myocardial remodeling involving changes in the ECM often advances to cause dysfunction and heart failure. In patients who experience progressive LV remodeling, morbidity and mortality are increased (13–15). Alterations in myocardial tissue structure can occur following MI with the development of cardiomyopathy and myocardial hypertrophy (7,16–20). Although the initiating stimulus differs between these three etiologies and the cellular remodeling processes are distinctly different, the end result is the development of cardiac dysfunction and the onset of heart failure.

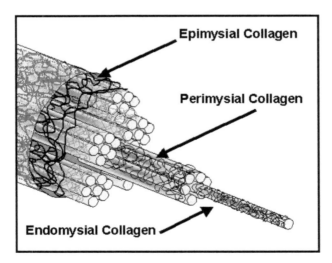

Figure 1   Schematic of the myocardial collagen network. The highly organized collagen network is composed of three units: the endomysial collagen, the perimysial collagen, and the epimysial collagen. *Source*: From Refs. 1,9,10.

Dilated Cardiomyopathy

Remodeling in dilated cardiomyopathy (DCM) is characterized by an increase in the ratio of LV chamber radius to wall thickness. This process increases myocardial wall stress and eventually results in progressive LV chamber dilation (21). The ECM undergoes structural realignment, loss of myocyte connections, degradation of the normal collagen fibrillar weave, and abnormalities in collagen crosslinking (7,8). Markers of collagen degradation are increased in DCM patients compared to age-matched control subjects (22). Thus, it is likely that changes in ECM structure and function are largely responsible for the structural remodeling of the LV seen in patients with DCM.

Left Ventricular Hypertrophy

Hypertensive heart disease is often characterized by alterations in normal LV filling properties that in aggregate have been referred to as diastolic dysfunction. Although the etiology of diastolic dysfunction is multifactorial, common structural changes occur within the ECM (15). Increases in collagen content, as well as alterations in collagen crosslinking, have been observed in association with LV hypertrophy (LVH) (23), and there is an evidence of biochemical modification of the ECM in this condition (3,24).

These changes in ECM structure are, in turn, related to abnormalities in cardiac function. For example, disruption of advanced glycated end products in collagen directly affects diastolic properties (3,24).

There are two distinct patterns of LVH that occur in response to persistent load: POH and volume overload hypertrophy (VOH). The fundamental driving force for this adaptive hypertrophy is changes in LV wall stress. However, the myocardial adaptation by which the hypertrophic response is manifested is different in the two overload states. POH results in what is termed a concentric hypertrophy whereby wall thickness is increased and LV diameter remains the same or is decreased. VOH results in an eccentric hypertrophy whereby wall thickness is disproportional to the magnitude of LV dimensional increase (25). In turn, the cardiac myocyte itself takes on two distinct shapes. As myocytes undergo a concentric hypertrophic response, they add sarcomeres in parallel because of an increase in width. In contrast, with eccentric hypertrophy, sarcomeres undergo a series of additions owing to the large increase in myocyte length (26). In both of these hypertrophic remodeling processes, significant myocardial matrix remodeling occurs and facilitates changes in myocyte shape. In POH, accumulation of myocardial fibrillar collagen influences passive compliance properties of the LV and results in diastolic dysfunction (7,16,27). In contrast to patterns of collagen accumulation in POH, LV remodeling with volume overload is characterized by excessive ECM degradation, disruption of normal fibrillar collagen, and LV dilation (23,27–29).

Post-MI

LV remodeling post-MI is characterized by nonuniform changes in LV myocardial wall geometry (7,17,30). After undergoing severe cellular loss, the nonviable infarct region becomes a thinned, fibrous scar (7,17,18,30). The myocytes in the border zone are susceptible to infarct expansion and undergo hypertrophy, realignment, and slippage (30,31). Gross LV myocardial geometry changes from an elliptical to a spherical shape, and there is a concomitant fall in LV performance (7,17,30,31).

Of importance is that, while LV pump function is commonly depressed with post-MI remodeling, contractility in viable isolated myocytes may be preserved (32). These observations emphasize the ability of the ECM to translate myocyte shortening into overall ejection. Moreover, these observations indicate that maladaptive post-MI remodeling is due in part to changes in the ECM as well as in viable myocytes. Therefore, it is essential to identify the cascade of events that give rise to LV remodeling post-MI. As post-MI remodeling is the most dynamic and best studied of the three etiologies discussed, this chapter will present post-MI remodeling as a foundation of what is known in this area.

MI AND REMODELING

MI can lead to a number of structural changes (i.e., remodeling) in the both infarcted and noninfarcted regions of the myocardium. Postinfarction remodeling consists of two distinct phases. The first phase, or early remodeling, takes place within the first 72 hours of postinfarction and is characterized by dilation of infarcted myocardium (33). The second phase, known as late remodeling, involves global LV remodeling.
   Structural remodeling after an MI is a complex and dynamic process. There has been considerable attention focused on how the myocardial ECM contributes to this remodeling process. Specifically, the induction of matrix metalloproteinases (MMPs) has been observed in animal models and in patients with post-MI, and the role of these proteolytic enzymes and of the tissue inhibitors of matrix metallo-proteinases (TIMPs) that regulate them has been studied extensively in recent years. In consideration of the divergent physical and biochemical pathways involved in LV remodeling, defining the molecular triggers of MMP and TIMP expression and targeting the upstream mechanisms responsible may prove to be an important therapeutic option for the attenuation of LV remodeling after MI.

MATRIX MMPs

The MMPs are a family of more than 20 species of zinc-dependent proteases that are essential for normal tissue remodeling in processes such as bone growth, wound healing, and reproduction (34–37). MMPs are responsible for turnover of the ECM, which in turn facilitates tissue remodeling. MMPs are synthesized as inactive zymo-gens and are secreted into the extracellular space as proenzymes (34–37). The pro-MMP binds specific ECM proteins and remains enzymatically quiescent until the propeptide domain is cleaved. Cleavage results in an exposure of the zinc-active site in the catalytic domain and subsequent activation of the enzyme (38). Disruption of the cysteine–zinc interaction in the MMP active site is essential in the activation of MMPs, and this "cysteine switch" hypothesis of MMP activation is likely the mechanism by which most MMPs are activated in vivo (38). As opposed to a spora-dic distribution of pro-MMPs throughout the ECM, there is a region- and type-specific allotment where some are bound to matrix proteins, whereas others reside in the plasma membrane. Moreover, a large reservoir of recruitable MMPs exists, which upon activation can result in a rapid surge of ECM proteolytic activity (Fig. 2).

MMP Classification
MMPs are classified into subgroups according to substrate specificity and/or struc-ture (34–37). MMPs that have been identified in the myocardium are listed in

Table 1 Matrix Metalloproteinases Identified in Human Myocardium

Class	Name	Number	Substrate
Collagenases	Interstitial collagenase	MMP-1	Collagen I, II, III, VII, gelatin, proteoglycan, glycoprotein
	Collagenase 2[a]	MMP-8[a]	Collagen I, II, III, aggrecan
	Collagenase 3	MMP-13	Collagen I, II, III, gelatin, proteoglycan, pro-MMP-1
Gelatinases	Gelatinase A	MMP-2	Gelatins, collagen I, IV, V, VII, elastin, fibronectin, laminin, proteoglycan
	Gelatinase B	MMP-9	Gelatin, collagen IV, V, XIV, elastin, proteoglycan, glycoprotein
Stromelysins	Stromelysin-1	MMP-3	Fibronectin, laminin, collagen III, IV, IX, pro-MMP-1, -7, -9
Matrilysin	Matrilysin[a]	MMP-7[a]	Fibronectin, laminin, elastin, gelatin, collagen I, IV
Membrane-type MMPs	MT1-MMP	MMP-14	Collagen I, II, III, fibronectin, laminin-1, glycoprotein, proteoglycan, pro-MMP-2, -13

[a]May exist in the myocardium in states of inflammation and wound healing.

Table 1. Interstitial collagenase (MMP-1) and collagenase-3 (MMP-13) possess high substrate specificity for fibrillar collagens, whereas the gelatinases (MMP-2 and MMP-9) demonstrate high substrate affinity for basement membrane proteins (34–37). The substrate portfolio for stromelysin (MMP-3) includes important myocardial ECM proteins such as aggrecan, fibronectin, and fibrillar collagens (34–37). Stromelysins participate in the MMP activational cascade (38). Earlier in vitro studies have demonstrated that MMP-3 can proteolytically process pro-MMP species (38,39). For example, Murphy et al. reported a 12-fold increase in the conversion of pro-MMP-1 to active MMP-1 in the presence of MMP-3 (39). In addition, other MMPs such as MMP-1, MMP-2, and the membrane-type MMPs (MT-MMPs) can also activate pro-MMPs (34–38,40,41). MT-MMPs contain a transmembrane domain and are likely activated intracellularly through a proprotein convertase pathway (34–38,40,41). Thus, unlike the secretable MMPs, MT-MMPs are inserted into the cell membrane already activated. It has been demonstrated that MT1-MMP degrades fibrillar collagens and a wide range of ECM components as well as proteolytically processes, pro-MMP-2 and -13 (34–38,40,41). There is an emerging evidence to suggest that MMPs can also degrade nonmatrix substrates, in turn affecting cell proliferation, migration, and apoptosis (42).

MMP Activation/Inhibition

Because of the class size and substrate specificity of the MMPs, activation of a few selective enzymes can launch a cascade of proteolytic activity. For example, pro-MMPs are activated not only by MMP-1, MMP-2, and the MT-MMPs, but also by the serine proteases chymotrypsin, trypsin, and plasmin (34,35,38). Thus, the regulation of these upstream processes and proteolytic cascades may directly affect MMP activity. Because MMPs degrade various components of the ECM, it is

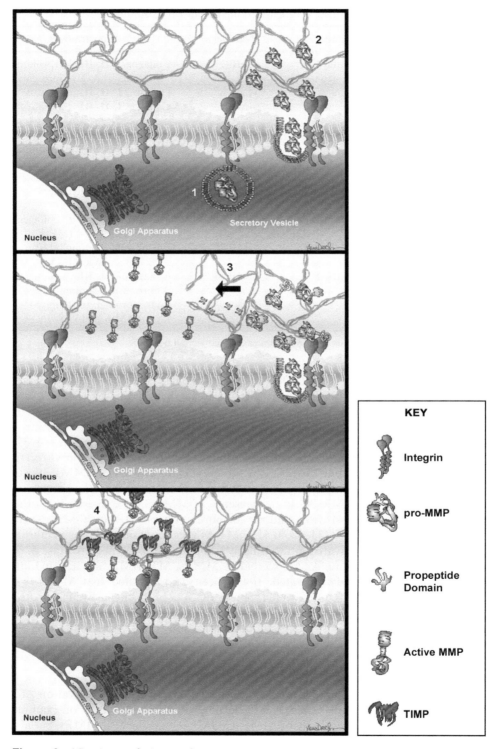

Figure 2 (*Caption on facing page*)

important that MMPs be tightly controlled to prevent excessive tissue degradation. Therefore, a group of endogenous proteins, the TIMPs, exists to regulate the activity of the MMPs. Currently, there are four known TIMP species (43–45). TIMPs bind to the active site of MMPs thereby blocking access to extracellular matrix substrates. The MMP/TIMP complex is formed in a stoichiometric 1:1 molar ratio and forms an important endogenous system for regulating MMP activity in vivo (43–45). The importance of inhibitory control of MMPs within the myocardium has been demonstrated through the genetic deletion of TIMP-1 expression (46). In TIMP-1 deficient mice, both LV and myocardial interstitial remodeling occurred in the absence of a pathological stimulus. LV end-diastolic volume, mass, and wall stress were increased at four months of age in TIMP-1 deficient mice when compared to age-matched controls (46). Additionally, TIMP-1 null mice exhibited a reduced myocardial fibrillar collagen content compared to wild-type animals (46). Although the TIMPs are expressed in an array of cells, TIMP-4 shows a high level of expression in human myocardial tissue (44). The following section assesses MMP and TIMP profiles in relation to LV remodeling post-MI.

LV Remodeling and MMPs in Post-MI Patients

Several studies have demonstrated increased plasma levels of MMPs in patients' early post-MI (47–49). For example, increased peripheral blood levels of MMP-2, -9, and TIMP-1 have been reported in patients of acute coronary syndromes (47–49). In these past observational reports, however, the temporal relationship between MMP and TIMP release into the plasma for the onset of myocardial ischemia/infarction remains unknown. Accordingly, a recent study focused on serial changes in plasma MMP levels following MI induced by alcohol ablation in patients with hypertrophic obstructive cardiomyopathy. In this study, a time-dependent release of MMPs post-MI was observed (50). As depicted in Figure 3, MMP-9 levels increased and remained elevated for up to 50 hours following alcohol-induced MI (50). While serial measurements do not reflect direct activity, they may serve as surrogate markers for relative activation of MMPs post-MI. Such studies may be useful in developing prognostic and diagnostic therapies.

LV Remodeling and MMPs: Animal Models of MI Demonstrating Cause and Effect

A cause–effect relationship between myocardial MMP activation and remodeling has been established through experimental studies utilizing genetic modifications and pharmacological MMP inhibition (51–53). These studies have shown that

---

Figure 2 (*Facing page*) Schematic of MMP secretion, activation, and inhibition. (**1**) After translation, secretory vesicles containing inactive or pro-MMPs are trafficked to the cell membrane. (**2**) The vesicle fuses with the cell membrane, thereby releasing the pro-MMP into the extracellular space. (**3**) When prompted by a stimulus, the pro-MMPs are activated by serine proteases such as chymotrypsin, trypsin, and plasmin or other MMPs, such as MMP-3 and MT1-MMP, by removal of the propeptide domain. After activation, the MMPs are able to degrade components of the extracellular matrix. (**4**) One point of inhibitory control is the family of TIMPs. Four identified TIMPs are able to bind noncovalently to the catalytic domain of the MMP and prevent its activity. Discordance in the MMP/TIMP ratio has been shown to be linked to pathologic myocardial remodeling. *Abbreviations*: MMP, matrix metalloproteinase; TIMP, inhibitors of metalloproteinase. (*See color insert.*)

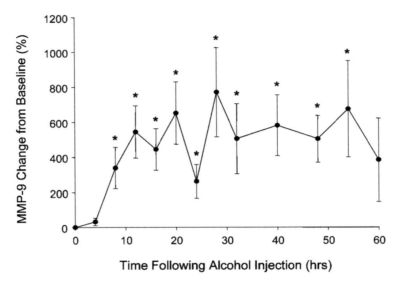

Figure 3 Time-dependent release of MMPs post-MI induced by alcohol ablation. A significant increase in plasma MMP-9 levels occurred following alcohol injection and appeared to plateau for up to 50 hours following injection. $^*p < 0.05$ versus time 0; baseline values. *Abbreviations*: MMP, matrix metalloproteinase; MI, myocardial infarction. *Source*: From Ref. 44.

altering the expression of certain MMP genes affects tissue remodeling within the MI region as well as influences the degree of global post-MI remodeling (51,52). Transgenic mouse models have been used to inspect the remodeling process after experimental MI. In mice devoid of the MMP-9 gene the degree of LV dilation was attenuated in post-MI (52). In a separate study, MMP-9 deficient mice exhibited a decreased rate of rupture after MI (54). Similarly, a study performed by Hayashidani et al. used an MMP-2 null mouse line to study the effects of MI on LV rupture and late remodeling (55). Mortality in MMP-2 null mice was significantly reduced when compared to wild-type controls after MI (55). In addition, a decrease in LV dilation and an increase in fractional shortening occurred in MMP-2 null mice (55). A study conducted by Creemers et al. used TIMP-1 deficient mice to determine the role of TIMP-1 in post-MI remodeling (56). In TIMP-1 deficient mice, LV end-diastolic volume and end-diastolic pressure were elevated when compared to control mice' post-MI (56). Myocyte cross-sectional area and LV mass increased indicating an amplified hypertrophic response in TIMP-1 null mice (56). Together, these data show that adverse LV remodeling post-MI was exacerbated by deletion of the TIMP-1 gene (56). This study demonstrated the importance of the role that TIMPs play in local endogenous control of MMP activity within the myocardium.

Cause–effect relationships between MMP activation and adverse LV remodeling have also been established through administration of pharmalogic MMP inhibitors post-MI (57–59). For example, utilizing a global MMP inhibition strategy in the setting of post-MI remodeling, myocardial MMP, and TIMP levels were evaluated (57). Levels of MMP-2 protein were decreased in the viable myocardium with pharmacological MMP inhibition when compared to MI only values (57). Also, TIMP-1 levels within the viable myocardium were increased in the MMP inhibitor group. Taken together, these results suggested that pharmacological inhibition altered endogenous MMP/TIMP profiles, which would favor ECM stabilization post-MI. Also, this past study as well as others (57–59), demonstrated that MMP inhibitors reduce the degree of LV dilation post-MI. As shown in Figure 4, the basis for this improvement in LV geometry post-MI with MMP inhibition is a reduction in the degree of infarct expansion. These past studies clearly established that increased MMP activity contributes, at least in part, to the adverse remodeling process post-MI.

A Unique MMP/TIMP Profile Following MI

A unique temporal profile of MMPs and TIMPs has been reported in the post-MI animal model (53,57,60). As is shown in Figure 5, Wilson et al. described a regional imbalance of MMPs and TIMPs following changes in the remote, transition, and infarct regions post-MI (60). Time dependent changes in myocardial MMP and TIMP levels following MI induction in rats have also been observed and show that differential profiles of MMPs are released throughout LV remodeling post-MI (53). MMPs-2, -9, and -13 were elevated early in the post-MI period, whereas a more gradual elevation of MMP-14 (MT1-MMP) followed 16 weeks post-MI. Furthermore, throughout the time course measured, TIMP-1 and -2 increased whereas TIMP-4 decreased (53). Therefore, changes in MMP and TIMP levels are not a global process. Rather, there is a region, time, and type-specific activational process. Targeting these certain species warrant further study.

The primary approaches used in the past for measuring MMP activation have been through collection of myocardial samples and the use of in vitro assay systems (50,53,57,60). These systems, however, do not provide a comprehensive measurement of MMP activity because endogenous regulatory mechanisms are removed. A microdialysis technique used previously provided a reliable way to interrogate the interstitium (61). By infusing a fluorogenic substrate specific for MMPs, the extracellular space was flooded and MMP activity directly measured (62). This approach provided a relative index of MMP activity within the interstitial space in vivo and has aided in the understanding of the activational patterns of MMPs during ischemia. A study conducted by Etoh et al. showed that interstitial MMP-2 and -9 activity began to increase

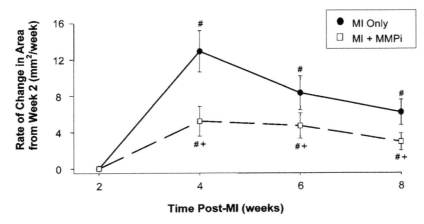

Figure 4 MMP inhibition attenuates post-MI remodeling. The time-dependent change in regional MI size, computed relative to week 2 values, was significantly lower in MI + MMPi group than MI-only group. $^{+}p < 0.05$ versus MI only, $^{\#}p < 0.05$ versus week 2. *Abbreviations*: MMP, matrix metalloproteinase; MMPi, MMP inhibition; MI, myocardial infarction. *Source*: From Ref. 50.

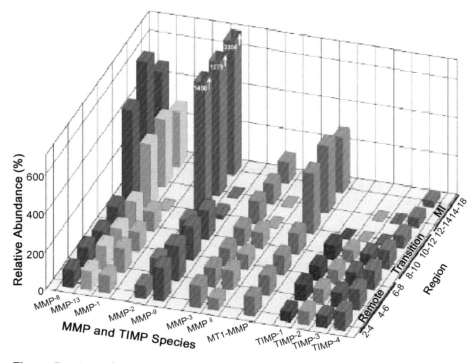

Figure 5 Three-dimensional histogram demonstrating regional distribution of MMPs and TIMPs after MI. Distinct region-specific changes in MMP profiles occurred moving from remote to MI regions. Certain MMPs, such as MMP-8 and -13, increased moving into transition and MI regions, whereas MMP-1 and -9 were undetectable within MI region. Levels of all TIMPs fell moving into MI region. *Abbreviations*: MMPs, matrix metalloproteinases; MI, myocardial infarction; TIMPs, tissue inhibitors of metalloproteinases. *Source*: From Ref. 51. (*See color insert.*)

as early as 10 minutes postocclusion of the coronary artery in the MI region with no change in the remote myocardium (62). As seen in Figure 6, MMP-2 and -9 activity remained elevated throughout the remainder of the ischemic period suggesting that MMP release and activation within the ischemic myocardial interstitium in the early post-MI period is related to subsequent remodeling of this region (62).

A loss of TIMP-mediated control has also been reported in LV remodeling following MI (53,57,60). In the rat MI model, MMP mRNA levels increased early post-MI, but were not associated with a concomitant increase in TIMP mRNA levels (53). In an in vitro system of ischemia and reperfusion, TIMP-1 expression was reduced in the early reperfusion period (63). These findings suggest a loss of endogenous MMP inhibitory control that occurs early in the post-MI period (63).

Identifying proteolytic systems, which promulgate regional geometry changes, offers greater potential for direct therapeutic targeting. However, developing specific targets will be predicated on recognition of the phases of post-MI remodeling. The early phase of post-MI remodeling occurs immediately following an acute myocardial insult. Localized MMP activity contributes to infarct scar formation, and healing is accompanied by neutrophil infiltration into the infarct area (7,17,18). The later phase in post-MI remodeling, however, leads to rampant MMP activity in the border region of the infarct scar which contributes to adverse remodeling and progressive infarct

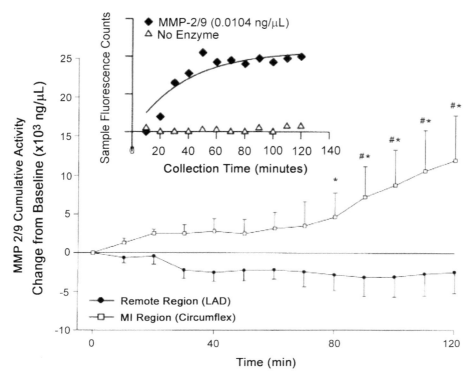

Figure 6 In vivo MMP activity during ischemia. Myocardial interstitial MMP activity was determined in vivo using a MMP-2/9 fluorogenic substrate and microdialysis. The cumulative change in dialysate fluorescence was computed as a function of time after coronary occlusion. The specificity of the MMP substrate was confirmed by in vitro assays using purified MMP-2/9 enzyme (*inset*). Increased fluorescence was observed in microdialysate collected from the MI region, whereas there was no change observed in the remote region served by the LAD coronary artery. With longer periods of coronary occlusion, cumulative MMP-2/9 activity within the myocardial interstitium increased. By two hours of coronary occlusion, MMP-2/9 activity was increased by nearly twofold from values at one hour. $^{\#}p < 0.05$ versus baseline; $^{*}p < 0.05$ versus remote region. *Abbreviations*: LAD, left anterior descending; MMP, matrix metalloproteinase; MI, myocardial infarction. *Source*: From Ref. 52.

scar expansion (57). Thus, targeting the MMP system post-MI must account for both the critical healing phase and the adverse post-MI remodeling phase. To determine the ideal time for MMP inhibitor deployment in the setting of LV remodeling post-MI, it is necessary to couple upstream cues that instigate adverse remodeling with improved imaging modalities.

Future Clinical Directions and Therapeutic Strategies

Heart failure is a clinical syndrome that can develop as a result of a variety of different pathological conditions. We have described earlier separate conditions (DCM, LVH because of overload, and MI) that involved myocardial remodeling that results in heart failure. While these conditions may ultimately give rise to a similar constellation of symptoms, the myocardial remodeling process that occurs post-MI is distinctly different than with the other conditions. In post-MI remodeling, structural alterations within the ECM occur, which result in mechanical disadvantage of surviving myocytes. In general, the factors that drive ECM remodeling are numerous. In the post-MI heart, however, a specific pattern of MMPs and TIMPs appear to occur (Fig. 5). Therefore, understanding the cellular and molecular triggers, which in turn give rise to changes in the ECM, could present opportunities to modify the remodeling process. Information gained from studies investigating this area likely hold therapeutic promise.

Modifying MMPs Post-MI

Animal studies have clearly demonstrated the modification of progressive LV dilation and dysfunction post-MI through the use of MMP inhibition (51,53,57,64). The clinical application of broad-spectrum MMP inhibitors, however, may be problematic (65,66). Specifically, broad-spectrum MMP inhibition has been associated with musculo-skeletal side effects, which suggests that broad-spectrum inhibition may interfere with normal tissue turnover (65,66). Therefore, the specific MMP types causative in the adverse LV remodeling process must be identified. MMP-1 is reduced in the myocardium of patients with LV failure, and experimental studies suggest that this MMP may not be causative to pathological LV remodeling (67). A study by Lindsey et al. utilized selective MMP inhibition, which effectively spared MMP-1 in a rabbit model of post-MI ventricular remodeling and reported that progressive LV dilation was prevented (58). Additionally, a study by Yarbrough et al. reported that selective inhibition, which spared MMP-1 and -7, favorably influenced LV remodeling after an MI (59).

Thus, a proof-of-concept has been established in which selective MMP inhibition is a practical pathway to modify LV remodeling post-MI. Along with selectivity, optimal timing must also be considered for interventions in post-MI remodeling. The acute, early phase of remodeling immediately following MI is essential for infarct scar formation and immune cell infiltration, and it is important to recognize the necessity for the wound-healing response post-MI. For example, past clinical studies that interfered with the early acute inflammatory response were associated with adverse outcomes (68). Differentiation of this acute critical healing phase from the adverse remodeling phase is essential. Thus, in targeting the MMP system, proteolytic cascades as well as their initiating events and temporal profiles must be carefully considered in properly deploying MMP inhibition.

SUMMARY: A NEED FOR FURTHER CLINICAL AND BASIC RESEARCH

Developing pharmacologic strategies that interfere with upstream signaling cascades involved in MMP transcription may improve our understanding of the complex myocardial remodeling process and the specific role of MMPs. For example, cytokine interruption, as with tumor necrosis factor-α (TNF-α) neutralizing proteins, may be a useful pharmacological tool in identifying the signaling pathways obligatory for MMP species induction (69). Ideally, through targeting specific bioactive molecules or blocking transcriptional binding sites, MMPs implicated in adverse remodeling could be inhibited, whereas necessary basal expression of beneficial MMP levels could continue. Gene delivery systems provide potential for local modification of MMPs and TIMPs. As genes can be selectively introduced into the myocardium in the form of naked deoxyribonucleic acid or using adenoviral vectors, gene delivery holds therapeutic promise for several heart failure processes (70–73). Emerging evidence also suggests that nonmatrix substrates of MMPs influence cell function. For example, the ability of

MMP-7 to cleave multiple nonmatrix substrates, such as cell surface bound Fas-ligand, β4 integrin, and pro-TNF-α, renders it significant in numerous pathways (42). Thus, more clinically applicable strategies specific to each disease state and time-course can be realized through acknowledging and targeting upstream signals of MMPs as well as their various nonmatrix substrates. In consideration of the divergent physical and biochemical pathways involved in LV remodeling, defining the molecular triggers of MMP and TIMP expression and targeting the upstream mechanisms responsible may prove to be an important therapeutic paradigm for heart failure treatment.

ABBREVIATIONS

DCM	Dilated cardiomyopathy
ECM	Extracellular matrix
LV	Left ventricle (ventricular)
LVH	Left ventricular hypertrophy
mRNA	Messenger ribonucleic acid
MI	Myocardial infarction
MMP	Matrix metalloproteinase
MT-MMP	Membrane-type MMP
TIMP	Tissue inhibitor of MMP
TNF-α	Tumor necrosis factor – alpha
POH	Pressure overload hypertrophy
VOH	Volume overload hypertrophy

REFERENCES

1. Weber KT, Sun Y, Tyagi SC, Cleutjens JP. Collagen network of the myocardium: function, structural remodeling and regulatory mechanisms. J Mol Cell Cardiol 1994; 26:279–292.
2. Geiger B, Bershadsky A, Pankov R, Yamada KM. Transmembrane crosstalk between the extracellular matrix–cytoskeleton crosstalk. Nat Rev Mol Cell Biol 2001; (11):793–805.
3. Candido R, Forbes JM, Thomas MC, Thallas V, Dean RG, Burns WC, Tikellis C, Ritchie RH, Twigg SM, Cooper ME, Burrell LM. A breaker of advanced glycation end products attenuates diabetes-induced myocardial structural changes. Circ Res 2003; 92:785–792.
4. Spinale FG, Zellner JL, Johnson WS, Eble DM, Munyer PD. Cellular and extracellular remodeling with the development and recovery from tachycardia-induced cardiomyopathy: changes in fibrillar collagen, myocyte adhesion capacity and proteoglycans. J Mol Cell Cardiol 1996; 28(8):1591–1608.
5. Spinale FG, Tomita M, Zellner JL, Cook JC, Crawford FA, Zile MR. Collagen remodeling and changes in LV function during development and recovery from supraventricular tachycardia. Am J Physiol 1991; 261(2 Pt 2):H308–H318.
6. Baicu CF, Stroud JD, Livesay VA, Hapke E, Holder J, Spinale FG, Zile MR. Changes in extracellular collagen matrix alter myocardial systolic performance. Am J Physiol Heart Circ Physiol 2003; 284(1):H122–H132.
7. Weber KT, Anversa P, Armstrong PW, Brilla CG, Burnett JC, Cruickshank JM, Devereux RB, Giles TD, Korsgaard N, Leier CV, Mendelsohn FAO, Motz WH, Mulvany MJ, Strauer BE. Remodeling and reparation of the cardiovascular system. J Am Coll Cardiol 1992; 20:3–16.

8. Factor SM. Role of extracellular matrix in dilated cardiomyopathy. Heart Fail 1994; 260–268.
9. Yurchenco PD, Schittny JC. Molecular architecture of basement membranes. FASEB J 1990; 4(6):1577–1590.
10. Kalluri R. Basement membranes: structure, assembly and role in tumour angiogenesis. Nat Rev Cancer 2003; 3(6):422–433.
11. Caulfield JB, Borg TK. The collagen network of the heart. Lab Invest 1979; 40(3):364–372.
12. Campbell SE. Collagen matrix in the heart. In: Eghbali-Webb M, ed. Molecular Biology of Collagen Matrix in the Heart. New York: Springer-Verlag, 1995:1–21.
13. White HD, Norris RM, Brown MA, Brandt PW, Whitlock RM, Wild CJ. Left ventricular end-systolic volume as the major determinant of survival after recovery from myocardial infarction. Circulation 1987; 76:44–51.
14. Douglas PS, Morrow R, Ioli A, Reichek N. Left ventricular shape, afterload and survival in idiopathic dilated cardiomyopathy. J Am Coll Cardiol 1989; 12:311–315.
15. Poole-Wilson PA. Relation of pathophysiologic mechanisms to outcome in heart failure. J Am Coll Cardiol 1993; 22:22A–29A.
16. Cohn JN, Ferrari R, Sharpe N. Cardiac remodeling–concepts and clinical implications: a consensus paper from an international forum on cardiac remodeling. J Am Coll Cardiol 2000; 35:569–582.
17. St. John Sutton MG, Sharpe N. Left ventricular remodeling after myocardial infarction: Pathophysiology and therapy. Circulation 2000; 101:2981–2988.
18. Anversa P, Olivetti G, Capasso JM. Cellular basis of ventricular remodeling after myocardial infarction. Am J Cardiol 1991; 68:7D–16D.
19. Francis GS. Pathophysiology of chronic heart failure. Am J Med 2001; 110:37S–46S.
20. Sackner-Bernstein JD. The myocardial matrix and the development and progression of ventricular remodeling. Curr Cardiol Rep 2000; 2:112–119.
21. Wynne J, Braunwald E. The cardiomyopathies and myocardities. In: Braunwald E, ed. Heart Disease. Philadelphia: WB Saunders, 1997:1404–1463.
22. Schwartzkopff B, Fassbach M, Pelzer B, Brehm M, Strauer BE. Elevated serum markers of collagen degradation in patients with mild to moderate dilated cardiomyopathy. Euro J Heart Fail 2002; 4:439–444.
23. Laviades C, Varo N, Fernandez J, Mayor G, Gil MJ, Monreal I, Diez J. Abnormalities of the extracellular degradation of collagen type I in essential hypertension. Circulation 1998; 98:535–540.
24. Vaitkevicius PV, Lane M, Spurgeon H, Ingram DK, Roth GS, Egan JJ, Vasan S, Wagle DR, Ulrich P, Brines M, Wuerth JP, Cerami A, Lakatta EG. A cross-link breaker has sustained effects on arterial and ventricular properties in older rhesus monkeys. P Natl Acad Sci 2001; 98:1171–1175.
25. Cotran, S. et al. Cardiac hypertrophy: pathophysiology and progression to failure. Robbins pathologic basis of disease (6th ed.). 1999: 547–549.
26. Pokharel S, Sharma UC, Pinto YM. Left ventricular hypertrophy: virtuous intentions, malign consequences. Int J Biochem Cell Biol 2003; 35(6):802–806.
27. Grossman W, Jones D, McLaurin LP. Wall stress and patterns of hypertrophy in the human left ventricle. J Clin Invest 1975; 56:56–64.
28. Nagatomo Y, Carabello BA, Coker ML, McDermott PJ, Nemoto S, Hamawaki M, Spinale FG. Differential effects of pressure or volume overload on myocardial MMP levels and inhibitory control. Am J Physiol Heart Circ Physiol 2000; 278(1):H151–H161.
29. Spinale FG, Ishihra K, Zile M, DeFryte G, Crawford FA, Carabello BA. The structural basis for changes in left ventricular function and geometry due to chronic mitral regurgitation and following correction of volume overload. J Thorac Cardiovasc Surg 1993; 106:1147–1157.
30. Pfeffer MA. Left ventricular remodeling after acute myocardial infarction. Ann Rev Med 1995; 46:455–466.

31. Olivetti G, Capasso JM, Sonnenblick EH, Anversa P. Side-to-side slippage of myocytes participates in ventricular wall remodeling acutely after myocardial infarction in rats. Circ Res 1990; 67:23–34.
32. Prahash AJC, Gupta S, Anand IS. Myocyte response to β-adrenergic stimulation is preserved in the noninfarcted myocardium of globally dysfunctional rat hearts after myocardial infarction. Circulation 2000; 102:1840–1846.
33. Erlebacher JA, Weiss JL, Weisfeldt ML, Bulkley BH. Early dilation of the infarcted segment in acute transmural myocardial infarction: role of infarct expansion in acute left ventricular enlargement. J Am Coll Cardiol 1984; 4(2):201–208.
34. Parsons SL, Watson SA, Brown PD, Collins HM, Steele RJC. Matrix metalloproteinases. Brit J Surg 1997; 84:160–166.
35. Nagase H, Woessner JF. Matrix metalloproteinases. J Biol Chem 1999; 274:21, 491–421, 494.
36. Woessner JF. Matrix metalloproteinases and their inhibitors in connective tissue remodeling. FASEB 1991; J5:2145–2154.
37. Vu TH, Werb Z. Matrix metalloproteinases: effectors of development and normal physiology. Genes Dev 2000; 14:2123–2133.
38. Nagase H. Activation mechanisms of matrix metalloproteinases. Biol Chem 1997; 378: 151–160.
39. Murphy G, Cockett MI, Stephens PE, Smith BJ, Docherty AJ. Stromelysin is an activator of procollagenase. A study with natural and recombinant enzymes. Biochem J 1987; 248:265–268.
40. Hernandez-Barrantes S, Bernardo M, Toth M, Fridman R. Regulation of membrane type-matrix metalloproteinases. Semin Cancer Biol 2002; 12:131–138.
41. Pei D, Weiss SJ. Transmembrane-deletion mutants of the membrane-type matrix metalloproteinase-1 process progelatinase A and express intrinsic matrix-degrading activity. J Biol Chem 1996; 271:9135–9140.
42. McCawley LJ, Matrisian LM. Matrix metalloproteinases: they're not just for matrix anymore! Curr Opin Cell Biol 2001; 13:534–540.
43. Edwards DR, Beaudry PP, Laing TD, Kowal V, Leco KJ, Leco PA, Lim MS. The roles of tissue inhibitors of metalloproteinases in tissue remodeling and cell growth. Int J Obes 1996; 20:S9–S15.
44. Green J, Wang M, Liu YE, Raymond LA, Rosen C, Shi YE. Molecular cloning and characterization of human tissue inhibitor of metalloproteinase 4. J Biol Chem 1996; 271:30375–30380.
45. Brew K, Dinakarpandian D, Nagase H. Tissue inhibitors of metalloproteinases: evolution, structure and function. Biochem et Biophy Acta 2000; 1477:267–283.
46. Roten L, Nemoto S, Simsic J, Coker ML, Rao V, Baicu S, Defreyte G, Soloway PJ, Zile MR, Spinale FG. Effects of gene deletion of tissue inhibitor of the matrix metalloproteinase-type 1 (TIMP-1) on left ventricular geometry and function in mice. J Mol Cell Cardiol 2000; 32:109–120.
47. Kai H, Ikeda H, Yusakawa H, Kai M, Seki Y, Kuwahara F, Ueno T, Sugi K, Imaizumi T. Peripheral blood levels of matrix metalloproteinases-2 and -9 are elevated in patients with acute coronary syndromes. J Am Coll Cardiol 1998; 32:368–372.
48. Hojo Y, Ikeda U, Ueno S, Arakawa H, Shimada K. Expression of matrix metalloproteinases in patients with acute myocardial infarction. Jpn Circ J 2001; 65:71–75.
49. Inokubo Y, Hanada H, Ishizaka H, Fukushi T, Kamada T, Okumura K. Plasma levels of matrix metalloproteinase-9 and tissue inhibitor of metalloproteinase-1 are increased in the coronary circulation in patients with acute coronary syndrome. Am Heart J 2001; 141:211–217.
50. Bradham WS, Gunasinghe H, Holder JR, Multani M, Killip D, Anderson M, Meyer D, Spencer WH, Torre-Amione G, Spinale FG. Release of matrix metalloproteinases following alcohol septal ablation in hypertrophic obstructive cardiomyopathy. J Am Coll Cardiol 2002; 40:2165–2173.

51. Rhode LE, Ducharme A, Arroyo LH, Aikawa M, Sukhova GH, Lopez-Anaya A, McClure KF, Mitchell PG, Libby P, Lee RT. Matrix metalloproteinase inhibition attenuates early left ventricular enlargement after experimental myocardial infarction in mice. Circulation 1999; 99:3063–3070.

52. Ducharme A, Frantz S, Aikawa M, Rabkin E, Lindsey M, Rohde LE, Schoen FJ, Kelly RA, Werb Z, Libby P, Lee RT. Targeted deletion of matrix metalloproteinase-9 attenuates left ventricular enlargement and collagen accumulation after experimental myocardial infarction. J Clin Invest 2000; 106:55–62.

53. Peterson TJ, Li H, Dillon L, Bryant JW. Evolution of matrix metalloproteinase and tissue inhibitor expression during heart failure progression in the infarcted rat. Cardiovasc Res 2000; 46:307–315.

54. Heymans S, Luttun A, Nuyens D, Theilmeier G, Creemers E, Moons L, Dyspersin GD, Cleutjens JP, Shipley M, Angellilo A, et al. Inhibition of plasminogen activators or matrix metalloproteinases prevents cardiac rupture but impairs therapeutic angiogenesis and causes cardiac failure. Nat Med 1999; 5(10):1135–1142.

55. Hayashidani S, Tsutsui H, Ikeuchi M, Shiomi T, Matsusaka H, Kubota T, Imanaka-Yoshida K, Itoh T, Takeshita A. Targeted deletion of MMP-2 attenuates early LV rupture and late remodeling after experimental myocardial infarction. Am J Physiol Heart Circ Physiol 2003; 285(3):H1229–H1235.

56. Creemers EE, Davis JN, Parkhurst AM, Leenders P, Dowdy KB, Hapke E, Hauet AM, Escobar PG, Cleutjens JP, Smits JF, Daemen MJ, Zile MR, Spinale FG. Deficiency of TIMP-1 exacerbates LV remodeling after myocardial infarction in mice. Am J Physiol Heart Circ Physiol 2003; 284(1):H364–H471.

57. Mukherjee R, Brinsa TA, Dowdy KB, Scott AA, Baskin JM, Deschamps AM, Lowry AS, Escobar P, Lucas DG, Yarbrough WM, Zile MR, Spinale FG. Myocardial infarct expansion and matrix metalloproteinase inhibition. Circulation 2003; 107:618–625.

58. Lindsey ML, Gannon J, Aikawa M, Schoen FJ, Rabkin E, Lopresti-Morrow L, Crawford J, Black S, Libby P, Mitchell PG, Lee RT. Selective matrix metalloproteinase inhibition reduces left ventricular remodeling but does not inhibit angiogenesis after myocardial infarction. Circulation 2002; 105:753–758.

59. Yarbrough WM, Mukherjee R, Escobar GP, Mingoia JT, Sample JA, Hendrick JW, Dowdy KB, McLean JE, Lowry AS, O'Neill TP, Spinale FG. Selective targeting and timing of matrix metalloproteinase inhibition in post-myocardial infarction remodeling. Circulation 2003; 108(14):1753–1759.

60. Wilson EM, Moainie SL, Baskin JM, Lowry AS, Deschamps AM, Mukherjee R, Guy TS, St. John-Sutton MG, Gorman JH, Edmunds LH, Gorman RC, Spinale FG. Region- and type-specific induction of matrix metalloproteinases in post-myocardial infarction remodeling. Circulation 2003; 107:2857–2863.

61. Ungerstedt U. Microdialysis–principles and applications for studies in animals and man. J Intern Med 1991; 230:365–373.

62. Etoh T, Joffs C, Deschamps AM, Davis J, Dowdy K, Hendrick J, Baicu S, Mukherjee R, Manhaini M, Spinale FG. Myocardial and interstitial matrix metalloproteinase activity after acute myocardial infarction in pigs. Am J Physiol Heart Circ Physiol 2001; 281(3):H987–H994.

63. Baghelai K, Marktanner R, Dattilo JB, Dattilo MP, Jakoi ER, Yager DR, Makhoul RG Wechsler AS. Decreased expression of tissue inhibitor of metalloproteinase 1 in stunned myocardium. J Surg Res 1998; 77:35–39.

64. Yarbrough WM, Mukherjee R, Brinsa TA, Dowdy KB, Scott AA, Escobar GP, Johns C, Lucas DG, Crawford FA Jr, Spinale FG. Matrix metalloproteinase inhibition modifies left ventricular remodeling after myocardial infarction in pigs. J Thorac Cardiovasc Surg 2003; 125:602–610.

65. Rasmussen HS, McCann PP. Matrix metalloproteinase inhibition as a novel anticancer strategy: a review with special focus on batimastat and marimastat. Pharmacol Ther 1997; 75:69–75.

66. Hidalgo M, Eckhardt SG. Development of matrix metalloproteinase inhibitors in cancer therapy. J Natl Cancer Inst 2001; 93:178–193.

67. Spinale FG, Coker ML, Heung LJ, Bond BR, Gunasinghe HR, Etoh T, Goldberg AT, Zellner JL, Crumbley AJ. A matrix metalloproteinase induction/activation system exists in the human left ventricular myocardium and is upregulated in heart failure. Circulation 2000; 102:1944–1949.

68. Silverman HS, Pfeifer MP. Relation between use of anti-inflammatory agents and left ventricular free wall rupture during acute myocardial infarction. Am J Cardiol 1987; 59:363–364.

69. Deswal A, Bozkurt B, Seta Y, Parilti-Eiswirth S, Hayes FA, Blosch C, Mann DL. Safety and efficacy of a soluble P75 tumor necrosis factor receptor (Enbrel, Etanercept) in patients with advanced heart failure. Circulation 1999; 99:3224–3226.

70. Baliga RR, Narula J. Pharmacogenomics of congestive heart failure. Med Clin of North Amer 2003; 87:569–578.

71. Miake J, Marbán E, Nuss HB. Biological pacemaker created by gene transfer. Nature 2002; 419:132–133.

72. Agata J, Chao L, Chao J. Kallikrein gene delivery improves cardiac reserve and attenuates remodeling after myocardial infarction. Hypertension 2002; 40:653–659.

73. Jeong JO, Byun J, Jeon E-S, Gwon H-C, Lim YS, Park J, Yeo S-J, Lee YJ, Kim S, Kim D-K. Improved expression by cytomegalovirus promoter/enhancer and behavior of vascular endothelial growth factor gene after myocardial injection of naked DNA. Exper and Molec Med 2002; 34:278–284.

4

Cardiac Myocyte Structural Remodeling

A. Martin Gerdes and Maurice S. Holder
Cardiovascular Research Institute, University of South Dakota and Sioux Valley Hospital and Health Systems, Sioux Falls, South Dakota, U.S.A.

INTRODUCTION

The arrangement of myocytes within the ventricular wall, as well as the entire orientation of the muscle mass 'of the heart, provides the structural basis for ventricular contractile function. Beyond the normal excitation–contraction process, both intracellular signal transduction pathways and mechanical tethering via the cytoskeletal–integrin–extracellular matrix are among supportive mechanisms by which cardiac functional state can be adjusted and maintained. For instance, when any type of load is continuously applied to the heart, a remodeling process occurs during which permanent and widespread alterations in myocyte structure is evidenced along with significant alterations in normal cardiac physiology. This myocyte structural remodeling underlies the advancement of many common abnormalities of the heart, and is a hallmark feature of cardiac hypertrophy and congestive heart failure (CHF).

Although it is uncontested that cardiac hypertrophy involves structural remodeling of myocytes, in the progression to failure there are time-dependent functional limitations in this compensatory process to support systemic demands. In cardiac failure, the nature of the structural remodeling of myocytes bears important clinical consequences as the remodeling of myocytes is thought to compensate for a heart, the performance of which is inadequate and is now characterized by a dilated, relatively thin-walled left ventricle (1). Increasing evidence, however, suggests that maladaptive remodeling may actually lead to pump dysfunction. Intrinsic to this situation is an increase in chamber diameter accompanied by inadequate wall thickening leading to elevated systolic and diastolic wall stress as predicted by the Laplace equation. According to this equation (in simplified form), $\sigma = PR/W$, where σ is wall stress, P is pressure, R is radius, and W is wall thickness. Thus, increased end diastolic pressure ($\uparrow$P) and chamber dilatation ($\uparrow$R) coupled with a "relatively" thinner wall ($\downarrow$W) leads to a dramatic increase in diastolic wall stress in CHF. As systolic blood pressure tends to be in the normal range in CHF, the major contributors to elevated systolic wall stress are largely anatomical ($\uparrow$ chamber diameter or R coupled with a reduction in "relative" wall thickness or $\downarrow$W). Here it should be pointed out that wall thickness is typically normal or increased if hypertension was present prior to the onset of CHF. Consequently, the adjective "relative" should not be omitted. It should also be pointed out that myocytes within the wall of the failing heart are not only enlarged (2), but often surrounded by increased levels of collagen (3). The extracellular matrix, as discussed elsewhere in this book, is also relevant to the remodeling process. The primary objective of this chapter is to provide evidence for a clear understanding of how alterations in myocyte shape contributes to chamber remodeling in heart disease. An understanding of how and when changes in myocyte dimensions occur in the progression of heart diseases to congestive failure may provide important targets for new therapies.

TECHNICAL CONSIDERATIONS IN ASSESSMENT OF MYOCYTE SHAPE ALTERATIONS

The spatial arrangement of myocytes within the ventricular wall is such that cell length contributes to chamber circumference while cell diameter primarily contributes to wall thickness. A large body of data on cardiac myocyte shape alterations in heart disease has appeared in the literature over many decades.

In particular, a classic study by Roberts and Wearns (4) merits discussion. To this day, this study represents the most accurate and comprehensive study of its kind conducted in humans. A total of 77 hearts were obtained from children, normal adults, and patients with cardiac hypertrophy and atrophy. Great care was taken to minimize sources of error in collection of morphometric data (e.g., tissue shrinkage, sectioning angle). Hearts were obtained so quickly after death (less than one hour postmortem) that rhythmic contractions were established by physiologic solutions used for tissue preparation. Basic information regarding myocyte size alterations during growth, maturation, cardiac hypertrophy, and atrophy were characterized along with associated changes in capillarization. The investigators also recognized that myocytes are elliptical in cross-section and, consequently reported values for mean diameter based on the major and minor transverse axes. Their value of 14 µm for mean myocyte diameter from normal human hearts is very similar to the mean value we observed in three normal human hearts (17 µm) using more sophisticated technology (5). Clearly, a high level of care must be taken in morphometric investigations to ascertain meaningful and widely representative results as to the behavior of the human heart.

In much of the recent literature, characterization of myocyte size typically involves H&E sections with minimal attention to errors from tissue shrinkage or sectioning angle. With this approach, it is not unusual to see discrepancies of more than 10 folds between parameters that should change in parallel during cardiac hypertrophy [e.g., left-ventricular (LV) myocyte cross-sectional area and mass in compensated hypertrophy from hypertension]. We have found that H&E processing of isolated myocytes produces a very large and variable amount of shrinkage (A.M. Gerdes, unpublished data). This is probably why whole tissue sections typically show considerable spaces between myocytes. Indeed, it would be virtually impossible to discern myocyte boundaries consistently in standard 5 µm H&E sections without considerable shrinkage, because this is difficult even in optimally fixed 1 µm plastic sections of myocardium. Additional considerations with this and other approaches to quantitate myocyte shape alterations include: (i) variation in sectioning angle, which leads to overestimation of cross-sectional area; (ii) variation in contractile phase of whole tissue samples (e.g., contracted cells are thicker); (iii) measurements are often made through the nucleus only, which provides a large overestimation of mean cross-sectional area; and (iv) cross-sectional area data alone may not provide an adequate understanding of the degree of cellular hypertrophy. These inadequacies are best overcome by collecting comprehensive data on myocyte volume, cross-sectional area (cellular analog of wall thickness), and length (cellular analog of chamber dimension) using standardized methods of proven reliability (discussed below). Changes in myocyte volume should correlate reasonably well with alterations in heart mass unless there is extensive fibrosis or alteration in myocyte number. Myocyte volume is a critical parameter needed to estimate myocyte number. With recent reports suggesting myocyte loss from apoptosis (6) and the possibility of myocyte hyperplasia in heart failure (7,8), such data are becoming more important.

Reliable and comprehensive methods to assess alterations in myocyte dimensions have been documented (9). In that study, an excellent correlation was found between three independent methods of measuring myocyte dimensions. An unusual feature of the study was that appropriate corrections were applied for all known sources of error with the morphometric measurements. Validation of volume measurements of isolated myocytes using a Coulter Channelyzer marked a major technical advance as this method is considerably faster and more objective than traditional approaches. A very important, and often overlooked feature of the Coulter Channelyzer–isolated myocyte approach is that all data can be readily compared. For instance, it was shown that dimensions of cardiac myocytes from normal adult mammals (e.g., humans, rats, cats, guinea pigs, and hamsters) are virtually indistinguishable (Fig. 1) (9–11). It would not be possible to make such comparisons with other technical approaches, such as the H&E sectioning method mentioned above, where cross-sectional area values from comparable animals can differ by several folds between labs, and between different studies from the same lab (not to mention the inability of this approach to provide meaningful data on volume and length). When assessing isolated myocyte volume by Coulter Channelyzer, it is important to use high quality preps that yield a distinct second peak. An example of a myocyte volume Coulter Histogram is shown in Figure 2. The first sharp peak contains debris

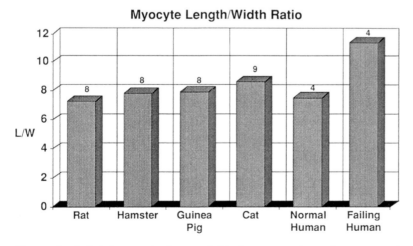

Figure 1 Isolated myocyte data available from normal rats, hamsters, guinea pigs, cats, and humans suggest that myocyte length/width ratio is regulated within a rather narrow range of approximately seven to nine in mammals. A hallmark feature of CHF is a substantial increase in myocyte length/width ratio, as indicated by human data shown here. Values are means with number of individuals per group indicated above bars. *Abbreviation*: CHF, congestive heart failure.

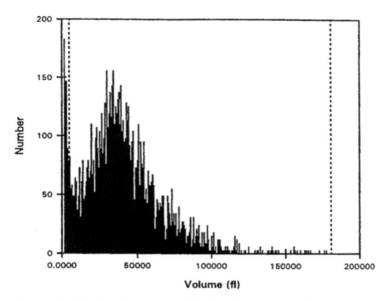

Figure 2 Typical histogram of isolated myocyte volumes generated by a Z2 Coulter Channelyzer. The first sharp peak is from debris and is excluded by the lower threshold indicated by the dotted line. The second dotted line indicates an assigned upper threshold. There is an insignificant number of counts above this line and they are likely to be cell clumps (e.g., 4× the median value). Cardiac myocytes are in the broad, normally distributed peak. The median value is used to represent cell volume. A shape factor of 1.5, representing a spherical cell, is built into the denominator of the equation in the software. Consequently, the median value should be adjusted to the appropriate shape factor for elongated myocytes (1.05) as indicated. Cells should be pipetted vigorously before each run to minimize clumps. Histograms from poor cell isolations often do not show a normal distribution (most likely because of myocyte fragments) and should not be used.

and it can be readily realized from calculations using myocyte length and width that these counts register well below those of the smallest myocytes within a given prep. Additionally, the Coulter Channelyzer software has a spherical shape factor (1.5) built into the formula, so this should be replaced by using a shape factor more representative of elongated cardiac myocytes (1.05). This entails multiplying myocyte volume by 1.43 (1.5/1.05) to replace the shape factor in the denominator of the equation (9). We report median values for myocyte volume as this minimizes the effect of myocyte clumps on the right side of the histogram (typically myocyte clumps represent very few counts if cells are vigorously pipetted).

ALTERATIONS IN MYOCYTE SHAPE IN HYPERTROPHY AND FAILURE

It was hypothesized and promoted in earlier years by Grossman et al. (12), and Linzbach (13) that alterations in myocyte shape, during cardiac hypertrophy, parallel changes in ventricular anatomy. Thus, in the presence of hypertension and other conditions in which there are sustained elevations in ventricular afterload concentric enlargement of the ventricle is present. This is characterized by an increase in wall thickness and myocyte cross-sectional area during the developmental and compensated phases of the disease process. The cellular changes derived from in vivo studies have invariably been supported by comprehensive data from studies using isolated myocytes [reviewed in Refs. (2,14)]. Increased volume load as is present with physiological growth (15), hyperthyroidism (16), and the opening of an aortocaval fistula (17) leads to proportional growth of myocyte length and diameter (translating to approximately two-thirds of the increase being due to cross-sectional area and the rest from increased length).

Comprehensive data collected from many mammalian species have suggested that myocyte length:width ratio (L/W) is normally regulated within a rather narrow range (approximately 7–9) in the normal heart (Fig. 1). Myocyte L/W declines slightly in situations of concentric hypertrophy, but remains relatively normal during volume overloading because of proportional myocyte growth. A deviation from this shape was noted when it was observed from human-isolated myocyte samples from the LV-free wall that myocyte L/W ratio increased dramatically in CHF owing to ischemic and dilated cardiomyopathy (11,18,19) (Fig. 1). Earlier studies of post-myocardial infarction (MI) hearts showed that remodeling of the spared LV plus septum in rats was due to an increase in myocyte length only with no change in cross-sectional area (20). Comparatively, it appears that cross-sectional area is normal in humans with heart failure and no prior history of hypertension. By using the same isolated myocyte techniques to examine post-MI remodeling in sheep, Kramer et al. (21) have observed more extensive remodeling in the peri-infarct region compared to remote spared myocardium. Again, the predominant change was lengthening of myocytes. As an increase in wall thickness (myocyte diameter) could help lower wall stress, it is possible that myocyte transverse growth is arrested under these conditions while series sarcomere addition is accentuated. The myocyte shape change occurring in the spared LV myocardium after MI is illustrated in Figure 3.

Data from patients with hypertension have suggested that myocyte cross-sectional area is almost double its normal value but does not change with progression to failure where cell lengthening is the predominant cellular change (22) (Fig. 3). It is not clear if transverse growth is arrested in this case or has simply reached an upper limit. Consistent with this possibility is the fact that cell volume would have to increase proportionately to surface membrane available for diffusion of nutrients and metabolites to be maintained in the face of the high metabolic rate of cardiac myocytes. This may provide considerable limitations to the growth response involved in remodeling. The landmark study by Page and McCallister (24) in aortic banded and hyperthyroid animals with modest hypertrophy supports this concept. They showed that cardiac myocytes are able to maintain a normal surface-to-volume relationship by increasing T tubular surface area disproportionately more than sarcolemmal surface area. It is not known whether this compensatory mechanism remains intact during the progression to failure.

Changes in myocyte shape have been temporally characterized in female spontaneously hypertensive heart failure (SHHF) rats over their life span (23) (Fig. 4). In these animals, early onset hypertension is superimposed on physiological growth during maturation. Myocyte cross-sectional area is about twice the normal values upon reaching adulthood, but the length is normal. After a period of compensated hypertrophy, it appears that length begins to increase between 6 and 12 months of age in these rats. Excessive addition of sarcomeres in series continues at a very slow rate until animals develop overt signs of CHF at about 24 months of age. Such data indicate that the onset of myocyte lengthening, without transverse growth, may be the critical early cellular event marking the transition of failure. It is likely that this slow, insidious process may occur over decades in human hypertensives.

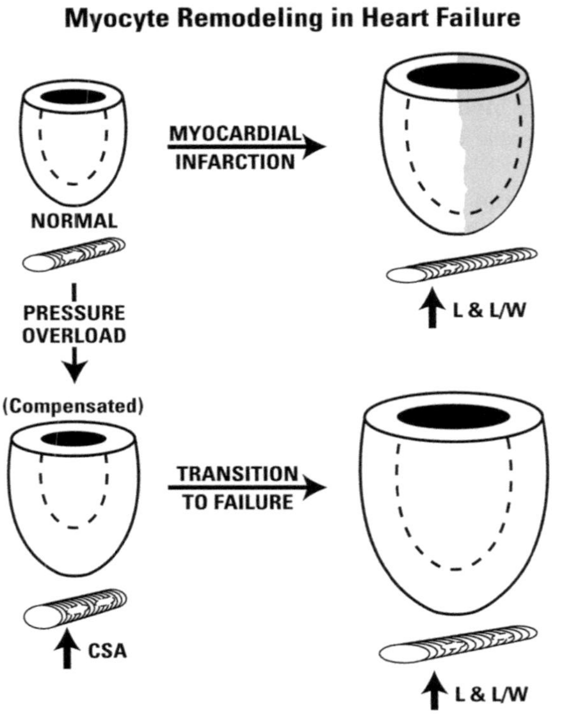

Myocyte Remodeling in Heart Failure

Figure 3 General changes in cardiac myocyte shape in progression to CHF are illustrated. Myocytes become more elongated and the ratio of myocyte L/W increases after myocardial infarction. There is little change in myocyte CSA. With pressure overload, CSA increases as the wall becomes thicker during the compensatory phase of hypertrophy. There is no further change in CSA with the transition to failure, but myocyte L and L/W increase substantially. *Abbreviations*: CHF, congestive heart failure; L/W, length to width, CSA, cross-sectional area. *Source*: From Ref. 2.

Interestingly, this cellular process may occur very rapidly in ischemic heart disease, but appears to progress much slower in hypertension. In both cases, the process begins well before the onset of symptomatic failure.

Important gender differences may predispose males to more rapid progression of remodeling; however, the general characteristics of the process appear to be the same. For instance, it appears that males, who typically have larger hearts than females, have a similar number of myocytes, but they are substantially larger than those from normal females (25). This is likely a result of the natural coexistent feature of larger body mass seen in males rather than a result of hormonal differences between males and females. Lean female SHHF rats have also been shown to develop failure six months later than lean SHHF males. From recent studies it appears that females, who have smaller myocytes initially, have more adaptive reserve for myocyte hypertrophy and are able to delay the onset of failure as a result of this gender difference in their heart composition (15,25). When females finally developed CHF, myocyte dimensions were similar to males who develop CHF earlier. With the onset of CHF, ejection fraction was better preserved in females versus males as is also the case in humans. It is likely that similar gender differences in myocyte size exist in humans but this is currently unknown.

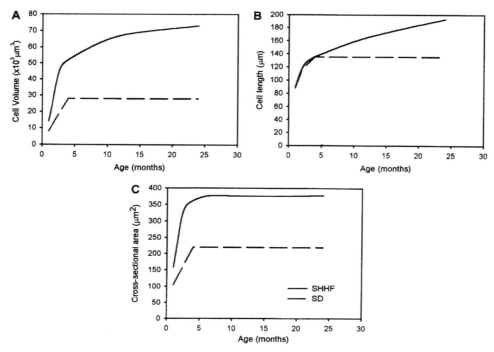

Figure 4 Temporal changes in LV myocyte shape in lean female SHHF rats over their life-time. The postnatal growth and maturation phase (zero to four months) is characterized by hypertension superimposed on normal physiological growth. Note, cell length is normal but hypertension has led to concentric hypertrophy at the myocyte level (e.g., increased myocyte volume owing to growth in cross-sectional area only). Over the remaining lifetime of the animal, there is a slow, progressive increase in myocyte length only because of series addition of new sarcomeres. The dashed line indicates the normal growth pattern found in female Sprague Dawley rats. *Abbreviation*: SHHF, spontaneously hypertensive heart failure. *Source*: From Ref. 23.

In my previous experience collecting isolated myocyte size data from the transplant program in Tampa, Florida, I was struck by the remarkable degree of heterogeneity in myocyte size from humans regardless of gender. Although mean LV myocyte size in humans was similar to that in other mammals, the range in mean values among patients was much greater than typically observed in animal studies. This should not be too surprising as animals are often selected from a defined colony with similar genetics. There was one failing human heart, in particular, that was worth noting (unpublished observation by author). Heart size was doubled in this patient with idiopathic dilated cardiomyopathy; yet myocyte volume was well below normal. It appeared that this patient might have had more than double the normal number of cardiac myocytes because there was minimal myocardial fibrosis. We believe the difference was genetic, as we have never observed a single bit of evidence supporting myocyte hyperplasia in adult myocytes in over 20 years of observations in thousands of animals with overloads of all types. Despite the potential survival advantage of having more cardiac myocytes, the patient had a dilated, failing heart. Of note, LV myocyte cross-sectional area was less than half normal so the myocyte LW ratio was well in the range observed in CHF (e.g., >10 in this case). There are clearly instances where a massive loss of cardiac myocytes may have been the primary underlying basis for pump dysfunction in chronic CHF. Indeed, we have recently confirmed that this is the case in BIO-TO2 cardiomyopathic hamsters (unpublished observation by the author). Nonetheless, we have observed that increased myocyte LW ratio is ubiquitous in CHF of any etiology. Our work also suggests that interventions to increase myocyte number, if possible and successful, may do little to improve heart function in the absence of improved myocyte and chamber geometry.

MOLECULAR MECHANISMS OF MYOCYTE LENGTHENING AND TRANSVERSE GROWTH

It is now clear that myocyte lengthening in the absence of transverse growth is a key early event in the transition to CHF. Many signaling pathways have been implicated in hypertrophic myocyte growth and excellent reviews address this topic in detail (26–29). In most cases, it is still not clear whether the primary stimulus for myocyte hypertrophy accompanying cardiovascular diseases is mechanical stretch or concurring increase in neurohormonal activation that activate the signaling pathways. Specific signaling pathways implicated in myocyte hypertrophy include protein kinase C, calcineurin, phosphoinositide-3 OH kinase (PI-3K), and G protein coupled receptor pathways. Additionally, signaling molecules, such as cardiotrophin-1 (30) and MEK 5 (31) have been linked to myocyte lengthening. Further work is needed, however, to determine if these or other signaling molecules are common triggers of pathological myocyte lengthening in progression to failure in humans and other animal models. Our findings regarding myocyte shape alterations in heart failure suggest that this avenue may eventually produce promising new targets for therapy. However, signal transduction related to myocyte hypertrophy is a very challenging field of research and is complicated by extensive redundancy and cross talk between the different signaling systems.

MYOCYTE SLIPPAGE

The phrase "myocyte slippage" is frequently used, but seldom adequately defined in the literature related to chamber dilatation in heart failure. This confusing phrase could be interpreted in several ways: (i) lateral side-to-side slipping of myocytes past one another in a transverse plane after disruption of endomysial collagen tethering (3); (ii) slippage of sheets of myocytes along fascial planes; or (iii) longitudinal slippage of myocytes past one another. The first situation, lateral side-to-side slippage, is what is most commonly intended by this phrase. The hypothesis of lateral side-to-side slippage, as first described in the English literature by Linzbach (13), implies that disruption of endomysial collagen struts may lead to this phenomenon. This hypothesis seems to ignore the fact that considerable branching occurs between myocytes and should impede this process. Surprisingly, illustrations from Linzbach's paper showed no lateral connections between myocytes when explaining slippage but showed extensive myocyte branching when explaining another concept. Scanning electron microscopy of human myocardial specimens by Yamamoto et al. (32) suggests that concentric hypertrophy leads to an increase in the number of these myocyte branches while eccentric hypertrophy leads to a reduction. Nonetheless, it seems difficult to envision a mechanism for

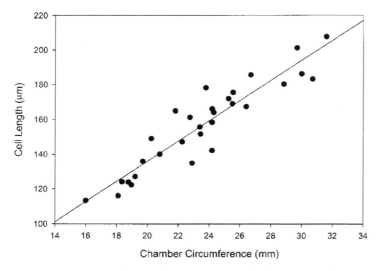

Figure 5 Myocyte lengthening was correlated with progressive chamber dilatation (chamber circumference) in SHHF from two months to 24 months of age (CHF stage). There is an approximate 1:1 relationship between cell length and chamber circumference, suggesting that cell lengthening can account for all chamber dilatation. In this experiment, myocytes were isolated from fixed tissue using KOH. This allowed simultaneous determination of cell length, chamber circumference, and sarcomere length (similar in all preps). These data suggest that myocyte slippage is not a necessary component of chamber dilatation in this model of CHF. *Abbreviations*: SHHF, spontaneously hypertensive heart failure; CHF, congestive heart failure. *Source*: From Ref. 33.

lateral side-to-side slippage considering the anatomical constraints. For the first time, a method accounting for myocyte length, sarcomere length, and chamber circumference simultaneously was used to demonstrate that myocyte lengthening from series sarcomere formation can account for all of the LV chamber dilatation in progression to failure in SHHF rats (33) (Fig. 5). In that study, there was an approximate doubling of myocyte length and chamber circumference over the rats' adult life (e.g., a 1:1 correlation from two to 24 months of age). Of note, this excellent linear correlation did not address the fact that myocytes are three-dimensional structures and change direction across the wall in a manner resembling a Japanese fan (34). In heart failure, myocytes also appear to increase their major transverse diameter, which tends to run circumferentially, and reduce their minor diameter, which tends to run transmurally (5). Thus, it appears that myocyte remodeling alone, without slippage, may sufficiently account for chamber dilatation in progression to failure. The extent to which myocyte lengthening of the spared segment in post-MI remodeling accounts for progressive chamber dilatation is not clear at this time, but this cellular change is clearly a major contributor. The concept that myocyte slippage contributes to chamber dilation in CHF has been repeated so often, it is largely accepted as fact. Nonetheless, I am not aware of any truly convincing data supporting this concept in a direct manner. For instance, a widely cited paper estimated changes in myocyte number across the wall using a complex formula involving myocyte patterns, cross-sectional area, and myocyte surface-to-volume ratio (35).

As we believe that it is virtually impossible to prove myocyte slippage with any current technology, our approach was to accurately account for everything, but myocyte slippage. At least in chronic hypertension progressing to failure, there is no need to invoke this concept to explain the cellular basis of chamber dilatation. Further work is needed to determine if this is also the case in chamber dilatation from other types of CHF. Finally, it should also be realized that extensive myocyte loss and replacement fibrosis could, theoretically, lead to chronic pump failure independent of myocyte lengthening but this has not been demonstrated to date.

SUMMARY

Changes in cardiac myocyte shape play a major role in progressive remodeling leading to CHF. Specifically, an increase in myocyte length from series sarcomere formation in the absence of myocyte transverse growth is likely the most important cellular change leading to cardiac dilatation in CHF. Furthermore, this cellular change alone may account for chamber dilatation in CHF. Meaningful improvements in the maladaptive chamber geometry present in CHF are not likely to occur without reversal of the change in cardiac myocyte shape. We have demonstrated pharmacologically in the SHHF animal model of CHF that it is possible to reverse myocyte shape and chamber remodeling nearly back to normal even when therapy is initiated near the onset of CHF (36). There is good reason to hope that this may some day become a reality for patients with CHF.

REFERENCES

1. Katz AM. Regression of left ventricular hypertrophy: new hope for dying hearts. Circulation 1998; 98:623–624.
2. Gerdes AM. Cardiac myocyte remodeling in hypertrophy and progression to failure. J Cardiac Failure 2002; 8:S264–S268.
3. Weber KT. Cardiac interstitium in health and disease: the fibrillar collagen network. J Am Coll Cardiol 1989; 13:1637–1652.
4. Roberts JT, Wearns JT. Quantitative changes in the capillary–muscle relationship in human hearts during normal growth and hypertrophy. Am Heart J 1941; 21:617–633.
5. Gerdes AM, Kellerman SE, Malec KB, Schocken DD. Transverse shape characteristics of cardiac myocytes from rats and humans. Cardioscience 1994; 5:31–36.
6. Olivetti G, Abbi R, Quaini F, Kajstura J, Cheng W, Nitahara JA, Quaini E, Di Loreto C, Beltrami CA, Krajewski S, Reed JC, Anversa P. Apoptosis in the failing human heart. N Engl J Med 1997; 336:1131–1141.
7. Beltrami AP, Urbanek K, Kajstura J, Yan SM, Finato N, Bussani R, Nadal-Ginard B, Silvestri F, Leri A, Beltrami CA, Anversa P. Evidence that human cardiac myocytes divide after myocardial infarction. N Engl J Med 2001; 344:1750–1757.
8. Kajstura J, Leri A, Finato N, Di Loreto C, Beltrami CA, Anversa P. Myocyte proliferation in end-stage cardiac failure in humans. Proc Natl Acad Sci 1998; 95:8801–8805.
9. Gerdes AM, Moore JA, Hines JM, Kirkland PA, Bishop SP. Regional differences in myocyte size in normal rat heart. Anat Rec 1986; 215:420–426.
10. Campbell SE, Gerdes AM, Smith TD. Comparison of regional differences in cardiac myocyte dimensions in rats, hamsters, and guinea pigs. Anat Rec 1987; 219:53–59.
11. Gerdes AM, Kellerman SE, Moore JA, Clark LC, Reaves PY, Malec KB, Muffly KE, McKeown PP, Schocken DD. Structural remodeling of cardiac myocytes from patients with chronic ischemic heart disease. Circulation 1992; 86:426–430.
12. Grossman W, Jones D, McLaurin LP. Wall stress and patterns of hypertrophy in the human left ventricle. J Clin Invest 1975; 56:56–64.

13. Linzbach AJ. Heart failure from the point of view of quantitative anatomy. Am J Cardiol 1960; 5:370–382.
14. Gerdes AM. The use of isolated myocytes to evaluate myocardial remodeling. TCM 1992; 2:152–155.
15. Bai S, Campbell SE, Moore JA, Morales MC, Gerdes AM. Influence of age, growth, and sex on cardiac myocyte size and number in rats. Anat Rec 1990; 226:207–212.
16. Gerdes AM, Moore JA, Hines JM. Regional changes in myocyte size and number in propranolol-treated hyperthyroid rats. Lab Invest 1987; 57:708–713.
17. Liu Z, Hilbelink DR, Gerdes AM. Regional changes in hemodynamics and cardiac myocyte size in rats with aortocaval fistulas II. Long-term effects. Circ Res 1991; 69:59–65.
18. Gerdes AM, Kellerman SE, Schocken DD. Implications of cardiomyocyte remodeling in heart dysfunction. In: Dhalla NS, Beamish RE, Takeda N, Nagano N, eds. The Failing Heart. New York: Raven Press, 1995:197–205.
19. Zafeiridis A, Jeevanandam V, Houser SR, Margulies KB. Regression of cellular hypertrophy after left ventricular assist device support. Circulation 1998; 98:656–662.
20. Zimmer HG, Gerdes AM, Lortet S, Mall G. Changes in heart function and cardiac cell size in rats with chronic myocardial infarctions. J Mol Cell Cardiol 1990; 22:1231–1243.
21. Kramer CM, Rogers WJ, Park CS, Seibel PS, Schaffer A, Theobald TM, Reichek N, Onodera T, Gerdes AM. Regional myocyte hypertrophy parallels regional myocardial dysfunction during post-infarction remodeling. J Mol Cell Cardiol 1998; 30:1773–1778.
22. Gerdes AM. Chronic ischemic heart disease. In: Weber KT, ed. Wound Healing Responses in Cardiovascular Disease. Mt Kisco, NY: Futura Pub. Co., 1995:61–66.
23. Onodera T, Tamura T, Said S, McCune SA, Gerdes AM. Maladaptive remodeling of cardiac myocyte shape begins long before failure in hypertension. Hypertension 1998; 32:753–757.
24. Page E, McCallister LP. Quantitative electron microscopic description of heart muscle cells. Application to normal, hypertrophied, and thyroxin-stimulated hearts. Am J Cardiol 1973; 31:172–181.
25. Tamura T, Said S, Gerdes AM. Gender-related differences in myocyte remodeling in progression to heart failure. Hypertension 1999; 33:676–680.
26. Frey N, Olson EN. Cardiac hypertrophy: the good the bad and the ugly. Ann Rev Physiol 2003; 65:45–79.
27. Molkentin JD, Dorn GW II. Cytoplasmic signaling pathways that regulate cardiac hypertrophy. Ann Rev Physiol 2001; 63:391–426.
28. Clerk AM, Sugden PH. Stimulation of p38 mitogen activated protein kinase pathway in neonatal rat ventricular myocytes by the G protein coupled receptor agonists endothelin I– and Phenylephrine: a role in cardiac myocyte hypertrophy? J cell Biol 1998; 142: 523–535.
29. Sadoshima J, Izumo S. Rapamycin selectively inhibits Angiotensin II induced increase in protein synthesis in cardiac myocytes in vitro. Potential role of 70 kD s6 in Angiotensin II induced cardiac hypertrophy. Circ Res 1995; 77:1040–1052.
30. Wollert KC, Taga T, Saito M, Narazaki M, Kishimoto T, Glembotski CC, Vernallis AB, Heath JK, Pennica D, Wood WI, Chien KR. Cardiotropin-1 activates a distinct form of cardiac muscle hypertrophy: assembly of sarcomeric units in series via gp 130/leukemia inhibitory factor receptor dependent pathways. J Biol Chem 1996; 271:9535–9545.
31. Nicol RL, Frey N, Pearson G, Cobb M, Richardson J, Olson EN. Activated MEK5 induces serial assembly of sarcomeres and eccentric cardiac hypertrophy. EMBO J 2001; 20:2757–2767.
32. Yamamoto S, James TN, Sawada K, Okabe M, Kawamura K. Generation of new intercellular junctions between cardiocytes. A possible mechanism compensating for mechanical overload in the hypertrophied human adult myocardium. Circ Res 1996; 78:362–370.
33. Tamura T, Onodera T, Said S, Gerdes AM. Correlation of myocyte lengthening to chamber dilation in the spontaneously hypertensive heart failure (SHHF) rat. J Mol Cell Cardiol 1998; 30:2175–2181.

34. Streeter DD, Hanna WT. Engineering mechanics for successive states in canine left ventricular myocardium. II. Fiber angle and sarcomere length. Circ Res 1973; 33:656–664.
35. Olivetti G, Capasso JM, Sonnenblick EH, Anversa P. Side-to-side slippage of myocytes participates in ventricular wall remodeling acutely after myocardial infarction in rats. Circ Res 1990; 67:23–34.
36. Tamura T, Said S, Harris J, Lu W, Gerdes AM. Reverse remodeling of cardiac myocyte hypertrophy in hypertension and failure by targeting of the renin-angiotensin system. Circulation 2000; 102:253–259.

5

Myofilament Remodeling in the Failing Human Heart

Peter VanBuren and Martin M. LeWinter
Department of Medicine and Molecular Physiology and Biophysics,
University of Vermont, Burlington, Vermont, U.S.A.

INTRODUCTION

The occurrence of myofilament remodeling in heart failure was first suggested over 40 years ago by the landmark observation of Gordon and Alpert that myofibrillar ATPase activity is markedly reduced in tissues obtained from failing human myocardium (1). This observation was subsequently confirmed by others (2–4). This finding indicates that, independent of numerous other functional changes, failing myocardium has an intrinsically reduced rate of cross-bridge formation, which contributes to impaired contractility and reduced power generation. An additional and favorable functional effect of slowed myofibrillar ATPase activity is an alteration in the mechano-energetics of contraction consisting of increased economy and efficiency of contraction, i.e., less chemical energy (ATP) is used to generate a unit of mechanical output (5–7).

Subsequently, numerous studies have been intended for understanding the modulation of cross-bridge cycling and the myofibrillar ATPase activity and explaining the mechanism or mechanisms of their depression in failing myocardium. Based on these studies, it is clear that the alterations in a number of myofilament proteins can influence the myofibrillar ATPase rate and remodeling, i.e., a change in the composition of these proteins, is indeed an important cause of this fundamental derangement. Figure 1 illustrates the composition and structural relationship of the myofilament's thick and thin filaments. This review will focus on changes in the thick filament (myosin) and the thin filament (actin, tropomyosin and the troponin complex) that contribute to the contractile defect in acquired heart failure.

THE THICK FILAMENT

Myosin as the Molecular Motor

The fundamental contractile unit of myofibril is the sarcomere in which inter-digitated thick and thin filaments slide past one another to give rise to muscle

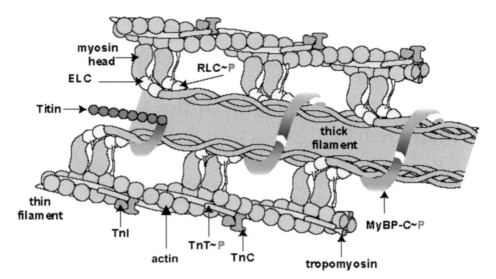

Figure 1 Myofilament's thick and thin filaments. Myosin cross-bridge heads project from the thick filament to interact with actin on the thin filament. The alpha-helical neck of the myosin cross-bridge is supported by ELC and RLC. The other thick filament–associated proteins include MyBP-C and titin. The thin filament contains actin, tropomyosin and the three troponin subunits TnC, TnI, and TnT. Proteins with phosphorylation sites are indicated (~P). See text for discussion. *Abbreviations*: ELC, essential light chains; RLC, regulatory light chains; MyBP-C, myosin binding protein C; TnC, troponin C; TnI, troponin I; TnT, troponin T. (*See color insert.*)

contraction. The thick filament consists primarily of cardiac myosin, and also includes components of myosin binding protein C and titin. Each myosin molecule is composed of two heavy chains and four light chains. Projections from the thick filament comprise the myosin head, a globular domain containing both the actin binding interface and the ATP hydrolysis site that has been aptly termed as the motor domain. The myosin head is attached to the base of the thick filament by a long alpha-helical neck region, which is supported by the regulatory and essential light chains. (As discussed below, phosphorylation of the regulatory light chain is one way to modulate myofibrillar ATPase activity). The neck region has been termed the lever arm as it is hypothesized that conformational changes in the motor domain are amplified through movement of the lever arm thus translating power to the base of the thick filament. Power generation by the myofilament is determined by the force generated as the myosin cross-bridge attaches and undergoes a conformational change during its cycle and the kinetic rate transitions of cross-bridge attachment and detachment.

Myosin Isoform Performance and Expression

Relatively soon after the discovery of depressed myofibrillar ATPase activity in failing myocardium the existence of the two major isoforms of myosin heavy chain (MHC) was recognized (8,9), and designated as α- and β-MHC (also referred to as V1 and V3 myosin based on the electrophoretic mobility of whole myosin). The two isoforms differ markedly in their intrinsic ATPase rate, with α-MHC being two to three times faster than β-MHC (10). In small mammals (rabbits, rodents), β-MHC is the predominant fetal form. In contrast, α-MHC is the predominant form in adult rodents, while there is about 2:1 ratio of β- to α-MHC in adult rabbits. In response to hemodynamic overload and hypothyroidism, small mammals rapidly increase the proportion of β-MHC present in the left ventricle (7,8). Thus, for example, Dahl salt-sensitive rats fed a high-salt diet demonstrate a shift from ~10% to ~75% β-MHC as they develop heart failure. Hypothyroidism causes a rapid shift to essentially 100% β-MHC in small mammals (7,9), while hyperthyroidism causes a similarly rapid shift to 100% α-MHC (11).

The Role of Myosin Isoforms in Human Heart Failure

When first discovered, this alteration in myosin isoforms offered an attractive explanation for slowing of myofibrillar ATPase activity in failing myocardium. However, myosin isoform shifting does not fully explain observed alterations in myocardial and ventricular mechano-energetics in small mammals (5,7). Moreover, it was reported that normal adult human myocardium contains exclusively β-MHC isoform (2,3), precluding a shift in this direction in failing myocardium and apparently invalidating this as a mechanism of depressed ATPase activity. Further supporting the lack of a role for myosin isoform shifting in failing myocardium is the fact that actomyosin ATPase activity is normal in failing myocardium, i.e., the intrinsic ATPase activity of myosin is unchanged (3,12). In view of these findings, attention shifted away from MHC isoform shifting to other mechanisms of depressed myofibrillar ATPase activity in failing myocardium (as discussed below).

Recently, however, there has been a revival of interest in the possibility that myosin isoform shifting is a mechanism of contractile abnormalities in failing myocardium. Careful high resolution gel electrophoresis studies show that normal human myocardium does contain small amounts of α-MHC, ranging from few to about 7% of total MHC (13–15). Interestingly, the contribution of mRNA for α-MHC is much larger in normal myocardium, amounting to ~35% of total (16). Moreover, assessed at either the mRNA or protein level, α-MHC is down-regulated in failing myocardium to levels near zero, with a corresponding relative increase in β-MHC (13–16). Thus, human failing myocardium demonstrates a similar modulation in myosin isoforms expression as in small mammals, albeit to a much smaller degree.

Molecular Performance of Human Failing and Non-failing Cardiac Myosin

A key question in failing human myocardium is whether the MHC isoform changes present at the protein level are functionally significant. In addition to the small magnitude, the normal actomyosin ATPase rate noted above argues against this. To more directly test this at a molecular level, we recently performed studies in which the ability of myosin isolated from end-stage failing hearts to propel actin and generate force in vitro was compared with myosin from control hearts (15). In these studies, a motility assay was employed in which myosin molecules are adhered to a microscope coverslip. Fluorescent-labeled unregulated actin filaments are propelled by the myosin-coated surface in the presence of ATP. The resulting motion is quantified and is analogous to slack shortening in muscle fibers. Isometric force generation was also assessed by the addition of α-actinin in increasing concentrations to the medium (15). α-Actinin is bound to the coverslip and avidly binds to cardiac actin, thus retarding actin filament movement. The α-actinin concentration at which all motion ceases is thus a sensitive index of isometric force. The results of these studies (Fig. 2) demonstrate that the motility and force produced by myosin from failing human myocardium is indistinguishable from that produced by myosin from normal myocardium in vitro, and argue against a significant functional role for MHC isoform shifting. From the same tissue (15), we confirmed that there is a small percent of α-MHC in control myocardium and that there is a downregulation in failing myocardium. To further assess the functional consequences of small myosin isoform shifts (15), we used cardiac myosin isolated from rabbits to compare the in vitro behavior of 100% β-MHC (simulating failing human myocardium) with a mixture of 90% β-MHC:10% α-MHC (simulating normal human myocardium). As with human myocardium, there was no difference in the in vitro behavior of simulated normal and simulated failing MHC composition. In summary, these studies argue strongly that the small MHC isoform shifts in failing human myocardium do not significantly alter the fundamental features of cross-bridge cycling. However, it is important to remember that the in vitro methods we employed lack the normal structural components of the sarcomere. To completely elucidate the possible functional significance of small MHC isoform shifts in failing human myocardium, it will be important that additional studies be performed using preparations that retain normal sarcomeric structure. The potential alterations in thick filament composition associated with myocardial failure are summarized in Table 1.

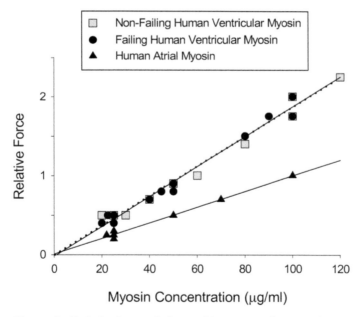

Figure 2   Relative isometric force of human cardiac myosin as a function of myosin surface concentration in the motility assay. Myosin isolated from failing ventricles generates the same force as myosin from non-failing ventricles. Ventricular tissue is comprised of predominantly V3 cardiac myosin (see text). Human atrial tissue contains approximately 70% V1 myosin and 30% V3 myosin. As V1 myosin is known to generate less force than V3 myosin (10), it is not surprising that myosin from atrial tissue generates less force than myosin from ventricular tissue. *Source*: From Ref. 15.

Table 1 Modulation of Contractile Protein Function

Protein	Isoform variation	Phosphorylation state
Myosin heavy chain	Yes[a,b]	No
Regulatory light chain (LC2)	No	Yes
Actin	Yes[b]	No
Troponin C	No	No
Troponin I	Yes[b]	Yes[c]
Troponin T	Yes[a,b]	Yes[c]
Tropomyosin	Yes[b]	No

[a]Abnormal variation reported in failing myocardium.
[b]Normal variation in fetal and/or adult life.
[c]Data reported suggestive of alterations in failing myocardium.

THE THIN FILAMENT

Evidence of Thin Filament-Mediated Alterations in Myofibrillar Performance in Human Heart Failure

As discussed above maximal myofibrillar ATPase activity is depressed by nearly 30% in human heart failure (1–4), which is not the case with myosin ATPase (2,3,12). Furthermore, Hasenfuss et al. (5,6) demonstrated that the isometric force–time-integral (cross-bridge force × attachment time) in failing human hearts is substantially increased, well out of proportion to any change in MHC isoform composition (Fig 3). When both mechanical and enzymatic data are considered, they suggest that the regulated interaction of myosin with the thin filament is altered in heart failure. In other words, the thin filament proteins may be directly involved in modulation of cross-bridge kinetics in human heart failure, resulting in a decrease in ATPase and an increase in the force–time integral. Below we present evidence that the regulatory proteins can significantly affect actomyosin's mechanical performance.

Current Concepts of Actomyosin Regulation

Thin Filament Structure

The thin filament is comprised of a double strand of filamentous actin. Tropomyosin (Tm) lies in the groove between these actin strands and spans seven actin monomers (17). The C-terminus of Tm overlaps the N-terminus of the adjacent Tm by 10 to 15 amino acids. Troponin (Tn) is an asymmetric molecule with a globular head and a long tail (18). Troponin contains three subunits: the calcium binding subunit (TnC), the inhibitory peptide (TnI), and the tropomyosin binding subunit (TnT). TnT comprises the tail of the troponin complex, with its N-terminal segment (amino acids 1–151) bound to the C-terminus of Tm and weakly to the N-terminus of the adjacent Tm (19). Crystallographic X-ray diffraction (21) and biochemical studies (22) reveal that the TnT tail segment binds to Tm both in the presence and absence of Ca^{2+} and, thus tethers the troponin complex to the thin filament during muscle activation. The C-terminal segment of TnT binds to TnI and TnC (23). In the absence of Ca^{2+}, TnI binds to actin and is weakly bound to TnC. In the presence of Ca^{2+}, TnC binds Ca^{2+} and undergoes a conformational change allowing strong binding to occur between TnC and TnI (24). With Ca^{2+} binding, TnI dissociates

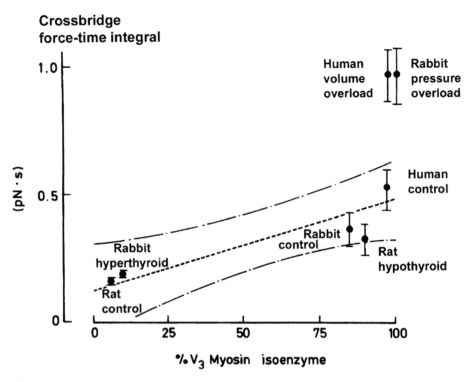

Figure 3 Myosin force–time-integral as a function of myosin isoform expression in human heart failure, animal models of myocardial failure, and differing thyroid states. While there is a linear correlation of the force–time-integral with the changes of myosin isoform expression associated with differing thyroid states, the increase in force–time-integral in both rabbit and human myocardial failure is out of proportion to the change in myosin isoform expression. These data support the concept of a thin filament–mediated defect in human heart failure (see text for discussion). *Source*: From Ref. 5.

from actin (20) and Tm movement on the thin filament occurs (25,26), allowing myosin to bind to actin. The potential alterations in thin filament composition associated with myocardial failure are summarized in Table 1.

Thin Filament Activation

In 1993, McKillop and Geeves proposed a three-state model of thin filament activation in which the thin filament is blocked, closed, or open (27). In the absence of Ca^{2+}, the thin filament is said to be in a "blocked state" as myosin binding is prohibited by Tm (28). In the "closed state," myosin is capable of binding to the thin filament in a weakly bound, nonforce generating state. Myosin is then able to undergo transition to a strongly bound, force generating state. This transition facilitates movement of Tm on the thin filament exposing other myosin strong-binding sites and leading to full thin filament activation, termed the "open state." We and others have demonstrated that myosin strong-binding plays a significant role in thin filament activation (24,29–33). The relative role of myosin strong-binding and calcium activation of the thin filament is illustrated in Figure 4. Thus, the thin filament appears to be only partially activated by Ca^{2+} binding; cooperative myosin strong

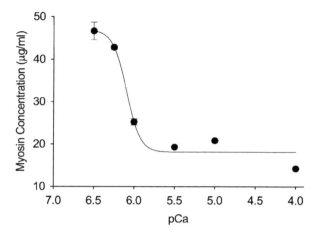

Figure 4 The relative role of myosin strong binding and calcium activation of the thin filament as determined in the in vitro motility assay. The concentration of myosin required to half-maximally activate the thin filament is an inverse function of the calcium concentration in the motility solution. Specifically, greater amounts of myosin are required to activate the thin filament at lower calcium concentrations than at higher calcium concentrations. While myosin strong binding is required to activate the thin filament at all calcium levels, the relative role of myosin strong binding appears to be the greatest at submaximal calcium. *Source*: From Ref. 37.

binding is required to switch the thin filament into the fully activated state (29,34). Cooperative activation is currently thought to be mediated along the thin filament through the overlap of adjacent Tms (35). Tn facilitates allosteric cooperativity of the thin filament, possibly through the N-terminal binding of TnT to the Tm overlap segment (20). Furthermore, there is direct evidence that the kinetics of the myosin cross-bridge can be modulated as a function of thin filament activation (36,37), giving credence to the concept that alterations in thin filament proteins may modulate contractile function in myocardial failure.

Alterations in Thin Filament Activation in Heart Failure

In several animal models and in human heart failure, altered myocardial Ca^{2+} sensitivity and/or altered maximal activation have been reported. Calcium sensitivity appears to be altered in both idiopathic dilated cardiomyopathy (DCM) and ischemic cardiomyopathy (ICM). In a rat myocardial infarction model, Ca^{2+} sensitivity of the noninfarcted myocardium is reduced (38). In contrast, both in a dog pacing heart failure model and human DCM, Ca^{2+} sensitivity was found to be increased (39,40). The authors of the latter studies postulated that increased Ca^{2+} sensitivity is related to reduced protein kinase A (PKA) activity.

Other investigators using both DCM and ICM human tissue have reported no change in peak tension or Ca^{2+} sensitivity under baseline conditions, but a differential response to protein kinase C (PKC) stimulation in failing myocardium, suggesting a possible thin filament based modulation of myocardial performance. Specifically, myofilament activation can be modulated in both failing and nonfailing human myocardium by stretch (Frank-Starling relation), and α-adrenergic and endothelin stimulation (41), all of which are postulated to occur by increasing Ca^{2+} responsiveness of the thin filament (42). These studies indirectly implicate a thin filament mediated alteration in mechanical performance in the failing heart.

Changes in Thin Filament Protein Expression in Human Heart Failure

Changes in Actin Isoform

The human heart predominantly expresses the α-cardiac and α-skeletal actin isoforms. There is a large shift in actin isoform expression (α-cardiac to α-skeletal) in the developing human heart (43). In addition, increased expression of α-skeletal actin mRNA has been reported in a rat myocardial infarction model (44,45). However, no changes in actin isoform expression levels have been demonstrated in end-stage human cardiomyopathy tissue compared to age matched controls (43,46). Interestingly, an increase in α smooth muscle actin (a fetal isoform) has been detected in human hibernating myocardium (47). While actin isoforms do not appear to exhibit altered mechanical effects in an unregulated system (48), actin isoform shifts in regulated systems may affect the kinetics of the cross-bridge cycle (49).

Changes in Regulatory Protein Expression

There is only one isoform of cardiac TnC. While Tm, TnI, and TnT manifest developmental isoform variation (50–54), no isoform shifts of either TnI or Tm at the protein level have been detected to date in human or experimental heart failure (50,53,54). In ICM, TnI degradation, as demonstrated in rat models of ischemia (55,56) and patients with unstable ischemic syndromes (57), could potentially play a role. In contrast, there are now a number of reports of altered TnT isoform expression in cardiac failure (4,58–61).

TnT Isoform Shifts

In several animal models and in various forms of human heart failure a TnT isoform shift has been described (4,39,58,60,62–64). In most cases there is expression of a fetal isoform in the disease state (4,39,58,60,62–65). Isoform variation of TnT is the result of alternative splicing of a single gene product, specifically 15 and/or 30 nt exons in the glutamate rich, highly charged N-terminal segment (58,66–68). Anderson et al. first reported a TnT isoform shift in human heart failure (both DCM and ICM), consisting of a modest increase in expression of a lower molecular weight, fetal isoform (TnT$_4$) compared with normal human myocardium ($12.0 \pm 4.9\%$ vs. $4.4 \pm 2.6\%$ p < 0.004). The increased expression of TnT$_4$ was correlated with a reduction in maximum Ca^{2+} activated myofibrillar ATPase (62). Subsequent studies have confirmed a TnT isoform shift in human heart failure (62,65), including ICM (39); however, others have failed to document this finding consistently (69). In patients with congenital heart disease, the degree of clinical heart failure correlates with the percent of TnT$_4$ isoform expression (64). We have demonstrated TnT isoform shifts in rabbit and rat models of hypertrophy and failure (7,61). In the rat, these shifts were temporally correlated with a reduction in maximum Ca^{2+} activated myofibrillar ATPase.

Two recent studies have directly examined the effect of troponin T isoforms on thin filament function. We assessed the two bovine adult cardiac troponin T isoforms (similar to that expressed in human heart failure) in the in vitro motility assay. These experiments did not demonstrate a significant difference in calcium:force or calcium:velocity relations (Fig. 5) (70). A subsequent study by Potter and colleagues demonstrated a small shift in calcium sensitivity for force in skinned fibers in which troponin complexes containing the different troponin T isoforms were exchanged (71). However, in light of the small troponin T isoform shift in human heart failure (62) and the modest functional differences of the two isoforms, it is unlikely that a troponin T isoform shift represents the primary cause of myofibrillar dysfunction in human heart failure. Thus,

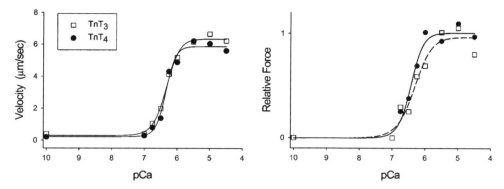

Figure 5 pCa:velocity and pCa:force relations for thin filaments reconstituted with the two adult beef cardiac troponin T isoforms (TnT_3 and TnT_4). An increase in TnT_4 expression has been demonstrated in human cardiac failure. No difference in function was detected for thin filaments containing all TnT_3 when compared to TnT_4. A small but nonsignificant increase in calcium sensitivity for force was detected for TnT_4. *Source*: From Ref. 70.

troponin T isoform shifting may simply reflect general activation of a more fetal gene program in myocardial failure (72).

Phosphorylation of Contractile Proteins: Possible Effect on Function

Factors other than isoform shifts may directly contribute to alterations in Ca^{2+} sensitivity and ATPase activity in human heart failure (42,69,73). An important potential contributor is contractile protein phosphorylation, which can be altered as a result of changes in activity of kinases and/or phosphatases.

Protein Kinase A

c-AMP–dependent PKA phosphorylates TnI and myosin binding protein C (C-protein) (74–79). Phosphorylation of TnI has been considered to be much more important functionally, and occurs at two adjacent serine residues near the N-terminus (75,76,78). TnI phosphorylation reduces the pCa_{50} for force and ATPase activity in skinned fibers, but does not alter maximum Ca^{2+} activated force and ATPase (74–79). Whether PKA phosphorylation influences unloaded shortening velocity or actomyosin kinetics is controversial (80–85). PKA phosphorylation of TnI has been shown to reduce TnC calcium affinity (86) and is felt to be the primary effect of PKA on thin filament function. Such a mechanism could account for PKA-mediated decreased calcium sensitivity for force and ATPase without affecting on maximal force or ATPase activity in skinned preparations (74–79). However, others have demonstrated an increase in myosin cross-bridge cycling rate (84,87), which suggests that PKA may also directly affect myosin cross-bridge kinetics.

Evidence for decreased PKA activity in heart failure is observed in failing dog and human myocardium, where an increase in pCa_{50}, but no difference in maximum Ca^{2+} activated tension in skinned preparations was reported (39,40). Treatment with the catalytic subunit of PKA normalized the tension-pCa relation. Bartel et al. (88) reported reduced in vitro TnI phosphorylation by isoproterenol in failing human myocardium implicating a PKA mediated defect. Moreover, TnI phosphorylation appears to be reduced in human heart failure (73,88). Consistent with these findings is the fact that PKA expression is reduced in human myocardial failure (89). These studies in aggregate provide compelling evidence that PKA activity is likely diminished in heart failure and likely contributes to altered myofilament function.

Protein Kinase C

PKC phosphorylates TnI (primarily at different sites than PKA), TnT and LC_2 (90–92). Phosphorylation of TnI/TnT by PKC decreases maximum Ca^{2+} stimulated actomyosin ATPase activity without changing pCa_{50} (92). Thus the effects of PKC activation on maximal ATPase are distinctly different from those produced by PKA. The PKC phosphorylatable sites on TnI are Ser 43, Ser 45 and Thr 143 (93–95). Ser 43, in the N-terminal region, appears to play a role in the binding of TnI to TnT as well as to TnC (93). Noland and Kuo (92) suggest that phosphorylation of Thr 143 and Ser 43/45 are the sites primarily responsible for the decrease in actomyosin ATPase activity with no change in pCa_{50}. In contrast, Endoh and Blinks (96) reported that α-adrenergic stimulation (presumably phosphorylating via a PKC mechanism) increases the myofilament sensitivity with Ca^{2+}. PKC dependent phosphorylation of TnT has recently been shown to inhibit TnT binding to Tm and thus inhibit ATPase activity (97). The net effect of nonspecific PKC activation is depression of maximal myofibrillar ATPase and contractile performance (92,98). As is illustrated in Figure 6, we recently demonstrated altered thin filament function in the Dahl salt sensitive rat model of myocardial failure (99).

Specifically, a decrease in thin filament calcium sensitivity is observed in failing thin filaments, which is ameliorated with treatment by the endothelin receptor blocker bosentan suggesting a PKC mediated mechanism. To date, preliminary human data indicate that PKC phosphorylation of the thin filament may be a key mechanism in the contractile deficit of the myofilament (100).

In contrast to PKC effects on the thin filament, PKC phosphorylation of myosin LC_2 increases actomyosin ATPase activity (90,91). LC_2 is also phosphorylated by Ca^{2+}/calmodulin dependent myosin LC kinase (MLCK). Phosphorylation

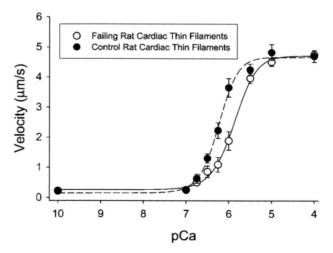

Figure 6 pCa:velocity relations for native thin filaments isolated from failing and normal Dahl salt sensitive rat myocardium. Thin filaments isolated from failing ventricles demonstrate a decrease in calcium sensitivity when compared to control ventricles. In this model of failure, a PKC-mediated modulation of thin filament function has been implicated. *Abbreviation*: PKC, protein kinase C. *Source*: From Ref. 99.

of LC_2 appears to increase maximal ATPase activity (91). The relative physiologic importance of PKC versus MLCK phosphorylation of cardiac LC_2 is uncertain. Normal myosin ATPase and contractile function in failing human myocardium argues against a role for LC_2 phosphorylation in depressed cross-bridge kinetics in human heart failure. In addition, recent evidence demonstrates that phosphorylation of the thick filament protein, myosin binding protein C, by either PKA or PKC alters the myosin cross-bridge orientation in α-MHC, but not β-MHC (101). The fact that no effect on β-MHC was observed would suggest that myosin binding protein C phosphorylation in humans has little, if any, effect.

SUMMARY

There is good evidence that the myofilament is integrally involved in the contractile deficit of human myocardial failure. To date no specific culprit deficit has been identified. Isoform changes both at the thick and thin filament level have not been large enough to suspect that these are major factors in myofilament remodeling. There is mounting evidence to support the hypothesis that myofilament protein phosphorylation may be a key mechanism in the contractile deficit of heart failure. Serum angiotensin II and norepinephrine are markedly elevated in heart failure. Direct myocardial stimulation by these signaling proteins is known to activate intracellular signaling of PKC and PKA. The fact that angiotensin II and β-adrenergic blockade dramatically improve mortality in heart failure underscores the potential importance of blocking PKC and PKA signaling in the adverse remodeling myofilament by protein phosphorylation.

REFERENCES

1. Alpert NR, Gordon MS. Myofibrillar adenosine triphosphatase activity in congestive heart failure. Am J Physiol 1962; 202:940–946.
2. Pagani ED, Alousi AA, Grant AM, Older TM, Dziuban SWJ, Allen PD. Changes in myofibrillar content and Mg-ATPase activity in ventricular tissues from patients with heart failure caused by coronary artery disease, cardiomyopathy, or mitral valve insufficiency. Circ Res 1988; 63:380–385.
3. Alousi AA, Grant AM, Etzler JR, Cofer BR, Van DB, Melvin D. Reduced cardiac myofibrillar Mg-ATPase activity without changes in myosin isozymes in patients with end-stage heart failure. Mol Cell Biochem 1990; 96:79–88.
4. Anderson PA, Malouf NN, Oakeley AE, Pagani ED, Allen PD. Troponin T isoform expression in humans. A comparison among normal and failing adult heart, fetal heart, and adult and fetal skeletal muscle. Circ Res 1991; 69:1226–1233.
5. Hasenfuss G, Mulieri LA, Blanchard EM, Holubarsch C, Leavitt BJ, Ittleman F, Alpert NR. Energetics of isometric force development in control and volume-overload human myocardium. Comparison with animal species. Circ Res 1991; 68:836–846.
6. Hasenfuss G, Mulieri LA, Leavitt BJ, Allen PD, Haeberle JR, Alpert NR. Alteration of contractile function and excitation-contraction coupling in dilated cardiomyopathy. Circ Res 1992; 70:1225–1232.
7. Kameyama T, Chen Z, Bell SP, VanBuren P, Maughan D, LeWinter MM. Mechano-energetic alterations during the transition from cardiac hypertrophy to failure in Dahl salt-sensitive rats. Circulation 1998; 98:2919–2929.
8. Alpert NR, Mulieri LA. Increased myothermal economy of isometric force generation in compensated cardiac hypertrophy induced by pulmonary artery constriction in the

rabbit. A characterization of heat liberation in normal and hypertrophied right ventricular papillary muscles. Circ Res 1982; 50:491–500.

9. Holubarsch C, Goulette RP, Litten RZ, Martin BJ, Mulieri LA, Alpert NR. The economy of isometric force development, myosin isoenzyme pattern and myofibrillar ATPase activity in normal and hypothyroid rat myocardium. Circ Res 1985; 56:78–86.

10. VanBuren P, Harris DE, Alpert NR, Warshaw DM. Cardiac V1 and V3 myosins differ in their hydrolytic and mechanical activities in vitro. Circ Res 1995; 77:439–444.

11. Goto Y, Slinker BK, LeWinter MM. Decreased contractile efficiency and increased nonmechanical energy cost in hyperthyroid rabbit heart. Relation between O2 consumption and systolic pressure-volume area or force-time integral. Circ Res 1990; 66:999–1011.

12. Nguyen TT, Hayes E, Mulieri LA, Leavitt BJ, ter Keurs H, Alpert NR, Warshaw DM. Maximal actomyosin ATPase activity and in vitro myosin motility are unaltered in human mitral regurgitation heart failure. Circ Res 1996; 79:222–226.

13. Miyata S, Minobe W, Bristow MR, Leinwand LA. Myosin heavy chain isoform expression in the failing and nonfailing human heart. Circ Res 2000; 86:386–390.

14. Reiser PJ, Portman MA, Ning XH, Schomisch MC. Human cardiac myosin heavy chain isoforms in fetal and failing adult atria and ventricles. Am J Physiol Heart Circ Physiol 2001; 280:H1814–H1820.

15. Noguchi T, Camp P, Alix SL, Gorga JA, Begin KJ, Leavitt BJ, Ittleman FP, Alpert NR, LeWinter MM, VanBuren P. Myosin from failing and non-failing human ventricles exhibit similar contractile properties. J Mol Cell Cardiol 2003; 35:91–97.

16. Lowes BD, Minobe W, Abraham WT, Rizeq MN, Bohlmeyer TJ, Quaife RA, Roden RL, Dutcher DL, Robertson AD, Voelkel NF, Badesch DB, Groves BM, Gilbert EM, Bristow MR. Changes in gene expression in the intact human heart. Downregulation of alpha-myosin heavy chain in hypertrophied, failing ventricular myocardium. J Clin Invest 1997; 100:2315–2324.

17. Nakajima-Taniguchi C, Matsui H, Nagata S, Kishimoto T, Yamauchi-Takihara K. Novel missense mutation in alpha-tropomyosin gene found in Japanese patients with hypertrophic cardiomyopathy. J Mol Cell Cardiol 1995; 27:2053–2058.

18. Zot AS, Potter JD. Structural aspects of troponin-tropomyosin regulation of skeletal muscle contraction. Annu Rev Biophys Biophys Chem 1987; 16:535–559.

19. Brisson JR, Golosinska K, Smillie LB, Sykes BD. Interaction of tropomyosin and troponin T: a proton nuclear magnetic resonance study. Biochemistry 1986; 25:4548–4555.

20. Arner A, Strauss JD, Svensson C, Ruegg JC. Effects of troponin-I extraction with vanadate and of the calcium sensitizer EMD 53998 on the rate of force generation in skinned cardiac muscle. J Mol Cell Cardiol 1995; 27:615–623.

21. Krumholz HM, Baker DW, Ashton CM, Dunbar SB, Friesinger GC, Havranek EP, Hlatky MA, Konstam M, Ordin DL, Pina IL, Pitt B, Spertus JA. Evaluating quality of care for patients with heart failure [published erratum appears in Circulation 2000 Jun 27;101(25):2995]. Circulation 2000; 101:E122–E140.

22. Inoko M, Kihara Y, Sasayama S. Neurohumoral factors during transition from left ventricular hypertrophy to failure in Dahl salt-sensitive rats. Biochem Biophys Res Commun 1995; 206:814–820.

23. Pearlstone JR, Smillie LB. Troponin T fragments: physical properties and binding to troponin C. Can J Biochem 1978; 56:521–527.

24. Tobacman LS. Thin filament-mediated regulation of cardiac contraction. Annu Rev Physiol 1996; 58:447–481.

25. Lehman W, Craig R, Vibert P. Ca(2+)-induced tropomyosin movement in Limulus thin filaments revealed by three-dimensional reconstruction. Nature 1994; 368:65–67.

26. Wakabayashi T, Huxley HE, Amos LA, Klug A. Three-dimensional image reconstruction of actin-tropomyosin complex and actin-tropomyosin-troponin T-troponin I complex. J Mol Biol 1975; 93:477–497.

27. Apstein CS. Increased glycolytic substrate protection improves ischemic cardiac dysfunction and reduces injury. Am Heart J 2000; 139:S107–S114.

28. Lorenz M, Poole KJ, Popp D, Rosenbaum G, Holmes KC. An atomic model of the unregulated thin filament obtained by X-ray fiber diffraction on oriented actin-tropomyosin gels. J Mol Biol 1995; 246:108–119.

29. VanBuren P, Palmiter KA, Warshaw DM. Tropomyosin directly modulates actomyosin mechanical performance at the level of a single actin filament. Proc Natl Acad Sci USA 1999; 96:12,488–12,493.

30. Grabarek Z, Grabarek J, Leavis PC, Gergely J. Cooperative binding to the Ca^{2+}-specific sites of troponin C in regulated actin and actomyosin. J Biol Chem 1983; 258:14,098–14,102.

31. Goldman YE, Hibberd MG, Trentham DR. Relaxation of rabbit psoas muscle fibres from rigor by photochemical generation of adenosine-5'-triphosphate. J Physiol (Lond) 1984; 354:577–604.

32. Millar NC, Homsher E. The effect of phosphate and calcium on force generation in glycerinated rabbit skeletal muscle fibers. A steady-state and transient kinetic study. J Biol Chem 1990; 265:20,234–20,240.

33. Swartz DR, Moss RL, Greaser ML. Calcium alone does not fully activate the thin filament for S1 binding to rigor myofibrils. Biophys J 1996; 71:1891–1904.

34. Holmes KC. The actomyosin interaction and its control by tropomyosin. Biophys J 1995; 68:2S–5S.

35. Chalovich JM. Actin mediated regulation of muscle contraction. Pharmacol Ther 1992; 55:95–148.

36. Brenner B. Effect of Ca^{2+} on cross-bridge turnover kinetics in skinned single rabbit psoas fibers: implications for regulation of muscle contraction. Proc Natl Acad Sci USA 1988; 85:3265–3269.

37. Gorga JA, Fishbaugher DE, VanBuren P. Activation of the calcium regulated thin filament by myosin strong-binding. Biophys J 2003; 85:2484–2491.

38. Li P, Hofmann PA, Li B, Malhotra A, Cheng W, Sonnenblick EH, Meggs LG, Anversa P. Myocardial infarction alters myofilament calcium sensitivity and mechanical behavior of myocytes. Am J Physiol 1997; 272:H360–H370.

39. Wolff MR, Buck SH, Stoker SW, Greaser ML, Mentzer RM. Myofibrillar calcium sensitivity of isometric tension is increased in human dilated cardiomyopathies: role of altered beta-adrenergically mediated protein phosphorylation. J Clin Invest 1996; 98:167–176.

40. Wolff MR, Whitesell LF, Moss RL. Calcium sensitivity of isometric tension is increased in canine experimental heart failure. Circ Res 1995; 76:781–789.

41. Pieske B, Schlotthauer K, Schattmann J, Beyersdorf F, Martin J, Just H, Hasenfuss G. Ca(2+)-dependent and Ca(2+)-independent regulation of contractility in isolated human myocardium. Basic Res Cardiol 1997; 92(suppl 1):75–86.

42. Holubarsch C, Ruf T, Goldstein DJ, Ashton RC, Nickl W, Pieske B, Pioch K, Ludemann J, Wiesner S, Hasenfuss G, Posival H, Just H, Burkhoff D. Existence of the Frank-Starling mechanism in the failing human heart. Investigations on the organ, tissue, and sarcomere levels. Circulation 1996; 94:683–689.

43. Boheler KR, Carrier L, de IB, Allen PD, Komajda M, Mercadier JJ, Schwartz K. Skeletal actin mRNA increases in the human heart during ontogenic development and is the major isoform of control and failing adult hearts. J Clin Invest 1991; 88:323–330.

44. Meggs LG, Tillotson J, Huang H, Sonnenblick EH, Capasso JM, Anversa P. Noncoordinate regulation of alpha-1 adrenoreceptor coupling and reexpression of alpha skeletal actin in myocardial infarction-induced left ventricular failure in rats. J Clin Invest 1990; 86:1451–1458.

45. Hanatani A, Yoshiyama M, Kim S, Omura T, Toda I, Akioka K, Teragaki M, Takeuchi K, Iwao H, Takeda T. Inhibition by angiotensin II type 1 receptor antagonist of cardiac phenotypic modulation after myocardial infarction. J Mol Cell Cardiol 1995; 27:1905–1914.

46. Schwartz K, Carrier L, Lompre AM, Mercadier JJ, Boheler KR. Contractile proteins and sarcoplasmic reticulum calcium-ATPase gene expression in the hypertrophied and failing heart. Basic Res Cardiol 1992; 87(suppl 1):285–290.

47. Ausma J, Schaart G, Thone F, Shivalkar B, Flameng W, Depre C, Vanoverschelde JL, Ramaekers F, Borgers M. Chronic ischemic viable myocardium in man: aspects of dedifferentiation. Cardiovas Pathol 1995; 4:29–37.

48. Harris DE, Warshaw DM. Smooth and skeletal muscle actin are mechanically indistinguishable in the in vitro motility assay. Circ Res 1993; 72:219–224.

49. Hewett TE, Grupp IL, Grupp G, Robbins J. Alpha-skeletal actin is associated with increased contractility in the mouse heart. Circ Res 1994; 74:740–746.

50. Gao L, Kennedy JM, Solaro RJ. Differential expression of TnI and TnT isoforms in rabbit heart during the perinatal period and during cardiovascular stress. J Mol Cell Cardiol 1995; 27:541–550.

51. McAuliffe JJ, Gao LZ, Solaro RJ. Changes in myofibrillar activation and troponin C Ca^{2+} binding associated with troponin T isoform switching in developing rabbit heart. Circ Res 1990; 66:1204–1216.

52. Ausoni S, De NC, Moretti P, Gorza L, Schiaffino S. Developmental expression of rat cardiac troponin I mRNA. Development 1991; 112:1041–1051.

53. Sasse S, Brand NJ, Kyprianou P, Dhoot GK, Wade R, Arai M, Periasamy M, Yacoub MH, Barton PJ. Troponin I gene expression during human cardiac development and in end-stage heart failure. Circ Res 1993; 72:932–938.

54. Hunkeler NM, Kullman J, Murphy AM. Troponin I isoform expression in human heart. Circ Res 1991; 69:1409–1414.

55. Van EJ, Powers F, Law W, Larue C, Hodges RS, Solaro RJ. Breakdown and release of myofilament proteins during ischemia and ischemia/reperfusion in rat hearts: identification of degradation products and effects on the pCa-force relation. Circ Res 1998; 82:261–271.

56. Gao WD, Atar D, Liu Y, Perez NG, Murphy AM, Marban E. Role of troponin I proteolysis in the pathogenesis of stunned myocardium. Circ Res 1997; 80:393–399.

57. Antman EM, Tanasijevic MJ, Thompson B, Schactman M, McCabe CH, Cannon CP, Fischer GA, Fung AY, Thompson C, Wybenga D, Braunwald E. Cardiac-specific troponin I levels to predict the risk of mortality in patients with acute coronary syndromes. N Engl J Med 1996; 335:1342–1349.

58. Anderson PA, Greig A, Mark TM, Malouf NN, Oakeley AE, Ungerleider RM, Allen PD, Kay BK. Molecular basis of human cardiac troponin T isoforms expressed in the developing, adult, and failing heart. Circ Res 1995; 76:681–686.

59. Harris DE, Warshaw DM. Smooth and skeletal muscle myosin both exhibit low duty cycles at zero load in vitro. J Biol Chem 1993; 268:14,764–14,768.

60. Gulati J, Akella AB, Nikolic SD, Starc V, Siri F. Shifts in contractile regulatory protein subunits troponin T and troponin I in cardiac hypertrophy. Biochem Biophys Res Commun 1994; 202:384–390.

61. Chen Z, Higashiyama A, Yaku H, Bell S, Fabian J, Watkins MW, Schneider DJ, Maughan DW, LeWinter MM. Altered expression of troponin T isoforms in mild left ventricular hypertrophy in the rabbit. J Mol Cell Cardiol 1997; 29:2345–2354.

62. Anderson PA, Malouf NN, Oakeley AE, Pagani ED, Allen PD. Troponin T isoform expression in the normal and failing human left ventricle: a correlation with myofibrillar ATPase activity. Basic Res Cardiol 1992; 87(suppl 1):117–127.

63. Kihara Y, Sasayama S. Transition from compensatory hypertrophy to dilated failing left ventricle in Dahl-Iwai salt-sensitive rats. Am J Hypertens 1997; 10:78S–82S.

64. Saba Z, Nassar R, Ungerleider RM, Oakeley AE, Anderson PA. Cardiac troponin T isoform expression correlates with pathophysiological descriptors in patients who underwent corrective surgery for congenital heart disease. Circulation 1996; 94:472–476.

65. Molina MI, Kropp KE, Gulick J, Robbins J. The sequence of an embryonic myosin heavy chain gene and isolation of its corresponding cDNA. J Biol Chem 1987; 262: 6478–6488.

66. Greig A, Hirschberg Y, Anderson PA, Hainsworth C, Malouf NN, Oakeley AE, Kay BK. Molecular basis of cardiac troponin T isoform heterogeneity in rabbit heart. Circ Res 1994; 74:41–47.

67. Gautel M, Furst DO, Cocco A, Schiaffino S. Isoform transitions of the myosin binding protein C family in developing human and mouse muscles: lack of isoform transcomplementation in cardiac muscle. Circ Res 1998; 82:124–129.

68. Shao Q, Ren B, Zarain-Herzberg A, Ganguly PK, Dhalla NS. Captopril treatment improves the sarcoplasmic reticular Ca(2+) transport in heart failure due to myocardial infarction. J Mol Cell Cardiol 1999; 31:1663–1672.

69. Solaro RJ, Powers FM, Gao L, Gwathmey JK. Control of myofilament activation in heart failure. Circulation 1993; 87(suppl VII):38–43.

70. VanBuren P, Alix SL, Gorga JA, Begin KJ, LeWinter MM, Alpert NR. Cardiac troponin T isoforms demonstrate similar effects on mechanical performance in a regulated contractile system. Am J Physiol Heart Circ Physiol 2002; 282:H1665–H1671.

71. Gomes AV, Guzman G, Zhao J, Potter JD. Cardiac troponin T isoforms affect the Ca^{2+} sensitivity and inhibition of force development. Insights into the role of troponin T isoforms in the heart. J Biol Chem 2002; 277:35,341–35,349.

72. Schwartz K, Chassagne C, Boheler KR. The molecular biology of heart failure. J Am Coll Cardiol 1993; 22:30A–33A.

73. Bodor GS, Oakeley AE, Allen PD, Crimmins DL, Ladenson JH, Anderson PA. Troponin I phosphorylation in the normal and failing adult human heart. Circulation 1997; 96:1495–1500.

74. Wattanapermpool J, Guo X, Solaro RJ. The unique amino-terminal peptide of cardiac troponin I regulates myofibrillar activity only when it is phosphorylated. J Mol Cell Cardiol 1995; 27:1383–1391.

75. Garvey JL, Kranias EG, Solaro RJ. Phosphorylation of C-protein, troponin I and phospholamban in isolated rabbit hearts. Biochem J 1988; 249:709–714.

76. Holroyde MJ, Small DA, Howe E, Solaro RJ. Isolation of cardiac myofibrils and myosin light chains with in vivo levels of light chain phosphorylation. Biochim Biophys Acta 1979; 587:628–637.

77. Zhang R, Zhao J, Potter JD. Phosphorylation of both serine residues in cardiac troponin I is required to decrease the Ca^{2+} affinity of cardiac troponin C. J Biol Chem 1995; 270:30,773–30,780.

78. Zhang R, Zhao J, Mandveno A, Potter JD. Cardiac troponin I phosphorylation increases the rate of cardiac muscle relaxation. Circ Res 1995; 76:1028–1035.

79. Mope L, McClellan GB, Winegrad S. Calcium sensitivity of the contractile system and phosphorylation of troponin in hyperpermeable cardiac cells. J Gen Physiol 1980; 75:271–282.

80. Herron TJ, Korte FS, McDonald KS. Power output is increased after phosphorylation of myofibrillar proteins in rat skinned cardiac myocytes. Circ Res 2001; 89:1184–1190.

81. Hofmann PA, Lange JH. Effects of phosphorylation of troponin I and C protein on isometric tension and velocity of unloaded shortening in skinned single cardiac myocytes from rats. Circ Res 1994; 74:718–726.

82. Janssen PM, de Tombe PP. Protein kinase A does not alter unloaded velocity of sarcomere shortening in skinned rat cardiac trabeculae. Am J Physiol 1997; 273:H2415–H2422.

83. Strang KT, Sweitzer NK, Greaser ML, Moss RL. Beta-adrenergic receptor stimulation increases unloaded shortening velocity of skinned single ventricular myocytes from rats. Circ Res 1994; 74:542–549.

84. Hoh JF, Rossmanith GH, Kwan LJ, Hamilton AM. Adrenaline increases the rate of cycling of crossbridges in rat cardiac muscle as measured by pseudo-random binary noise-modulated perturbation analysis. Circ Res 1988; 62:452–461.

85. de Tombe P, ter Keurs H. Lack of effect of isoproterenol on unloaded velocity of sarcomere shortening in rat cardiac trabeculae. Circ Res 1991; 68:382–391.

86. Robertson SP, Johnson JD, Holroyde MJ, Kranias EG, Potter JD, Solaro RJ. The effect of troponin I phosphorylation on the Ca^{2+}-binding properties of the Ca^{2+}-regulatory site of bovine cardiac troponin. J Biol Chem 1982; 257:260–263.

87. Kentish JC, McCloskey DT, Layland J, Palmer S, Leiden JM, Martin AF, Solaro RJ. Phosphorylation of troponin I by protein kinase A accelerates relaxation and cross-bridge cycle kinetics in mouse ventricular muscle. Circ Res 2001; 88:1059–1065.

88. Bartel S, Stein B, Eschenhagen T, Mende U, Neumann J, Schmitz W, Krause EG, Karczewski P, Scholz H. Protein phosphorylation in isolated trabeculae from nonfailing and failing human hearts. Mol Cell Biochem 1996; 157:171–179.

89. Zakhary DR, Moravec CS, Bond M. Regulation of PKA binding to AKAPs in the heart: alterations in human heart failure. Circulation 2000; 101:1459–1464.

90. Venema RC, Raynor RL, Noland TAJ, Kuo JF. Role of protein kinase C in the phosphorylation of cardiac myosin light chain 2. Biochem J 1993; 294:401–406.

91. Noland TAJ, Kuo JF. Phosphorylation of cardiac myosin light chain 2 by protein kinase C and myosin light chain kinase increases Ca(2+)-stimulated actomyosin MgATPase activity. Biochem Biophys Res Commun 1993; 193:254–260.

92. Noland TAJ, Kuo JF. Protein kinase C phosphorylation of cardiac troponin I or troponin T inhibits Ca2(+)-stimulated actomyosin MgATPase activity. J Biol Chem 1991; 266:4974–4978.

93. Noland TAJ, Raynor RL, Kuo JF. Identification of sites phosphorylated in bovine cardiac troponin I and troponin T by protein kinase C and comparative substrate activity of synthetic peptides containing the phosphorylation sites. J Biol Chem 1989; 264: 20,778–20,785.

94. Clement O, Puceat M, Walsh MP, Vassort G. Protein kinase C enhances myosin light-chain kinase effects on force development and ATPase activity in rat single skinned cardiac cells. Biochem J 1992; 285:311–317.

95. Venema RC, Kuo JF. Protein kinase C-mediated phosphorylation of troponin I and C-protein in isolated myocardial cells is associated with inhibition of myofibrillar actomyosin MgATPase. J Biol Chem 1993; 268:2705–2711.

96. Endoh M, Blinks JR. Actions of sympathomimetic amines on the Ca^{2+} transients and contractions of rabbit myocardium: reciprocal changes in myofibrillar responsiveness to Ca^{2+} mediated through alpha- and beta-adrenoceptors. Circ Res 1988; 62:247–265.

97. Noland TAJ, Kuo JF. Protein kinase C phosphorylation of cardiac troponin T decreases Ca(2+)-dependent actomyosin MgATPase activity and troponin T binding to tropomyosin-F-actin complex. Biochem J 1992; 288:123–129.

98. Watson JE, Karmazyn M. Concentration-dependent effects of protein kinase C-activating and—nonactivating phorbol esters on myocardial contractility, coronary resistance, energy metabolism, prostacyclin synthesis, and ultrastructure in isolated rat hearts. Effects of amiloride. Circ Res 1991; 69:1114–1131.

99. Noguchi T, Kihara Y, Begin KJ, Gorga JA, Palmiter KA, LeWinter MM, VanBuren P. Altered myocardial thin-filament function in the failing Dahl salt-sensitive rat heart: amelioration by endothelin blockade. Circulation 2003; 107:630–635.

100. Noguchi T, LeWinter MM, VanBuren P. Thin filament based modulation of contractile performance in human heart failure; a potential role of PKC mediated phosphorylation. J Card Fail 2002; 8:S8.

101. Weisberg A, Winegrad S. Relation between crossbridge structure and actomyosin ATPase activity in rat heart. Circ Res 1998; 83:60–72.

6
Alterations in Cardiac Myocyte Cytoskeleton

Henk L. Granzier
Washington State University, Pullman, Washington, U.S.A.

INTRODUCTION

The cytoskeleton is an intricate network of filaments and accessory proteins that allows cardiac myocytes to withstand the large mechanical stresses that are experienced during each heart beat and maintains the structural integrity of the myocyte. The cytoskeleton is comprised of three main types of filaments: microtubules, intermediate filaments, and actin filaments (also known as microfilaments). In addition, the cytoskeleton consists of membrane associated proteins that link the cytoskeleton to the extracellular matrix and of the giant protein titin together with many titin-binding proteins constitutes a sarcomeric skeleton. The cytoskeleton is shown schematically in Figure 1. The cytoskeleton supports the fragile cell membranes, positions organelles, provides for intracellular transport, organizes myofilaments, and provides mechanical strength and structural integrity to the cardiac myocyte. The cytoskeleton also senses biomechanical stresses and strains and can respond to these by triggering signaling pathways that are involved in remodeling. The various cytoskeletal filaments are discussed below with a special emphasis on recent progress in understanding the structure and function of titin in cardiac function and disease.

MICROTUBULES

These are stiff cylindrical structures composed of tubulin subunits (a heterodimer of α-tubulin and β-tubulin) that polymerize head-to-tail to form 13 parallel protofilaments that surround a hollow core (1). In addition, the cell contains soluble tubulin monomers that are in equilibrium with polymerized tubulin (microtubules are highly dynamic and continuously undergo cycles of polymerization followed by disassembly). A well-established role of microtubules is to provide tracks along which motor molecules transport cargo throughout the cell (1).

In earlier studies Rappaport and Samuel (2) and Schaper et al. (3) showed that, in cardiac myocytes from normal hearts, microtubules form a fine network that is most prominent in the perinuclear regions and tends to be more sparse elsewhere.

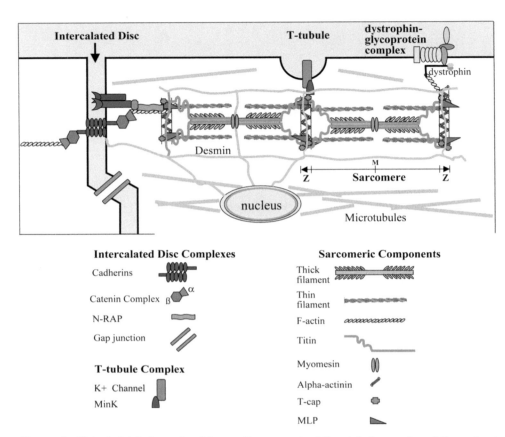

Figure 1 Cytoskeletal elements of the cardiac myocyte. Microtubules are found throughout the cytoplasm and are most dense near the nuclei. Desmin forms intermediate filaments that connect myofibrils laterally at the level of the Z-disk and M-line and that connect successive Z-disks longitudinally. They also link myofibrils to nuclei and cell membranes. F-actin filaments are part of the sarcomeric thin filaments and constitute various linking systems that connect Z-disks to the plasma membrane and intercalated disk. They are linked to the cadherins of the intercalated disks by interacting with catenins and to the DGC via binding to dystrophin. Titin spans half-sarcomeres, with its terminal ends overlapping in the Z-disk and at the M-line. Titin is associated with the other parts of the cytoskeleton at various locations. T-cap interacts with the Z-disk end of titin and with MLP, forming part of an MLP-dependent stretch sensing complex. MLP interacts with N-RAP, a protein that may be involved in mediating interactions between myofibrils and the cell membrane at adherens junctions through its possible interactions with cadherin-based protein complexes and/or integrin-associated vinculin. T-cap also interacts with the potassium channel subunit, minK, at T-tubules forming a complex that may be involved in the stretch-dependent regulation of potassium flux. *Abbreviations*: DGC, dystrophin–glycoprotein complex; MLP, muscle LIM protein. *Source*: From Ref. 101. (*See color insert.*)

In the early work it was also noted by Samuel et al. (4) that in a rat hypertrophy model the microtubular network was increased at the onset of hypertrophy and then normalized. Schaper et al. (3,5) examined microtubules in end-stage heart failure patients with dilated cardiomyopathy (DCM) and found more abundant microtubules in failing human myocardium. The role that microtubules might play in hypertrophy and cardiac failure has been investigated by Cooper et al. (6–9). Using a feline right ventricular banding model of hypertrophy, these authors revealed an increase in density of microtubules as well as in the total amount of tubulin (6,7). These changes were accompanied by contractile abnormalities of cardiac myocytes (reduction in extent of unloaded shortening) that were normalized by colchicine (a drug that depolymerizes microtubules). Notably, the authors showed that contractile abnormalities can be induced in control myocytes by promoting tubulin polymerization with Taxol® (6). Supportive results have been reported with different models, including a pig heart failure model (10), a guinea pig hypertrophy model (11), and a dog model of left ventricular (LV) hypertrophy (9). These findings have led to the proposal that microtubular networks can constitute an internal viscous load which leads to contractile dysfunction and that can ultimately contribute to heart failure. This mechanism, however, is unlikely to be universal because several groups have reported no effect of either depolymerizing microtubular networks (using colchicne) or polymerizing tubulin (using taxol) on passive behavior or contractile function during hypertrophy and heart failure (12–14). As suggested by Dr. ter Keurs (15) it is possible that the increased density of microtubular networks reflects mainly cellular growth (hypertrophy) and increased protein turnover, processes that require enhanced microtubular-based intracellular transport systems, and the increase in viscous load is a secondary effect that occurs in some circumstances. Considering that the increase in microtubular networks appears to be one of the hallmarks of human heart failure (5,16,17), its role in remodeling and contractile dysfunction during heart failure warrants further investigation.

INTERMEDIATE FILAMENTS

The main intermediate filament type in cardiac myocytes is desmin (18,19). Desmin monomers polymerize to long rope-like filaments (diameter of ~12 nm) that are likely to provide mechanical integrity and strength to the cardiac myocyte (1,18). Desmin filaments surround myofibrils and laterally connect adjacent Z-disks and M-lines, ensuring thereby that during contraction sarcomeres of adjacent myofibrils remain in register. In addition to connecting Z-disks of adjacent myofibrils, desmin filaments also connect successive Z-disks longitudinally (20). These longitudinally oriented desmin filaments are likely to bear force in highly stretched myocytes (21,22). Force levels are only significant at sarcomere lengths (SLs) that exceed the physiological range and desmin filaments might thereby provide a safety mechanism against over-stretch of sarcomeres (21,22). Desmin filaments also form physical links between the myofibrils, mitochondria, nuclei and the sarcolemma; this cytoplasmic network maintains the spatial relationship between the contractile apparatus and other organelles of the myocyte. Desmin is especially abundant at the intercalated disks (23,24) and contributes to a strong linking system that connects adjacent cells. Desmin constitutes ~2% of total protein in the heart (~0.35% in skeletal muscle) (19) a high level probably reflects the need for protection from the high stresses that are exerted on the myocytes with every heart beat. αB-crystallin is a small heat shock protein that colocalizes with desmin filaments and that prevents their irreversible aggregation (25,26).

Changes in the expression and distribution of desmin have been reported in many cardiac diseases. Progressive increases in both desmin protein levels and the desmin filament density in cardiac myocytes have been observed in the progression of chronic pressure overload cardiac hypertrophy and failure in guinea pigs (11,12).

Desmin was among the six most upregulated genes in an extensive study of familial hypertropic cardiomyopathic hearts (27). Several studies on human hearts have shown that in idiopathic hypertrophy and failure, desmin protein expression is upregulated and desmin filaments are disorganized (5,28). The effects of upregulation of desmin on structure and function of the heart has been studied in transgenic mouse lines in which wild-type desmin is overexpressed specifically in the heart (29). Desmin is elevated at the Z-disk and intercalated disk, but otherwise no structural abnormalities were noted. Isolated myocyte mechanics, isolated heart mechanics, and in vivo hemodynamics and echocardiography were all normal (29). Thus, it appears that upregulation of desmin expression reported in hypertrophy may reflect a need to counteract elevated wall stress (5,28), but is not in itself detrimental to the heart (25,29).

Several studies have revealed that missense mutations of the desmin gene (25,30,31) are associated with general myopathy involving both striated and smooth muscles. These desmin-related diseases are characterized by abnormal desmin aggregation and often have as main cause of death cardiomyopathy. Clinical phenotypes vary and the age of onset and disease progression are also variable. Proposed disease mechanisms include accumulation of misfolded desmin into toxic insoluble aggregates that ultimately destroy the cell, and loss of desmin function (31). Surprisingly, gene targeting studies have shown that mice can do without desmin during embryonic development and desmin-null animals are viable and fertile (32,33), suggesting limited functions for desmin during early stages of life (at least in the mouse).

Desmin-null mice ultimately do develop cardiac hypertrophy that is followed by the development of DCM with abnormalities that include myofibrillar misalignments and mitochondrial abnormalities (31), suggesting a more critical function of desmin later in life. Overall, evidence suggests that upregulation of wild-type desmin itself does not negatively impact the heart, that downregulation or absence of desmin results in cardiomyopathy later in life without apparent effects on development and viability, whereas misfolding of mutated desmin has a varied and often severe disease phenotype.

ACTIN FILAMENTS

Actin filaments (also known as F-actin) consist of actin monomers that polymerized head-to-tail to form slender filaments. In the sarcomere, F-actin filaments are associated with regulatory proteins to form thin filaments, and are crosslinked by α-actinin in the Z-disk (1). During contraction myosin interacts with actin and the resulting force is transmitted through the thin filament to the Z-disk. Actin filaments also link the terminal half-sarcomere of the cell to the intercalated disk, by binding to catenins which in turn bind to the cytoplasmic domain of caherins which mediate cell–cell attachment (34) (Fig. 1). Thus, actin filaments transmit force between adjacent sarcomeres and neighboring myocytes. Actin filaments also link to the sarcolemma in structural complexes known as costamers (present at the level of the Z-disk) by binding to the aminoterminus of dystrophin. Dystrophin is part of a complex that consists of peripheral and integral proteins, known as dystrophin–glycoprotein complex (DGC) (35,36). This complex forms a structural linkage between the cytoskeleton and the extracellular matrix, and provides mechanical support to the plasma membrane during contraction. Multiple mutations in the genes encoding DGC proteins have been shown to underlie DCM (36–38). That linking of the DGC to the actin-based cytoskeleton is important and is also shown by actin mutations (in subdomains 1 and 3) that reduce actin's ability to bind dystrophin and which give rise to DCM (39). These DCM causing mutations in actin are thought to result in disease by causing force transmission abnormalities. Actin mutations that lead to impaired myosin binding and a reduction in active force generation have also been reported and these result in familial hypertrophic cardiomyopathy (FHC) (40). Many of the genes implicated in FHC encoded proteins are involved in generation of force (myosin, cardiac troponin T, tropmysoin, MyBP-C, etc.) and it appears that the reduction in force generation stimulates hypertrophy as a compensatory response (37,38). Defects in force transmission, as for example actin mutations that affect dystrophin binding, cannot be compensated by hypertrophy (this increases the stress on linkages and may actually enhance force transmission deficits). Instead, the heart appears to respond to defects in force transmission by the serial addition of sarcomeres, which may be an attempt to lower active force (according to the Frank-Starling mechanism) and reduces stress on force transmission sites. However, this process results in chamber dilatation and ultimately DCM. Thus, actin performs critically important roles in force generation, with mutations that impact this role resulting in FHC, and in force transmission, with mutations that cause defects in force transmission giving rise to DCM.

TITIN

The main component of the sarcomeric cytoskeleton is the giant protein titin. Single titin molecules anchor in the Z-disk and extend all the way to the M-line region of the sarcomere (Fig. 2). Successive titin molecules are arranged head-to-head and tail-to-tail, providing a continuous filament along the full length of the myofibril (41). The majority of titin's I-band region is extensible and functions as a molecular spring that when extended develops passive force (41). Titin's passive force is largely responsible for myocardial passive stiffness within the physiological SL range, and together with collagen it determines the upper limit of the physiological SL range of the heart (42). Because each half thick filament (edge of A-band to M-line) contains its own set of titin filaments connecting it to the nearest Z-disk, titin's force also maintains the central position of the A-band within the sarcomere (43,44), ensuring efficient muscle contraction. Titin is also responsible for the so-called restoring force that is generated when sarocomeres shorten to below their slack length ($\sim$1.9 μm). This is made possible by the fact that the extensible region of titin is held away from the Z-disk by titin's inextensible near Z-disk region (this region binds strongly to the thin filament and can therefore withstand compressive forces) (45–47). As a result, when the sarcomere shortens to below the slack length, the thick filament moves into titin's near Z-disk region (48,49) and titin's extensible region is stretched in a direction opposite to that when the sarcomere is elongated above the slack length, generating the restoring force that pushes the Z-disks apart, towards their slack length position. Titin's restoring force is likely to be a factor in setting during systole, the lower SL limit, and in the elastic recoil that drives early diastolic filling (48). Thus, titin is important in maintaining the structural integrity of the sarcomere, setting the physiological SL range, and in diastolic filling.

Cardiac Titin Isoforms

There is only a single titin gene (located on chromosome 2) and titin isoforms derive from alternative splicing (50). In humans, the titin gene contains 363 exons that code

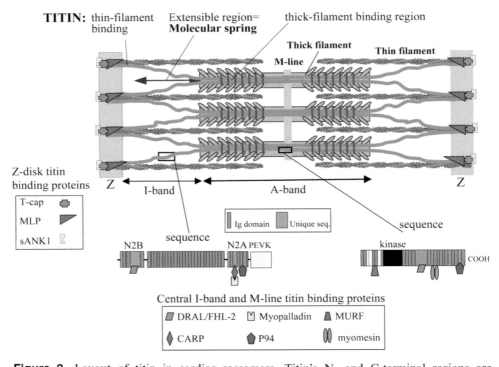

Figure 2 Layout of titin in cardiac sarcomere. Titin's N- and C-terminal regions are embedded in the Z-disk and M-line, respectively (73,102). Titin's I-band region is extensible and functions as a molecular spring that contains three distinct elements: tandem Ig segments (formed by serially linked Ig domains, shown as red rectangles), the PEVK segment (yellow), and the unique sequence (blue) that is part of the cardiac specific N2B element. Two main classes of cardiac titin isoforms exist: 1) The N2B isoform with a short spring that consists of tandem Ig segments with ~40 Ig domains, the N2B element and a short PEVK (~180 residues). 2) The N2BA isoform has a much longer spring, consisting of ~60 Ig domains, the N2B element, a ~800 residue proline (P), glutamate (E), valine (V) and lysine (K) rich domain (PEVK), and in addition the N2A element. (The shown sequence is part of the I-band sequence of N2BA cardiac titin.) Outside the molecular spring region the isoforms are largely identical in sequence. Titin-binding ligands located in Z-disk, I-band and M-line are also shown. The N-terminal region of titin interacts with T-cap and MLP, forming part of a MLP-dependent stretch sensing complex, as well as s-ANK1 that may have a structural role. Titin's N2B element (found in both N2B and N2BA isoforms) interacts with DRAL/FLH-2, a protein involved in recruiting multiple metabolic enzymes to the I-band. Titin's N2A element (only present in N2BA titin) interacts with CARP, a potential regulator of gene expression during stretch-induced myocyte hypertrophy, and the calpain protease p94. CARP also interacts with the muscle-specific nuclear and sarcomeric protein, myopalladin. The M-line region of titin contains a kinase domain of unknown function, and a MURF-interacting region both of which may be involved in signaling. Additionally, titin's M-line region interacts with myomesin, a structural protein. (Vertical gray lines in A-band region denote MyBP-C. Note that the figure is not to scale.) *Abbreviation*: s-ANK, small ankyrin-1. *Source*: From Ref. 103. (*See color insert.*)

for a total of about 38,138 amino acid residues (50). Exons 1 to ~251 are Z-disk and I-band exons, of which exons 45 to 224 are extensively differentially spliced. The multiple splice pathways in the I-band encoding region of the titin gene give rise to cardiac isoforms with distinct spring compositions: the "small" 2970 kDa cardiac isoform known as N2B titin (so named because it contains the N2B element) and the large so-called N2BA titin isoform (name reflects the presence of both N2B and N2A elements). A given sarcomere stretch results in a fractional extension of titin's extensible segment (end-to-end length divided by the contour length) that is much higher for the N2B isoform (because of its shorter extensible segment) than for the N2BA isoform (51). Thus, cardiac myocytes that express high levels of N2B titin will have higher passive stiffness than those that express N2BA titin.

Large mammals co-express N2B and N2BA titins, with co-expression occurring at the level of the half-sarcomere (52). Each half thick filament binds six titin molecules (53) and this value appears to be constant despite widely varying co-expression ratios in different species (51). Titin's constant stoichiometry (six titins/half thick filament) may reflect the constant requirement for functions performed by titin's inextensible regions, such as thick-filament length control and construction and maintenance of Z-disks and M-lines (54) (note that the inextensible regions of isoforms are largely identical (55)). Thus, varying the co-expression ratio of isoforms while keeping the stoichiometry constant does not impact critical functions performed by titin's inextensible region, but allows the development of graded passive force levels, in between those of N2B- and N2BA-pure myocytes (52). In the adult, the co-expression ratio of titin isoforms varies across the LV wall [in pig (51) and dog (56) the N2BA/N2B expression ratio is ~30% higher in subendocardium than in subepicardium] and titin-based passive varies accordingly. A prominent example of varying the co-expression ratio of isoforms is during fetal and neonatal heart development. Recent work revealed an expression of fetal N2BA titin isoforms, which is characterized by additional spring elements both in the tandem Ig and PEVK region of the molecule (57). The fetal N2BA isoform predominates in fetal myocardium and gradually disappears during postnatal development with a time course that varies (from days to weeks) in different species (57).

As expected, passive myocardium is much more compliant in the neonate than in the adult. Thus, functional transitions and diastolic filling behavior during development of the heart appear to be controlled by the expression ratio of titin isoforms.

Titin Adaptations During Disease

To understand how titin responds to chronic mechanical challenge of the heart, a canine tachycardia-induced model of DCM has been used, in which rapid pacing results in chamber dilation and elevated chamber stiffness (56). Two weeks of pacing gives rise to an exaggerated transmural titin isoform ratio gradient (56) and four weeks of pacing results in elevated N2B (stiff) titin expression and downregulation of N2BA (compliant) titin, accompanied by increased titin-based passive stiffness (58). A recent study of the spontaneously hypertensive rat model (SHR) has shown a reduced expression of N2BA titin in response to pressure overload, consistent with elevated passive stiffness of SHR cardiac myocytes (59). The adjustment in cardiac isoform expression is not restricted to animal models, but also occurs in patients with coronary artery disease (CAD). Interestingly, in CAD an increase of the more compliant N2BA cardiac titin isoform occurs (60). The role of titin in heart failure patients was recently investigated in patients with end-stage heart failure because of nonischemic DCM (61). Results revealed small N2B (stiff) and large N2BA (compliant) cardiac titin isoforms with a mean N2BA:N2B expression ratio that was significantly increased in heart failure patients. Mechanical measurements on LV muscle strips dissected from these hearts revealed that passive muscle stiffness was significantly reduced in patients with high

N2BA:N2B expression ratio (Fig. 3A). Clinical correlations support the relevance of these changes for LV function. LV function was assessed by invasive hemodynamics and Doppler echocardiography. A positive correlation between the N2BA:N2B titin isoform ratio and DT of mitral E velocity, a wave transit time, and EDV/EDP ratio was found. Thus, in end-stage failing hearts the more compliant N2BA isoform comprised a greater percentage of titin and changes in titin isoform expression in heart failure patients with DCM significantly impacts diastolic filling by lowering myocardial stiffness (61).

The reduction in passive stiffness seen in heart failure patients is the reverse of the process that occurs during normal fetal and neonatal heart development where titin expression switches from a predominance of compliant fetal N2BA titin to the more stiff N2B isoform, giving rise to increased passive stiffness (Fig. 3B) (57). Considering that reactivation of fetal gene expression occurs in many myocardial disease states (62) it was studied whether expression of fetal cardiac titin occurs in DCM patients. The recently developed titin exon microarray was used to examine titin at the transcript level and myocardium from DCM patients was compared with

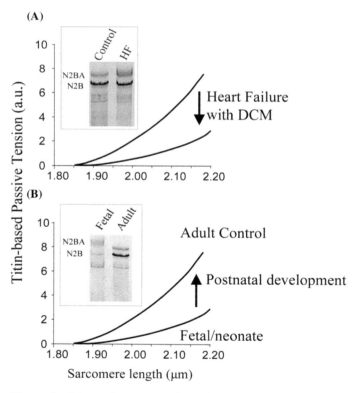

Figure 3 Schematic representations of adjustments in titin expression that decreases titin-based passive myocardial stiffness during DCM heart failure (**A**) and increases titin-based passive myocardial stiffness during normal neonatal development (**B**). Titin gels are shown in the inset. They reveal that normal adult myocardium (control in **A**) expresses more N2B titin than N2BA titin and that N2BA titin in upregulated in myocardium of HF patients. In fetal myocardium (inset of **B**) N2BA titin also dominates. *Abbreviations*: DCM, dilated cardiomyopathy; HF, heart failure. *Source*: From Refs. 61,57.

that of fetal and adult controls. This analysis indicated that the examined DCM hearts did not express fetal cardiac titin (57). Upregulated N2BA expression is instead likely because of upregulation of adult N2BA-type isoforms.

Titin's size and complex sequence makes it a prominent target for post-translational modifications and damage. Several studies have reported proteolysis of titin in end-stage heart failure (63,64). This may be because of heightened proteolytic activity and/or enhanced proteolytic sensitivity of titin. Modified titins may also result from, for example, action of free radicals, glycation reactions, and X-ray radiation [because of its extremely large size, radiation preferentially damages titin (65)]. Chemo- and radio-therapy for breast cancers and mediastinal lymphomas are known to result in secondary progressive cardiomyopathies (66) and whether damage to the titin filament is involved in this pathology warrants investigation.

Hereditary Titin Diseases

Because of its enormous size and multiple functions, titin is a prominent target for mutations that give rise to muscle disease. Earlier linkage studies in families with DCM identified titin as a candidate gene, but sequencing of the cardiac specific exon 49 (N2B element) failed to detect mutations (67). The skeletal muscular dystrophy tibialis muscular dystrophy (TMD) is a genetic muscle disease of dominant inheritance (common in Finland, but also found elsewhere), that has also been linked to titin (68). Sequencing of titin exons that code for skeletal muscle specific spring elements, however, failed to detect disease-causing mutations. Thus, earlier work supports the notion that titin is a candidate gene for hereditary muscle disease, but sequencing a small subset of all titin exons was not sufficient to identify the mutations involved.

The recent completion of titin's genomic structure (50) has facilitated the search for titin mutations [primer pairs and PCR conditions are now available to survey all 363 titin exons, see Ref. (69)]. This has enabled identification of several exons with mutations. Sequencing the entire titin gene in DCM families (69) revealed two DCM-causing mutations in titin, i.e., in exons 18 and 326. The titin mutation in exon 326 is predicted to cause a frameshift resulting in truncated titin ($\sim$2000 kDa). Western blot analysis revealed that a truncated titin is indeed expressed in skeletal muscle but of a size smaller than expected ($\sim$1100 kDa), suggesting that the truncated protein is very sensitive to proteolysis and is cleaved. Exon 18 encodes an Ig domain and the identified mutation in this exon is predicted to disrupt its native structure (69), underscoring the importance of maintaining Ig domains in their folded state. Sequencing of the entire titin gene in TMD families (70) revealed a mutation in exon 363. This exon codes for the most C-terminal Ig domain of titin and its mutation is also predicted to destabilize the folded state. Finally, a recent large sequencing project (71) analyzed patients with a variety of cardiac diseases and identified in HCM patients mutations in exons 2 and 14 (Z-disk) and 49 (N2B element).

Protein–protein interaction studies showed that the two identified missense mutations in exons 2 and 14 lower titin's affinity for binding to T-cap and α-actinin, respectively, implicating impairment/perturbation of the interaction between Z-disk titin and its ligands as a potential disease mechanism (71). Although, these sequencing studies are recent and few, and thus the frequency of genetic titin diseases remains to be established, the work by Kimura and colleagues (71) raises the possibility that up to several percent of all cardiac diseases include titin mutations.

Most titin mutations identified so far are in regions of titin that are expressed in all striated muscle types. Surprisingly, different muscles are not affected equally by the mutations. For example, the mutation of exon 362 in TMD is expressed in all muscles, but the mutation affects selectively only the extensor muscles of the frontal leg compartment, whereas adjacent muscles are spared (68). Similarly, the two mutations identified in DCM are expressed in all titin isoforms, but there is no clinically detectable phenotype in skeletal muscle (69). These muscle-specific phenotypes are puzzling and underscore the need to more deeply study the titin's multiple functions as well as their relative importance in different muscle types under both normal and pathological conditions.

Titin-Based Protein Complexes as Biomechanical Sensors

Many titin-binding proteins are part of structural complexes and biomechanical sensing pathways (Fig. 2). Protein-binding sites in titin are found in and near the Z-disk, in the central I-band region, and in the M-line region of the molecule. The titin-capping protein T-cap (or telethonin (72)) interacts with titin's N-terminus (73) and functions as an adaptor protein that links signaling and structural molecules to titin. T-cap interacts with the cytoplasmic domains of two membrane associated proteins: the potassium channel subunit minK/isk (74), and the small ankyrin-1 (sANK1) (75), found in the T-tubules and the sarcoplasmic reticulum (SR), respectively. Possibly, T-cap's interaction with minK anchors the T-tubules close to the Z-disk region of the sarcomere and may regulate potassium channel function in response to myocyte stretch. sANK1 is a transmembrane protein of the SR and its interaction with titin's N-terminus may position the SR around the Z-disk of each sarcomere (75). A further role for titin in organizing the SR is suggested by the interaction between titin's near Z-disk domains Z9–Z10 and obscurin (76). Obscurin contains signaling domains (50,76,77), and interacts with the ankyrin isoform 1.5; this ankyrin isoform appears to link the SR to the sarcomere and to regulate ryanodine receptor distribution in the SR (78). These interactions with titin's Z-disk region may be involved in positioning the SR and T-tubular membrane systems in close proximity to the I-band region of the sarcomere. Furthermore, it also ensures that these membrane systems move with the Z-disk, thereby preventing excessive strains, which possibly could tear membranes when the sarcomeres shorten during contraction. Thus, titin maintains the central position of the A-band in the sarcomere (see above) as well as the organization of the SR and T-tubular systems.

T-cap also interacts with muscle LIM protein (MLP), an essential nuclear regulator of myogenic differentiation (79). Interestingly, stretching of cultured cardiac myocytes induces expression of the well known stretch response markers, brain natriuretic peptide and atrial natriuretic factor, but this response is absent in cardiac myocytes from a MLP KO mouse model (80), suggesting that MLP could be involved in stretch sensing. The impaired stretch sensing of the MLP KO mouse model could be at the basis of the DCM phenotype of these animals, a conclusion supported by the finding of a MLP mutation in a subset of human patients with DCM (80).

The N2B and N2A elements, located in the central I-band region, constitute hotspots for interactions with signaling molecules. The cardiac-specific N2B element interacts with a member of the LIM protein family known as DRAL/FHL-2 (81). DRAL/FHL-2 in turn binds the metabolic enzymes creatine kinase, adenylate kinase, and phosphofructokinase (81). Thus, the N2B element may play a role in the compartmentalization of metabolic enzymes, ensuring generation of ATP close to the overlap region of the sarcomere, where high levels of ATP are consumed during contraction. The N2B element also interacts with αB-crystallin (82), a member of the small heat shock protein family, which functions as chaperones that maintain the folded state of proteins. In heart muscle, αB-crystallin participates in the ischemic stress signaling response and possibly protects titin from structural damage under conditions of heightened vulnerability.

The N2A element is found in both skeletal muscle titins and cardiac N2BA titins. Two of its Ig domains (Ig82/83) interact with the calpain protease P94, also known as calpain-3 (83). P94 is involved in protein degradation, and binding to titin is thought to regulate P94 activity and fine-tune its functions (83). A tyrosine-rich binding motif between Ig80 and Ig81 interacts with a conserved motif present in the three homologous muscle ankyrin repeat proteins: cardiac ankyrin repeat protein (CARP), and the two closely related proteins, ankrd2 (or Arpp) and diabetes ankyrin repeat protein (DARP) (84). CARP also interacts with myopalladin, a ~145 kDa protein found in the sarcomere and nucleus (85). All three ankyrin repeat proteins were identified previously by their cytokine-like induction following cardiac injury and muscle denervation (CARP) (86–88), skeletal muscle stretch (ankrd2/Arpp) (89), or during recovery after metabolic challenge (DARP) (90). Furthermore, CARP, ankrd2, and DARP are all up-regulated in myocardium of heart failure patients with DCM (61). A recent study of a muscular dystrophy mouse model revealed dysregulation of the N2A protein complex with CARP up-regulation as a possible primary event (91). Overall these findings suggest that ankyrin repeat proteins are part of muscle stress response pathways.

All I-band ligands of titin and their associated binding partners are also found in the nucleus where they participate in transcriptional and cell cycle regulation. It seems to be likely that this dual localization (I-band and nucleus) reflects a dual function for these proteins: being part of a titin-based stretch sensing complex in the I-band and regulating transcription in the nucleus. Furthermore, such dual localization may also provide a communication pathway between the I-band and nucleus that links stretch sensing to gene expression. The molecular mechanisms of I-band based signaling have been studied in the naturally occurring *mdm* (muscular dystrophy with myositis) mouse. In this mouse four small titin exons coding for 83 amino acids are excised from the N2A element, resulting in death at about six to eight weeks of age (92). A gene expression study (91) revealed striking early changes in a subset of genes, including MLP, CARP, and MURF-1 (a M-line based protein, see below), showing that the deletion of this small portion of the N2A element simultaneously affects titin ligands in the Z-disk (MLP), I-band (CARP) and M-line (MURF-1).

The M-line region of titin contains a conserved serine/threonine kinase domain, phosphorylation motifs (93), a second binding site for p94 (94), a second binding site for DRAL/FHL-2 (81), possibly additional binding sites for T-cap and osbcurin (76,95), and a binding site for the RING finger protein MURF-1 (96). Analogous to the Z-disk protein T-cap, MURF-1 is a multifunctional adaptor protein that may function in the regulation of gene expression (MURF-1 interacts with the steroid regulated transcriptional activator GMEB1, see Ref. (97)) and in protein turnover by acting as an ubiquitin ligase (98). MURF-1 interacts with the two titin Ig domains, A169/170, located at the M-line periphery, and can form heterodimers with the homologous proteins MURF-2 and MURF-3 (96). Both MURF-2 and MURF-3 also bind microtubules and this binding seems to regulate microtubular stability (99,100).

Titin is the only protein that spans the half-sarcomeric distance and this together with the many titin-based protein complexes makes titin an ideal candidate to function as a sarcomere stretch sensor that underlies length-dependent signaling processes in the cardiac myocyte. As discussed above, the last few years has seen great progress in establishing many of the players involved in titin-based sensing and signaling, and protein interaction hotspots have been identified at titin's Z-disk, central I-band, and M-line regions. The I-band region of titin is independent of actomyosin-based force and mechanical input to the I-band sensing complex is likely to be purely titin-based. The Z-disk based titin ligands may receive biomechanical input not just from titin, but also from actomyosin-based tension (communicated to the Z-disk via the thin filaments), and tensions that are communicated via the sarcolemma. Thus, titin is likely to integrate a variety of biomechanical signals.

SUMMARY

Titin-based stiffness adjustments occur in various muscle diseases and several disease-causing titin mutations have been discovered. Early indications are that the frequency of genetic titin diseases may be relatively high. As titin's size is enormous, multiple biological functions can be assigned to titin. In addition to titin's well-established role in determining passive muscle stiffness, recent evidence suggests a role in protein metabolism (by regulating ubiquination and calpain activity), compartmentalization of metabolic enzymes (binding of DRAL/FHL-2), binding of chaperones (αB-crystallin), and positioning of the membrane systems of the T-tubules and SR (binding of sANK1 and obscurin). Titin is the only protein that spans the half-sarcomeric distance and, therefore, titin might act as a sarcomere stretch sensor that underlies length-dependent signaling processes in the cardiac myocyte. The last few years has seen great progress in establishing many of the players involved in titin-based sensing and signaling, and protein interaction hotspots have been identified at titin's Z-disk, central I-band, and M-line regions. The stage has been set for gaining a deeper understanding of the role of titin in muscle function and disease.

ACKNOWLEDGMENTS

Supported by grants from the National Institutes of Health HL-61487/62881.

REFERENCES

1. Alberts B, Johnson A, Lewis J, Raff M, Roberts K, Walter P. In: Molecular Biology of the Cell. New York: Garland Science, 2002:907–982.
2. Rappaport L, Samuel JL. Microtubules in cardiac myocytes. Int Rev Cytol 1988; 113: 101–143.
3. Schaper J, Froede R, Hein S, Buck A, Hashizume H, Speiser B, Friedl A, Bleese N. Impairment of the myocardial ultrastructure and changes of the cytoskeleton in dilated cardiomyopathy. Circulation 1991; 83:504–514.
4. Samuel JL, Marotte F, Delcayre C, Rappaport L. Microtubule reorganization is related to rate of heart myocyte hypertrophy in rat. Am J Physiol 1986; 251:H1118–H1125.
5. Hein S, Kostin S, Heling A, Maeno Y, Schaper J. The role of the cytoskeleton in heart failure. Cardiovasc Res 2000; 45:273–278.
6. Tsutsui H, Tagawa H, Kent RL, McCollam PL, Ishihara K, Nagatsu M, Cooper G. Role of microtubules in contractile dysfunction of hypertrophied cardiocytes. Circulation 1994; 90:533–555.
7. Tsutsui H, Ishihara K, Cooper G. Cytoskeletal role in the contractile dysfunction of hypertrophied myocardium. Science 1993; 260:682–687.
8. Ishibashi Y, Takahashi M, Isomatsu Y, Qiao F, Iijima Y, Shiraishi H, Simsic JM, Baicu CF, Robbins J, Zile MR, Cooper G. Role of microtubules versus myosin heavy

chain isoforms in contractile dysfunction of hypertrophied murine cardiocytes. Am J Physiol Heart Circ Physiol 2003; 285:H1270–H1285.

9. Tagawa H, Koide M, Sato H, Zile MR, Carabello BA, Cooper G. Cytoskeletal role in the transition from compensated to decompensated hypertrophy during adult canine left ventricular pressure overloading. Circ Res 1998; 82:751–761.

10. Eble DM, Spinale FG. Contractile and cytoskeletal content, structure, and mRNA levels with tachycardia-induced cardiomyopathy. Am J Physiol 1995; 268:H2426–H2439.

11. Wang X, Li F, Campbell SE, Gerdes AM. Chronic pressure overload cardiac hypertrophy and failure in guinea pigs: II. Cytoskeletal remodeling. J Mol Cell Cardiol 1999; 31:319–331.

12. Collins JF, Pawloski-Dahm C, Davis MG, Ball N, Dorn GW II, Walsh RA. The role of the cytoskeleton in left ventricular pressure overload hypertrophy and failure. J Mol Cell Cardiol 1996; 28:1435–1443.

13. Bailey BA, Dipla K, Li S, Houser SR. Cellular basis of contractile derangements of hypertrophied feline ventricular myocytes. J Mol Cell Cardiol 1997; 29:1823–1835.

14. de Tombe PP. Altered contractile function in heart failure. Cardiovasc Res 1998; 37: 367–380.

15. ter Keurs HE. Microtubules in cardiac hypertrophy: a mechanical role in decompensation? Circ Res 1998; 82:828–831.

16. Zile MR, Green GR, Schuyler GT, Aurigemma GP, Miller DC, Cooper G. Cardiocyte cytoskeleton in patients with left ventricular pressure overload hypertrophy. J Am Coll Cardiol 2001; 37:1080–1084.

17. Aquila-Pastir LA, DiPaola NR, Matteo RG, Smedira NG, McCarthy PM, Moravec CS. Quantitation and distribution of beta-tubulin in human cardiac myocytes. J Mol Cell Cardiol 2002; 34:1513–1523.

18. Lazarides E. Intermediate filaments as mechanical integrators of cellular space. Nature 1980; 283:249–256.

19. Price MG. Molecular analysis of intermediate filament cytoskeleton—a putative load-bearing structure. Am J Physiol 1984; 246:H566–H572.

20. Wang K, Ramirez-Mitchell R. A network of transverse and longitudinal intermediate filaments is associated with sarcomeres of adult vertebrate skeletal muscle. J Cell Biol 1983; 96:562–570.

21. Granzier HL, Irving TC. Passive tension in cardiac muscle: contribution of collagen, titin, microtubules, and intermediate filaments. Biophys J 1995; 68:1027–1044.

22. Wang K, McCarter R, Wright J, Beverly J, Ramirez-Mitchell R. Viscoelasticity of the sarcomere matrix of skeletal muscles. The titin-myosin composite filament is a dual-stage molecular spring. Biophys J 1993; 64:1161–1177.

23. Price MG, Sanger JW. Intermediate filaments in striated muscle. A review of structural studies in embryonic and adult skeletal and cardiac muscle. Cell Muscle Motil 1983; 3:1–40.

24. Price MG, Lazarides E. Expression of intermediate filament-associated proteins paranemin and synemin in chicken development. J Cell Biol 1983; 97:1860–1874.

25. Wang X, Osinska H, Gerdes AM, Robbins J. Desmin filaments and cardiac disease: establishing causality. J Card Fail 2002; 8:S287–S292.

26. Wang X, Klevitsky R, Huang W, Glasford J, Li F, Robbins J. AlphaB-crystallin modulates protein aggregation of abnormal desmin. Circ Res 2003; 93:998–1005.

27. Hwang DM, et al. A genome-based resource for molecular cardiovascular medicine: toward a compendium of cardiovascular genes. Circulation 1997; 96:4146–4203.

28. Heling A, Zimmermann R, Kostin S, Maeno Y, Hein S, Devaux B, Bauer E, Klovekorn WP, Schlepper M, Schaper W, Schaper J. Increased expression of cytoskeletal, linkage, and extracellular proteins in failing human myocardium. Circ Res 2000; 86:846–853.

29. Wang X, Osinska H, Dorn GW II, Nieman M, Lorenz JN, Gerdes AM, Witt S, Kimball T, Gulick J, Robbins J. Mouse model of desmin-related cardiomyopathy. Circulation 2001; 103:2402–2407.

30. Li D, Tapscoft T, Gonzalez O, Burch PE, Quinones MA, Zoghbi WA, Hill R, Bachinski LL, Mann DL, Roberts R. Desmin mutation responsible for idiopathic dilated cardiomyopathy. Circulation 1999; 100:461–464.

31. Goldfarb LG, Vicart P, Goebel HH, Dalakas MC. Desmin myopathy. Brain 2004; 127:723–734.

32. Milner DJ, Weitzer G, Tran D, Bradley A, Capetanaki Y. Disruption of muscle architecture and myocardial degeneration in mice lacking desmin. J Cell Biol 1996; 134: 1255–1270.

33. Li Z, Mericskay M, Agbulut O, Butler-Browne G, Carlsson L, Thornell LE, Babinet C, Paulin D. Desmin is essential for the tensile strength and integrity of myofibrils but not for myogenic commitment, differentiation, and fusion of skeletal muscle. J Cell Biol 1997; 139:129–144.

34. Borg TK, Goldsmith EC, Price R, Carver W, Terracio L, Samarel AM. Specialization at the Z line of cardiac myocytes. Cardiovasc Res 2000; 46:277–285.

35. Cohn RD, Campbell KP. Molecular basis of muscular dystrophies. Muscle Nerve 2000; 23:1456–1471.

36. Durbeej M, Campbell KP. Muscular dystrophies involving the dystrophin–glycoprotein complex: an overview of current mouse models. Curr Opin Genet Dev 2002; 12: 349–361.

37. Towbin JA, Bowles NE. The failing heart. Nature 2002; 415:227–233.

38. Seidman JG, Seidman C. The genetic basis for cardiomyopathy: from mutation identification to mechanistic paradigms. Cell 2001; 104:557–567.

39. Olson TM, Michels VV, Thibodeau SN, Tai YS, Keating MT. Actin mutations in dilated cardiomyopathy, a heritable form of heart failure. Science 1998; 280:750–752.

40. Mogensen J, Klausen IC, Pedersen AK, Egeblad H, Bross P, Kruse TA, Gregersen N, Hansen PS, Baandrup U, Borglum AD. Alpha-cardiac actin is a novel disease gene in familial hypertrophic cardiomyopathy. J Clin Invest 1999; 103:R39–R43.

41. Granzier H, Labeit S. Cardiac titin: an adjustable multi-functional spring. J Physiol 2002; 541:335–342.

42. Wu Y, Cazorla O, Labeit D, Labeit S, Granzier H. Changes in titin and collagen underlie diastolic stiffness diversity of cardiac muscle. J Mol Cell Cardiol 2000; 32:2151–2162.

43. Horowits R, Podolsky RJ. The positional stability of thick filaments in activated skeletal muscle depends on sarcomere length: evidence for the role of titin filaments. J Cell Biol 1987; 105:2217–2223.

44. Horowits R, Podolsky RJ. Thick filament movement and isometric tension in activated skeletal muscle. Biophys J 1988; 54:165–171.

45. Granzier H, Kellermayer M, Helmes M, Trombitas K. Titin elasticity and mechanism of passive force development in rat cardiac myocytes probed by thin-filament extraction. Biophys J 1997; 73:2043–2053.

46. Trombitas K, Greaser M, Labeit S, Jin JP, Kellermayer M, Helmes M, Granzier H. Titin extensibility in situ: entropic elasticity of permanently folded and permanently unfolded molecular segments. J Cell Biol 1998; 140:853–859.

47. Linke WA, Ivemeyer M, Labeit S, Hinssen H, Ruegg JC, Gautel M. Actin–titin interaction in cardiac myofibrils: probing a physiological role. Biophys J 1997; 73:905–919.

48. Helmes M, Trombitas K, Granzier H. Titin develops restoring force in rat cardiac myocytes. Circ Res 1996; 79:619–626.

49. Granzier H, Helmes M, Cazorla O, McNabb M, Labeit D, Wu Y, Yamasaki R, Redkar A, Kellermayer M, Labeit S, Trombitas K. Mechanical properties of titin isoforms. Adv Exp Med Biol 2000; 481:283–300; discussion 300–304.

50. Bang ML, Centner T, Fornoff F, Geach AJ, Gotthardt M, McNabb M, Witt CC, Labeit D, Gregorio CC, Granzier H, Labeit S. The complete gene sequence of titin, expression of an unusual approximately 700-kDa titin isoform, and its interaction with obscurin identify a novel Z-line to I-band linking system. Circ Res 2001; 89:1065–1072.

51. Cazorla O, Freiburg A, Helmes M, Centner T, McNabb M, Wu Y, Trombitas K, Labeit S, Granzier H. Differential expression of cardiac titin isoforms and modulation of cellular stiffness. Circ Res 2000; 86:59–67.

52. Trombitas K, Wu Y, Labeit D, Labeit S, Granzier H. Cardiac titin isoforms are coexpressed in the half-sarcomere and extend independently. Am J Physiol Heart Circ Physiol 2001; 281:H1793–H1799.

53. Liversage AD, Holmes D, Knight PJ, Tskhovrebova L, Trinick J. Titin and the sarcomere symmetry paradox. J Mol Biol 2001; 305:401–409.

54. Gregorio CC, Granzier H, Sorimachi H, Labeit S, Muscle assembly: a titanic achievement? Curr Opin Cell Biol 1999; 11:18–25.

55. Labeit S, Kolmerer B. Titins: giant proteins in charge of muscle ultrastructure and elasticity. Science 1995; 270:293–296.

56. Bell SP, Nyland L, Tischler MD, McNabb M, Granzier H, LeWinter MM. Alterations in the determinants of diastolic suction during pacing tachycardia. Circ Res 2000; 87:235–240.

57. Lahmers S, Wu Y, Call DR, Labeit S, Granzier H. Developmental control of titin isoform expression and passive stiffness in fetal and neonatal myocardium. Circ Res 2004; 94:505–513.

58. Wu Y, Bell SP, Trombitas K, Witt CC, Labeit S, LeWinter MM, Granzier H. Changes in titin isoform expression in pacing-induced cardiac failure give rise to increased passive muscle stiffness. Circulation 2002; 106:1384–1389.

59. Warren CM, Jordan MC, Roos KP, Krzesinski PR, Greaser ML. Titin isoform expression in normal and hypertensive myocardium. Cardiovasc Res 2003; 59:86–94.

60. Neagoe C, Kulke M, del Monte F, Gwathmey JK, de Tombe PP, Hajjar RJ, Linke WA. Titin isoform switch in ischemic human heart disease. Circulation 2002; 106:1333–1341.

61. Nagueh SF, Shah G, Wu Y, Guillermo TA, King NMP, Lahmers S, Witt C, Becker K, Labeit S, Granzier H. Altered titin expression, myocardial stiffness, and left ventricular function in patients with dilated cardiomyopathy. Circulation 2004; 110:115–162.

62. Colucci WS. Molecular and cellular mechanisms of myocardial failure. Am J Cardiol 1997; 80:15L–25L.

63. Hein S, Scholz D, Fujitani N, Rennollet H, Brand T, Friedl A, Schaper J. Altered expression of titin and contractile proteins in failing human myocardium. J Mol Cell Cardiol 1994; 26:1291–1306.

64. Morano I, Hadicke K, Grom S, Koch A, Schwinger RH, Bohm M, Bartel S, Erdmann E, Krause EG. Titin, myosin light chains and C-protein in the developing and failing human heart. J Mol Cell Cardiol 1994; 26:361–368.

65. Horowits R, Kempner ES, Bisher ME, Podolsky RJ. A physiological role for titin and nebulin in skeletal muscle. Nature 1986; 323:160–164.

66. Shusterman S, Meadows AT. Long term survivors of childhood leukemia. Curr Opin Hematol 2000; 7:217–222.

67. Siu BL, Niimura H, Osborne JA, Fatkin D, MacRae C, Solomon S, Benson DW, Seidman JG, Seidman CE. Familial dilated cardiomyopathy locus maps to chromosome 2q31. Circulation 1999; 99:1022–1026.

68. Udd B, Haravuori H, Kalimo H, Partanen J, Pulkkinen L, Paetau A, Peltonen L, Somer H. Tibial muscular dystrophy—from clinical description to linkage on chromosome 2q31. Neuromuscul Disord 1998; 8:327–332.

69. Gerull B, Gramlich M, Atherton J, McNabb M, Trombitas K, Sasse-Klaassen S, Seidman JG, Seidman C, Granzier H, Labeit S, Frenneaux M, Thierfelder L. Mutations of TTN, encoding the giant muscle filament titin, cause familial dilated cardiomyopathy. Nat Genet 2002; 30:201–204.

70. Hackman P, Vihola A, Haravuori H, Marchand S, Sarparanta J, De Seze J, Labeit S, Witt C, Peltonen L, Richard I, Udd B. Tibial muscular dystrophy is a titinopathy caused by mutations in TTN, the gene encoding the giant skeletal-muscle protein titin. Am J Hum Genet 2002; 71:492–500.

71. Itoh-Satoh M, Hayashi T, Nishi H, Koga Y, Arimura T, Koyanagi T, Takahashi M, Hohda S, Ueda K, Nouchi T, Hiroe M, Marumo F, Imaizumi T, Yasunami M, Kimura A. Titin mutations as the molecular basis for dilated cardiomyopathy. Biochem Biophys Res Commun 2002:291:385–393.

72. Mues A, van der Ven PF, Young P, Furst DO, Gautel M. Two immunoglobulin-like domains of the Z-disk portion of titin interact in a conformation-dependent way with telethonin. FEBS Lett 1998; 428:111–114.

73. Gregorio CC, Trombitas K, Centner T, Kolmerer B, Stier G, Kunke K, Suzuki K, Obermayr F, Herrmann B, Granzier H, Sorimachi H, Labeit S. The NH2 terminus of titin spans the Z-disc: its interaction with a novel 19-kD ligand (T-cap) is required for sarcomeric integrity. J Cell Biol 1998; 143:1013–1027.

74. Furukawa T, Ono Y, Tsuchiya H, Katayama Y, Bang ML, Labeit D, Labeit S, Inagaki N, Gregorio CC. Specific interaction of the potassium channel beta-subunit minK with the sarcomeric protein T-cap suggests a T-tubule-myofibril linking system. J Mol Biol 2001; 313:775–784.

75. Kontrogianni-Konstantopoulos A, Bloch RJ. The hydrophilic domain of small ankyrin-1 interacts with the two N-terminal immunoglobulin domains of titin. J Biol Chem 2003; 278:3985–3991.

76. Young P, Ehler E, Gautel M. Obscurin, a giant sarcomeric Rho guanine nucleotide exchange factor protein involved in sarcomere assembly. J Cell Biol 2001; 154:123–136.

77. Russell MW, Raeker MO, Korytkowski KA, Sonneman KJ. Identification, tissue expression and chromosomal localization of human Obscurin-MLCK, a member of the titin and Dbl families of myosin light chain kinases. Gene 2002; 282:237–246.

78. Bagnato P, Barone V, Giacomello E, Rossi D, Sorrentino V. Binding of an ankyrin-1 isoform to obscurin suggests a molecular link between the sarcoplasmic reticulum and myofibrils in striated muscles. J Cell Biol 2003; 160:245–253.

79. Arber S, Halder G, Caroni P. Muscle LIM protein, a novel essential regulator of myogenesis, promotes myogenic differentiation. Cell 1994; 79:221–231.

80. Knoll R, Hoshijima M, Hoffman HM, Person V, Lorenzen-Schmidt I, Bang ML, Hayashi T, Shiga N, Yasukawa H, Schaper W, McKenna W, Yokoyama M, Schork NJ, Omens JH, McCulloh AD, Kimura W, Gregorio CC, Poller W, Schaper J, Schultheiss HP, Chien KR. The cardiac mechanical stretch sensor machinery involves a Z disc complex that is defective in a subset of human dilated cardiomyopathy. Cell 2002; 111:943–955.

81. Lange S, Auerbach D, McLoughlin P, Perriard E, Schafer BW, Perriard JC, Ehler E. Subcellular targeting of metabolic enzymes to titin in heart muscle may be mediated by DRAL/FHL-2. J Cell Sci 2002; 115:4925–4936.

82. Bullard B, Ferguson C, Minajeva A, Leake MC, Gautel M, Labeit D, Ding L, Labeit S, Horwitz J, Leonard KR, Linke WA. Association of the chaperone alphaB-crystallin with titin in heart muscle. J Biol Chem 2004; 279:7917–7924.

83. Ono Y, Kakinuma K, Torii F, Irie A, Nakagawa K, Labeit S, Abe K, Suzuki K, Sorimachi H. Possible regulation of the conventional calpain system by skeletal muscle-specific calpain, p94. J Biol Chem 2004; 279(4):2761–2771. Epub 2003 Nov 1.

84. Miller MK, Bang ML, Witt C, Labeit D, Trombitas K, Watanabe K, Granzier H, McElhinny AS, Gregorio CC, Labeit S. The muscle ankyrin repeat proteins: CARP, ankrd2/Arpp and DARP as a family of titin filament based stress response molecules. J Mol Biol 2003; 333:951–964.

85. Bang ML, Mudry RE, McElhinny AS, Trombitas K, Geach AJ, Yamasaki R, Sorimachi H, Granzier H, Gregorio CC, Labeit S. Myopalladin, a novel 145-kilodalton sarcomeric protein with multiple roles in Z-disc and I-band protein assemblies. J Cell Biol 2001; 153:413–427.

86. Nakada C, Oka A, Nonaka I, Sato K, Mori S, Ito H, Moriyama M. Cardiac ankyrin repeat protein is preferentially induced in atrophic myofibers of congenital myopathy and spinal muscular atrophy. Pathol Int 2003; 53:653–658.

87. Tsukamoto Y, Senda T, Nakano T, Nakada C, Hida T, Ishiguro N, Kondo G, Baba T, Sato K, Osaki M, Mori S, Ito H, Moriyama M. Arpp, a new homolog of carp, is preferentially expressed in type 1 skeletal muscle fibers and is markedly induced by denervation. Lab Invest 2002; 82:645–655.

88. Kuo H, Chen J, Ruiz-Lozano P, Zou Y, Nemer M, Chien KR. Control of segmental expression of the cardiac-restricted ankyrin repeat protein gene by distinct regulatory pathways in murine cardiogenesis. Development 1999; 126:4223–4234.

89. Kemp TJ, Sadusky TJ, Saltisi F, Carey N, Moss J, Yang SY, Sassoon DA, Goldspink G, Coulton GR. Identification of Ankrd2, a novel skeletal muscle gene coding for a stretch-responsive ankyrin-repeat protein. Genomics 2000; 66:229–241.

90. Ikeda K, Emoto N, Matsuo M, Yokoyama M, Molecular identification and characterization of a novel nuclear protein whose expression is up-regulated in insulin-resistant animals. J Biol Chem 2003; 278:3514–3520.

91. Witt CC, Ono Y, Puschmann E, McNabb M, Wu Y, Gotthardt M, Witt SH, Haak M, Labeit D, Gregorio CC, Sorimachi H, Granzier H, Labeit S. Induction and myofibrillar targeting of CARP, and suppression of the Nkx2.5 pathway in the MDM mouse with impaired titin-based signaling. J Mol Biol 2004; 336:145–154.

92. Garvey SM, Rajan C, Lerner AP, Frankel WN, Cox GA. The muscular dystrophy with myositis (mdm) mouse mutation disrupts a skeletal muscle-specific domain of titin. Genomics 2002; 79:146–149.

93. Gautel M, Leonard K, Labeit S. Phosphorylation of KSP motifs in the C-terminal region of titin in differentiating myoblasts. EMBO J 1993; 12:3827–3834.

94. Kinbara K, Sorimachi H, Ishiura S, Suzuki K. Muscle-specific calpain, p94, interacts with the extreme C-terminal region of connectin, a unique region flanked by two immunoglobulin C2 motifs. Arch Biochem Biophys 1997; 342:99–107.

95. Mayans O, van der Ven PF, Wilm M, Mues A, Young P, Furst DO, Wilmanns M, Gautel M. Structural basis for activation of the titin kinase domain during myofibrillogenesis. Nature 1998; 395:863–869.

96. Centner T, Yano J, Kimura E, McElhinny AS, Pelin K, Witt CC, Bang ML, Trombitas K, Granzier H, Gregorio CC, Sorimachi H, Labeit S. Identification of muscle specific ring finger proteins as potential regulators of the titin kinase domain. J Mol Biol 2001; 306:717–726.

97. McElhinny AS, Kakinuma K, Sorimachi H, Labeit S, Gregorio CC. Muscle-specific RING finger-1 interacts with titin to regulate sarcomeric M-line and thick filament structure and may have nuclear functions via its interaction with glucocorticoid modulatory element binding protein-1. J Cell Biol 2002; 157:125–136.

98. Bodine SC, et al. Identification of ubiquitin ligases required for skeletal muscle atrophy. Science 2001; 294:1704–1708.

99. Spencer JA, Eliazer S, Ilaria RL Jr, Richardson JA, Olson EN. Regulation of microtubule dynamics and myogenic differentiation by MURF, a striated muscle RING-finger protein. J Cell Biol 2000; 150:771–784.

100. Pizon V, Iakovenko A, Van Der Ven PF, Kelly R, Fatu C, Furst DO, Karsenti E, Gautel M. Transient association of titin and myosin with microtubules in nascent myofibrils directed by the MURF2 RING-finger protein. J Cell Sci 2002; 115:4469–4482.

101. Miller MK, Granzier H, Ehler E, Gregorio CC. The sensitive giant: the role of titin-based stretch sensing complexes in the heart. Trends Cell Biol 2004; 14:119–126.

102. Obermann WM, Gautel M, Steiner F, van der Ven PF, Weber K, Furst DO. The structure of the sarcomeric M band: localization of defined domains of myomesin, M-protein, and the 250-kD carboxy-terminal region of titin by immunoelectron microscopy. J Cell Biol 1996; 134:1441–1453.

103. Granzier HL, Labeit S. The giant protein titin: a major player in myocardial mechanics, signaling, and disease. Circ Res 2004; 94:284–295.

7

Importance of Myocyte Loss and Regeneration During the Cardiac Remodeling Process

Shaila Garg and Jagat Narula

Departments of Internal Medicine and Cardiology, Drexel University College of Medicine, Philadelphia, Pennsylvania, U.S.A.

The international forum on cardiac remodeling has defined cardiac remodeling as "alterations in the genome expression, molecules, cells and interstitium that are manifested clinically as changes in the size, shape and function of the heart after cardiac injury" (1). Despite the multiple etiologies possible for cardiac injury, the remodeling process seems to be the final common pathway leading to heart failure. The changes that occur within the myocardium during the remodeling process involve myocytes, interstitium, fibroblasts, collagen, and coronary vasculature. Myocytes have deservingly attracted much attention in the remodeling process in view of their contractile activity and numeric contribution to the cardiac structure. Myocytes undergo multiple changes during the remodeling process, including hypertrophy, loss, and regeneration.

HEART MUSCLE CELL LOSS IN MYOCARDIAL REMODELING

Myocyte loss not only happens at the time of the initial insult but also continues thereafter as part of the remodeling process. The three major structural counterparts of the maladaptive processes of ventricular decompensation are fibrosis, cellular degeneration represented by impairment of myocyte ultrastructure, and actual myocyte loss. The latter contributes to mural thinning, cavitary dilatation, depressed ventricular contractile performance, and an increase in diastolic stress (2–4). Diffuse myocyte death appears to have a worse effect on cardiac hemodynamics than segmental loss (2). Experimental studies as well as observational evidence in humans indicate that over 40% of myocytes need to die as a result of acute myocardial infarction to cause overt heart failure. On the other hand, conditions associated with diffuse myocyte loss, such as systemic hypertension, pacing-induced heart failure, and diffuse nonocclusive coronary artery disease, can lead to heart failure even when less than 20% of cells are lost.

Several mechanisms have been postulated for ongoing myocyte loss; the most accepted one being increased wall stress leading to energy imbalance and ischemia. Myocardial hypertrophy that develops during remodeling also leads to relative ischemia, which leads to a vicious cycle of increased wall thickness, stress, and ischemia. In addition, reactive and replacement fibrosis associated with remodeling also leads to a reduction in capillary density in the myocardium. Numerous factors present in failing myocardium, such as angiotensin II, catecholamines, cytokines, reactive oxygen species, mechanical stress, and natriuretic peptide can induce myocyte death. Myocyte hypertrophy occurring as a part of the remodeling process can eventually lead to cell death. In a hypertrophied cell, the nucleus/cell volume ratio is not able to take care of all the transcriptional activity required, resulting in cell exhaustion and death.

Cardiomyocytes, for a long time, were considered to be capable of dying only by necrosis. However, over the last few years, it has been increasingly recognized that cell death in the failing heart may occur by caspase-dependent and -independent programmed cell death known as apoptosis, as well as by autophagy associated with ubiquitinated protein accumulation (Fig. 1).

Cell Death by Necrosis

Myocyte death by necrosis is associated with the disruption of cell membrane and loss of cell contents (5). Causes of necrosis in failing heart may include ongoing ischemia, immunoinflammatory diseases, and chemical insults such as alcohol and doxorubicin toxicity. Following an inciting agent, several biochemical pathways mediate cell death, including adenosine triphosphate (ATP) depletion, oxidative stress, and intracellular increase in calcium. These changes lead to the activation of phospholipases, proteases, ATPases, and endonucleases, and eventually mitochondrial and cell membrane damage (6). The loss of cell membrane integrity defines death by necrosis and precedes influx of extracellular fluid, cellular swelling, and release of proteolytic enzymes that cause further cell disruption and induction of inflammation. Necrotic cells are characterized by an overt breach in the plasma membrane, dilatation of mitochondria with large amorphous densities, intracellular accumulation of fluffy material representing denatured protein, and breakdown of nuclear DNA as karyolysis, pyknosis, or karyorrhexis.

Monoclonal antibodies to heavy chains of myosin, labeled with appropriate radiotracers, are commonly used for in vivo detection of the extent of necrotic myocyte cell death in heart failure. Myocytes with rupture of the sarcolemmal membrane allow antimyosin to enter the cells and bind to myofibrillar myosin, whereas nonnecrotic cells with intact sarcolemma do not accumulate the antibody (7–10). Necrosis not only accounts for cell loss but also leads to the generation of fibrosis, which in itself is an important constituent of remodeling (11).

Cell Death by Apoptosis

Apoptosis also leads to heart muscle cell loss. In acute myocardial infarction, there is necrosis of myocytes within the infarct zone. However, both immediately and later in the postinfarct course, there is evidence of apoptosis in the peri-infarct zone and in the myocardium remote from the infarct site. The degree of late apoptosis at both these sites correlates with macroscopic signs of postinfarction ventricular remodeling, such as increases in left ventricular dimensions, free wall thickness, diameter-

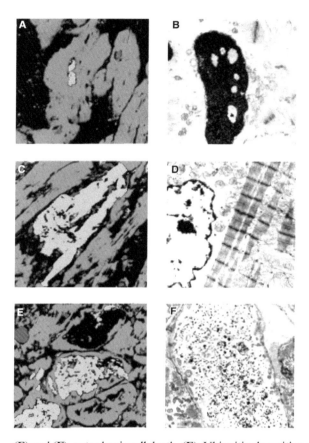

Figure 1 Different types of cell death in the myocardium. (**A**), (**C**), and (**E**): Confocal microscopic pictures (*dark grey is actin*); (**B**), (**D**), and (**F**): Electron microscopic pictures (*all bars* = 2 μm). (**A**) and (**B**): Apoptotic cell death demonstrated by TUNEL method. (**A**), nuclei with DNA fragmentation are light grey. (**B**), nuclei show condensed chromatin. (**C**) and (**D**), necrotic cell death. (**C**), single cell necrosis demonstrated by C9-labeling (*seen as light grey*). (**D**), nuclei are electron-lucent with clumped chromatin, mitochondria are damaged with flocculent densities. (**E**) and (**F**), autophagic cell death. (**E**), Ubiquitin deposition (*seen as light grey*). (**F**), ultrastructural appearance with numerous autophagic vacuoles. *Abbreviations*: TUNEL, terminal deoxynucleotidyl transferase biotin-dUTP nick end labeling. *Source*: From Ref. 24. (*See color insert.*)

to-wall thickness ratio, and mass. Apoptosis rate has been reported to be higher in patients with persistent infarct-related artery occlusion (12,13), and it correlates with the severity of clinical manifestations and rapidity of progression of heart failure in patients with ischemic heart disease (14). Unlike necrosis, apoptosis is not associated with activation of an inflammatory reaction, vascular proliferation, or collagen deposition.

Apoptosis is a genetically programmed, active-energy requiring process that is mediated through the activation of proteolytic caspases, which eventually leads to cytoplasmic proteolysis and DNA fragmentation (15). The initiation of the apoptotic process in end-stage heart failure is evidenced by the release of cytochrome c from the mitochondrial compartment, and its association with upregulation and activation of caspases 8, 9, and 3 (16).

The inducers of apoptosis lead to the activation of caspase 3 either by release of cytochrome c or by activation of downstream executionary caspases. Activation of caspase 3 has been found to contribute to the impairment of systolic function by damaging myofilaments, α-actin, α-actinin, and troponin T (17). However, due to gradual loss of DNAases, nuclei remain intact and these cells may exist in a state of suspended animation. This is relevant for development of therapeutic interventions that may allow reverse remodeling and improvement in heart function.

It has been noted in various experimental models that attenuation of the apoptotic process in the overloaded myocardium is associated with positive effects on hemodynamic performance and remodeling. Long-term antihypertensive treatment with an angiotensin-converting enzyme (ACE) inhibitor blocks apoptosis of cardiomyocytes in adult rats with genetic hypertension, and this therapy is associated with a reduction in heart failure and reverse remodeling (18). Activation of apoptotic and necrotic processes is also inhibited by insulin-like growth factor-1 (IGF-1); transgenic mice overexpressing human IGF-1β in their myocytes were able to prevent activation of cell death in the viable myocardium after infarction, limiting ventricular dilation, myocardial loading, and cardiac hypertrophy (10,19). β-Receptor agonism has been established to induce myocyte apoptosis in vitro, and treatment with beta-blockers significantly reduces apoptosis in experimental heart failure models (20). Moreover, direct inhibition of caspases attenuates the development of cardiomyopathy in transgenic mice (Fig. 2) (21).

DNA fragmentation is the hallmark of apoptosis and can be demonstrated in tissue samples by DNA gel electrophoresis. The TUNEL (terminal deoxynucleotidyl transferase biotin-dUTP nick end labeling) technique, however, is the most

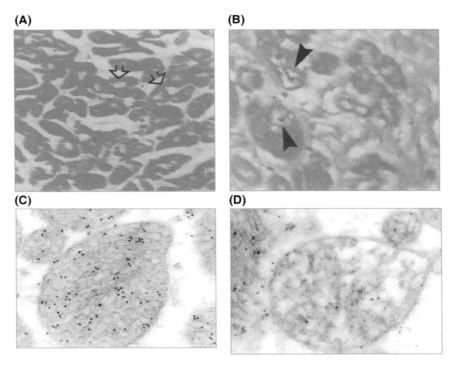

Figure 2 Histologic and ultrastructural evidence of apoptosis in heart failure. A histologic specimen from a victim of a motor vehicle accident shows absence of TUNEL staining (*clear nuclear regions, open arrows*) (**A**). Two myocytes in the cardiomyopathic heart demonstrate TUNEL-positive nuclei (*blue-black color, arrowheads*) (**B**). The ultrastructural examination with immunoelectron microscopy reveals cytochrome c (immunogold particles) localized in the mitochondrial compartment in the normal heart (**C**). On the other hand, cytochrome c is released from the mitochondrial to the cytoplasmic compartment in a cardiomyopathic heart, which is seen sprinkled on to the background of contractile proteins (**D**). (*See color insert.*)

commonly used method for identification of apoptosis in tissue samples. The histochemical technique is used here to identify DNA fragmentation in the nuclei of individual cells. Electron microscope pictures of apoptotic cells reveal nuclear condensation with intact mitochondria and sarcolemma.

Autophagic Cell Death

Autophagic machinery with the ubiquitin/protein degradation system is highly conserved through evolution (22). Ubiquitin-mediated proteolysis has an important part to play in a variety of cellular processes such as regulation of cell cycle and division, differentiation and development, and cellular response to stress and extracellular effectors. This is functionally analogous to autophagy, which ensures turnover of cellular organelles (23). Autophagic cell death may be secondary to stress and oxidative damage to the cell proteins. Degradation of proteins via this pathway involves labeling of lysine residues of the substrate proteins with ubiquitin, followed by degradation of labeled proteins by the 26S proteosome complex with release of free and reusable ubiquitin. Ubiquitin is a 76-residue polypeptide and its conjugation to the protein is a three-step cascade mechanism. First, in an ATP-requiring reaction, a high-energy thiol-ester intermediate is generated by the ubiquitin-activating enzyme E1. One of the several ubiquitin-conjugating enzymes E2 then transfers the activated ubiquitin moiety from E1 to the ubiquitin-ligase E3. By successively adding activated ubiquitin molecules to lysine residues on the previously conjugated ubiquitin molecule, a polyubiquitin chain is synthesized. The E3 also serve as the specific recognition factors of the system. After protein degradation, short peptides and ubiquitin are released. Addition of four or more ubiquitin molecules leads a protein to destruction.

Shorter chains, however, may contribute to survival functions including gene transcription and DNA repair. Also, ubiquitination is not an irreversible process and ubiquitin can be removed from the proteins by enzymes like isopeptidase T and the ubiquitin-fusion degradation system. Provided ATP resources are not exhausted, storage of ubiquitin/protein complexes is a slow process that ultimately results in loss of nucleus and death of the myocyte.

A recent study of 19 explanted hearts, from patients with idiopathic dilated cardiomyopathy, demonstrated different stages of ubiquitination ranging from the deposition of small nuclear or cytosolic aggregates to large accumulations. The exclusive accumulation of ubiquitin in myocytes positive for monodansylcadaverine was also observed, establishing a link between ubiquitin accumulations and autophagy (24). Autophagic cell death was found to be connected with major defects in the ubiquitin/proteosome cascade. There was an upregulation of ubiquitin conjugation pathway and reduction in deubiquitinating enzymes. It has been postulated that downregulation of lysosomal proteolytic enzyme cathepsin D accompanied by the loss of deubiquitinating enzymes leads to accumulation of polyubiquitinated proteins which are responsible for autophagic cell death (24). Autophagic cell death is recognized by the presence of multiple ubiquitin accumulations and loss of nuclei. Ultrastructurally, these cells show numerous autophagic vacuoles with loss of contractile and nuclear material (Fig. 3).

Relative Contribution of Different Modes of Cell Death

The rate of myocyte loss by diverse processes is highly variable suggesting that different modes of death may be interchangeable depending on several highly

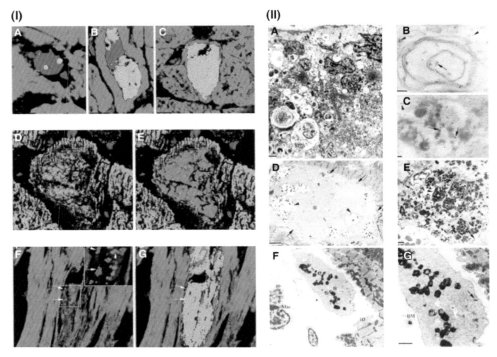

Figure 3 (**I**) Immunoconfocal patterns of ubiquitin-accumulations, sarcomeric proteins, and signs of autophagy in myocytes with ubiquitin-inclusions (*ubiquitin seen as light grey*). (**A**), Only two small dots of ubiquitin signal in a myocyte nucleus. (**B**), Punctate nuclear and massive cytoplasmic ubiquitin labeling in one myocyte. (**C**), Massive deposition of ubiquitin in a myocyte without a nucleus (*dark grey is actin*). (**D**) and (**E**), Double labeling for ubiquitin and myosin showing colocalization of these proteins and that ubiquitinated myosin lacks a typical cross-striated pattern (*dark grey is myosin*). (**F**) and (**G**), Double labeling for ubiquitin (*light grey*) with monodansylcadaverine [*dark grey shown with arrows in* (***F***)]. Shown with arrows in (**F**) are the autophagic vacuoles. (**G**), Double labeling for ubiquitin and monodansylcadaverine showing colocalization of these two signals. (**II**) Ultrastructural features of autophagic myocytes in patients with DCM. (**A**), Extensive cytoplasmic accumulations of autophagic vacuoles and myelin figures. Severe disintegration of the nuclei into 7 particles and only a narrow rim of the contractile material confined to the cell periphery is seen (arrows). (**B**) and (**C**), Immunogold electron micrographs of autophagic myocytes show that ubiquitin is localized in the cytosol (*arrowhead*) and within a vacuolar-lamellar structure (*arrow*) corresponding to those shown in (**A**) with arrowheads. Similar distribution of ubiquitin was found in vacuolar-lysosomal structures (**C**) corresponding to those shown with arrowheads in (**D**) and (**E**). (**D**), Myocyte (*asterisk*) with numerous autophagic vacuoles in the absence of contractile filaments and a nucleus. (**E**), Myocyte (*asterisk*) displaying electron dense autophagic vacuoles. Myocytes in (**D**) and (**E**) show still existing cell-to-cell contacts (*arrow*) at the intercalated disk (ID) region of the neighboring myocyte. (**F**), Very small myocyte (*asterisk*) detached at the ID from an apparently normal myocyte. Macrophage (Mac) is seen in close vicinity with the degenerating myocyte. (**G**), Enlargement of the central part of the degenerating cell (*asterisk*) shown in (**F**) displaying sarcoplasmic reticulum (SR) structures, remnants of Z-disks (*arrow*), and a basement membrane (BM). Scale bars = 5 μm in (**A**) and (**D**), 2 μm in (**E**) through (**G**), 250 nm in (**B**) and (**C**). *Source*: From Ref. 24. (*See color insert.*)

modifiable intra- and extracellular conditions. Apoptosis is an energy-requiring process, and exhaustion of energy during the process may lead to cell death by necrosis. On the other hand, an early necrotic stimulus may divert to apoptotic pathways, if energy is restored (as by reperfusion in the setting of acute myocardial infarction). Presence of multiple morphologies and, more so, hybrid morphologies between various types of cell death (particularly between necrosis and apoptosis) may consolidate the concept of interchangeability between the various forms of cell death (25). Cellular triggers, classically known to cause one form of cell death, sometimes are seen to mediate other forms of cell death, depending on the factors such as nature and intensity of the stressor, degree of ATP depletion, amount of poly-ADP-ribose polymerase (PARP) cleavage, or the rate of mitochondrial permeability transition (26–29). In addition to apoptosis and necrosis, autophagy may contribute to cell death (24). Ubiquitination and autophagy can modulate apoptosis by influencing degradation of caspases and various regulator proteins. Therefore, it is difficult to assign any quantitative figures to relative contribution of each type of cell death, as the proportion of cells dying through each mechanism may differ at various stages in the natural history of heart failure; and therapeutic interventions may also trigger the balance from one to another.

MYOCYTE REGENERATION AND REMODELING

As discussed, myocyte loss is an inexorable process in the failing heart. However, even in this setting, the tissue sections performed during routine histologic examination show a significant amount of intact myocardium, which may exceed in absolute volume than the muscle mass present in control hearts (30–34). This may occur as a result of reactive hypertrophy and to a small extent by proliferation of the residual myocytes. Over the last few years, several workers have tried to study the effector pathways involved in the translation of mechanical signals generated by hemodynamic overload into molecular events resulting in myocyte hypertrophy and proliferation (7,8,35,36). The overload generated by the loss of cells, the hemodynamic alterations produced by the combination of the changes in cardiac anatomy, and the depression in myocardial performance initiate reactive growth processes in the myocardium. In the initial stages, the increase in myocyte volume may be able to normalize the alterations in systolic and diastolic wall stresses. Pathologists have been generally unable to detect mitotic figures in the myocyte nuclei, and therefore myocytes have been considered to be terminally differentiated cells. However, recently, some evidence of myocyte proliferation has been observed following cell injury, which may be secondary to cellular replication, stem cell immigration, or revitalization of partially injured cells (37).

Myocyte Replication

The entire machinery required for the entry of myocytes into cell cycle is present at baseline and it is upregulated in the failing heart (38). Induction of late growth-related genes and activation of cyclins occurs in acute cardiac decompensation following myocardial infarction or coronary constriction in animal models (38,39). This is associated with high levels of DNA replication, karyokinesis, and cytokinesis. Increase in the number of myocytes has been noticed on quantitative studies in severe myocardial hypertrophy (40). Some investigators have demonstrated that human ventricular myocytes are able to synthesize DNA and re-enter the mitotic cycle even in the absence of physiological load (41).

Study of the explanted hearts from patients with ischemic and idiopathic dilated cardiomyopathy, using confocal microscopy, has demonstrated that of a million myocytes, only 14 were in mitosis under normal circumstance and this number increased ten times in the failing myocardium (42). Of note is the fact that no evidence of apoptosis was found in the dividing cells. In the patients dying 4–12 days after acute myocardial infarction, myocytes in cell division cycle, identified by labeling with the nuclear antigen Ki-67, were observed in 4% of peri-infarct myocytes and 1% of myocytes in the distant regions (43). Curiously, no myocyte regeneration is evident in the infarct zone, which thins out and gets fibrosed during postmyocardial infarction remodeling (Figs. 5 and 6).

Stem Cell Migration in Failing Myocardium

The origin of the replicating myocytes in failing myocardium is not yet clear. As opposed to increase in cell number related to dividing myocytes, myocardial renewal may be secondary to immigration of a differentiated progeny of stem cells under appropriate circumstances. It remains to be clarified whether the precursor cells are cardiac stem cells that had accumulated in the heart early during development or are the progeny of hematopoietic stem cells that transfer to myocardium later in life.

Hematopoietic stem cells appear to have the capacity to sense injury to distant target organs, which enable them to migrate to the site of damage and undergo differentiation (44,45). These events may promote structural and functional alterations in the target organ. This has been noted in several organ systems including the nervous system and liver. Recent studies have proposed that bone marrow cells when injected into myocardium can acquire characteristics of cardiac stem cells and differentiate to

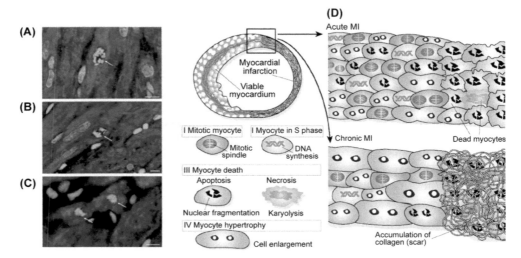

Figure 4 Evidence of mitotic activity in the cardiomyocytes following myocardial infarction in humans. Myocyte division a–c, Stages of mitosis in human myocytes: (**A**) metaphase chromosomes, (**B**) karyokinesis, and (**C**) cytokinesis are shown by fluorescence of propidium iodide (*seen here as light grey*). Dark grey reflects cardiac myosin antibody staining of myocyte cytoplasm. Arrows indicate mitosis. (**D**) Changes in myocyte growth in acute and chronic infarcts. MI, myocardial infarct. *Source*: From Ref. 37. (*See color insert.*)

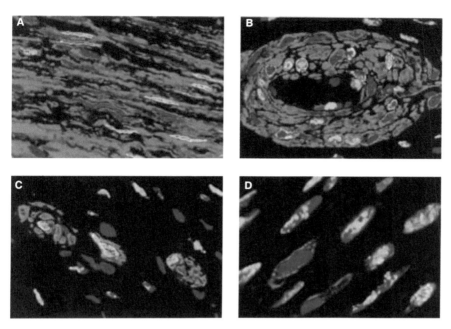

Figure 5 Regeneration of myocytes and vessels following cytokine-induced stem cell mobilization to the infarcted heart in mice. Ki-67 (**A** and **B**) and BrdUrd (**C** and **D**) labeling of myocytes (**A**), smooth muscle cells (**B** and **C**), and endothelial cells (EC) (**D**) in the forming myocardium. Myocytes are stained by cardiac myosin (**A**), SMC by a -smooth muscle actin (**B** and **C**), and EC by factor VIII (**D**). Bright fluorescence of nuclei (seen here as light grey) reflects the combination of propidium iodide (PI) and Ki-67 (**A** and **B**) or PI and BrdUrd (**C** and **D**). *Source*: From Ref. 47. (*See color insert.*)

contribute to the regeneration of myocardium (46–48). Differentiation of bone marrow cells into myocytes and coronary vessels has been observed after direct injection into the peri-infarct zone. Migration of stem cells into the infarcted portions of the heart can be facilitated by the use of cytokines in experimental models. The migration is accompanied by significant myocardial repair and recovery in ventricular function. Cytokine-induced stem cell transfer has been reproduced in infarct zone in a mice model, which contributed to myocyte and vascular regeneration, decreased apoptosis of hypertrophic myocytes in the peri-infarct region, long-term salvage of viable myocardium, reduction in collagen deposition, and sustained improvement in cardiac function. These concepts have formed the basis of emerging interventions to replete myocyte loss after myocardial infarction and to reverse remodeling in the failing heart.

CONCLUSIONS

Myocyte loss and regeneration, though present in healthy myocardium, are significantly amplified in the failing heart. Together with interstial changes and fibrosis, contractile dysfunction of myocytes is responsible for the phenotype of remodeling. The significance of understanding various mechanisms for cell death lies in the ability to prevent or reverse cell death. The concept of cellular proliferation, although yet not fully accepted, offers newer vistas for recovery of functional myocardium. Cell death and proliferation may vary depending on the mode of stress and the stage of the disease. Efforts are being made to prevent the cell loss, augment cell proliferation and to reverse remodeling.

REFERENCES

1. Jay N. Cohn, Roberto Ferrari, Norman Sharpe and on Behalf of an International Forum on Cardiac Remodeling. Cardiac remodeling—concepts and clinical implications: a consensus paper from an international forum on cardiac remodeling. JACC 2000; 35:569–582.
2. Anversa P, Zhang X, Li P, Capasso JM. Chronic coronary artery constriction leads to moderate myocyte loss and left ventricular dysfunction and failure in rats. J Clin Invest 1992; 89:618–629.
3. McKay RG, Pfeffer MA, Pasternak RC, Markis JE, Come GC, Nakao C, Alderman JD, Ferguson JJ, Safian RD, Grossman W. Left ventricular remodeling after myocardial infarction: a corollary to infarct expansion. Circulation 1986; 74:693–702.
4. Olivetti G, Capasso JM, Meggs LG, Sonnenblick EH, Anversa P. Cellular basis of chronic ventricular remodeling after myocardial infarction in rats. Circ Res 1991; 68: 856–869.
5. Narula J, Hofstra L. Imaging myocardial necrosis and apoptosis. In: Dilsizian V, Narula J, eds. Series Ed Braunwald E. Atlas of Nuclear Cardiology. Philadelphia: Curr Med 2003:197–216.
6. Kumar V. Cellular Pathology I: Cell injury and cell death. In: Cotran RS, Kumar V, Collins T, eds. Robbin's Pathologic Basis of Disease. 6th ed. Philadelphia: WB Saunders, 1999:8.
7. Anversa P, Kajstura J, Cheng W, Reiss K, Cigola E, Olivetti G. Insulin like growth factor-1 and myocyte growth: the danger of a dogma. Part 1. Postnatal myocardial development: Normal growth. Cardiovasc Res 1996; 32:219–225.
8. Anversa P, Kajstura J, Cheng W, Reiss K, Cigola E, Olivetti G. Insulin like growth factor-1 and myocyte growth: the danger of a dogma. Part 2. Induced myocardial growth: pathological hypertrophy. Cardiovasc Res 1996; 32:495–502.
9. Benjamin IJ, Jalil JE, Tan LB, Cho K, Weber KT, Clark WA. Isoproterenol-induced myocardial fibrosis in relation to myocyte necrosis. Circ Res 1989; 67:657–670.
10. Li Q, Li B, Wang X, Leri A, Jana KP, Liu Y, Kajstura J, Baserga R, Anversa P. Overexpression of insulin like growth factor-1 in mice protects from myocyte death after infarction, attenuating ventricular dilatation, wall stress and cardiac hypertrophy. J Clin Invest 1997; 100:1991–1999.
11. Darzynkiewiez Z, Juan G, Li X, Gorczyca W, Murakami T, Traganos F. Cytometry in cell necrobiology. Analysis of apoptosis, accidental cell death and necrosis. Cytometry 1997; 27:1–20.
12. Baldi A, Abbate A, Bussani R, Patti R, Melfi R, Angelini A, Dobrina A, Rossiello R, Silvestri F, Baldi F, Di Sciascio G. Apoptosis and post-infarction left ventricular remodeling. J Mol Cell Cardiol 2002; 34:165–174.
13. Abbate A, Biondi-Zoccai GGL, Bussani R, Dobrina A, Camilot D, Feroce F, Rossiello R, Baldi F, Silvestri F, Biasucci LM, Baldi A. Increased myocardial apoptosis in patients with unfavorable left ventricular remodeling and early symptomatic post-infarction heart failure. J Am Coll Cardiol 2003; 41:753–760.
14. Saraste A, Pulkki K, Kallajoki M, Heikkila P, Laine P, Mattila S, Neiminen NS, Parvinen M, Voipio-Pulkki LM. Cardiomyocyte apoptosis and progression of heart failure to transplantation. Eur J Clin Invest 1999; 29:380–386.
15. Stegh AH, Peter ME. Apoptosis and caspases. Cardiol Clinics 2001; 19:13–29.
16. Narula J, Pandey P, Arbustini E, Haider N, Narula N, Kolodgie FD, Dal Bello B, Semigran MJ, Bielsa-Masdeu A, Dec GW, Israels S, Ballester M, Virmani R, Saxena S,

Kharbanda S. Apoptosis in heart failure: release of cytochrome-c and activation of caspase-3 in human cardiomyopathy. Proc Natl Acad Sci USA 1999; 96:8144–8149.

17. Communal C, Sumandea M, de Tombe P, Narula J, Solaro RJ, Hajjar RJ. Functional consequences of caspase activation in cardiac myocytes. Proc Natl Acad Sci 2002; 99: 6252–6256.

18. Diez J, Panizo A, Hernandez M, Vega F, Sola I, Fortuno MA, Pardo J. Cardiomyocyte apoptosis and cardiac angiotensin-converting enzyme in spontaneously hypertensive rats. Hypertension 1997; 30:1029–1034.

19. Buerke M, Murohara T, Skurk C, Nuss C, Tomaselli K, Lefer AM. Cardioprotective effect of insulin-like growth factor 1 in myocardial ischemia followed by reperfusion. Proc Natl Acad Sci USA 1995; 92:8031–8035.

20. Sabbah HN, Sharov VG, Gupta RC, Todor A, Singh V, Goldstein S. Chronic therapy with metoprolol attenuates cardiomyocyte apoptosis in dogs with heart failure. J Am Coll Cardiol 2000; 36:1698–1705.

21. Wenker D, Chandra M, Nguyen K, Miao W, Garantziotis S, Factor SM, Shirani J, Armstrong RC, Kitsis RN. A mechanistic role for cardiac myocyte apoptosis in heart failure. J Clin Invest 2003; 111:1497–1504.

22. Glickman MH, Ciechanover A. The ubiquitin-proteasome proteolytic pathway: destruction for the sake of construction. Physiol Rev 2002; 82:373–428.

23. Bursch W, Hochegger K, Török L, Marian B, Ellinger A, Hermann RS. Autophagic and apoptotic types of programmed cell death exhibit different fates of cytoskeletal filaments. J Cell Sci 2000; 113:1189–1198.

24. Kostin S, Pool L, Elsässer A, Hein S, Drexler HCA, Arnon E, Hayakawa Y, Zimmermann R, Bauer E, Klövekorn W-P, Schaper J. Myocytes die by multiple mechanisms in failing human hearts. Circ Res 2003; 92:715–724.

25. Sperandio S, de Belle I, Bredesen DE. An alternate non apoptotic form of cell death. Proc Natl Acad Sci USA 2000; 97:14376–14381.

26. Fischer S, Maclean AA, Liu M, Cardella JA, Slutsky AS, Suga M, Moreira JF, Keshavjee S. Dynamic changes in apoptotic and necrotic cell death correlate with severity of ischemia-reperfusion injury in lung transplantation. Am J Respir Crit Care Med 2000; 162: 1932–1939.

27. Lieberthal W, Menza SA, Levine JS. Graded ATP depletion can cause necrosis or apoptosis of cultured mouse proximal tubular cells. Am J Physiol Renal Physiol 1998; 274: F315–F327.

28. Los M, Mozoluk M, Ferrari D, Stepczynska A, Stroh C, Renz A, Herceg Z, Wang ZQ, Schulze-Osthoff K. Activation and caspase-mediated inhibition of PARP: a molecular switch between fibroblast necrosis and apoptosis in death receptor signaling. Mol Biol Cell 2002; 13:978–988.

29. Lemasters JJV. Necrapoptosis and the mitochondrial permeability transition: shared pathways to necrosis and apoptosis. Am J Physiol Gastrointest Liver Physiol 1999; 276:G1–G6.

30. Arbustini E, Pozzi R, Grasso M, Gavazzi A, Diegoli M, Bramerio M, Specchia G. Pathologic substrates and clinical correlates of coronary artery disease and chronic congestive heart failure requiring cardiac transplantation. Coron Artery Dis 1991; 2:605–612.

31. Beltrami CA, Finato N, Rocco M, Feruglio GA, Puricelli C, Cigola E, Sonnenblick EH, Olivetti G, Anversa P. The cellular basis of dilated cardiomyopathy in humans. J Mol Cell Cardiol 1995; 27:291–305.

32. Buja LM, Willerson JT. The role of coronary artery lesions in ischemic heart disease: insight from recent clinicopathologic, coronary angiographic and experimental studies. Hum Pathol 1987; 18:451–461.

33. Roberts WC. The coronary arteries and left ventricle in clinically isolated angina pectoris: a necropsy analysis. Circulation 1976; 54:388–390.

34. Beltrami CA, Finato N, Rocco M, Feruglio GA, Puricelli C, Cigola E, Quaini G, Sonnenblick EH, Olivetti G, Anversa P. Structural basis of end-stage failure in ischemic cardiomyopathy in humans. Circulation 1994; 89:151–163.

35. Anversa P, Kajstura J. Ventricular myocytes are not terminally differentiated in the adult mammalian heart. Circ Res 1998; 83:1–14.

36. Chien KR, Knowlton KU, Zhu H, Chien S. Regulation of cardiac gene expression during myocardial growth and hypertrophy: molecular studies of an adaptive physiologic response. FASEB J 1991; 55:3037–3046.

37. Anversa P, Nadal-Ginard B. Myocyte renewal and ventricular remodeling. Nature 220; 415:240–243.

38. Reiss K, Cheng W, Giordano A, De Luca A, Li B, Kajstura J, Anversa P. Myocardial infarction is coupled with the activation of cyclins and cyclin dependent kinases in myocytes. Exp Cell Res 1996; 225:44–54.

39. Reiss K, Kajstura J, Capasso JM, Marino TA, Anversa P. Impairment of myocyte contractility following coronary artery narrowing is associated with activation of the myocyte IGF-1 autocrine system, enhanced expression of late growth related genes, DNA-synthesis and myocyte nuclear mitotic division in rats. Exp Cell Res 1993; 207: 348–360.

40. Astorri E, Bolognesi R, Colla B, Chizzola A, Visioli O. Left ventricular hypertrophy: a cytometric study on 42 human hearts. J Mol Cell Cardiol 1977; 9:763–775.

41. Beltrami CA, Di Loreto C, Finato N, Rocco M, Artico D, Cigola E, Gambert SR, Olivetti G, Kajstura J, Anversa P. Proliferating cell nuclear antigen (PCNA), DNA synthesis and mitosis in myocytes following cardiac transplantation in man. J Mol Cell Cardiol 1997; 29:2789–2802.

42. Kajstura J, Leri A, Finato N, Di Loreto C, Beltrami CA, Anversa P. Myocyte proliferation in end-stage cardiac failure in humans. Proc Natl Acad Sci USA 1998; 95:8801–8805.

43. Beltrami AP, Urbanek K, Kajstura Jan, Yan SM, Finato N, Bussani R, Nadal-Ginard B, Silvestri F, Leri A, Beltrami CA, Anversa P. Evidence that human cardiac myocytes divide after myocardial infarction. N Engl J Med 2001; 344:1750–1757.

44. Eglitis MA, Mezey E. Hematopoietic cells differentiate into both microglia and macroglia in the brains of adult mice. Proc Natl Acad Sci USA 1997; 94:4080–4085.

45. Theise ND, Nimmakayalu M, Gardner R, Illei PB, Morgan G, Teperman L, Henegariu O, Krause DS. Liver from bone marrow in humans. Hepatology 2000; 32:11–16.

46. Orlic D, Kajstura J, Chimenti S, Jakoniuk I, Anderson S, Li B, Pickel J, McKay R, Nadal-Ginard B, Bodine D, Leri A, Anversa P. Bone marrow cells regenerate infarcted myocardium. [Letter]. Nature 2001; 410(6829):701–705.

47. Orlic D, Kajstura J, Chimenti S, Limana F, Jakoniuk I, Quaini F, Nadal-Ginard B, Bodine DM, Leri A, Anversa P. Mobilized bone marrow cells repair the infarcted heart improving function and survival. Proc Natl Acad Sci USA 2001; 98:10344–10349.

48. Kocher AA, Schuster MD, Szabolcs MJ, Takuma S, Burkhoff D, Wang J, Homma S, Edwards NM, Itescu S. Neovascularization of ischemic myocardium by human bone-marrow-derived angioblasts prevents cardiomyocyte apoptosis, reduces remodeling and improves cardiac function. Nature Med 2001; 7:430–436.

8

Remodeling from Compensated Hypertrophy to Heart Failure

Functional-Structural Correlations of Remodeling in the Human Heart

Stefan Hein
Department of Cardiac Surgery, Kerckhoff-Clinic, Bad Nauheim, Hessen, Germany

Jutta Schaper
Department of Experimental Cardiology, Max-Planck-Institute, Bad Nauheim, Hessen, Germany

The classical definition of remodeling as "genome expression resulting in molecular, cellular, and interstitial changes" was established by Cohn et al. (1) in 2000. Accordingly, in this chapter we will try to identify the cellular and interstitial changes that are characteristic of the remodeling process, which eventually lead to failure of the hypertrophied human heart.

Experimental animal models have frequently been used to define the processes leading to heart failure under pressure or volume overload conditions [for review see (2)]. Despite their obvious usefulness, these animal experiments are not able to completely mimic the situation in the human heart, where mechanical overload, beginning at a very low degree, is a chronic situation present over years or decades. In contrast to the experimental condition in animals, left ventricular (LV) systolic function in the human heart is maintained at a constant level for a longer period of time (many years), even when valve stenosis is pronounced. The duration of animal experiments is usually only weeks or months and they comprise an abrupt induction of overload in completely normal hearts without allowing for chronic structural decompensation to occur.

Since the introduction of cardiac surgery and transplantation studies in human hearts have been optimized. Nowadays, during open-heart surgical procedures or cardiological interventions, tissue samples can be removed from living human hearts and immediately preserved in liquid nitrogen or buffered glutaraldehyde. Immediate fixation offers the unique advantage that any autolytic alterations of the myocardium are avoided. Moreover, during cardiac transplantation procedures, numerous tissue samples from whole hearts can be obtained and optimally preserved for structural and molecular biological analyses. In addition, the technical progress in histology by the introduction of laser confocal microscopy and the commercial availability of a large variety of antibodies further facilitates studies in human hearts. Accordingly, this chapter will describe the structural alterations that occur in the human heart during the transition from compensated hypertrophy to cardiac decompensation and it will correlate these findings with clinical functional data.

GENERAL MORPHOLOGICAL CONSIDERATIONS

Myocyte Structure and the Interstitium

Normal myocardium consists of 90% myocytes and 10% of interstitial tissue (3). Myocytes are about 80 to 120 μm long and 15 to 20 μm thick; the normal cross-sectional area is 400 to 420 μm^2 (4). The interstitial tissue represents the scaffold, which provides stability to the continuously contracting myocyte system. It consists of the extracellular matrix (ECM), fibroblasts, macrophages, and blood vessels; 75% of all cell nuclei are in the interstitial space. The ECM includes fibronectin, various collagen isoforms, the proteoglycans, and laminin (5,6).

In human hearts, the nucleus/cell volume ratio is 1.0%, and 95% of all myocytes are mononucleated (3). Myocytes consist of 22% mitochondria, 56% myofilaments, and 12% cytoplasm containing cellular organelles such as sarcoplasmic reticulum (SR), the T-tubular system, ribosomes, Golgi apparatus, endoplasmic reticulum and the intercalated disc (7,8). This structural composition provides a high-performance contractile machinery with sufficient energy supply and fine-tuned impulse conduction, which guarantees its incessant activity during the lifetime of an individual.

Myocyte Degeneration

The term "degeneration" has almost been deleted from the literature (9). However, it will be used here because it seems to be very appropriate for summarizing those structural changes in myocytes that indicate structural deterioration. Furthermore, the term "degeneration" was chosen because it characterizes involvement of all cellular organelles in a chronic and most probably slow process of degradation. It finally results in cellular sequestration, atrophy, and cell death followed by replacement fibrosis (4).

Characteristic degenerative changes include the reduction of the contractile material and disturbance of the morphological integrity of the remaining sarcomeres, aberrations in size and shape of mitochondria and nuclei, and an increase in the amount of free cytoplasm, of prominent SR and of the T-tubular system (10). These changes can be identified by electron microscopy as well as confocal microscopy using immunofluorescent staining methods (Fig. 1). The latter permits identification of proteins that cannot be separately identified in the electron microscope, such as the different components of the Z-line, e.g., titin, α-actinin, telethonin, muscle lim protein (MLP), or of the cytoskeleton, the intercalated disc, or of the sarcolemma (Fig. 2). Combined, these morphological methods allow for a rather comprehensive analysis of cellular and interstitial genome alterations.

The Problem of Cell Death

There is no doubt that in hypertrophied hearts myocyte loss occurs during the transition to heart failure (11). Three different forms of cell death should be considered: apoptosis, oncosis, and ubiquitin-related autophagic cell death.

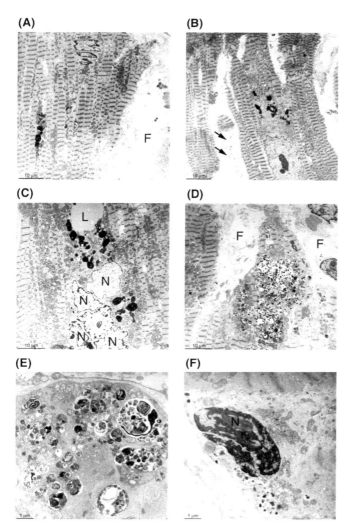

Figure 1 Myocyte degeneration evaluated by electron microscopy: (**A**) Human myocardium with normally structured myocytes but increased interstitial tissue (*right side of the figure*) indicating fibrosis (F). (**B**) Beginning of myocyte degeneration with increased cytoplasmic space and lack of sarcomeres (*center*). Note the presence of fibrosis and the small dying myocyte with only a few sarcomeres and absence of nucleus (*large arrows*). (**C**) Degenerating myocyte with nuclear fragmentation (N) and a lipid vacuole (L). (**D**) Accumulation of autophagic vacuoles in a degenerating myocyte. Note the "tapering out" of the myocyte at the top of the figure and the fibrotic tissue (F) surrounding the cell. (**E**) Enlargement of autophagic vacuoles revealing the typical structure in a sarcomere-free area of a myocyte. (**F**) Typical appearance of an apoptotic nucleus showing condensed chromatin. N-nucleus.

Apoptosis, the suicidal cell death, is characterized by fragmentation of DNA into regular 185 kB pieces that can be identified by the TUNEL method and other techniques. It is a programmed, ATP- and caspase-dependent type of cell death. Structurally, the most obvious alteration is dense condensation of nuclear chromatin in the presence of intact mitochondria and sarcolemma (12) (Fig. 1F).

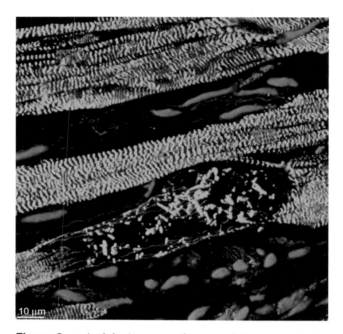

Figure 2 α-Actinin (*green, nuclei are red*) for myocyte degeneration (note the severe disorganization of the sarcomeric structure in the myocyte in the lower part of the picture. (*See color insert.*)

Oncosis is the accidental mode of cell death, predominantly caused by ischemia with random fragmentation of the nuclear DNA resulting in clumping of the chromatin. The mitochondria appear swollen and cleared, the sarcolemma is leaky, and the cell is edematous. Oncosis may be identified using ultrastructural analysis (13,14). Furthermore, since the leaky cell membrane permits the permeation of proteins including complement 9 (C9) into the cells, it can be monitored by confocal microscopy using immunofluorescence staining with an antibody against C9 (15) (Fig. 3).

The protein turnover, which is a tightly regulated process in any tissue, is highly ensured by the ubiquitin–proteasomal system (16,17). Ubiquitin-related autophagic cell death is known to occur in neurodegenerative conditions such as Parkinson's or Alzheimer's disease (18,19) and it has recently been found in cardiomyopathic human hearts as well (20,21). The preliminary stage consists of deposition of ubiquitin/protein complexes, while the final phase includes fragmentation and disappearance of the nucleus (22,23) (Fig. 4). Autophagic vacuoles are hallmarks of cellular degeneration (Figs. 1D,E). In an extensive analysis of the different steps of the autophagic cell death cascade, our group identified a significant defect of the deubiquitinating enzyme systems as a possible cause of excessive ubiquitin/protein storage and later cell death (21).

The term "necrosis" should be reserved for degrading processes occurring after cell death, regardless of the pathogenetic mechanism of cell death.

Fibrosis

Fibrosis is an excess of any fibrous tissue (9), which consists of fibroblasts (myofibroblasts in some cases), macrophages, mast cells, and occasionally lymphocytes and the

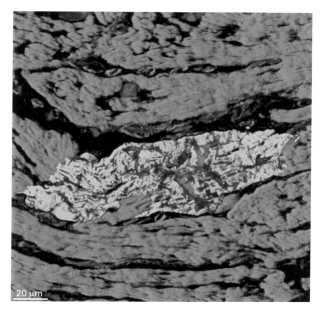

Figure 3 Oncotic cell death characterized by C9 deposition (*green*) in hypertrophied myocardium (*myocytes are red, nuclei blue*). (*See color insert.*)

ECM. All cells in the ECM are able to produce cytokines, which in turn stimulate the synthesis of matrix proteins, matrix metalloproteinases (MMPs), tissue inhibitors of MMPs (TIMPs), growth factors, etc. (24–26). Fibroblasts show an increased rate of proliferation in chronic cardiac hypertrophy (27). The rate of fibrosis is increased in

(A)   **(B)**

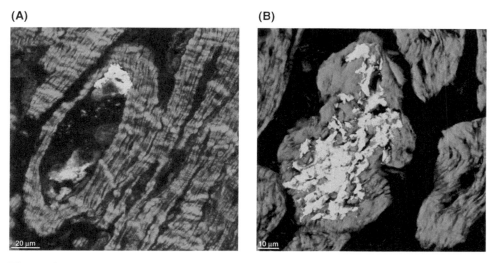

Figure 4 Autophagic cell death identified by ubiquitin labeling (*green*) in hypertrophied myocardium (*cardiomyocytes are red, nuclei blue*): (**A**) Moderate accumulation of ubiquitin/protein complexes in a severely degenerated cell with loss of sarcomeres. This cell is still viable because it contains a nucleus. (**B**) Significant accumulation of ubiquitin/protein conjugates in a myocyte doomed to die because the nucleus is lacking. (*See color insert.*)

(A)   **(B)**

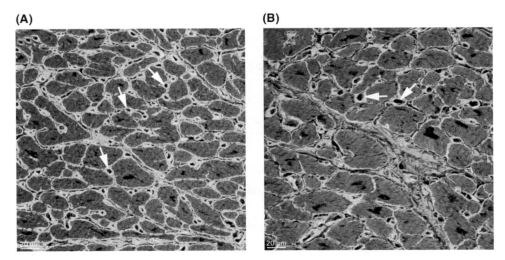

Figure 5   Fibronectin (*green*) for determination of fibrosis. Myocytes are red and black spots are nuclei: (**A**) Normal myocardium shows fine septa of ECM between cardiomyocytes. (**B**) In hypertrophied myocardium, the myocytes are enlarged and the amount of replacement fibrosis is increased. Note the distinct labeling of capillaries (*arrows*) and their reduced number in B. *Abbreviation*: ECM, extra cellular matrix. (*See color insert.*)

hypertrophied hearts and it consists of fibronectin, fibrillar collagen (type I and III,) and nonfibrillar collagen isoforms (collagen IV and VI), the proteoglycans, laminin, and other proteins (Fig. 5). The amount of collagen I, because of its immense tensile strength, determines the stiffness of fibrotic tissue (28). Fibrosis may be focal or generally distributed. In hypertrophied hearts, the predilection site of fibrosis is the subendocardium from where it spreads into the mid- and subepimyocardial layers at later stages.

"Reactive fibrosis" is defined as a reaction of fibroblasts to increased cytokine production, e.g., in hypertension or in cardiomyopathies (9). Usually it is localized in the perivascular area and occasionally it is found in the interstitium as well. "Replacement fibrosis" is focal and caused by the loss of myocytes. In hypertrophied hearts, both modes of fibrosis occur simultaneously. Increased rates of fibrosis are accompanied by loss of microvascular elements, which increases the perfusion distance for oxygen and metabolites, and causes further myocyte loss (Figs. 6A,B) (29). Furthermore, angiotensin II and transforming growth factor-β_1 (TGF-β_1) are profibrotic cytokines that are increased during the development of fibrosis (30–32) (Fig. 6E).

STUDIES IN PATIENTS

We investigated a well-defined cohort of patients with isolated aortic valve stenosis, who were divided into groups according to their functional capacity defined by changes of ejection fraction (EF): Differentiation was made between one group with well preserved systolic function (EF>50%, group I), one with moderately reduced function (EF: 50–30%, group II), and one with severely reduced function (EF < 30%, group III). Data from patients undergoing transplantation because of overt heart

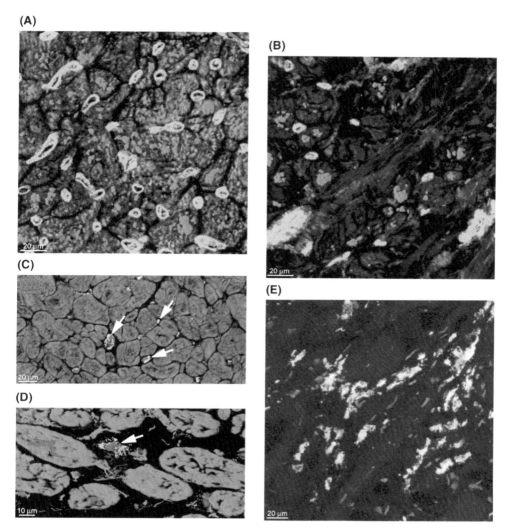

Figure 6 Changes in the microvasculature and the extracellular space. (**A**) and (**B**): Capillaries in human myocardium labeled by CD 31 (*green*). Lipofuscin (*red*) is abundant, myocytes are greenish: (**A**) Myocardium from a patient from group I shows a slightly reduced number of capillaries. (**B**) Tissue from group III exhibits only a few myocytes and a reduced number of capillaries. (**C**) Several capillaries are positive for ACE (*arrows, green, nuclei are blue*). (**D**) Occurrence of TGF-β₁ (*green, arrow*) in a macrophage within fibrotic tissue (*black area*). (**E**) An increased number of macrophages (*CD 68, green, nuclei are red*) is typical of fibrotic tissue in hypertrophied myocardium. *Abbreviation*: ACE, angiotensin-converting enzyme. (*See color insert.*)

failure due to dilated cardiomyopathy (DCM) are reported for comparison. Systolic and diastolic performance was determined by echocardiography one day before aortic valve replacement (AVR). LV subvalvular septum myectomies were harvested at the time of AVR, and samples were investigated using immunohistochemical and morphometrical methods. Informed consent was given in all cases. In the present text we have tried to avoid the term "compensation." In fact, the term "compensated" hypertrophy should be reserved for that type of hypertrophy, where despite an increase in LV-mass, the functional potential of the whole organ is unchanged or even elevated, such as during adolescence or endurance training in the adult. There is a balance between chamber size and wall thickness in these hearts, wall stress is kept constant, and the diastolic parameters are normal. The term "congestive heart failure" is more suitable to describe both diastolic and systolic dysfunction of failing hearts (33–35).

To facilitate the differentiation between normal human hearts and those with different degrees of hypertrophy and functional limitations, methods such as pulse Doppler echocardiography and magnetic resonance imaging (MRI) can be used to set standards in defining more precisely flow velocity and local wall motion patterns at varying time points of the cardiac cycle (36–39).

Group I with Preserved Cardiac Function

All data described in the following text are summarized in Tables 1 and 2.

The myocytes in hearts from group I patients showed hypertrophy but appeared more or less structurally normal with only 5% of all cells exhibiting slight degenerative changes. Myocyte hypertrophy was evident. The most obvious structural divergence was the significantly increased degree of fibrosis (29%). The prevailing type of fibrosis was "reactive fibrosis" indicating stimulation of the fibroblasts, most probably by cytokines because of an increased LV wall tension. "Replacement fibrosis" was found as well and appears to reflect the loss of myocytes. Capillary density was reduced but the ratio of angiotensin-converting enzyme (ACE) containing capillaries was significantly increased. TGF-β_1 as measured by quantitative confocal microscopy was elevated but did not reach statistical significance (4).

All three different modes of cell death were observed.

Table 1 Morphometric Data of Normal and Diseased Human Myocardium

	Control	Group I	Group II	Group III	DCM
Myocytes (n/mm^2)	380 ± 20	366 ± 45	290 ± 21^a	216 ± 32^a	240 ± 12^a
Cross-sectional area (μm^2)	424 ± 37	561 ± 91	658 ± 84^a	593 ± 64^a	543 ± 81
Myocytes <300 μm^2 (%)	2.5 ± 2	10.5 ± 3^a	6.2 ± 4	$17. \pm 5^a$	19 ± 4^a
Longitudinal cell area (μm^2)	1397 ± 76	2268 ± 203	2488 ± 481^a	2397 ± 559^a	–
Nuclear area (μm^2)	71 ± 7	80 ± 6	87 ± 5	103 ± 5^a	127 ± 12^a
Nuclear/myocyte area	0.05	0.04	0.04	0.04	–
Ki-67 (n/mm^2)	0.11 ± 0.09	0.25 ± 0.09	0.55 ± 0.23	0.3 ± 0.13	0.27 ± 0.08
Ubiquitin positive (‰)	0.5 ± 0.01	0.5 ± 0.03	1 ± 0.4^a	6 ± 0.5^a	0.8 ± 0.02
C9 positive (‰)	0	3 ± 0.5^a	5 ± 0.8^a	4 ± 0.3^a	0.6 ± 0.01
Apoptosis (‰)	0	0.02 ± 0.002	0.01 ± 0.003	0	0.02 ± 0.05
Myocyte degeneration (%)	0.5 ± 0.4	5.1 ± 4.1	11.4 ± 5.3^a	15.9 ± 8.7^a	12.4 ± 3^a
Fibrosis (%)	11.6 ± 1.2	29.5 ± 2.9^a	27.9 ± 1.5^a	$38.8 \pm 2.4^{a,b,c}$	28 ± 10^a
Capillary density (n/mm^2)	1874 ± 262	1413 ± 151	1322 ± 203^a	1198 ± 157^a	1212 ± 103^a

$^a p < 0.05$ vs. control, $^b p < 0.05$ vs. group I, $^c p < 0.05$ vs. group II.

Table 2 Summary of Clinical Data

	Control	Group I	Group II	Group III	DCM
Diastolic function	Normal	Impaired	Impaired	Severely impaired	Disturbed
Systolic function	Normal	Normal hypercontractile	Depressed	Severely depressed	Severely depressed
NYHA Class	–	2.3 ± 0.8	2.5 ± 0.7	3 ± 0.5	3 ± 1
Ejection fraction (%)	61 ± 8	59 ± 8	43 ± 5	24 ± 5	16 ± 5
RWT (2xPWT/ LVEDD)	0.42 ± 0.01	0.63 ± 0.14	0.5 ± 0.18	0.45 ± 0.1	0.34 ± 0.6
Rel LV-mass (g/m^2 BSA)	93 ± 11	137 ± 26	131 ± 57	153 ± 35	189 ± 63
LVEDP (mmHg)	8 ± 1	15 ± 5	18 ± 5	24 ± 5	28 ± 3
LVESP (mmHg)	130 ± 17	191 ± 25	180 ± 15	156 ± 24	98 ± 8
Recovery	–	Complete	Complete/ Incomplete	Incomplete	Transplantation

Abbreviations: NYHA, New York Heart Association; RWT, Relative wall thickness; Rel LV-mass, Relative left ventricular mass; LVEDP, Left ventricular end-diastolic pressure; LVESP, Left ventricular end-systolic pressure.

LV mass calculated with the Devereux formula increased from 93 g/m^2 body surface area (BSA) in controls to 137 g/m^2. Relative wall thickness (RWT) changed from 0.42 in controls to 0.63 in group I, identifying concentric hypertrophy. In addition, LV chamber dimensions were significantly reduced in group I: Left ventricular end-systolic diameter (LVESD) was 45 mm in control and 35 mm in group I, and left ventricular end-diastolic diameter (LVEDD) was 42 mm versus 55 mm. Despite preserved LV systolic function, patients of group I were in New York Heart Association (NYHA) class 2.3 with typical symptoms of congestive heart failure. Together with the rise in left ventricular end-diastolic pressure (LVEDP) to 15 mmHg (controls: 8 mmHg), which represents a rightward shift of the pressure volume loop of diastolic filling, and a decrease in the E/A ratio of mitral flow pattern to below 1, the syndrome of diastolic heart failure (DHF) was diagnosed in these patients (40,41). Disturbed diastolic function is a very sensitive parameter of regional ischemia, since actin-myosin detachment, Ca^{2+} dissociation from Troponin C, as well as SERCA activity are ATP dependent processes.

Several findings indicated the presence of subendocardial ischemia in patients with pressure overload hypertrophy: (1) Increased oxygen demand of the hypertrophied ventricle and increased LVEDP reduce coronary flow reserve (42). (2) The reduction of capillary density from 1874 n/mm^2 in controls to 1418 n/mm^2 in group I leads to an increased diffusion distance. (3) Myocyte death in group I and II was caused mainly by oncosis, which is the ischemic type of cell death. These findings confirm earlier reports on the occurrence of ischemic foci in the subendocardium of hypertrophied myocardium.

The presence of ischemia was also reported in studies using phosphorous-magnetic resonance spectroscopy evaluating high energy phosphates during stress and at rest; they showed a significant drop of the PCr/ATP ratio in pressure overload hypertrophied LV compared to normal controls and endurance trained individuals (43,44).

We propose that significant myocyte hypertrophy and the huge accumulation of fibrosis are the morphological correlate of the disturbed diastolic filling pattern of the hypertrophied left ventricle characterized by impaired relaxation and increased passive stiffness.

It is concluded that in group I diastolic dysfunction is present, which originates from myocyte hypertrophy and an elevated rate of fibrosis while the systolic function is normal to hypercontractile.

Group II with Moderately Reduced Function

In this second group of patients the degree of fibrosis was similar to that of the first group (27%), but replacement fibrosis occurred more frequently than in group I. Capillary density was significantly reduced as compared to control myocardium.

The number of ACE positive microvessels and the fibroblast content of TGF-β_1 were similar to group I. Myocyte hypertrophy was more pronounced. Myocyte degeneration steeply increased to 11% of all myocytes and was significantly more severe. The number of C9 positive myocytes was increased in comparison to group I, indicating an aggravation of the ischemic condition.

Thus, it seems that the differences between the first and the second group rest on the increased severity of myocyte hypertrophy and damage including occurrence of cell death, i.e., worsening of the structural composition of the myocytes rather than that of the interstitium.

This morphological situation is reflected in the clinical parameters. Comparable to the unchanged rate of hypertrophy and fibrosis, LVEDP shows only an insignificant increase. On the other hand, the elevated number of myocytes with loss of contractile material is the cause of the decrease in systolic function, which can be seen from the drop in EF and the increase in LVSP. Furthermore, in this group a few patients showed episodes of LV failure.

It is concluded that in group II, patients' hearts are characterized by myocyte hypertrophy, degeneration, and cell death as well as fibrosis, which are interpreted as the morphological correlates of diastolic as well as systolic dysfunction.

Group III with Severely Reduced Function

In the third group of patients, the rate of fibrosis was significantly higher than in group II and the number of proliferating fibroblasts was increased. Capillary density was significantly reduced. The number of ACE positive capillaries was unchanged as compared to that of the other groups, but in the interstitium TGF-β_1 was significantly elevated. Most myocytes were hypertrophied with a mean cross-sectional area of about 593 μm^2, but the number of small myocytes with a cross-sectional area $<300\,\mu m^2$ had sharply increased (17%). This indicates that myocyte hypertrophy was much more pronounced than reflected by the mean values given here. The number of myocytes showing degenerative changes was elevated (16%). These morphological alterations indicate progressive degeneration which is also reflected in the increased number of cells undergoing autophagic cell death. The rate of ischemic cell death was unchanged compared to group II, apoptosis was minimal.

From these data it is evident that in patients of group III with significantly decreased EF, both, the cellular and the interstitial components of the myocardium underwent progressive remodeling, which finally resulted in significant deterioration of cardiac function.

From a clinical point of view, it is important to note that in these patients with severely impaired systolic function, left ventricular end-systolic pressure (LVESP) and the pressure gradient declined. The incidence of chronic or paroxysmal atrial fibrillation was higher than 50%, concomitant episodes of LV-decompensation with severe pulmonary congestion (i.e., NYHA class IV) and hospitalization were far more frequent than in group II (17% vs. 60%). While chamber dimensions were smaller in group I and II than in controls, in group III it reached normal values but LV-mass levels showed a further increase up to 153 g/m^2.

It is concluded that in these patients both, diastolic and systolic dysfunction is prominent and that the morphological equivalent is represented by impaired myocyte integrity, decreased capillary density and a high rate of fibrosis.

Overt Heart Failure

In order to test the hypothesis that structural alterations in failing hearts are similar qualitatively but quantitatively different, we also studied hearts from patients with DCM undergoing transplantation because of intractable heart failure. These patients represent the final stage of remodeling where any therapeutic measure is ineffective and replacement of the damaged heart is the sole solution to ensure survival of the patient.

Myocardium from those patients showed a significant degree of fibrosis which, however, varied widely from patient to patient, in dependence of the duration of the disease (3,10). The myocytes were enlarged and showed degeneration. The appearance of degenerated myocytes was similar to that observed in aortic stenosis (AS) patients but it occurred less frequently. The rate of all three different modes of cell death was lower than in patients with AS (21). These findings indicate that both, the cellular and the interstitial components of the myocardium are affected in end-stage heart failure and that this group of patients represent the "burnt-out" phase of the remodeling process.

Nuclear Changes

The nucleus is especially important for survival of the cell and its alterations are therefore discussed separately here. In the course of development of hypertrophy from the stage of well- preserved function to final complete heart failure, the nucleus enlarges in parallel with the increase in cell size. However, in hypertrophied myocytes the ratio between the nuclear area and the longitudinal cell surface area was 20% smaller than in normal myocytes (0.05 in controls, 0.04 in all groups of hypertrophy) (3,4). A certain number of myocyte nuclei showed a positive reaction with the proliferation marker Ki-67 indicating DNA repair or replication processes leading to polyploidy in the absence of mitotic cell division (4) (Fig. 7A).

The observation of Ki-67 positivity was confirmed by measurements of the DNA content in myocyte nuclei, which revealed a steady and significant increase during transition from one phase of hypertrophy to the next. The DNA concentration per nuclear area was unchanged compared to control, indicating that the increase in nuclear size was accompanied by an increase in its DNA (4) as shown by other groups as well (45). This, however, may be insufficient as compensatory

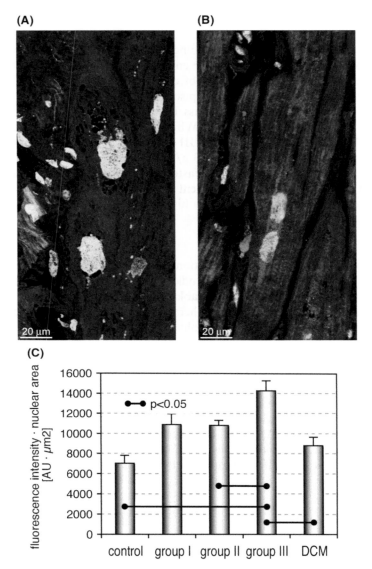

Figure 7 Nuclear staining: **(A)** Feulgen staining for DNA content (*yellow*), myocytes are reddish. **(B)** Sc-35 (*green*) in nearly black myocytes. Lipofuscin is red in **(A)** and **(B)**. **(C)**: Quantitative evaluation of the Sc-35 content in myocyte nuclei of controls and patients with aortic stenosis or DCM. Whereas a significant increase in Sc-35 is present in hypertrophied hearts, especially in group III, in patients with overt heart failure the Sc-35 content is significantly reduced. *Abbreviation*: DCM, dilated cardiomyopathy. (*See color insert.*)

mechanism because the ratio between the nuclear/myocyte area was diminished, i.e., the nuclei are not enlarged enough to adapt to the increase in cell volume (Table 1).

The presence of the non-snRNP factor Sc-35, a factor required for the first step of splicing and spliceosome assembly, indicates ongoing transcription (46). It was significantly increased in the enlarged nuclei but its concentration was similar to the control situation (4) (Fig. 7B and C). Therefore, it may be assumed that transcription occurs in cardiomyocytes as long as they contain a nucleus, even in the presence of significant degenerative alterations. This finding is interpreted to indicate the possibility of reversal of injury and recovery on the cellular level of damaged myocytes as soon as mechanical overload has been eliminated, i.e., after valve replacement and abolishment of the pressure gradient.

In patients with overt heart failure due to DCM, the Sc-35 content did not differ from control myocytes despite the increase in nuclear size; it was significantly reduced as compared to group III of the hypertrophied hearts (Fig. 7C). It may be assumed that in earlier stages of the disease, probably decades earlier, the Sc-35 content was elevated comparable to the situation of hypertrophied hearts, but that in later stages the capability for adaptation is exhausted and transcription cannot be maintained at an elevated level. This, however, is necessary for the preservation of the structural integrity of enlarged myocytes.

GENERAL CLINICAL CONSIDERATIONS

Structure–Function Relationships

The first mechanism of adaptation in LV human myocardium exposed to chronic pressure overload is concentric hypertrophy with an increase in LV mass and RWT, thereby procuring compensation for excessive afterload with hypernormal systolic function. These pathophysiological changes are accompanied (or caused) by substantial cellular and interstitial remodeling, which finally leads to deleterious consequences for the entire organ.

We found that diastolic dysfunction is the first clinical symptom of ongoing remodeling, while systolic function still appeared to be normal (group I). This is evident from the elevated LVEDP and the disturbance of relaxation determined by the echocardiographic E/A ratio and the increase in left atrial (LA) size (Table 2). Consequently, a strong correlation was found to exist between LVEDP and the degree of fibrosis (Fig. 8A). This can easily be explained by the fact that increased fibrillar collagen content is one of the major determinants of the rise in passive stiffness, which leads to a leftward shift of the diastolic pressure–volume loop and a concomitant elevation of LVEDP (25). Furthermore, this increased LVEDP causes elevation of LA pressure and size, which in turn is also strongly correlated with the degree of LV fibrosis in all patient groups (Fig. 8B).

In groups II and III, diastolic as well as systolic function is disturbed (Table 2). This is evident not only from the strong correlation between LVEDP and fibrosis, but also from the relationship between EF and fibrosis (Fig. 8C). The structural correlate of disturbed function, therefore, is represented not only by the persistence of an increased rate of fibrosis, (which is even further increased in group III) but also by an additional cellular factor, i.e., the occurrence of myocyte degeneration and cell death. As discussed in detail in the foregoing text, in group II, myocyte degeneration and cell death are prominent and the "reactive" phenotype of fibrosis changed to "replacement fibrosis" because of myocyte loss. This is reflected by the direct relationship between fibrosis and myocyte degeneration (Fig. 8D): It is evident that an increased rate of myocyte degeneration is accompanied by an elevated degree of fibrosis thereby reducing both, diastolic and systolic function.

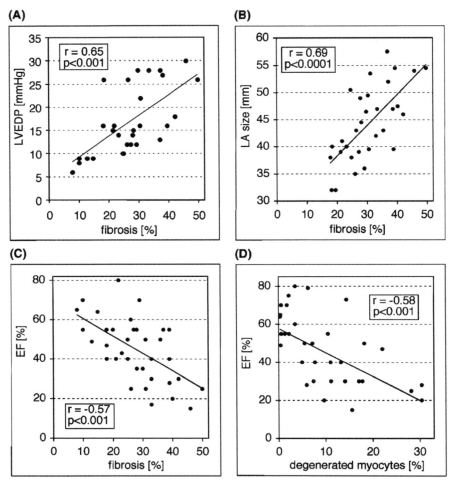

Figure 8 Correlation of clinical and morphological findings: (**A**) and (**B**) Elevated LV-fibrosis is correlated to increased LVEDP and LA-dimension; these are symptoms of disturbed LV-diastolic function. (**C**) Fibrosis also influences LV-systolic performance, (**D**) An increase in myocyte degeneration also reduces LV-EF. *Abbreviations*: LV, left ventricular; LVEDP, left ventricular end-diastolic pressure; LA, left atrial; EF, ejection fraction.

The Problem of Reversibility of Structural/Functional Changes

The present data show that a strong structure/function relationship exists and that the probability of functional recovery may be deduced from the structural alterations that are present. In LV myocardium with monocausal pathology of chronic pressure overload, structural remodeling is strongly dependent on the degree and duration of increased afterload, which directly determines the degree of structural remodeling during recovery. The postoperative re-remodeling process, however, can only achieve the level of complete restitution if the number of viable myocytes is above a critical level. A high degree of replacement fibrosis is not compatible with functional recovery because of the loss of myocytes, which implies a partial loss of the contractile machinery. Scar tissue cannot contract and many of the myocytes still present most probably will not contribute to contraction because of their loss of sarcomeres and separation/isolation by fibrotic tissue. When a high rate of replacement fibrosis is present implying myocyte loss, functional recovery will be incomplete. The data presented here are in accordance with those published earlier by Krayenbühl et al. (47) who described an increase of fibrosis and myocyte degeneration in a selected group of patients with AS with normal EF but elevated LVEDP, which corresponds to group I of the patients described in our study. In these patients, incomplete structural recovery occurred during the first two postoperative years and fibrosis was still augmented at seven years after surgery. The authors concluded that reversal of diastolic dysfunction in AS takes years because interstitial fibrosis regresses very slowly and that systolic overload persists due to a reduced volume fraction of myofibrils (47,48).

In hearts from patients suffering from other pathologies, such as DCM, the postmyocarditis situation, or ischemic heart disease, the potential of functional recovery is much more difficult to define. In these hearts, the alterations are very heterogenous in character and severity and they also vary in regional distribution. Functional and structural recovery under these conditions are mostly examined in patients before and after left ventricular assist device (LVAD) support (49–51). These studies have great difficulties to define structural recovery. Nevertheless, reverse remodeling of the ECM combined with restitution of injured myocytes has been occasionally reported in patients after LVAD support (52). It appears therefore from these studies that there exists also a close correlation between structural and functional recovery (53,54).

Successful weaning from LVAD (bridge to recovery) was furthermore strongly correlated with the age of the patients and the duration of heart failure, i.e., young patients with short duration of heart failure will profit the most from this procedure. However, a permanent recovery depends not only on a shorter history but also on the speed of restoration of normal heart function on mechanical support (55–57). This indicates and confirms the deleterious effects of an abnormal hemodynamic situation on myocardial structural components. The major goal in patients treated with a LVAD, however, is to prevent death by acute heart failure, with the perspectives of heart transplantation, bridge to bridge or bridge to recovery. In this context, functional recovery is usually incomplete, but it ensures survival and avoids heart transplantation.

In patients with aortic valve stenosis, however, LVAD is usually not applied but operative correction of the valve defect is carried out, even under emergency conditions.

In conclusion: Progression of structural remodeling to the final stage of heart failure in the chronically pressure-overloaded hypertrophied human heart occurs by a significant increase in fibrosis as the primary event. This is followed by myocyte degeneration and cell death. Diastolic dysfunction mainly originates from the augmentation of fibrosis although subendocardial ischemia contributes by damaging myocytes. Myocyte deficiencies are the main cause of systolic dysfunction. The extent of diastolic as well as systolic dysfunction, present at different degrees at the time of valve replacement, will determine the course of postoperative recovery after pathological afterload has been corrected. After a period of re-remodeling, when first cellular hypertrophy and thereafter fibrosis will have largely disappeared (48), patients in group I will achieve completely normal heart function. Recovery in group III patients, however, will be incomplete because of the irreversible loss of parts of the contractile machinery by cell death and its replacement by fibrosis (4). The final conclusion of this work, therefore, is the recommendation to alleviate AS in earlier stages of the disease by operative AVR.

REFERENCES

1. Cohn JN, Ferrari R, Sharpe N. Cardiac remodeling-concepts and clinical implications: A consensus paper from an international forum on cardiac remodeling [Review]. J Am Coll Cardiol 2000; 35:569–582.

2. Swynghedauw B. Molecular mechanisms of myocardial remodeling. Physiological Reviews 1999; 79:215–262.

3. Scholz D, Diener W, Schaper J. Altered nucleus/cytoplasm relationship and degenerative structural changes in human dilated cardiomyopathy. Cardioscience 1994; 5:127–138.

4. Hein S, Arnon E, Kostin S, Schonburg M, Elsasser A, Polyakova V, Bauer EP, Klovekorn WP, Schaper J. Progression from compensated hypertrophy to failure in the pressure-overloaded human heart–Structural deterioration and compensatory mechanisms. Circulation 2003; 107:984–991.

5. Speiser B, Riess CF, Schaper J. The extracellular matrix in human myocardium: Part I: Collagen I, III, VI, and IV. Cardioscience 1991; 2:225–232.

6. Speiser B, Weihrauch D, Riess CF, Schaper J. The extracellular matrix in human cardiac tissue. Part II: vimentin, laminin, and fibronectin. Cardioscience 1992; 3:41–49.

7. Schaper J, Meiser E, Stämmler G. Ultrastructural morphometric analysis of myocardium from dogs, rats, hamsters, mice and from human hearts. Circ Res 1985; 56:377–391.

8. Barth E, Stämmler G, Speiser B, Schaper J. Ultrastructural quantitation of mitochondria and myofilaments in cardiac muscle from 10 different animal species including man. J Mol Cell Cardiol 1992; 24:669–681.

9. Majno G, Joris I. Cells, Tissues, and Disease. Principles of general pathology Blackwell Science: Cambridge, Mass 1996.

10. Schaper J, Froede R, Hein S, Buck A, Hashizume H, Speiser B, Friedl A, Bleese N. Impairment of the myocardial ultrastructure and changes of the cytoskeleton in dilated cardiomyopathy. Circulation 1991; 83:504–514.

11. Olivetti G, Melissari M, Capasso JM, Anversa P. Cardiomyopathy of the aging human heart (myocyte loss and reactive cellular hypertrophy). Circ Res 1991; 68:1560–1568.

12. Majno G, Joris I. Apoptosis, oncosis, and necrosis. An overview of cell death. Am J Pathol 1995; 146:3–15.

13. Jennings RB, Sommers HM, Herdson PB, Kaltenbach JP. Ischemic injury of myocardium. Ann N Y Acad Sci 1969; 156:61.

14. Schaper J, Mulch J, Winkler B, Schaper W. Ultrastructural, functional, and biochemical criteria for estimation of reversibility of ischemic injury: A study on the effects of global ischemia on the isolated dog heart. J Mol Cell Cardiol 1979; 11:521–541.

15. Ferreira MAS, Owen HE, Howie AJ. High prevalence of acute myocardial damage in a hospital necropsy series, shown by C9 immunohistology. J Clin Pathol 1998; 51:548–551.

16. Ciechanover A, Orian A, Schwartz AL. The ubiquitin-mediated proteolytic pathway: Mode of action and clinical implications. J Cell Biochem, suppl 2000; 34:40–51.

17. Glickman MH, Ciechanover A. The ubiquitin-proteasome proteolytic pathway: destruction for the sake of construction. Physiol Rev 2002; 82:373–428.

18. Alves-Rodrigues A, Gregori L, Figueiredo-Pereira ME. Ubiquitin, cellular inclusions and their role in neurodegeneration. Trends Neurosci 1998; 21:516–520.

19. Jellinger KA. Cell death mechanisms in Parkinson's disease. J Neural Transm 2000; 107:1–29.

20. Knaapen MW, Davies MJ, De Bie M, Haven AJ, Martinet W, Kockx MM. Apoptotic versus autophagic cell death in heart failure. Cardiovasc Res 2001; 51:304–312.

21. Kostin S, Pool L, Elsässer H, Hein S, Drexler H, Arnon E, Hayakawa Y, Zimmermann R, Bauer E, Klövekorn W-P, Schaper J. Myocytes die by multiple mechanisms in failing human hearts. Circ Res 2003; 92:715–724.

22. Klionsky DJ, Emr SD. Autophagy as a regulated pathway of cellular degradation. Science 2000; 290:1717–1720.

23. Bursch W, Hochegger K, Török L, Marian B, Ellinger A, Hermann RS. Autophagic and apoptotic types of programmed cell death exhibit different fates of cytoskeletal filaments. J Cell Sci 2000; 113:1189–1198.

24. Weber KT. Cardiac interstitium in health and disease: the fibrillar collagen network. J Am Coll Cardiol 1989; 13:1637–1652.

25. Weber KT, Sun Y, Tyagi SC, Cleutjens JP. Collagen network of the myocardium: function, structural remodeling and regulatory mechanisms. J Mol Cell Cardiol 1994; 26: 279–292.

26. Brilla CG, Maisch B, Zhou G, Weber KT. Hormonal regulation of cardiac fibroblast function. Eur Heart J 1995; 16:45–50.

27. Polyakova V, Hein S, Kostin S, Zeigelhöfer T, Schaper J. Remodeling of the extracellular matrix in the progression from hypertrophy to heart failure. Circulation 108 (Suppl IV): 216 (abstr), 2003.

28. Hein S, Schaper J. The extracellular matrix in normal and diseased myocardium. J Nucl Cardiol 2001; 8:188–196.

29. Sabbah HN, sharov VG, Lesch M, Goldstein S. Progression of heart failure: a role for interstitial fibrosis. Mol Cell Biochem 1995; 147:29–34.

30. Weber KT, Swamynathan SK, Guntaka RV, Sun Y. Angiotensin II and extracellular matrix homeostasis. Int J Biochem Cell Bio 1999; 31:395–403.

31. Deten A, Holzl A, Leicht M, Barth W, Zimmer HG. Changes in extracellular matrix and in transforming growth factor beta isoforms after coronary artery ligation in rats. J Mol Cell Cardiol 2001; 33:1191–1207.

32. Schultz J, Witt SA, Glascock BJ, Nieman ML, Reiser PJ, Nix SL, Kimball TR, Doetschman T. TGF-beta1 mediates the hypertrophic cardiomyocyte growth induced by angiotensin II. J Clin Invest 2002; 109:787–796.

33. Derumeaux G, Mulder P, Richard V, Chagraoui A, Nafeh C, Bauer F, Henry JP, Thuillez C. Tissue Doppler imaging differentiates physiological from pathological pressure-overload left ventricular hypertrophy in rats. Circulation 2002; 105:1602–1608.

34. Angeja BG, Grossman W. Evaluation and management of diastolic heart failure. Circulation 2003; 107:659–663.

35. Zile MR, Gaasch WH, Carroll JD, Feldman MD, Aurigemma GP, Schaer GL, Ghali JK, Liebson PR. Heart failure with a normal ejection fraction: is measurement of diastolic function necessary to make the diagnosis of diastolic heart failure? Circulation 2001; 104:779–782.

36. Stuber M, Scheidegger MB, Fischer SE, Nagel E, Steinemann F, Hess OM, Boesiger P. Alterations in the local myocardial motion pattern in patients suffering from pressure overload due to aortic stenosis. Circulation 1999; 100:361–368.

37. Diamant M, Lamb HJ, Groeneveld Y, Endert EL, Smit JWA, Bax JJ, Romijn JA, de Roos A, Radder JK. Diastolic dysfunction is associated with altered myocardial metabolism in asymptomatic normotensive patients with well-controlled type 2 diabetes mellitus. J Am Coll Cardiol 2003; 42:328–335.

38. Ennis DB, Epstein FH, Kellmann P, Fananapazir L, McVeigh ER, Arai AE. Assessment of regional systolic and diastolic dysfunction in familial hypertrophic cardiomyopathy using MR tagging. Mag Res in Med 2003; 50:638–642.

39. Schillaci G, Pasqualini L, Verdecchia P, Vaudo G, Marchesi S, Porcellati C, de Simone G, Mannarino E. Prognostic significance of left ventricular diastolic dysfunction in essential hypertension. J Am Coll Cardiol 2002; 39:2005–2011.

40. Zile MR, Brutsaert DL. New concepts in diastolic dysfunction and diastolic heart failure: Part II – Causal mechanisms and treatment. Circulation 2002; 105:1503–1508.

41. Zile MR, Brutsaert DL. New concepts in diastolic dysfunction and diastolic heart failure: Part I: diagnosis, prognosis, and measurements of diastolic function. Circulation 2002; 105:1387–1393.

42. Julius BK, Spillmann M, Vassalli G, Villari B, Eberli FR, Hess OM. Angina pectoris in patients with aortic stenosis and normal coronary arteries. Mechanisms and pathophysiological concepts. [comment]. Circulation 1997; 95:892–898.

43. Lamb HJ, Beyerbacht HP, van der Laarse A, Stoel BC, Doornbos J, van der Wall EE, de Roos A. Diastolic dysfunction in hypertensive heart disease is associated with altered myocardial metabolism. Circulation 1999; 99:2261–2267.

44. Pluim BM, Lamb HJ, Kayser HW, Leujes F, Beyerbacht HP, Zwinderman AH, van der Laarse A, Vliegen HW, de Roos A, van der Wall EE. Functional and metabolic evaluation of the athlete's heart by magnetic resonance imaging and dobutamine stress magnetic resonance spectroscopy. Circulation 1998; 97:666–672.

45. Soonpaa MH, Field I. Survey of studies examining mammalian cardiomyocyte DNA synthesis. Circ Res 1998; 83:15–26.

46. Fu XD, Maniatis T. The 35-kDa mammalian splicing factor SC35 mediates specific interactions between U1 and U2 small nuclear ribonucleoprotein particles at the $3'$ splice site. Proc Natl Acad Sci USA 1992; 89:1725–1729.

47. Krayenbuehl HP, Hess OM, Monrad ES, Schneider J, Mall G, Turina M. Left ventricular myocardial structure in aortic valve disease before, intermediate, and late after aortic valve replacement. Circulation 1989; 79:744–755.

48. Villari B, Vassalli G, Monrad ES, Chiariello M, Turina M, Hess OM. Normalization of diastolic dysfunction in aortic stenosis late after valve replacement. Circulation 1995; 91:2353–2358.

49. Young JB. Healing the heart with ventricular assist device therapy: mechanisms of cardiac recovery. Ann Thorac Surg 2001; 71:S210–S219.

50. Mann DL, Willerson JT. Left ventricular assist devices and the failing heart: a bridge to recovery, a permanent assist device, or a bridge too far? Circulation 1998; 98:2367–2369.

51. Kumpati GS, McCarthy PM, Hoercher KJ. Left ventricular assist device bridge to recovery: a review of the current status. Ann Thorac Surg 2001; 71:S103–S108; discussion S114–S115.

52. McCarthy PM, Nakatani S, Vargo R, Kottke-Marchant K, Harasaki H, James KB, Savage RM, Thomas JD. Structural and left ventricular histologic changes after implantable LVAD insertion. Ann Thorac Surg 1995; 59:609–613.

53. Frazier OH, Benedict CR, Radovancevic B, Bick RJ, Capek P, Springer WE, Macris MP, Delgado R, Buja LM. Improved left ventricular function after chronic left ventricular unloading. Ann Thorac Surg 1996; 62:675–681; discussion 681–682.

54. Barbone A, Oz MC, Burkhoff D, Holmes JW. Normalized diastolic properties after left ventricular assist result from reverse remodeling of chamber geometry. Circulation 2001; 104:I229–I232.

55. Mancini DM, Beniaminovitz A, Levin H, Catanese K, Flannery M, DiTullio M, Savin S, Cordisco ME, Rose E, Oz M. Low incidence of myocardial recovery after left ventricular assist device implantation in patients with chronic heart failure. Circulation 1998; 98:2383–2389.

56. Martin J, Sarai K, Schindler M, van de Loo A, Yoshitake M, Beyersdorf F. MEDOS HIA-VAD biventricular assist device for bridge to recovery in fulminant myocarditis. Ann Thorac Surg 1997; 63:1145–1146.

57. Hetzer R, Muller JH, Weng YG, Loebe M, Wallukat G. Midterm follow-up of patients who underwent removal of a left ventricular assist device after cardiac recovery from end-stage dilated cardiomyopathy. J Thorac Cardio Surg 2000; 120:843–853.

9

Myocardial Remodeling: Physiological and Pathological

Jeffrey S. Borer and Karl H. Schuleri
*Division of Cardiovascular Pathophysiology and The Howard Gilman Institute for
Valvular Heart Diseases, Weill Medical College of Cornell University,
New York, New York, U.S.A.*

INTRODUCTION

"Physiological" vs. "Pathological" Remodeling: Definitions

In response to changes in hemodynamic stresses and other environmental alterations the heart alters its size, shape, and cell biology. These changes generally enable the heart to adapt functionally to its new environment and to maintain performance measures within normal limits, at least for a limited time. Though the cardiomyocyte is a "fully differentiated" cell and, thus, has been thought to be incapable of a hyperplastic response, recent evidence suggests that, in certain situations, this may not be true; more importantly, the cardiomyocyte has been long known to be capable of net synthesis of a variety of its components, resulting in cellular hypertrophy. Other myocardial cell lines (fibroblasts, endothelium, various leukocytes) can produce hyperplasia by cell division and may also be capable of hypertrophy. Atrophy may also be possible for some or all normal cell lines, as suggested by recent data from echocardiography acquired in the microgravity environment as part of the NASA space exploration program (1,2).

Environmental adaptations can be "physiological," i.e., generated solely by hemodynamic stresses and dietary variations that are considered within the normal range of human activity, e.g., athletic training. Alternatively, adaptations can be "pathological," generated by stimuli or boundary conditions outside the range of normal human activity and specifically encompassing a variety of disease processes. These processes can directly alter cell biology (e.g., ischemia, infection) or molecular biology (e.g., genetically determined cardiomyopathy), or they can alter myocardial biology secondarily by generating abnormal exogenous stresses (e.g., valvular heart diseases). A "hybrid" of direct and secondary effects is also possible: patients with prior myocardial infarction, having suffered dramatic and irreversible alteration in the intrinsic biology of a portion of myocardium, may consequently subject the uninfarcted and, possibly, nonischemic myocardium to abnormal mechanical stresses despite putatively normal hemodynamic profiles. Similarly, hypertension causes pressure loading of the left ventricular (LV) myocardium but simultaneously imposes a pressure load directly on the walls of the coronary arteries, causing or potentiating arterial injury and occlusive plaque formation that can confound the effects of LV pressure loading with the ischemia of coronary occlusive disease. As subsequently discussed, pathological remodeling is very complex, resulting from a variety of disease processes that are commonly associated. This chapter will review the myocardial trophic response to physiological stress and will then discuss myocardial responses to prototypical examples of pathological remodeling caused by the well-defined and relatively pure hemodynamic alterations of LV pressure loading [aortic stenosis (AS) and hypertension] and of LV volume loading [aortic regurgitation (AR)]. Cellular and molecular mechanisms believed to be responsible for remodeling of the primary myocardial cell lines and their products is reviewed briefly, specifically as they relate to alterations in the quantity or quality of the major myocardial protein components.

Importantly, components of the remodeling process resulting from pressure and volume loading are likely to affect all forms of pathological remodeling to a greater or lesser extent. For example, as stated above, mechanical analysis suggests that the myocardial loads of AR are similar to those in the noninfarcted myocardium of patients who have suffered large myocardial infarctions and have subsequently developed heart failure (3). Thus, pathological remodeling occurring in patients with intrinsic myocardial diseases (caused by ischemia, viral infection, genetic variants, etc.) can be understood, in part, as an adaptive or compensatory response to normal exogenous loading conditions by myocardium that cannot generate or transmit contractile force normally.

Force Generation vs. Force Transmission

The function of the heart is to pump blood by actively generating contractile force. Therefore, remodeling of the cardiomyocyte, the cell specifically containing the contractile elements, is of central interest. Indeed, in patients subjected to the isolated, pure volume loading of AR, loss of contractility is the best predictor of outcome in unoperated patients (4,5) as well as in those who undergo aortic valve replacement (5,6). As a corollary, the lusiotropic characteristics of the heart can result in abnormalities in diastolic chamber filling leading to clinically important hemodynamic disturbances. The cardiomyocyte is responsible for active myocardial relaxation, as well as active contraction, and has some intrinsic elastic properties that affect other aspects of diastolic function.

However, the myocardium also comprises fibroblasts, endothelial cells, macrophages, and monocytes/lymphocytes. All are involved in the remodeling response. Most prominent among these noncontractile cell lines is the fibroblast, which produces the extracellular matrix (ECM) and also some of the metalloproteases (MMP) that modulate the response of the myocardium to its environment (7). The ECM is the "scaffold" on which the cardiomyocytes are organized. For this reason alone, the effects of stresses on fibroblast remodeling intrinsically are of interest. Moreover, the fibrous proteins synthesized by the fibroblast are primary determinants of the passive diastolic properties of the heart (8,9). Perhaps most importantly, during the past decade, with the identification of well-documented clinical examples (10–12), force transmission has gained increasing attention as a critical contributor to myocardial contractility and mechanical performance. Force transmission is heavily dependent on the fibroblast and the ECM. Thus, while systolic ejection requires force generation by the contractile elements of the cardiomyocyte, unless the generated force is transmitted to the ECM, contractility and ejection will be deficient. Moreover, for optimal mechanical performance of the cardiac chambers, the force must be transmitted by the ECM in a directionally efficient manner, i.e., according to specific vectors. Force transmission to the ECM is mediated by a series of proteins that reach and pierce the sarcolemma to connect the actin cytoskeleton of the cardiomyocyte, itself in direct contact with the sarcomere, with the ECM.

The data supporting the key role of force transmission in determining myocardial contractility have their origin in studies of skeletal muscle, specifically in patients with certain forms of muscular dystrophy (10–12). When these genetically transmitted diseases are manifest, in addition to skeletal muscle weakness, the myocardium is dysfunctional, expressing the phenotype of congestive cardiomyopathy with heart failure. Affected patients have point mutations that result in absence of the dystrophins, a series of proteins, which form a part of the chain that connects the contractile element with the ECM. When the dystrophins are absent, even though cardiomyocyte force generation is normal in in vitro preparations, force transmission is deficient and LV performance is deranged. Indeed, paucity of dystrophins accounts for both skeletal muscle weakness and dilated, hypocontractile cardiomyopathy. Thus, normal force transmission is critical for the capacity of the myocardium to subserve its pump function.

The dystrophins do not bind directly to the ECM. Rather, at the sarcolemma, the dystrophins are bound to a glycoprotein complex, the latter predominantly comprising fibronectin (10–14), a normal ECM component. If the actin cytoskeleton is structured abnormally, if actin interactions with the dystrophins are defective, if dystrophins are deficient or qualitatively abnormal, if the glycoprotein complex is qualitatively or quantitatively abnormal affecting its interaction with the dystrophins or with the ECM, then force transmission is defective and heart failure results. Thus, normal myocyte–matrix interaction must be reasonably preserved if contractility and cardiac mechanical function are to be normal.

In theory, force transmission can be deranged both by quantitative subnormality and abnormality of normal connecting proteins, as well as by qualitative alterations of the connectors or the ECM or cardiomyocyte/fibroblast integrins with which they interact. Increasing evidence suggests that in valvular diseases and, by extrapolation, in settings associated with similar loading abnormalities, hyperproduction of selected normal ECM proteins may be particularly important in causing contractile deficits.

In addition to mediating the physical interaction of the sarcomere and the ECM, cardiomyocyte structural proteins are believed to transmit mechanical stresses to the nucleus. This may be important in triggering variations in gene expression that result in the alterations in cellular synthetic and degradative processes defining the specific form of remodeling. The recent description of "stress responsive" genomic elements, promoters that can be activated in response to specific mechanical stresses, provides the molecular basis for the remodeling process (15).

Endothelial cells line the cardiac chambers and the myocardial blood vessels. Their specific role in myocardial remodeling is not clear, but these cells can produce autocrine or paracrine factors in response to stress. Such factors may well affect local myocardial responses to environmental variations. Leukocytes may be even more importantly involved in remodeling as the producers of cytokines and other cell signaling molecules that modulate the molecular response to environmental variations (16).

PHYSIOLOGICAL REMODELING

The prototypical and most dramatic example of physiological remodeling is the response to intensive athletic training. Undoubtedly, more modest cardiac remodeling, perhaps imperceptible with current measuring tools, occurs in response to less intense variations in human activity. Considerable literature has now defined the training response, providing a useful comparator for the effects of disease processes.

Changes in cardiac size in response to participation in sports have long been recognized. However, for many years, the nature of these adaptations was poorly understood. For example, as recently as four decades ago, in his definitive monograph, *"Diseases of the Heart,"* Charles K. Friedberg summarized existing knowledge by noting that, "The so-called 'athlete's heart', i.e., cardiac dilatation and hypertrophy formerly attributed to athletic activity, is now believed to be the consequence of independent rheumatic, congenital or syphilitic heart disease. Transient cardiac enlargement occurs occasionally during strenuous physical exertion, but this represents reversible compensatory dilatation which disappears promptly with cessation of the activity." (17). References to support these statements had been published 30 to 40 years earlier. At about the time of Friedberg's observations, when I was an undergraduate, I overheard a young faculty member/long-distance runner describing the discovery of his enlarged "athletic" heart on chest X-ray and the reassurance by a famous Boston cardiologist that he need not worry—he would live at least until age 60! The true form and extent of cardiac remodeling in response to physical exercise, and its functional concomitants, were not credibly defined until the advent of echocardiography. The stimulus to the first echocardiographic surveys was the sudden death of a basketball player of the University of Maryland in the mid-1970s. A NIH team took the then relatively new tool to the University, studied many team members and determined that a variety of remodeling patterns were present in healthy athletes with no apparent underlying structural or other heart disease (18). When supplemented by subsequent studies by members of this team and others (19–21), a pattern emerged that now is well recognized and has been frequently reviewed (22,23).

Athletic training generally comprises a mixture of static (isometric) and dynamic (isotonic) exercise. The former involves the achievement of external work by marked force generation by skeletal muscles associated with pronounced increases in intramuscular pressure with limited or absent muscle shortening compared with the precontraction state; the latter is characterized by changes in skeletal muscle length with relatively modest force generation (24). Athletic activities characterized predominantly by dynamic exercise (e.g., long-distance running and other endurance-related contests) generally involve frequently repeated contractions during an extended interval; the exercising muscles require oxygen and nutrients, and hence blood flow, far in excess of that required in the resting state. Heart rate, peripheral resistance, cardiac output, and blood volume vary to meet these needs, with resulting venous return to the heart increased relative to resting levels and to that achieved during nonathletic activities. Athletic activities characterized predominantly by static exercise (for example, weight lifting or shot-putting) are marked by substantial increases in peripheral vascular resistance; requisite increases in muscle flow and cardiac output must be achieved against markedly abnormal pressure loads. These hemodynamic variations mirror to some limited extent those associated with volume loading of regurgitant valve diseases (dynamic exercise) and the pressure loading of AS or hypertension (static exercise), respectively. The heart is genetically programmed to maintain normal wall tension and, hence, myocardial energy requirements, as long as possible. This end can be achieved in the face of abnormal loads only with an increase in myocardial mass. When faced with volume loads, as in dynamic exercise, the LV increases mass eccentrically, with increase in LV wall surface area to accommodate the enlarged chamber volume, and mildly increased wall thicknesses that generally remain within the normal range (23). When faced with pressure loads, as in static exercise, the LV increases mass concentrically (25), characteristically involving wall thickening to values that commonly exceed the normal limits. However, in contrast to the effects of chronic volume and pressure loading achieved by valvular diseases or hypertension, LV myocardial contractility and chamber function in athletes remains within normal limits (though measurement is confounded by variations in peripheral vascular dynamics, skeletal muscle efficiency, autonomic activity, etc.) (24). Furthermore, again in contrast with disease states, experimental evidence suggests that, in the otherwise normal heart, the coronary vascular bed increases in size to accommodate exercise-induced myocardial hypertrophy, precluding the development of ischemia (24,26).

The cellular/molecular basis for these manifestations of physiological remodeling is incompletely defined, but experimental data suggest the involvement of several cellular/molecular variations. Thus, in rats exposed to treadmill (i.e., dynamic) exercise for 13 weeks, myocardial α-myosin heavy chain content was augmented compared with nonexercised controls (a finding not observed in animals with remodeling associated with myocardial infarction) and, significantly, there was no evidence of apoptosis, a common finding in postinfarction remodeling (27). In vitro data obtained by passive stretch of cardiomyocyte cultures, perhaps also most relevant to the alterations of dynamic exercise, suggest enhanced expression of genes coding for several other proteins, including the contractile protein, β-myosin heavy chain, and skeletal α-actin (28). Though firm conclusions are not possible given the differences in the model systems employed, the disparate result in these two studies suggest a modulation of the direct effects of mechanical stress and strain by neurohumoral/hormonal/cytokine systems in vivo. However, the relevance of these particular data to humans is unclear, since humans, unlike rats, show little evidence of myosin isoform shifts in the various settings that have been studied. In addition, as discussed below, cellular/molecular responses differ in experimental volume and pressure loading, suggesting the importance of the specific characteristics of the mechanical perturbation in determining the cellular responses. Nonetheless, the experimental data suggest important alterations in contractile proteins, the predominant cardiomyocyte protein components, when exercise is intense. Myocardial ECM is also affected by training. After 10 weeks of treadmill exercise, normal age-related diminution of expression of the genes coding for the two principal myocardial fibrillar collagens, isoforms I and III, was significantly attenuated (29); simultaneously, the normal age-related increase in mature collagen cross-linking (also increased after myocardial infarction) was likewise attenuated (29), i.e., training prevented structural changes plausibly associated with enhanced myocardial stiffness, diastolic dysfunction, and alteration in force transmission.

The morphologic changes observed with intensive physical training are often associated with electrocardiographic alterations (Fig. 1). The latter may be attributable to the well-documented increase in vagal tone associated with athletic training; the sympathetic nervous system response also contributes to the concomitant reduction in resting heart rate and blood pressure, and reduction in the response of these variables to exercise. However, it is possible that the cellular response to exercise involves remodeling of ion channels and/or connexins, altering the characteristics

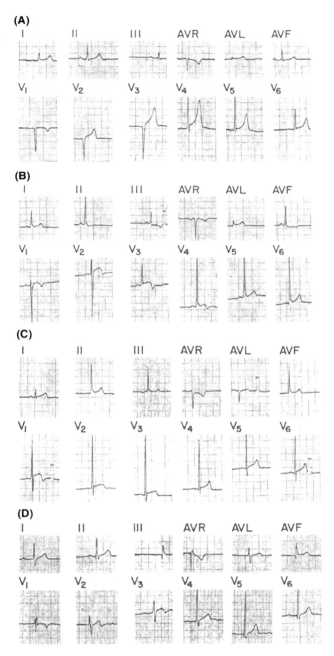

Figure 1 Electrocardiographic variations commonly observed in intensely trained athletes. (**A**) Sinus bradycardia and prominent Q waves in leads V1 to V3 accompanied by tall, peaked T waves in mid and lateral precordial leads. (**B**) Supernormal voltage and ST-segment elevation in precordial leads. There is also terminal P-wave inversion in leads V1 to V4. (**C**) Tall R waves in right precordial leads along with evidence of generalized ST-segment elevation. (**D**) First degree A-V block, a right ventricular conduction abnormality and prominent lateral precordial voltage. Not illustrated are sinus bradycardia and intermittent 4:2 2° Wenckebach A-V block that were also recorded in this patient. *Abbreviation*: A-V, atrio-ventricular. *Source*: From Ref. 60.

of membrane depolarization and transmission of electrical signals among cells. Although this hypothesis has not been tested in the context of exercise, evaluation of pressure loaded human hearts (hypertension and aortic stenosis) (30,31) and experimental studies in volume loading (AR) (32) have demonstrated significant variation in connexin density, potentially altering the function of membrane gap junctions that transmit electrical impulses among cardiomyocytes. The relation of these changes to clinically evident arrhythmias and sudden death in patients with overload lesions is not clear, but it is a plausible target for future study.

The foregoing discussion reviews the effects of pressure and volume loading on cardiac morphology and cell biology. Another factor, gravity, may have as great or greater impact on cardiac form and function, but the characteristics and magnitude of this influence are poorly understood because their understanding requires studies in gravity's absence. Spaceflight opportunities for scientific missions have been limited and sample sizes employed to elucidate the effects of gravity on cardiac remodeling have been small. Nonetheless, the potential importance of this factor has been recently suggested by analysis of echocardiography obtained in astronauts undertaking four- to five-month spaceflight missions (1) (Table 1). In contrast to the small nominal increase in LV mass seen after one to two weeks of microgravity exposure [and associated with a small decrease in LV systolic volume and increase in left ventricular ejection fraction (LVEF)], these longer duration flights were associated with a 14% reduction in LV mass (from 219 g pre-flight to 192 g immediately upon landing, with little change several days later), a 7% increase in LV end-systolic

Table 1　Changes in Left Ventricular Chamber Mass/Geometry with Microgravity Comparison of Percent Change from Pre-flight to Landing Day Between Short- and Long-Duration Flights

	Short duration ($n = 13$) Δ %	Long duration ($n = 13$) Δ %
LVDD	-4.00 ± 0.02	-1.30 ± 0.01
LVDS	-7.00 ± 0.02	7.40 ± 0.05[a]
FS	7.00 ± 0.03	-11.00 ± 0.07[b]
LVPWD	9.00 ± 0.03	-14.20 ± 0.01[b]
LVPWS	2.00 ± 0.02	-14.00 ± 0.05[a]
Mass	3.00 ± 0.07	-1.20 ± 0.08
LVVD	-8.00 ± 0.04	-3.50 ± 0.06
LVVS	-12.00 ± 0.06	39.00 ± 0.24[b]
EF	6.00 ± 0.02	-10.50 ± 0.03[b]
E/A ratio	5.00 ± 0.09	10.30 ± 0.18
HR	15.70 ± 0.09	6.00 ± 0.10
SDP	6.96 ± 2.47	8.93 ± 3.12
DBP	6.64 ± 4.12	16.60 ± 3.90
SV	-5.00 ± 0.03	-17.40 ± 0.05
CO	-2.30 ± 0.07	-12.20 ± 0.09

[a]$P < 0.05$ and [b]$P < 0.01$ between short- and long-duration flights.

Abbreviations: CO, cardiac output; DBP, diastolic blood pressure; EF, ejection fraction; FS, percent fractional shortening; HR, heart rate; LVDD, left ventricular diameter at end diastole; LVDS, left ventricular diameter at peak systole; LVPWD, left ventricular posterior wall thickness at end diastole; LVPWS, left ventricular posterior wall thickness at peak systole; LVVD, left ventricular volume at end diastole; LVVS, left ventricular volume at end systole; Mass, left ventricular mass; SBP, systolic blood pressure; SV, stroke volume.

Source: From Ref. 1.

volume and an 11% decrease in LVEF. This apparent atrophy may result from reduction in demand for cardiac output from muscles that have been unloaded by microgravity, or from the known microgravity-related reduction in blood volume. However, a primary influence of gravity on myocardial cell biology must also be considered, involving either variation in cell water or, more likely, changes in intracellular protein content. Similarly, prolongation in electrocardiographic QTc interval duration seen with relatively long-duration spaceflight (2) suggests that gravity may significantly modulate sarcolemma ion channel expression.

PATHOLOGICAL REMODELING

Left Ventricular Pressure Loading—Aortic Stenosis and Hypertension

The remodeling that occurs in chronic AS can be considered to be the central characteristic of this disease. Heart failure and death do not result from the deformed valve per se, but, rather from abnormal myocardial loads (stresses and strains) caused by the hemodynamic effects of the deformed valves and the resulting adaptive and, ultimately, maladaptive cellular and molecular myocardial responses. This is true for hypertension as well, but the latter may be modified to the extent that the sequelae of hypertension include not only heart failure but also vascular complications such as myocardial infarction and stroke, due to direct effects of hypertension on vascular biology. For purposes of this discussion, only the effects of hypertension directly on the myocardium will be considered.

Chamber Sizes and Wall Thicknesses

Pressure loading due to AS typically results in concentric hypertrophy, i.e., LV wall thickening without increase in chamber volume. However, some latitude exists in the myocardial response to pressure loading, presumably based on genetic variation: concentric hypertrophy occurs in 95% of patients with hemodynamically severe AS (33). Wall thickening can be quite marked and with it, diastolic characteristics can be dramatically affected. When contractile compensatory capacity is exceeded, dilatation may occur, but this is a late and relatively uncommon phenomenon that is almost invariably associated with subnormal LV systolic performance (33). After valve replacement, reverse remodeling continues during an interval of approximately six months, although return to normal LV is not a uniform finding. Little change occurs thereafter; in one early series, no change in wall thicknesses or chamber volume, or in associated LV fractional shortening, was apparent between six months and three years after valve replacement (33). Compared with the effects of volume loading (below), not only is the gross morphological response to pressure loading different, but also the kinetics of the relevant cellular metabolic responses differ, since recovery after removal of pathological volume loading continues over the course of several years (34) (Fig. 2A). The functional response to relief of pressure loading depends on the preoperative effects of the pressure loading; among patients with supernormal LVEF at rest before valve replacement, this parameter tends to diminish toward normal; among those with subnormal ejection fraction at rest before operation, the postoperative ejection fraction generally improves, commonly reaching the normal range (35). Relief of pressure loading uniformly improves LVEF response to exercise (35) (Fig. 2B).

(A)

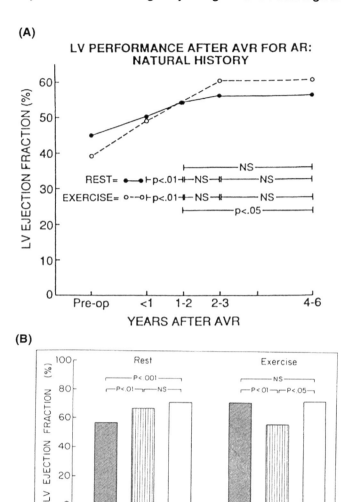

(B)

Figure 2 Variation in left ventricular performance with relief of pathological volume (AR) (**A**) and pressure (AS) (**B**) loading. *Abbreviations*: AR, aortic regurgitation; AS, aortic stenosis, AVR, aortic valve replacement; LV, left ventricular; NS, not significant. *Source*: From Refs. 34, 35.

Cellular and Molecular Responses: The Cardiomyocyte

From catheter-based endomyocardial biopsies, it has long been known that cardio-myocytes are markedly hypertrophied in patients with pressure loading and, specifically, in patients with AS, in whom myocyte diameter is approximately twice that of normal (8,36). Consistent with this finding, experimental data uniformly indicate that, with interposition of pressure loads, contractile protein synthesis by the cardiomyocyte increases markedly (37). For example, in the mouse, banding the aorta to simulate AS or hypertension resulted in a 25% increase in heart-to-body weight ratio after seven days, compared with sham controls; of particular note, banded animals manifested a fourfold greater increase in β-myosin heavy chain content (normally comprising >50% of cardiomyocyte protein) than in sham animals; there was a parallel increase in β-myosin heavy chain gene expression, indicating that the protein increase was primarily due to abnormal synthesis rate (37). In a parallel rabbit model isolating the right ventricular (RV) myocardium, RV pressure overload by pulmonary artery constriction resulted in similar upregulation of β-myosin heavy chain synthesis; in this study, however, α-myosin heavy chain synthesis was not increased, suggesting that pressure loading results selectively in β-chain upregulation (38). Unlike research into the effects of volume loading (below) in the area of pressure loading, little information is available about contractile protein synthesis rates beyond one week, or about associated degradation rates. However, the rate and magnitude of wall thickening during pressure loading has led to the assumption that the predominant factor underlying hypertrophy is abnormal protein synthesis rate. Signaling pathways and modulators involved in cardiomyocyte hypertrophy are not fully understood, but several appear to be involved, including modules of the mitogen activated protein kinase (MAPK) superfamily, specific G proteins, GTPases, transforming growth factor β (TGFβ), fibroblast growth factor (FGF), insulin-like growth factor I receptor pathway, protein kinase C, and calcineurin, among other pathways and factors (39). It is noteworthy that, when assessed in vitro, the hypertrophied LV myocytes manifest normal percentage shortening with electrical stimulation, but maximal velocity of shortening is only about half normal, suggesting that eventual development of hypocontractility in the pressure overloaded heart may be due, in part, to transcriptional upregulation of β-myosin heavy chain and its intrinsic contractile capacities (37). In addition, cellular remodeling in response to chronic pressure loading involves reduction in the density of connexins (30,31), which comprise the gap junctions facilitating intercellular transmission of membrane electrical impulses. [These changes are directionally opposite to the alterations in connexins observed in chronic volume loading (32).] Although speculation, as above, has suggested functional consequences of changes in metabolism of specific cardiomyocyte-derived proteins, the results of these changes are not well understood. In addition, histopathological analyses of human biopsies taken during cardiac surgery for AS have indicated that cardiomyocyte hypertrophy is associated with abnormal cell degeneration, mainly by autophagy and oncosis, with modest but clearly apparent cardiomyocyte apoptosis (9). Recent data from experimental studies in hypertensive models similarly implicate apoptosis as an important component of the remodeling process, leading to reduced cardiomyocyte cell number as surviving cells hypertrophy (40,41).

Cellular and Molecular Responses: The Fibroblast

The importance of the fibroblast response to pressure overload was suggested at least three decades ago from surgical specimens demonstrating substantial fibrosis in patients with AS (36,42). Soon thereafter, catheter-based endomyocardial biopsies examined by light microscopy and painstaking morphometric techniques in patients with and without heart valve disease, indicated that interstitial fibrosis was sevenfold more prominent in patients with AS than in controls with no clinically apparent LV disease, a more pronounced increase in ECM than the corresponding increase in myocyte mass inferred from the same biopsies (8). This finding has been confirmed and extended recently in biopsies from humans undergoing valve replacement for AS, evaluated with standard microscopic morphometry enhanced by immunohistochemical staining and supplemented by electron microscopy, DNA and apoptosis assays (9). Abnormal collagen synthesis is well described in AS (43,44); content of other ECM components, including fibronectin (9), is also increased, but little information is available about the relative magnitude of hyperproduction of collagen and noncollagen ECM elements. Myocardial stiffness, a measure of LV diastolic performance, once thought to result from increased myocyte mass in AS, is now known to relate directly to ECM variations (8). Ultrastructural studies of cardiac autopsy specimens obtained within eight hours of death from patients with and without hypertension (45) have extended these findings [and other earlier findings (46–48)], reporting abnormal variation in organization of ECM collagen, abnormally dense weave of collagen encasing myocytes, and abnormal ECM interconnections among myocytes, all suggesting the basis for abnormalities in myocyte-generated force transmission and, in addition, for subnormal myocyte diastolic lengthening necessary to maximize contractile force ("Frank–Starling effect"). The cellular and molecular basis of ECM alterations from pressure loading is not fully understood. However, recent observations in experimental volume loading suggest that, at least in part, fibrosis is a primary response to the specific characteristics of mechanical stresses and strains impacting the fibroblast (14), probably acting via specific stress-responsive genomic elements (15). Genetically determined modulators of this process are probably important as well, as suggested by the finding that serum biomarkers of net collagen synthesis varies in hypertensive patients as a function of angiotensin II type I (AT_1) polymorphism, and that these patterns of gene expression (Fig. 3) predict responses of AT_1-receptor blocker therapy on LV chamber stiffness (49). Importantly, the ECM response to pressure loading is at least partially reversible with removal of the inciting stresses, though the time course of reversal is relatively long (50).

Left Ventricular Volume Loading—Aortic Regurgitation

Typically, the LV increases in diameter gradually during the course of AR, while wall thicknesses (septum and free wall) remain normal (eccentric hypertrophy).

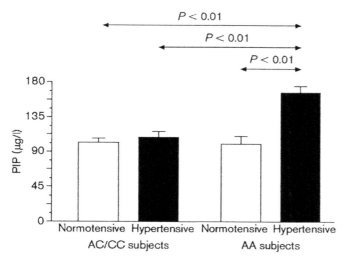

Figure 3 Serum concentration of the carboxy-terminal propeptide of procollagen type I (PIP) measured in normotensive subjects and hypertensive patients classified according to their genotypes for the A1166C polymorphism of the angiotensin II type I receptor gene. *Source*: From Ref. 49.

The rate of remodeling depends on the acuity of the regurgitant lesion; in human disease, AR generally progresses gradually, allowing sufficient time for hypertrophy to enable functional adaptation of the myocardium. When AR occurs acutely, the contractile capacity is overwhelmed until adaptive hypertrophy is adequate for compensation. However, if the acute event is sustained, experimental data indicate that ventricular hypertrophy reaches a "plateau" phase within three months, after which further increases in LV mass are relatively slow (51). The functional concomitants of the geometric changes that occur in the plateau phase generally include maintenance of systolic performance at values achieved prior to chronic volume loading, along with enhancement of diastolic performance (enhanced relaxation and/or passive elastic properties) for the first six months, followed by reduction at 12 months and then a plateau (52). Once volume loading is relieved by valve replacement, reverse remodeling occurs. This process is relatively slow; complete remodeling, with maximization of left ventricular performance, requires three years (34).

Cellular and Molecular Responses: The Cardiomyocyte

Experimental data indicate that, when AR is first created, myocytes respond with abnormally rapid synthesis of normal contractile proteins for approximately one week, combined with a subnormal degradation rate (51). By one month, synthesis rate is subnormal; however, LV mass continues to increase, albeit at a somewhat slower rate than that achieved during the first week, solely attributable to markedly subnormal contractile and total protein degradation rate. Thereafter, myocardial mass increases only very slightly during the next two years. Individual myofibrillar proteins mirror this overall pattern: myosin heavy chain, myosin light chains 1 and 2, and actin manifest supernormal synthesis rates with relatively normal degradation rates through three days of AR, but by one month, synthesis rates are normal or reveal a tendency to subnormality, while degradation rates fall; synthesis and degradation of α-actinin and desmin are normal or tend to be subnormal at three days; both become subnormal by one month (51,53). Preliminary data indicate that, when heart failure supervenes in AR, both synthesis and degradation rates are markedly subnormal (53). The importance of these changes in the pathophysiology of the heart failure (that inevitably develops if volume loading is maintained) is not clear. The possible functional significance of myofibrillar protein senescence in AR has not been evaluated. Nonetheless, it is noteworthy that the pattern of myocyte-derived protein metabolism appears to differ in volume versus pressure loading. Moreover, despite important differences in the hemodynamic stresses impacting on the myocardium, volume loading due to mitral regurgitation results in a relatively similar response pattern to that in AR, also differing from AS (54).

Connexins, important in mediating intercellular electrical signals, are also abnormally expressed in AR, with increased connexin43 density apparent in chronic experimental AR (32), possibly providing one of the bases for arrhythmia in this setting.

Cellular and Molecular Responses: The Fibroblast

As previously noted, the fibroblast synthesizes most of the components of the ECM, as well as some of the MMP responsible for modulating the ECM response to environmental changes (7). Volume loading is associated with myocardial fibrosis, indicating hyperactivity of the fibroblast. Indeed, in patients with AR who have undergone myocardial biopsy at catheterization or during valve replacement surgery, abnormal quantities of ECM, as myocardial fibrosis, are uniformly observed (8,14,37,42,50). Experimental AR causes similar changes (55). In addition, consistent with the clinical observations in patients with evidence of decompensation and about to undergo surgery, fibrosis is most exuberant in experimental animals with AR and congestive heart failure, but is also present in the absence of heart failure, albeit generally with more modest intensity. However, until recently, it was unclear whether these changes were secondary to myocyte damage, possibly associated with inflammation, or were primary responses to the mechanical effects of volume loading. Experimental data now indicate that, in AR, fibrosis occurs in the absence of evidence of inflammation (14,56) and that, when cardiac fibroblasts are subjected in cell culture to the mechanical strains of AR, ECM alteration is virtually identical with that observed in the diseased state in vivo; thus, the LV wall strain of volume loading directly causes fibrosis, and the magnitude of the change appears to be directly related to the magnitude of strain (14,56). The fibrotic lesion of volume loading comprises qualitatively abnormal fibrous tissue: the ECM is particularly rich in noncollagen matrix elements, among which fibronectin (the predominant component of the glycoprotein associated complex connecting dystrophins and ECM) is most prominent (56). Northern analyses of gene expression indicate that other glycoproteins are also abnormally produced; preliminary experimental data indicate that proteoglycans are not abnormally expressed, but that integrins are altered. [However, the biology of AR and fibroblast strain is not yet completely characterized: experimentally, multiple genes are up and downregulated, the functional importance of which is as yet unclear (57).] Despite fibrosis, myocardial collagen content per unit myocardium remains normal in experimental AR (58); fibroblasts cultured from animals with AR utilize normal quantities of proline, the collagen precursor, and manifest normal expression of genes coding for collagen isoforms as well as normal synthesis rates of collagen types I and III, most commonly found in myocardial ECM (56). Preliminary data suggest that similar changes are produced by normal human cardiac fibroblasts when they are subjected to the strains of AR in cell culture (59).

ECM remodeling induced by volume loading is modulated by abnormal activity of MMP and subnormal activity of tissue inhibitors (TIMPs) of these enzymes. Preliminary data from fibroblast cell cultures indicate that myocardial collagenase activity and content are both modestly abnormal in AR for the most prominently expressed collagen isoforms (59), and that the corresponding TIMPs are subnormally expressed. In contrast, fibronectinase activity is normal. Although these data indicate that collagen degradation is abnormal in volume loading, further research is needed to determine whether this results in a qualitative abnormality of collagen structure or in the distribution of specific isoforms within the ECM collagen, potentially altering the force-transmitting and scaffold function of the ECM. The latter may be particularly important: relative collagen deficiency may fail to provide adequate resistance to volume loading in AR, allowing a progressive increase in the mechanical factors generating pathologic remodeling.

Remodeling of the ECM in volume loading involves a series of cellular and molecular changes that transduce the mechanical effects of the hemodynamic abnormality. The initiating event may take any one of several forms, but an attractive hypothesis is that mechanical strain stimulates the interaction of a fibroblast integrin with extracellular fibronectin (59), followed by a cascade of intracellular reactions that culminate in activation of multiple genes. Fibronectin is only one of the abnormally expressed ECM genes already identified in AR; comprehensive elucidation of the pathobiology of volume loading requires evaluation of many genes and their products. However, given the prominence of fibronectin upregulation among the ECM changes in AR, some insight may be gained from studies of the fibronectin gene and its product.

Preliminary data indicate that the MAPK pathway is central to the transduction cascade: both the c-jun-N-terminal-kinase (JNK) module and the p38-mitogen activated kinase module of the MAPK pathway are upregulated by AR, although the extracellular response kinase (ERK) module, activated by various stressors in other systems, is not (59). Confirmation has been provided by preliminary assessment with known selective inhibitors of these modules. Activation of these modules leads to downward modulation of the formation of c-jun/activating transcription factor (ATF)-2 complex formation with the cyclic adenosine monophosphate (c-AMP) response element (CRE) of the fibronectin promoter of the fibronectin gene.

CONCLUSION

The heart responds to changes in its exogenous environment, similar to changes in its internal milieu, with alterations in gross morphology (shape, chamber volumes, wall thicknesses). These responses result from alterations in the cellular and molecular biology of all myocardial cell lines, stimulated by environmental variations, which can affect both the qualitative and quantitative characteristics of myocardial component tissues. Since function follows structure, remodeling can result in useful adaptation to the new environment. Clinically, this may entail adaptation of the normal heart to the demands of abnormal activities that involve stresses and strains of volume and pressure loading. When global stresses and strains are particularly intense in specific disease states such as regurgitant and stenotic valvular diseases and hypertension, or when regional stresses and strains are heightened by loss of myocardial tissue (ischemic injury), or when intrinsic myocardial disease processes render myocardial functional reserve inadequate for adaptation to normal stresses and strains, remodeling can be maladaptive. Current data indicate that the pattern of remodeling can vary significantly with the form of environmental change. A full understanding of the various patterns of remodeling at the cellular and molecular levels is expected to enable therapeutic strategies to minimize maladaptation and to promote cardiac health.

REFERENCES

1. Martin DS, South DA, Wood ML, Bungo MW, Meck JV. Comparison of echocardiographic changes after short and long duration space flight. Aviation Space and Environ Med 2002; 73:532–536.
2. D'Aunno DS, Dougherty AH, DeBlock HF, Meck JV. Effect of short- and long-duration spaceflight on QTc intervals in healthy astronauts. Am J Cardiol 2003; 91:494–497.
3. Pfeffer MA. Cardiac remodeling and its prevention. In: Colucci WS, Braunwald E, eds. Atlas of Heart Diseases; Heart Failure: Cardiac Function and Dysfunction. Vol. IV, Philadelphia: Current Medicine, 1995:5.1–5.14.
4. Borer JS, Hochreiter C, Herrold EM, Supino P, Aschermann M, Wencker D, Devereux RB, Roman MJ, Szulc M, Kligfield P, Isom OW. Prediction of indications for valve replacement among asymptomatic or minimally symptomatic patients with chronic aortic regurgitation and normal left ventricular performance. Circulation 1998; 97:525–534.

5. Borer JS, Herrold EM, Hochreiter CA, Supino PG, Yin A, Krieger K, Isom OW. Aortic regurgitation: selection of asymptomatic patients for valve surgery. Adv Cardiol 2002; 39:74–85.

6. Borer JS, Supino P, Hochreiter C, Herrold EM, Yin A, Krieger K, Isom OW. Valve surgery in the asymptomatic patient with aortic regurgitation: current indications and the effect of change rates in objective measures. Adv Cardiol 2004; 41:36–47.

7. Li YY, McTiernan CF, Feldman AM. Interplay of matrix metalloproteinases, tissue inhibitors of metalloproteinases and their regulators in cardiac matrix remodeling. Cardiovas Res 2000; 46:214–224.

8. Hess OM, Schneider J, Koch R, Bamert C, Grimm J, Krayenbuehl HP. Diastolic function and myocardial structure in patients with myocardial hypertrophy. Circulation 1981; 63:360–371.

9. Hein S, Arnon E, Kostin S, Schonburg M, Elsasser A, Polyakova V, Bauer EP, Klovekorn W-P, Schaper J. Progression from compensated hypertrophy to failure in the pressure-overloaded human heart: structural deterioration and compensatory mechanisms. Circulation 2003; 107:984–991.

10. Towbin JA. Role of cytoskeletal proteins in cardiomyopathies. Curr Opin Cell Biol 1998; 10:131–139.

11. Bowles NE, Bowles KR, Towbin JA. The "final common pathway" hypothesis and inherited cardiovascular disease: the role of cytoskeletal proteins in dilated cardiomyopathy. Herz 2000; 25:168–175.

12. Leiden J. The genetics of dilated cardiomyopathy: emerging clues to the puzzle. N Engl J Med 1997; 337:1080–1081.

13. Ahumada G, Saffitz J. Fibronectin in the rat heart: link between cardiac myocytes and collagen. J Histochem Cytochem 1984; 32:383–388.

14. Borer JS, Truter SL, Herrold EM, Supino PG, Carter JN, Gupta A. The cellular and molecular basis of heart failure in regurgitant valvular diseases: the myocardial extracellular matrix as a building block for future therapy. Adv Cardiol 2002; 39:7–14.

15. Gimbroni M, Najel T, Topper J. Biochemical activation: an emerging paradigm in endothelial adhesion biology. J Clin Invest 1997; 99:1809–1813.

16. Weihrauch D, Arras M, Zimmermann R, Schaper J. Importance of monocytes/macrophages and fibroblasts for healing of micronecroses in porcine myocardium. Molecular and Cellular Biochemistry 1995; 147:13–19.

17. Friedberg CK. Diseases of the Heart, 3rd ed. Philadelphia: Saunders, 1966:1698.

18. Morganroth J, Maron BJ, Henry WL, Epstein SE. Comparative left ventricular dimensions in trained athletes. Ann Int Med 1975; 82:521–529.

19. Blomqvist CG, Saltin B. Cardiovascular adaptations to physical training. Ann Rev Physiol 1983; 45:169–189.

20. Convertino VA, Brock PJ, Keil LC, Bernauer EM, Greenleaf JE. Exercise-training induced hypervolemia: role of plasma albumin, renin, and vasopressin. J Appl Physiol 1980; 48:665–669.

21. Mitchell JH, Haskell WL, Raven PB. Classification of sports. Medical Science of Sports and Exercise 1994; 26:S242–S245.

22. Crawford MH. Physiological consequences of systematic training. Cardio Clin 1992; 10:209–218.

23. Maron BJ. Structural features of the athletic heart as defined by echocardiography. J Am Coll Cardio 1986; 7:190–203.

24. Gallagher KM, Raven PB, Mitchell JH. Classification of sports and the athlete's heart. In: Williams RA, ed. The Athlete and Heart Disease: Diagnosis, Evaluation and Management. Philadelphia: Lippincott Williams and Wilkins, 1999:9–22.

25. Longhurst JC, Kelly AR, Gonyea WJ, Mitchell JH. Echocardiographic left ventricular masses in distance runners and weight lifters. Journal of Applied Physiology: Respiratory, Environmental and Exercise Physiology 1980; 48:154–162.

26. Schaible TF, Scheuer J. Cardiac function in hypertrophied hearts from chronically exercised female rats. J Appl Physiol 1981; 50:1140–1145.
27. Jin H, Yang R, Li W, Lu H, Ryan AM, Ogasawara AK, van Peborgh J, Paoni NF. Effects of exercise training on cardiac function, gene expression, and apoptosis in rats. Am J Physiol: Heart and Circulatory Physiology 2000; 279:H2994–H3002.
28. Yamazaki T, Komuro I, Yazaki Y. Molecular mechanism of cardiac cellular hypertrophy by mechanical stress. J Mol Cell Cardiol 1995; 27:133–140.
29. Thomas DP, Zimerman SD, Hansen TR, Martin DT, McCormick RJ. Collagen gene expression in rat left ventricle: interactive effect of age and exercise training. J Appl Physiol 2000; 29:1462–1466.
30. Peters NS, Green CR, Poole-Wilson PA, Severs NJ. Reduced content of connexin43 gap junctions in ventricular myocardium from hypertrophied and ischemic human hearts. Circulation 1993; 88:864–875.
31. Peters N. New insights into myocardial arrhythmogenesis: distribution of gap-junctional coupling in normal, ischaemic and hypertrophied human hearts. Clin Sci 1996; 90: 447–452.
32. Goldfine SM, Walcott B, Brink PR, Magid NM, Borer JS. Myocardial connexin43 expression in left ventricular hypertrophy resulting from aortic regurgitation. Cardiovas Pathol 1999; 8:1–6.
33. Henry WL, Bonow RO, Borer JS, Kent KM, Ware JH, Redwood DR, Itscoitz SB, McIntosh CL, Morrow AG, Epstein SE. Evaluation of valve replacement in patients with aortic stenosis. Circulation 1980; 61:814–825.
34. Borer JS, Herrold EM, Hochreiter C, Roman M, Supino P, Devereux RB, Kligfield P, Nawaz H. Natural history of left ventricular performance at rest and during exercise after aortic valve replacement for aortic regurgitation. Circulation 1991; 84(Suppl III):III-133–III-139.
35. Borer JS, Jason M, Devereux RB, Fisher J, Green MV, Bacharach SL, Pickering T, Laragh JH. Function of the hypertrophied left ventricle at rest and exercise: hypertension and aortic stenosis. Am J Med 1983; 75(Suppl III):34–39.
36. Maron BJ, Ferrans VJ, Roberts WC. Myocardial ultrastructure in patients with chronic aortic valve disease. Am J Cardiol 1975; 35:725–731.
37. Dorn GW, Robbins J, Ball N, Walsh RA. Myosin heavy chain regulation and myocyte contractile depression after LV hypertrophy in aortic-banded mice. Am J Physiol: Heart and Circulatory Physiology 1994; 267:H400–H405.
38. Nagai R, Pritzl N, Low RB, Stirewalt WS, Zak R, Alpert NR, Litten RZ. Myosin isozyme synthesis and mRNA levels in pressure overloaded rabbit hearts. Circ Res 1987; 60:692–699.
39. Molkentin JD, Dorn GW. Cytoplasmic signaling pathways that regulate cardiac hypertrophy. Ann Rev Physiol 2001; 63:391–426.
40. Fortuno MA, Ravassa S, Fortuno A, Zalba G, Diez J. Cardiomyocyte apoptotic cell death in arterial hypertension: mechanisms and potential management. Hypertension 2001; 38:1406–1412.
41. Gonzalez A, Fortuno MA, Querejeta R, Ravassa S, Lopez B, Lopez N, Diez J. Cardiomyocyte apoptosis in hypertensive cardiomyopathy. Cardiovas Res 2003; 59:549–562.
42. Schwarz F, Flaming W, Schaper J, Langebartels F, Sesto M, Hehrlein F, Schlepper M. Myocardial structure and function in patients with aortic valve disease and their relation to post-operative results. Am J Cardiol 1978; 441:661–668.
43. Weber KT, Janicki JS, Shroff SG, Pick R, Chen RM, Bashey RI. Collagen remodeling of the pressure overloaded, hypertrophied non-human primate myocardium. Circ Res 1988; 62:757–765.
44. Weber KT, Brilla CG. Pathological hypertrophy and the cardiac interstitium. Circulation 1991; 83:1849–1865.
45. Rossi MA. Pathologic fibrosis and connective tissue matrix in left ventricular hypertrophy due to chronic arterial hypertension in humans. J Hyp 1998; 16:1031–1041.

46. Weber KT, Sun Y, Tyagi SC, Cleutjens JPM. Collagen network of the myocardium: function, structural remodeling and regulatory mechanisms. J Mol Cell Cardiol 1994; 26:279–292.

47. Robinson TF, Cohen-Gould L, Factor SM. The skeletal framework of mammalian heart muscle: arrangement of inter- and pericellular connective tissue structures. Laboratory Investigation 1983; 49:482–498.

48. Eobinson TF, Factor SM, Capasso JM, Wittenberg IA, Blumenfeld OO, Seiffer S. Morphology, composition and function of struts between cardiac myocytes of rat and hamster. Cell and Tissue Research 1987; 249:247–255.

49. Diez J, Laviades C, Orbe J, Zalba G, Lopez B, Gonzalez A, Mayor G, Paramo JA, Beloqui O. The A1166C polymorphism of the AT_1 receptor gene is associated with collagen type I synthesis and myocardial stiffness in hypertensives. J Hyper 2003; 21:2085–2092.

50. Krayenbuehl HP, Hess OM, Monrad ES, Schneider J, Mall G, Turina M. Left ventricular myocardial structure in aortic valve disease before, intermediate and late after aortic valve replacement. Circulation 1989; 79:744–755.

51. Magid NM, Borer JS, Young MS, Wallerson DC, De Monteiro C. Suppression of protein degradation in progressive cardiac hypertrophy of chronic aortic regurgitation. Circulation 1993; 87:1249–1257.

52. Magid NM, Wallerson DC, Borer JS, Mukherjee A, Young MS, Devereux RB, Carter JN. Left ventricular diastolic and systolic performance during chronic experimental aortic regurgitation. Am J Physiol 1992; 263(Heart Circ Physiol):H226–H233.

53. Magid NM, Wallerson DC, Borer JS. Myofibrillar protein turnover in cardiac hypertrophy due to aortic regurgitation. Cardiol 1993; 82:20–29.

54. Borer JS, Herrold EM, Hochreiter CA, Truter SL, Carter JN, Goldfine SM. Pathophysiology of heart failure in regurgitant valvular diseases: relation to ventricular dysfunction and clinical debility. In: Sanders MR, Kostis JB, eds. Molecular Cardiology in Clinical Practice. Boston: Kluwer Academic, 1999: 43–56.

55. Liu S-K, Magid NM, Fox PR, Goldfine SM, Borer JS. Fibrosis, myocyte degeneration and heart failure in chronic experimental aortic regurgitation. Cardiol 1998; 90:101–109.

56. Borer JS, Truter SL, Herrold EM, Falcone DJ, Pena M, Carter JN, Dumlao T, Lee J, Supino PG. Myocardial fibrosis in chronic aortic regurgitation: molecular and cellular response to volume overload. Circulation 2002; 105:1837–1842.

57. Truter SL, Goldin D, Kolesar J, Dumlao TF, Borer JS. Abnormal gene expression of cardiac fibroblasts in experimental aortic regurgitation. Am J Ther 2000; 7:237–243.

58. Goldfine SM, Pena M, Magid NM, Liu S-K, Borer JS. Myocardial collagen in cardiac hypertrophy resulting from chronic aortic regurgitation. Am J Ther 1998; 5:139–146.

59. Borer JS, Truter SL, Gupta A, Herrold EM, Carter J, Lee E, Pitlor L. Heart failure in aortic regurgitation: the role of primary fibrosis and its cellular and molecular pathophysiology. Adv Cardiol 2004;41, in press.

60. Lichtman J, O'Rourke RA, Klein A, Karliner JS. Electrocardiogram of the athlete: alterations simulating those of organic heart disease. Arch Int Med 1973; 132:763–770.

10

Experimental Animal Models of Cardiac Remodeling

Elaine J. Tanhehco and Hani N. Sabbah
Henry Ford Health System, Detroit, Michigan, U.S.A.

INTRODUCTION

Cardiac remodeling is a compensatory physiologic response to an event or condition that compromises cardiac function. Triggers for remodeling include infarction, hypertension, wall stress, inflammation, pressure overload, and volume overload (Table 1). Alterations in myocardial structure can occur as quickly as within a few hours of injury, and may progress over months and years. The definition of cardiac remodeling encompasses the global, cellular, and genetic changes that lead to alterations in the ventricular shape and function (Table 2). While initially beneficial, these changes over time (months to years) can impair myocardial function to the point of chronic intractable heart failure, one of the most common causes of cardiovascular mortality. Therefore, cardiac remodeling is generally considered detrimental. The adverse effect of remodeling on the clinical course has stimulated intense interest in trying to define the mechanisms involved, and in the treatments that can prevent or reverse the remodeling process.

The hallmarks of cardiac remodeling are manifested as chamber dilation, increase in ventricular sphericity, and the development of interstitial and perivascular fibrosis. Increased sphericity is positively associated with mitral regurgitation and a poor prognosis in heart failure patients (1,2). Ventricular dilation mainly results from cardiomyocyte hypertrophy and lengthening, and to a lesser extent from increases in the ventricular mass.

Cell death via necrosis and apoptosis also contributes to cardiac remodeling through the loss of viable functional cardiac units. Reactive interstitial fibrosis decreases compliance, impedes oxygen diffusion to cardiomyocytes, and may alter the transmission of force from individual cell to global chamber contraction, all of which hinder cardiac function. Upregulation and activation of matrix metalloproteinases (MMPs) has been proposed as one mechanism that underlies the development of interstitial and perivascular fibrosis in myocardial tissue (3,4). The importance of cardiomyocyte hypertrophy is reflected in the fact that despite ongoing cardiomyocyte loss, ventricular enlargement prevails during the course of remodeling. Progressive left-ventricular dilation, a hallmark of ongoing ventricular remodeling, can lead

Table 1 Stimuli for Cardiac Remodeling

Myocardial infarction
Pressure overload (stenosis, hypertension)
Volume overload (valvular regurgitation)
Inflammation/infection
Genetic cardiomyopathy
Toxic cardiomyopathy
Neurohumoral activation
Nutritional deficiency
Idiopathic dilated cardiomyopathy

to increased wall stress and myocardial stretch; both of these can lead to increased oxygen consumption, an undesirable consequence in the setting of heart failure.

Activation of neurohormonal systems importantly contributes to many of the critical aspects of cardiac remodeling. Circulating levels of norepinephrine have been directly correlated with prognosis of heart failure patients (5), while brain natriuretic peptide (BNP) is quickly emerging as a clinically useful marker for this purpose (6). Other neurohormones associated with cardiac remodeling include renin, aldosterone, angiotensin II (AII) (7), atrial natriuretic peptide (ANP) (6), and tumor necrosis factor-α (TNF-α) (8); all the five hormones have been found to be elevated in patients with heart failure. Norepinephrine and ANP trigger apoptosis of cardio-myocytes (9,10), while AII causes cellular necrosis and stimulates fibrosis (11,12). Norepinephrine and AII also induce cardiomyocyte hypertrophy (13,14). Over-expression of TNF-α has been associated with cardiac dysfunction, left-ventricular dilation (15), cardiomyocyte hypertrophy (16), and an increase in collagen deposi-tion in the interstitial compartment (17).

Both β-adrenergic receptor blockers and angiotensin converting enzyme (ACE) inhibitors have been shown to attenuate cardiac remodeling and ameliorate heart failure, confirming the importance of norepinephrine and AII in the pathophysio-logy of this disease (18,19).

Modification of the expression of certain cardiac genes associated with the fetal phenotype, namely ANP, BNP, α-myosin heavy chain (MHC), and β-MHC, is increasingly being recognized as another aspect involved in the progression of heart failure. The exact purposes for these changes in gene expression remains under inves-tigation. It has been proposed that the activation of natriuretic peptides may have a

Table 2 Characteristics of Cardiac Remodeling

Ventricular dilation
Ventricular wall thinning
Infarct expansion
Ventricular sphericity
Interstitial fibrosis
Cardiomyocyte hypertrophy
Cardiomyocyte apoptosis
Cardiomyocyte necrosis

pressure–volume compensatory role in heart failure (20). The failing heart also exhibits a decrease in the ratio of α-MHC to β-MHC (21,22). Since α-MHC possesses two to three times more rapid, actin-activated ATPase activity than β-MHC (23), alterations in the α-MHC to β-MHC ratio could adversely impact contractile function and may contribute to the decline in ejection fraction associated with heart failure.

A variety of whole animal models have been developed to elucidate the pathophysiology of cardiac remodeling. The majority of these models are inextricably linked to the study of heart failure and generally utilize the development of overt heart failure as an endpoint. In these studies, the defining feature of heart failure is a significant decrease in ejection fraction. A subset of studies also examines the parameters of compensated cardiac remodeling, in order to determine which factors and at what point cardiac remodeling undergoes transitions into decompensated heart failure. In humans, however, cardiac remodeling is usually not discovered until clinical symptoms of heart failure appear. Since remodeling is stimulated by a myriad of factors and conditions, it is important to examine a variety of models in which remodeling is induced by several different means. The following review will concentrate on three major models of remodeling followed by an overview of less common models. The three major models that will be discussed are the chronic rapid pacing model, the myocardial infarction (MI) model, and intracoronary microembolization model.

CHRONIC RAPID PACING

Rapid ventricular pacing causes cardiac enlargement and failure within a matter of weeks, and is one of the most widely used methods to induce myocardial remodeling. The severity of cardiac dysfunction directly correlates with the rate of pacing. Rapid pacing can be extremely detrimental and in some cases results in death in less than six weeks (24). This problem can be circumvented by pacing at a lower rate for the first few weeks, followed by a higher rate of pacing until clinical heart failure ensues. The contractile dysfunction evident in this model has been proposed to be due to exhaustion of the Frank–Starling mechanism (25). Alternatively, heart failure in this model may be due to the depletion of myocardial energy stores, as result of high demands for oxygen. Similar to heart failure patients, increased circulating levels of neurohormones, including norepinephrine, AII, and renin are observed in paced animals (26–34). Although chamber dilation occurs, this is not accompanied by an increase in ventricular mass (24,35). Most of the effects of rapid pacing on myocardial function, remodeling, and neurohumoral activation are reversible upon cessation of the pacing stimulus (24,36). Though the rapid pacing model is one of the most popular in the study of cardiac remodeling, it does not entirely replicate the human situation. Discrepancies between this model and what is observed in humans include chamber dilation without increase in ventricular mass, a lack of definite interstitial fibrosis development, and cardiomyocyte hypertrophy (in dogs). Also, remodeling generally progressively worsens despite removal of the initiating stimulus in humans, rather than reverses, as in rapidly paced animals.

Rapid Pacing in Dogs

Rapid ventricular pacing is primarily employed in dogs paced at 210–260 beats per minute, through the left or right ventricle, for three to six weeks (24,26,27,35,37–41).

This technique produces chamber dilation within two weeks, and causes a decrease in the ejection fraction and cardiac output within three weeks (24,26). Deterioration of systolic function is evident prior to diastolic dysfunction, which sets in upon the appearance of clinical heart failure (edema, lethargy, weight gain, etc.) (37). Pacing also leads to increases in left-ventricular end-diastolic pressure (36), mean arterial pressure, pulmonary artery pressure, and wall stress (35,38). When pacing is suspended, cardiac output returns to normal within two weeks and posterior wall function improves (24). Isolated cardiomyocytes from paced hearts also exhibit impaired contraction (39). These cells show abnormal calcium handling, which is apparent by a decrease in the time taken to decline from peak intracellular calcium. This kind of cell behavior correlates with a prolonged contraction and relaxation (40). In addition, there is evidence of endothelial dysfunction in the coronary vessels of paced hearts, which may adversely affect coronary circulation (41).

Chronic rapid pacing in dogs elicits augmented early sympathetic tone and decreased parasympathetic activity (27). Increases in plasma norepinephrine and renin plateau are observed after approximately three weeks (26,28,29), along with a corresponding decrease in the myocardial tissue norepinephrine content (26). The increase in plasma norepinephrine occurs regardless of whether the paced animal is compensated or decompensated (26). In contrast, plasma AII levels have been noted to rise only when overt heart failure ensues (30). There is also an increase of both the circulating ANP and BNP, in this model (28–30,37,38). The hemodynamic and renal responses to the exogenous administration of ANP to paced dogs is blunted, suggesting that increased ANP production may not be an important compensatory mechanism to counteract the initial stages of heart failure (38). As with the functional changes induced by pacing, neurohumoral activation is reversible, with plasma renin levels falling 40% within a couple days and circulating ANP dropping 60% within a few hours of stopping pacing (28).

Global remodeling occurs in rapidly paced canines within two to three weeks, which is demonstrated by the left-ventricular dilation and increases in right and left-ventricular volumes (24). Unlike hemodynamic dysfunction and neurohumoral activation, chamber dilation persists even after pacing is terminated (24). Since pacing does not produce an increase in ventricular mass (24,35), chamber dilation is most probably attributable to ventricular wall thinning (24). Reports have been conflicting on whether focal necrosis and interstitial and replacement fibrosis could develop in this model (24,26,35). This may be due to differences in the rigor of various pacing protocols and/or the point at which these observations were made, i.e., if during the pacing or the post pacing recovery period. Isolated cardiomyocytes derived from paced hearts exhibit increased length, width, and volume (35,39). Transmural cell loss also occurs, as determined by the histological analysis (35). Cellular hypertrophy may compensate for cell loss, but its occurrence is controversial (35). Diffuse apoptosis has also been detected in this model (42–44). TUNEL (Terminal deoxynucleotidyl Transferase Biotin-dUTP Nick End Labeling)-positive cells are observed in the ventricles and atria, with a higher incidence of apoptosis occurring in the ventricles (42). The prevalence of apoptosis in cardiomyocytes is markedly elevated when compared to that of noncardiomyocytes (fibroblasts, endothelial cells) (42,43). Decreased expression of the antiapoptotic gene Bcl-2 and an increased expression of the proapoptotic genes Fas, caspase-3, and Bax have been demonstrated in paced cardiomyocytes (42,44). It has been suggested that the leakage of cytochrome c from mitochondria into the cytosol of cardiomyocytes triggers apoptosis in this model (43).

Rapid Pacing in Other Species

Pig

Unlike most canine models, rapid pacing in pigs is accomplished via electrode implantation in the left, right, or both the atria, instead of using the ventricle. The effects of chronic rapid pacing in pigs are similar to those seen in dogs in terms of hemodynamic function (36,45), neurohumoral activation (42–l45), cardiomyocyte injury and loss (33,45), and reversibility upon cessation of pacing (36). In contrast to the dog model, myocardial mass and stiffness increase *after* termination of pacing in pigs (36,45). Collagen production is augmented throughout the myocardium, albeit unevenly, with more detected in the subendocardium when compared to the subepicardial and subendocardial regions (46). Broad inhibition of MMP activity decreases left-ventricular end-diastolic pressure, wall stress, plasma norepinephrine levels, and attenuates the increase in collagen content (47). These results suggest a critical role for MMPs in the rapid pacing pig model of cardiac remodeling.

Rabbit

Rapid chronic pacing has also been performed in rabbits, and features changes that are similar to the myocardial remodeling and neurohumoral activation seen in dogs and pigs (48). In addition to this, myocardium from paced rabbits exhibits significant increases in the rate of MHC protein synthesis, without any change in the MHC mRNA expression and left-ventricular (LV) MHC protein accumulation, thereby suggesting abnormalities in contractile protein turnover (48). Though not widely employed, rabbits may be an attractive alternative for rapid pacing studies due to their low-cost and ease of handling and maintenance as compared to larger animals.

MYOCARDIAL INFARCTION

Remodeling as a result of myocardial infarction is an important cause of heart failure in patients; it is not surprising, therefore, that induction of myocardial infarction has been used in a variety of animal models to study the evolution of cardiac remodeling and heart failure. Myocardial infarction is usually produced via partial to complete occlusion of a coronary artery without reperfusion. Unlike pacing-induced cardiac remodeling, infarction causes irreversible damage to the myocardial tissue.

Infarction-induced remodeling has been most widely characterized in the rat, though it has also been studied in mice, sheep, pigs, and dogs.

Infarction by Coronary Artery Ligation

Rat

Permanent coronary artery occlusion in rats leads to heart failure within three to six weeks, as evidenced by a systolic and diastolic dysfunction, decrease in $+dP/dt$, wall thickening, left-ventricular diastolic filling, and increases in left-ventricular end-diastolic pressure (49). When challenged, isolated infarcted hearts exhibit an attenuated Frank–Starling response (50). Despite the global decline in pump function, cardiomyocytes isolated from infarcted hearts do not show any changes in the percent and velocity of shortening when compared to normal cells (50). In addition, no changes in the amplitude and velocity of the rise and decrease in intracellular calcium are evident, indicating that calcium handling remains normal in these hearts (50).

Increased neurohumoral activation in the rat myocardial infarct model is detected in the plasma, as well as in the cardiac tissue. Circulating levels of ANP (51), TNF-α (50), BNP (50), and endothelin (52) have all been shown to be elevated after infarction, although the circulating plasma renin activity is usually not increased in this model. Examination of left-ventricular tissue has been shown to demonstrate increased ACE activity that is inversely associated with left-ventricular pressure (53). Plasma norepinephrine is also increased in this model; however, observations regarding alterations in cardiac tissue production of norepinephrine vary. Some report an increase in right-ventricular norepinephrine content (54), whereas others claim an overall decrease in cardiac norepinephrine (51). These discrepancies may be due to differences in the breeds and sex of the rats studied. Turnover of norepinephrine was also found to be increased in the left ventricle and was associated with an increase in left-ventricular weight (54).

Adverse changes in left-ventricular remodeling include ventricular dilatation and development of the interstitial fibrosis. Upon microscopic examination, cardiomyocytes from infarcted hearts exhibit eccentric hypertrophy (50). A direct relationship exists between an increase in average myocyte length and the expansion of left-ventricular volume (50). Increases in collagenase and gelatinase activity may account for the interstitial fibrosis observed in infarcted hearts (55), since both types of enzymes have been proposed to be involved in development of interstitial fibrosis. Collagenase activity peaks within seven days after infarction, suggesting that the triggers for fibrosis may be activated within a short period of time (55). Interestingly, expression of the tissue inhibitor of metalloproteinase (TIMP) is also increased after infarction. However, this may be due to increased collagenase activity, since these enzymes negatively regulate each other (55). Treatment with ACE inhibitors decreases infarct size and collagen volume fraction, thereby emphasizing the role of activation of the renin–angiotensin system on the development of interstitial fibrosis, and its possible relationship to infarct expansion (56). In addition, it has been shown that inhibition of collagen synthesis improves left-ventricular function and decreases left-ventricular enlargement (57).

Apoptosis occurs rapidly in the rat infarct model, with significant increases apparent 24 hours after infarction in the infarct area and border zones. At four weeks after infarction, apoptotic activity declines and levels off (58). Measurement of the expression of apoptotic genes reveals decreased expression Bcl-2 and increased expression of Bax (59).

Several advantages of this model include ease of care for rats, cost, and simplicity of the ligation procedure. However, rats differ from humans in terms of electrophysiology, coronary circulation, and cardiac protein isoforms (60,61). Therefore, data obtained from this model must be interpreted with caution when determining its clinical significance.

Mouse

Myocardial infarction-induced remodeling has also been studied in mice. Effects are similar to that seen in rats, including the left-ventricular dysfunction, chamber dilation, and increase in wall thickness and weight (62). Increased cardiac ANP expression has been noted in this model, suggesting reactivation of the fetal gene program. Apoptosis appears in the area at risk within hours, and in the infarcted area seems to level off within seven days after ischemia (63); however, apoptotic cardiomyocytes are evident in areas remote from the infarct, months after injury (62).

Sheep and Pig

The human-like coronary vasculature of sheep makes this species useful for study of the pathophysiology of ischemia-induced remodeling (64,65). In order to create a suitable amount of injury, usually the homonymous artery and its diagonal branch are both ligated approximately 40% of the distance, from the apex to the base of the heart. As with rats, coronary artery ligation leads to chamber enlargement, however, without significant decreases in ejection fraction or fractional shortening (66). The severe degree of ischemia used in the ovine model of myocardial infarction can precipitate an abundance of arrhythmias, leading to a high rate of mortality. In an effort to reduce the mortality associated with this model, Kim et al. (66) employed sequential ligation instead of simultaneous ligation that was performed in previous models. In this method, the homonymous artery is ligated first, which is followed by ligation of the diagonal branch one hour later. This method still produced a high incidence of arrhythmias, but accomplished its goal of allowing a greater rate of survival.

As with sheep, pigs also possess human-like coronary vasculature (67). However, only the left circumflex artery needs to be occluded to produce a sufficient amount of injury. Clinical heart failure (as defined by weight loss, ascites, decreased activity, and significantly increased left-ventricular end-diastolic pressure) evolves in 30% to 40% of pigs within four weeks (68). Both compensated and decompensated animals experience ventricular dilatation, increases in left-ventricular mass, and decreases in left-ventricular ejection fraction (68). However, these changes are significantly greater in decompensated versus compensated pigs. Inhibition of MMP activity in the porcine infarct model attenuates infarct expansion, decreases the left-ventricular end-diastolic dimension, and improves the wall-thinning index as well, suggesting an important role for MMPs in porcine ventricular remodeling (69).

Infarction Via Transmyocardial Shock

Apart from direct coronary artery ligation, regional infarcts can be created in dogs by repetitive transmyocardial direct current shock applied externally to the chest cavity, therefore eliminating the need for open-heart surgery (70). The procedure results in 15% to 35% necrosis of the left ventricle, as well as an atrioventricular block (70). The incidence of mortality associated with this model is modest with approximately 70% of animals surviving the procedure (70). Direct current shock leads to acute depression of cardiac function and increased norepinephrine production, and heart failure takes months to develop (70,71). In a number of cases, however, this method does not ever lead to decompensated heart failure (70). In contrast to the functional decline, remodeling becomes evident much more quickly. Left-ventricular mass increases and LV dilation occurs one week after the direct current shock and progressively rises for up to 12 months (70–72).

CORONARY MICROEMBOLIZATION

In contrast to the regional myocardial infarct models discussed previously, an alternate method to produce remodeling and heart failure involves the creation of dispersed microinfarcts by injection of microspheres directly into the coronary circulation. Coronary microembolization has primarily been employed in dogs, though attempts have also been made to apply this technique in calves (73). Previous models using a bolus injection of glass beads or microspheres into the left circumflex artery of dogs precipitated a high incidence of mortality (74,75). Another approach entailing slow injection of latex microspheres over the course of four to six hours decreased mortality, but resulted in only moderate damage (73). In order to circumvent these issues, Sabbah et al. (76) developed a canine model whereby the heart failure is produced by three to nine sequential intracoronary microembolizations with polystyrene microspheres, over a course of one to three weeks. Intracoronary microembolizations are continued until the target ejection fraction is reached (usually below 35%). A key feature of this model is that the deterioration of function and remodeling continues to progress for several months, after the coronary microembolizations are suspended (76). The model manifests many of the features of hemodynamic dysfunction of heart failure in humans, including profound systolic and diastolic left-ventricular dysfunction, increased LV filling pressures, increased systemic vascular resistance, and decreased cardiac output (76,77). Isolated cardiomyocytes derived from the hearts of these animals also exhibit intrinsic impairment of contractility. Cardiac dysfunction is accompanied by neurohumoral activation, including sustained elevation of plasma norepinephrine, increased levels of plasma angiotensin-II, ANP, endothelin-1, and TNF-α. The model also exhibits down-regulation of cardiac β-adrenergic receptors (78), development of chronic ventricular arrhythmias (79), third heart sound (80), and exercise intolerance (81).

Coronary microembolization induces global cardiac remodeling, which is evident due to the left-ventricular dilation, chamber enlargement, and development of interstitial fibrosis (82–84) (Fig. 1). Increased chamber sphericity also develops and has been directly linked to the functional mitral regurgitation seen in these animals (84) (Fig. 2). At the cellular level, cardiomyocyte hypertrophy is apparent (85). The extent of cellular hypertrophy exceeds in the left-ventricular regions bordering scarred tissue than the left-ventricular regions that are remotely located from any scarred tissue (86,87).

Apoptosis is also seen; with the highest incidence of apoptotic events found in myocardial regions bordering scarred tissue of old infarcts (88). These regions show a greater extent of cardiomyocyte degeneration (89) and an increased lactate dehydrogenase activity (90). Regions bordering old infarcts exhibit considerable interstitial fibrosis associated with reduced capillary density and increased oxygen diffusion distance between cardiomyocytes (91). The gelatinases, MMP-2 and MMP-9, increase two- to four-fold in this model and may contribute to the development of interstitial fibrosis (92). It is suggested that the fibrous areas of left ventricle regions bordering old infarcts may be subject to chronic hypoxia, which in turn may trigger cardiomyocyte apoptosis (93). In addition, a decrease in the ratio of cardiac α-MHC to β-MHC has been observed in microembolized dogs, suggesting reinduction of the fetal gene program.

Many clinical therapies used to treat heart failure patients have been shown to ameliorate cardiac function and remodeling, in this model. A few of these treatments include ACE inhibitors, aldosterone receptor antagonists, and β-blockers (94,95). In addition, all prevent LV sphericity and enlargement, interstitial fibrosis, myocyte hypertrophy, and apoptosis. Table 3 shows the effects of two clinical treatments, enalapril (ACE inhibitor) and metoprolol (β-blocker), on hemodynamic measurements in dogs with microembolization-induced heart failure. As in humans, these drugs prevent the decline in cardiac function and LV enlargement associated with heart failure.

(A) **(B)** **(C)**

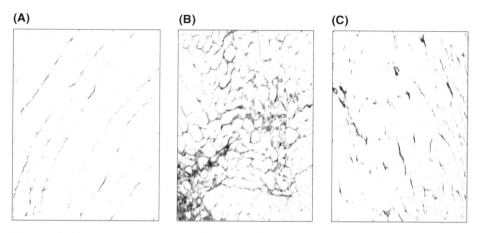

Figure 1 Representative staining for interstitial fibrosis with collagen-specific picosirius red. Interstitial fibrosis is indicated by dark purple staining. Magnification 200X. (**A**) Left-ventricular (LV) section from normal canine myocardium. (**B**) LV tissue section from micorem-bolization-induced failing canine myocardium. (**C**) LV tissue section from a canine treated with eprosartan, an angiotensin-1 receptor antagonist. Eprosartan ameliorates LV dysfunction asso-ciated with heart failure in the canine coronary microembolization model. *Source:* From Ref. 95. (*See color insert.*)

OTHER MODELS OF REMODELING

In addition to the major models of cardiac remodeling reviewed previously in this discussion, a host of other, albeit less common, animal models have been reported that may be useful toward study of remodeling (Table 4). Since cardiac remodeling can be triggered by a variety of stimuli, the development of a variety of models to explore this phenomenon would be advantageous in understanding its underlying mechanisms. Though these may not all represent major pathways by which remodel-ing occurs in humans, a variety of factors most likely contribute to remodeling and render the information gleaned from these models are potentially useful.

End-systole **End-diastole**

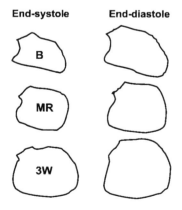

Figure 2 End-systolic (*left*) and end-diastolic (*right*) left-ventricular silhouettes from a study dog, traced from a ventriculogram. The silhouettes depict the changes of global left-ventricular shape at baseline (*B*), at the onset of mitral regurgitation (*MR*) and at 3 weeks (*3W*) after the onset of functional mitral regurgitation.

Table 3 Hemodynamic Effects of Enalapril and Metoprolol in Dogs with Microembolization-Induced Heart Failure

	Pre	Post
Control		
HR (beats per minute)	79 ± 7	97 ± 8
LVEDP (mmHg)	17 ± 3	20 ± 5
Stroke volume (mL)	36 ± 3	29 ± 2
LVEF (%)	36 ± 1	26 ± 1
LVESV (mL)	39 ± 4	57 ± 6
LVEDV (mL)	61 ± 6	78 ± 8
Wall stress (g/cm^2)	74 ± 12	80 ± 21
Enalapril		
HR (beats per minute)	87 ± 5	81 ± 4
LVEDP (mmHg)	21 ± 4	16 ± 4
Stroke volume (mL)	35 ± 3	39 ± 3
LVEF (%)	35 ± 1	38 ± 3
LVESV (mL)	40 ± 4	40 ± 3
LVEDV (mL)	61 ± 6	65 ± 5
Wall stress (g/cm^2)	79 ± 16	71 ± 19
Metoprolol		
HR (beats per minute)	95 ± 8	83 ± 6
LVEDP (mmHg)	18 ± 3	10 ± 2
Stroke volume (mL)	35 ± 3	40 ± 3
LVEF (%)	35 ± 1	40 ± 3
LVESV (mL)	41 ± 3	41 ± 3
LVEDV (mL)	64 ± 3	69 ± 4
Wall stress (g/cm^2)	78 ± 13	43 ± 8

Abbreviations: HR, heart rate; LVEDP, left-ventricular end-diastolic pressure; LVEF, left-ventricular ejection fraction; LVESV, left-ventricular end-systolic volume; LVEDV, left-ventricular end-diastolic volume.

Volume and Pressure Overload

Chronic volume and pressure overload, achieved by a variety of surgical methods, leads to cardiac remodeling, which is most notably evidenced by cardiac hypertrophy. Banding of the ascending aorta results in left-ventricular pressure overload, whereas right-ventricular overload is accomplished by banding the pulmonary artery. One technique to induce overload involves banding rodents while they are still young, so that aortic or arterial stenosis develops as they grow (96–98). Animals can also be banded as adults, which may be more convenient in the case of large animals such as sheep and pigs (99–102). Two other methods that produce volume overload include chordal rupture that causes mitral regurgitation and surgical manipulation that creates an aortocaval fistula (103–108). These various methods of creating overload have been performed in a wide range of animals including rats, mice, guinea pigs, rabbits, dogs, cats, ferrets, pigs, and sheep. In most cases, chronic overload leads to a derangement of pump function; however, hypertrophy can occur in the absence of ventricular dysfunction (96).

While the exact physiological consequences of overloading vary between species and the different techniques used to create an overload, the general characteristics associated with cardiac remodeling, as seen in the previously mentioned models,

Table 4 Summary of Animal Models of Remodeling

Experimental method	Species
Chronic rapid pacing	Dog, pig, rabbit
Myocardial infarction	Rat, mouse, sheep, dog
Coronary microembolization	Dog, calf, pig
Volume/pressure overload	
Aortic stenosis	Rat, sheep, pig, rabbit, dog, guinea pig
Pulmonary artery stenosis	Rat, mouse, cat, ferret
Mitral regurgitation	Dog
Aortocaval fistula	Rat, dog
Genetic	Rat, hamster, cow, turkey, cat, mouse (transgenic)
Viral myocarditis	Mouse

have also been observed in overload models. Chronic volume and pressure overload causes ventricular hypertrophy and chamber dilation, as well as increases in cell volume, cellular hyperplasia, and capillary proliferation (103–108). Volume overload also leads to an increase in cardiac mass (103,105,106). Neurohumoral activation occurs with chronic pressure overload as evidenced by elevations in plasma renin activity and plasma aldosterone, ANP, and BNP (99). In addition to the increase in ANP expression, β-MHC expression has been noted to increase in the aortic banded rats, suggesting reinduction of the fetal gene program in this setting (97). Finally, apoptosis has been observed in volume-overloaded animals, peaking within days of initiation of overload, while cardiac hypertrophy continues to progress for weeks thereafter (109).

Efforts to identify factors that delineate the transition from compensated to a decompensated heart failure have been made in volume and pressure overload models. In pressure overload models, a decline in hemodynamic function has been found to correlate with decreases in Ca^{2+}-ATPase mRNA expression, glucose uptake, coronary effluent adenosine concentration, and increased cardiomyocyte microtubule density, compared to corresponding compensated animals (97,110–112). It is also interesting to note that pressure-overloaded mast cell deficient mice do not develop decompensated heart failure, suggesting the involvement of an immunological component in the shift from compensated to decompensated heart failure (113). In addition, volume-overloaded rats exhibit increases in cardiac monocyte chemoattractant protein-1 production (114). Recognizing the factors that distinguish compensated from decompensated cardiac remodeling carries tremendous clinical implications, both for the diagnosis as well as the development of new therapies.

Cardiomyopathies

Thus far, all of the models discussed in this chapter have involved some type of surgical or mechanical intervention to induce cardiac remodeling. Though these techniques do produce clinically applicable data, it should be taken into account that remodeling can also occur as a result of congenital defects or infection. Several animal models of cardiac remodeling have emerged due to spontaneous genetic defects. Advances in genetic engineering have also made it possible to generate animals with inborn cardiomyopathies [see Ref. (115) for review]. For example, mice overexpressing the G-protein subunit, $G_{s\alpha}$, exhibit decreased left-ventricular ejection fraction, and increased left-ventricular end-diastolic dimension (116), while mutations in the α-MHC gene result in hypertrophic cardiomyopathy (117). In addition, cardiac remodeling can be brought about by inoculating mice with a virus that causes myocarditis, which manifests itself through increases in right and left ventricle cavity dimensions, elevated circulating cytokines, decreases in right- and left-ventricular wall thickness, and eventually culminates in heart failure (118,119).

The more rare and unusual models include cats, turkeys, and cows with congential cardiomyopathies (120–122, see Table 4 for more details). They exhibit many of the symptoms of human cardiac remodeling, such as hypertrophy and the development of interstitial fibrosis (120,122). A much more common practice in the investigation of cardiac remodeling and heart failure is the use of spontaneously hypertensive rats. These animals develop heart failure after a sustained period of hypertrophy (123). In addition, spontaneously hypertensive rats show progressive increases in plasma renin activity, aldosterone, and ANP levels as they age (124). Histologic evaluation of their hearts reveals visible fibrosis (123). Spontaneously hypertensive rats also exhibit a decrease in the ratio of β-MHC to α-MHC and increased ANP expression as they transition into heart failure (123,125). Strains of cardiomyopathic hamsters have also been characterized (126–129). In these animals, cardiac interstitial fibrosis and hypertrophy develop within a year, leading to reduced cardiovascular function (124). Increased norepinephrine turnover also occurs as cardiomyopathic hamsters enter decompensated heart failure (130). Abnormalities in calcium homeostasis have been suggested to underlie cardiomyopathy in these animals (129–131).

CONCLUSION

Remodeling of the heart involves a characteristic series of progressive structural and functional changes that ultimately result in the development of heart failure. Studying the pathophysiology of cardiac remodeling in human patients is difficult due to the presence of diverse underlying etiologies, comorbid conditions that may influence the process, the effects of concurrent therapy, and difficulties inherent in sampling human tissue. In addition, the time course may be quite prolonged as occurs in patients who undergo remodeling post-MI. Animal models allow for thorough investigation of the mechanisms of remodeling under controlled conditions (Table 5). As discussed in this chapter, numerous models exist that address the various etiologies by which remodeling is initiated. Characterizing the various mechanisms by which

Table 5 Summary of Remodeling Triggers and Corresponding Experimental Techniques

Physiological insult	Model/technique
Reversible	Rapid pacing
Infarction	Coronary artery ligation
	Transmyocardial shock
	Coronary microembolization
Hypertension	Aortic banding
	Pulmonary artery banding
	Aortalcaval fistula
	Genetic abnormality

cardiac remodeling can occur is important for both understanding the pathophysiology of remodeling/heart failure, as well as for the design of future therapies.

REFERENCES

1. Nass O, Rosman H, Al-Khaled N, Shimoyama H, Alam M, Sabbah HN, Goldstein S. Relation of left ventricular chamber shape in patients with low ($\leq 40\%$) ejection fraction to severity of functional mitral regurgitation. Am J Cardiol 1995; 76:402–404.
2. Sabbah HM, Rosman H, Kono T, Alam M, Khaja F, Goldstien S. On the mechanism of functional mitral regurgitation. Am J Cardiol 1993; 72:1074–1076.
3. Li YY, Feng YQ, Kadokami T, McTiernan CF, Feldman AM. Modulation of matrix metalloproteinase activities remodels myocardial extracellular matrix in TNFα transgenic mice. Circulation 1999; 100(Suppl I):I752.
4. Li YY, McTiernan CF, Feldman AM. Interplay of matrix metalloproteinases, tissue inhibitors of metalloproteinases and their regulators in cardiac matrix remodeling. Cardiovasc Res 2000; 46:214–224.
5. Cohn JN, Levine TB, Olivari MT, Garberg V, Lura D, Francis GS, Simon AB, Rector T. Plasma norepinephrine as a guide to prognosis in patients with chronic congestive heart failure. N Engl J Med 1984; 211:819–823.
6. Tsutamoto T, Wada A, Maeda K, Hisanaga T, Maeda Y, Fukai D, Ohnishi M, Sugimoto Y, Kinoshita M. Attenuation of compensation of endogenous cardiac natriuretic peptide system in chronic heart failure: prognostic role of plasma brain natriuretic peptide concentration in patients with chronic symptomatic left ventricular dysfunction. Circulation 1997; 96:509–516.
7. Francis GS. Neuroendocrine manifestations of congestive heart failure. Am J Cardiol 1988; 62:9A–13A.
8. Levine B, Kalman J, Mayer L, Fillit HM, Packer M. Elevated circulating levels of tumor necrosis factor in severe chronic heart failure. N Engl J Med 1990; 323: 236–241.
9. Communal C, Singh K, Pimentel DR, Colucci WS. Norepinephrine stimulates apoptosis in adult rat ventricular myocytes by activation of the beta-adrenergic pathway. Circulation 1998; 98:1329–1334.
10. Wu CF, Bishopric NH, Pratt RE. Atrial natriuretic peptide induces apoptosis in neonatal rat cardiac myocytes. J Biol Chem 1997; 272:14860–14866.
11. Tan LB, Jalil JE, Pick R, Janicki JS, Weber KT. Cardiac myocyte necrosis induced by angiotensin II. Circ Res 1991; 69:1185–1195.
12. Weber KT, Brilla CG. Pathological hypertrophy and cardiac interstitium. Fibrosis and renin–angiotensin–aldosterone system. Circulation 1991; 83:1849–1865.
13. Ruzicka M, Yuan B, Harmsen E, Leenen FH. The renin–angiotensin system and volume overload-induced cardiac hypertrophy in rats. Effects of angiotensin converting enzyme inhibitor versus angiotensin II receptor blocker. Circulation 1993; 87:921–930.
14. Simpson P. Norepinephrine-stimulated hypertrophy of cultured rat myocardial cells is an alpha 1 adrenergic response. J Clin Invest 1983; 72:732–738.
15. Sivasubramanian N, Coker ML, Kurrelmeyer KM, MacLellan WR, DeMayo FJ, Spinale FG, Mann DL. Left ventricular remodeling in transgenic mice with cardiac restricted overexpression of tumor necrosis factor. Circulation 2001; 104:826–831.
16. Janczewski AM, Kadokami T, Lemster B, Frye CS, McTiernan CF, Feldman AM. Morphological and functional changes in cardiac myocytes isolated from mice overexpressing TNF-alpha. Am J Physiol Heart Circ Physiol 2003; 284:H960–969.
17. Li YY, Feng YQ, Kadokami T, McTiernan CF, Draviam R, Watkins SC, Feldman AM. Myocardial extracellular matrix remodeling in transgenic mice overexpressing

tumor necrosis factor alpha can be modulated by anti-tumor necrosis factor alpha therapy. Proc Natl Acad Sci USA 2000; 97:12746–12751.

18. Sleight P. Angiotensin II and trials of cardiovascular outcomes. Am J Cardiol 2002; 89:11A–16A.

19. Sabbah HN. The cellular and physiologic effects of beta blockers in heart failure. Clin Cardiol 1999; 22(Suppl 5):V16–V20.

20. Walther T, Schultheiss H-P, Tschöpe C, Stepan H. Natriuretic peptide system in fetal heart and circulation. J Hyperten 2002; 20:785–791.

21. Lompre AM, Nadal-Ginard B, Madhavi V. Expression of cardiac alpha and beta myosin heavy chain genes is developmentally and hormonally regulated. J Biol Chem 1984; 259:6437–6446.

22. Reiser PJ, Portman MA, Ning X, Moravec CS. Human cardiac myosin heavy chain isoforms in fetal and failing adult atria and ventricles. Am J Physiol Heart Circ Physiol 2001; 280:H1818–H1820.

23. Litten RZ, Martin BF, Low RB, Alpert NR. Altered myosin isozyme pattern from pressure-overload and thyroxic hypertrophied rabbit hearts. Circ Res 1982; 50:856–864.

24. Wilson JR, Douglas P, Hickey WF, Lanoce V, Ferraro N, Muhammad A, Reichek N. Experimental congestive heart failure produced by rapid ventricular pacing in the dog: cardiac effects. Circulation 1987; 75:857–867.

25. Komamura K, Shannon RP, Ihara T , Shen YT, Mirsky I, Bishop SP, Vatner SF. Exhaustion of Frank–Starling mechanism in conscious dogs with heart failure. Am J Physiol 1993; 265:H1119–1131.

26. Armstrong PW, Stopps TP, Ford SE, DeBold AJ. Rapid ventricular pacing in the dog: pathophysiologic studies of heart failure. Circulation 1986; 5:1075–1084.

27. Eaton GM, Cody RJ, Nunziata E, Binkley PF. Early left ventricular dysfunction elicits activation of sympathetic drive and attenuation of parasympathetic tone in the paced canine model of congestive heart failure. Circulation 1995; 92:555–561.

28. Travill CM, Williams TDM, Pate P, Song G, Chalmers J, Lightman SL, Sutton R, Noble MIM. Haemodynamic and neurohumoral response in heart failure produced by rapid ventricular pacing. Cardiovasc Res 1992; 26:783–790.

29. Spinale FG, Holzgrefe HH, Mukherjee R, Barry Hird R, Walker JD, Arnim-Barker A, Powell JR, Koster WH. Angiotensin-converting enzyme inhibition and the progression of congestive cardiomyopathy. Effects on left ventricular and myocyte structure and function. Circulation 1995; 92:562–578.

30. Luchner A, Stevens TL, Borgeson DD, Redfield MM, Bailey JE, Sandberg SM, Heulbein DM, Burnett JC. Angiotensin II in the evolution of experimental heart failure. Hypertension 1996; 28:472–477.

31. Krombach RS, Clair MJ, Hendrick JW, Houck WV, Zellner JL, Kribbs SB, Whitebread S, Mukherjee R, deGasparo M, Spinale FG. Angiotensin converting enzyme inhibition, AT1 receptor inhibition, and combination therapy with pacing induced heart failure: effects on left ventricular performance and regional blood flow patterns. Cardiovasc Res 1998; 38:631–645.

32. Burchell SA, Spinale FG, Crawford FA, Tanaka R, Zile MR. Effects of chronic tachycardia-induced cardiomyopathy on the beta-adrenergic receptor system. J Thorac Cardiovasc Surg 1992; 104:1006–1012.

33. Eble DM, Spinale FG. Contractile and cytoskeletal content, structure, and mRNA levels with tachycardia-induced cardiomyopathy. Am J Physiol Heart Circ Physiol 1995; 37:H2426–H2439.

34. Klinge R, Hystad M, Kjekshus J, Karlberg BE, Djoseland O, Aakvaag A, Hall C. An experimental study of cardiac natriuretic peptides as markers of development of congestive heart failure. Scand J Clin Lab Invest 1998; 58:683–691.

35. Kajstura J, Zhang X, Liu Y, Szoke E, Cheng W, Olivetti G, Hintze TH, Anversa P. The cellular basis of pacing-induced dilated cardiomyopathy. Myocyte cell loss and myocyte cellular reactive hypertrophy. Circulation 1995; 92:2306–2317.

36. Spinale FG, Tomita M, Zellner JL, Cook JC, Crawford FA, Zile MR. Collagen remodeling and changes in LV function during development and recovery from supraventricular tachycardia. Am J Physiol Heart Circ Physiol 1991; 261:H308–H318.

37. Williams RE, Kaas DA, Kawagoe Y, Pak P, Tunin RS, Shah R, Hwang A, Feldman AM. Endomyocardial gene expression during development of pacing tachycardia-induced heart failure in the dog. Circ Res 1994; 75:615–623.

38. Riegger GAJ, Elsner D, Kromer EP, Daffner C, Forssmann WG, Muders F, Pashcer EW, Kochsiek K. Atrial natriuretic peptide in congestive heart failure in the dog: plasma levels, cyclic guanosine monophosphate, ultrastructure of atrial myoendocrine, and hemodynamic, hormonal, and renal effects. Circulation 1988; 77:398–406.

39. Ravens U, Davia K, Davies CH, O'Gara P, Drake-Holland AJ, Hynd JW, Noble MI, Harding SE. Tachycardia-induced failure alters contractile properties of canine ventricular myocytes. Cardiovasc Res 1996; 32:613–621.

40. Perreault CL, Shannon RP, Komamura K, Vatner SF, Morgan JP. Abnormalities in intracellular calcium regulation and contractile function in myocardium from dogs with pacing-induced heart failure. J Clin Invest 1992; 89:932–938.

41. Wang J, Seyedi N, Xu XB, Wolin MS, Hintze TH. Defective endothelium-mediated control of coronary circulation in conscious dogs after heart failure. Am J Physiol 1994; 266:H670–680.

42. Heinke MY, Yao M, Chang D, Einstein R, dos Remedios CG. Apoptosis of ventricular and atrial myocytes from pacing-induced canine heart failure. Cardiovasc Res 2001; 49:127–134.

43. Cesselli D, Jakoniuk I, Barlucchi L, Beltrami AP, Hintze TH, Nadal-Ginard B, Kajstura J, Leri A, Anversa P. Oxidative stress-mediated cardiac cell death is a major determinant of ventricular dysfunction and failure in dog dilated cardiomyopathy. Circ Res 2001; 89:279–286.

44. Leri A, Liu Y, Malholtra A, Li A, Stiegler P, Claudio PP, Giordano A, Kajstura J, Hintze TH, Anversa P. Pacing-induced heart failure in dogs enhances the expression of p53-dependent genes in ventricular myocytes. Circulation 1998; 97:194–203.

45. Spinale FG, Hendrick DA, Crawford FA, Smith AC, Hamada Y, Carabello BA. Chronic supraventricular tachycardia causes ventricular dysfunction and subendocardial injury in swine. Am J Physiol Heart Circ Physiol 1990; 28:H218–H229.

46. Spinale FG, Zellner JL, Johnson WS, Eble DM, Munyer PD. Cellular and extracellular remodeling with the development and recovery from tachycardia-induced cardiomyopathy: changes in fibrillar collagen, myocyte adhesion capacity and proteoglycans. J Mol Cell Cardiol 1996; 28:1591–1608.

47. King MK, Coker ML, Goldberg A, McElmurray JH, Gunasinghe HR, Mukherjee R, Zile MR, O'Neill TP, Spinale FG. Selective matrix metalloproteinase inhibition with developing heart failure. Effects on left ventricular function and structure. Circ Res 2003; 92:177–185.

48. Eble DM, Walker JD, Mukherkjee R, Samarel AM, Spinale FG. Myosin heavy chain synthesis is increased in a rabbit model of heart failure. Am J Physiol Heart Circ Physiol 1997; 41:H969–H978.

49. Litwin SE, Katz SE, Morgan JP, Douglas PS. Serial echocardiographic assessment of left ventricular geometry and function after large myocardial infarction in the rat. Circulation 1994; 89:345–354.

50. Anand IS. Ventricular remodeling without cellular contractile dysfunction. J Card Failure 2002; 8:S401–S408.

51. Hodsman GP, Kohzuki M, Howes LG, Sumithran E, Tsunoda K, Johnston CI. Neurohumoral responses to chronic myocardial infarction in rats. Circulation 1988; 78:376–381.

52. Teerlink JR, Löffler B-M, Hess P, Maire J-P, Clozel M, Clozel J-P. Role of endothelin in the maintenance of blood pressure in conscious rats with chronic heart failure. Acute effects of the endothelin receptor antagonist Ro 47–0203 (Bosentan). Circulation 1994; 90:2510–2518.

53. Pinto YM, de Smet BG, van Gilst WH, Scholtens E, Monnink S, de Graeff PA, Wesseling H. Selective and time related activation of the cardiac renin–angiotensin system after experimental heart failure: relation to ventricular function and morphology. Cardiovasc Res 1993; 27:1933–1938.

54. Ganguly PK, Dhalla KS, Shao Q, Beamish RE, Dhalla NS. Differential changes in sympathetic activity in left and right ventricles in congestive heart failure after myocardial infarction. Am Heart J 1997; 133:340–345.

55. Cleutjens JP, Kandala JC, Guarda E, Guntaka RV, Weber KT. Regulation of collagen degradation in the rat myocardium after infarction. J Mol Cell Cardiol 1995; 27: 1281–1292.

56. DeCarvalho Frimm C, Sun Y, Weber KT. Angiotensin II receptor blockade and myocardial fibrosis of the infarcted rat heart. J Lab Clin Med 1997; 129:439–446.

57. Nwogu JI, Geenen D, Bean M, Brenner MC, Huang X, Butterick PM. Inhibition of collagen synthesis with prolyl 4-hydroxylase inhibitor improves left ventricular function and alters the pattern of left ventricular dilation after myocardial infarction. Circulation 2001; 104:2216–2221.

58. Palojoki E, Saraste A, Eriksson A, Pulkki K, Kallajoki M, Voipio-Pulkki L-M, Tikkanen I. Cardiomyocyte apoptosis and ventricular remodeling after myocardial infarction in rats. Am J Physiol Heart Circ Physiol 2001; 280:H2726–H2731.

59. Gupta S, Prahash AJ, Anand IS. Myocyte contractile function is intact in the post-infarct remodeled rat heart despite molecular alterations. Cardiovasc Res 2000; 48:77–88.

60. Baker HJ, Lindsey JR, Weisbroth. The Laboratory Rat. Vol. II. Research applications. New York: Academic Press 1980.

61. Gross DR. Animal Models in Cardiovascular Research. Dordrecht, The Netherlands: Kluwer, 1994.

62. Sam F, Sawyer DB, Chang DL-F, Eberli FR, Ngoy S, Jain M, Amin J, Apstein CS, Colucci WS. Progressive left ventricular remodeling and apoptosis late after myocardial infarction in mouse heart. Am J Physiol Heart Circ Physiol 2000; 279:H422–H428.

63. Bialik S, Geenen DL, Sasson IE, Cheng R, Horner JW, Evans SM, Lord EM, Koch CJ, Kitis RN. Myocyte apoptosis during acute myocardial infarction in the mouse localizes to hypoxic regions but occurs independently of p53. J Clin Invest 1997; 100:1363–1372.

64. Millner RW, Mann JM, Pearson I, Pepper JR. Experimental model of left ventricular failure. Ann Thorac Surg 1991; 52:78–83.

65. Frink RJ, Merrick B. The sheep heart: Coronary and conduction system anatomy with special reference to the presence of an oscordis. Anat Rec 1974; 179:189–200.

66. Kim WG, Park JJ, Oh SI. Chronic heart failure model with sequential ligation of the homonymous artery and its diagonal branch in the sheep. ASAIO J 2001; 47:667–672.

67. Maxwell MP, Hearse DJ, Yellon DM. Species variation in the coronary collateral circulation during regional myocardial ischemia: a critical determinant of the rate of evolution and extent of myocardial infarction. Cardiovasc Surg 1987; 21:737–746.

68. Zhang J, Wilke N, Wang Y, Zhang Y, Wang C, Eijgelshoven MHJ, Cho YK, Murakami Y, Ugurbil K, Bache RJ. Functional and bioenergetic consequences of post-infarction left ventricular remodeling in a new porcine model. MRI and ^{31}P-MRS study. Circulation 1996; 94:1089–1100.

69. Mukherjee R, Brinsa TA, Dowdy KB, Scott AA, Baskin JM, Deschamps AM, Lowry AS, Escobar GP, Lucas DG, Yarbrough WM, Zile MR, Spinale FG. Myocardial infarct expansion and matrix metalloproteinase inhibition. Circulation 2003; 107:618–625.

70. Carlyle PF, Cohn JN. A nonsurgical canine model of chronic left ventricular myocardial dysfunction. Am J Physiol Heart Circ Physiol 1983; 13:H769–H774.

71. McDonald KM, Francis GS, Carlyle PF, Hauer K, Matthews J, Hunter DW, Cohn JN. Hemodynamic, left ventricular structural and hormonal changes after discrete myocardial damage in the dog. J Am Coll Cardiol 1992; 19:460–467.

72. McDonald KM, Garr M, Carlyle PF, Francis GS, Hauer K, Hunter DW, Parish T, Stillman A, Cohn JN. Relative effects of α_1-adrenoceptor blockade converting enzyme inhibitor therapy, and angiotensin II subtype 1 receptor blockade on ventricular remodeling in the dog. Circulation 1994; 90:3034–3046.

73. Weber KT, Malinin TI, Dennison BH, Fuqua JM Jr, Speaker DM, Hastings FW. Experimental myocardial ischemia and infarction. Production of diffuse myocardial lesions on unanesthetized calves. Am J Cardiol 1972; 29:793–802.

74. Franciosa JA, Heckel R, Limas C, Cohn JN. Progressive myocardial dysfunction associated with increased vascular resistance. Am J Physiol Heart Circ Physiol 1980; 239:H477–H482.

75. Smiseth OA, Lindal S, Mjos OD, Vik-Mo H, Jorgensen L. Progression of myocardial damage following coronary microembolization in dogs. Acta Pathol Microbiol Immunol Scand 1983; 91:115–124.

76. Sabbah HN, Stein PD, Kong T, Gheorghiade T, Levine B, Jafri S, Hawkins ET, Goldstein S. A canine model of chronic heart failure produced by multiple sequential coronary microembolization. Am J Physiol Heart Circ Physiol 1991; 29:H1379–H1384.

77. Kono T, Sabbah HN, Rosman H, Alam M, Stein PD, Goldstein S. Left atrial contribution to ventricular filling during the course of evolving heart failure. Circulation 1992; 86:1317–1322.

78. Gengo PG, Sabbah HN, Steffen RP, Sharpe JK, Kono T, Stein PD, Goldstein S. Myocardial beta adrenoceptor and voltage sensitive calcium channel changes in a canine model of chronic heart failure. J Mol Cell Cardiol 1992; 12:1361–1369.

79. Sabbah HN, Goldberg AD, Schoels W, Kono T, Brachmann J, Goldstein S. Spontaneous and inducible ventricular arrhythmias in a canine model of chronic heart failure: Relation to hemodynamics and sympathoadrenergic activation. Eur Heart J 1992; 13:1562–1572.

80. Kono T, Rosman H, Alam M, Stein PD, Sabbah HN. Hemodynamic correlates of the third heart sound during the evolution of chronic heart failure. J Am Coll Cardiol 1993; 21:419–423.

81. Sabbah HN, Hansen-Smith F, Sharov VG, Kono T, Lesch M, Gengo PJ, Steffen RP, Levine TB, Goldstein S. Decreased proportion of type-I myofibers in skeletal muscle of dogs with chronic heart failure. Circulation 1993; 87:1729–1737.

82. Sabbah HN, Sharov VG, Lesch M, Goldstein S. Progression of heart failure: a role for interstitial fibrosis. Mol Cell Biochem 1995; 147:29–34.

83. Sabbah HN, Goldstein S. Ventricular remodeling: consequences and therapy. Eur Heart J 1993; 14(Supplement C):24–29.

84. Sabbah HN, Kono T, Stein PD, Mancini GBJ, Goldstein S. Left ventricular shape changes during the course of evolving heart failure. Am J Physiol Heart Circ Physiol 1992; 32:H266–H270.

85. Sharov VG, Sabbah HN, Ali AS, Shimoyama H, Lesch M, Goldstein S. Abnormalities of cardiocytes in regions bordering fibrous scars of dogs with heart failure. Int J Cardiol 1997; 60:273–279.

86. Shevlyagin S, Sharov VG, Sahan G, Silverman N, Lesch M, Goldstein S. Disproportionate increase in cardiocyte hypertrophy in myocardial regions bordering old infarctions: studies in patients and dogs with chronic heart failure. Circulation 1995; 92:3811.

87. Ali AS, Sharov VG, Lesch M, Sabbah HN, Goldstein S. Nonhomogeneous distribution of cardiocyte size in the border zone of old infarcts in dogs with chronic heart failure [abstr]. J Heart Failure 1996; 3:234.

88. Sharov VG, Sabbah HN, Shimoyama H, Goussev AV, Lesch M, Goldstein S. Evidence of cardiocyte apoptosis in myocardium of dogs with chronic heart failure. Am J Pathol 1996; 148:141–149.

89. Sharov VG, Sabbah HN, Kono T, Ali AS, Shimoyama H, Lesch M, Goldstein S. Ultrastructural abnormalities of cardiomyocytes in border zone of old infarction: studies in dogs with chronic heart failure. FASEB J 1993; 7:A112.

90. Shimoyama H, Sabbah HN, Sharov VG, Cook J, Lesch M, Goldstein S. Accumulation of interstitial collagen in the failing left ventricular myocardium is associated with increased anaerobic metabolism. J Am Coll Cardiol 1994; 98A (Special issue).

91. Sabbah HN, Sharov VG, Cook JM, Shimoyama H, Lesch M, Goldstein S. Accumulation of collagen in the cardiac interstitium of dogs with chronic heart failure is associated with decreased capillary density and increased oxygen diffusion distance. J Am Coll Cardiol 1995; 222A (Special issue).

92. Kawamoto RM, Suzuki G, Morita H, O'Neill TP, Sabbah HN. Matrix metalloproteinase 2 and 9 activity is increased in dogs with progressive left ventricular failure. Circulation 2001; 102(suppl II):105770A.

93. Tanaka M, Ito H, Adachi S, Akimoto H, Nishikawa T, Kasajima T, Marumo F, Hiroe M. Hypoxia induces apoptosis with enhanced expression of Fas antigen messenger RNA in cultured neonatal rat cardiomyocytes. Circ Res 1994; 75:426–433.

94. Sabbah HN, Shimoyama H, Kono T, Gupta RC, Sharov VG, Scicli G, Levine B, Goldstein S. Effects of long-term monotherapy with enalapril, metoprolol, and digoxin on the progression of left ventricular dysfunction and dilation in dogs with reduced ejection fraction. Circulation 1994; 89:2852–2859.

95. Suzuki G, Mishima T, Tanhehco EJ, Sharov VG, Todor A, Rastogi S, Gupta RC, Chaudhry PA, Anagnostopoulos PA, Nass O, Goldstein S, Sabbah HN. Effects of the AT_1-receptor antagonist eprosartan on the progression of left ventricular dysfunction in dogs with heart failure. Br J Pharmacol 2003; 138:301–309.

96. Julian FJ, Morgan DL, Moss RL, Gonzalez M, Dwivedi P. Myocyte growth without physiological impairment in gradually induced rat cardiac hypertrophy. Circ Res 1981; 49:1300–1310.

97. Feldman AM, Wienberg EO, Ray PE, Lorell BH. Selective changes in cardiac gene expression during compensated hypertrophy and the transition to cardiac decompensation in rats with chronic aortic banding. Circ Res 1993; 73:184–192.

98. Weinberg EO, Schoen FJ, George D, Kagaya Y, Douglas PS, Litwin SE, Schunkert H, Benedict CR, Lorell BH. Angiotensin-converting enzyme inhibition prolongs survival and modifies the transition to heart failure in rats with pressure overload hypertrophy due to ascending aortic stenosis. Circulation 1994; 90:1410–1422.

99. Charles CJ, Kaaja RJ, Espiner EA, Nicholls MG, Pemberton CJ, Richards AM, Yandle TG. Natriuretic peptides in sheep with pressure overload left ventricular hypertrophy. Clin Exp Hypertens 1996; 18:1051–1071.

100. Carroll SM, Nimmo LE, Knoepfler PS, White FC, Bloor CM. Gene expression in a swine model of right ventricular hypertrophy: intercellular adhesion molecule, vascular endothelial growth factor and plasminogen activators are upregulated during pressure overload. J Mol Cell Cardiol 1995; 27:1427–1441.

101. Tagawa H, Koide M, Sato H, Cooper G IV. Cytoskeletal role in the contractile dysfunction of cardiocytes from hypertrophied and failing right ventricular myocardium. Proc Asso Am Physicians 1996; 108:218–229.

102. Koide M, Nagatsu M, Zile MR, Hamawaki M, Swindle MM, Keech G, DeFreyte G, Tagawa H, Cooper G IV, Carabello BA. Premorbid determinants of left ventricular dysfunction in novel model of gradually induced pressure overload in the adult canine. Circulation 1997; 95:1601–1610.

103. Kleaveland JP, Kussmaul WG, Vinciguerra T, Diters R, Carabello BA. Volume overload hypertrophy in a closed-chest model of mitral regurgitation. Am J Physiol Heart Circ Physiol 1988; 254:H1034–H1041.

104. Spinale FG, Ishihara K, Zile M, DeFryte G, Crawford FA, Carabello BA. Structural basis for changes in left ventricular function and geometry because of chronic mitral regurgitation and after correction of volume overload. Thorac Cardiovasc Surg 1993; 106:1147–1157.

105. Liu Z, Hilbelink DR, Crockett WB, Gerdes AM. Regional changes in hemodynamics and cardiac myocyte size in rats with aortocaval fistulas. 1. Developing and established hypertrophy. Circ Res 1991; 69:52–58.

106. Liu Z, Hilbelink DR, Gerdes AM. Regional changes in hemodynamics and cardiac myocyte size in rats with aortocaval fistulas. 2. Long-term effects. Circ Res 1991; 69:59–65.

107. Newman WH, Webb JG. A differential inotropic responsiveness to isoprenaline and ouabain in dogs with heart failure. Cardiovasc Res 1980; 14:530–536.

108. Pinsky WW, Lewis RM, Hartley CJ, Entman ML. Permanent changes of ventricular contractility and compliance in chronic volume overload. Am J Physiol Heart Circ Physiol 1979; 237:H575–H583.

109. Teiger E, Than VD, Richard L, Wisnewsky C, Tea BS, Gaboury L, Tremblay J, Schwartz K, Hamet P. Apoptosis in pressure overload-induced heart hypertrophy in the rat. J Clin Invest 1996; 97:2891–2897.

110. Friehs I, Moran AM, Stamm C, Colan SD, Takeuchi K, Cao-Danh H, Rader CM, McGowan FX, del Nido PJ. Impaired glucose transporter activity in pressure-overload hypertrophy is an early indicator of progression to failure. Circulation 1999; 100: II187–II193.

111. Meyer TE, Chung ES, Perlini S, Norton GR, Woodiwiss AJ, Lorbar M, Fenton RA, Dobson JG Jr. Antiadrenergic effects of adenosine in pressure overload hypertrophy. Hypertension 2001; 37:862–868.

112. Tagawa H, Koide M, Sato H, Zile MR, Carabello BA, Cooper G IV. Cytoskleleton role in the transition from compensated to decompensated hypertrophy during adult canine left ventricular pressure overloading. Circ Res 1998; 82:751–761.

113. Hara M, Ono K, Hwang MW, Iwasaki A, Okada M, Nakatani K, Sasayama S, Matsumori A. Evidence for a role of mast cells in the evolution to congestive heart failure. J Exp Med 2002; 195:375–381.

114. Behr TM, Wang X, Aiyar N, Coatney RW, Li X, Koster P, Angermann CE, Ohlstein E, Feuerstein GZ, Winaver J. Monocyte chemoattractant protein-1 is upregulated in rats with volume-overload congestive heart failure. Circulation 2000; 102:1315–1322.

115. Hasenfuss G. Animal models of human cardiovascular disease, heart failure and hypertrophy. Cardiovasc Res 1998; 39:60–76.

116. Iwase M, Uechi M, Vatner DE, Asai K, Shannon RP, Kudej RK, Wagner TE, Wight DC, Patrick TA, Ishikawa Y, Homcy CJ, Vatner SF. Cardiomyopathy induced by cardiac G$_s$ alpha overexpression. Am J Physiol Heart Circ Physiol 1997; 272:H585–589.

117. Geisterfer-Lowrance AA, Christe M, Conner DA, Ingwall JS, Schoen FJ, Seidman CE, Seidman JG. A mouse model of familial hypertrophic cardiomyopathy. Science 1996; 272:731–734.

118. Matsumori A, Kawai C. An experimental model for congestive heart failure after encephalomyocarditis virus myocarditis in mice. Circulation 1982; 65:1230–1235.

119. Lane JR, Neumann DA, Lafond-Walker A, Herskowitz A, Rose NR. Role of IL-1 and tumor necrosis factor in coxsackie virus-induced autoimmune myocarditis. J Immunol 1993; 151:1682–1690.

120. Fox PR, Liu S-K, Maron BJ. Echocardiographic assessment of spontaneously occurring feline hypertrophic cardiomyopathy. An animal model of human disease. Circulation 1995; 92:2645–2651.

121. Gruver EJ, Glass MG, Marsh JD, Gwathmey JK. An animal model of dilated cardiomyopathy: characterization of dihydropyridine receptors and contractile performance. Am J Physiol Heart Circ Physiol 1993; 265:H1704–H1711.

122. Eschenhagen T, Diederich M, Kluge SH, Magnussen O, Mene U, Müller F, Schmitz W, Scholz H, Weil J, Sent U, Schaad A, Scholtysik G, Wüthrich A, Gaillard C. Bovine hereditary cardiomyopathy: an animal model of human dilated cardiomyopathy. J Mol Cell Cardiol 1995; 27:357–370.

123. Boluyt MO, O'Neill L, Meredith AL, Bing OHL, Brooks WW, Conrad CH, Crow MT, Lakatta EG. Alterations in cardiac gene expression during the transition from stable hypertrophy to heart failure. Marked upregulation of genes encoding extracellular matrix components. Circ Res 1994; 75:23–32.

124. Holycross BJ, Summers BM, Dunn RB, McCune SA. Plasma renin activity in heart failure-prone SHHF/Mccp-facp rats. Am J Physiol Heart Circ Physiol 1997:H228–H233.

125. Li Z, Bing OH, Long X, Robinson KG, Lakatta EG. Increased cardiomyocyte apoptosis during the transition to heart failure in the spontaneously hypertensive rat. Am J Physiol Heart Circ Physiol 1997; 272:H2313–H2319.

126. Gertz EW. Cardiomyopathic Syrian hamster: a possible model of human disease. Prog Exp Tumor Res 1972; 16:242–260.

127. Forman R, Parmley WW, Sonnenblick EH. Myocardial contractility in relation to hypertrophy and failure in myopathic Syrian hamster. J Mol Cell Cardiol 1972; 4:203–211.

128. Sole MJ, Lo CM, Laird CW, Sonnenblick EH, Wurtman RJ. Norepinephrine turnover in the heart and spleen of the cardiomyopathic Syrian hamster. Circ Res 1975; 37:855–862.

129. Iannini JP, Spinale FG. The identification of contributory mechanisms for the development and progression of congestive heart failure in animal models. J Heart Lung Transplant 1996; 15:1138–1150.

130. Capasso JM, Sonnenblick EH, Anversa P. Chronic calcium channel blockade prevents the progression of myocardial contractile and electrical dysfunction in the cardiomyopathic Syrian hamster. Circ Res 1990; 67:1381–1391.

131. Rouleau JL, Chuck LH, Hollosi G, Kidd P, Sievers RE, Wikman–Coffelt J, Parmley WW. Verapamil preserves myocardial contractility in the hereditary cardiomyopathy of the Syrian hamster. Circ Res 1982; 50:405–412.

11

The Use of Genetically Modified Mice in the Investigation of Cardiac Remodeling

Daniel J. Lips
Department of Cardiology, Heart Lung Center Utrecht, Utrecht, The Netherlands

Rutger J. Hassink and Aart Brutel de la Riviere
Department of Cardio-Thoracic Surgery, Heart Lung Center Utrecht, Utrecht, The Netherlands

Pieter A. Doevendans
Interuniversity Cardiology Institute The Netherlands, Utrecht, The Netherlands and Department of Cardiology, Heart Lung Center Utrecht, The Netherlands

INTRODUCTION

Cardiac remodeling describes the adaptive response of the myocardium to pathophysiological changes. The remodeling process can be measured at different levels. Firstly, the composition of myocardial tissue will change in response to injury. Concomitantly, the left ventricular contractile function and relaxation pattern adapt. The mechanism of remodeling, albeit adaptive or maladaptive, can be studied at the organ, cellular, or molecular level. To unravel the importance of various signaling pathways and specific proteins, often gene modification studies are being performed in rodents, predominantly mice (1). In mice, the plain effect of transgenesis (adding genes) or gene targeting (replacing genes) on left ventricular morphology and function can be assessed (2–4). Specific alterations in individual genes result in (A) the absence of a selected protein (i.e., knockout); (B) the abundance of a selected protein (i.e., overexpression); (C) truncated and/or dysfunctional proteins with a potential dominant-negative effect or (D) the introduction of constitutively active proteins (dominant-positive). Because single genes are targeted, it is possible to elucidate the specific functions of the corresponding protein in intracellular signaling pathways. All aspects of cardiac remodeling can be investigated through this approach. In addition, surgical interventions can be applied to induce pathology, including aortic banding, myocardial infarction (MI) or ischemia/reperfusion (I/R). By systematic application of these surgical interventions to various recently generated mouse lines, several crucial pathways have been recognized and subsequently unraveled step-by-step (5).

The importance of genes involved in postischemic cardiac remodeling, hypertrophy, and heart failure, and pro- and antiapoptotic pathways have been studied.

A fascinating observation is the close relationship between various intracellular signaling events during distinct pathological conditions. For instance the family of mitogen-activated protein kinases (MAPK) was described as a group of crucial molecules for the development of hypertrophy in the heart (6–8). More recent studies outlined their protective role in ischemia-induced apoptosis (9,10). The opposite is true for the Akt protein. Initially identified as a prosurvival protein (11), a potential role in ischemic heart disease and even dilated cardiomyopathy has been proposed recently (12). The conclusion could be drawn that intracellular pathways result in distinct cellular outcomes depending on the specific pathophysiological situation. Moreover, the myocyte response is not solely determined by activation of a single pathway, but by the relative activity of various simultaneous-acting signaling molecules. Part of the complexity involved could be elucidated by using more advanced genetic models in the adult mouse, where genes can be activated or repressed at will of the investigator. For example, the tamoxifen-induced cardiomyocyte-specific and temporally regulated gene-expression in a Mer-Cre-Mer based mouse model can be used to delete a floxed gene (13). This is an example of an inducible, tissue-specific knockout approach. Here, we will focus on the most recent information available on cardiac remodeling. The more important aspect of remodeling is cardiac function, as the morphological changes are less crucial for the outcome or prognosis compared to functional changes in man and mouse alike.

CARDIAC REMODELING AND MYOCARDIAL ISCHEMIA

Cardiac remodeling is defined by the structural and concomitant functional alterations in the heart following an ischemic event. The ischemia-induced architectural remodeling encompasses ventricular dilatation, myocardial hypertrophy, deposition of collagen, and apoptosis in the area-at-risk (for ischemia) and the remote myocardium. Cardiac remodeling in man begins classically within days following the ischemic event and it continues indefinitely. The process is characterized by an enlargement of ventricular dimensions or volumes (end-diastolic and -systolic dimensions) combined with a decrease in ejection fraction and increase in left ventricular mass. Even nowadays, in the era of aggressive medical therapy for ischemic heart disease and during acute MI, cardiac remodeling remains important for long-term cardiac function (14). Current therapy in acute MI in humans can attenuate ventricular volume enlargement and dilation at one month following the ischemic event, coinciding with improvement of left ventricular geometry and cardiac function as defined by ejection fraction (14). More importantly, cardiac death and hospitalization for clinical heart failure are still significantly higher in patients with left ventricular remodeling following ischemia as compared to patients without (15). Thorough investigation into the prevention of adverse effects of cardiac remodeling has important implications in daily clinical practice.

Cardiac remodeling has been investigated extensively in animal (Table 1) (16–18) and human studies, using imaging techniques as echocardiography (19), computed tomography (CT) (20), radionuclide ventriculography (21), and recently, magnetic resonance imaging (MRI) (14,22). With the development of genetically altered mice, an intense interest arose for murine models of I/R and MI (16,23). Through genetic engineering, it is possible to elucidate the spectrum of specific functions of single proteins in the intracellular signaling pathways involved in ischemia (–reperfusion) injury. Mouse models of ischemia constitute the occlusion of the left

Table 1 Mouse Models Used in Cardiac Remodeling

Molecular target	Genetic modification	Phenotype	References
Cardiac remodeling			
AT1a receptor	Knockout	Less LV remodeling, improved survival	29
AT2 receptor	Overexpression	Less LV remodeling, improved survival	31
	Knockout	No different cardiac remodeling	27
IGF-1	Knockout	Attenuation DNA synthesis, higher apoptosis rates	33
	Overexpression	Smaller infarctions, lower apoptosis rates	58
VEGF	Knockout	Deterioration cardiac function, impaired angiogenesis	34
FGF-1	Overexpression	Delayed infarct development, prosurvival cellular signaling	36
MMP-9	Knockout	Attenuated LV enlargement, augmented collagen deposition	41
TIMP-1	Knockout	Loss of fibrillar collagen	44
Apoptosis and hypertrophy			
Fas	Non-functional	Smaller infarction, lower apoptosis rate	59
Caspase-3	Overexpression	Larger infarction, higher apoptosis rate	75
Bax	Knockout	Smaller infarction, lower apoptosis rate	54
Mst1	Overexpression	Higher apoptosis rate, abolishment cardiac hypertrophy	123
	Dominant-negative overexpression	Smaller infarction, lower apoptosis rate	
A1 adenosine receptor	Overexpression	Smaller infarction, lower apoptosis rate	76
Caspase-1	Knockout	Smaller infarction, lower apoptosis rate	74
TNF receptor	Knockout	Larger infarction, higher apoptosis rate	65
TNF-α	Knockout	Smaller infarction	67
gp130	Overexpression	Cardiac hypertrophy	89

(*Continued*)

Table 1 Mouse Models Used in Cardiac Remodeling (*Continued*)

Molecular target	Genetic modification	Phenotype	References
	Dominant-negative overexpression	Attenuated hypertrophic response	90
	Knockout	Progressive pressure-induced cardiomyopathy, augmented apoptosis rates	127
STAT3	Overexpression	Cardiac hypertrophy	93
MEKK1	Knockout	Abolishment of cardiac hypertrophy, attenuated hypertrophic response	94,101
MKP-1	Overexpression	Diminished developmental myocardial growth, attenuated hypertrophic response	97
MEK1	Overexpression	Concentric cardiac hypertrophy	06,86
		Smaller infarction, lower apoptosis rate	
MEK5	Overexpression	Eccentric cardiac hypertrophy, sudden death	98
JNK	Inducible overexpression of MKK7D	Progressive cardiomyopathy	13
p38	Inducible overexpression of MKK3bE or MKK6bE	Restrictive cardiomyopathy, heterogeneous mycoyte atrophy	07
p38α	Dominant-negative overexpression	Hypertrophic cardiomyopathy	105
Calcineurin	Overexpression	Cardiac hypertrophy	107
		Sudden cardiac death	108
		Smaller infarction, lower apoptosis rate	109
	Dominant-negative overexpression	Attenuated adrenergic-stimulated hypertrophic response	111
Calcineurin Aβ	Knockout	Larger infarction, higher apoptosis rate	82

(*Continued*)

Table 1 Mouse Models Used in Cardiac Remodeling (*Continued*)

Molecular target	Genetic modification	Phenotype	References
NFAT3c	Knockout	Abolishment of cardiac hypertrophy, attenuated hypertrophic response	113
MCIP-1	Overexpression	Attenuated hypertrophic response	116
	Knockout	Exacerbated hypertrophic response	118
GSK-3β	Overexpression	Abolishment of cardiac hypertrophy	119
Cain/cabin	Overexpression	Attenuated hypertrophic response	120
A-kinase anchoring protein 79	Overexpression	Attenuated hypertrophic response	120

Mouse models in cardiac remodeling are as discussed in the text and figures. Abbreviations used are explained in the text. Models are listed in order of appearance in the text.

anterior descending coronary artery permanently (MI) or temporarily (I/R), allowing the assessment of the cardiac remodeling process. The fruit of these experimental studies was the revelation of important roles for the renin-angiotensin-system (RAS), androgenic hormones, growth factors, and tissue-matrix components in both acute and chronic remodeling (Fig. 1).

The importance of the renin-angiotensin axis, with its effector protein angiotensin II (Ang II), originates from its roles in promotion of cardiomyocyte hypertrophy and induction of fibrosis (24). Inhibition of Ang II reduces the amount of collagen deposition and improves cardiac function during the remodeling phase, while the opposite is true for expression of Ang II. All the components necessary to generate Ang II are present in the myocardium, and cardiac Ang II formation appears to be regulated independently from the systemically circulating RAS (25). Ang II receptors are predominantly located on cardiomyocytes instead of fibroblasts (26). Chimeric mice with cardiomyocytes baring the angiotensin (AT) 1A receptor and cardiomyocytes carrying the endogenous AT1A promoter driving the *lacZ* gene, showed communication between cardiomyocytes and fibroblast upon Ang II stimulation, leading to proliferation and collagen deposition of the latter (26). Most biological effects of Ang II are mediated through the AT1, instead of AT2, receptor (27). Activation of this G-protein coupled receptor initiates the expression of growth-related genes through MAPK and the family of Janus kinase/signal transducer and activator of transcription (JAK/STAT) proteins dependent pathways (28). Activation of the AT2 receptor has a negative regulatory effect on these signaling cascades.

Genetically altered mouse models have been the key in elucidating the specific roles for the RAS components, for example, as in the case of the Ang II receptors.

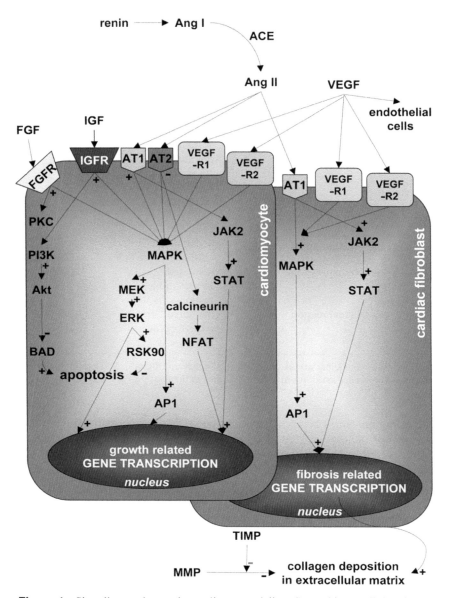

Figure 1 Signaling pathways in cardiac remodeling. Several intracellular signaling pathways involved in the cellular adaptation of cardiac remodeling are depicted in the figure. The purpose of the figure is not to be complete, but to give an overview of the signaling cascades discussed in the text. Activation or inhibition of single molecules are indicated by (+) and (−) respectively.

Gene-targeting studies revealed the different effects of AT receptor 1 and 2 stimulation. As mentioned above, activation of the AT1 receptor stimulates MAPK and JAK/STAT activity, while AT2 receptors effects in inactivation of both signaling cascades (29–31). Knocking out the AT1 a receptor led to less left ventricular remodeling and improved survival in long-term follow-up after MI, while AT2 receptor deficiency resulted in increased death shortly after acute MI (29,30). Consequently,

AT2 receptor overexpression showed preservation of left ventricular performance during post-MI remodeling reducing the early mortality rate after MI (25,31). Cardiomyocyte-specific transgenesis of the AT1 receptor results in a lethal phenotype, with gross malformations of the heart and early death within one week postnatal (32). The presented data evidently show the discrepancies between AT receptors and their effects on cardiac remodeling.

Growth factors stimulate various cell types into hyperplastic or hypertrophic responses, and initiate angiogenesis. These cellular functions enable growth factors to protect the heart against deleterious remodeling and impaired cardiac performance. An important role for these factors could have been anticipated as ischemia results in the loss of significant amounts of cardiomyocytes, providing a trigger for regeneration and growth. Insulin-like growth factor (IGF) is an endogenous protector of cardiomyocytes and inducer of prosurvival processes ex vivo and in vivo. In IGF-1 knockout mice myocardial ischemia resulted in the attenuation of DNA synthesis and an augmentation of apoptosis rates (33). Mice lacking isoforms of the vascular endothelial growth factor (VEGF) showed impaired angiogenesis and worsened cardiac performance following myocardial ischemia (34,35). The fibroblast growth factor (FGF) is produced by cardiomyocytes and fibroblasts, and mainly stimulates cellular proliferation and deposition of collagen. Cardiac-specific over-expression of human FGF-1 delayed infarct development because of a constitutive higher expression of prosurvival ERK1/2 signaling pathways (36). Furthermore, FGF-1 overexpression led to significant increments in collagen deposition following myocardial ischemia (37). FGF-2 confers increased resistance to ischemic injury (38). An isolated mouse heart model showed the cardioprotective effect of FGF-2 over-expression in the heart toward ischemia, as myocyte viability was protected and lactate dehydrogenase release inhibited. In mice deficient for nitric oxide synthase-2 (NOS2), FGF-2 did not attenuate ischemia-induced cardiac dysfunction (39). FGF-2 may therefore protect the heart from ischemia-induced cardiac dysfunction by stimulating nitric oxide production. The signaling cascades of the MAPK family have been proven important in the cellular action of growth factors, as all factors stimulate MAPK activity. On top of that, both IGF and FGF activate Akt through PI3K, which is another pathway besides the MAPK cascade counteracting cell death. Conclusively, these studies demonstrate the importance of growth factors during the remodeling phase as they mediate proliferative processes.

Adequate control of collagen organization is reflected by the main physiological functions of the interstitial collagen network: to retain tissue integrity and cardiac pump function (40). Both the RAS and growth factors are involved in the induction of fibrosis, and highlight the extracellular collagen organization as an important feature of cardiac remodeling. The families of serine and matrix metalloproteinases (MMP) belong to the only proteinase enzymes capable of degrading the extracellular collagen matrix. These MMPs are prominently expressed after myocardial ischemia (40). Targeted deletion of MMP-9 attenuated left ventricular enlargement and increased myocardial collagen content following infarction (41,42). Knockout of MMP-2 resulted in comparable infarct sizes as in wildtype mice, but MMP-2 null mice exerted less ventricular dilation and better cardiac performance (43). MMP-2 was also considered to be involved in early cardiac rupture following ischemia, as the incidence of ventricular rupture was decreased in the knockout mice. Incidence of cardiac rupture decreased subsequently. Moreover, uncontrolled MMP activity as seen by tissue-inhibitor of MMP (TIMP)-1 deficient mice, demonstrated amplified adverse left ventricular remodeling (i.e., significant loss of fibrillar collagen) following MI and thereby emphasized the importance of local endogenous control of cardiac MMP activity (44).

Yet, the important aspect of inflammation and neutrophil infiltration has not been discussed. Infiltration of the ischemic area by polymorphic neutrophils greatly influences infarct size and long-term remodeling and is highly regulated by, for instance, endothelial cell intercellular adhesion molecule-1 (ICAM-1). In ICAM-1 deficient mice, the infarct area is significantly attenuated with marked reduction in I/R-injury serum parameters (45,46). P-selection is another molecule mediating early neutrophil adhesion in the coronary arteries. ICAM-1/P-selection double knockout mice show decreased myocardial neutrophils infiltration upon ischemia. Deficiency of neutrophil cell markers such as CD18 does result in myocardial protection against ischemia. In CD18 knockout mice, the rate of myocardial neutrophil infiltration was reduced in combination with smaller infarct sizes (46).

The discussed mouse models have taught us much about the process of cardiac remodeling and have provided us with molecular tools to intervene in the deleterious disease state. However, all involved factors center around cell death, which is the main item of ischemia-induced remodeling. Cell death leads to loss of viable myocardium, responsively hypertrophic growth and ventricular dilation. The destructed cells and extracellular matrix are removed by unfavorable inflammation and replaced with large amounts of noncontractile collagen. The most gain could be obtained by intervening in the development of cell death and limiting the myocardial damage. The relation between myocardial ischemia and cardiomyocyte death is the subject of the next section.

MYOCARDIAL ISCHEMIA AND APOPTOSIS

Myocardial ischemia will lead to the loss of viable cardiomyocytes by three different mechanisms: necrosis, apoptosis, and autophagic cell death. Necrosis is well characterized by cell swelling, disruption of cell organelles and cell membranes, and an inflammatory tissue reaction. In apoptosis (i.e., programmed cell death), the cell undergoes shrinkage, nuclear chromatin condensation, and formation of apoptotic bodies, which are removed by macrophages without eliciting inflammation (47). In contrast to necrosis, apoptosis is an energy-requiring and multiprotein-involving process (Fig. 2). Apoptosis can be assessed by several methods, like electron microscopy, terminal deoxynucleotidyl transferase-mediated dUTP nick end labeling (TUNEL) or in situ end labeling (ISEL) methods for identification of DNA fragments, Annexin-V staining (48), or by the appearance of DNA laddering in gel electrophoresis. Technical and experimental difficulties, however, have to be taken into account, as overestimation of the number of apoptotic cardiomyocytes (false-positives) and misinterpretation of electron microscopic photos are pitfalls in these studies. Autophagic cell death is another form of programmed cell death that seems to be important in heart failure (49). Like apoptosis, autophagic cell death is regulated and is associated with DNA fragmentation. Unlike apoptosis, it is caspase-independent and morphologically resembles necrosis.

Ischemia-induced apoptosis is characterized by cytochrome C release from the mitochondria into the cytosol and the activation of the caspase-cascade by the death-receptor pathway (Fig. 2). Cytochrome C release following opening of the mitochondrial permeability transition (MPT) pores is under control of members of the Bcl-2 family, constituting the antiapoptotic Bcl-2 and Bcl-xL, and proapoptotic members

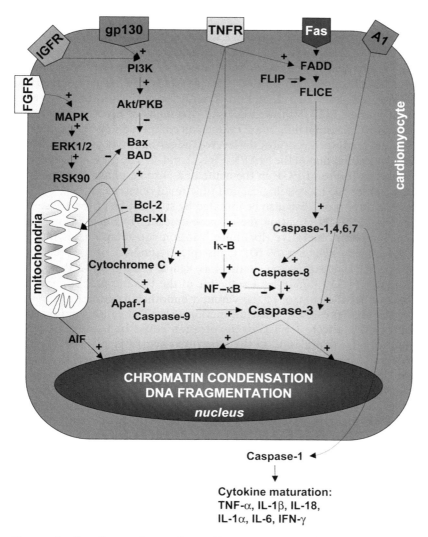

Figure 2 Signaling pathways in cardiomyocyte apoptosis. Several intracellular signaling pathways involved in cardiomyocyte apoptosis are depicted in the figure. The purpose of the figure is not to be complete, but to give an overview of the signaling cascades discussed in the text. Activation or inhibition of single molecules are indicated by (+) and (−) respectively.

Bad and Bax (50,51). In mice, overexpression of the antiapoptotic human Bcl-2 gene rendered the heart more resistant to apoptotic cell death and functional deterioration following coronary artery occlusion (52,53). Homozygotic Bax ablation in mice significantly reduced apoptotic markers (54). Bax ablation even led to reduced damage to mitochondria, the source of cytochrome C. Bax and BAD activity is under control of Akt, the serine-threonine kinase that inhibits cardiomyocyte apoptosis in vitro and in vivo. Transgenic mice overexpressing a constitutively active mutant of Akt in the heart showed a 50% reduction of infarct size in combination with concentric hypertrophy and preserved systolic function (55). The growth promoting peptide IGF-1 induces antiapoptotic signals in cardiomyocytes through PI3K-dependent

Akt activation (Fig. 2) in vivo (56,57). IGF-1 deficient mice develop increased amounts of cardiomyocyte apoptosis one week post-MI, affecting cardiac remodeling by thinning of the ventricular walls (33). The IGF-1 protective effect on apoptosis was confirmed by the report that overexpression of the peptide in mice with chronic MI, reduced myocyte death by blocking apoptosis rates (58). This resulted in preserved wall thickness and decreased ventricular dilatation. Transgenic mice with cardiac-specific overexpression of human FGF-1 showed delayed first signs of myocardial damage and postponed reach of maximal infarct area extension compared to wildtype animals (36). FGF-1 provides cardioprotection at the level of the cardiac myocyte and is at least partially mediated via activation of the MAPK ERK-1 and -2. The presence of growth-promoting peptides as IGF and FGF in the acute and chronic phase of myocardial ischemia is beneficial for the morphological and functional outcome.

Cardiomyocyte apoptosis can also be initiated by stimulation of the cell surface death-receptor (Fas, Fig. 2) activating the caspase-cascade. In mice lacking functional Fas, as in lymphoproliferative (lpr) mice, cardiomyocyte apoptosis rate was reduced upon myocardial ischemia (59,60). Consequently, infarct sizes were smaller in these mice. Specific overexpression of the Fas ligand (FasL, a member of the tumor necrosis factor (TNF) family, and activator of the Fas receptor) on vascular endothelium cells under the control of the vascular endothelial cadherin promoter, reduced infarct size and preserved cardiac function (61). The cause for this paradoxical cardioprotective effect of FasL overexpression could be because of a 54% reduction of neutrophil accumulation in the reperfused heart tissue (61). This highlights the important aspect of inflammation in I/R-induced myocardial damage. TNF itself may have antiapoptotic and subsequent beneficial effects on the cell survival of myocytes injured by ischemia (62,63). This was shown by TNF-α overexpression in mice, resulting in the activation of NF-κB, a mediator of antiapoptotic pathways (64). TNF-α exerts its effect by binding to two cell-surface receptors, TNFR1 and -2. In TNF-receptor (TNFR)-1 and -2 double knockout mice post-infarction remodeling was characterized by significantly greater infarct areas and accelerated rates and extent of apoptosis, though single TNFR-1 or -2 gene targeting did not result in increased infarct size or cell death (65). Isolated mouse heart studies in mice lacking the TNF-α gene demonstrated loss of the cardioprotective effect of ischemic preconditioning (IP) in the investigated genotype (66). These data suggest that the presence and stimulation of TNFRs gives rise to one or more cytoprotective signals that prevent and/or delay the development of cardiac myocyte apoptosis after acute ischemic injury. However, in large contrast to the above-mentioned TNF-α overexpressing mice, TNF-α knockout mice showed significantly reduced infarct sizes following ischemia compared to wildtype littermates (67). Furthermore, antagonizing TNF-α with soluble TNFRs decreased the size of infarction and the amount of apoptosis following MI (68). In IP studies, the use of TNF-α antibodies or soluble TNFRs reduced infarct extent significantly too (68–70). Moreover, mice overexpressing TNF-α specifically in the heart developed chronic heart failure mediated by severe leukocyte infiltration (71). The presumed deleterious effect of TNF-α in cardiac remodeling could be because of its contributions in promoting leukocyte infiltration of the myocardium. Currently available data support the concept that TNF-α is involved in ischemia-induced injury and that a discrepancy exists between ligand and receptor in cardiac remodeling. This discrepancy could be because of the form in which TNF-α is expressed, i.e., in the circulating (cleavable) or the transmembrane form (72). Further studies in genetically engineered mice will be necessary to address this interesting question. However, currently, clinical trials are already in progress to investigate the possible protective effect of TNF-antibodies during myocardial ischemia.

Repressor domains and other inhibitory proteins like FLIP [Fas-associated death domain protein-like interleukin 1β-converting enzyme (FLICE)/caspase-8-inhibitory protein], are able to block apoptosis via the death receptors (Fig. 2). Subsequently, the caspase-cascade is activated. Caspases are the specialized cysteine-dependent proteases that cleave major structural proteins of the cytoplasm and nucleus (51). The generation of caspase-knockout mice by homologous recombination provided the possibility to study the role of individual caspase proteins (73). The proapototic active caspase-1 is known to be important for the production of various cytokines such as interferon (IFN)γ and TNF-α, which contribute to the cardiac remodeling process after MI. Targeted deletion of caspase-1 (interleukin-1β (IL-1β)-converting enzyme) protects the heart from early postinfarct mortality and ventricular dilation, partly through a reduction in the rate of apoptosis in the remodeling remote myocardium (74). Cardiac-overexpression of the main effector caspase, caspase-3, increased infarct size and rendered mice more susceptible to postischemic mortality (75). It should be noted that caspase-3 transgenic mice already show depressed cardiac performance and myofibrillar and nuclear damage. However, caspase-3 overexpression itself induced no full-blown apoptotic response in cardiomyocytes. Transgenic mice have clarified that mammalian sterile 20-like kinase Mst1, a prominent myelin basic protein kinase activated by proapoptotic stimuli in cardiac myocytes, is an efficient mediator of apoptosis through the activation of caspase-3. Overexpression of the dominant negative Mst1 transgene significantly decreased the rate of apoptosis and caspase-3 activity in mice following I/R. The activity of caspase-3 and induction of apoptosis could also be reduced by overexpressing the cardiac A1 adenosine receptors (76). In most studies mentioned above, as in the Bax null mice study, caspase-3 activity was inhibited and the rate of TUNEL-positive cardiomyocytes reduced following myocardial ischemia (54). Blockade of the apoptotic process by a caspase-inhibitor during the subacute stage of MI has recently been shown to have beneficial effects on apoptosis rates and cardiac remodeling in rats (77). The discussed studies present evident genetic proof defining caspase-3 as an important determinant in I/R induced apoptotic signaling.

Deprivation of serum and glucose during ischemia in vivo, results in cardiomyocyte cell death (78). However, whether apoptosis is triggered predominantly during ischemia or reperfusion remains a controversial issue. The total amount of cell death in MI is significantly greater compared to ischemic events with eventual reperfusion. Prolongation of ischemia-time proportionally increases apoptosis (and necrosis) rates with estimates running from 3% to 12% of myocyte nuclei, and apoptosis has been observed solely in hypoxic regions of the myocardium within 48 hours of the ischemic event (10,79–81). On the other hand, initial signs of apoptosis are first detected within minutes of reperfusion and heightened rates of cardiomyocyte apoptosis are continuously present in the border zones of the infarct-area and the remote myocardium over a time-period of 12 weeks following MI (82,83). The increments in apoptosis following myocardial ischemia could be owing to other causes than the ischemia itself, for instance, because of the development of heart failure. However, its presence in this extended period designates *apoptosis* to be *a key player* in *post-infarction cardiac remodeling*. Similar findings have been reported from human studies where the presence of high grades of apoptosis have been observed in human hearts of patients dying within 12 to 62 days following acute MI (84). The rate of apoptosis is enhanced enormously by myocardial reperfusion (85). Essentials for the survival of viable cells are provided as reperfusion restores oxygen and glucose supplies, thereby also delivering the required energy for apoptosis and restarting or accelerating the apoptotic process compared with the situation in continuously ischemic myocardium.

As apoptosis is an active process, it could in principle be inhibited. This makes apoptotic cell death an interesting potential therapeutic target. Moreover, current clinical therapeutic techniques, including thrombolytic therapy and primary percutaneous transluminal coronary angioplasty (PTCA) or on-pump coronary artery bypass grafting (CABG), increase the relevance of IR injury and provide a therapeutic portal to the affected myocardium. A study using real-time imaging of apoptosis in the beating murine heart showed first signs of cardiomyocyte apoptosis within minutes after reperfusion, while during ischemia hardly any apoptotic cell could be found (48). Apoptosis was detected by Annexin-V binding to phosphatidylserine (PS), which is rapidly translocated from the inner to the outer leaflet of the cell membrane following activation of the cell-death program. Maximal Annexin-V binding was reached within 20 to 25 minutes of reperfusion in the mid-myocardium. Therefore, the cell-death program, involving cytochrome-c release from the mitochondria and activation of the caspase-cascade, functions within minutes. Additionally, as solely cardiomyocytes were found to be apoptotic within the time-scale investigated, these cells seem to be more prone to I/R injury than fibroblast or other cell types within the myocardium. Successful rescue of cardiomyocytes affected by caspase-inhibition proved the ability to inhibit apoptosis in cells which otherwise would have died (48). Although in this study caspase-inhibitors were administered before the ischemic event, the results stress the possible important aspect of reperfusion-induced apoptosis inhibition. One could furthermore argue that apoptosis is a biological process to dispose the myocardium of affected and injured cells. Inhibition of apoptosis could leave the heart with inefficient cardiomyocytes and thereby affect cardiac performance. However, studies genetically blocking I/R-induced apoptosis show preservation of cardiac function in mice following severe ischemic events (10). These results suggest that it would be worthwhile to therapeutically inhibit apoptosis following I/R-injury regarding long-term follow up, although the time-window of opportunity is narrow.

MYOCARDIAL HYPERTROPHY

The hypertrophic response of cardiomyocytes has been studied extensively in the past decade using genetically modified mouse models and in vitro models. Cardiomyocyte hypertrophic growth in vivo is triggered by various stimuli such as ventricular pressure- or volume-overload, IR, or drug induced (86). Furthermore, analogous to human hypertrophic cardiomyopathies, the possibility arose to evoke hypertrophy by engineering mutations in the sarcomeric genes of mice. Many aspects of cardiac hypertrophic growth have been investigated with these techniques. Hypertrophy is characterized by cardiomyocyte growth, myofibrillar disarray, fibrosis, apoptosis, arrhythmias, elevated filling and end-diastolic pressures, decline in systolic function, cellular and mitochondrial energy inefficiency, alterations in calcium handling, and eventually transition toward heart failure (86). Moreover, the single most powerful predictor for the development of heart failure is the presence of left ventricular hypertrophy (87). For this reason, much effort is currently undertaken to investigate the etiology of molecular hypertrophic mechanisms and to unravel the signaling pathways underlying this affliction (Fig. 3).

To date, more than 100 genetically altered mouse models of hypertrophic cardiomyopathy, myocardial hypertrophy and heart failure have been engineered (88).

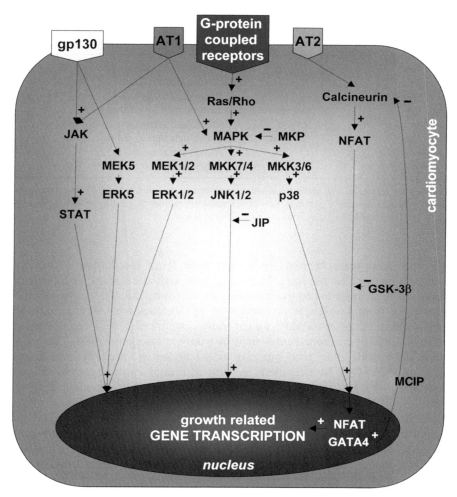

Figure 3 Signaling pathways in cardiomyocyte hypertrophy. Several intracellular signaling pathways involved in cardiomyocyte hypertrophy are depicted in the figure. The purpose of the figure is not to be complete, but to give an overview of the signaling cascades discussed in the text. Activation or inhibition of single molecules are indicated by (+) and (−) respectively.

Because of the variety of genes involved, numerous molecular mechanisms in the development of hypertrophy have been investigated (Table 2). Proposed molecular mechanisms in the development of hypertrophy include abnormalities in contractile, cytoskeletal, and intracellular calcium (Ca^{2+})-regulatory proteins, alterations in excitation-contraction coupling, signal transduction system, cardiac metabolism or myocyte apoptosis, and many others (88). All hypertrophic stimuli (i.e., myocyte stretch, hemodynamic stress, ischemia, hormones, vasoactive peptides, or neurotransmitters) converge in several intracellular signaling cascades mediating these extrinsic and intrinsic growth signals into coordinated alterations of genetic profiles, and increments in the overall rate of RNA transcription and protein synthesis.

An important signaling pathway in cardiomyocyte hypertrophy is initiated by the gp130 cytokine receptor. Transgenic mice expressing constitutively active gp130

Table 2 Mechanism in Hypertrophy

Extracellular causes	Intracellular causes
Pressure overload	Alterations in molecules involved in:
Volume overload	Contractile apparatus
Ischemia–reperfusion	Cytoskeletal structure
Hormones	Intracellular Ca^{2+}-handling
Vasoactive peptides	Excitation-contraction coupling
Neurotransmitters	Intracellular signal transduction
	Cardiac metabolism
	Cardiomyocyte apoptosis

These are the proposed extracellular and intracellular causes for hypertrophic growth. These extrinsic and intrinsic growth signals eventually converge into coordinated alterations of genetic profiles, and increments in the overall rate of RNA transcription and protein synthesis.

protein in the heart were created by mating mice from interleukin (IL)-6 and IL-6-receptor transgenic mouse lines. The continuous activation of the gp130 signaling pathways in these mice led to overt cardiac hypertrophy (89). In transgenic mice overexpressing a dominant-negative mutant of gp130, pressure-overload induced smaller increments in heart-to-body weight ratios, ventricular wall thickness and cross-sectional areas of cardiomyocytes than in wildtype littermates (90). The gp130 protein-induced hypertrophic gene program could be antagonized by monoclonal anti-gp130 antibodies (91). Investigations in cultured murine cardiomyocytes infected with adenoviruses suggested that mainly the STAT3-dependent signaling pathway downstream of gp130 promoted cardiac myocyte hypertrophy (92). Pressure-overload activated ERKs and STAT3 in the heart of wildtype mice, whereas pressure overload-induced activation of STAT3, but not that of ERKs, was suppressed in transgenic mice overexpressing dominant-negative gp130 (90). Moreover, transgenic mice with cardiac-specific overexpression of the STAT3 gene manifested myocardial hypertrophy at the age of 12 weeks (93). These results suggest that gp130 plays a critical role in stimulus-induced cardiac hypertrophy possibly through the STAT3 signaling pathway.

ERK belongs to another important pathway through which extracellular stimuli induce hypertrophy. Hypertrophic stimuli induce the activation of the MAPK superfamily cascades. MAPKs are a widely distributed group of intracellular proteins composed of three terminal MAPK branches: (i) the extracellular signal-regulated kinases (ERKs), (ii) c-Jun NH(2)-terminal kinases (JNKs), and (iii) the stress-induced p38 MAPKs. Hypertrophic stimuli induce the activation of MAPKs through G-protein coupled receptors (94) and low molecular weight GTP binding proteins Ras and Rho depending pathways (Fig. 3) (95,96). A family of MAPK phosphatases (MKPs) acts as the critical counteracting factors of p38, JNK and ERK (97). Selective ERK1/2 stimulation by cardiac specific overexpression of MEK1 (endogenous kinase activator of ERK1/2) demonstrated concentric hypertrophy without signs of cardiomyopathy combined with an improved cardiac function (6). On the contrary, cardiac-specific expression of activated the MEK5–ERK5 pathway in transgenic mice resulted in eccentric cardiac hypertrophy that progressed to dilated cardiomyopathy and sudden death (98).

JNK MAPK contributes to the pressure-overload and catecholamine-induced hypertrophic response (95,99). Prevention of JNK activity through JNK-interacting protein 1 (JIP-1), a cytosolic scaffolding protein, results in reduced cellular growth in response to G-protein coupled receptor agonists (100). Pressure-overload however caused significant levels of cardiac hypertrophy coinciding with signs of clinical heart failure and higher mortality in mice deficient in mitogen-activated protein kinase kinase kinase (MEKK1; mediates JNK activation) (101). Direct JNK-gene overexpression or deficiency in cardiomyocytes has not been published until very recently. However, through an ingenious mouse model using tamoxifen-inducible cardiac-specific-gene expression, a constitutively activated upstream activator of JNK was investigated in the adult mouse heart (13). Prolonged activation of JNK resulted in progressive cardiomyopathy adding molecular proof for a regulatory role of JNK in maladaptive cardiac growth.

Stress-induced p38 MAPK activity increases significantly in mouse hearts after chronic transverse aortic constriction, coinciding with the onset of ventricular hypertrophy. Adenoviral overexpression of wild type p38β or a dominant-negative p38β variant, respectively, enhanced and suppressed the hypertrophic response following aortic constriction (102). P38 activity has also been associated with catecholamine-induced hypertrophy (103). Transgenic male and female mice with fourfold phospholamban (PLB) overexpression exhibited enhanced circulating catecholamines concurrent with p38 MAPK activation levels (103). However, in vivo cardiac overexpression of p38 MAPK activation resulted in varying degrees of myocyte atrophy. This was shown using a transgenic gene-switch strategy with activated mutants of well-established upstream activators of p38, namely MKK3bE and MKK6bE (7). The gene-switch strategy is based on cre/loxP-mediated DNA recombination to restrict transgene expression to ventricular myocytes and avoid potentially adverse effects of transgene expression on the survival of the founder transgenic animals (104). Moreover, dominant-negative p38α transgenic mice showed enhanced cardiac hypertrophy following both pressure-overload and catecholamine drug infusion (105). Further investigations in these mice revealed an enhancement of calcineurin–NFAT signaling following reduced p38 activity, providing a possible explanation for the contradicting results in the studies presented. However, probably both p38 and JNK MAPK activation are directly associated with a failing cardiac phenotype. In longstanding hypertrophy, only p38 and JNK MAPKs are activated, associated with progressive deterioration to maladaptive chronic hypertrophy and congestive heart failure (99,106).

The calcium/calmodulin-dependent protein phosphatase calcineurin is important in cardiac hypertrophy in response to numerous stimuli. Transgenic mice overexpressing the activated form of calcineurin exhibited a severe form of cardiac hypertrophy with concurrent transition toward apoptosis-independent heart failure and sudden death because of lethal arrhythmias (107–109). Besides the ability of calcineurin itself to provoke hypertrophic growth of the heart, it has been demonstrated to play significant roles in pressure-overload–induced and isoproterenol-induced cardiac hypertrophy as observed in transgenic mice overexpressing a dominant negative mutant of calcineurin (110,111). Calcineurin was shown to dephosphorylate the transcription factor NFAT3, enabling it to translocate to the nucleus, where NFAT interacted mainly with transcription factor GATA4, resulting in synergistic activation of cardiac transcription (107). NFAT3 appeared to be required for calcineurin-mediated hypertrophic signaling based on reduced cardiac growth upon calcineurin stimulation in NFAT3 knockout mice (112,113). Transgenic mice expressing the activated form of calcineurin showed interaction with other developmental pathways by inactivation of p38 and increased expression of the dual specificity phosphatase MAPK phosphatase-1 (MKP-1) (114). Calcineurin hypertrophic signaling was furthermore interconnected with PKCα, theta, and JNK in the heart (115).

Endogenous myocyte-enriched calcineurin-interacting protein (MCIP) counteracts calcineurin through a negative feedback loop. Cardiac-specific expression of human MCIP1 in mice inhibited cardiac hypertrophy, provoked by constitutively active calcineurin (116). Even hypertrophy induced by catecholamines, exercise training or pressure-overload was attenuated, without deterioration of ventricular performance (116,117). Even so, the lack of MCIP1 in homozygous knockout mice exacerbated the hypertrophic response to activated calcineurin (118). Another endogenous inhibiting protein of the calcineurin-dependent hypertrophic signaling pathway is glycogen synthase kinase (GSK)-3β. Transgenic mice that express a constitutively active form of GSK-3β showed a significant ability to attenuate the hypertrophic response to calcineurin activation (119). GSK-3β mice expressing the calcineurin inhibitory domains of Cain/Cabin-1 and A-kinase anchoring protein 79 demonstrated reduced cardiac calcineurin activity and reduced hypertrophy in response to catecholamine infusion and pressure overload (120). Conclusively, the calcium/calmodulin-dependent protein phosphatase calcineurin is an important regulator of cardiac hypertrophy induced by various stimuli as shown by multiple studies in genetically engineered mice.

Apoptosis and Its Relation to Cardiac Hypertrophy

Cardiac hypertrophy might be compensatory to the loss of viable cells after myocardial ischemia (121). Cardiomyopathy in transgenic mice is related to increased apoptosis rates. Nix/Bnip3L, one of the mitochondrial death proteins, was shown to be upregulated in cardiac hypertrophy, and also to play a major role in the apoptotic process (122). As reported before, the cardiac-specific overexpression of Mst1 resulted in activation of caspases and increase of apoptosis. Interestingly, Mst1 also prevents cardiac myocyte elongation and hypertrophy despite increased wall stress (123). Mst1 may inhibit (unknown) signaling molecules causing hypertrophy.

However, several molecular effectors in hypertrophic signaling have been proven important in antiapoptotic signaling also, as part of the second leg of cardiomyocyte survival pathways. For instance, the MEK1–ERK1/2 signaling pathway stimulates cardiac hypertrophic growth associated with augmented cardiac function (i.e., adaptive hypertrophy) combined with partial resistance to apoptosis (10,124). The gp130 protein, another potent cardiac survival factor, mediates cardiotrophin-1 (CT-1)-induced cardiac hypertrophy and is capable of inhibiting cardiomyocyte apoptosis also (125,126). In gp130 knockout mice, the application of pressure overload by aortic banding resulted in massive apoptosis rates plus reduced cardiac hypertrophy (127). To date, the role of the calcineurin-NFAT pathway in pro- or antiapoptotic responses is not clear yet (128). For instance, adrenergic stimulation led to calcineurin-mediated cardiomyocyte apoptosis (129), while other investigations found myocardial protection against I/R induced apoptosis through calcineurin in vitro and in vivo (82,109). It seems that NFAT activity is the critical component mediating effects of calcineurin stimulation resulting in the activation of pro- or antiapoptotic pathways in cardiomyocytes (82,130). Selective NFAT inhibition during phenylephrine stimulation prevented calcineurin-mediated hypertrophy but resulted in increased cardiomyocyte apoptosis (130). Therefore, certain prohypertrophic and antiapoptotic pathways can come together as a common survival pathway, suggesting interplay between cellular pathways related to adaptive myocyte responses.

HEMODYNAMICS

The use of genetically engineered mice rendered vital information about the precise role various proteins play in the diversity of intracellular signaling pathways in the event of cellular hypertrophy or apoptosis. The link between the genotype of the mouse and the concurrent phenotype is investigated with the use of sophisticated molecular and cellular techniques. An equally important aspect of the resultant phenotype is cardiac function. Genetic engineering provides the possibility to probe the precise mechanism in which specific proteins exert their effects on the plain cardiac function or the process of remodeling. The small size of the murine heart and the rapid pace at which it contracts rendered difficulties in measuring cardiac performance. Various techniques have been used to assess murine cardiac function, such as transthoracic ultrasonography (131), Langendorff perfusion systems (132), aortic flow probes (37), and micromanometers (133). The mentioned techniques have been and are being used extensively in mouse studies, notwithstanding the relative inaccuracy and the lack of total cardiac performance assessment. This could theoretically be obtained through measuring pressure and volume simultaneously. The development of new cardiac function assessment techniques such as MRI (134) and conductance-micromanometers (135) provide new tools for accurate and total cardiac performance assessment.

The concept of simultaneous in vivo left ventricular pressure and volume measurements has been developed by Baan et al. in the early 1980s (136). The volume-signal is derived in accordance with Ohm's Law, and depends on the varying amount of ventricular blood volume related to the cardiac cycle. An electrical current is produced by an intraventricular catheter to measure conductance, which is directly related to the varying blood volume. The pressure and volume signals as produced by the conductance-micromanometer catheter are displayed in a two-dimensional diagram. This presentation results in pressure–volume loops accurately representing cardiac performance, which can be analyzed meticulously. The pressure–volume technique originates from large animal and human studies, but has recently been miniaturized and made suitable for mouse studies (135). Pressure–volume assessment protocols have been developed (137) and hemodynamic studies have been successfully performed in genetically engineered mice (10,138). Figure 4 shows pressure–volume (PV) loops derived in *Erk2* heterozygous null mice, MEK1 transgenic mice and their respective littermates, one week following I/R injury. PV-loops show the dramatic decrease in cardiac function in Erk and MEK wildtype mice following I/R. Indications for functional deterioration are the rightward shift of the PV-loops, the decrease in systolic pressure and stroke volume (therefore, attenuation of stroke work) and the increase in diastolic pressure. The MEK1 transgenic phenotype protects the heart against IR injury, thereby maintaining cardiac function at the normal baseline level.

The use of MRI in mice is another relatively new concept in the studies of murine hemodynamics (134). The noninvasive technique of MRI enables the investigator to study in vivo murine cardiac metabolism, morphology, and function under (patho) physiological conditions. To quantify left ventricular function high-resolution (HR) MRI is used. To date, MRI studies have been successfully implemented in cardiac phenotyping. Evaluation of cardiac mass and function in post-MI remodeling was performed by MRI studies in mice overexpressing the AT2 receptor (Fig. 5) (25). Serial MRI studies in transgenic mice overexpressing TNF-α demonstrated increased ventricular mass and deteriorating cardiac function over time in this murine model of dilated cardiomyopathy (139).

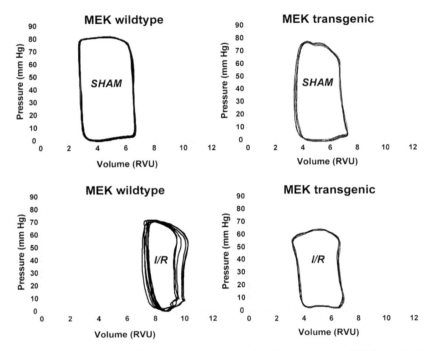

Figure 4 Pressure–volume loops derived in mice. Representative PV-loops derived in different genotypes following an ischemia–reperfusion experiment. Seven days following the ischemic event cardiac function was measured in MEK1 transgenic mice and their respective littermates. Deterioration of cardiac function following IR in wildtype mice and Erk2 heterozygous mice is evident, as indicated by the changes in ejection fraction, systolic and diastolic pressures, and stroke work. MEK1 transgenic mice were protected against the deleterious effects of the ischemic event and maintained cardiac function at normal level. *Source*: From Ref. 10.

To stress the importance of accurate phenotyping of cardiac performance in (genetically altered) mice, some relevant studies are discussed in more detail. In accordance with the general consensus concerning hypertrophy is the increase in cardiac mass, a compensating mechanism to withstand augmented hemodynamic stress. Cardiac hypertrophy can be defined as the increase in myocardial mass in an effort to alleviate the elevation in wall stress according to Laplace's principle. Studies using a heart size-independent analysis for in vivo murine cardiac function determination suggested that the development of cardiac hypertrophy is associated with a heightened contractile state, perhaps as an early compensatory response to pressure overload (140). The systolic wall stress increases significantly in the early response to pressure-overload concomitant with ejection fraction and fractional shortening attenuation. In time, a gradual normalization of wall stress and cardiac performance is seen with serial measurements (141). Pharmacological inhibition of hypertrophic growth in pressure-overloaded mice showed maintenance of normal LV size and systolic function as measured by echocardiography (142), without reducing left ventricular wall stress (143). Moreover, two genetically altered mouse models [mice with myocardial expression of a carboxyl terminal peptide of $G\alpha_q$ (TgGqI) that specifically inhibits G_q-mediated signaling and genetically altered mice that lack endogenous norepinephrine and epinephrine created by disruption of the dopamine β-hydroxylase gene (Dbh$^{-/-}$)] have a blunt hypertrophic response to pressure-overload and showed inadequacy to normalize

(A)

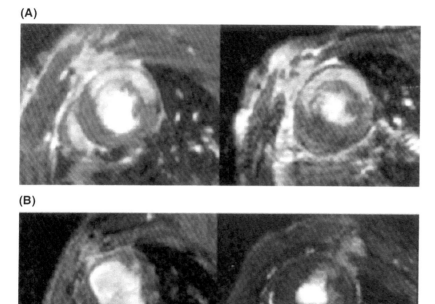

(B)

Figure 5 Representative images of Gadolinium-enhanced MRI recordings in mice. (**A**) MR images one day postmyocardial infarction in wildtype (*left*) and transgenic mice (*right*). In this end-systolic short-axis image are the infarct areas hyper-enhanced by the Gadolinium. Furthermore, note the increased end-systolic cavity area in WT mouse compared with TG mouse despite similar infarct sizes. (**B**). MR images 28 days post myocardial infarction in wild-type (*left*) and transgenic (*right*) mice. Note the marked anteroseptal thinning and markedly increased end-systolic cavity in wildtype mouse compared with the transgenic mouse. *Abbreviation*: MRI, magnetic resonance imaging; TG, transgenic; WT, wild type; *Source*: From Ref. 25.

the twofold increased wall stress. Despite the increase in wall stress cardiac function as measured by serial echocardiography showed little deterioration in either of the pressure-overloaded genetic mouse strains (144). In contrast, wild-type mice with similar pressure overload showed a significant increase in chamber dimensions and progressive deterioration in cardiac function. These data suggest that under conditions of pressure-overload, the development of cardiac hypertrophy and normalization of wall stress may not be necessary to preserve cardiac function, as previously hypothesized. The data provided suggest that cardiac hypertrophic growth is not mandatory to maintain ventricular wall stress at normal levels, as augmented wall stress is not a predictor for deterioration of cardiac function.

CONCLUSION

Genetic modification of murine DNA provides the possibility to elucidate the spectrum of specific functions of a single molecule, and its corresponding intracellular signaling pathways in specific pathophysiological situations as for instance myocardial ischemia. Cardiac remodeling is defined by the ischemia-induced structural and concomitantly functional alterations in the heart. Murine models of I/R and MI mimic the pathophysiological changes observed in man. Cell death, deposition of collagen, and myocardial hypertrophy are the characteristics of cardiac remodeling in man and mouse alike. The predominant factor in determining the severity of the resultant cardiac phenotype is ventricular function (Fig. 6), which is best assessed by the use of in vivo pressure–volume measurements in mice. The use of genetically modified murine models in I/R studies led to a wealth of knowledge about the molecular mechanisms involved in these processes. Despite the data available, intracellular signaling is still a mystery as we lack the tools to study several pathways simultaneously and maybe even more importantly the temporal changes in signaling cascade interactions. Further studies should more intensively incorporate the crossbreeding of different engineered genotypes to intervene at multiple steps in signaling cascades, or in parallel cascades. The use of inducible gene-alteration in mice will

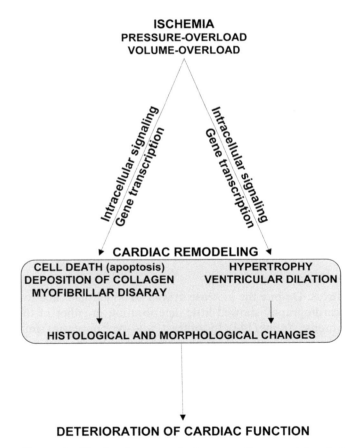

Figure 6   The development of adverse cardiac remodeling in human heart disease. Presented is a scheme of the development of deteriorating cardiac function, mediated by the characteristics of cardiac remodeling. Cardiac remodeling is initiated by several pathophysiological situations as myocardial ischemia. Activated molecular pathways and newly expressed gene transcription profiles are the basis of cellular, histological, and morphological alterations leading to the development of clinical heart disease in the patient.

better control the influence of genomics on the resultant phenotype and the circumstances involved. This way, genetic studies in mice could accelerate the understanding of cardiac remodeling, and result in useful therapeutical tools in the clinic.

REFERENCES

1. Bueno OF, van Rooij E, Molkentin JD, Doevendans PA, De Windt LJ. Calcineurin and hypertrophic heart disease: novel insights and remaining questions. Cardiovasc Res 2002; 53:806–821.
2. Doevendans PA, Hunter JJ, Lembo G. Strategies for studying cardiovascular diseases in transgenic mice and gene-targeted mice. In: Monastersky GM RJ, ed. Strategies in Transgenic Animal Science. Washington: American Society for Microbiology, 1995: 107–144.
3. Doevendans PA, Daemen M, de Muinck E, Smits JF. Cardiovascular phenotyping in mice. Cardiovasc Res 1998; 39:34–49.
4. Kubalak S, Doevendans PA, Rockman H. Molecular analysis of cardiac muscle diseases based on mouse genetics. In: Adolph KW, ed. Human Molecular Genetics. Orlando: Academic Press, 1996:470–487.
5. Bueno OF, van Rooij E, Lips DJ, Doevendans PA, De Windt LJ. Cardiac hypertrophic signaling: the Good, the Bad and the Ugly. In: Doevendans PA, Kaab S, eds. Cardiovascular Genomics: New Insights Into Pathophysiology. Dordrecht: Kluwer, 2002:131–155.
6. Bueno OF, De Windt LJ, Tymitz KM, Witt SA, Kimball TR, Klevitsky R, Hewett TE, Jones SP, Lefer DJ, Peng CF, Kitsis RN, Molkentin JD. The MEK1-ERK1/2 signaling pathway promotes compensated cardiac hypertrophy in transgenic mice. Embo J 2000; 19:6341–6350.
7. Liao P, Georgakopoulos D, Kovacs A, Zheng M, Lerner D, Pu H, Saffitz J, Chien KR, Xiao RP, Kass DA, Wang Y. The in vivo role of p38 MAP kinases in cardiac remodeling and restrictive cardiomyopathy. Proc Natl Acad Sci USA 2001; 98:12283–12288.
8. Zhang D, Gaussin V, Taffet GE, Belaguli NS, Yamada M, Schwartz RJ, Michael LH, Overbeek PA, Schneider MD. TAK1 is activated in the myocardium after pressure overload and is sufficient to provoke heart failure in transgenic mice. Nat Med 6: 556–563.
9. Eefting F, Rensing B, Wigman J, Pannekoek WJ, Liu WM, Cramer MJ, Lips DJ, Doevendans PA. Apoptosis in reperfusion injury. Cardiovasc Res 2003; 61:414–426.
10. Lips DJ, Bueno OF, Wilkins BJ, Lorenz JN, Meloche S, Pages G, De Windt LJ, Doevendans PA, Molkentin JD. The MEK1-ERK2 signaling pathway protects the myocardium from ischemic damage in vivo. Circulation 2004; 109:1938–1941.
11. Camper-Kirby D, Welch S, Walker A, Shiraishi I, Setchell KD, Schaefer JE, Kajstura J, Anversa P, Sussman MA. Myocardial Akt activation and gender: increased nuclear activity in females versus males. Circ Res 2001; 88:1020–1027.
12. Haq S, Choukroun G, Lim H, Tymitz KM, del Monte F, Gwathmey J, Grazette L, Michael A, Hajjar R, Force T, Molkentin JD. Differential activation of signal transduction pathways in human hearts with hypertrophy versus advanced heart failure. Circulation 2001; 103:670–677.
13. Petrich BG, Molkentin JD, Wang Y. Temporal activation of c-Jun N-terminal kinase in adult transgenic heart via cre-loxP-mediated DNA recombination. FASEB J 2003; 17:749–751.
14. Bellenger NG, Swinburn JM, Rajappan K, Lahiri A, Senior R, Pennell DJ. Cardiac remodelling in the era of aggressive medical therapy: does it still exist? Int J Cardiol 2002; 83:217–225.

15. Bolognese L, Neskovic AN, Parodi G, Crisano G, Buonamici P, Santoro GM, Antoniucci D. Left ventricular remodeling after primary coronary angioplasty: patterns of left ventricular dilation and long-term prognostic implications. Circulation 2002; 106:2351–2357.

16. Lutgens E, Daemen MJ, de Muinck ED, Debets J, Leenders P, Smits JF. Chronic myocardial infarction in the mouse: cardiac structural and functional changes. Cardiovasc Res 1999; 41:586–593.

17. Eaton LW, Bulkley BH. Expansion of acute myocardial infarction: its relationship to infarct morphology in a canine model. Circ Res 1981; 49:80–88.

18. Holmes JW, Yamashita H, Waldman LK, Covell JW. Scar remodeling and transmural deformation after infarction in the pig. Circulation 1994; 90:411–420.

19. Eaton LW, Weiss JL, Bulkley BH, Garrison JB, Weisfeldt ML. Regional cardiac dilatation after acute myocardial infarction: recognition by two-dimensional echocardiography. N Engl J Med 1979; 300:57–62.

20. Rumberger JA, Behrenbeck T, Breen JR, Reed JE, Gersh BJ. Nonparallel changes in global left ventricular chamber volume and muscle mass during the first year after transmural myocardial infarction in humans. J Am Coll Cardiol 1993; 21:673–682.

21. Gaudron P, Eilles C, Kugler I, Ertl G. Progressive left ventricular dysfunction and remodeling after myocardial infarction. Potential mechanisms and early predictors. Circulation 1993; 87:755–763.

22. Nahrendorf M, Wiesmann F, Hiller KH, Han H, Hu K, Waller C, Ruff J, Haase A, Ertl G, Bauer WR. In vivo assessment of cardiac remodeling after myocardial infarction in rats by cine-magnetic resonance imaging. J Cardiovasc Magn Reson 2000; 2:171–180.

23. Michael LH, Entman ML, Hartley CJ, Youker KA, Zhu J, Hall SR, Hawkins HK, Berens K, Ballantyne CM. Myocardial ischemia and reperfusion: a murine model. Am J Physiol Heart Circ Physiol 1995; 269:H2147–54.

24. Booz GW, Baker KM. Molecular signalling mechanisms controlling growth and function of cardiac fibroblasts. Cardiovasc Res 1995; 30:537–543.

25. Yang Z, Bove CM, French BA, Epstein FH, Berr SS, DiMaria JM, Gibson JJ, Carey RM, Kramer CM. Angiotensin II type 2 receptor overexpression preserves left ventricular function after myocardial infarction. Circulation 2002; 106:106–111.

26. Matsusaka T, Katori H, Inagami T, Fogo A, Ichikawa I. Communication between myocytes and fibroblasts in cardiac remodeling in angiotensin chimeric mice. J Clin Invest 1999; 103:1451–1458.

27. Xu J, Carretero OA, Liu YH, Shesely EG, Yang F, Kapke A, Yang XP. Role of AT2 receptors in the cardioprotective effect of AT1 antagonists in mice. Hypertension 2002; 40:244–250.

28. Bader M. Role of the local renin-angiotensin system in cardiac damage: a minireview focussing on transgenic animal models. J Mol Cell Cardiol 2002; 34:1455–1462.

29. Harada K, Sugaya T, Murakami K, Yazaki Y, Komuro I. Angiotensin II type 1A receptor knockout mice display less left ventricular remodeling and improved survival after myocardial infarction. Circulation 1999; 100:2093–2099.

30. Adachi Y, Saito Y, Kishimoto I, Harada M, Kuwahara K, Takahashi N, Kawakami R, Nakanishi M, Nakagawa Y, Tanimoto K, Saitoh Y, Yasuno S, Usami S, Iwai M, Horiuchi M, Nakao K. Angiotensin II type 2 receptor deficiency exacerbates heart failure and reduces survival after acute myocardial infarction in mice. Circulation 2003; 107:2406–2408.

31. Oishi Y, Ozono R, Yano Y, Teranishi Y, Akishita M, Horiuchi M, Oshima T, Kambe M. Cardioprotective role of AT2 receptor in postinfarction left ventricular remodeling. Hypertension 2003; 41:814–818.

32. Hein L, Stevens ME, Barsh GS, Pratt RE, Kobilka BK, Dzau VJ. Overexpression of angiotensin AT1 receptor transgene in the mouse myocardium produces a lethal phenotype associated with myocyte hyperplasia and heart block. Proc Natl Acad Sci USA 1997; 94:6391–6396.

33. Palmen M, Daemen MJ, Bronsaer R, Dassen WR, Zandbergen HR, Kockx M, Smits JF, van der Zee R, Doevendans PA. Cardiac remodeling after myocardial infarction is impaired in IGF-1 deficient mice. Cardiovasc Res 2001; 50:516–524.

34. Carmeliet P, Ng YS, Nuyens D, Theilmeier G, Brusselmans K, Cornelissen I, Ehler E, Kakkar VV, Stalmans I, Mattot V, Perriard JC, Dewerchin M, Flameng W, Nagy A, Lupu F, Moons L, Collen D, D'Amore PA, Shima DT. Impaired myocardial angiogenesis and ischemic cardiomyopathy in mice lacking the vascular endothelial growth factor isoforms VEGF164 and VEGF188. Nat Med 1999; 5:495–502.

35. Bellomo D, Headrick JP, Silins GU, Paterson CA, Thomas PS, Gartside M, Mould A, Cahill MM, Tonks ID, Grimmond SM, Townson S, Wells C, Little M, Cummings MC, Hayward NK, Kay GF. Mice lacking the vascular endothelial growth factor-B gene (Vegfb) have smaller hearts, dysfunctional coronary vasculature, and impaired recovery from cardiac ischem. Circ Res 2000; 86:E29–35.

36. Buehler A, Martire A, Strohm C, Wolfram S, Fernandez B, Palmen M, Wehrens XH, Doevendans PA, Franz WM, Schaper W, Zimmermann R. Angiogenesis-independent cardioprotection in FGF-1 transgenic mice. Cardiovasc Res 2002; 55:768–777.

37. Palmen M, Daemen MJ, Buehler A, Bronsaer RJ, Zimmermann R, Smits JF, Doevendans PA. Impaired cardiac remodeling and function after myocardial infarction in FGF-1 transgenic mice. Circulation 1999; 100:250.

38. Sheikh F, Sontag DP, Fandrich RR, Kardami E, Cattini PA. Overexpression of FGF-2 increases cardiac myocyte viability after injury in isolated mouse hearts. Am J Physiol Heart Circ Physiol 2001; 280:H1039–H1050.

39. Hampton TG, Amende I, Fong J, Laubach VE, Li J, Metais C, Simons M. Basic FGF reduces stunning via a NOS2-dependent pathway in coronary-perfused mouse hearts. Am J Physiol Heart Circ Physiol 2000; 279:H260–8.

40. Cleutjens JP, Creemers EE. Integration of concepts: cardiac extracellular matrix remodeling after myocardial infarction. J Card Fail 2002; 8:S344–S348.

41. Ducharme A, Frantz S, Aikawa M, Rabkin E, Lindsey M, Rohde LE, Schoen FJ, Kelly RA, Werb Z, Libby P, Lee RT. Targeted deletion of matrix metalloproteinase-9 attenuates left ventricular enlargement and collagen accumulation after experimental myocardial infarction. J Clin Invest 2000; 106:55–92.

42. Heymans S, Luttun A, Nuycns D, Theilmeier G, Creemers EE, Moons L, Dyspersin GD, Cleutjens JP, Shipley M, Angellilo A, Levi M, Nube O, Baker A, Keshet E, Lupu F, Herbert JM, Smits JF, Shapiro SD, Baes M, Borgers M, Collen D, Daemen MJ, Carmeliet P. Inhibition of plasminogen activators or matrix metalloproteinases prevents cardiac rupture but impairs therapeutic angiogenesis and causes cardiac failure. Nat Med 1999; 5:1135–42.

43. Hayashidani S, Tsutsui H, Jkeuchi M, Shiomi T, Matsusaka H, Kubota T, Imanaka-Yoshida K, Itoh T, Takeshita A. Targeted deletion of MMP-2 attenuates early LV rupture and late remodeling after experimental myocardial infarction. Am J Physiol Heart Circ Physiol 2003; 285:H1229–35.

44. Creemers EE, Davis JN, Parkhurst AM, Leenders P, Dowdy KB, Hapke E, Hauet AM, Escobar PG, Cleutjens JP, Smits JF, Daemen MJ, Zile MR, Spinale FG. Deficiency of TIMP-1 exacerbates LV remodeling after myocardial infarction in mice. Am J Physiol Heart Circ Physiol 2003; 284:H364–71.

45. Metzler B, Mair J, Lercher A, Schaber C, Hintringer F, Pachinger O, Xu Q. Mouse model of myocardial remodelling after ischemia: role of intercellular adhesion molecule-1. Cardiovasc Res 2001; 49:399–407.

46. Palazzo AJ, Jones SP, Girod WG, Anderson DC, Granger DN, Lefer DJ. Myocardial ischemia-reperfusion injury in CD18- and ICAM-1-deficient mice. Am J Physiol 1998; 275:H2300–H2307.

47. Majno G, Joris I. Apoptosis, oncosis, and necrosis. An overview of cell death. Am J Pathol 1995; 146:3–15.

48. Dumont EA, Reutelingsperger CP, Smits JF, Daernen MJ, Doevendans PA, Wellens HJ, Hofstra L. Real-time imaging of apoptotic cell-membrane changes at the single-cell level in the beating murine heart. Nat Med 2001; 7:1352–5.

49. Knaapen MW, Davies MJ, De Bie M, Haven AJ, Martinet W, Kockx MM. Apoptotic versus autophagic cell death in heart failure. Cardiovasc Res 2001; 51:304–312.

50. Li P, Nijhawan D, Budihardjo I, Srinivasula SM, Ahmad M, Alnemri ES, Wang X. Cytochrome c and dATP-dependent formation of Apaf-l/caspase-9 complex initiates an apoptotic protease cascade. Cell 1997; 91:479–89.

51. Borutaite V, Brown GC. Mitochondria in apoptosis of ischemic heart. FEBS Lett 2003; 541:1–5.

52. Brocheriou V, Hagege AA, Oubenaissa A, Lambert M, Mallet VO, Duriez M, Wassef M, Kahn A, Menascbe P, Gilgenkrantz H. Cardiac functional improvement by a human Bcl-2 transgene in a mouse model of ischemia/reperfusion injury. J Gene Med 2000; 2:326–33.

53. Chen Z, Chua CC, Ho YS, Hamdy RC, Chua BH. Overexpression of Bcl-2 attenuates apoptosis and protects against myocardial I/R injury in transgenic mice. Am J Physiol Heart Circ Physiol 2001; 280:H2313–H2320.

54. Hochhauser E, Kivity S, Offen D, Maulik N, Otani H, Barhum Y, Pannet H, Shneyvays V, Shainberg A, Goldshtaub V, Tobar A, Vidne BA. Bax ablation protects against myocardial ischemia-reperfusion injury in transgenic mice. Am J Physiol Heart Circ Physiol 2003; 284:H2351–9.

55. Matsui T, Li L, Wu JC, Cook SA, Nagoshi T, Picard MH, Liao R, Rosenzweig A. Phenotypic spectrum caused by transgenic overexpression of activated Akt in the heart. J Biol Chem 2002; 277:22896–901.

56. Kuwahara K, Saito Y, Kishimoto I, Miyamoto Y, Harada M, Ogawa E, Hamanaka I, Kajiyama N, Takahashi N, Izumi T, Kawakami R, Nakao K. Cardiotrophin-1 phosphorylates akt and BAD, and prolongs cell survival via a PI3K-dependent pathway in cardiac myocytes. J Mol Cell Cardiol 2000; 32:1385–94.

57. Wu W, Lee WL, Wu YY, Chen D, Liu TJ, Jang A, Sharma PM, Wang PH. Expression of constitutively active phosphatidylinositol 3-kinase inhibits activation of caspase 3 and apoptosis of cardiac muscle cells. J Biol Chem 2000; 275:40113–9.

58. Li Q, Li B, Wang X, Leri A, Jana KP, Liu Y, Kajstura J, Baserga R, Anversa P. Overexpression of insulin-like growth factor-1 in mice protects from myocyte death after infarction, attenuating ventricular dilation, wall stress, and cardiac hypertrophy. J Clin Invest 1997; 100:1991–1999.

59. Lee P, Sata M, Lefer DJ, Factor SM, Walsh K, Kitsis RN. Fas pathway is a critical mediator of cardiac myocyte death and MI during ischemia-reperfusion in vivo. Am J Physiol Heart Circ Physiol 2003; 284:H456–H463.

60. Jeremias I, Kupatt C, Martin-Villalba A, Habazettl H, Schenkel J, Boekstegers P, Debatin KM. Involvement of CD95/Apo1/Fas in cell death after myocardial ischemia. Circulation 2000; 102:915–20.

61. Yang J, Jones SP, Suhara T, Greer JJ, Ware PD, Nguyen NP, Perlman H, Nelson DP, Lefer DJ, Walsh K. Endothelial cell overexpression of fas ligand attenuates ischemia-reperfusion injury in the heart. J Biol Chem 2003; 278:15185–91.

62. Beg AA, Baltimore D. An essential role for NF-kappaB in preventing TNF-alpha-induced cell death. Science 1996; 274:782–784.

63. Grag AK, Aggarwal BB. Reactive oxygen intermediates in TNF signaling. Mol Immunol 2002; 39:509–517.

64. Kubota T, Miyagishima M, Frye CS, Alber SM, Bounoutas GS, Kadokami T, Watkins SC, McTiernan CF, Feldman AM. Overexpression of tumor necrosis factor- alpha activates both anti- and pro-apoptotic pathways in the myocardium. J Mol Cell Cardiol 2001; 33:1331–44.

65. Kurrelmeyer KM, Michael LH, Baumgarten G, Taffet GE, Peschon JJ, Sivasubramanian N, Entman ML, Mann DL. Endogenous tumor necrosis factor protects the adult

cardiac myocyte against ischemic-induced apoptosis in a murine model of acute myocardial infarction. Proc Natl Acad Sci USA 2000; 97:5456–61.

66. Smith RM, Suleman N, McCarthy J, Sack MN. Classic ischemic but not pharmacologic preconditioning is abrogated following genetic ablation of the TNFalpha gene. Cardiovasc Res 2002; 55:553–560.

67. Maekawa N, Wada H, Kanda T, Niwa T, Yarnada Y, Saito K, Fujiwara H, Sekikawa K, Seishima M. Improved myocardial ischemia/reperfusion injury in mice lacking tumor necrosis factor-alpha. J Am Coll Cardiol 2002; 39:1229–35.

68. Sugano M, Koyanagi M, Tsuchida K, Hata T, Makino N. In vivo gene transfer of soluble TNF-alpha receptor 1 alleviates myocardial infarction. FASEB J 2002; 16: 1421–1422.

69. Yamashita N, Hoshida S, Otsu K, Taniguchi N, Kuzuya T, Hori M. The involvement of cytokines in the second window of ischaemic preconditioning. Br J Pharmacol 2000; 131:415–422.

70. Belosjorow S, Bolle I, Duschin A, Heusch G, Schulz R. TNF-alpha antibodies are as effective as ischemic preconditioning in reducing infarct size in rabbits. Am J Physiol Heart Circ Physiol 2003; 284:H927–H930.

71. Graciano AL, Bryant DD, White DJ, Horton J, Bowles NE, Giroir BP. Targeted disruption of ICAM-1, P-selectin genes improves cardiac function and survival in TNF-alpha transgenic mice. Am J Physiol Heart Circ Physiol 2001; 280:H1464–H1471.

72. Dibbs ZL, Diwan A, Nemoto S, DeFreitas G, Abdellatif M, Carabello BA, Spinale FG, Feuerstein G, Sivasubramanian N, Mann DL. Targeted overexpression of transmembrane tumor necrosis factor provokes a concentric cardiac hypertrophic phenotype. Circulation 2003; 108:1002–8.

73. Zheng TS, Hunot S, Kuida K, Flavell RA. Caspase knockouts: matters of life and death. Cell Death Differ 1999; 6:1043–1053.

74. Frantz S, Ducharme A, Sawyer D, Rohde LE, Kobzik L, Fukazawa R, Tracey D, Allen H, Lee RT, Kelly RA. Targeted deletion of caspase-1 reduces early mortality and left ventricular dilatation following myocardial infarction. J Mol Cell Cardiol 2003; 35:685–94.

75. Condorelli G, Roncarati R, Ross JJ, Pisani A, Stassi G, Todaro M, Trocha S, Drusco A, Gu Y, Russo MA, Frati G, Jones SP, Lcfcr DJ, Napoli C, Croce CM. Heart-targeted overexpression of caspase3 in mice increases infarct size and depresses cardiac function. Proc Natl Acad Sci USA 2001; 98:9977–82.

76. Regan SE, Broad M, Byford AM, Lankford AR, Cerniway RJ, Mayo MW, Matherne GP. A1 adenosine receptor overexpression attenuates ischemia-reperfusion-induced apoptosis and caspase 3 activity. Am J Physiol Heart Circ Physiol 2003; 284:H859–66.

77. Hayakawa K, Takernura G, Kanoh M, Li Y, Koda M, Kawase Y, Maruyama R, Okada H, Minatoguchi S, Fujiwara T, Fujiwara H. Inhibition of granulation tissue cell apoptosis during the subacute stage of myocardial infarction improves cardiac remodeling and dysfunction at the chronic stage. Circulation 2003; 108:104–9.

78. Bialik S, Cryns VL, Drincic A, Miyata S, Wollowick AL, Srinivasan A, Kitsis RN. The mitochondrial apoptotic pathway is activated by serum and glucose deprivation in cardiac myocytes. Circ Res 1999; 85:403–14.

79. Fliss H, Gattinger D. Apoptosis in ischemic and reperfused rat myocardium. Circ Res 1996; 79:949–956.

80. Bialik S, Geenen DL, Sasson IE, Cheng R, Horncr JW, Evans SM, Lord EM, Loch CJ, Kitsis RN. Myocyte apoptosis during acute myocardial infarction in the mouse localizes to hypoxic regions but occurs independently of p53. J Clin Invest. 1997; 100(6): 1363–1372.

81. Kajstura J, Cheng W, Reiss K, Clark WA, Sonnenblick EH, Krajewski S, Reed JC, Olivetti G, Anversa P. Apoptotic and necrotic myocyte cell deaths are independent contributing variables of infarct size in rats. Lab Invest 1996; 74:86–107.

82. Bueno OF, Lips PJ, Kaiser RA, Wilkins BJ, Dai YS, Glascock BJ, Kimball TR, Aronow BJ, Doevendans PA, Molkentin JD. Calcineurin Abeta gene targeting predisposes the myocardium to stress-induced apoptosis and dysfunction. Circ Res 2004; 94:98–99.

83. Palojoki E, Saraste A, Eriksson A, Pulkki K, Kallajoki M, Voipio-Pulkki LM, Tikkanen I. Cardiomyocyte apoptosis and ventricular remodeling after myocardial infarction in rats. Am J Physiol Heart Circ Physiol 2001; 280:H2726–31.

84. Baldi A, Abbate A, Bussani R, Patti G, Melfi R, Angelini A, Dobrina A, Rossiello R, Silvestri F, Baldi F, Di Sciascio G. Apoptosis and post-infarction left ventricular remodeling. J Mol Cell Cardiol 2002; 34:165–74.

85. Gottlieb RA, Burleson KO, Kloner RA, Babior BM, Engler RL. Reperfusion injury induces apoptosis in rabbit cardiomyocytes. J Clin Invest 1994; 94:1621–1628.

86. Lips DJ, Van Kraaij DL, De Windt LJ, Doevendans PA. Molecular determinants of myocardial hypertrophy and failure: alternative pathways for beneficial and maladaptive hypertrophy. Eur Heart J 2003; 24:883–896.

87. Maron BJ. Hypertrophic cardiomyopathy. Lancet 1997; 350:127–133.

88. Chu G, Haghighi K, Kranias EG. From mouse to man: understanding heart failure through genetically altered mouse models. J Card Fail 2002; 8:S432–S439.

89. Hirota H, Yoshida K, Kishimoto T, Taga T. Continuous activation of gp130, a signal-transducing receptor component for interleukin 6-related cytokines, causes myocardial hypertrophy in mice. Proc Natl Acad Sci USA 1995; 92:4862–4866.

90. Uozumi H, Hiroi Y, Zou Y, Takimoto E, Toko H, Niu P, Shimoyama M, Yazaki Y, Nagai R, Komuro I. gp130 plays a critical role in pressure overload-induced cardiac hypertrophy. J Biol Chem 2001; 276:23115–9.

91. Wollert KC, Taga T, Saito M, Narazaki M, Kishimoto T, Glembotski CC, Vernallis AB, Heath JK, Pennica D, Wood WI, Chien KR. Cardiotrophin-1 activates a distinct form of cardiac muscle cell hypertrophy. Assembly of sarcomeric units in series VLA gpl30/leukemia inhibitory factor receptor-dependent pathways. J Biol Chem 1996; 271:9535–45.

92. Kunisada K, Tone E, Fujio Y, Matsui H, Yamauchi-Takihara K, Kishimoto T. Activation of gp130 transduces hypertrophic signals via STAT3 in cardiac myocytes. Circulation 1998; 98:346–352.

93. Kunisada K, Negoro S, Tone E, Funamoto M, Osugi T, Yarnada S, Okabe M, Kishimoto T, Yamaucbi-Takihara K. Signal transducer and activator of transcription 3 in the heart transduces not only a hypertropbic signal but a protective signal against doxorubicin-induced cardiomyopathy. Proc Natl Acad Sci USA 2000; 97:315–9.

94. Minamino T, Yujiri T, Terada N, et al. MEKK1 is essential for cardiac hypertrophy and dysfunction induced by Gq. Proc Natl Acad Sci USA 2002; 99:3866–3871.

95. Ramirez MT, Sah VP, Zhao XL, Hunter JJ, Chien KR, Brown JH. The MEKK-JNK pathway is stimulated by alpha1-adrenergic receptor and ras activation and is associated with in vitro and in vivo cardiac hypertrophy. J Biol Chem 1997; 272:14,057–14,061.

96. Thorburn J, Xu S, Thorburn A. MAP kinase- and Rho-dependent signals interact to regulate gene expression but not actin morphology in cardiac muscle cells. EMBO J 1997; 16:1888–1900.

97. Bueno OF, De Windt LJ, Lim HW, Tymitz KM, Witt SA, Kimball TR, Molkentin JD. The dual-specificity phosphatase MKP-1 limits the cardiac hypertrophic response in vitro and in vivo. Circ Res 2001; 88:88–96.

98. Nicol RL, Frey N, Pearson G, Cobb M, Richardson J, Olson EN. Activated MEK5 induces serial assembly of sarcomeres and eccentric cardiac hypertrophy. Embo J 2001; 20:2757–2767.

99. Esposito G, Prasad SV, Rapacciuolo A, Mao L, Koch WJ, Rockman HA. Cardiac overexpression of a G(q) inhibitor blocks induction of extracellular signal-regulated kinase and c-Jun NH(2)-terminal kinase activity in in vivo pressure overload. Circulation 2001; 103:1453–1458.

100. Finn SG, Dickens M, Fuller SJ. c-Jun N-terminal kinase-interacting protein 1 inhibits gene expression in response to hypertrophic agonists in neonatal rat ventricular myocytes. Biochem J 2001; 358:489–495.

101. Sadoshima J, Montagne O, Wang Q, Yang G, Warden J, Liu J, Takagi G, Karoor V, Hong C, Johnson GL, Vatner DE, Vatner SF. The MEKK1-JNK pathway plays a protective role in pressure overload but does not mediate cardiac hypertrophy. J Clin Invest 2002; 110:271–9.

102. Wang Y, Huang S, Sah VP, Ross JJ, Brown JH, Han J, Chien KR. Cardiac muscle cell hypertrophy and apoptosis induced by distinct members of the p38 mitogen-activated protein kinase family. J Biol Chem 1998; 273:2161–8.

103. Dash R, Schmidt AG, Pathak A, Gerst MJ, Biniakiewicz D, Kadambi VJ, Hoit BD, Abraham WT, Kranias EG. Differential regulation of p38 mitogen-activated protein kinase mediates gender-dependent catecholamine-induced hypertrophy. Cardiovasc Res 2003; 57:704–14.

104. Rajewsky K, Gu H, Kuhn R, Betz UA, Muller W, Rocs J, Schwenk F. Conditional gene targeting. J Clin Invest 1996; 98:600–3.

105. Braz JC, Bueno OF, Liang Q, Wilkins BJ, Dai Y-S, Parsons S, Braunwart J, Glascock BJ, Klevitsky R, Kimball TF, Hewett TE, Molkentin JD. Targeted inhibition of p38 MAPK promotes hypertrophic cardiomyopathy through upregulation of calcineurin NFAT signaling. J Clin Invest 2003; 111:1475–86.

106. Hayashida W, Kihara Y, Yasaka A, Inagaki K, Iwanaga Y, Sasayama S. Stage-specific differential activation of mitogen-activated protein kinases in hypertrophied and failing rat hearts. J Mol Cell Cardiol 2001; 33:733–744.

107. Molkentin JD, Lu JR, Antos CL, Markham B, Richardson J, Robbins J, Grant SR, Olson EN. A calcineurin-dependent transcriptional pathway for cardiac hypertrophy. Cell 1998; 93:215–28.

108. Dong D, Duan Y, Guo J, Roach DE, Swirp SL, Wang L, Lees-Miller JP, Sheldon RS, Molkentin JD, Duff HJ. Overexpression of calcineurin in mouse causes sudden cardiac death associated with decreased density of K+ channels. Cardiovasc Res 2003; 57: 320–32.

109. De Windt LJ, Lim HW, Taigen T, Wencker D, Condorclli G, Dorn GWn, Kitsis RN, Molkentin JD. Calcineurin-mediated hypertrophy protects cardiomyocytes from apoptosis in vitro and in vivo: An apoptosis-independent model of dilated heart failure. Circ Res 2000; 86:255–63.

110. Zou Y, Hiroi Y, Uozumi H, Takimoto E, Toko H, Zhu W, Kudoh S, Mizukami M, Shimoyama M, Shibasaki F, Nagai R, Yazaki Y, Komuro I. Calcineurin plays a critical role in the development of pressure overload-induced cardiac hypertrophy. Circulation 2001; 104:97–101.

111. Zou Y, Yao A, Zhu W, Kudoh S, Hiroi Y, Shimoyama M, Uozumi H, Kohmoto O, Takahashi T, Shibasaki F, Nagai R, Yazaki Y, Komuro I. Isoproterenol activates extracellular signal-regulated protein kinases in cardiomyocytes through calcineurin. Circulation 2001; 104:102–8.

112. Van Rooij E, Doevendans PA, de Theije CC, Babiker FA, Molkentin JD, de Windt LJ. Requirement of nuclear factor of activated T-cells in calcineurin-mediated cardiomyocyte hypertrophy. J Biol Chem 2002; 277:48,617–48,626.

113. Wilkins BJ, De Windt LJ, Bueno OF, Braz JC, Glascock BJ, Kimball TF, Molkentin JD. Targeted disruption of NFATc3, but not NFATc4, reveals an intrinsic defect in calcineurin-mediated cardiac hypertrophic growth. Mol Cell Biol 2002; 22:7603–13.

114. Lim HW, New L, Han J, Molkentin JD. Calcineurin enhances MAPK phosphatase-1 expression and p38 MAPK inactivation in cardiac myocytes. J Biol Chem 2001; 276:15,913–15,919.

115. De Windt LJ, Lim HW, Haq S, Force T, Molkentin JD. Calcineurin promotes protein kinase C and c-Jun NH2-terminal kinase activation in the heart. Cross-talk between cardiac hypertrophic signaling pathways. J Biol Chem 2000; 275:13,571–13,579.

116. Rothermel BA, McKinsey TA, Vega RB, Nicol RL, Mammen P, Yang J, Antos CL, Shelton JM, Bassel-Duby R, Olson EN, Williams RS. Myocyte-enriched calcineurin-interacting protein, MCIPl, inhibits cardiac hypertrophy in vivo. Proc Natl Acad Sci USA 2001; 98:3328–33.

117. Hill JA, Rothermel B, Yoo KD, Cabuay B, Demetroulis E, Weiss RM, Kutschke W, Bassel-Duby R, Williams RS. Targeted inhibition of calcineurin in pressure-overload cardiac hypertrophy. Preservation of systolic function. J Biol Chem 2002; 277:10251–5.

118. Vega RB, Rothermel BA, Weinheimer CJ, Kovacs A, Naseem RH, Bassel-Duby R, Williams RS, Olson EN. Dual roles of modulatory calcineurin-interacting protein 1 in cardiac hypertrophy. Proc Natl Acad Sci USA 2003; 100:669–74.

119. Antos CL, McKinsey TA, Frey N, Kutschke W, McAnally J, Shelton JM, Richardson JM, Hill JA, Olson EN. Activated glycogen synthase-3 beta suppresses cardiac hypertrophy in vivo. PNAS 2002; 99:907–12.

120. De Windt LJ, Lim HW, Bueno OF, Liang Q, Delling U, Braz JC, Glascock BJ, Kimball TF, del Monte F, Hajjar RJ, Molkentin JD. Targeted inhibition of calcineurin attenuates cardiac hypertrophy in vivo. Proc Natl Acad Sci USA 2001; 98:3322–7.

121. Adams JW, Sakata Y, Davis MG, Sah VP, Wang Y, Liggett SB, Chien KR, Brown JH, Dorn GWn. Enhanced Galphaq signaling: a common pathway mediates cardiac hypertrophy and apoptotic heart failure. Proc Natl Acad Sci USA 1998; 95:10140–5.

122. Yussrnan MG, Toyokawa T, Odley A, Lynch RA, Wu G, Colbert MC, Aronow BJ, Lorenz JN, Dorn GWn. Mitochondrial death protein Nix is induced in cardiac hypertrophy and triggers apoptotic cardiomyopathy. Nat Med 2002; 8:725–30.

123. Yamamoto S, Yang G, Zablocki D, Liu J, Hong C, Kim SJ, Soler S, Odashima M, Thaisz J, Yehia G, Molina CA, Yatani A, Vatner DE, Vatner SF, Sadoshima J. Activation of Mstl causes dilated cardiomyopathy by stimulating apoptosis without compensatory ventricular myocyte hypertrophy. J Clin Invest 2003; 11:1463–74.

124. Bueno OF, Molkentin JD. Involvement of extracellular signal-regulated kinases 1/2 in cardiac hypertrophy and cell death. Circ Res 2002; 91:776–781.

125. Pennica D, King KL, Shaw KJ, Luis E, Rullamas J, Luoh SM, Darbonne WC, Knutzon DS, Yen R, Chien KR, Baker JB, Wood WI. Expression cloning of cardiotrophin 1, a cytokine that induces cardiac myocyte hypertrophy. PNAS 1995; 92:1142–6.

126. Sheng Z, Knowlton K, Chen J, Hoshijima M, Brown JH, Chien KR. Cardiotrophin 1 (CT-1) inhibition of cardiac myocyte apoptosis via a mitogen-activated protein kinase-dependent pathway. Divergence from downstream CT-1 signals for myocardial cell hypertrophy. J Biol Chem 1997; 272:5783–5791.

127. Hirota H, Chen J, Betz UA, Rajewsky K, Gu Y, Ross JJ, Mullex W, Chien KR. Loss of a gp130 cardiac muscle cell survival pathway is a critical event in the onset of heart failure during biomechanical stress. Cell 1999; 97:189–98.

128. Lotem J, Kama R, Sachs L. Suppression or induction of apoptosis by opposing pathways downstream from calcium-activated calcineurin. Proc Natl Acad Sci USA 1999; 96:12,016–12,020.

129. Saito S, Hiroi Y, Zonu Y, Aikawa R, Toko H, Shibasaki F, Yazaki Y, Nagai R, Komuro I. beta-Adrenergic pathway induces apoptosis through calcincurin activation in cardiac myocytes. J Biol Chem 2000; 275:34528–33.

130. Pu WT, Ma Q, Izumo S. NFAT transcription factors are critical survival factors that inhibit cardiomyocyte apoptosis during phenylephrine stimulation in vitro. Circ Res 2003; 92:725–731.

131. Tanaka N, Dalton N, Mao L, et al. Transthoracic echocardiography in models of cardiac disease in the mouse. Circulation 1996; 94:1109–1117.

132. De Windt LJ, Willems J, Reneman RS, Van der Vusse GJ, Arts T, Van Bilsen M. An improved isolated, left ventricular ejecting, murine heart model. Functional and metabolic evaluation. Pflugers Arch 1999; 437:182–190.

133. Rockman HA, Hamilton RA, Jones LR, Milano CA, Mao L, Lefkowitz RJ. Enhanced myocardial relaxation in vivo in transgenic mice overexpressing the beta2-adrenergic

receptor is associated with reduced phospholamban protein. J Clin Invest 1996; 97:1618–1623.

134. Chacko VP, Aresta F, Chacko SM, Weiss RG. MRI/MRS assessment of in vivo murine cardiac metabolism, morphology, and function at physiological heart rates. Am J Physiol Heart Circ Physiol 2000; 279:H2218–H2224.

135. Georgakopoulos D, Mitzner WA, Chen CH, Byrne BJ, Millar HD, Hare JM, Kass DA. In vivo murine left ventricular pressure-volume relations by miniaturized conductance naicromanometry. Am J Physiol 1998; 274:H1416–22.

136. Baan J, van der Velde ET, de Bruin HG, Smeenk GJ, Koops J, van Dijk AD, Tenmrnerman D, Senden J, Buis B. Continuous measurement of left ventricular volume in animals and humans by conductance catheter. Circulation 1984; 70:812–23.

137. Lips DJ, Van der Nagel T, Steendijk P, Palmen M, Janssen B, Van Dantzig J-M, De Windt LJ, Doevendans PA. Left ventricular pressure-volume measurements in mice: A comparative study between a closed-chest versus open-chest approach. Basic Res Cardiol 2004; 99:551–559.

138. Georgakopoulos D, Christe ME, Giewat M, Seidman CM, Seidman JG, Kass DA. The pathogenesis of familial hypertrophic cardiomyopathy: early and evolving effects from an alpha-cardiac myosin heavy chain missense mutation. Nat Med 1999; 5:327–330.

139. Franco F, Thomas GD, Giroir B, Bryant D, Bullock MC, Chwialkowski MC, Victor RG, Peshock RM. Magnetic resonance imaging and invasive evaluation of development of heart failure in transgenic mice with myocardial expression of tumor necrosis factor-alpha. Circulation 1999; 99:448–54.

140. Takaoka H, Esposito G, Mao L, Suga H, Rockman HA. Heart size-independent analysis of myocardial function in murine pressure overload hypertrophy. Am J Physiol Heart Circ Physiol 2002; 282:H2190–H2197.

141. Nakamura A, Rokosh DG, Paccanaro M, Yee RR, Simpson PC, Grossman W, Foster E. LV systolic performance improves with development of hypertrophy after transverse aortic constriction in mice. Am J Physiol Heart Circ Physiol 2001; 281:H1104–12.

142. Hill JA, Karimi M, Kutschke W, Davisson RL, Zimmerman K, Wang Z, Kerber RE, Weiss RM. Cardiac hypertrophy is not a required compensatory response to short-term pressure overload. Circulation 2000; 101:2863–9.

143. Kai T, Ishikawa K. Lisinopril reduces left ventricular hypertrophy and cardiac polyamine concentrations without a reduction in left ventricular wall stress in transgenic Tsukuba hypertensive mice. Hypertens Res 2000; 23:625–631.

144. Esposito G, Rapacciuolo A, Naga Prasad SV, Takaoka H, Thomas SA, Koch WJ, Rockman HA. Genetic alterations that inhibit in vivo pressure-overload hypertrophy prevent cardiac dysfunction despite increased wall stress. Circulation 2002; 105:85–92.

12

The Renin–Angiotensin–Aldosterone System in Cardiac Remodeling

Barry Greenberg

Advanced Heart Failure Treatment Program, University of California, San Diego, California, U.S.A.

INTRODUCTION

The concept that the renin–angiotensin–aldosterone system (RAAS) plays a major role in determining cardiovascular (CV) structure and function, in both health and disease, is now firmly ensconced in the thinking of clinicians and researchers alike. Widespread recognition of its importance, new information about how the RAAS is regulated, the signaling pathways through which it alters cell functions and the ways in which it affects the CV system continue to emerge. As an indication of the ongoing intense interest in this area, over 3500 new citations for articles describing various aspects of the RAAS were identified on a Medline search extending from 2000 through 2004. Many of the articles contained novel information about previously unrecognized and/or incompletely understood aspects of this complex system. As a result of the ongoing assessment of the RAAS, traditional concepts defining its scope and function have had to be re-examined and (sometimes substantially) modified. Thus, the RAAS as we know it today, is very different from the system that was conceptualized 20–30 years ago when the first therapeutic agents designed to inhibit the effects of Angiotensin (Ang) II were introduced into clinical practice for the treatment of hypertension and heart failure.

As our view of the RAAS has evolved over time, it is not at all surprising that concepts regarding the role, pathways, and importance of this system in the process of cardiac remodeling also need to be modified. Moreover, our understanding of the mechanisms through which drugs that inhibit the RAAS affect cardiac remodeling, has also been transformed by the results of recent investigations. The goal of this chapter is to provide an overview of the role of the RAAS in cardiac remodeling, and to describe the effects (and mechanisms as we currently understand them) of drugs that block the RAAS in this process. Throughout the chapter, emphasis will be placed on new, emerging, and controversial aspects of the RAAS.

THE CLASSICAL SYSTEMIC AND THE LOCAL TISSUE-BASED RENIN–ANGIOTENSIN–ALDOSTERONE SYSTEMS

The initial description of the RAAS depicted a cascade of events that occurred mainly within the blood stream and which culminated in the production of Ang II (1). In this pathway (Fig. 1) the precursor molecule angiotensinogen is released from the liver into the circulation where it is enzymatically degraded by kidney-derived plasma renin activity (PRA) to form Ang I, an inactive decapeptide. When Ang I comes into contact with angiotensin-converting enzyme (ACE) at the endothelial cell surface, two additional amino acids are removed to form Ang II. These effector molecules are distributed via the circulation to tissues throughout the body, where they activate specific receptors on a wide variety of cells to produce changes in organ structure and function. Most of these effects are initiated by the binding of Ang II to its Type 1 (AT_1) receptor. Ang II has protean effects on the CV system that include peripheral, renal and glomerular efferent arterial vasoconstriction, salt and water retention, release of norepinephrine from sympathetic nerve endings, vascular hypertrophy, and endothelial cell dysfunction. In addition, Ang II has potent growth promoting effects on cardiac myocytes and fibroblasts that are directly related to the remodeling process.

Over time, however, it became clear that not all the effects of RAAS on the CV system could be easily related to the classical circulatory system. For instance, although the circulatory RAAS may be activated in the immediate postmyocardial infarction (MI) period (particularly when extensive myocardial injury results in cardiac dysfunction), it soon returns to baseline as hemodynamic perturbations begin to stabilize (2–5). Despite the relative quiescence of the circulatory RAAS, however, extensive changes in cardiac structure and function progress over time. Drugs that block the RAAS during this period, inhibit remodeling, delay progression to heart failure and improve survival in both experimental animal models and in human patients (6–10). Moreover, these effects transcend the ability of the drugs to alter the load on the heart. How then to explain the favorable effects of inhibiting the RAAS on cardiac remodeling and the natural history of diseases that cause myocardial damage?

Evidence that the genes for all components of the RAAS are expressed by cardiac cells (11–16) and that renin activity is taken up from the coronary circulation

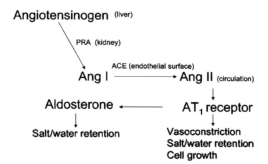

Figure 1 The classical renin-angiotensin aldosterone system. As first conceived, this system was contained largely within the circulation. Angiotensin II, the main effector molecule, was distributed by the circulatory system throughout the body where it interacted with its, Type 1 receptor to bring about changes in cell structure and function. *Abbreviations*: PRA, plasma renin activity; Ang, angiotensin; ACE, angiotensin converting enzyme; AT_1, Type1 Angiotensin II receptor.

(17), indicates the presence of a localized tissue-based RAAS in the heart. Moreover, it is now recognized that alternative pathways for the conversion of Ang I to Ang II exist in tissue. The best characterized and the most important of these pathways involves the enzyme chymase that is released by mast cells within the heart (18). The importance of this alternative pathway is supported by evidence that the majority of Ang II in the heart is generated locally from the cardiac RAAS rather than being transported to the heart by the circulatory RAAS (19,20).

The cardiac RAAS is regulated independent of the circulatory RAAS. Whereas the latter is mainly responsive to systemic factors related to changes in blood pressure and intravascular volume, the tissue-based RAAS responds to local factors such as myocardial stretch and growth factors that are present in the heart. Post-MI (and in other settings in which cardiac injury results in remodeling), the cardiac RAAS is upregulated and Ang II levels in the heart are increased (21–26). The Ang II that is generated and released directly within the heart, then functions in an autocrine/paracrine manner to influence cardiac structure and function (19,27). The significance of the cardiac RAAS in promoting remodeling is supported by the evidence that Ang II levels in the interstitial fluid of the heart are much higher than levels in plasma (28), and by the close correlation between structural changes in the heart and the extent of cardiac RAAS activation (16,21).

The use of cardiac-specific transgenic approaches has helped confirm the critical role of the cardiac RAAS in the remodeling process. Transgenic mice overexpressing angiotensinogen in cardiomyocytes have been shown to develop cardiac hypertrophy without fibrosis, despite the presence of normal blood pressure. Moreover, this effect could be prevented by the use of an angiotensin-converting enzyme inhibitor or by Ang II type 1 receptor blockade (29). In transgenic mice, in which the AT_1 receptor was under the control of the mouse alpha-myosin heavy chain promoter, cardiomyocyte specific overexpression of the AT_1 receptor also induced significant remodeling characterized by cardiac hypertrophy and fibrosis along with increased expression of atrial natriuretic factor (ANF) in the ventricle (30). These changes occurred despite the absence of any change in systolic blood pressure or heart rate and they resulted in premature death secondary to heart failure. Since most effects of Ang II in promoting cardiac remodeling are mediated through its AT_1 receptor, the role of the RAAS in post-MI cardiac remodeling has been studied in knockout mice that are null for the AT_1 receptor (31). When these mice undergo coronary artery ligation to induce a large MI, they develop significantly less left ventricular (LV) dilatation, less fibrosis in noninfarcted segments of myocardium, and better LV systolic function over time than wild type controls despite the fact that infarct sizes in the two study groups were equal. Moreover, post-MI survival is significantly improved in AT_1 knockout mice compared to wild-type controls (31).

CARDIAC REMODELING AND THE EFFECTS OF ANGIOTENSIN II IN THE HEART

Cardiac remodeling has been defined as "genome expression, molecular, cellular and interstitial changes that are manifested clinically as changes in size, shape and function of the heart after cardiac injury" (32). In the post-MI heart, there is evidence of cardiac myocyte hypertrophy and lengthening, ongoing myocyte loss due to apoptosis, and restructuring of the extracellular matrix (ECM). The latter leads to substantial increases in the amount of fibrous tissue, particularly in noninfarcted segments of the myocardium (33–42). As the heart remodels, cardiac myocytes demonstrate altered expression of genes that encode structural and functional proteins including changes in the relative abundance of sarcomeric proteins involved in contractile function such as the myosin heavy chain isoforms, proteins that help regulate calcium flux such as SERCA (Sarcoplasmic Endoplasmic Reticulum Calcium) and phospholamban, and markers of hypertrophy such as (ANF) (43–46). The net effect is a change in cardiac myocyte phenotype. Fibroblasts within the heart are stimulated to replicate, migrate, and to produce ECM proteins and "secondary" growth factors that advance the remodeling process through autocrine/paracrine effects (25,33,38,39,41,42,47–53).

Ang II promotes cardiac remodeling by several mechanisms. The peptide is a potent vasoconstrictor and it also enhances salt and water retention (both directly and by stimulating aldosterone release). The resultant elevations in peripheral vascular resistance and intravascular volume increase the pressure and volume load on the LV. Although the patterns of cell growth differ according to the nature of the load, increases in pressure and volume stimulate the development of myocardial hypertrophy. With pressure load, hypertrophy is concentric (increases in muscle mass but not chamber volume) whereas with volume load it is eccentric (increases in both mass and volume). Ang II also causes cardiac myocyte loss due to its direct toxic effects and by initiating programmed cell death or apoptosis (54). Myocyte loss, in turn, stimulates remodeling by increasing the load on remaining viable cells. In addition to altering load and causing myocyte loss, Ang II has direct growth promoting effects on cells within the heart. Thus, Ang II has been shown to cause cardiac myocyte hypertrophy and to stimulate changes in cardiac fibroblasts that play an important role in the remodeling process (55–59). The effects on fibroblasts include replication, migration, production of ECM proteins and production of substances such as the tissue inhibitors of metalloproteinase activity (TIMPs) that regulate the balance between ECM production and degradation (60). Finally, there is considerable evidence that many effects of Ang II may not be direct, but rather are related to the release of secondary growth factors that are then responsible for initiating changes in cell structure and function (47–52).

EFFECTS OF ANGIOTENSIN II ON CARDIAC FIBROBLASTS

It is well-recognized that the ECM component of the heart plays a vital role in maintaining cardiac structure and that changes in the amount and/or composition of the ECM can greatly influence cardiac function. Increased amounts of interstitial fibrosis adversely affect the ability of the heart to relax and fill at normal pressures (i.e., diastolic dysfunction). It also impairs transmission and coordination of the shortening of individual myocytes into synchronized chamber contraction (i.e., systolic dysfunction). Cardiac fibroblasts play a crucial role in formation and maintenance of the ECM. They produce ECM proteins such as fibronectin and various collagens and they regulate the breakdown of fibrous tissue through production of matrix metalloproteinases and TIMPs (34,60–65). They also produce growth factors that act in an autocrine and paracrine manner to stimulate cardiac myocyte hypertrophy (48,52,66).

Ang II stimulates fibroblast functions involved in post-MI remodeling including replication, migration and production of ECM proteins and secondary growth factors. In cardiac fibroblasts (and myocytes) most effects of Ang II are mediated through the AT_1 receptor (34,57,59,63,67). The AT_1 receptor is a member of the

7 transmembrane G-protein coupled receptor class that is traditionally associated with promotion of cell growth. Interestingly, the AT_1 receptor is more abundant on cardiac fibroblasts than on cardiac myocytes (68). In the post-MI heart where extensive ECM remodeling occurs, the density of AT_1 receptors on cardiac fibroblasts is increased (24–26,69). In studies performed in cultured neonatal rat cardiac fibroblasts, it was shown that the proinflammatory cytokines tumor necrosis factor-alpha (TNF-α) and interleukin-1beta (IL-1β) are potent inducers of AT_1 receptor upregulation, whereas other factors known to be present in the remodeling heart fail to increase AT_1 receptor expression (Fig. 2) (70). This effect of the proinflammatory cytokines appears to involve activation of the transcription factor nuclear factor-κB (NF-κB) since AT_1 receptor upregulation can be selectively blocked by interfering with NF-κB dissociation from IκB, a regulatory protein that restricts nuclear translocation which is essential to activate gene expression (71). As shown in Figure 3, there are strong temporal and spatial associations between the appearance of these cytokines in the post-MI heart and increases in AT_1 receptor density that are consistent with the possibility that the relationship may be causal in nature. Although the significance of increased AT_1 receptor density on cardiac fibroblasts during remodeling is uncertain, there is evidence that proinflammatory cytokine induced AT_1 upregulation enhances Ang II stimulated proline incorporation and production of TIMP-1 (Fig. 4) (60). These findings strongly suggest that upregulation of the AT_1 receptor contributes to the remodeling process by increasing the responsiveness of cardiac fibroblasts to the pro-fibrotic effects of Ang II (60,70).

EFFECTS OF ANGIOTENSIN II ON CARDIAC MYOCYTES

In contrast to cardiac fibroblasts, AT_1 receptors appear to be sparsely distributed on cardiac myocytes (47,68). These findings raise the question of whether Ang II causes

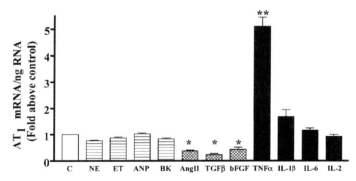

Figure 2 Cytokine induction of the AT_1 receptor gene in cardiac fibroblasts. Results of quantitative RT-PCR analysis of RNA from neonatal rat cardiac fibroblasts that had been exposed for 24 hours to a series of growth factors are depicted. A significant, nearly fivefold increase was noted in response to TNF-α. Subsequent experiments demonstrated a similar response to IL-1β after 12 hours of cytokine stimulation (71). *Abbreviations*: C, control; NE, norepinephrine; ET, endothelin; ANP, atrial natriuretic peptide; BK, bradykinin; TGFβ, transforming growth factor beta; bFGF, basic fibroblast growth factor; TNF-α, tumor necrosis factor alpha; IL-1β, interleukin 1 beta; IL-6, interleukin 6; IL-2, interleukin 2. $^*p \leq 0.05$; $^{**}p \leq 0.01$. *Source*: From Ref. 70.

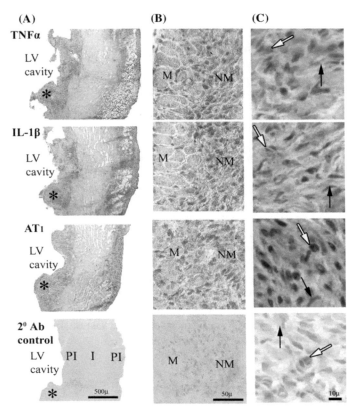

Figure 3 TNF-α, IL-1β and AT₁ receptor immunostaining is associated with non-myocytes in the post-MI heart. Immunostaining of the infarct region is shown in three magnifications as indicated by the scale bars. (**A**) Accumulation of TNF-α and IL-1β and increased density of the AT₁ receptor are shown at the border of the infarct that involves the free wall of the left ventricle. (**B**) Labeling of the proteins is associated only with the non-myocytes [*area magnified from (**A**)]. (**C**) Immunostaining is associated with macrophages (*white arrow*) and fibroblasts (*black arrow*). Immunostaining using only a secondary antibody results in absence of staining. *Abbreviations*: TNF-α, tumor necrosis factor alpha; IL-1β, interleukin 1 beta; AT₁, Type1 Angiotensin II receptor; Ab, antibody; LV, left ventricle; PI, peri-infarction zone; I, infarct zone; M, myocytes; NM, non-myocytes. (*See color insert.*)

cardiac myocyte hypertrophy by direct effects on the cell or whether other mechanisms are involved. Increased levels of Ang II in the circulation can elevate arterial pressure thereby increasing load on the heart. This increase in systolic wall stress is a potent stimulus for the development of concentric hypertrophy of the LV. However, as mentioned previously the effects of inhibiting the RAAS on LV hypertrophy transcend reductions in blood pressure. Several laboratories have shown that Ang II activation of the AT₁ receptor on cardiac fibroblasts results in the production of a variety of secondary growth factors including interleukin-6 (IL-6), leukemia inhibitory factor-1, transforming growth factor-beta and endothelin-1 (ET-1), all of which are believed to play a role in promoting cardiac remodeling (47,47,8,51,72–79). Evidence that Ang II induces hypertrophy of cardiac myocytes only when these cells are cocultured with cardiac fibroblasts, or when a conditioned medium from fibroblasts stimulated by Ang II is added to myocyte culture (Fig. 5), supports the notion that production of

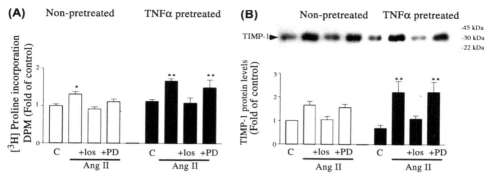

Figure 4 Effects of cytokine-induced increases in AT_1 receptor density on cardiac fibroblast functions. (**A**) shows the effects of cytokine induced AT_1 receptor upregulation in cardiac fibroblasts on [^{3}H] Proline incorporation as an indicator of extracellular matrix protein production. Upregulation of the AT_1 receptor two to threefold by prior exposure to TNF-α more than doubled the response to Ang II. In both untreated control and pretreated cells the effects of Ang II could be blocked by losartan indicating that the effect was mediated through the AT_1 receptor. (**B**) shows the effects of cytokine induced AT_1 receptor upregulation in cardiac fibroblasts on TIMP-1 production. As with proline incorporation, cytokine pre-treatment designed to upregulate the AT_1 receptor enhanced the response to Ang II. *Abbreviations*: TIMP-1, tissue inhibitor of metalloproteinase-1; los, losartan; PD, PD 123319; Ang II, angiotensin II. $^*p \le 0.05$; $^{**}p \le 0.01$. *Source*: Adapted from Ref. 60.

growth factors from fibroblasts is a critical component of the remodeling process (47,48) and that Ang II-mediated cardiac myocyte hypertrophy may not be a direct effect of the peptide (42,47,48,52).

CROSS-TALK BETWEEN THE RAAS AND OTHER SYSTEMS

Within the complex milieu of the heart in which there are numerous growth factors and cytokines, it is not at all surprising that there is a considerable cross-talk between various agents and that many of these interactions are believed to play an important role in cardiac remodeling (80). Some of the interactions between systems have been recognized for many years now whereas others have only recently been discovered. The ability of Ang II to enhance sympathetic nervous system activity by promoting presynaptic release of catecholamines and, conversely, evidence that catecholamines stimulate production and release of PRA from the kidney are good examples of the reciprocal nature of the extensive cross-talk between systems. Another example discussed earlier in this chapter is the effect of proinflammatory cytokines in upregulating AT_1 receptor density on cardiac fibroblasts, a phenotypic change that renders these cells more responsive to the pro-fibrotic effects of Ang II (60,70).

Based on the paucity of AT_1 receptors on cardiac myocytes (compared to their abundance on cardiac fibroblasts) (47,68) and evidence that Ang II stimulation of cardiac fibroblasts increases production of several important secondary growth factors, it is quite possible that Ang II stimulation of cardiac hypertrophy is dependent on interactions between fibroblasts and myocytes within the heart (42,47,48,52). Interestingly, myocytes may also contribute to the production of ECM proteins by cardiac fibroblasts. There is evidence that cultured cardiac fibroblasts express genes

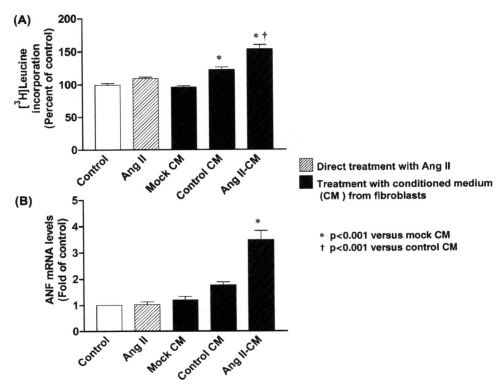

Figure 5 Angiotensin II stimulation of cardiac myocyte hypertrophy is mediated through the release of secondary factors from cardiac fibroblasts. In these experiments myocyte hypertrophy was assessed by measuring [³H]Leucine incorporation (**A**) and ANF gene expression (**B**). Direct exposure of neonatal rat cardiac myocytes to Ang II failed to increase either leucine incorporation of ANF gene expression. Compared to mock culture media, exposure of myocytes to culture media from adult rat cardiac fibroblasts significantly increased both measures of hypertrophy, indicating the release of growth factors from the fibroblasts. However, culture media taken from fibroblasts that had been stimulated with Ang II demonstrated a significantly greater increase in both leucine incorporation and ANF gene expression. *Abbreviations*: ANF, atrial natriuretic factor; Ang II, angiotensin II; CM, culture media.

for collagen production only when they are cocultured with cardiac myocytes, or culture media from these cells is added to the fibroblasts (81). The mechanism involved appears to be related to Ang II stimulation of the growth factor transforming growth factor-beta (TGF-β) in the myocytes, which in turn stimulates IL-6 in the fibroblasts. This cytokine, in particular, appears to be responsible for the effects of Ang II in promoting collagen gene expression.

ADDITIONAL COMPONENTS OF THE RENIN–ANGIOTENSIN– ALDOSTERONE SYSTEM

Aldosterone

Aldosterone is a mineralocorticoid that plays a major role in maintaining circulatory homeostasis by enhancing sodium and water retention in the kidney. This agent, however, has numerous other CV effects that could influence cardiac remodeling.

These include effects on endothelial cell function, vascular compliance and arterial pressure that would tend to promote remodeling by increasing load on the heart. In addition, cardiac fibroblasts possess mineralocorticoid receptors (82) whose activation by aldosterone stimulates remodeling, particularly the enhanced production of ECM proteins (63). Increased cardiac fibrosis is known to be an important component of cardiac hypertrophy and remodeling, and the amount of fibrosis in the remodeling heart can be extensive. Although fibrosis was traditionally thought to develop predominantly as replacement for myocytes that have undergone cell death, it is now apparent that cardiac fibrosis may develop as an independent process. The latter point was convincingly demonstrated in patients with evidence of extensive remodeling post-MI who presented with end-stage heart failure. In these patients, approximately two-third of the fibrosis found in the heart was located in areas that were distant from previous MI sites (36). Since fibrosis is known to be an important determinant of both diastolic and systolic functions of the heart (37,83–86), the presence of diffuse fibrosis in noninfarcted myocardium was considered to be the major cause of dysfunction of these severely failing hearts. Interstitial fibrosis is also an important structural substrate for the development of both conduction abnormalities that lead to dyssynchronous myocardial contraction, and cardiac arrhythmias. The former impair pump efficiency while the latter leads to further deterioration in cardiac function and can result in sudden cardiac death. In experimental animal models, both administration of aldosterone and overexpression of the mineralocorticoid receptor have been shown to promote cardiac fibrosis (87,88).

Conversely, blockade of aldosterone inhibits fibrosis and improves cardiac function (89,90). In the post-MI rat model, the addition of the aldosterone blocker eplerenone inhibited remodeling and it significantly potentiated the effects of ACE inhibition on LV fibrosis and cardiac hypertrophy (91).

The production and release of aldosterone from the adrenal gland has long been considered to be under the exclusive control of Ang II. There is evidence, however, that the levels of circulating aldosterone are regulated by additional factors including potassium, Adrenocorticotropic hormone (ACTH), catecholamines, potassium serotonin, ET and nitric oxide (NO) (92). The dissociation between Ang II levels and regulation of aldosterone has been demonstrated convincingly in patients with heart failure who were treated with either an ACE inhibitor, an Ang receptor blocker (ARB), or the combination of agents in the Randomized Evaluation of Strategies for Left Ventricular Dysfunction (RESOLVD) pilot study (93). Although higher doses of the drugs reduced serum aldosterone levels at 17 weeks after the initiation of therapy, by 43 weeks, the levels had returned to or exceeded baseline levels in most cases. Thus, therapies designed to treat RAAS activation in heart failure (though effective in inhibiting remodeling and improving the clinical course) are insufficient to suppress the effects of aldosterone.

Based on these considerations it is not surprising that blockade of aldosterone using receptor antagonists has beneficial effects on cardiac remodeling in patients with hypertension and with heart failure. In the 4E-Left Ventricular Hypertrophy Study, administration of eplerenone, an aldosterone receptor antagonist, to hypertensive patients with left ventricular hypertrophy (LVH) over a 9 month period, reduced blood pressure and LV mass to a similar extent as an ACE inhibitor (94). In the Randomized Aldactone Evaluation Study (RALES) trial, patients with advanced heart failure due to LV systolic dysfunction (who were already receiving an ACE inhibitor in the vast majority of cases) were treated in a double-blinded manner with the aldosterone antagonist spironolactone or placebo (95). Patients who received spironolactone experienced a 30% reduction in all-cause mortality—the primary endpoint in the study—and also a significant reduction in hospitalizations. In these patients, determination of cardiac collagen turnover by measuring serum levels of procollagen type III N-terminal peptide demonstrated that high levels of turnover were associated with a poor outcome. Patients randomized to placebo, continued to increase or had no change in procollagen type III N-terminal peptide after six months of followup. Patients who received spironolactone, however, demonstrated significant reduction in this marker of collagen turnover during that period (96). These findings are consistent with the concept that aldosterone receptor blockade limits excessive ECM turnover in heart failure patients and that this effect benefits the clinical course of heart failure patients.

Cardiac remodeling post-MI is one of the most important causes of progressive LV dysfunction leading to heart failure in the United States and other developed nations. The effect of aldosterone blockade in patients with post-MI LV dysfunction was evaluated in the Eplerenone Post-Acute Myocardial Infarction Heart Failure Efficacy and Survival Study Investigators (EPHESUS) trial (97). In this study, 3313 patients with evidence of depressed ejection fraction following an acute MI were randomized to receive either eplerenone or placebo in addition to optimal medical therapy which included the use of ACE inhibitors in 87% and beta blockers in 75%. During a mean followup of 16 months, eplerenone reduced all-cause mortality by 15% ($P = 0.008$), cardiovascular mortality by 17% ($P = 0.005$), and the combined endpoint of death from cardiovascular causes or hospitalization for cardiovascular events by 13% ($P = 0.002$). There was also a 21% reduction ($P = 0.03$) in the rate of sudden cardiac death. Although the effects of eplerenone on post-MI cardiac remodeling was not assessed in the EPHESUS trial, there is evidence that addition of an aldosterone receptor antagonist to an ACE inhibitor can prevent remodeling better than an ACE inhibitor alone (98).

The AT$_2$ Receptor

Most physiologic and pathophysiologic effects of Ang II are mediated through the interaction of this peptide with its AT$_1$ receptor. However, Ang II can also bind to its Type 2 (AT$_2$) receptor (99). Both the AT$_1$ and AT$_2$ receptor are 7 transmembrane domain G-protein coupled receptors of the class A rhodopsin-like family. There is, however, only ~30% sequence homology between them. These receptors have similar affinities for Ang II that are in the nanomolar range. The AT$_2$ receptor is highly expressed in developing fetuses, but its expression declines rapidly after birth (100). In the adult, it is expressed in very low abundance in the CV system (101). Under pathologic conditions, however, expression levels within the CV system may increase substantially (102–105). Since neither the signaling pathways nor the consequences on cell structure and function of AT$_2$ receptor activation have been fully delineated, its importance in maintaining CV homeostasis and its role in the pathogenesis of disease remain speculative and may be context specific (106).

Signaling pathways associated with the AT$_2$ receptor include activation of protein phosphatases (with resultant protein dephosphorylation), activation of phospholipase A$_2$, regulation of the bradykinin-NO-cGMP system, and sphingolipid-derived ceramide formation (107). Consistent with evidence indicating an association between the AT$_2$ receptor and protein phosphatase activity, initial reports of the effects of AT$_2$ receptor activation suggested that it mediated anti-growth effects in cardiac myocytes and fibroblasts (108,109). However, experiments in mice which were null for the AT_2 receptor gene, provided conflicting results with regard to its impact on the remodeling process (108,109). Of interest, is the observation that transgenic mice with ventricular-specific increases in AT_2 receptor density develop a phenotype of dilated cardiomyopathy characterized by ventricular dilatation, wall thinning, increased fibrosis and depressed contractile function (110). Although the mechanism for these abnormalities in ventricular structure and function are uncertain, increased apoptosis within the ventricular myocardium may be involved. Moreover, this response appeared to be "dose related," since the severity of the phenotype and extent of apoptosis is directly related to the level of AT_2 receptor gene expression. Transgenic animals also demonstrated evidence of activation of protein kinase C alpha and beta (PKC-α and -β), both of which have been implicated in cardiac remodeling and in the development of heart failure.

Another mechanism through which the AT_2 receptor might influence the response to Ang II is through formation of structural heterodimers with the AT_1 receptor (111). In this case, the interposition of the AT_2 molecule to the usual AT_1 homodimer appears to inhibit AT_1 receptor signaling and thus serves as a dominant negative or antagonist of AT_1 receptor function (112). Since both types of Ang receptors are expressed in tissue, the impact of this effect on AT_1 receptor signaling might well be related to the relative abundance of the two receptor subtypes. Thus, in settings where there is evidence of selective down regulation of AT_2 receptor gene expression, the change of ratio between the receptors could result in a lower likelihood of heterodimerization, an effect that would tend to favor an enhanced response of the AT_1 receptor to Ang II. Conversely, situations which result in an increase in the ratio of AT_2 to AT_1 would be expected to decrease AT_1 mediated effects of Ang II.

Although evidence that signaling from the AT_2 receptor plays a role in the development of cardiac hypertrophy in humans is limited, results from a recent study evaluating the association between a common intronic polymorphism of the AT_2 gene (-1332 G/A) and the presence of LV hypertrophy in hypertensive patients, provide some insights into this issue (112). The AT_2 gene consists of three exons and two introns, with the entire reading frame for the AT_2 receptor located on exon 3. The polymorphism site is located just downstream from exon 2 in a region that is important for transcriptional activity. Individuals with the G allele lack exon 2 of the AT_2 gene and this leads to less effective transcription. The investigators found an association between the G allele and LV hypertrophy, despite the fact that patients had been treated for their hypertension, suggesting that in the setting of hypertension the AT_2 receptor may mediate anti-hypertrophic effects.

Alternative Pathway Involving ACE2 and Ang-(1–7)

Recent evidence suggests that an alternative pathway of the RAAS involving ACE2 and Ang-(1–7) may play a role in regulating cardiac remodeling. ACE2 is a homologue of ACE (113,114). Human ACE2 cDNA predicts an 805 amino acid protein that has a 42% homology with the N-terminal catalytic domain of ACE. Although first identified in the heart, kidney and testes (113,114), the distribution of ACE2 now appears to be considerably more widespread than originally believed (115).

As shown in Figure 6, ACE2 functions as a carboxypeptidase that cleaves a single peptide from either the decapeptide Ang I to form Ang-(1–9), or the octapeptide Ang II to form Ang-(1–7). Evidence that the catalytic activity of ACE2 for Ang II is substantially greater than for Ang I suggests that its primary role is to convert Ang II to Ang-(1–7) (116). In contrast to ACE, ACE2 does not convert Ang I to Ang II nor

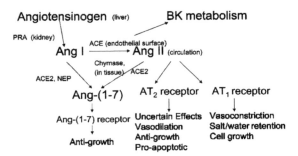

Figure 6 The evolving renin–angiotensin–aldosterone system. Over the past two decades there have been major advances in understanding the scope of this system. These include recognition of alternative tissue based pathways that generate angiotensin II from angiotensin I and the presence of the AT_2 receptor. In addition, ACE2, a homologue of ACE, provides an alternative pathway for processing angiotensin peptides. Angiotensin-(1–7), a by product of this pathway has effects on cells that appear to oppose those of Ang II and, thus, may play a role in modulating the effects of the renin angiotensin system during cardiac remodeling. *Abbreviations*: PRA, plasma renin activity; BK, bradykinin; Ang, angiotensin; ACE, angiotensin converting enzyme; NEP, neutral endopeptidase; AT_1, Type 1 angiotensin II receptor; AT_2, Type 2 angiotensin II receptor.

does it play a role in the metabolism of bradykinin (BK) (114,116). The activity of ACE2 is not affected by ACE inhibitors (113,114).

There is evidence that ACE2 is involved in the regulation of blood pressure and that it may help protect against increased blood pressure. In three different rat models of hypertension, the ACE2 gene maps to a quantitative trait locus on the X chromosome and in all three of these strains of rat, ACE2 levels are reduced (117). ACE2 has also been implicated in regulating cardiac function. Mice that are null for the ACE2 gene develop LV dilatation in association with severe cardiac contractile abnormalities (117). These abnormalities are both gender- and time-dependent since they are more severe in males and they progress with age. Evidence that cardiac dysfunction can be prevented in ACE2 null mice by the ablation of ACE expression suggests that the balance between ACE and ACE2 plays a critical role in regulating cardiac function (117).

Although the mechanism(s) by which ACE2 may exert these beneficial effects on the CV system is not yet defined, regulation of Ang II levels seems likely to be involved. Increases in ACE2 activity would be expected to reduce Ang II levels by both increasing its degradation and by reducing availability of Ang I, its immediate precursor. An additional mechanism might be through the production of Ang-(1–7), a peptide which has the property of being able to inhibit the C-terminal active site of ACE (118) and thus prevent conversion of Ang I to Ang II.

In addition to regulating Ang II levels, ACE2 may also affect cardiac remodeling by increasing levels of Ang-(1–7). This heptapeptide can be generated from either Ang II (by ACE2 activity) or from Ang I (in a single step via neutral endopeptidase activity or in two steps via successive ACE2 and ACE activity). Ang-(1–7) is believed to have beneficial effects on cardiac structure and function by virtue of its ability to inhibit the pressor, proliferative and cell growth–promoting effects of Ang II (119,120). In vascular smooth muscle cells (VSMCs), Ang-(1–7) inhibits Ang II-mediated cell growth and infusion of Ang-(1–7) after vascular injury inhibits neointimal growth through a mechanism independent of effects on heart rate or blood pressure (121). In the rat coronary ligation model of MI, Ang-(1–7) administration helps preserve cardiac function and limit increases in cardiac myocyte diameter (122). In the same experimental model, the administration of an ARB inhibits development of post-MI cardiac hypertrophy and this effect is associated with increases in cardiac ACE2 gene expression as well as in circulating Ang-(1–7) levels (123).

The pathways through which Ang-(1–7) favorably affects cellular structure and function have also been evaluated. In cultured rat aortic VSMCs, the anti-proliferative effects of Ang-(1–7) involve release of prostacyclin and prostacyclin-mediated production of cAMP, activation of cAMP-dependent protein kinase, and attenuation of ERK1 and ERK2 phosphorylation (124). Ang-(1–7) has also been shown to inhibit Ang II effects by stimulating production of NO and vasodilatory prostaglandins (118,124–126). As mentioned in the preceding paragraph, Ang-(1–7) might also counter Ang II effects by virtue of its ability to inhibit ACE activity (118), an effect that would reduce the production of Ang II as well as augment levels of anti-growth factors such as BK by blocking their breakdown. Although Ang-(1–7) might interfere with Ang II signaling through the AT_1 receptor by competing for receptor binding sites or by inducing AT_1 receptor internalization (127), the concentrations required for these effects are substantially higher than are required for the effects on cell functions making it unlikely that interference with Ang II binding is the major pathway involved. Thus, the available evidence suggests that Ang-(1–7) inhibits Ang II cellular effects by mechanisms that vary according to species, cell type and the particular function involved.

Ang-(1–7) can be generated locally in the myocardium of various species (28,128). It is formed from Ang I or Ang II in the interstitium of the canine LV and the concentration of Ang-(1–7) immunoreactivity is increased in the rat heart following induction of a large MI due to coronary artery ligation (129). In the failing human heart, Ang-(1–7) forming activity related to both neutral endopeptidase (NEP) and ACE2 is increased (130). In this setting, NEP has a preference for Ang I while ACE2 appears to have substrate preference for Ang II (130). Evidence that the major pathway for the generation of Ang-(1–7) in the human heart depends on the availability of Ang II as a substrate, suggests the importance of ACE2 in this process (130). Neither localization within the myocardium nor the cell type involved in the generation of Ang-(1–7) is known with certainty. In the post-MI rat heart, Ang-(1–7) immunoreactivity is associated with cardiac myocytes but not with interstitial cells or blood vessels and the most intense signal is noted at four weeks in the zone surrounding the replacement scar. Although the increase in Ang-(1–7) in the post-MI heart suggests that ACE2 is likely to be increased in this setting (131) in a manner analogous to what has been reported with ACE (21,132), whether or not this actually occurs is uncertain. In one report neither ACE nor ACE2 mRNA levels were increased in noninfarcted segments of LV in rats four weeks post-MI (123).

The results summarized above indicate that while some effects of Ang-(1–7) could be related to either inhibition of Ang II binding to the AT_1 receptor or inhibition of ACE activity (thereby reducing both Ang II generation and BK breakdown), most available information is consistent with the possibility of this heptapeptide having direct effects on cell functions. There has been, however, uncertainty regarding the identity and role of the specific cell surface receptor involved in mediating Ang-(1–7) effects. Santos et al. (133) reported the existence of a distinct Ang-(1–7) receptor encoded by the Mas protooncogene. Studies done in Mas null mice demonstrated that the absence of this G-protein coupled receptor abolishes the binding of Ang-(1–7) to kidney cells as well as the anti-diuretic effects of Ang-(1–7) after administration of an acute water load. The aortas of Mas deficient mice also fail to relax in response to Ang-(1–7). Transfection of Chinese hamster ovary cells with the Mas protoocogene increased arachidonic acid release. These findings suggest that Mas may be the functional receptor for Ang-(1–7) and that it mediates many of the functional effects of this peptide that have been noted in recent studies.

EFFECTS OF INHIBITING THE RAAS ON CARDIAC REMODELING

Some of the first evidence that blocking the RAAS could favorably affect cardiac remodeling, came from seminal studies performed by Mark and Janice Pfeffer and their colleagues (7,134). These investigators initially showed that the administration of the ACE inhibitor captopril to rats following an MI induced by coronary artery ligation substantially inhibited the remodeling process. They further demonstrated that the favorable effects of ACE inhibition on structural changes in the post-MI heart were associated with improved cardiac function and reduced mortality (7). Similar beneficial effects on post-MI remodeling in this and other experimental animal models have since appeared in the medical literature (135,136). Moreover, ACE inhibitors have been shown to inhibit remodeling that develops in response to other processes that stress the heart including pressure and/or volume overload and in genetic models of cardiomyopathy (137–139).

The mechanisms responsible for the favorable effects of ACE inhibition are less clear. Reduction in the hemodynamic load on the heart, related to a decrease in Ang II- mediated vasoconstriction, does not appear to be the major factor since there is evidence that the inhibition in remodeling with ACE inhibitors significantly exceeds that seen in control groups treated with vasodilating drugs administered to reduce arterial pressure to a similar range (140,141). Given the growth promoting effects of Ang II, it is logical to assume that protection against remodeling is related to a reduction in the production of this effector molecule. This interpretation is supported by studies demonstrating that remodeling that occurs post-MI as well as in response to increased load on the heart is inhibited by ARBs (142–144), agents that compete with Ang II for occupancy of the AT_1 receptor. In addition, there is evidence that post-MI remodeling is diminished in mice that are null for the AT_{1A} receptor (31). The situation, however, is complicated by the fact that ACE inhibition has other effects including blocking the breakdown of biologically active molecules such as BK. Since BK has been shown to have anti-growth properties, it is plausible that (at least some of) the effects on remodeling could be related to the increase in BK levels with ACE inhibition. Evidence that administration of a BK blocker in association with an ACE inhibitor diminishes protection against remodeling in some experimental models favors this possibility (145–148). Further support for this concept comes from studies evaluating remodeling in mice that are null for the BK B_2 receptor gene (149,150).

The effects of ACE inhibition on cardiac remodeling that have been reported in various experimental animal models have been replicated in patients with LV dysfunction. Extending the work done in the post-MI rat model, Pfeffer et al. demonstrated that ACE inhibition could favorably affect ventricular enlargement in patients with post-MI LV dysfunction (151). These encouraging findings provided the mechanistic justification for the Survival and Ventricular Enlargement (SAVE) study (8) and other trials (10,152–154) which tested the hypothesis that ACE inhibition could improve survival of heart failure and post-MI patients with a depressed ejection fraction (EF). The results of these studies have shown quite conclusively that in patients with post-MI dysfunction ACE inhibitors reduce mortality in the range of 20% to 25%. They also significantly reduce the rate of progression of heart failure in this population. Results from the SAVE trial also provided further important information that there was a strong association between the extent of cardiac remodeling in the first year post-MI (as assessed by an increase in LV volumes) and subsequent cardiovascular events (155). Consistent with the concept that ACE inhibitors favorably affect the clinical course of patients with post-MI LV dysfunction, captopril was shown to significantly inhibit increases in LV volume compared to the control group in the SAVE study.

Once heart failure develops, the remodeling process is by no means quiescent as evidenced by changes that occurred in the control population of the Studies of Left Ventricular Dysfunction (SOLVD) echocardiographic substudy (9). These patients, who were representative of the SOLVD population in general and were included in the trial based on the presence of symptomatic heart failure and evidence of LV ejection fraction ≤ 0.35, already demonstrated considerable LV dilatation and hypertrophy at the time of randomization in the study. Over the first 12 months, the placebo group experienced further significant increases in both volumes and mass. In contrast, patients randomized to the ACE inhibitor enalapril demonstrated significant attenuation of these adverse structural changes. The beneficial effects of ACE inhibition in preventing cardiac remodeling were related to clinical improvements since patients in SOLVD who received enalapril experienced improved survival, lower rates of hospitalization for heart failure, and reduced likelihood of progression from asymptomatic LV dysfunction to symptomatic heart failure.

Based on results demonstrating highly favorable effects of ACE inhibitors on the clinical course (including an overall 20–25% reduction in mortality) of post-MI LV dysfunction or heart failure, these drugs have become a cornerstone of therapy for patients with these conditions. The power of these drugs to affect clinical outcomes is so persuasive that administration of ACE inhibitors to heart failure patients has been adopted as a quality indicator to assess the adequacy of medical management. Although ACE inhibitors remain the first line of therapy for patients with post-MI LV dysfunction and heart failure, there is evidence that ARBs also have beneficial effects on the clinical course of these populations. The clinical efficacy of this approach has now been shown in the VALIANT study in which valsartan, an ARB, was found to have comparable effects on mortality and morbidity as the ACE inhibitor captopril in MI survivors with LV dysfunction (156). ARBs have also been shown to favorably affect the course of heart failure patients who were not receiving ACE inhibitors in a subgroup of the population in the Val-HeFT study and in the much larger population included in the CHARM-Alternative study (157,158). The favorable effects of ARB treatment on cardiac remodeling had already previously been demonstrated in the RESOLVD Pilot Study (93).

CONCLUSIONS

The role of the RAAS in the development of cardiac remodeling is now well established based on studies done in isolated, experimental animal models and human patients who had either remodeled their heart or were at risk of doing so. The clinical correlate of this investigative work is that the use of drugs such as ACE Inhibitors or ARBs that are designed to block the effects of RAAS activation are now the cornerstone of therapy for a variety of CV diseases that result in cardiac remodeling. What is not so clear, however, is exactly how these drugs bring about their beneficial effects on the remodeling process and clinical course. In particular, the role of the AT_2 receptor and the ACE2/Ang-(1–7) pathway require more work in order to understand how they influence the development of remodeling in various pathological settings. Although we can only anticipate what insights ongoing studies in these areas will bring, based on the trajectory of discovery in this area it is likely to be an interesting experience for clinicians and researchers alike.

REFERENCES

1. Reid IA, Morris BJ, Ganong WF. The renin-angiotensin system. Ann Rev Physiol 1978; 40:377–410.
2. Hodsman GP, Kohzuki M, Howes LG, Sumithran E, Tsunoda K, Johnston CI. Neurohumoral responses to chronic myocardial infarction in rats. Circulation 1988; 78(2):376–381.
3. Francis GS, Benedict C, Johnstone DE, Kirlin PC, Nicklas J, Liang CS, Kubo SH, Rudin-Toretsky E, Yusuf S. Comparison of neuroendocrine activation in patients with left ventricular dysfunction with and without congestive heart failure. A substudy of the Studies of Left Ventricular Dysfunction (SOLVD). Circulation 1990; 82(5):1724–1729.
4. Michel JB, Lattion AL, Salzmann JL, Cerol ML, Philippe M, Camilleri JP, Corvol P. Hormonal and cardiac effects of converting enzyme inhibition in rat myocardial infarction. Circ Res 1988; 62(4):641–650.
5. White M, Rouleau JL, Hall C, Arnold M, Harel F, Sirois P, Greaves S, Solomon S, Ajani U, Glynn R. Changes in vasoconstrictive hormones, natriuretic peptides, and left ventricular remodeling soon after anterior myocardial infarction. Am Heart J 2001; 142(6):1056–1064.
6. Ambrose J, Pribnow DG, Giraud GD, Perkins KD, Muldoon L, Greenberg BH. Angiotensin type 1 receptor antagonism with irbesartan inhibits ventricular hypertrophy and improves diastolic function in the remodeling post-myocardial infarction ventricle. J Cardiovasc Pharmacol 1999; 33(3):433–439.
7. Pfeffer JM, Pfeffer MA, Braunwald E. Influence of chronic captopril therapy on the infarcted left ventricle of the rat. Circ Res 1985; 57(1):84–95.
8. Pfeffer MA, Braunwald E, Moye LA, Basta L, Brown EJ Jr, Cuddy TE, Davis BR, Geltman EM, Goldman S, Flaker GC. Effect of captopril on mortality and morbidity in patients with left ventricular dysfunction after myocardial infarction. Results of the survival and ventricular enlargement trial. The SAVE Investigators [see comments]. N Engl J Med 1992; 327(10):669–677.
9. Greenberg B, Quinones MA, Koilpillai C, Limacher M, Shindler D, Benedict C, Shelton B. Effects of long-term enalapril therapy on cardiac structure and function in patients with left ventricular dysfunction. Results of the SOLVD echocardiography substudy. Circulation 1995; 91(10):2573–2581.
10. Effect of enalapril on survival in patients with reduced left ventricular ejection fractions and congestive heart failure. The SOLVD Investigators. N Engl J Med 1991; 325(5): 293–302.
11. Paul M, Wagner J, Dzau VJ. Gene expression of the renin-angiotensin system in human tissues. Quantitative analysis by the polymerase chain reaction. J Clin Invest 1993; 91(5):2058–2064.
12. Dostal DE, Baker KM. Angiotensin II stimulation of left ventricular hypertrophy in adult rat heart. Mediation by the AT1 receptor. Am J Hypertens 1992; 5(5 Pt 1): 276–280.

13. Katwa LC, Ratajska A, Cleutjens JP, Sun Y, Zhou G, Lee SJ, Weber KT. Angiotensin converting enzyme and kininase-II-like activities in cultured valvular interstitial cells of the rat heart. Cardiovasc Res 1995; 29(1):57–64.

14. Endo-Mochizuki Y, Mochizuki N, Sawa H, Takada A, Okamoto H, Kawaguchi H, Nagashima K, Kitabatake A. Expression of renin and angiotensin-converting enzyme in human hearts. Heart Vessels 1995; 10(6):285–293.

15. Zhang X, Dostal DE, Reiss K, Cheng W, Kajstura J, Li P, Huang H, Sonnenblick EH, Meggs LG, Baker KM. Identification and activation of autocrine renin-angiotensin system in adult ventricular myocytes. Am J Physiol 1995; 269(5 Pt 2):H1791–H1802.

16. Serneri GG, Boddi M, Cecioni I, Vanni S, Coppo M, Papa ML, Bandinelli B, Bertolozzi I, Polidori G, Toscano T. Cardiac angiotensin II formation in the clinical course of heart failure and its relationship with left ventricular function. Circ Res 2001; 88(9):961–968.

17. Muller DN, Fischli W, Clozel JP, Hilgers KF, Bohlender J, Menard J, Busjahn A, Ganten D, Luft FC. Local angiotensin II generation in the rat heart: role of renin uptake. Circ Res 1998; 82(1):13–20.

18. Urata H, Boehm KD, Philip A, Kinoshita A, Gabrovsek J, Bumpus FM, Husain A. Cellular localization and regional distribution of an angiotensin II-forming chymase in the heart. J Clin Invest 1993; 91(4):1269–1281.

19. Dostal DE, Baker KM. The cardiac renin-angiotensin system: conceptual, or a regulator of cardiac function? Circ Res 1999; 85(7):643–650.

20. van Kats JP, Danser AH, van Meegen JR, Sassen LM, Verdouw PD, Schalekamp MA. Angiotensin production by the heart: a quantitative study in pigs with the use of radiolabeled angiotensin infusions. Circulation 1998; 98(1):73–81.

21. Hirsch AT, Talsness CE, Schunkert H, Paul M, Dzau VJ. Tissue-specific activation of cardiac angiotensin converting enzyme in experimental heart failure. Circ Res 1991; 69(2):475–482.

22. Lindpaintner K, Lu W, Neidermajer N, Schieffer B, Just H, Ganten D, Drexler H. Selective activation of cardiac angiotensinogen gene expression in post- infarction ventricular remodeling in the rat. J Mol Cell Cardiol 1993; 25(2):133–143.

23. Meggs LG, Coupet J, Huang H, Cheng W, Li P, Capasso JM, Homcy CJ, Anversa P. Regulation of angiotensin II receptors on ventricular myocytes after myocardial infarction in rats. Circ Res 1993; 72(6):1149–1162.

24. Nio Y, Matsubara H, Murasawa S, Kanasaki M, Inada M. Regulation of gene transcription of angiotensin II receptor subtypes in myocardial infarction. J Clin Invest 1995; 95(1):46–54.

25. Sun Y, Cleutjens JP, Diaz-Arias AA, Weber KT. Cardiac angiotensin converting enzyme and myocardial fibrosis in the rat. Cardiovasc Res 1994; 28(9):1423–1432.

26. Lefroy DC, Wharton J, Crake T, Knock GA, Rutherford RA, Suzuki T, Morgan K, Polak JM, Poole-Wilson PA. Regional changes in angiotensin II receptor density after experimental myocardial infarction. J Mol Cell Cardiol 1996; 28(2):429–440.

27. De Mello WC, Danser AH. Angiotensin II and the heart: on the intracrine renin-angiotensin system. Hypertension 2000; 35(6):1183–1188.

28. Wei CC, Ferrario CM, Brosnihan KB, Farrell DM, Bradley WE, Jaffa AA, Dell'Italia LJ. Angiotensin peptides modulate bradykinin levels in the interstitium of the dog heart in vivo. J Pharmacol Exp Ther 2002; 1(300):324–329.

29. Mazzolai L, Nussberger J, Aubert JF, Brunner DB, Gabbiani G, Brunner HR, Pedrazzini T. Blood pressure-independent cardiac hypertrophy induced by locally activated renin-angiotensin system. Hypertension 1998; 31(6):1324–1330.

30. Paradis P, Dali-Youcef N, Paradis FW, Thibault G, Nemer M. Overexpression of angiotensin II type I receptor in cardiomyocytes induces cardiac hypertrophy and remodeling. Proc Natl Acad Sci USA 2000; 97(2):931–936.

31. Harada K, Sugaya T, Murakami K, Yazaki Y, Komuro I. Angiotensin II type 1A receptor knockout mice display less left ventricular remodeling and improved survival after myocardial infarction. Circulation 1999; 100(20):2093–2099.

32. Cohn JN, Ferrari R, Sharpe N. Cardiac remodeling–concepts and clinical implications: a consensus paper from an international forum on cardiac remodeling. Behalf of an International Forum on Cardiac Remodeling. J Am Coll Cardiol 2000; 35(3):569–582.

33. Anversa P, Beghi C, Kikkawa Y, Olivetti G. Myocardial infarction in rats. Infarct size, myocyte hypertrophy, and capillary growth. Circ Res 1986; 58(1):26–37.

34. Weber KT. Extracellular matrix remodeling in heart failure: a role for de novo angiotensin II generation. Circulation 1997; 96(11):4065–4082.

35. Lutgens E, Daemen MJ, de Muinck ED, Debets J, Leenders P, Smits JF. Chronic myocardial infarction in the mouse: cardiac structural and functional changes [see comments]. Cardiovasc Res 1999; 41(3):586–593.

36. Beltrami CA, Finato N, Rocco M, Feruglio GA, Puricelli C, Cigola E, Quaini F, Sonnenblick EH, Olivetti G, Anversa P, et al. Structural basis of end-stage failure in ischemic cardiomyopathy in humans. Circulation 1994; 89(1):151–163.

37. Litwin SE, Litwin CM, Raya TE, Warner AL, Goldman S. Contractility and stiffness of noninfarcted myocardium after coronary ligation in rats. Effects of chronic angiotensin converting enzyme inhibition. Circulation 1991; 83(3):1028–1037.

38. van Krimpen C, Smits JF, Cleutjens JP, Debets JJ, Schoemaker RG, Struyker Boudier HA, Bosman FT, Daemen MJ. DNA synthesis in the noninfarcted cardiac interstitium after left coronary artery ligation in the rat: effects of captopril. J Mol Cell Cardiol 1991; 23(11):1245–1253.

39. Cleutjens JP, Kandala JC, Guarda E, Guntaka RV, Weber KT. Regulation of collagen degradation in the rat myocardium after infarction. J Mol Cell Cardiol 1995; 27(6):1281–1292.

40. Sun Y, Weber KT. Infarct scar: a dynamic tissue. Cardiovasc Res 2000; 46(2):250–256.

41. Woodiwiss AJ, Tsotetsi OJ, Sprott S, Lancaster EJ, Mela T, Chung ES, Meyer TE, Norton GR. Reduction in myocardial collagen cross-linking parallels left ventricular dilatation in rat models of systolic chamber dysfunction. Circulation 2001; 103(1):155–160.

42. Manabe I, Shindo T, Nagai R. Gene expression in fibroblasts and fibrosis: involvement in cardiac hypertrophy. Circ Res 2002; 91(12):1103–1113.

43. Mercadier JJ, Lompre AM, Wisnewsky C, Samuel JL, Bercovici J, Swynghedauw B, Schwartz K. Myosin isoenzyme changes in several models of rat cardiac hypertrophy. Circ Res 1981; 49(2):525–532.

44. Hunter JJ, Chien KR. Signaling pathways for cardiac hypertrophy and failure. N Engl J Med 1999; 341(17):1276–1283.

45. Braunwald E, Bristow MR. Congestive heart failure: fifty years of progress. Circulation 2000; 102(20 Suppl 4):IV14–IV23.

46. Razeghi P, Young ME, Alcorn JL, Moravec CS, Frazier OH, Taegtmeyer H. Metabolic gene expression in fetal and failing human heart. Circulation 2001; 104(24):2923–2931.

47. Gray MO, Long CS, Kalinyak JE, Li HT, Karliner JS. Angiotensin II stimulates cardiac myocyte hypertrophy via paracrine release of TGF-beta 1 and endothelin-1 from fibroblasts. Cardiovasc Res 1998; 40(2):352–363.

48. Harada M, Itoh H, Nakagawa O, Ogawa Y, Miyamoto Y, Kuwahara K, Ogawa E, Igaki T, Yamashita J, Masuda I. Significance of ventricular myocytes and nonmyocytes interaction during cardiocyte hypertrophy: evidence for endothelin-1 as a paracrine hypertrophic factor from cardiac nonmyocytes. Circulation 1997; 96(10):3737–3744.

49. Cheng TH, Cheng PY, Shih NL, Chen IB, Wang DL, Chen JJ. Involvement of reactive oxygen species in angiotensin II-induced endothelin-1 gene expression in rat cardiac fibroblasts. J Am Cardiol 2003; 42(10):1845–1854.

50. ItO H, Hirata Y, Adachi S, Tanaka M, Tsujino M, Koike A, Nogami A, Murumo F, Hiroe M. Endothelin-1 is an autocrine/paracrine factor in the mechanism of angiotensin II-induced hypertrophy in cultured rat cardiomyocytes. J Clin Invest 1993; 92(1):398–403.

51. Campbell SE, Katwa LC. Angiotensin II stimulated expression of transforming growth factor- beta1 in cardiac fibroblasts and myofibroblasts. J Mol Cell Cardiol 1997; 29(7):1947–1958.

52. Kim NN, Villarreal FJ, Printz MP, Lee AA, Dillmann WH. Trophic effects of angiotensin II on neonatal rat cardiac myocytes are mediated by cardiac fibroblasts. Am J Physiol 1995; 269(3 Pt 1):E426–E437.

53. Bouzegrhane F, Thibault G. Is angiotensin II a proliferative factor of cardiac fibroblasts? Cardiovasc Res 2002; 53(2):304–312.

54. Horiuchi M, Hayashida W, Kambe T, Yamada T, Dzau VJ. Angiotensin type 2 receptor dephosphorylates Bcl-2 by activating mitogen-activated protein kinase phosphatase-1 and induces apoptosis. J Biol Chem 1997; 272(30):19022–19026.

55. Sadoshima J, Izumo S. Molecular characterization of angiotensin II–induced hypertrophy of cardiac myocytes and hyperplasia of cardiac fibroblasts. Critical role of the AT1 receptor subtype. Circ Res 1993; 73(3):413–423.

56. Baker KM, Aceto JF. Angiotensin II stimulation of protein synthesis and cell growth in chick heart cells. Am J Physiol 1990; 259(2 Pt 2):H610–H618.

57. Schorb W, Booz GW, Dostal DE, Conrad KM, Chang KC, Baker KM. Angiotensin II is mitogenic in neonatal rat cardiac fibroblasts. Circ Res 1993; 72(6):1245–1254.

58. Crawford DC, Chobanian AV, Brecher P. Angiotensin II induces fibronectin expression associated with cardiac fibrosis in the rat. Circ Res 1994; 74(4):727–739.

59. Sun Y, Ramires FJ, Zhou G, Ganjam VK, Weber KT. Fibrous tissue and angiotensin II. J Mol Cell Cardiol 1997; 29(8):2001–2012.

60. Peng J, Gurantz D, Tran V, Cowling RT, Greenberg BH. Tumor necrosis factor-alpha-induced AT1 receptor upregulation enhances angiotensin II-mediated cardiac fibroblast responses that favor fibrosis. Circ Res 2002; 91(12):1119–1126.

61. Spinale FG, Coker ML, Krombach SR, Mukherjee R, Hallak H, Houck WV. Matrix metalloproteinase inhibition during the development of congestive heart failure: effects on left ventricular dimensions and function. Circ Res 1999; 85(4):364–376.

62. Siwik DA, Pagano PJ, Colucci WS. Oxidative stress regulates collagen synthesis and matrix metalloproteinase activity in cardiac fibroblasts. Am J Physiol Cell Physiol 2001; 280(1):C53–C60.

63. Brilla CG, Zhou G, Matsubara L, Weber KT. Collagen metabolism in cultured adult rat cardiac fibroblasts: response to angiotensin II and aldosterone. J Mol Cell Cardiol 1994; 26(7):809–820.

64. Carver W, Nagpal ML, Nachtigal M, Borg TK, Terracio L. Collagen expression in mechanically stimulated cardiac fibroblasts. Circ Res 1991; 69(1):116–122.

65. Eghbali M, Blumenfeld OO, Seifter S, Buttrick PM, Leinwand LA, Robinson TF, Zern MA, Giambrone MA. Localization of types I, III and IV collagen mRNAs in rat heart cells by in situ hybridization. J Mol Cell Cardiol 1989; 21(1):103–113.

66. Yue P, Massie BM, Simpson PC, Long CS. Cytokine expression increases in non-myocytes from rats with postinfarction heart failure. Am J Physiol 1998; 275(1 Pt 2): H250–H258.

67. Brilla CG, Zhou G, Rupp H, Maisch B, Weber KT. Role of angiotensin II and prostaglandin E2 in regulating cardiac fibroblast collagen turnover. Am J Cardiol 1995; 76(13):8D–13D.

68. Villarreal FJ, Kim NN, Ungab GD, Printz MP, Dillmann WH. Identification of functional angiotensin II receptors on rat cardiac fibroblasts. Circulation 1993; 88(6):2849–2861.

69. Sun Y, Weber KT. Angiotensin II receptor binding following myocardial infarction in the rat. Cardiovasc Res 1994; 28(11):1623–1628.

70. Gurantz D, Cowling RT, Villarreal FJ, Greenberg BH. Tumor necrosis factor-alpha upregulates angiotensin II type 1 receptors on cardiac fibroblasts. Circ Res 1999; 85(3):272–279.

71. Cowling RT, Gurantz D, Peng J, Dillmann WH, Greenberg BH. Transcription factor NF-kappa B is necessary for up-regulation of type 1 angiotensin II receptor mRNA

in rat cardiac fibroblasts treated with tumor necrosis factor-alpha or interleukin-1 beta. J Biol Chem 2002; 277(8):5719–5724.

72. Fujisaki H, Ito H, Hirata Y, Tanaka M, Hata M, Lin M, Adachi S, Akimoto H, Marumo F, Hiroe M. Natriuretic peptides inhibit angiotensin II-induced proliferation of rat cardiac fibroblasts by blocking endothelin-1 gene expression. J Clin Invest 1995; 96(2):1059–1065.

73. Sano M, Fukuda K, Sato T, Kawaguchi H, Suematsu M, Matsuda S, Koyasu S, Matsui H, Yamauchi-Takihara K, Harada M. ERK and p38 MAPK, but not NF-kappaB, are critically involved in reactive oxygen species-mediated induction of IL-6 by angiotensin II in cardiac fibroblasts. Circ Res 2001; 89(8):661–669.

74. Sano M, Fukuda K, Kodama H, Pan J, Saito M, Matsuzaki J, Takahashi T, Makino S, Kato T, Ogawa S. Interleukin-6 family of cytokines mediate angiotensin II-induced cardiac hypertrophy in rodent cardiomyocytes. J Biol Chem 2000; 275(38):29717–29723.

75. Piacentini L, Gray M, Honbo NY, Chentoufi J, Bergman M, Karliner JS. Endothelin-1 stimulates cardiac fibroblast proliferation through activation of protein kinase C. J Mol Cell Cardiol 2000; 32(4):565–576.

76. Wang F, Trial J, Diwan A, Gao F, Birdsall H, Entman M, Hornsby P, Sivasubramaniam N, Mann D. Regulation of cardiac fibroblast cellular function by leukemia inhibitory factor. J Mol Cell Cardiol 2002; 34(10):1309–1316.

77. Yue TL, Gu JL, Wang C, Reith AD, Lee JC, Mirabile RC, Kreutz R, Wang Y, Maleeff B, Parsons AA, Ohlstein EH. Extracellular signal-regulated kinase plays an essential role in hypertrophic agonists, endothelin-1 and phenylephrine-induced cardiomyocyte hypertrophy. J Biol Chem 2000; 275(48):37895–37901.

78. Kodama H, Fukuda K, Pan J, Makino S, Baba A, Hori S, Ogawa S. Leukemia inhibitory factor, a potent cardiac hypertrophic cytokine, activates the JAK/STAT pathway in rat cardiomyocytes. Circ Res 1997; 81(5):656–663.

79. Murata M, Fukuda K, Ishida H, Miyoshi S, Koura T, Kodama H, Nakazawa HK, Ogawa S. Leukemia inhibitory factor, a potent cardiac hypertrophic cytokine, enhances L-type Ca^{2+} current and $(Ca^{2+})i$ transient in cardiomyocytes. J Mol Cell Cardiol 1999; 31(1):237–245.

80. Dostal DE. Regulation of cardiac collagen: angiotensin and cross-talk with local growth factors. Hypertension 2001; 37(3):841–844.

81. Sarkar S, Vellaichamy E, Young D, Sen S. Influence of cytokines and growth factors in Ang II-mediated collagen upregulation by fibroblasts in rats: role of myocytes. Am J Physiol Heart Circ Physiol 2004.

82. Arriza JL, Weinberger C, Cerelli G, Glaser TM, Handelin BL, Housman DE, Evans RM. Cloning of human mineralocorticoid receptor complementary DNA: structural and functional kinship with the glucocorticoid receptor. Science 1987; 237(4812): 268–275.

83. Carroll EP, Janicki JS, Pick R, Weber KT. Myocardial stiffness and reparative fibrosis following coronary embolisation in the rat. Cardiovasc Res 1989; 23(8):655–661.

84. Jalil JE, Doering CW, Janicki JS, Pick R, Shroff SG, Weber KT. Fibrillar collagen and myocardial stiffness in the intact hypertrophied rat left ventricle. Circ Res 1989; 64(6):1041–1050.

85. Ohkubo N, Matsubara H, Nozawa Y, Mori Y, Murasawa S, Kijima K, Maruyama K, Masaki H, Tsutumi Y, Shibazaki Y. Angiotensin type 2 receptors are reexpressed by cardiac fibroblasts from failing myopathic hamster hearts and inhibit cell growth and fibrillar collagen metabolism. Circulation 1997; 96(11):3954–3962.

86. Volders PG, Willems IE, Cleutjens JP, Arends JW, Havenith MG, Daemen MJ. Interstitial collagen is increased in the noninfarcted human myocardium after myocardial infarction. J Mol Cell Cardiol 1993; 25(11):1317–1323.

87. Ramires FJ, Sun Y, Weber KT. Myocardial fibrosis associated with aldosterone or angiotensin II administration: attenuation by calcium channel blockade. J Mol Cell Cardiol 1998; 30(3):475–483.

88. Le Menuet D, Isnard R, Bichara M, Viengchareun S, Muffat-Joly M, Walker F, Zennaro MC, Lombes M. Alteration of cardiac and renal functions in transgenic mice overexpressing human mineralocorticoid receptor. J Biol Chem 2001; 276(42): 38911–38920.

89. Zannad F, Dousset B, Alla F. Treatment of congestive heart failure: interfering the aldosterone-cardiac extracellular matrix relationship. Hypertension 2001; 38(5): 1227–1232.

90. Delyani JA, Robinson EL, Rudolph AE. Effect of a selective aldosterone receptor antagonist in myocardial infarction. Am J Physiol Heart Circ Physiol 2001; 281(2): H647–H654.

91. Fraccarollo D, Galuppo P, Hildemann S, Christ M, Ertl G, Bauersachs J. Additive improvement of left ventricular remodeling and neurohormonal activation by aldosterone receptor blockade with eplerenone and ACE inhibition in rats with myocardial infarction. J Am Coll Cardiol 2003; 42(9):1666–1673.

92. Lombes M, Farman N, Bonvalet JP, Zennaro MC. Identification and role of aldosterone receptors in the cardiovascular system. Ann Endocrinol (Paris) 2000; 61(1):41–46.

93. McKelvie RS, Yusuf S, Pericak D, Avezum A, Burns RJ, Probstfield J, Tsuyuki RT, White M, Rouleau J, Latini R. Comparison of candesartan, enalapril, and their combination in congestive heart failure: randomized evaluation of strategies for left ventricular dysfunction (RESOLVD) pilot study. The RESOLVD pilot study investigators. Circulation 1999; 100(10):1056–1064.

94. Pitt B, Reichek N, Willenbrock R, Zannad F, Phillips RA, Roniker B, Kleiman J, Krause S, Burns D, Williams GH. Effects of eplerenone, enalapril, and eplerenone/enalapril in patients with essential hypertension and left ventricular hypertrophy: the 4E-left ventricular hypertrophy study. Circulation 2003; 108(15):1831–1838.

95. Pitt B, Zannad F, Remme WJ, Cody R, Castaigne A, Perez A, Palensky J, Wittes J. The effect of spironolactone on morbidity and mortality in patients with severe heart failure. Randomized Aldactone Evaluation Study Investigators. N Engl J Med 1999; 341(10):709–717.

96. Zannad F, Alla F, Dousset B, Perez A, Pitt B. Limitation of excessive extracellular matrix turnover may contribute to survival benefit of spironolactone therapy in patients with congestive heart failure: insights from the randomized aldactone evaluation study (RALES). Rales Investigators. Circulation 2000; 102(22):2700–2706.

97. Pitt B, Remme W, Zannad F, Neaton J, Martinez F, Roniker B, Bittman R, Hurley S, Kleiman J, Gatlin M. Eplerenone post-acute myocardial infarction heart failure efficacy and survival study investigators. Eplerenone, a selective aldosterone blocker, in patients with left ventricular dysfunction after myocardial infarction. N Engl J Med 2003; 348(14):1309–1321.

98. Hayashi M, Tsutamoto T, Wada A, Tsutsui T, Ishii C, Ohno K, Fujii M, Taniguchi A, Hamatani T, Nozato Y. Immediate administration of mineralocorticoid receptor antagonist spironolactone prevents post-infarct left ventricular remodeling associated with suppression of a marker of myocardial collagen synthesis in patients with first anterior acute myocardial infarction. Circulation 2003; 107(20):2559–2565.

99. Mukoyama M, Nakajima M, Horiuchi M, Sasamura H, Pratt RE, Dzau VJ. Expression cloning of type 2 angiotensin II receptor reveals a unique class of seven-transmembrane receptors. J Biol Chem 1993; 268(33):24539–24542.

100. Shanmugam S, Lenkei ZG, Gasc JM, Corvol PL, Llorens-Cortes CM. Ontogeny of angiotensin II type 2 (AT2) receptor mRNA in the rat. Kidney Int 1995; 47(4): 1095–1100.

101. Wang ZQ, Moore AF, Ozono R, Siragy HM, Carey RM. Immunolocalization of subtype 2 angiotensin II (AT2) receptor protein in rat heart. Hypertension 1998; 32(1): 78–83.

102. Adachi Y, Saito Y, Kishimoto I, Harada M, Kuwahara K, Takahashi N, Kawakami R, Nakanishi M, Nakagawa Y, Tanimoto K. Angiotensin II type 2 receptor deficiency

exacerbates heart failure and reduces survival after acute myocardial infarction in mice. Circulation 2003; 107(19):2406–2408.

103. Lopez JJ, Lorell BH, Ingelfinger JR, Weinberg EO, Schunkert H, Diamant D, Tang SS. Distribution and function of cardiac angiotensin AT1- and AT2-receptor subtypes in hypertrophied rat hearts. Am J Physiol 1994; 267(2 Pt 2):H844–H852.

104. Tsutsumi Y, Matsubara H, Ohkubo N, Mori Y, Nozawa Y, Murasawa S, Kijima K, Maruyama K, Masaki H, Moriguchi Y. Angiotensin II type 2 receptor is upregulated in human heart with interstitial fibrosis, and cardiac fibroblasts are the major cell type for its expression. Circ Res 1998; 83(10):1035–1046.

105. Wharton J, Morgan K, Rutherford RA, Catravas JD, Chester A, Whitehead BF, De Leval MR, Yacoub MH, Polak JM. Differential distribution of angiotensin AT2 receptors in the normal and failing human heart. J Pharmacol Exp Ther 1998; 284(1): 323–336.

106. Booz GW. Cardiac angiotensin AT2 receptor: what exactly does it do? Hypertension 2004; 43(6):1162–1163.

107. Berry C, Touyz R, Dominiczak AF, Webb RC, Johns DG. Angiotensin receptors: signaling, vascular pathophysiology, and interactions with ceramide. Am J Physiol Heart Circ Physiol 2001; 281(6):H2337–H2365.

108. Bartunek J, Weinberg EO, Tajima M, Rohrbach S, Lorell BH. Angiotensin II type 2 receptor blockade amplifies the early signals of cardiac growth response to angiotensin II in hypertrophied hearts. Circulation 1999; 99(1):22–25.

109. van Kesteren CA, van Heugten HA, Lamers JM, Saxena PR, Schalekamp MA, Danser AH. Angiotensin II-mediated growth and antigrowth effects in cultured neonatal rat cardiac myocytes and fibroblasts. J Mol Cell Cardiol 1997; 29(8):2147–2157.

110. Yan X, Price RL, Nakayama M, Ito K, Schuldt AJ, Manning WJ, Sanbe A, Borg TK, Robbins J, Lorell BH. Ventricular-specific expression of angiotensin II type 2 receptors causes dilated cardiomyopathy and heart failure in transgenic mice. Am J Physiol Heart Circ Physiol 2003; 285(5):H2179–H2187.

111. Breitwieser GE. G protein-coupled receptor oligomerization: implications for G protein activation and cell signaling. Circ Res 2004; 94(1):17–27.

112. Alfakih K, Maqbool A, Sivananthan M, Walters K, Bainbridge G, Ridgway J, Balmforth AJ, Hall AS. Left ventricle mass index and the common, functional, X-linked angiotensin II type-2 receptor gene polymorphism (-1332 G/A) in patients with systemic hypertension. Hypertension 2004; 43(6):1189–1194.

113. Tipnis SR, Hooper NM, Hyde R, Karran E, Christie G, Turner AJ. A human homolog of angiotensin-converting enzyme–Cloning and functional expression as a captopril-insensitive carboxypeptidase. Journal of Biological Chemistry 2000; 275(43):33238–33243.

114. Donoghue M, Hsieh F, Baronas E, Godbout K, Gosselin M, Stagliano N, Donovan M, Woolf B, Robison K, Jeyaseelan R, Breitbart RE, Acton S. A novel angiotensin-converting enzyme-related carboxypeptidase (ACE2) converts angiotensin I to angiotensin 1–9. Circ Res 2000; 87(5):E1–E9.

115. Harmer D, Gilbert M, Borman R, Clark KL. Quantitative mRNA expression profiling of ACE 2, a novel homologue of angiotensin converting enzyme. FEBS Lett 2002; 532(1-2):107–110.

116. Vickers C, Hales P, Kaushik V, Dick L, Gavin J, Tang J, Godbout K, Parsons T, Baronas E, Hsieh F. Hydrolysis of biological peptides by human angiotensin-converting enzyme-related carboxypeptidase. J Biol Chem 2002; 277(17):14838–14843.

117. Crackower MA, Sarao R, Oudit GY, Yagil C, Kozieradzki I, Scanga SE, Oliveira-dos-Santos AJ, da Costa J, Zhang L, Pei Y. Angiotensin-converting enzyme 2 is an essential regulator of heart function. Nature 2002; 417(6891):822–828.

118. Li P, Chappell MC, Ferrario CM, Brosnihan KB. Angiotensin-(1–7) augments bradykinin-induced vasodilation by competing with ACE and releasing nitric oxide. Hypertension 1997; 29(1 Pt 2):394–400.

119. Freeman EJ, Chisolm GM, Ferrario CM, Tallant EA. Angiotensin-(1–7) inhibits vascular smooth muscle cell growth. Hypertension 1996; 28(1):104–108.

120. Zhu Z, Zhong J, Zhu S, Liu D, Van Der GM, Tepel M. Angiotensin-(1–7) inhibits angiotensin II-induced signal transduction. J Cardiovasc Pharmacol 2002; 40(5):693–700.

121. Strawn WB, Ferrario CM, Tallant EA. Angiotensin-(1–7) reduces smooth muscle growth after vascular injury. Hypertension 1999; 33(1 Pt 2):207–211.

122. Loot AE, Roks AJ, Henning RH, Tio RA, Suurmeijer AJ, Boomsma F, van Gilst WH. Angiotensin-(1–7) attenuates the development of heart failure after myocardial infarction in rats. Circulation 2002; 105(13):1548–1550.

123. Ishiyama Y, Gallagher PE, Averill DB, Tallant EA, Brosnihan KB, Ferrario CM. Upregulation of angiotensin-converting enzyme 2 after myocardial infarction by blockade of angiotensin II receptors. Hypertension 2004; 43(5):970–976.

124. Tallant EA, Clark MA. Molecular mechanisms of inhibition of vascular growth by angiotensin-(1–7). Hypertension 2003; 42(4):574–579.

125. Brosnihan KB, Li P, Ferrario CM. Angiotensin-(1-7) dilates canine coronary arteries through kinins and nitric oxide. Hypertension 1996; 27(3 Pt 2):523–528.

126. Muthalif MM, Benter IF, Uddin MR, Harper JL, Malik KU. Signal transduction mechanisms involved in angiotensin-(1–7)-stimulated arachidonic acid release and prostanoid synthesis in rabbit aortic smooth muscle cells. J Pharmacol Exp Ther 1998; 284(1):388–398.

127. Clark MA, Diz DI, Tallant EA. Angiotensin-(1–7) downregulates the angiotensin II type 1 receptor in vascular smooth muscle cells. Hypertension 2001; 37(4):1141–1146.

128. Zisman LS, Meixell GE, Bristow MR, Canver CC. Angiotensin-(1–7) formation in the intact human heart: in vivo dependence on angiotensin II as substrate. Circulation 2003; 108(14):1679–1681.

129. Averill DB, Ishiyama Y, Chappell MC, Ferrario CM. Cardiac angiotensin-(1–7) in ischemic cardiomyopathy. Circulation 2003; 108(17):2141–2146.

130. Zisman LS, Keller RS, Weaver B, Lin Q, Speth R, Bristow MR, Canver CC. Increased angiotensin-(1–7)-forming activity in failing human heart ventricles: evidence for upregulation of the angiotensin-converting enzyme Homologue ACE2. Circulation 2003; 108(14):1707–1712.

131. Burrell LM, Johnston CI, Tikellis C, Cooper ME. ACE2, a new regulator of the renin-angiotensin system. Trends Endocrinol Metab 2004; 15(4):166–169.

132. Duncan AM, Burrell LM, Kladis A, Campbell DJ. Effects of angiotensin-converting enzyme inhibition on angiotensin and bradykinin peptides in rats with myocardial infarction. J Cardiovasc Pharmacol 1996; 28(6):746–754.

133. Santos RAS, Silva ACSE, Maric C, Silva DMR, Machado RP, de Buhr I. Angiotensin-(1–7) is an endogenous ligand for the G protein-coupled receptor Mas. Proceedings of the National Academy of Sciences of the United States of America 2003; 100(14):8258–8263.

134. Pfeffer MA, Pfeffer JM, Steinberg C, Finn P. Survival after an experimental myocardial infarction: beneficial effects of long-term therapy with captopril. Circulation 1985; 72(2):406–412.

135. Gay RG. Early and late effects of captopril treatment after large myocardial infarction in rats. J Am Coll Cardiol 1990; 16(4):967–977.

136. McDonald KM, Rector T, Carlyle PF, Francis GS, Cohn JN. Angiotensin-converting enzyme inhibition and beta-adrenoceptor blockade regress established ventricular remodeling in a canine model of discrete myocardial damage. J Am Coll Cardiol 1994; 24(7):1762–1768.

137. Qing G, Garcia R. Chronic captopril and losartan (DuP 753) administration in rats with high-output heart failure. Am J Physiol 1992; 263(3 Pt 2):H833–H840.

138. Weinberg EO, Schoen FJ, George D, Kagaya Y, Douglas PS, Litwin SE, Schunkert H, Benedict CR, Lorell BH. Angiotensin-converting enzyme inhibition prolongs survival and modifies the transition to heart failure in rats with pressure overload hypertrophy due to ascending aortic stenosis. Circulation 1994; 90(3):1410–1422.

139. Haleen SJ, Weishaar RE, Overhiser RW, Bousley RF, Keiser JA, Rapundalo SR, Taylor DG. Effects of quinapril, a new angiotensin converting enzyme inhibitor, on left ventricular failure and survival in the cardiomyopathic hamster. Hemodynamic, morphological, and biochemical correlates. Circ Res 1991; 68(5):1302–1312.

140. Kim S, Ohta K, Hamaguchi A, Yukimura T, Miura K, Iwao H. Angiotensin II induces cardiac phenotypic modulation and remodeling in vivo in rats. Hypertension 1995; 25(6):1252–1259.

141. Raya TE, Gay RG, Aguirre M, Goldman S. Importance of venodilatation in prevention of left ventricular dilatation after chronic large myocardial infarction in rats: a comparison of captopril and hydralazine. Circ Res 1989; 64(2):330–337.

142. Schieffer B, Wirger A, Meybrunn M, Seitz S, Holtz J, Riede UN, Drexler H. Comparative Effects of Chronic Angiotensin-Converting Enzyme-Inhibition and Angiotensin-Ii Type-1 Receptor Blockade on Cardiac Remodeling After Myocardial-Infarction in the Rat. Circulation 1994; 89(5):2273–2282.

143. Kojima M, Shiojima I, Yamazaki T, Komuro I, Zou Z, Wang Y, Mizuno T, Ueki K, Tobe K, Kadowaki T. Angiotensin II receptor antagonist TCV-116 induces regression of hypertensive left ventricular hypertrophy in vivo and inhibits the intracellular signaling pathway of stretch-mediated cardiomyocyte hypertrophy in vitro. Circulation 1994; 89(5):2204–2211.

144. Ruzicka M, Yuan B, Harmsen E, Leenen FH. The renin-angiotensin system and volume overload-induced cardiac hypertrophy in rats. Effects of angiotensin converting enzyme inhibitor versus angiotensin II receptor blocker. Circulation 1993; 87(3):921–930.

145. Farhy RD, Ho KL, Carretero OA, Scicli AG. Kinins mediate the antiproliferative effect of ramipril in rat carotid artery. Biochem Biophys Res Commun 1992; 182(1):283–288.

146. Linz W, Scholkens BA. A specific B2-bradykinin receptor antagonist HOE 140 abolishes the antihypertrophic effect of ramipril. Br J Pharmacol 1992; 105(4):771–772.

147. McDonald KM, Mock J, D'Aloia A, Parrish T, Hauer K, Francis G, Stillman A, Cohn JN. Bradykinin antagonism inhibits the antigrowth effect of converting enzyme inhibition in the dog myocardium after discrete transmural myocardial necrosis. Circulation 1995; 91(7):2043–2048.

148. Liu YH, Yang XP, Sharov VG, Nass O, Sabbah HN, Peterson E, Carretero OA. Effects of angiotensin-converting enzyme inhibitors and angiotensin II type 1 receptor antagonists in rats with heart failure. Role of kinins and angiotensin II type 2 receptors. J Clin Invest 1997; 99(8):1926–1935.

149. Emanueli C, Maestri R, Corradi D, Marchione R, Minasi A, Tozzi MG, Salis MB, Straino S, Capogrossi MC, Olivetti G. Dilated and failing cardiomyopathy in bradykinin B(2) receptor knockout mice. Circulation 1999; 100(23):2359–2365.

150. Yang XP, Liu YH, Mehta D, Cavasin MA, Shesely E, Xu J, Liu F, Carretero OA. Diminished cardioprotective response to inhibition of angiotensin-converting enzyme and angiotensin II type 1 receptor in B(2) kinin receptor gene knockout mice. Circ Res 2001; 88(10):1072–1079.

151. Pfeffer MA, Lamas GA, Vaughan DE, Parisi AF, Braunwald E. Effect of captopril on progressive ventricular dilatation after anterior myocardial infarction. N Engl J Med 1988; 319(2):80–86.

152. Effects of enalapril on mortality in severe congestive heart failure. Results of the Cooperative North Scandinavian Enalapril Survival Study (CONSENSUS). The CONSENSUS Trial Study Group. N Engl J Med 1987; 316(23):1429–1435.

153. Ambrosioni E, Borghi C, Magnani B. The effect of the angiotensin-converting-enzyme inhibitor zofenopril on mortality and morbidity after anterior myocardial infarction. The Survival of Myocardial Infarction Long-Term Evaluation (SMILE) Study Investigators. N Engl J Med 1995; 332(2):80–85.

154. Effect of ramipril on mortality and morbidity of survivors of acute myocardial infarction with clinical evidence of heart failure. The Acute Infarction Ramipril Efficacy (AIRE) Study Investigators. Lancet 1993; 342(8875):821–828.

155. St John SM, Pfeffer MA, Plappert T, Rouleau JL, Moye LA, Dagenais GR, Lamas GA, Klein M, Sussex B, Goldman S. Quantitative two-dimensional echocardiographic measurements are major predictors of adverse cardiovascular events after acute myocardial infarction. The protective effects of captopril. Circulation 1994; 89(1):68–75.

156. Pfeffer MA, McMurray JJ, Velazquez EJ, Rouleau JL, Kober L, Maggioni AP, Solomon SD, Swedberg K, Van de Werf F, White H. Valsartan in Acute Myocardial Infarction Trial Investigators. Valsartan, captopril, or both in myocardial infarction complicated by heart failure, left ventricular dysfunction, or both. N Engl J Med 2003; 349(20):1893–1906.

157. Maggioni AP, Anand I, Gottlieb SO, Latini R, Tognoni G, Cohn JN. Effects of valsartan on morbidity and mortality in patients with heart failure not receiving angiotensin-converting enzyme inhibitors. J Am Coll Cardiol 2002; 40(8):1414–1421.

158. Granger CB, McMurray JJ, Yusuf S, Held P, Michelson EL, Olofsson B, Ostergren J, Pfeffer MA, Swedberg K. CHARM Investigators and Effects of candesartan in patients with chronic heart failure and reduced left-ventricular systolic function intolerant to angiotensin-converting-enzyme inhibitors: the CHARM-Alternative trial. Lancet 2003; 362(9386):772–776.

13

Adrenergic Receptor Signaling and Cardiac Remodeling

Gerald W. Dorn II and Lynne E. Wagoner
Department of Internal Medicine, University of Cincinnati, Cincinnati, Ohio, U.S.A.

CARDIAC REMODELING AS AN ADRENERGIC RECEPTOR-MEDIATED RESPONSE

Hypertrophy is the common chronic adaptive response of the adult heart to injury or abnormal hemodynamic load and is physically manifested as increased cardiomyocyte size and protein content. Results from the Framingham Study show that myocardial hypertrophy predisposes to early death (1). However, the broad categorization of enlarged hearts as "hypertrophied" is a generalization that ignores diverse physiologic or pathologic stimuli for hypertrophy and fails to distinguish between critical differences in etiology, mechanism, and prognosis. For example, physiologic hypertrophy that develops in highly trained athletes is not associated with either altered myocardial protein or genetic makeup, or with any predisposition to heart failure or death. In contrast, pathologic hypertrophy, the precursor to myocardial remodeling, is associated with specific changes in gene and protein expression that recapitulate an embryonic phenotype (2,3). Although pathologic hypertrophy initially compensates for increased hemodynamic loading by normalizing wall stress, these functional benefits are transitory and the hypertrophied heart ultimately dilates and fails in a poorly understood process commonly referred to as decompensation (4–6).

Physiologic and molecular studies have identified the stimuli and functional consequences of pathologic myocardial hypertrophy and have defined a host of accompanying changes in gene and gene product expression. Ongoing investigations into the mechanisms by which the heart translates a mechanical stimulus of hemodynamic overload into a biochemical signal(s) for hypertrophy have implicated a number of distinct signaling events, but a complete and coherent biochemical mechanism describing the development and decompensation of myocardial hypertrophy has not yet been achieved. It is increasingly clear that hypertrophy signaling occurs through multiple parallel pathways with critical kinases and phosphatases acting as biochemical transducers (7). Taken together, the accumulated data on hypertrophy signaling and the obvious distinction between physiologic and pathologic forms of myocardial hypertrophy support the existence of unique processes for each. Hence, the notion that cardiac endocrine/autocrine mechanisms regulate pathologic cardiomyocyte growth has gained acceptance, based initially on detailed investigations of cardio-myocyte hypertrophy signaling in cultures of spontaneously beating neonatal ventri-cular myocytes. Simpson and colleagues observed that the alpha adrenergic agonist norepinephrine (NE), but not the beta adrenergic agonist isoproterenol, increased cardiomyocyte cell size in a dose-dependent manner. Thus, a direct hypertrophic effect was demonstrated for stimulating adrenergic receptors (α_1AR) (8,9). Using variations of this tissue culture model, hypertrophic signaling by phenylephrine and other receptor agonists with similar signaling profiles [angiotensin II, endothelin, and prostaglandin (PG) $F_2\alpha$] have since been demonstrated (10–13). The potential relevance of these secreted factors in pathologic cardiomyocyte hypertrophy is sup-ported by the observations that angiotensin II and endothelin-1 can be released from mechanically deformed cardiomyocytes (14–16) and that plasma levels of many of these factors are measurably increased in heart failure (17–19). This notion is further supported by in vivo infusion studies where angiotensin II has cardiotrophic effects (20,21). However, traditional physiological and pharmacological experiments have not been able to distinguish between direct cardiotrophic effects of infused agonists and indirect effects secondary to increased blood pressure and cardiac mechanical load. Therefore, the most definitive characterizations of signaling mechanisms for in vivo myocardial hypertrophy have been produced by genetic manipulation of G-protein-coupled receptors and G-proteins themselves in experimental systems.

Gq Signaling in Cardiac Hypertrophy

Gq is the heterotrimeric G-protein that transduces signals from ligand-occupied heptahelical membrane receptors for angiotensin II, endothelin-1, alpha-adrenergic agonists, and other hormones known to provoke a hypertrophy response in cardio-myocytes (Fig. 1). As noted above, of the many candidate signaling pathways for cardiac hypertrophy, special importance has been assigned to receptors that signal via Gq. When transfected into cultured neonatal rat cardiomyocytes, wild-type Gαq stimulates cardiomyocyte hypertrophy similar to that caused by these agonists or mechanical stress (22,23). When Gαq activity is inhibited on the other hand, agonist-mediated cardiomyocyte hypertrophy is blunted (22). It is, perhaps, not surprising therefore that experimental and clinical studies show that pharmacological anta-gonism of Gq-coupled receptors regresses hypertrophy and improves prognosis in heart failure (24–26).

Transgenic Analysis of In Vivo Myocardial Gαq/PLC/PKC Signaling

In vivo analysis of cardiac Gq/PLC/PKC signaling has been greatly advanced by utilization of transgenic techniques to overexpress Gq-coupled receptors, Gαq or its dominant negative inhibitor, or the downstream Gq effector, PKC, in the mouse heart. The hallmark transgenic experiment, which established that activation of a cardio-myocyte Gq-coupled receptor was sufficient to stimulate cardiac hypertrophy, was from Lefkowitz and coworkers who overexpressed a mutant constitutively activated (CA) α_{1B} AR in the mouse heart (27). This experimental design, i.e., overexpression of a mutant receptor that activated downstream signaling pathways in the absence of a bound agonist, represented an attempt to reproduce the effects of chronic α adre-nergic stimulation of cardiomyocytes without the confounding extracardiac effects of adrenergic agonist. Three-fold overexpression of the CA α_{1B}AR caused cardiac hyper-trophy with some characteristics of pressure overload hypertrophy. Conversely, in vivo

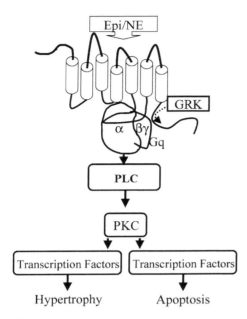

Figure 1 Cardiac function of myocardial PKC isoforms as determined by in vivo over-expression or translocation modification models. *Abbreviations*: NE, norepinephrine; GRK, G-protein receptor kinase; PLC, phospholipase C; PKC, protein kinase C.

inhibition of Gq signaling by deactivation (RGS4, 28) or competition (Gq "minigene"; 29) prevents or substantially attenuates the normal myocardial hypertrophic response to pressure overload. Finally, whereas ablation of the Gαq gene alone has no detectable cardiac effects (30), simultaneous ablation of Gα11 with conditional, cardiac-specific ablation of Gαq resulted in absence of pressure overload hypertrophy after transverse aortic banding, thus demonstrating an absolute requirement for Gq/11 signaling in reactive myocardial hypertrophy (31).

The hypothesis that ligand-independent activation of Gq signaling would, by itself, be sufficient to remodel the heart was tested by overexpression of the alpha subunit of the Gq α-subunit (32). Mice expressing a four- or fivefold excess of wild-type Gαq exhibited phenotypes of cardiac hypertrophy, mild contractile dysfunction without overt heart failure, and unresponsiveness to beta-adrenergic stimulation. Further increasing Gαq expression to eight times that of normal resulted in progressive ventricular dilation and development of a fibrotic cardiomyopathy. Gαq-mediated hypertrophy was similar to pressure overload hypertrophy in terms of the extent of cardiac hypertrophy, the pattern of fetal gene expression, and the increase in cardiomyocyte cross-sectional area (33,34). However, Gαq overexpressors exhibited some atypical features as well, such as eccentric ventricular remodeling, resting sinus bradycardia, and left ventricular systolic dysfunction at matched (atrial-paced) heart rates.

To determine how enhanced signaling through Gαq might modify the hypertrophic response to pressure overload, Gαq overexpressors are subjected to transverse aortic banding (34). Serially assessed for three weeks, pressure-overloaded Gαq overexpressors rapidly developed eccentric hypertrophy with progressively declining ventricular function, ultimately resulting in overt functional decompensation and pulmonary edema. This rapid progression to dilated cardiomyopathy in aortic banded Gαq overexpressors was analogous to that seen more slowly in mice expressing Gαq at higher levels in the heart (32), and was subsequently determined to be associated with increased cardiomocyte apoptosis (23). Indeed, cultured neonatal rat cardiac myocytes infected with adenovirus encoding a CA form of Gαq, briefly hypertrophied, but subsequently underwent apoptosis (35). The most dramatic in vivo correlate of Gαq -mediated cardiomyocyte apoptosis is the lethal peripartum cardiomyopathy typically observed during the first week after delivery (see below). Thus, the Gαq transgenic mice develop cardiac hypertrophy that predisposed to cardiomyocyte apoptosis under a variety of extrinsic or intrinsic stresses, which establishes a biological link between cardiac hypertrophy and programmed cell death, and suggests a cellular mechanism for cardiac remodeling.

APOPTOSIS AS A MECHANISM FOR CARDIAC HYPERTROPHY DECOMPENSATION

The transition from compensated hypertrophy to decompensated hypertrophy and dilated cardiomyopathy is classically associated with loss of cardiomyocytes and their replacement with fibrotic tissue. Studies over the past decade have increasingly supported an important pathophysiological role for cardiomyocyte apoptosis in heart failure decompensation. Since apoptotic cell death is a "voluntary" process (suicide), as opposed to necrotic cell death, which is imposed from without (murder), a tantalizing potential exists for prevention of hypertrophy decompensation by inhibition of cardiomyocyte apoptosis.

An important role for apoptotic cardiomyocyte loss in the pathophysiology of heart failure is suggested by studies of human heart failure syndromes. Although these data are necessarily associative and are limited primarily to tissue samples obtained firom explanted end-stage hearts of patients undergoing cardiac transplantation, the common finding of increased cardiomyocyte apoptosis in severe heart failure among virtually all studies cannot be ignored. Whereas cardiomyocyte apoptosis, most frequently measured as nuclear labeling by terminal deoxynucleotidyl transferase-mediated dUTP nick end labeling (the TUNEL assay), is extremely rare in normal hearts (~0.001%), its frequency is increased up to several hundred–fold in end-stage heart failure, to between 0.1% and 0.3% in most reports (36,37). Although this appears to be a very low incidence of a process that is postulated to play a major role in heart failure progression, these types of studies generate only a "snapshot" view of a chronic process. If one estimates the total loss of cardiomyocytes by integrating the frequency of apoptosis over time, and assuming that TUNEL labeling can detect an apoptosing cell for 24 hours, a sustained rate of only 0.2% TUNEL positivity (loss of 0.2% of cardiomyocytes/day) would, over one year, cause the death of more than half of the total cardiomyocytes (38). Furthermore, apoptosis in human heart failure appears to be a regulated process in that it is more common in male than female patients (39), may occur at a higher rate in dilated than in ischemic cardiomyopathy (36), and occurs in a regional distribution in infarcted hearts and those with arrhythmogemc right ventricular dysplasia (40,41). Thus, numerous clinical studies have proven that the incidence of cardiomyocyte apoptosis is strikingly increased in several human heart failure syndromes, and support an important role for apoptotic cell death in the progression of heart failure. Unfortunately, the critical mechanisms that modulate and stimulate apoptosis in human heart failure remain unknown, despite intriguing observations by some investigators

of regulated expression of pro- and antiapoptotic proteins in the failing human myocardium (37,42). More mechanistic data have been obtained in physiologically or
genetically modified animal models.

Since angiotensin II is widely considered to be a critically important stimulus for
pathological hypertrophy, it is especially interesting that it has also been implicated in
transducing cardiomyocyte apoptosis. Piero Anversa first demonstrated that angiotensin II, via activation of AT1 receptors, causes apoptosis of cultured neonatal rat
cardiomyocytes (43,44). Conversely, in vivo inhibition of the renin–angiotensin system
(RAS), through administration of angiotensin converting enzyme inhibitors, effectively blocks myocardial apoptosis in spontaneously hypertensive rats and dogs with
heart failure caused by sequential coronary microembolizations (45,46). The components of a likely RAS-coupled apoptosis-signaling pathway have been delineated in
a series of recent publications (47–50). Briefly, heart failure is associated with increased
expression of angiotensinogen, renin, and AT1 receptors, which, at least in part, is
mediated by p53. In contrast, IGF-1 overexpression attenuates p53 transcriptional
activity and diminishes the content of p53-responsive proapoptotic genes such as
angiotensinogen, AT1, and Bax. Hence, apoptosis in heart failure is promoted by regulated expression of p53-responsive genes encoding members of the RAS, and the antiapoptotic effects of IGF-1 appear to be mediated, in part, by inhibition of the RAS.

A key observation of genetic mouse models of apoptosis is that programmed
cardiomyocyte death is sufficient, by itself, to cause heart failure. The first genetically
modified mouse model in which this was clearly the case was the Gq transgenic mouse,
which exhibits a unique apoptotic peripartum cardiomyopathy (23). Whereas mice
expressing the alpha subunit of the Gq signaling protein develop a baseline hypertrophy with the molecular, cellular, and functional characteristics of *nonfailing* pressure
overload hypertrophy, shortly after giving birth, Gq overexpressors develop massive
cardiomyocyte apoptosis that leads to lethal heart failure. This model demonstrated
that apoptosis could cause rapidly progressive heart failure, and suggested a connection between hypertrophy and apoptosis signaling. Another useful genetic model of
apoptosis was a transgenic mouse expressing a caspase-8-FK506 binding protein
chimera, in which administration of FK1012 dimerizes the fusion protein, thereby activating caspase-8 and downstream caspases (51). This was rapidly lethal, again demonstrating that massive acute apoptosis can cause an aggressive heart failure syndrome.
Perhaps of more relevance to human heart failure however is the fact that mice
expressing the fusion protein demonstrated chronic, indolent cardiomyocyte apoptosis even in the absence of the dimerizing agent, and developed a dilated cardiomyopathy, showing that sustained, low grade apoptosis in the absence of any external
stress can cause heart failure. Finally, a mouse lacking the cytokine receptor, gp130,
demonstrated that absence of a critical cardiomyocyte survival pathway and the
consequent dis-inhibition of apoptosis can result in lethal dilated cardiomyopathy
after induction of surgical pressure overload (transverse aortic constriction) (52).
Not only is this a highly relevant model because it perturbs a cytokine receptor,
but it further links hypertrophy and apoptosis, in that, gp130 knockout mice fail to
hypertrophy to pressure overload. In this case, failure to undergo adaptive hypertrophy is associated with induction of apoptosis, suggesting that adaptive hypertrophy
requires antiapoptotic factors such as gp130. Indeed, the growth factor IGF-1, which
promotes "adaptive" hypertrophy/hyperplasia, also protects from apoptosis (53).
Thus, pathological hypertrophy in response to pressure overload, postinfarction
remodeling, or Gq-overexpression is associated with increased apoptosis, whereas
adaptive hypertrophy appears to be associated with protection from apoptosis.

Mechanisms for cardiomyocyte apoptosis have been defined by in vivo and in vitro expression of Gαq, which provokes spontaneous apoptosis at high signaling activities, but at lower signaling intensities only predisposes to apoptosis stimulated by agonists or physiological stress (23,35,53,54). Thus, Gαq signaling not only causes apoptosis outright, but also lowers the threshold for apoptosis stimulated by other hypertrophic agonists. A pivotal role for mitochondria in Gαq-mediated cardiomyocyte apoptosis was established using adenoviral infection of wild-type (i.e., not mutationally activated) or CA (GTPase-deficient) Gαq in cultured neonatal rat cardiac myocytes (23,35). Mitochondrial damage, measured as cytochrome c release, electron microscopic abnormalities, and loss of the mitochondrial membrane potential, correlated with Gαq-mediated apoptosis. In these studies, pharmacological inhibition of caspases demonstrated the necessity for caspase activation in Gαq-stimulated apoptosis, but not for mitochondrial damage, suggesting that caspase activation was downstream of mitochondrial injury. Finally, stabilization of the mitochondrial permeability transition pore with bongkrekic acid blocked not only Gαq-induced mitochondrial damage, but also caspase activation, proving that caspases are activated as a consequence of mitochondrial injury in Gαq-mediated apoptosis.

In myocardial disease, special importance has been assigned to this mitochondrial pathway, initiated by release of cytochrome c from damaged mitochondria (35,55,56). In the presence of ATP, cytochrome c, apoptotic protein-activating factor-1 (APAF-1), and the cysteine protease caspase-9 form a complex that activates executioner caspases such as caspase-3 (57,58). These proteases cleave numerous terminal proteins, including the DNA repair enzyme poly (ADP-ribose) polymerase (PARP), the nuclear envelope protein lamin, focal adhesion kinase (FAK), and protein kinase C (PKC) δ. Ultimately, the characteristic cellular features of apoptosis appear, including condensation and fragmentation of nuclear chromatin with internucleosomal cleavage of DNA, compaction of cytoplasmic organelles, dilation of the endoplasmic reticulum, cellular shrinkage, and membrane blebbing. In cardiac myocytes, what is much less clear is what happens prior to mitochondrial cytochrome c release, and how, at a molecular level, cytochrome c release is accomplished. Clearly, members of the Bcl-2 (*B*-cell *l*ymphoma) family of apoptosis regulating proteins play important roles in both promoting and preventing mitochondrial pathway apoptosis. Bcl-2, the prototypical member of this family, and closely related Bcl-xl are both potent antiapoptotic (i.e., cell survival) factors located on the outer mitochondrial membrane. Other Bcl-2 family members such as Bax and Bad are proapoptotic and translocate to the mitochondria early in the apoptotic process. In general, all Bcl-2 related proteins dimerize and it is speculated that Bcl-2 and Bcl-xl may protect against Bad and Bax by heterodimerizing with them (59,60). This notion is consistent with the observation that Bcl family proteins share homology with bacterial toxins that can form ion conductive pores, although such pores have never been demonstrated in intact cells and pores of that size would be too small to accommodate cytochrome c (61,62). Nevertheless, the notion that a normal balance between anti- and proapoptotic Bcl-2 related proteins is critical to preventing cell apoptosis is widely accepted and there are numerous examples of conditions in which expression of Bcl-2 related proteins is perturbed, including human and experimental heart failure (42,45,47,50,63,64). The likely mechanism for Gq-induced mitochondrial injury was revealed after transcriptome analysis of Gq transgenic mouse hearts uncovered a striking upregulation of several apoptotic effector genes, leading to the cloning of the mitochondrial death protein Nix (also known as Bnip3L), and its dominant inhibitory truncated splice isoform, sNix. Transgenic mouse models demonstrated that Nix expression was sufficient to cause a lethal apoptotic cardiomyopathy, and that inhibition of Nix protected the Gq mouse from peripartum apoptosis, ventricular decompensation, and death (65). These studies suggest that "compensated" hypertrophy carries within it the genetic seeds of apoptotic destruction, and establish one known molecular mechanism for ventricular remodeling.

PKC SIGNALING IN CARDIAC HYPERTROPHY

A major downstream effector of Gq/PLC signaling is the PKC family of ubiquitous serine–threonine kinases (66). PKC has long been considered to be a pathophysiologic mediator of cardiac hypertrophy and/or failure, based on observed associations between PKC activity and various pathological myocardial conditions. In cardiac tissue, PKC enzymatic activity is increased after ischemia and acute or chronic pressure overload, and in these conditions PKC is postulated to mediate ischemic preconditioning and to transduce hypertrophy signaling, respectively (67–69). However, the PKC family consists of at least 10 isoenzymes encoded by different genes and exhibit distinct patterns of tissue-specific expression and agonist-mediated activation. The heterogeneity of PKC isoform expression and differences in PKC isoform regulation and activation in the heart have complicated attempts to precisely define the role of PKC in adaptive cardiac responses and remodeling. Based on enzymatic properties, PKC isoforms are classified as being "conventional" (cPKC) or calcium-dependent, "novel" (nPKC) or calcium-independent, and "atypical" (aPKC), which are activated by lipids other than diacylglycerol (DAG). Mammalian myocardial tissue typically contains one or two cPKCs (PKCα and/or PKCβ), at least two nPKCs (PKCδ and PKCε), and one or more aPKCs (PKCζ and λ). An important feature of cardiac PKC isoforms is that, when activated, they translocate to distinct subcellular sites. For example, in cultured rat cardiac myocytes, stimulation of α1 ARs is associated with translocation of PKC β1 from cytosol to nucleus, PKCβII from fibrillar structures to perinucleus and sarcolemma, and PKCε from nucleus and cytosol to myofibrils (70,71). This differential subcellular compartmentalization of activated PKC isoforms implies distinct substrates, and therefore, unique cellular functions for each isoform (Fig. 2) (72,73).

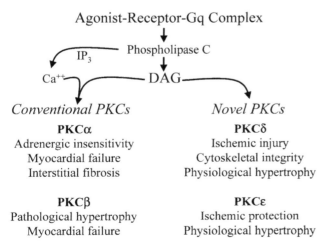

Agonist-Receptor-Gq Complex

Phospholipase C

IP_3

Ca^{++} — DAG

Conventional PKCs

PKCα
Adrenergic insensitivity
Myocardial failure
Interstitial fibrosis

PKCβ
Pathological hypertrophy
Myocardial failure

Novel PKCs

PKCδ
Ischemic injury
Cytoskeletal integrity
Physiological hypertrophy

PKCε
Ischemic protection
Physiological hypertrophy

Figure 2   Schematic depiction of Receptor/Gq/PKC signaling pathway. *Abbreviation*: PKC, protein kinase C.

The mechanism for subcellular translocation and activation of PKC isoforms involves binding to anchoring proteins termed RACKs (receptors for activated C kinases) (71,74). Each PKC isoform or group of related isoforms bind to a specific RACK through unique binding domains, and interference of PKC–RACK binding with peptide analogs of the RACK binding domain can inhibit translocation of PKC isoforms. Recently, the introduction of isoform-selective PKC translocation modulating peptides into cultured neonatal cardiomyocytes and transgenic mouse hearts has helped define roles for PKCε in cardiomyocyte contractile rate (75) and ischemic preconditioning (69), and cPKC in regulation of L-type calcium channels (76). This experimental approach takes advantage of the observation that unstimulated PKC exists as an inactive folded protein where the substrate binding site is occupied by the pseudosubstrate domain, thereby maintaining the enzyme in its catalytically inactive state. In the presence of phospholipid such as DAG, the PKC protein unfolds and exposes the substrate and RACK binding sites. Subcellular localization of membrane bound RACKs, which exhibit specificity for different PKC isoforms, determines the subcellular translocation of activated PKC isoforms, and therefore the likelihood that an activated PKC isoform will have access to a particular set of substrate proteins. RACK binding domains have been identified within the C_2 domain of conventional PKCs (i.e., cardiac PKC α, β) and the V_1 domain of novel PKCs (i.e., cardiac PKC δ, ε) and aPKCs (i.e., cardiac PKCζ, λ), and protein fragments or small peptides corresponding to a portion of the RACK binding domain are effective, isoform-specific inhibitors of PKC translocation, and hence activation.

Using this approach, a PKCβ C_2 domain peptide has inhibited phorbol ester attenuation of isoproterenol-stimulated calcium channel activity, suggesting that a cPKC (PKCα or β) mediates phorbol 12-myristate 13-acetate (PMA)-induced inhibition of this channel (76). A PKCε V_1 fragment (144 amino acids) or the eight amino acid PKCε RACK binding site peptide attenuated PMA or NE-dependent negative chronotropy (75), and prevented ischemic preconditioning in cultured neonatal cardiac myocytes (69). A PKCδ inhibitor peptide had no effects in these studies and the effects of aPKCs are relatively unexplored in cardiomyocytes. Thus, the introduction of peptides corresponding to specific PKC isoform RACK binding sites was an effective technique for inhibiting individual PKC isoforms in cardiomyocytes, and has laid the groundwork for parallel experiments in transgenic mice.

The advantage of manipulating in vivo cardiac PKC translocation, rather than the level (higher or lower) of PKC expression, is that proper stoichiometry between PKC isoforms and their respective substrates is maintained, thereby ensuring normal enzyme specificity, which has not been the case with standard transgenic overexpression of cardiac PKC isoforms (77). Activation of PKCε resulted in myocardial hypertrophy with normal function, i.e., "physiological hypertrophy," which appeared to result from cardiomyocyte hyperplasia rather than cell enlargement (78). The reciprocal experiment achieved the opposite result; inhibition of PKCε at the highest levels of inhibitor peptide expression caused a dilated cardiomyopathy phenotype with cardiomyocyte hypoplasia (78). Since the Gq model is associated with increased PKCε activity, the translocation modifier mice were crossed with the Gq mouse to attempt a "rescue." Unexpectedly, further increasing PKCε translocation by co-expression of ψεRACK with Gαq improved left ventricular function, whereas "normalization" of PKCε activity, i.e., restoration of a normal proportion of cytosolic to particulate PKCε by co-expression of low levels of εPKC inhibitor peptide with Gαq, resulted in a lethal dilated cardiomyopathy (79). These studies demonstrate that *PKCε* activation in the heart, which has long been assumed to be pathologic because of associations with pressure overload and ischemia (80,81), is in fact a beneficial compensatory event in cardiac hypertrophy. The notion that PKCε is a beneficial isoform in the heart has been supported by conventional transgenic overexpression studies (82).

Parallel studies have targeted PKCδ and PKCα using an identical peptide translocation activator/inhibitor approach. As with PKCε, selective activation of PKCδ resulted in "hypertrophy" phenotype, demonstrating parallel functions of PKCs ε and δ in myocardial growth signaling (83). Strikingly, selective inhibition of PKCδ by its truncated first variable region (δV1) resulted in a restrictive myopathic phenotype resembling the myofibrillar cardiomyopathies described with desmin and α-B-crystallin mutations in humans and mouse models (84). Finally, modulation of PKCα had no effects at all on myocardial growth or structure, but its activation blunted the response to β-ARs, an effect that, when superimposed on the Gq transgenic model, was lethal (85).

β-ADRENERGIC SIGNALING IN HEART FAILURE

One of the hallmark characteristics of cardiac remodeling is myocardial insensitivity to β-adrenergic agonists. ARs are members of the superfamily of seven transmembrane spanning (TMS) receptors and are activated by the endogenous catecholamines, epinephrine and NE. They represent the receptor component of the sympathetic nervous system, which serves a major homeostatic role in virtually all organ systems (86). Their structural features are similar to rhodopsin, with the β_2AR being considered prototypic. As shown in Figures 3, 4, and 5 below, the amino terminus is extracellular, there are three extracellular and three intracellular loops (ECL, ICL, respectively), and there is a carboxy terminus that is intracellular. Agonist and antagonist binding occur in the TMS regions, the former stabilizing a conformation of ICL2 and ICL3, and the proximal cytoplasmic tail, which promotes G-protein coupling. Post-translational modifications can include phosphorylation by various kinases in ICL3 and the carboxy terminus, palmitoylation of cysteines in the proximal portion of the carboxyterminus, and glycosylation of the amino terminus.

There are nine AR subtypes: β_1, β_2, β_3, α_{1A}, α_{1B}, α_{1D}, and α_{2A}, and α_{2B}, and α_{2C}. Classic signal transduction by these receptors is via their binding to one or more heterotrimeric G-proteins. Both the G_α and $G_{\beta\gamma}$ subunits of these G-proteins carry out signaling. βAR couple to $G_{\alpha s}$ and act to stimulate adenylyl cyclase, increasing cAMP and activating protein kinase A (PKA). The α_1AR couple to $G_{\alpha q}$ and stimulate phospholipase-C (PLC). α_2AR couple to $G_{\alpha i}$, inhibiting adenylyl cyclase, activating voltage-gated calcium channels, and inhibiting inwardly rectifying potassium channels. In the cardiomyocyte the dominant βAR is the β_1AR subtype, which stimulates cardiac inotropy and chronotropy upon binding to NE (from presynaptic nerves) or to circulating epinephrine. In humans, β_2AR also stimulate rate and contractility (87,88), although studies in knockout mice suggest that β_2AR have a minimal role in contractility (89). There is a lack of agreement on whether a functional β_3AR is expressed in the heart.

α_{2A} and α_{2C}AR in cardiac presynaptic nerves inhibit NE release, forming an autoinhibitory feedback loop. α_{2A}AR inhibit NE release in the setting of high frequency synaptic stimulation, while α_{2C}AR inhibit NE release during low-stimulation frequencies. α_{2A}/α_{2C} knockout mice develop a severe cardiomyopathy thought to be due to unregulated cardiac presynaptic NE release (90). These two α_2AR are localized to other presynaptic nerve terminals throughout the sympathetic nervous system as well, with the $\alpha_{2A}AR$ also found post synaptically. Within the central nervous system, $\alpha_{2A}ARs$ act to decrease sympathetic outflow and are the target of antihypertensive agents such as clonidine. The localization of α_{2B} subtype includes the central nervous system and peripheral vasculature, including the microvasculature of the heart, where it acts to vasoconstrict.

Agonist binding to ARs stabilizes the receptor in a so-called active conformation, which results in interactions with G-proteins and initiation of signaling to the effector. With continuous agonist exposure, some ARs undergo a decrease in coupling efficiency, defined as desensitization. Both β_1AR and β_2AR undergo a rapid agonist-promoted desensitization due to phosphorylation by G-protein-coupled receptor kinases, such as the βAR kinase (now termed GRK2, see below). GRK-phosphorylated receptor is then bound by β-arrestin, which evokes a partial uncoupling of the receptor from G-protein interaction. For β_1- and β_2ARs, phosphorylation by PKA also uncouples receptors. These phosphorylation events take place within moments of agonist occupancy and are thought to be the mechanisms by which receptor function is integrated within a cell that is receiving many signals. Some ARs also undergo agonist-promoted internalization, which serves as a mechanism for dephosphorylation, sorting of receptors to degradation pathways, or reinsertion into the membrane. After prolonged agonist exposure some of the ARs undergo processes leading to a decrease in overall expression. This phenomenon, termed downregulation, can be due to altered transcription, message stability, and protein degradation. The functional consequences of receptor uncoupling from phosphorylation, and loss of receptor expression, can result in marked agonist-promoted desensitization, as in heart failure (91).

Regulation of AR Function in Heart Failure

As discussed above, the sympathetic nervous system acts to increase cardiac output and maintain blood flow to critical regions through its cardiac and vascular distribution. Thus, under normal physiologic events, such as rising from the supine position, blood pressure is maintained, despite the potential for dependent pooling of intravascular volume. Similarly, during physical activity cardiac output/perfusion is regulated, in part, by the adrenergic system. During pathologic conditions, adaptive as well as maladaptive regulatory events can unfold. In acute events, such as abrupt hemodynamic shock in an otherwise normal individual, the sympathetic nervous system appears to be well suited for short-term maintenance of perfusion to vital organs. On the other hand, chronic activation by endogenous catecholamines or exogenous agonists can lead to events with a negative impact on end-organ function. This seems to be particularly so for the heart. Evidence from a number of studies over the last few decades indicates paradoxical AR events that represent both detrimental and beneficial effects.

For many years it has been known that infusions of catecholamines in animals can cause various cardiomyopathic features, including hypertrophy and failure. In transgenic studies, moderate overexpression of β_1AR in the mouse heart results in dilated cardiomyopathy (92). Interestingly, cardiac β_2AR overexpression is better tolerated: in "dose-ranging" studies of transgenic β_2AR overexpression marked overexpression or extended observation times were necessary to evoke pathologic features (93). Thus, even in the absence of a precipitating event, chronic adrenergic stimulation of the heart (particularly β_1AR) has the potential to cause cardiac remodeling. Further evidence from multiple clinical trials lends credence to the potential deleterious effects of adrenergic drive in chronic heart failure. High levels of circulating NE are negatively correlated with heart failure survival (94). In addition, some of the strongest evidence comes from the success of βAR antagonists (β-blockers) for treating heart failure (95). A developing paradigm suggests that regardless of the initial basis of failure (ischemia/infarction, hypertension, valvular, myocarditis, or "idiopathic"), the persistent stimulation of the βAR within the setting of a metabolically and physiologically compromised heart leads to progression (96).

Although reduced βAR expression and/or function in human heart failure may be a protective adaptation (91), βAR desensitization events can apparently have deleterious effects as well. Physiologically, patients with markedly desensitized βAR are unable to acutely increase cardiac output via the sympathetic system, and thus exhibit intolerance to acute exertion (a major symptom of the syndrome). Similarly, changes in preload, afterload, coronary perfusion, or the parasympathetic system are less tolerated owing to an ineffective sympathetic system. In addition, it appears that the antiapoptotic effects of β_2AR (97,98) are negated under these circumstances, since there is a global blunting of both βAR subtype functions. Interestingly, cardiac β_1AR expression and function are decreased in chronic heart failure, but β_2AR expression is unchanged (91). This has suggested the presence of an underlying mechanism to preserve β_2AR function. The mechanism(s) by which β-blockers exert their effects, which include a decrease in symptoms, an increase in LVEF and a decrease in mortality, is not likely to be simply an accentuation of the desensitization process already in place. Indeed, βAR function and/or expression actually increase during β-blocker therapy (95). The dosing strategy that has proven successful for β-blockers in heart failure is an initial low dose with a slow up-titration. Improvement is often not evident until several months of stable therapy. This suggests that β-blockers act to modify some of the biochemical remodeling that has occurred within the adrenergic system, shifting the function back toward "normal" while maintaining a low level of antagonism to protect against enhanced catecholamines. As the initial sympathetic stress on the heart is relieved, and improvement occurs, systemic catecholamines often decrease during β-blocker therapy, thus affecting one of the stimuli for the desensitization (99). The "rescue" of certain mouse failure phenotypes by inhibition of GRKs or expression of β_2AR is consistent with the notion that partial normalization of certain aspects of βAR signaling is beneficial in the failing heart (100,101).

Coding Polymorphisms of α_2- and βAR Polymorphisms and Their Signaling Phenotypes

Over the last ten years, a number of nonsynonymous polymorphisms of human α_{2A}, α_{2B}, α_{2C}, β_1-, and β_2AR have been identified in human subjects and phenotyped in transfected cells, transgenic mice, and heart failure patients (102). The locations of these SNPs are shown diagrammatically in Figures 3, 4, and 5, and the cellular and human phenotypes for each are individually described below.

(1) *The $\alpha_{2A}AR$ Lys251 has enhanced function.* The $\alpha_{2A}AR$ coding region has one variant with a frequency >1% (Fig. 3), which consists of a substitution of Lys for Arg in the third intracellular loop of the receptor (103). It is uncommon, with a frequency of ~4% in African Americans and ~0.4% in Caucasians. When expressed in COS-7 cells, agonist-promoted stimulation of [^{35}S]GTPγS was 40%

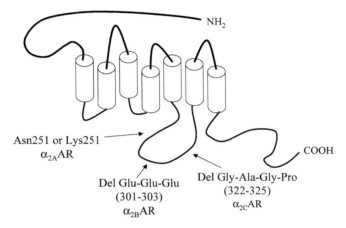

Figure 3 Schematic representation of the alpha adrenergic receptors and localization of known polymorphisms. The alpha-2A receptor has one known polymorphism at amino acid position 251 (aspartic acid or lysine). The alpha-2B receptor has one known polymorphism that is a deletion of amino acids 301–303. The alpha-2C receptor has one known polymorphisms that is a deletion of amino acid 322–325.

greater with the polymorphic Lys251 receptor compared to that of its allelic counterpart. Expression, ligand binding, and basal function did not differ between the two receptors. This enhanced function was true for several classes of agonists, including catecholamines, azepines, and imidazolines, albeit to different extents. In CHO cells, agonist-promoted inhibition of adenylyl cyclase was increased. The phenotypic difference was even greater when MAP-kinase stimulation was assessed, again with Lys251 function being greater than Arg251. Thus the α_{2A}Lys251 phenotype is a gain-of-function.

(2) *The α_{2B}Del301–303 displays altered GRK-mediated phosphorylation and desensitization.* One nonsynonymous polymorphic locus was detected in the

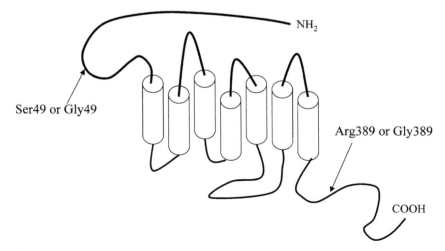

Figure 4 Schematic representation of the beta-1 adrenergic receptor and localization of known polymorphisms at amino acid positions 49 (serine or glycine) and 389 (arginine or glycine).

α_{2B}AR coding region, which consists of an in-frame deletion of nine nucleotides resulting in the loss of three consecutive Glu residues in the third intracellular loop (Fig. 3) (104). This polymorphism is relatively common, with frequencies of ∼30% in Caucasians and ∼12% in African Americans. In recombinantly expressed CHO cells, a small decrease in coupling efficiency was noted. The major phenotype, though, displayed a marked decrease in agonist-promoted phosphorylation and desensitization by GRK2. Since GRK2 phosphorylation is the major mediator of short-term agonist-promoted desensitization of α_{2B}AR, functional studies were carried out (inhibition of adenylyl cyclase) in the absence or presence of short-term pretreatment with agonist. The α_{2B}Del301–303 failed to undergo desensitization. Accordingly, the phenotype of the α_{2B}Del301–303 receptor displays the inability to undergo rapid agonist-promoted desensitization.

(3) *The α_{2C}Del322–325 displays impaired functional coupling.* Another in-frame deletion in ICL3 was noted in the coding region of the α_{2C}AR subtype, which results in a loss of the four amino acids Gly-Ala-Gly-Pro (Fig. 3) (105). The polymorphism is common in African Americans (∼40%), but is found in ∼4% of Caucasians. When expressed to equal levels in CHO cells, high-affinity agonist binding was moderately decreased, suggesting a receptor G-protein coupling defect. Indeed, in adenylyl cyclase inhibition studies, a marked decrease in function was observed. Decreased function was also observed for agonist-mediated stimulation of MAP-kinase and IP$_3$ production. Thus the α_{2C}Del322–325 phenotype is one of marked loss-of-function.

(4) *The Gly49 β_1AR is atypically glycosylated and undergoes enhanced down-regulation.* An SNP in the amino terminus of the β_1AR was found, which results in a substitution of Gly for Ser at amino acid 49 (Fig. 4) (106). Its allele frequency is ∼15% in both racial groups. In HEK-293 cells recombinantly expressing each receptor, we found no differences in agonist or antagonist binding affinities, or stimulation of adenylyl cyclase. Since the amino terminus in some ARs is involved in baseline and agonist-promoted trafficking, these parameters were explored. Initial studies showed that glycosylation patterns differed between the two receptors. Using various approaches, this was found to be due to altered N-linked glycosylation at amino acid 15. Agonist-promoted internalization of both receptors was similar, as was the rate of new receptor synthesis and membrane insertion. However, long-term exposure to agonist revealed that the less common Gly49 receptor underwent enhanced down-regulation compared to Ser49. Thus, the functional phenotype of the Gly49 polymorphism is one of enhanced long-term, agonist-promoted downregulation.

(5) *The Arg389 β_1AR displays enhanced function and undergoes increased desensitization.* Within the intracellular domain of the β_1AR, a polymorphic locus was identified where Arg or Gly can be present at amino acid 389, in a predicted small intracellular α-helix between TM7 and the membrane anchoring palmitoylation site (Fig. 4) (107). Gly389 is considered as the reference genotype, since all previous structure/function studies have been carried out with this receptor. In transfected cells, agonist-promoted [^{35}S]GTPγS production was increased with Arg389, as was agonist high-affinity binding, indicating enhanced agonist-receptor-G$_s$ interaction. In functional studies (stimulation of adenylyl cyclase), basal and, particularly, agonist stimulation of adenylyl cyclase was higher for Arg compared to Gly389.

(6) *The β_2AR position 16 and 27 polymorphisms alter downregulation.* Two common polymorphic loci are in the amino terminus of the human β_2AR, where Arg or Gly can be found at amino acid 16, and Gln or Glu at position 27 (Fig. 5) (108). At position 16, Gly is more common, but, again, because Arg was the only cloned human receptor prior to our study, and all the pharmacology had been carried

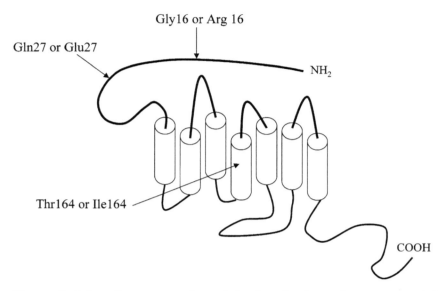

Figure 5 Schematic representation of the beta-2 adrenergic receptor and localization of known polymorphisms at amino acid positions 16 (arginine or glycine), 27 (glutamine or glutamic acid), and 164 (threonine or isoleucine).

out with this receptor, it is considered to be the reference allele in order to remain consistent with the term "wild-type" that has been associated with Arg16. In transfected cells, agonist and antagonist binding affinities were not different between all four possible combinations of the two polymorphic sites. Their presence in the amino terminus prompted studies of trafficking, where the extent of long-term agonist-promoted downregulation varied based on genotype. Arg16/Gln27 (reference genotype) underwent ~26% downregulation, compared to ~43% for Gly16/Gln27, ~0% for Arg16/Glu27 (a rare haplotype), and ~45% for Gly16/Gln27.

(7) *The β_2AR Ile164 function is impaired.* The polymorphism of the β_2AR where Ile is substituted for Thr in TM4 (Fig. 5) is uncommon, with a heterozygous frequency of 2% to 5% (109). (A homozygous Ile164 has never been reported.) Recombinantly expressed Ile164 receptor had an intrinsic decrease in affinity for some ligands, as well as a decrease in function of the agonist receptor–G_s complex. Consistent with this altered conformation, agonist-promoted internalization and GRK-mediated desensitization are decreased with the Ile164 β_2AR. In transgenic mice expressing equivalent expression of both receptors in the heart (~1 pmol/mg), baseline and agonist-stimulated contractility was reduced in the Ile164 mice (110). It should be noted that the Ile164 receptor has been associated with decreased survival in heart failure, and depressed exercise capacity in patients otherwise matched for clinical characteristics (see below).

Clinical Consequences of α_2- and βAR Polymorphisms in Human Heart Failure

Studies have been reported by a number of groups addressing whether these polymorphisms are associated with the risk of heart failure (or other cardiovascular diseases such as hypertension and myocardial infarction), or act to modify the disease (also described as association with "intermediate phenotypes" or "clinical subsets") (Table 1).

(1) *β_2AR polymorphisms.* Cardiac β_2AR have an inotropic and chronotropic effect and also appear to signal to anti-apoptotic events (93,97,98,101). Thus, the markedly uncoupled Ile164 variant might predispose individuals who have heart failure to a rapid decline in function and early death/transplant. In a study of 212 normals and 259 heart failure patients (idiopathic or ischemic) (111), there was no association between Ile164 and heart failure. However, those with failure who had Ile164 reached the death/transplant endpoint much earlier (adjusted relative risk = 4.81, $P < 0.001$). There was no significant confounding by demographic or clinical parameters, including etiology or baseline LVEF. Even before overt decompensation, patients with Ile164 exhibited diminished exercise capacity, a depressed cardiac output response and smaller alterations in systemic vascular resistance, the latter suggesting that the extracardiac (arteriolar) Ile164-β_2AR contributed to the altered physiology. In contrast, there was no apparent effect of the position 16/27 β_2AR polymorphisms on survival.

The β_2AR polymorphisms also impact exercise capacity, a critical determinant of prognosis in heart failure patients (112). Cardiopulmonary exercise testing was performed on 232 compensated heart failure patients with defined β_2AR genotype. Patients with the Ile164 polymorphism had a lower peak VO_2 than did patients with Thr164. Catheterization-based invasive exercise testing revealed depressed exercise-induced cardiac index, systemic vascular resistance, stroke volume, and VO_2 in patients with Ile164. The polymorphisms at position 16 also impacted exercise capacity. Arg16 patients had a higher peak VO_2 than Gly16 patients. Additionally, Arg16/Glu27 had the highest VO_2 and Gly16/Gln27 had the lowest, not confounded by baseline clinical characteristics, including β-blocker usage.

(2) *β_1AR polymorphisms.* A number of clinical studies have been carried out with the two nonsynonymous β_1AR polymorphisms at positions 49 and 389, primarily in hypertension, heart failure, and obesity. The gain-of-function Arg389 variant is associated with improved exercise capacity (113). Arg389 has been associated with a greater decrease in blood pressure in response to β-blocker treatment (114,115). Taken together, it is apparent that genetic variation of the β_1AR, which is also expressed in the vasculature and the kidney, may play a role in hypertension.

Additionally, in 244 heart failure patients, homozygous Arg was associated with an improvement in left ventricular function during treatment with carvedilol (8.7% ± 1.1%) compared to that in homozygous Gly389 patients (0.93% ± 1.7%, $P < 0.02$). No effect modification was observed based on age, sex, race, etiology of heart failure, baseline LVEF, or carvedilol dose (116). This is the first pharmocogenomic result

Table 1 Human Clinical Endpoints Observed from Adrenergic Receptor Polymorphisms

Receptor	Causation	Survival	Exercise capacity	Other
$\beta_1$389	+ (in combination with α_2c)	−	+	Role in HTN Carvedilol effectiveness
$\beta_2$16	−	−	+	−
$\beta_2$164	−	+	+	−
α_2c	+ (in combination with β_1AR)	?	?	−

published, showing differences in responsiveness to β-blocker treatment based on genetic variants.

(3) *$\alpha_{2C}AR$ polymorphisms and the two-gene α_{2C}/β_1AR interaction.* A recent study has examined the hypothesis that multiple polymorphisms of different genes that have the same net effect on receptor signaling could act synergistically to affect disease, i.e., chronically increased β_1AR drive predisposes to heart failure. As the β_1AR-Arg389 has enhanced signaling compared to Gly389 (107), the Arg variant has the potential to chronically stimulate the heart. Likewise, since the α_{2C}Del322–325 polymorphic receptor has depressed signaling (105), and presynaptic $\alpha_{2C}AR$ control NE release during low frequency stimulation (which we equate with chronic control of release), this variant should predispose to higher NE levels in the synapse. Accordingly, the combination of the β_1Arg389 and α_{2C}Del322–325 could represent greater risk than either alone. In a study of 159 nonischemic dilated cardiomyopathy patients and 189 controls (117), the $\alpha_{2C}AR$ revealed a higher allele frequency for African Americans with heart failure than for healthy African American controls (61.5% vs. 41.1%), and the hypothesis that α_{2C}Del322–325 and β_1Arg389 act together as a risk for heart failure was examined. The double homozygous combination revealed an adjusted odds ratio of 10.11, and the combination of the two genotypes had a multiplicative association (i.e., more than an additive effect) with the risk of heart failure. As pharmacogenomics becomes increasingly a clinical, rather than simply an investigative approach, these findings demonstrate the importance of assessing phenotype in the context of multi-gene variants within a similar functional group.

MODULATION OF RECEPTOR SIGNALING IN THE REMODELING HEART

It is interesting that, in comparison with the receptors described above, nonsynonymous genetic variation in the G-proteins that couple to them is an extremely rare event (118). Since there are literally hundreds of heptahelical receptors that couple to only a handful of G-proteins, these observations suggest that significant polymorphic variation in signaling protein pathways is possible for those gene products that exhibit a high degree of functional redundancy.

G-Protein Receptor Kinases as Disease Modifiers in Heart Failure

G-protein-coupled receptor signaling is, in addition to the presence of ligand, dependent upon the balance between receptor desensitization and resensitization. Desensitization, defined as the time-dependent loss of receptor responsiveness to agonist, is a critical feedback mechanism that protects receptors from overstimulation. G-protein receptor kinases were first identified in studies of the mechanisms of β2AR desensitization, when agonist-induced receptor phosphorylation during homologous desensitization was found in cells that lacked the cAMP-dependent protein kinase (119). The responsible serine–threonine kinase was purified and designated as β-adrenergic receptor kinase (β-ARK) (120), later renamed as GRK-2. Identification of the seven-member mammalian GRK family followed as the cDNAs were cloned (121). Structurally, GRKs have been divided into three subfamilies: GRK1 and 7 (retinal opsin kinases), GRK2 and 3 (a.k.a. β-ARK1 and 2), and GRK4, 5, and 6. All GRKs contain a central $\sim$265 amino acid kinase domain, preceded by a highly conserved $\sim$185 amino acid amino terminal regulatory domain that contains an RGS domain and is thought to be critical for receptor recognition. The variable C-terminal domain determines the subfamily: GRK2 and 3 contain a pleckstrin homology domain that is responsible for Gβγ binding, whereas GRK4, 5, and 6 lack this domain and instead have palmitoylated residues (GRK4 and 6) or a basic amino acid extension (GRK5) that likely mediate plasma membrane phospholipid interactions. Individual GRKs exhibit tissue-specific expression, with GRK5 having the highest expression in heart, followed by GRK2 and 3, and GRK 6 (122).

The primary function of GRKs is to phosphorylate agonist-occupied G-protein-coupled receptors on their third intracellular loop (i.e., α2aAR) or C-termini (i.e., β2AR). As described in detail in recent reviews (123,124), GRK-phosphorylated receptors uncouple from their G-protein effectors (desensitization) and are internalized into clathrin-coated pits for either degradation (downregulation) or recycling to cell membranes (resensitization). A secondary function of GRKs relates to their N-terminal RGS domains, through which they can regulate receptor-mediated signaling directly at the level of G-proteins. Indeed, GRK2 has recently been shown to directly interact with activated Gαq and inhibit its stimulation of PLC, and phosphorylation-independent "desensitization" of the parathyroid hormone receptor by overexpressed GRKs is indicative of RGS-mediated GAP activity of these proteins (125).

GRK activity can be modulated in two ways:inhibition of catalytic kinase activity or alteration in subcellular localization. For GRK2 family members, which are primarily cytosolic and translocate to cell membranes upon receptor activation, proper membrane targeting and subcellular localization depend upon Gβγ binding to residues 643–670 of the C-terminal pleckstrin homology domain. The currently accepted model describes agonist occupancy of receptor leading to activation of heterotrimeric G-proteins and dissociation of free Gα and Gβγ subunits. While Gα activates primary downstream effectors, Gβγ, localized at the plasma membrane, interacts with the extreme carboxyl terminus of the pleckstrin homology domain plus a portion of the remaining C-terminus of GRK2 family members, and recruits them to their membrane receptor substrates. These observations have led to the use of a C-terminal peptide, "β-ARKct" or "βARK minigene," to sequester free Gβγ and thereby inhibit GRK function.

Genetic mouse models support a critical regulatory function for GRKs in the heart. Targeted myocardial overexpression of GRK2 or GRK5 impaired βAR signaling and βAR agonist-mediated contractile function (126,127), whereas overexpression of the inhibitory βARKct "minigene" had opposing effects of enhancing βAR-mediated signaling and contractility (126). Importantly, homozygous GRK2 knockout mice were embryonic lethal (myocardial hypoplasia and death by embryonic day 15.5), but hemizygous GRK2 knockouts, with 50% diminished GRK2 expression, have helped to establish a dose–response relationship between GRK2 levels and myocardial contractile function (128,129). In the context of cardiac hypertrophy and failure, inhibition of GRK2 by myocardial-expressed βARKct restored βAR signaling and contractile function in pressure overload hypertrophy (100) and the dilated cardiomyopathy of the MLP (muscle LIM protein) knockout model (130). These studies show that forced regulation of GRK expression or function can profoundly affect myocardial function in cardiac hypertrophy and the progression to heart failure, and support the notion that functionally significant polymorphisms of cardiac-expressed GRKs can act as disease modifiers in heart failure.

To date, there has only been one published report of a human polymorphic GRK associated with cardiovascular disease, hypertension, and GRK4 (131). Since GRK4 is predominantly expressed in the genito-urinary system, and has not been detected in the heart or vasculature, the mechanism of this effect is unclear. On the other hand, regulated expression of cardiac-expressed GRKs has been reported in several human cardiovascular syndromes: GRK2 levels are reportedly increased in circulating lymphocytes of hypertensive patients, in whom resistance to β2AR-mediated peripheral vasodilation is an associated finding (132). Because β2AR levels in lymphocytes generally parallel receptor levels in vascular smooth muscle, this finding suggests a causal relationship between increased GRK2 and blunted β2AR responses. GRK2 is also upregulated in experimental (rat) myocardial ischemia, and is temporally associated with diminished β2AR responsiveness in this model (133). Finally, myocardial GRK activity (presumably GRK2 and GRK5) is increased in human heart failure, which is associated with diminished βAR responsiveness. Indeed, the increase in GRK levels appears to be inversely correlated with myocardial contractility, suggesting a potential therapeutic target (134–137).

CONCLUSIONS

A host of changes in gene and gene product expression have been demonstrated to occur as a result of pathologic myocardial hypertrophy. Clearly, hypertrophy signaling occurs through multiple parallel pathways with critical kinases and phosphatases acting as biochemical transducers. Such pathways are potential sites for pharmacologic treatment and/or regression of hypertrophy.

Experimental and clinical studies have shown that pharmacological antagonism of Gq-coupled receptors regresses hypertrophy and improves prognosis in heart failure. On the other hand, the Gαq transgenic mice develop cardiac hypertrophy, which predisposes to cardiomyocyte apoptosis establishing a biological link between cardiac hypertrophy and programmed cell death, and suggests a cellular mechanism for cardiac remodeling. Thus, the potential exists for prevention of hypertrophy decompensation by inhibition of myocyte apoptosis. For example, Nix expression causes a lethal apoptotic cardiomyopathy, and inhibition of Nix protects against apoptosis, ventricular decompensation, and death. Thus, "compensated" hypertrophy carries within it the genetic seeds of apoptotic destruction, and establishes one known molecular mechanism for ventricular remodeling.

Activation of PKCε and PKCδ, downstream effectors of Gq/PLC signaling, results in myocardial hypertrophy with normal function or cardiomyocyte hyperplasia, rather than cell enlargement. Thus, potential pharmacologic agents that activate PKCε and PKCδ activation in the heart may have beneficial compensatory effects in cardiac hypertrophy.

The AR system (beta and alpha) has been studied extensively in heart failure. β-blockers, a proven therapy for heart failure, act to modify the biochemical remodeling that has occurred within the adrenergic system. AR polymorphisms are common genetic variants shown to have a biologic role in the etiology and progression of heart failure. GRKs phosphorylate agonist-occupied G-protein-coupled receptors (such as the adrenergic receptors), allowing them to uncouple (desensitization), internalize (downregulation) or recycle back to cell membranes (resensitization). Studies have shown that forced regulation of GRK expression or function can profoundly affect myocardial function in cardiac hypertrophy and the progression to heart failure, and support the notion that functionally significant polymorphisms of cardiac-expressed GRKs can act as disease modifiers in heart failure. Pharmacogenomics will become an increasingly clinical approach to the management of heart failure patients, as the importance of assessing phenotype in the context of multi-gene variants within a similar functional group is determined.

REFERENCES

1. Levy D, Garrison RJ, Savage DD, Kannel WB, Castelli WP. Prognostic implications of echocardiographically determined left ventricular mass in the Framingham Heart Study. N Engl J Med 1990; 322:1561–1566.
2. Chien KR, Knowlton KU, Zhu H, Chien S. Regulation of cardiac gene expression during myocardial growth and hypertrophy: molecular studies of an adaptive physiologic response. FASEB J 1991; 5:3037–3046.
3. Izumo S, Nadal-Ginard B, Mahdavi V. Protooncogene induction and reprogramming of cardiac gene expression produced by pressure overload. Proc Natl Acad Sci USA 1988; 85:339–343.
4. Grossman W, Jones D, McLaurin LP. Wall stress and patterns of hypertrophy in the human left ventricle. J Clin Invest 1975; 56:56–64.
5. Meerson FZ, Kapelko VI. The effect of cardiac hyperfunction on its automaticity and reactivity to the chronotropic effect of the vagus. Cor Vasa 1965; 7:264–272.
6. Osler W. The Principles and Practice of Medicine. New York: D. Appleton & Co., 1892:634.
7. Molkentin JD, Dorn II GW. Cytoplasmic signaling pathways that regulate cardiac hypertrophy. Annu Rev Physiol 2001; 63:391–426.
8. Simpson P, McGrath A, Savion S. Myocyte hypertrophy in neonatal rat heart cultures and its regulation by serum and by catecholamines. Circ Res 1982; 51:787–801.
9. Simpson P. Norepinephrine-stimulated hypertrophy of cultured rat myocardial cells is an alpha 1 adrenergic response. J Clin Invest 1983; 72:732–738.
10. Adams JW, Migita DS, Yu MK, Young R, Hellickson MS, Castro-Vargas FE, Domingo JD, Lee PH, Bui JS, Henderson SA. Prostaglandin F2 alpha stimulates hypertrophic growth of cultured neonatal rat ventricular myocytes. J Biol Chem 1996; 271: 1179–1186.
11. Knowlton KU, Michel MC, Itani M, Shubeita HE, Ishihara K, Brown JH, Chien KR. The alpha 1A-adrenergic receptor subtype mediates biochemical, molecular, and morphologic features of cultured myocardial cell hypertrophy. J Biol Chem 1993; 268:15,374–15,380.
12. Sadoshima J, Izumo S. Molecular characterization of angiotensin II–induced hypertrophy of cardiac myocytes and hyperplasia of cardiac fibroblasts. Critical role of the AT1 receptor subtype. Circ Res 1993; 73:413–423.
13. Shubeita HE, McDonough PM, Harris AN, Knowlton KU, Glembotski CC, Brown JH, Chien KR. Endothelin induction of inositol phospholipid hydrolysis, sarcomere assembly, and cardiac gene expression in ventricular myocytes. A paracrine mechanism for myocardial cell hypertrophy. J Biol Chem 1990; 265:20,555–20,562.
14. Kojima M, Shiojima I, Yamazaki T, Komuro I, Zou Z, Wang Y, Mizuno T, Ueki K, Tobe K, Kadowaki T. Angiotensin II receptor antagonist TCV-116 induces regression of hypertensive left ventricular hypertrophy in vivo and inhibits the intracellular signaling pathway of stretch-mediated cardiomyocyte hypertrophy in vitro. Circulation 1994; 89:2204–2211.
15. Sadoshima J, Xu Y, Slayter HS, Izumo S. Autocrine release of angiotensin II mediates stretch-induced hypertrophy of cardiac myocytes in vitro. Cell 1993; 75:977–984.
16. Yamazaki T, Komuro I, Kudoh S, Zou Y, Shiojima I, Hiroi Y, Mizuno T, Maemura, K, Kurihara H, Aikawa R, et al. Endothelin-1 is involved in mechanical stress-induced cardiomyocyte hypertrophy. J Biol Chem 1996; 271:3221–3228.
17. Berger HJ, Zaret BL, Speroff L, Cohen LS, Wolfson S. Regional cardiac prostaglandin release during myocardial ischemia in anesthetized dogs. Circ Res 1976; 38:566–571.

18. Lai J, Jin H, Yang R, Winer J, Li W, Yen R, King KL, Zeigler F, Ko A, Cheng J, et al. Prostaglandin F2 alpha induces cardiac myocyte hypertrophy in vitro and cardiac growth in vivo. Am J Physiol 1996; 271:H2197–H2208.

19. Rockman HA, Koch WJ, Milano CA, Lefkowitz RJ. Myocardial beta-adrenergic receptor signaling in vivo: insights from transgenic mice. J Mol Med 1996; 74:489–495.

20. Baker KM, Chernin MI, Wixson SK, Aceto JF. Renin-angiotensin system involvement in pressure-overload cardiac hypertrophy in rats. Am J Physiol 1990; 259:H324–H332.

21. Dostal DE, Baker KM. Angiotensin II stimulation of left ventricular hypertrophy in adult rat heart. Mediation by the AT1 receptor. Am J Hypertens 1992; 5:276–280.

22. LaMorte VJ, Thorburn J, Absher D, Spiegel A, Brown JH, Chien KR, Feramisco JR, Knowlton KU. Gq- and ras-dependent pathways mediate hypertrophy of neonatal rat ventricular myocytes following alpha 1-adrenergic stimulation. J Biol Chem 1994; 269:13,490–13,496.

23. Adams JW, Sakata Y, Davis MG, Sah VP, Wang Y, Liggett SB, Chien KR, Brown JH, Dorn GW. Enhanced Galphaq signaling: a common pathway mediates cardiac hypertrophy and apoptotic heart failure. Proc Natl Acad Sci USA 1998; 95:10,140–10,145.

24. Dunn FG, Oigman W, Ventura HO, Messerli FH, Kobrin I, Frohlich ED. Enalapril improves systemic and renal hemodynamics and allows regression of left ventricular mass in essential hypertension. Am J Cardiol 1984; 53:105–108.

25. Garavaglia GE, Messerli FH, Nunez BD, Schmieder RE, Frohlich ED. Immediate and short-term cardiovascular effects of a new converting enzyme inhibitor (lisinopril) in essential hypertension. Am J Cardiol 1988; 62:912–916.

26. Nakashima Y, Fouad FM, Tarazi RC. Regression of left ventricular hypertrophy from systemic hypertension by enalapril. Am J Cardiol 1984; 53:1044–1049.

27. Milano CA, Dolber PC, Rockman HA, Bond RA, Venable ME, Allen LF, Lefkowitz RJ. Myocardial expression of a constitutively active alpha 1B-adrenergic receptor in transgenic mice induces cardiac hypertrophy. Proc Natl Acad Sci USA 1994; 91:10,109–10,113.

28. Rogers JH, Tamirisa P, Kovacs A, Weinheimer C, Courtois M, Blumer KJ, Kelly DP, Muslin AJ. RGS4 causes increased mortality and reduced cardiac hypertrophy in response to pressure overload. J Clin Invest 1999; 104:567–576.

29. Akhter SA, Luttrell LM, Rockman HA, Iaccarino G, Lefkowitz RJ, Koch WJ. Targeting the receptor-Gq interface to inhibit in vivo pressure overload myocardial hypertrophy. Science 1998; 280:574–577.

30. Offermanns S, Toombs CF, Hu YH, Simon MI. Defective platelet activation in G alpha(q)-deficient mice. Nature 1997; 389:183–186.

31. Wettschureck N, Rutten H, Zywietz A, Gehring D, Wilkie TM, Chen J, Chien KR, Offermanns S. Absence of pressure overload induced myocardial hypertrophy after conditional inactivation of Galphaq/Galpha11 in cardiomyocytes. Nat Med 2001; 7: 1236–1240.

32. D'Angelo DD, Sakata Y, Lorenz JN, Boivin GP, Walsh RA, Liggett SB, Dorn GW. Transgenic Galphaq overexpression induces cardiac contractile failure in mice. Proc Natl Acad Sci USA 1997; 94:8121–8126.

33. Dorn GW, Robbins J, Ball N, Walsh RA. Myosin heavy chain regulation and myocyte contractile depression after LV hypertrophy in aortic-banded mice. Am J Physiol 1994; 267:H400–H405.

34. Sakata Y, Hoit BD, Liggett SB, Walsh RA, Dorn GW. Decompensation of pressure-overload hypertrophy in G alpha q-overexpressing mice. Circulation 1998; 97: 1488–1495.

35. Adams JW, Pagel AL, Means CK, Oksenberg D, Armstrong RC, Brown JH. Cardiomyocyte apoptosis induced by Galphaq signaling is mediated by permeability transition pore formation and activation of the mitochondrial death pathway. Circ Res 2000; 87:1180–1187.

36. Narula J, Haider N, Virmani R, DiSalvo TG, Kolodgie FD, Hajjar RJ, Schmidt U, Semigran MJ, Dec GW, Khaw BA. Apoptosis in myocytes in end-stage heart failure. N Engl J Med 1996; 335:1182–1189.

37. Olivetti G, Abbi R, Quaini F, Kajstura J, Cheng W, Nitahara JA, Quaini E, Di Loreto C, Beltrami CA, Krajewski S, et al. Apoptosis in the failing human heart. N Engl J Med 1997; 336:1131–1141.

38. Colucci WS. Apoptosis in the heart. N Engl J Med 1996; 335:1224–1226.

39. Guerra S, Leri A, Wang X, Finato N, Di Loreto C, Beltrami CA, Kajstura J, Anversa P. Myocyte death in the failing human heart is gender dependent. Circ Res 1999; 85: 856–866.

40. Mallat Z, Tedgui A, Fontaliran F, Frank R, Durigon M, Fontaine G. Evidence of apoptosis in arrhythmogenic right ventricular dysplasia. N Engl J Med 1996; 335: 1190–1196.

41. Itoh G, Tamura J, Suzuki M, Suzuki Y, Ikeda H, Koike M, Nomura M, Jie T, Ito K. DNA fragmentation of human infarcted myocardial cells demonstrated by the nick end labeling method and DNA agarose gel electrophoresis. Am J Pathol 1995; 146: 1325–1331.

42. Misao J, Hayakawa Y, Ohno M, Kato S, Fujiwara T, Fujiwara H. Expression of bcl-2 protein, an inhibitor of apoptosis, and Bax, an accelerator of apoptosis, in ventricular myocytes of human hearts with myocardial infarction. Circulation 1996; 94:1506–1512.

43. Cigola E, Kajstura J, Li B, Meggs LG, Anversa P. Angiotensin II activates programmed myocyte cell death in vitro. Exp Cell Res 1997; 231:363–371.

44. Kajstura J, Cigola E, Malhotra A, Li P, Cheng W, Meggs LG, Anversa P. Angiotensin II induces apoptosis of adult ventricular myocytes in vitro. J Mol Cell Cardiol 1997; 29:859–870.

45. Fortuno MA, Ravassa S, Etayo JC, Diez J. Overexpression of Bax protein and enhanced apoptosis in the left ventricle of spontaneously hypertensive rats: effects of AT1 blockade with losartan. Hypertension 1998; 32:280–286.

46. Goussev A, Sharov VG, Shimoyama H, Tanimura M, Lesch M, Goldstein S, Sabbah HN. Effects of ACE inhibition on cardiomyocyte apoptosis in dogs with heart failure. Am J Physiol 1998; 275:H626–H631.

47. Leri A, Claudio PP, Li Q, Wang X, Reiss K, Wang S, Malhotra A, Kajstura J, Anversa P. Stretch-mediated release of angiotensin II induces myocyte apoptosis by activating p53 that enhances the local renin-angiotensin system and decreases the Bcl-2-to-Bax protein ratio in the cell. J Clin Invest 1998; 101:1326–1342.

48. Leri A, Liu Y, Wang X, Kajstura J, Malhotra A, Meggs LG, Anversa P. Overexpression of insulin-like growth factor-1 attenuates the myocyte renin-angiotensin system in transgenic mice. Circ Res 199; 84:752–762.

49. Leri A, Liu Y, Li B, Fiordaliso F, Malhotra A, Latini R, Kajstura J, Anversa P. Up-regulation of AT(1) and AT(2) receptors in postinfarcted hypertrophied myocytes and stretch-mediated apoptotic cell death. Am J Pathol 2000; 156:1663–1672.

50. Barlucchi L, Leri A, Dostal DE, Fiordaliso F, Tada H, Hintze TH, Kajstura J, Nadal-Ginard B, Anversa P. Canine ventricular myocytes possess a renin-angiotensin system that is upregulated with heart failure. Circ Res 2001; 88:298–304.

51. Wencker D, Chandra M, Nguyen K, Miao W, Garantziotis S, Factor SM, Shirani J, Armstrong RC, Kitsis RN. A mechanistic role for cardiac myocyte apoptosis in heart failure. J Clin Invest 2003; 111:1497–1504.

52. Hirota H, Chen J, Betz UA, Rajewsky K, Gu Y, Ross J, Jr., Muller W, Chien KR. Loss of a gp130 cardiac muscle cell survival pathway is a critical event in the onset of heart failure during biomechanical stress. Cell 1999; 97:189–198.

53. Li Q, Li B, Wang X, Leri A, Jana KP, Liu Y, Kajstura J, Baserga R, Anversa P. Overexpression of insulin-like growth factor-1 in mice protects from myocyte death after infarction, attenuating ventricular dilation, wall stress, and cardiac hypertrophy. J Clin Invest 1997; 100:1991–1999.

54. De Windt LJ, Lim HW, Taigen T, Wencker D, Condorelli G, Dorn GW, Kitsis RN, Molkentin JD. Calcineurin-mediated hypertrophy protects cardiomyocytes from apoptosis in vitro and in vivo: An apoptosis-independent model of dilated heart failure. Circ Res 2000; 86:255–263.

55. Bialik S, Cryns VL, Drincic A, Miyata S, Wollowick AL, Srinivasan A, Kitsis RN. The mitochondrial apoptotic pathway is activated by serum and glucose deprivation in cardiac myocytes. Circ Res 1999; 85:403–414.

56. Narula J, Pandey P, Arbustini E, Haider N, Narula N, Kolodgie FD, Dal Bello B, Semigran MJ, Bielsa-Masdeu A, Dec GW, et al. Apoptosis in heart failure: release of cytochrome c from mitochondria and activation of caspase-3 in human cardiomyopathy. Proc Natl Acad Sci USA 1999; 96:8144–8149.

57. Green DR, Reed JC. Mitochondria and apoptosis. Science 1998; 281:1309–1312.

58. O'Rourke B. Apoptosis: rekindling the mitochondrial fire. Circ Res 1999; 85:880–883.

59. Adams JM, Cory S. The Bcl-2 protein family: arbiters of cell survival. Science 1998; 281:1322–1326.

60. Pellegrini M, Strasser A. A portrait of the Bcl-2 protein family: life, death, and the whole picture. J Clin Immunol 1999; 19:365–377.

61. Kluck RM, Bossy-Wetzel E, Green DR, Newmeyer DD. The release of cytochrome c from mitochondria: a primary site for Bcl-2 regulation of apoptosis. Science 1997; 275:1132–1136.

62. Antonsson B, Conti F, Ciavatta A, Montessuit S, Lewis S, Marinou I, Bernasconi L, Bernard A, Mermod JJ, Mazzei G, et al. Inhibition of Bax channel-forming activity by Bcl-2. Science 1997; 277:370–372.

63. Cook SA, Sugden PH, Clerk A. Regulation of bcl-2 family proteins during development and in response to oxidative stress in cardiac myocytes: association with changes in mitochondrial membrane potential. Circ Res 1999; 85:940–949.

64. Fortuno MA, Zalba G, Ravassa S, E D'Elom, Beaumont FJ, Fortuno A, Diez J. p53-mediated upregulation of BAX gene transcription is not involved in Bax-alpha protein overexpression in the left ventricle of spontaneously hypertensive rats. Hypertension 1999; 33:1348–1352.

65. Yussman MG, Toyokawa T, Odley A, Lynch RA, Wu G, Colbert MC, Aronow BJ, Lorenz JN, Dorn GW. Mitochondrial death protein Nix is induced in cardiac hypertrophy and triggers apoptotic cardiomyopathy. Nat Med 2002; 8:725–730.

66. Nishizuka Y. Studies and perspectives of protein kinase C. Science 1986; 233:305–312.

67. Gu X, Bishop SP. Increased protein kinase C and isozyme redistribution in pressure-overload cardiac hypertrophy in the rat. Circ Res 1994; 75:926–931.

68. Schunkert H, Sadoshima J, Cornelius T, Kagaya Y, Weinberg EO, Izumo S, Riegger G, Lorell BH. Angiotensin II-induced growth responses in isolated adult rat hearts. Evidence for load-independent induction of cardiac protein synthesis by angiotensin II. Circ Res 1995; 76:489–497.

69. Gray MO, Karliner JS, Mochly-Rosen D. A selective epsilon-protein kinase C antagonist inhibits protection of cardiac myocytes from hypoxia-induced cell death. J Biol Chem 1997; 272:30,945–30,951.

70. Disatnik MH, Jones SN, Mochly-Rosen D. Stimulus-dependent subcellular localization of activated protein kinase C; a study with acidic fibroblast growth factor and transforming growth factor-beta 1 in cardiac myocytes. J Mol Cell Cardiol 1995; 27:2473–2481.

71. Disatnik MH, Buraggi G, Mochly-Rosen D. Localization of protein kinase C isozymes in cardiac myocytes. Exp Cell Res 1994; 210:287–297.

72. Hug H, Sarre TF. Protein kinase C isoenzymes: divergence in signal transduction? Biochem J 1993; 291(Pt 2):329–343.

73. Steinberg SF, Goldberg M, Rybin VO. Protein kinase C isoform diversity in the heart. J Mol Cell Cardiol 1995; 27:141–153.

74. Mochly-Rosen D. Localization of protein kinases by anchoring proteins: a theme in signal transduction. Science 1995; 268:247–251.
75. Johnson JA, Gray MO, Chen CH, Mochly-Rosen D. A protein kinase C translocation inhibitor as an isozyme-selective antagonist of cardiac function. J Biol Chem 1996; 271:24,962–24,966.
76. Zhang ZH, Johnson JA, Chen L, El Sherif N, Mochly-Rosen D, Boutjdir M. C2 region-derived peptides of beta-protein kinase C regulate cardiac Ca^{2+} channels. Circ Res 1997; 80:720–729.
77. Pass JM, Zheng Y, Wead WB, Zhang J, Li RC, Bolli R, Ping P. PKCepsilon activation induces dichotomous cardiac phenotypes and modulates PKCepsilon-RACK interactions and RACK expression. Am J Physiol Heart Circ Physiol 2001; 280(3): H946–H955.
78. Mochly-Rosen D, Wu G, Hahn H, Osinska H, Liron T, Lorenz JN, Yatani A, Robbins J, Dorn GW. Cardiotrophic effects of protein kinase C epsilon: analysis by in vivo modulation of PKCepsilon translocation. Circ Res 2000; 86:1173–1179.
79. Wu G, Toyokawa T, Hahn H, Dorn GW. Epsilon protein kinase C in pathological myocardial hypertrophy. Analysis by combined transgenic expression of translocation modifiers and Galphaq. J Biol Chem 2000; 275:29,927–29,930.
80. Gu X, Bishop SP. Increased protein kinase C and isozyme redistribution in pressure-overload cardiac hypertrophy in the rat. Circ Res 1994; 75:926–931.
81. Schunkert H, Sadoshima J, Cornelius T, Kagaya Y, Weinberg EO, Izumo S, Riegger G, Lorell BH. Angiotensin II-induced growth responses in isolated adult rat hearts. Evidence for load-independent induction of cardiac protein synthesis by angiotensin II. Circ Res 1995; 76:489–497.
82. Takeishi Y, Ping P, Bolli R, Kirkpatrick DL, Hoit BD, Walsh RA. Transgenic overexpression of constitutively active protein kinase C epsilon causes concentric cardiac hypertrophy. Circ Res 2000; 86:1218–1223.
83. Chen L, Huhn H, Wu G, Chen CH, Liron T, Schectman D, Cavallaro G, Banci L, Guo Y, Bolli R, et al. Opposing cardioprotective actions and parallel hypertrophic effects of delta PKC and epsilon PKC. Proc Natl Acad Sci USA 2001; 98:11,114–11,119.
84. Hahn HS, Yussman MG, Toyokawa T, Marreez Y, Barrett TJ, Hilty KC, Osinska H, Robbins J, Dorn GW. Ischemic protection and myofibrillar cardiomyopathy: dose-dependent effects of in vivo delta PKC inhibition. Circ Res 2002; 91:741–748.
85. Hahn HS, Marreez Y, Odley A, Sterbling A, Yussman MG, Hilty C, Bodi I, Liggett SB, Schwartz A, Dorn II GW. Protein kinase C alpha negatively regulates systolic and diastolic function in pathologic hypertrophy. Circ Res 2003; 93:1111–1119.
86. Hoffman BB. Catecholamines, sympathomimetic drugs, and adrenergic antagonists. In: Hardman JG, Limbird LE, eds. The Pharmacological Basis of Therapeutics. New York: McGraw-Hill, 2001:215–268.
87. Bristow MR, Ginsburg R, Umans V, Fowler M, Minobe W, Rasmussen R, Zera P, Menlove R, Shah P, Jamieson S. Beta 1- and beta 2-adrenergic-receptor subpopulations in nonfailing and failing human ventricular myocardium: coupling of both receptor subtypes to muscle contraction and selective beta 1-receptor down-regulation in heart failure. Circ Res 1986; 59:297–309.
88. del Monte F, Kaumann AJ, Poole-Wilson PA, Wynne DG, Pepper J, Harding SE. Coexistence of functioning beta 1- and beta 2-adrenoceptors in single myocytes from human ventricle. Circulation 1993; 88:854–863.
89. Rohrer DK, Desai KH, Jasper JR, Stevens ME, Regula DP Jr, Barsh GS, Bernstein D, Kobilka BK. Targeted disruption of the mouse beta1-adrenergic receptor gene: developmental and cardiovascular effects. Proc Natl Acad Sci USA 1996; 93:7375–7380.
90. Hein L, Altman JD, Kobilka BK. Two functionally distinct alpha2-adrenergic receptors regulate sympathetic neurotransmission. Nature 1999; 402:181–184.

91. Bristow MR, Hershberger RE, Port JD, Minobe W, Rasmussen R. Beta 1- and beta 2-adrenergic receptor-mediated adenylate cyclase stimulation in nonfailing and failing human ventricular myocardium. Mol Pharmacol 1989; 35:295–303.

92. Engelhardt S, Hein L, Wiesmann F, Lohse MJ. Progressive hypertrophy and heart failure in beta1-adrenergic receptor transgenic mice. Proc Natl Acad Sci USA 1996; 96:7059–7064.

93. Liggett SB, Tepe NM, Lorenz JN, Canning AM, Jantz TD, Mitarai S, Yatani A, Dorn GW. Early and delayed consequences of beta(2)-adrenergic receptor overexpression in mouse hearts: critical role for expression level. Circulation 2000; 101:1707–1714.

94. Cohn JN, Levine TB, Olivari MT, Garberg V, Lura D, Francis GS, Simon AB, Rector T. Plasma norepinephrine as a guide to prognosis in patients with chronic congestive heart failure. N Engl J Med 1984; 311:819–823.

95. Bristow MR. beta-adrenergic receptor blockade in chronic heart failure. Circulation 2000; 101:558–569.

96. Liggett SB. Beta-adrenergic receptors in the failing heart: the good, the bad, and the unknown. J Clin Invest 2001; 107:947–948.

97. Chesley A, Lundberg MS, Asai T, Xiao RP, Ohtani S, Lakatta EG, Crow MT. The beta(2)-adrenergic receptor delivers an antiapoptotic signal to cardiac myocytes through G(i)-dependent coupling to phosphatidylinositol 3'-kinase. Circ Res 2000; 87:1172–1179.

98. Zhu WZ, Zheng M, Koch WJ, Lefkowitz RJ, Kobilka BK, Xiao RP. Dual modulation of cell survival and cell death by beta(2)-adrenergic signaling in adult mouse cardiac myocytes. Proc Natl Acad Sci USA 2001; 98:1607–1612.

99. Van Campen LC, Visser FC, Visser CA. Ejection fraction improvement by beta-blocker treatment in patients with heart failure: an analysis of studies published in the literature. J Cardiovasc Pharmacol 1998; 32(suppl 1):S31–S35.

100. Rockman HA, Chien KR, Choi DJ, Iaccarino G, Hunter JJ, Ross J, Jr., Lefkowitz RJ, Koch WJ. Expression of a beta-adrenergic receptor kinase 1 inhibitor prevents the development of myocardial failure in gene-targeted mice. Proc Natl Acad Sci USA 1998; 95:7000–7005.

101. Dorn GW, Tepe NM, Lorenz JN, Koch WJ, Liggett SB. Low- high-level transgenic expression of beta2-adrenergic receptors differentially affect cardiac hypertrophy and function in Galphaq-overexpressing mice. Proc Natl Acad Sci USA 1999; 96:6400–6405.

102. Small KM, DW McGraw, Liggett SB. Pharmacology and physiology of human adrenergic receptor polymorphisms. Annu Rev Pharmacol Toxicol 2003; 43:381–411.

103. Small KM, Forbes SL, Brown KM, Liggett SB. An asn to lys polymorphism in the third intracellular loop of the human alpha 2A-adrenergic receptor imparts enhanced agonist-promoted Gi coupling. J Biol Chem 2000; 275:38,518–38,523.

104. Small KM, Brown KM, Forbes SL, Liggett SB. Polymorphic deletion of three intracellular acidic residues of the alpha 2B-adrenergic receptor decreases G protein-coupled receptor kinase-mediated phosphorylation and desensitization. J Biol Chem 2001; 276: 4917–4922.

105. Small KM, Forbes SL, Rahman FF, Bridges KM, Liggett SB. A four amino acid deletion polymorphism in the third intracellular loop of the human alpha 2C-adrenergic receptor confers impaired coupling to multiple effectors. J Biol Chem 2000; 275: 23,059–23,064.

106. Levin MC, Marullo S, Muntaner O, Andersson B, Magnusson Y. The myocardium-protective Gly-49 variant of the beta 1-adrenergic receptor exhibits constitutive activity and increased desensitization and down-regulation. J Biol Chem 2002; 277:30,429–30,435.

107. Mason DA, Moore JD, Green SA, Liggett SB. A gain-of-function polymorphism in a G-protein coupling domain of the human beta1-adrenergic receptor. J Biol Chem 1999; 274:12,670–12,674.

108. Green SA, Turki J, Innis M, Liggett SB. Amino-terminal polymorphisms of the human beta 2-adrenergic receptor impart distinct agonist-promoted regulatory properties. Biochemistry 1994; 33:9414–9419.

109. Green SA, Cole G, Jacinto M, Innis M, Liggett SB. A polymorphism of the human beta 2-adrenergic receptor within the fourth transmembrane domain alters ligand binding and functional properties of the receptor. J Biol Chem 1993; 268:23,116–23,121.

110. Turki J, Lorenz JN, Green SA, Donnelly ET, Jacinto M, Liggett SB. Myocardial signaling defects and impaired cardiac function of a human beta 2-adrenergic receptor polymorphism expressed in transgenic mice. Proc Natl Acad Sci USA 1996; 93:10,483–10,488.

111. Liggett SB, Wagoner LE, Craft LL, Hornung RW, Hoit BD, McIntosh TC, Walsh RA. The Ile164 beta2-adrenergic receptor polymorphism adversely affects the outcome of congestive heart failure. J Clin Invest 1998; 102:1534–1539.

112. Wagoner LE, Craft LC, Singh BK, Suresh DP, Zengel PW, McGuire N, Abraham WT, Chenier TC, Dorn GW, Liggett SB. Polymorphisms of the β2 Adrenergic Receptor determine exercise capacity in patients with heart failure. Circ Res 2000; 86:834–840.

113. Wagoner LE, Craft LL, Zengel P, N McGuire, Rathz DA, Dorn GW, Liggett SB. Polymorphisms of the beta1-adrenergic receptor predict exercise capacity in heart failure. Am Heart J 2002; 144:840–846.

114. Johnson JA, Zineh I, Puckett BJ, SP McGorray, Yarandi HN, Pauly DF. Beta 1-adrenergic receptor polymorphisms and antihypertensive response to metoprolol. Clin Pharmacol Ther 2003; 74:44–52.

115. Humma LM, Puckett BJ, Richardson HE, Terra SG, Andrisin TE, Lejeune BL, Wallace MR, Lweis JF, McNamara DM, Picoult-Newberg L. Effects of beta1-adrenoceptor genetic polymorphisms on resting hemodynamics in patients under-going diagnostic testing for ischemia. Am J Cardiol 2001; 88:1034–1037..

116. Perez JM, Rathz DA, Petrashevskaya NN, Hahn HS, Wagoner LE, Schwartz A, Dorn GW, Liggett SB, et al. Beta(1)-adrenergic receptor polymorphisms confer differential function and predisposition to heart failure. Nat Med 2003; 9:1300–1305.

117. Small KM, Wagoner LE, Levin AM, Kardia SL, Liggett SB. Synergistic polymorphisms of beta1- and alpha2C-adrenergic receptors and the risk of congestive heart failure. N Engl J Med 2002; 347:1135–1142.

118. Lynch RA, Wagoner L, Li S, Sparks L, Molkentin J, Gorn GW. Novel and nondetected human signaling protein polymorphisms. Physiol Genomics 2002; 10:159–168.

119. Strasser RH, Sibley DR, Lefkowitz RJ. A novel catecholamine-activated adenosine cyclic 3′,5′-phosphate independent pathway for beta-adrenergic receptor phosphorylation in wild-type and mutant S49 lymphoma cells: mechanism of homologous desensitization of adenylate cyclase. Biochemistry 1986; 25:1371–1377.

120. Benovic JL, Strasser RH, Caron MG, Lefkowitz RJ. Beta-adrenergic receptor kinase: identification of a novel protein kinase that phosphorylates the agonist-occupied form of the receptor. Proc Natl Acad Sci USA 1986; 83:2797–2801.

121. Benovic JL, DeBlasi A, Stone WC, Caron MG, Lefkowitz RJ. Beta-adrenergic receptor kinase: primary structure delineates a multigene family. Science 1989; 246:235–240.

122. Kunapuli P, Benovic JL. Cloning and expression of GRK5: a member of the G protein-coupled receptor kinase family. Proc Natl Acad Sci USA 1993; 90:5588–5592.

123. Pitcher JA, Freedman NJ, Lefkowitz RJ. G protein-coupled receptor kinases. Annu Rev Biochem 1998; 67:653–692.

124. Ferguson SS. Evolving concepts in G protein-coupled receptor endocytosis: the role in receptor desensitization and signaling. Pharmacol Rev 2001; 53:1–24.

125. Dicker F, Quitterer U, Winstel R, Honold K, Lohse MJ. Phosphorylation-independent inhibition of parathyroid hormone receptor signaling by G protein-coupled receptor kinases. Proc Natl Acad Sci USA 1999; 96:5476–5481.

126. Koch WJ, Rockman HA, Samama P, Hamilton RA, Bond RA, Milano CA, Lefkowitz RJ. Cardiac function in mice overexpressing the beta-adrenergic receptor kinase or a beta ARK inhibitor. Science 1995; 268:1350–1353.

127. Rockman HA, Choi DJ, Rahman NU, Akhter SA, Lefkowitz RJ, Koch WJ. Receptor-specific in vivo desensitization by the G protein-coupled receptor kinase-5 in transgenic mice. Proc Natl Acad Sci USA 1996; 93:9954–9959.

128. Jaber M, Koch WJ, Rockman H, Smith B, Bond RA, Sulik KK, Ross J Jr, Lefkowitz RJ, Caron MG, Giros B. Essential role of beta-adrenergic receptor kinase 1 in cardiac development and function. Proc Natl Acad Sci USA 1996; 93:12,974–12,979.

129. Rockman HA, Choi DJ, Akhter SA, Jaber M, Giros B, Lefkowitz RJ, Caron MG, Koch WJ. Control of myocardial contractile function by the level of beta-adrenergic receptor kinase 1 in gene-targeted mice. J Biol Chem 1998; 273:18,180–18,184.

130. Manning BS, Shotwell K, Mao L, Rockman HA, Koch WJ. Physiological induction of a beta-adrenergic receptor kinase inhibitor transgene preserves ss-adrenergic responsiveness in pressure-overload cardiac hypertrophy. Circulation 2000; 102:2751–2757.

131. Felder RA, Sanada H, Xu J, Yu PY, Wang Z, Watanabe H, Asico LD, Wang W, Zheng S, Yamaguchi I, Williams SM, Gainer J, Brown NJ, Hazen-Martin D, Wong LJ, Robillard JE, Carey RM, Eisner GM, Jose PA. G protein-coupled receptor kinase 4 gene variants in human essential hypertension. Proc Natl Acad Sci USA 2002; 99:3872–3877.

132. Gros R, Benovic JL, Tan CM, Feldman RD. G-protein-coupled receptor kinase activity is increased in hypertension. J Clin Invest 1997; 99:2087–2093.

133. Ungerer M, Kessebohm K, Kronsbein K, Lohse MJ, Richardt G. Activation of beta-adrenergic receptor kinase during myocardial ischemia. Circ Res 1996; 79:455–460.

134. Bristow MR, Hershberger RE, Port JD, Gilbert EM, Sandoval A, Rasmussen R, Cates AE, Feldman AM. Beta-adrenergic pathways in nonfailing and failing human ventricular myocardium. Circulation 1990; 82:I12–I25.

135. Bristow MR, Kantrowitz NE, Ginsburg R, Fowler MB. Beta-adrenergic function in heart muscle disease and heart failure. J Mol Cell Cardiol 1985; 17(suppl 2):41–52.

136. Ungerer M, Bohm M, Elce JS, Erdmann E, Lohse MJ. Altered expression of beta-adrenergic receptor kinase and beta 1-adrenergic receptors in the failing human heart. Circulation 1993; 87:454–463.

137. Ungerer M, Parruti G, Bohm M, Puzicha M, A DeBlasi, Erdmann E, Lohse MJ. Expression of beta-arrestins and beta-adrenergic receptor kinases in the failing human heart. Circ Res 1994; 74:206–213.

14

The Role of Inflammatory Mediators in Cardiac Remodeling

Gabor Szalai, Yasushi Sakata, Jana Burchfield, Shintaro Nemoto, and Douglas L. Mann
Winters Center for Heart Failure Research, The Cardiology Section, Department of Medicine, Houston Veterans Affairs Medical Center and Baylor College of Medicine, Houston, Texas, U.S.A.

INTRODUCTION

One of the major conceptual advances in our understanding of the pathogenesis of heart failure has been the insight that it may progress as the result of the sustained overexpression of biologically active molecules, such as norepinephrine and angiotensin II. By virtue of their deleterious effects, these factors are sufficient to contribute to disease progression by provoking worsening left ventricular (LV) remodeling and progressive LV dysfunction (1–3). Recently, a second class of biologically active molecules, termed cytokines, has also been identified in the setting of heart failure (4,5). Analogous to the situation with neurohormones, the overexpression of cytokines provokes worsening LV remodeling and progressive LV dysfunction (1–3, 6–8), suggesting that these molecules may play an important role in disease progression in heart failure. In the present review, we will discuss the evidence which suggests that inflammatory mediators play an important role in cardiac remodeling, as well as recent evidence which suggests that there are functionally significant interactions between the neurohormones and the proinflammatory cytokines that contribute to adverse cardiac remodeling.

OVERVIEW OF THE BIOLOGY OF PROINFLAMMATORY CYTOKINES

The term "cytokine" is applied to a group of relatively small molecular weight protein molecules (generally, 15–30 kDa) that are secreted by cells in response to a variety of inducing stimuli. Although cytokines are similar to polypeptide hormones in many respects, they can be produced by a variety of cell types in a number of different tissues, as opposed to being produced by a specific cell type in a specific organ, as is the case for polypeptide hormones. Whereas "proinflammatory cytokines" have traditionally been thought to be produced by the immune system, one of the important recent observations is that virtually all nucleated cell types within the myocardium, including cardiac myocytes themselves, are capable of synthesizing a portfolio of proinflammatory cytokines in response to virtually all forms of cardiac injury [reviewed in (9)]. Thus, from a conceptual standpoint, these molecules should be envisioned as proteins that are produced locally within the myocardium by "cardiocytes" (i.e., cells that reside within the myocardium), in response to one or more different forms of environmental stress. An important corollary of this statement is that the expression of these "stress activated" cytokines can occur in the complete absence of activation of the immune system. The biological properties of several important proinflammatory cytokines, including tumor necrosis factor (TNF) and members of the interleukin-1 (IL-1) family [including the recently described member IL-18 (10)], as well as those of the interleukin-6 (IL-6) family of cytokines has been reviewed recently (11,12).

EFFECT OF INFLAMMATORY MEDIATORS ON CARDIAC REMODELING

The first suggestion that inflammatory mediators played a role in LV remodeling was from a study in which human volunteers were administered endotoxin intravenously. These subjects developed a 20% increase in LV end-diastolic volume within five hours of endotoxin administration. Moreover, these effects appeared to be independent of discrete changes in LV loading conditions (13). Subsequent to these early human studies, experimental studies have consistently shown that sustained expression of proinflammatory cytokine leads to an increase in LV cavity dimension and LV wall thinning (6,8,14,15). While the mechanisms that are responsible for these potentially deleterious changes in cardiac structure are undoubtedly complex, the current literature suggests that inflammatory mediators produce important changes in the biology of cardiac myocytes and nonmyocytes, as well as changes in the myocardial extracellular matrix. Accordingly, in the following section, we will review the spectrum of deleterious biological effects that are exerted by inflammatory mediators in the adult mammalian heart.

Cardiac Myocyte Hypertrophy

As shown in the Table 1, inflammatory mediators have a number of important effects that contribute to LV remodeling, including myocyte hypertrophy, alterations

Table 1 Effects of Inflammatory Mediators on Left Ventricular Remodeling

Alterations in the biology of the myocyte
   Myocyte hypertrophy
   Contractile abnormalities
   Fetal gene expression
   Negative inotropic effects
Alteration in the extracellular matrix
   Degradation of the matrix
   Myocardial fibrosis
Progressive myocyte loss
   Necrosis
   Apoptosis

in fetal gene expression, negative inotropic effects, as well as progressive myocyte loss through apoptosis. Palmer et al. (16) demonstrated that incubation with interleukin-1β led to cell hypertrophy and reinitiated DNA synthesis in cultured neonatal cardiac myocytes. In keeping with these findings, TNF stimulation has been shown to provoke a modest increase in overall protein synthesis and sarcomeric protein synthesis in cultured adult cardiac myocytes (17). With respect to the mechanisms that are responsible for TNF–induced myocyte hypertrophy, a recent study suggests that reactive oxygen intermediates may be responsible for myocyte hypertrophy (18). In this study, TNF provoked the expression of reactive oxygen intermediates, as well as a significant increase in [^{3}H]leucine uptake and cardiac myocyte size that were sensitive to the antioxidant-butylated hydroxyanisole, suggesting that reactive oxygen intermediates were necessary for the effects of TNF on cardiac myocyte hypertrophy (18). Relevant to this discussion is the observation that TNF is sufficient to provoke cardiac myocyte hypertrophy in experimental models wherein TNF was infused at pathophysiologically relevant concentrations (8), as well as in transgenic mice that harbor cardiac-restricted overexpression of TNF (15).

IL-6 has also been implicated in hypertrophic growth in cardiac myocytes. Transgenic mice overexpressing both IL-6 and the IL-6 receptor (IL-6R) developed substantial cardiac hypertrophy, whereas transgenic mice overexpressing either IL-6 or IL-6R alone did not show detectable myocardial abnormalities. Furthermore, cultured neonatal cardiac myocytes enlarged following incubation with a combination of IL-6 and a soluble form of IL-6R, thus suggesting that the IL-6–induced activation of the gp130 signaling pathways leads to cardiac hypertrophy (19).

Cardiac Myocyte Apoptosis

The role of inflammatory mediators and programmed cell death has been examined in several studies. As an example, IL-1β–induced activation of nitric oxide (NO) has been implicated as a mediator of programmed cell death in neonatal cardiac myocytes (20). Incubation of rat neonatal cardiac myocytes with IL-1β and interferon-gamma for 48 hours resulted in an increase in the expression of inducible nitric oxide synthase (iNOS), nitrite production, and programmed cell death. Both the cytokine-induced nitrite accumulation and myocyte apoptosis could be completely prevented by the nonselective NOS inhibitor L-nitroarginine or the specific iNOS inhibitor 2-amino-5, 6-dihydro-6-methyl-4H-1, 3-thiazine, implicating a role for NO in programmed cell death. TNF-induced programmed cell death has been linked to activation of the type 1 TNF receptor (TNFR1) in adult rat cardiac myocytes. As shown in Figure 1, TNFR1-induced apoptosis may occur through at least two separate pathways that are coupled to distinct domains of TNFR1. Activation of an 80-amino acid region of TNFR1, termed the "death domain," has been shown to trigger apoptosis in certain cell types through the so-called extrinsic pathway (21,22). As shown in Figure 1, fas associated protein with death domain (FADD) recruits an interleukin converting enzyme (ICE)-like protease termed "MACH1" (for MORT1-associated CED homologue) (23)/"FLICE" (FADD-like *ICE*)/"Caspase 8" (24) to the death domain, with resultant activation of the apoptogenic death machinery. The apoptogenic signaling pathways, downstream from the receptor interacting protein (RIP), are comparatively less well understood, but appear to involve interaction with a protein termed "RAIDD" (RIP-associated ICH-1/CED3 protein with a death domain), which recruits ICH-1 (caspase 2) to the death domain complex (25), and is therefore capable of triggering a second apoptogenic

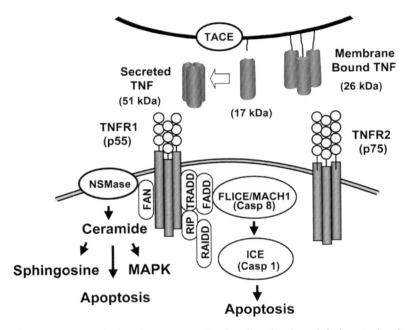

Figure 1 TNF-induced proapoptotic signaling in the adult heart. As shown, membrane-bound TNF is "shed" (cleaved) from cell membranes by TACE. Secreted TNF is then capable of binding to both the type 1 (TNFR1) and type 2 (TNFR2) TNF receptors on a variety of different cell types, including cardiac myocytes. TNF binding to TNFR1 activates two proapoptotic pathways: the intrinsic pathway mediated by activation of death domain proteins (TRADD, FADD, FLICE, ICE) or the neutral sphingomyelinase pathway (FAN, NSmase, Ceramide). *Abbreviations*: TNF, tumor necrosis factor; TNFR1, tumor necrosis factor receptor (type 1); TNFR2, tumor necrosis factor receptor (type 2); TACE, TNF-converting enzyme; FAN, factor associated with neutral sphingomyelinase; FADD, fas associated protein with death domain; TRADD, TNFRSF1A-associated via death domain; ICE, interleukin-1 converting enzyme; FLICE, fas associated protein with death domain like interleukin-1-converting enzyme.

signal from the death domain (26). TNF-induced activation of ICE-like proteases triggers programmed cell death through a pathway that involves increased mitochondrial permeability transition (27), presumably related to the proteolytic action of the upstream caspases, or alternatively through TNF-induced generation of reactive oxygen intermediates that may damage the mitochondria (28–30).

The second major pathway that links TNF to programmed cell death is the neutral sphingomyelinase pathway, which is also responsible for mediating the immediate negative inotropic effects of TNF (31). Recently, activation of the neutral sphingomyelinase pathway has been linked to the development of apoptosis in cultured adult cardiac myocytes (32). Although all of the steps that link TNF–induced activation of the neutral sphingomyelinase pathway with the induction of apoptosis have not been elucidated, the current literature suggests that occupancy of TNFR1 leads to recruitment of a protein termed "FAN" (factor associated with neutral sphingomyelinase) (33) to the neutral sphingomyelinase domain of TNFR1 (Fig. 1). FAN is then thought to interact directly with membrane bound neutral sphingomyelinase, with resultant generation of ceramide (34). Ceramide triggers programmed cell death through an as yet unknown pathway(s) that may involve the generation of reactive oxygen intermediates that are capable of causing mitochondrial damage and increased mitochondrial permeability transition, as was discussed above for the death domain (35–37). The importance of FAN in hypoxia/reoxygenation injury in rat cardiac myocytes has been show recently in a study in which a dominant-negative FAN construct significantly attenuated hypoxia/reoxygenation-induced cell death, whereas overexpression of wild-type FAN construct led to an increase in myocyte apoptosis (38).

Negative Inotropic Effects

The negative inotropic properties of TNF, the IL-1 family members [including the recently described member IL-18 (10)], as well as the interleukin-6 (IL-6) family of cytokines have been reviewed recently, and will only be discussed briefly herein [reviewed in (39)]. The extant literature suggests that TNF modulates myocardial function through at least two different pathways i.e., an immediate pathway that is manifest within minutes and is mediated by activation of the neutral sphingomyelinase pathway (31), and a delayed pathway that requires hours to days to develop and is mediated by NO (40,41). Several studies have shown that treatment with IL-1β will lead to a depression in cardiac myocyte contractility (42–46). Although the exact signal transduction pathways have not yet been identified for IL-1, the delayed nature of these IL-1β–mediated effects (47,48) suggests that they are mediated by NOS. Recently, it has been suggested that TNF and IL-1 may produce negative inotropic effects indirectly through activation and/or release of IL-18 which is a recently described member of the IL-1 family of cytokines (49). Relevant to the present discussion is the observation that specific blockade of IL-18 using neutralizing IL-18 binding protein leads to an improvement in myocardial contractility in atrial tissue that was subjected to ftlineischemia reperfusion injury (50). Although the signaling pathways that are responsible for the IL-18–induced negative inotropic effects have not been delineated thus far, it is likely that they will overlap those for IL-1, given that the IL-18 receptor complex utilizes components of the IL-1 signaling chain, including IL-1R-activating kinase (IRAK) and TNFR–associated factor-6 (TRAF-6) (49). IL-6 has been shown to decrease cardiac contractility via a NO–dependent pathway that is secondary to IL-6–induced phosphorylation of signal transducer and activator of transcription 3 (STAT3). In this study, IL-6 enhanced de novo synthesis of iNOS protein, increased NO production, and decreased rat cardiac myocyte contractility after two hours of incubation. The effects of IL-6 on iNOS production and myocyte contractility were blocked by genistein at concentrations that were sufficient to block IL-6–induced activation of STAT3. Taken together, these observations suggest that IL-6 is sufficient to produce negative inotropic effects through STAT3–mediated activation of iNOS (51).

The negative inotropic effects of TNF have been observed in vivo as well, both in studies in which rats were infused with pathophysiologically levels of TNF and in studies in which in transgenic mice with targeted overexpression of TNF (8,52,53). Franco et al. (52) used cine-magnetic resonance imaging to demonstrate that there was a significant increase in LV volume and a significant decrease in LV ejection fraction over time in transgenic mice with targeted overexpression of TNF. Importantly, these effects were shown to be dependent on gene dosage i.e., when the line of transgenic mice with high TNF expression (lineage 1) was compared to a transgenic line with lower myocardial TNF expression (lineage 2), there was a significantly

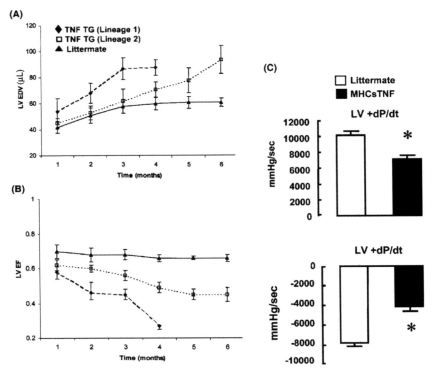

Figure 2 Effect of TNF on LV function, LV volume, and LV ejection fraction were serially examined by magnetic resonance imaging in two lines of transgenic mice (TNF TG) with high (*lineage 1*) and low (*lineage 2*) levels of myocardial TNF expression in comparison to age-matched littermate control mice: (**A**) Serial changes in LV volume in transgenic mice and littermate control mice. (**B**) Serial changes in LV ejection fraction in transgenic mice and littermate control mice. (**C**) LV contractility in mice with targeted overexpression of TNF (MHCsTNF) (53). For these studies, the animals were paced via the atrium to a heart rate at which positive dP/dt was maximal, as defined by examination of the force frequency curves for each animal, and peak positive and negative dP/dt assessed for MHCsTNF mice and littermate controls. *Abbreviations*: TNF, tumor necrosis factor; LV, left ventricular; TNF TG, tumor necrosis factor transgenic. *Source*: From Ref. 52.

greater increase in LV volume (Fig. 2A) and a significantly greater decrease in LV ejection fraction (Fig. 2B) in the transgenic mouse lines with higher TNF expression (52). In studies from laboratory, we measured LV function in line of transgenic mice with targeted overexpression of TNF (53) using Millar catheters. The animals were paced from the atrium to a heart rate at which dP/dt was maximal, as defined by the examination of the force frequency curves for each animal, and peak positive and negative dP/dt assessed for TNF transgenic mice and littermate controls. As shown in Figure 2C, there was a significant decrease in peak + dP/dt and peak − dP/dt in the TNF transgenic mice, consistent with the findings reported by Franco et al. (52).

Effects of Inflammatory Mediators on the Biology of Cardiac Fibroblasts

In addition to exerting effects on cardiac myocytes, proinflammatory cytokines can have a profound effect on a variety of different functions in cardiac fibroblasts. Two recent studies have shown that TNF inhibits collagen gene expression and/or collagen synthesis in cardiac fibroblasts (54,55). IL-1β has been shown to exert a potent antiproliferative effect on cardiac fibroblasts through upregulation of the transcriptional repressor yin yang-1 (YY1) (56). Although the effects of IL-6 have not been directly studied in cardiac fibroblasts, leukemia inhibitory factor (LIF) and cardiotrophin-1, both of which are members of the IL-6 family, have been shown to influence fibroblast function. Both CT-1 and LIF have been shown to stimulate fibroblast growth (57,58). LIF stimulation also inhibited the differentiation of cardiac fibroblasts into cardiac myofibroblasts and blunted the effects of tissue growth factor-β (TGF-β) on collagen synthesis (58). Thus, the results of this study suggest that the expression of LIF may serve as an autocrine/paracrine factor that dampens and negatively modulates ongoing remodeling of the extracellular matrix by preventing excessive fibrosis.

Effects of Inflammatory Mediators on the Extracellular Matrix

The expression of inflammatory mediators in the myocardium leads to disparate changes in the extracellular matrix that range from fibrillar collagen degradation, which would be expected to promote LV dilation, to excessive fibrillar collagen deposition, which would be expected to promote increased LV stiffness. As will be discussed below, the effects of inflammation of the myocardium are largely governed by the duration of inflammatory signaling.

Two lines of evidence suggest that TNF may promote LV remodeling through degradation of the extracellular matrix component of the myocardium. First, a recent experimental study (8) demonstrated that a systemic infusion of pathophysiologically relevant concentrations of TNF led to time-dependent increase in LV dimension (Fig. 3A) that was accompanied by progressive degradation of the extracellular matrix (Fig. 3C). Moreover, similar findings have been reported following a single infusion of TNF in dogs (14). Second, a recent study by Kubota et al. (15) showed that a transgenic mouse line that overexpressed TNF in the cardiac compartment, developed progressive LV dilation over a 24-week period of observation (Fig. 4). Similar findings have also been reported by Bryant et al. (6) and Sivasubramanian et al. (53), who observed identical findings with respect to LV dysfunction and LV dilation in transgenic mice with targeted overexpression of TNF in the heart. With respect to the mechanisms that are involved in TNF–induced LV dilation, it has been suggested that TNF–induced activation of matrix metalloproteinases (MMP) is responsible for this effect (53,59). As shown in Figures 5 and 6, respectively, there was a progressive loss of fibrillar collagen and increased MMP activation in the hearts of the transgenic mice overexpressing TNF in the cardiac compartment. The dissolution of the fibrillar collagen weave that surrounds the individual cardiac myocytes and links the myocytes together would be expected to allow for rearrangement ("slippage") of myofibrillar bundles within the ventricular wall (60). However, Figure 5 also shows that long-term stimulation (i.e., 8–12 weeks) with TNF resulted in an increase in fibrillar collagen content that was accompanied by decreased MMP activity (Fig. 6B), and increased expression of the tissue inhibitors of matrix metalloproteinases (TIMPs) (Fig. 6C). Taken together, these observations suggest that sustained myocardial inflammation provokes time-dependent changes in the balance between MMP activity and TIMP activity, i.e., during the early stages of inflammation, there is an increase in the ratio of MMP activity to TIMP levels that fosters LV dilation. However, with chronic inflammatory signaling there is a time-dependent

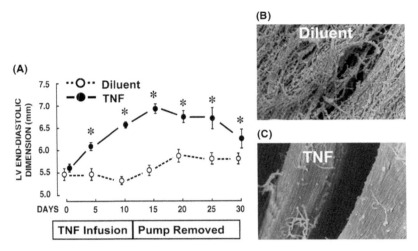

Figure 3 Effect of a continuous TNF infusion in vivo on LV structure and myocardial fibrillar collagen in rats. LV dimensions were determined in rats that underwent implantation of an intraperitoneal osmotic infusion pump that contained either diluent or TNF. The amount of TNF in the osmotic infusion pumps was titrated to achieve systemic levels that are observed in patients with heart failure (80–100 U/mL). After 15 days, the osmotic infusion pumps were removed and the animals were allowed to recover. LV dimensions were assessed serially by 2-D–directed M-mode echocardiography at baseline and every 5 days for a total of 30 days: (**A**) Serial changes in LV end-diastolic diameter before and after the osmotic infusion pumps were removed. (**B**) Scanning electron micrograph of a rat heart infused with diluent for 15 days, showing the presence of fibrillar collagen network investing individual cardiac myocytes. (**C**) Scanning electron micrograph of a rat heart infused with diluent for 15 days showing striking decrease in fibrillar collagen weave surrounding the cardiac myocytes. *Abbreviations*: TNF, tumor necrosis factor; LV, left ventricular. *Source*: From Ref. 8.

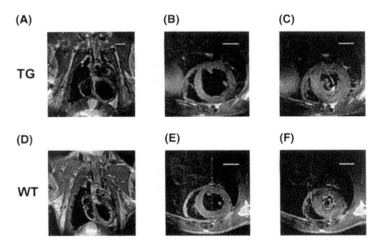

Figure 4 LV remodeling in a transgenic mouse model of TNF overexpression. TNF overexpression was targeted to the cardiac compartment of transgenic mice. Magnetic resonance images of the heart were obtained from 24-week old transgenic mice (*panels A–C*) and an age-matched control mouse (*panels D–F*). As shown, there was significant LV dilation in the animal harboring the TNF transgene in the cardiac compartment. *Abbreviations*: TNF, tumor necrosis factor; LV, left ventricular. *Source*: From Ref. 15.

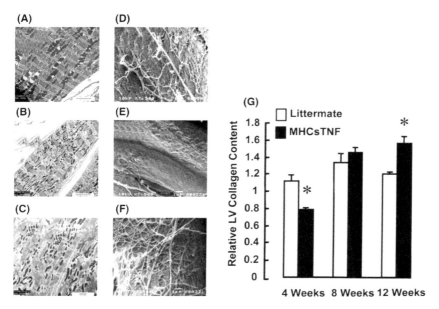

Figure 5 Effects of sustained proinflammatory cytokine expression on myocardial ultrastructure and collagen content. Panels (A)–(C): (A) representative transmission electron micrographs in littermate controls and (B) the TNF transgenic mice at four and (C) eight weeks of age. The transmission electron micrographs from the littermate control mice at four weeks (A) revealed a characteristic linear array of sarcomere and myofibril. In contrast, the myofibril in the four-week old TNF transgenic mice were less organized, with loss of sarcomeric registration observed in many of the sections (Fig. 3B). The ultrastructural abnormalities in the TNF transgenic mice were further exaggerated in the 12-week old TNF transgenic mice, which showed a significant loss of sarcomere registration and myofibril disarray (C). Panels (D)–(F): representative scanning electron micrographs in littermate controls (D) and the TNF transgenic mice a four (E) and eight weeks of age (F). (E) significant loss of fibrillar collagen in the TNF transgenic mice at four weeks of age when compared to age-matched littermate controls (D). However, as the TNF transgenic mice aged (12 weeks), there was an obvious increase in myocardial fibrillar collagen content. (Panel G): illustration of the myocardial collagen content as determined by picrosirius red staining. There was a loss of myocardial collagen content at four weeks of age in the TNF transgenic mice, that was later followed by a progressive increase in myocardial collagen content at 8 and 12 weeks of age. *Abbreviations*: TNF, tumor necrosis factor. *Source*: From Ref. 53.

increase in TIMP levels, with a resultant decrease in the ratio of MMP activity to TIMP activity, and a subsequent increase in myocardial fibrillar collagen content. Although the molecular mechanisms that are responsible for the transition between excessive degradation and excessive synthesis of the extracellular matrix are not known, these studies are consistent with parallel studies in experimental models of chronic injury/inflammation in the liver, lung, and kidney, wherein an initial increase in MMP expression is superseded by increased TIMP expression and increased expression of a number of fibrogenic cytokines, most notably TGF-β (61,62).

In contrast to the acute effects of myocardial inflammation which are capable of leading to degradation of fibrillar collagen and LV dilation, sustained myocardial inflammation leads to myocardial fibrosis. For example, three lines of transgenic

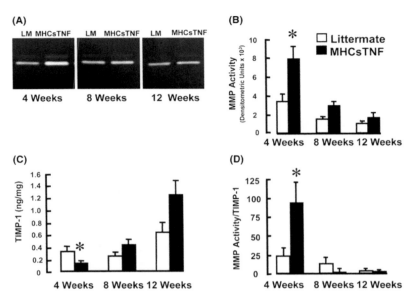

Figure 6 Effects of sustained proinflammatory cytokine expression on MMP activity and TIMP levels. (**A**) A zymogram of total MMP activity in the TNF transgenic mice (TNF-TG) and littermate (LM) control mice a 4, 8, and 12 weeks of age, (**B**) Summary of the results of group data for total MMP zymographic activity. MMP activity was significantly ($p < 0.001$) greater in the TNF transgenic mice at four weeks of age; however, MMP activity was no different from littermate control mice at 8 and 12 weeks of age. (**C**) Time-dependent changes in TIMP levels at 4, 8, and 12 weeks in the TNF transgenic and littermate control mice. At four weeks of age, TIMP-1 levels were significantly less in the TNF transgenic mice; however, TIMP-1 levels increased progressively in the TNF transgenic mice from 8 to 12 weeks of age. (**D**) Time-dependent changes in the ratio of MMP activity/TIMP levels in the TNF transgenic and littermate control mice. As shown, at four weeks of age the ratio of MMP activity/TIMP-1 levels was significantly greater in the TNF transgenic mice, thus favoring collagen degradation (Fig. 4D); however, the ratio of MMP activity/TIMP-1 decreased progressively from 8 to 12 weeks of age, thus favoring collagen accumulation (Fig. 4D). *Abbreviations*: TNF, tumor necrosis factor; TNF TG tumor necrosis factor Transgenic; TIMP, tissue inhibitors of matrix metalloproteinases; MMP, matrix metalloproteinase. *Source*: From Ref. 53.

mice with cardiac-restricted overexpression of TNF developed by three independent laboratories, all of them develop a common profibrotic phenotype as they age (6,15,53). Given that TNF inhibits collagen gene expression and/or collagen synthesis in cardiac fibroblasts in vitro (54,55), it is likely that the increased myocardial fibrosis in transgenic mice with targeted overexpression of TNF is mediated by one or more indirect effects of TNF. As an example, TNF stimulation has been shown to increase the density of angiotensin type I receptors (AT_1) on cardiac fibroblasts (63), as well as increase the sensitivity of these cells to the profibrotic actions of endogenous angiotensin II (55). A second potential explanation for the observed fibrosis in the TNF transgenic mice is that TNF increases the expression of TGF-β. As shown in Figure 7, TGF-β_1 and TGF-β_2 mRNA and protein levels were significantly increased in the hearts of the TNF transgenic mice (MHCsTNF) relative to littermate controls (53). Moreover, it bears emphasis that many of the profibrotic actions of angiotensin II in vitro (64) and in vivo (65) are mediated indirectly by

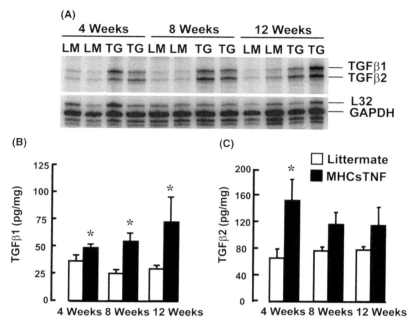

Figure 7 TGF-β_1 and TGFβ_2 levels in the mice with targeted overexpression of TNF (MHCsTNF) and littermate control mice. (**A**) Representative ribonuclease protection assay for TGF-β_1 and TGF-β_2 mRNA levels in the MHCsTNF and littermate controls at 4, 8 and 12 weeks of age. (**B**) The protein levels for TGF-β_1 and (**C**) The protein levels for TGF-β_2 in the MHCsTNF and littermate controls at 4, 8 and 12 weeks of age. *Abbreviations*: GAPDH, glyceraldehyde-3-phosphate dehydrogenase; LM, littermate control; TG, MHCsTNF transgenic; *, $p < 0.05$ compared to littermate control mice; TNF, tumor necrosis factor; TGF-β, tissue growth factor-β; mRNA, messenger ribonucleic acid. *Source*: From Ref. 53.

TGF-β. Thus, taken together, these latter two observations suggest that TGF-β may play an important role in the pathological myocardial fibrosis that develops following sustained myocardial inflammation.

INTERACTIONS BETWEEN THE RENIN-ANGIOTENSIN SYSTEM AND PROINFLAMMATORY CYTOKINES IN ADVERSE CARDIAC REMODELING

Although neurohormonal and cytokine systems have been regarded as functionally distinct biological systems, recent studies suggest that these two systems can cross-regulate each other, with the result that neurohormonal and cytokine systems may participate in positive feed forward loops that contribute to adverse cardiac remodeling. Whereas angiotensin II was traditionally viewed as a circulating neurohormone that stimulated the constriction of vascular smooth muscle cells, aldosterone release from the adrenal gland, sodium reabsorption in the renal tubule, and/or as a stimulus for growth of cardiac myocytes or fibroblasts (66), it is becoming increasingly apparent that angiotensin II provokes inflammatory responses in a variety of cell and tissue types. For example, angiotensin II activates a redox sensitive transcription factor termed "nuclear factor-kappa B (NF-κB)" (67) that is critical for initiating the coordinated expression of the classical components of the myocardial inflammatory response, including increased expression of proinflammatory cytokines, NO, chemokines, and cell adhesion molecules (68,69). Pathophysiologically relevant concentrations of angiotensin II are sufficient to provoke TNF mRNA and protein synthesis in the adult heart through a NF-κB–dependent pathway (70). Figure 8 shows that treatment with angiotensin II resulted in a rapid increase in TNF mRNA

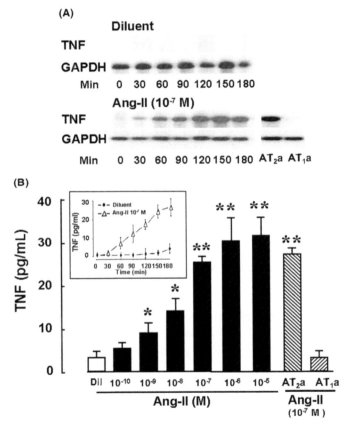

Figure 8 Angiotensin II–induced myocardial TNF biosynthesis in the adult heart: (A) TNF mRNA expression (RNase protection assay) was assessed ex vivo in diluent and angiotensin II–(10^{-7} M) treated (0–180 minutes), buffer-perfused Langendorff hearts, in the presence or absence of 10^{-6} M PD123319, an AT_2 receptor antagonist (AT_2a) or 10^{-6} M losartan, an AT_1 receptor antagonist (AT_1a). (B) Myocardial TNF protein production was assessed in the superfusates of the angiotensin II–treated hearts using ELISA, in the presence or absence of PD123319 (10^{-6} M) or losartan (10^{-6} M) pretreatment. The main panel of Figure 1B shows the dose-dependent effects of angiotensin II (10^{-10} M to 10^{-5} M), whereas the inset shows the time course (0–180 minutes) for TNF protein synthesis following stimulation with either diluent (solid circles) or 10^{-7} M angiotensin II (*open triangles*). *Abbreviations*: AT_1a, AT_1 receptor antagonist [losartan]; AT_2a, AT_2 receptor antagonist [PD123319]; *, $p < 0.05$; **, $p < 0.01$ compared to diluent treated hearts; TNF, tumor necrosis factor; mRNA, messenger ribonucleic acid. *Source*: From Ref. 73.

(Fig. 8A) and protein synthesis (Fig. 8B) in isolated buffer-perfused hearts. Stimulation of isolated adult cardiac myocytes with angiotensin II resulted in a threefold increase in TNF protein biosynthesis within one hour, and a 15-fold increase in TNF protein biosynthesis within 24 hours, suggesting that the increase in TNF biosynthesis in the intact heart was mediated, at least in part, at the level of the cardiac myocyte. The effects of angiotensin II on TNF mRNA and protein synthesis were mediated exclusively by the angiotensin type 1 receptor, insofar as pretreatment with the angiotensin type 1 receptor antagonist losartan completely abolished the effects of angiotensin II on TNF biosynthesis, whereas pretreatment with the angiotensin type 2 receptor receptor antagonist PD123319 had no effect on angiotensin II induced TNF biosynthesis (Fig. 8B). This study further showed that the effects of angiotensin II on TNF myocardial biosynthesis were dependent upon protein kinase C (PKC) mediated activation of NF-κB (70).

There is also increasing evidence that inflammatory mediators are capable of upregulating various components of the renin angiotensin system in a variety of

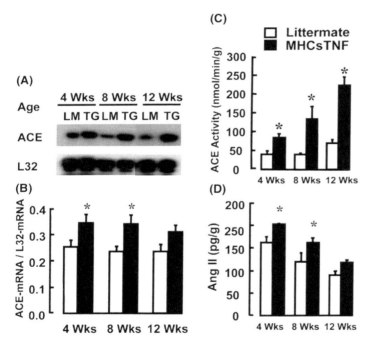

Figure 9 ACE mRNA, ACE activity, and angiotensin II peptide levels in mice with targeted overexpression of TNF, MHCsTNF, and littermate control mice: (**A**) Ribonuclease protection assay for ACE mRNA in the hearts of the 4-, 8-, and 12-week MHCsTNF transgenic and littermate control mice. (**B**) Group data in hearts from 4-, 8-, and 12-week-old MHCsTNF (n) and 4-, 8-, and 12-week-old littermate control mice (n). (**C**) ACE activity in the hearts from 4-, 8-, and 12-week-old MHCsTNF and the 4-, 8-, and 12-week old littermate control mice. (**D**) Group data for angiotensin II peptide levels in the hearts of the MHCsTNF and littermate control mice at 4-, 8-, and 12-weeks of age. *Abbreviations*: n, 7 hearts/time; mRNA, messenger ribonucleic acid; TNF, tumor necrosis factor; ACE, angiotensin converting enzyme; MHC, major histocompatibility complex. *, $p < 0.05$ versus age-matched control group by Tukey's test. *Source*: From Ref. 71.

mammalian tissues, including the heart. As a recent example, studies using transgenic mice with cardiac restricted overexpression of TNF have shown that targeted overexpression of TNF leads to an increase in angiotensin II peptide levels in the heart (71). This study serially examined several components of the renin angiotensin system, including angiotensinogen, renin, angiotensin converting enzyme (ACE), and angiotensin I and II peptide levels in a transgenic mouse line with cardiac restricted overexpression of TNF (MHCsTNF). There was a significant increase in ACE mRNA levels (Figs. 9A,B) and ACE activity (Fig. 9C), as well as increased angiotensin II peptide levels (Fig. 9D) in the hearts of the MHCsTNF mice relative to littermate controls. Importantly, the expression of renin and angiotensinogen was not increased in MHCsTNF mice compared with littermate controls. Thus, this study suggested that the increased levels of angiotensin II peptide in the MHCsTNF mice was principally the result of increased ACE activity, as opposed to increased activation of the more proximal components of the renin angiotensin system, namely renin and angiotensinogen. This study also showed that the activation of the renin angiotensin system was functionally significant in the TNF transgenic mice. That is, treatment of the MHCsTNF mice from four to eight weeks of age with losartan, significantly attenuated cardiac hypertrophy, myocardial fibrosis, and cardiac myocyte apoptosis in the MHCsTNF mice (71). Taken together, these observations suggest that interactions between the renin angiotensin system and inflammatory mediators may contribute to adverse cardiac remodeling in the adult mammalian heart. Although speculative, one potential reason for the so-called phenomenon of "neurohormonal escape," (72) in which there is progressive cardiac remodeling despite pharmacological blockade of renin angiotensin system, may relate to the redundancy that exists between cross-regulated biological systems, such as the renin angiotensin systems and proinflammatory cytokines.

CONCLUSION

In the foregoing chapter, we have reviewed the experimental literature which suggests that inflammatory mediators play an important role in adverse cardiac remodeling. As noted above, the extant literature suggests that pathophysiologically relevant concentrations of inflammatory mediators mimic many aspects of the heart failure phenotype in experimental animals, including LV dysfunction, LV dilation, activation of fetal gene expression, cardiac myocyte hypertrophy, and cardiac myocyte apoptosis (Table 1). Thus, analogous to the proposed role for neurohormones in cardiac remodeling, inflammatory mediators appear to represent a second distinct class of biologically active molecules that can contribute to adverse cardiac remodeling. Moreover, there is a growing body of evidence which suggests that many of the adverse effects that have been attributed to neurohormones may be mediated, at least in part, by inflammatory mediators.

ACKNOWLEDGMENTS

The author would like to thank Mary Helen Soliz for secretarial assistance. This material is the result of work supported with the use of resources and facilities at the Houston VA Medical Center, and by research funds from the N.I.H. (P50 HL-O6H and RO1 HL58081–01, RO1 HL61543–01, HL-42250–10/10).

REFERENCES

1. Teerlink JR, Pfeffer JM, Pfeffer MA. Progressive ventricular remodeling in response to diffuse isoproterenol-induced myocardial necrosis in rats. Circ Res 1994; 75:105–113.
2. Tan LB, Jalil JE, Pick R, Janicki JS, Weber KT. Cardiac myocyte necrosis induced by angiotensin II. Circ Res 1991; 69:1185–1195.
3. Mann DL, Kent RL, Parsons B, Cooper G IV. Adrenergic effects on the biology of the adult mammalian cardiocyte. Circulation 1992; 85:790–804.
4. Levine B, Kalman J, Mayer L, Fillit HM, Packer M. Elevated circulating levels of tumor necrosis factor in severe chronic heart failure. N Engl J Med 1990; 223:236–241.
5. Torre-Amione G, Kapadia S, Benedict CR, Oral H, Young JB, Mann DL. Proinflammatory cytokine levels in patients with depressed left ventricular ejection fraction: a report from the studies of left ventricular dysfunction (SOLVD). J Am Coll Cardiol 1996; 27:1201–1206.
6. Bryant D, Becker L, Richardson J, Shelton J, Franco F, Pechock RM, Thompson M, Giroir BP. Cardiac Failure in transgenic mice with myocardial expression of tumor necrosis factor-α (TNF). Circulation 1998; 97:1375–1381.
7. Kubota T, McNamara DM, Wang JJ, Trost M, McTiernan CF, Mann DL, Feldman AM. Effects of tumor necrosis factor gene polymorphisms on patients with congestive heart failure. Circulation 1998; 97:2499–2501.
8. Bozkurt B, Kribbs S, Clubb FJ Jr, Michael LH, Didenko VV, Hornsby PJ, Seta Y, Oral H, Spinale FG, Mann DL. Pathophysiologically relevant concentrations of tumor necrosis factor-α promote progressive left ventricular dysfunction and remodeling in rats. Circulation 1998; 97:1382–1391.
9. Mann DL. Stress-activated cytokines and the heart: from adaptation to maladaptation. Annu Rev Physiol 2003; 65:81–101.
10. Pomerantz BJ, Reznikov LL, Harken AH, Dinarello CA. Inhibition of caspase 1 reduces human myocardial ischemic dysfunction via inhibition of IL-18 and IL-1beta. Proc Natl Acad Sci U.S.A. 2001; 98:2871–2876.
11. Mann DL. Activation of inflammatory mediators in heart failure. In: Mann DL, ed. Heart Failure: A Companion to Braunwald's Heart Disease, Philadelphia: Saunders, 2003; 159–180.
12. Mann DL. Inflammatory Mediators and the Failing Heart: Past, Present, and the Foreseeable Future. Circ Res 2002; 91:988–998.
13. Suffredini AF, Fromm RE, Parker MM, Brenner M, Kovacs JA, Wesley RA, Parrillo JE. The cardiovascular response of normal humans to the administration of endotoxin. N Engl J Med 1989; 321:280–287.
14. Pagani FD, Baker LS, Hsi C, Knox M, Fink MP, Visner MS. Left ventricular systolic and diastolic dysfunction after infusion of tumor necrosis factor-α in conscious dogs. J Clin Invest 1992; 90:389–398.
15. Kubota T, McTiernan CF, Frye CS, Slawson SE, Koretsky AP, Demetris AJ, Feldman AM. Dilated cardiomyopathy in transgenic mice with cardiac specific overexpression of tumor necrosis factor-alpha. Circ Res 1997; 81:627–635.
16. Palmer JN, Hartogensis WE, Patten M, Fortuin FD, Long CS. Interleukin-1β induces cardiac myocyte growth but inhibits cardiac fibroblast proliferation in culture. J Clin Invest 1995; 95:2555–2564.
17. Yokoyama T, Nakano M, Bednarczyk JL, McIntyre BW, Entman ML, Mann DL. Tumor necrosis factor-α provokes a hypertrophic growth response in adult cardiac myocytes. Circulation 1997; 95:1247–1252.
18. Nakamura K, Fushimi K, Kouchi H, Mihara K, Miyazaki M, Ohe T, Namba M. Inhibitory effects of antioxidants on neonatal rat cardiac myocyte hypertrophy induced by tumor necrosis factor-alpha and angiotensin II. Circulation 1998; 98:794–799.
19. Hirota H, Yoshida K, Kishimoto T, Taga T. Continuous activation of gp130, a signal-transducing receptor component for interleukin 6-related cytokines, causes myocardial hypertrophy in mice. Proc Natl Acad Sci U.S.A. 1995; 92:4862–4866.

20. Arstall MA, Sawyer DB, Fukazawa R, Kelly RA. Cytokine-mediated apoptosis in cardiac myocytes: the role of inducible nitric oxide synthase induction and peroxynitrite generation [see comments]. Circ Res 1999; 85:829–840.

21. Tartaglia LA, Ayres TM, Wong GHW, Goeddel DV. A novel domain with the 55 kd TNF receptor signals cell death. Cell 1993; 74:845–853.

22. MacLellan WR, Schneider MD. Death by design: programmed cell death in cardiovascular biology and disease. Circ Res 1997; 81:137–144.

23. Boldin MP, Goncharov TM, Goltsev YV, Wallach D. Involvement of MACH, a novel MORT1/FADD-interacting protease, in fas/APO-1- and TNF receptor-induced cell death. Cell 1996; 85:803–815.

24. Muzio M, Chinnaiyan AM, Kischkel FC, O'Rourke K, Shevchenko A, Ni J, Scaffidi C, Bretz JD, Zhang M, Gentz R, Mann M, Krammer PH, Peter ME, Dixit VM. FLICE, a novel FADD-homologous ICE/CED-3-like protease, is recruited to the CD95 (Fas/APO-1) death-inducing signaling complex. Cell 1996; 85:817–827.

25. Duan H, Dixit VM. RAIDD is a new 'death' adaptor molecule. Nature 1997; 385:86–89.

26. Nagata S. Apoptosis by death factor. Cell 1997; 88:355–365.

27. Higuchi M, Aggarwal BB, Yeh ETH. Activation of CPP32-like proteases in tumor necrosis factor-induced apoptosis is dependent on mitochondrial function. J Clin Invest 1997; 99:1751–1758.

28. Goossens V, Grooten J, DeVos K, Fiers W. Direct evidence for tumor necrosis factor-induced mitochondrial reactive oxygen intermediates and their involvement in cytotoxicity. Proc Natl Acad Sci U.S.A. 1995; 92:8115–8119.

29. Albrecht H, Tschopp J, Jongeneel CV. Bcl-2 protects from oxidative damage and apoptotic cell death without interfering with activation of NF-κB by TNF. FEBS Lett 1994; 351:45–48.

30. Schulze-Osthoff K, Bakker AC, Vanhaesebroeck B, Beyaert R, Jacob WA, Fiers W. Cytotoxic activity of tumor necrosis factor is mediated by early damage of mitochondrial functions. J Biol Chem 1992; 267:5317–5323.

31. Oral H, Dorn GW, II, Mann DL. Sphingosine mediates the immediate negative inotropic effects of tumor necrosis factor-α in the adult mammalian cardiac myocyte. J Biol Chem 1997; 272:4836–4842.

32. Krown KA, Page MT, Nguyen C, Zechner D, Gutierrez V, Comstock KL, Glembotski CC, Quintana PJE, Sabbadini RA. Tumor necrosis factor alpha-induced apoptosis in cardiac myocytes: involvement of the sphingolipid signaling cascade in cardiac cell death. J Clin Invest 1996; 98:2854–2865.

33. Adam-Klages S, Adam D, Wiegmann K, Struve S, Kolanus W, Schneider-Mergener J, Kronke M. FAN, a novel WD-repeat protein couples the p55 TNF receptor to neutral sphingomyelinase. Cell 1996; 86:937–947.

34. Kolesnick RN. Sphingomyelin and derivatives as cellular signals. Prog Lipid Res 1991; 30:1–38.

35. Jarvis WD, Kolesnick RN, Fornari FA, Traylor RS, Gewirtz DA, Grant S. Induction of apoptotic DNA damage and cell death by activation of the sphingomyelin pathway. Proc Natl Acad Sci U.S.A. 1994; 91:73–77.

36. Pushkareva M, Obeid LM, Hannun YA. Ceramide: an endogenous regulator of apoptosis and growth suppression. Immunol Today 1995; 16:294–297.

37. Susin SA, Zamzami N, Castedo M, Daugas E, Wang H-G, Geley S, Fassy F, Reed JC, Kroemer G. The central executioner of apoptosis: Multiple connections between protease activation and mitochondria in Fas/APO-1CD95- and ceramide-induced apoptosis. J Exp Med 1997; 186:25–37.

38. O'Brien NW, Gellings NM, Guo M, Barlow SB, Glembotski CC, Sabbadini RA. Factor associated with neutral sphingomyelinase activation and its role in cardiac cell death. Circ Res 2003; 92:589–591.

39. Mann DL. Cytokines as mediators of disease progression in the failing heart. In: Hosenpud JD, Greenberg BH, eds. Congestive Heart Failure. Philadelphia: Lippincott Williams & Wilkins, 1999:213.

40. Gulick TS, Chung MK, Pieper SJ, Lange LG, Schreiner GF. Interleukin 1 and tumor necrosis factor inhibit cardiac myocyte β-adrenergic responsiveness. Proc Natl Acad Sci U.S.A. 1989; 86:6753–6757.

41. Balligand JL, Ungureanu D, Kelly RA, Kobzik L, Pimental D, Michel T, Smith TW. Abnormal contractile function due to induction of nitric oxide synthesis in rat cardiac myocytes follows exposure to activated macrophage-conditioned medium. J Clin Invest 1993; 91:2314–2319.

42. Combes A, Frye CS, Lemster BH, Brooks SS, Watkins SC, Feldman AM, McTiernan CF. Chronic exposure to interleukin 1beta induces a delayed and reversible alteration in excitation-contraction coupling of cultured cardiomyocytes. Pflugers Arch 2002; 445:246–256.

43. Bick RJ, Liao JP, King TW, LeMaistre A, McMillin JB, Buja LM. Temporal effects of cytokines on neonatal cardiac myocyte Ca2+ transients and adenylate cyclase activity. Am J Physiol 1997; 272:H1937–H1944.

44. Pinsky DJ, Cai B, Yang X, Rodriguez C, Sciacca RR, Cannon PJ. The lethal effects of cytokine-induced nitric oxide on cardiac myocytes are blocked by nitric oxide synthase antagonism or transforming growth factor beta. J Clin Invest 1995; 95:677–685.

45. Simms MG, Walley KR. Activated macrophages decrease rat cardiac myocyte contractility: importance of ICAM-1-dependent adhesion. Am J Physiol 1999; 277:H253–H260.

46. Weisensee D, Bereiter-Hahn J, Schoeppe W, Low-Friedrich I. Effects of cytokines on the contractility of cultured cardiac myocytes. Int J Immunopharmacol 1993; 15:581–587.

47. Hosenpud JD. The Effects of Interleukin-1 on Myocardial Function and Metabolism. Clin Immunol Immunopathol 1993; 68:175–180.

48. Schulz R, Panas DL, Catena R, Moncada S, Olley PM, Lopaschuk GD. The role of nitric oxide in cardiac depression induced by interleukin-1β and tumour necrosis factor-α. Br J Pharmacol 1995; 114:27–34.

49. Dinarello CA. Interleukin-18. Methods 1999; 19:121–132.

50. Pomerantz BJ, Reznikov LL, Harken AH, Dinarello CA. Inhibition of caspase 1 reduces human myocardial ischemic dysfunction via inhibition of IL-18 and IL-1beta. Proc Natl Acad Sci U.S.A. 2001; 98:2871–2876.

51. Yu X, Kennedy RH, Liu SJ. JAK2/STAT3, not ERK1/2, mediates interleukin-6-induced activation of inducible nitric-oxide synthase and decrease in contractility of adult ventricular myocytes. J Biol Chem 2003; 278:16304–16309.

52. Franco F, Thomas GD, Giroir BP, Bryant D, Bullock MC, Chwialkowski MC, Victor RG, Peshock RM. Magnetic resonance imaging and invasive evaluation of development of heart failure in transgenic mice with myocardial expression of tumor necrosis factor-alpha. Circulation 1999; 99:448–454.

53. Sivasubramanian N, Coker ML, Kurrelmeyer K, DeMayo F, Spinale F.G, Mann DL. Left ventricular remodeling in transgenic mice with cardiac restricted overexpression of tumor necrosis factor. Circulation 2001:826–831.

54. DA Siwik, DL Chang, WS Colucci. Interleukin-1beta and tumor necrosis factor-alpha decrease collagen synthesis and increase matrix metalloproteinase activity in cardiac fibroblasts in vitro. Circ Res 2000; 86:1259–1265.

55. Peng J, Gurantz D, Tran V, Cowling RT, Greenberg BH. Tumor necrosis factor-alpha-induced AT1 receptor upregulation enhances angiotensin II-mediated cardiac fibroblast responses that favor fibrosis. Circ Res 2002; 91:1119–1126.

56. Felgner PL, Ringold GM. Cationic liposome-mediated transfection. Nature 1989; 337:387–388.

57. Tsuruda T, Jougasaki M, Boerrigter G, Huntley BK, Chen HH, D'Assoro AB, Lee SC, Larsen AM, Cataliotti A, Burnett JC Jr. Cardiotrophin-1 stimulation of cardiac fibroblast

growth: roles for glycoprotein 130/leukemia inhibitory factor receptor and the endothelin type A receptor. Circ Res 2002; 90:128–134.

58. Wang F, Trial J, Diwan A, Gao F, Birdsall H, Entman M, Hornsby P, Sivasubramanian D, Mann D. Regulation of cardiac fibroblast cellular function by leukemia inhibitory factor. J Mol Cell Cardiol 2002; 34; 34:1309.

59. Li YY, Feng YQ, Kadokami T, McTiernan CF, Draviam R, Watkins SC, Feldman AM. Myocardial extracellular matrix remodeling in transgenic mice overexpressing tumor necrosis factor alpha can be modulated by anti- tumor necrosis factor alpha therapy. Proc Natl Acad Sci U.S.A. 2000; 97:12746–12751.

60. Weber KT. Cardiac Interstitium in Health and Disease: The Fibrillar Collagen Network. J Am Coll Cardiol 1989 June; 13(7):1637–1652.

61. Knittel T, Mehde M, Grundmann A, Saile B, Scharf JG, Ramadori G. Expression of matrix metalloproteinases and their inhibitors during hepatic tissue repair in the rat [In Process Citation]. Histochem Cell Biol 2000; 113:443–453.

62. Sime PJ, Marr RA, Gauldie D, Xing Z, Hewlett BR, Graham FL, Gauldie J. Transfer of tumor necrosis factor-alpha to rat lung induces severe pulmonary inflammation and patchy interstitial fibrogenesis with induction of transforming growth factor-beta1 and myofibroblasts. Am J Pathol 1998; 153:825–832.

63. Gurantz D, Cowling RT, Villarreal FJ, Greenberg BH. Tumor necrosis factor-alpha upregulates angiotensin II type 1 receptors on cardiac fibroblasts. Circ Res 1999; 85:272–279.

64. Gray MO, Long CS, Kalinyak JE, Li HT, Karliner JS. Angiotensin II stimulates cardiac myocyte hypertrophy via paracrine release of TGF-beta 1 and endothelin-1 from fibroblasts. Cardiovasc Res 1998; 40:352–363.

65. Schultz JJ, Witt SA, Glascock BJ, Nieman ML, Reiser PJ, Nix SL, Kimball TR, Doetschman T. TGF-beta1 mediates the hypertrophic cardiomyocyte growth induced by angiotensin II. J Clin Invest 2002; 109:787–796.

66. Dostal DE, Baker KM. The cardiac renin-angiotensin system: conceptual, or a regulator of cardiac function? Circ Res 1999; 85:643–650.

67. Brasier AR, Jamaluddin M, Han Y, Patterson C, Runge MS. Angiotensin II induces gene transcription through cell-type-dependent effects on the nuclear factor-kappaB (NF-κB) transcription factor. Mol Cell Biochem 2000; 212:155–169.

68. Hernandez-Presa M, Bustos C, Ortega M, Tunon J, Renedo G, Ruiz-Ortega M, Egido J. Angiotensin-converting enzyme inhibition prevents arterial nuclear factor-κB activation, monocyte chemoattractant protein-1 expression, and macrophage infiltration in a rabbit model of early accelerated atherosclerosis. Circulation 1997; 95:1532–1541.

69. Luft FC. Workshop: mechanisms and cardiovascular damage in hypertension. Hypertension 2001; 37:594–598.

70. Kalra D, Baumgarten G, Dibbs Z, Seta Y, Sivasubramanian N, Mann DL. Nitric Oxide Provokes Tumor Necrosis Factor-alpha Expression in Adult Feline Myocardium Through a cGMP-Dependent Pathway. Circulation 2000; 102:1302–1307.

71. Flesch M, Hoper A, Dell'Italia L, Evans K, Bond R, Peshock R, Diwan A, Brinsa TA, Wei CC, Sivasubramanian N, Spinale FG, Mann DL. Activation and functional significance of the renin-angiotensin system in mice with cardiac restricted overexpression of tumor necrosis factor. Circulation 2003; 108:598–604.

72. Francis GS, Cohn JN, Johnson G, Rector TS, Goldman S, Simon A. Plasma norepinephrine, plasma renin activity, and congestive heart failure. Circulation 1993; 87: VI-40–VI-48.

73. D Kalra, Sivasubramanian N, Mann DL. Angiotensin II induces tumor necrosis factor biosynthesis in the adult mammalian heart through a protein kinase C-dependent pathway. Circulation 2002; 105:2198–2205.

15

Oxidative Stress in Heart Failure: Impact on Cardiac Function and Ventricular Remodeling

Luciano C. Amado, Anastasios P. Saliaris, and Joshua M. Hare
Department of Medicine, Johns Hopkins Medical Institutions, Baltimore, Maryland, U.S.A.

INTRODUCTION

Despite several decades of advances in heart failure (HF) therapeutics resulting in substantial reductions in morbidity and mortality, symptomatic HF continues to be a major cause of mortality and hospitalization in the United States (1,2). Because of this, improved understanding of HF pathophysiology is crucial in developing novel and effective treatments for this syndrome. The process of ventricular remodeling is viewed as a central pathophysiologic mechanism driving the progression of HF. In this regard, HF is viewed as a syndrome characterized not only by hemodynamic disturbances, but also by local and systemic neurohormonal, molecular, and cellular abnormalities, which lead to changes in size, shape and function of the heart, together representing ventricular remodeling.

Clinical benefits obtained with neurohormonal antagonists (3,4) have prompted an ongoing search for other key pathophysiologic pathways in the failing cardiovascular system. As such, oxidative stress has emerged as a novel therapeutic target in the treatment of HF, and there is growing evidence implicating this biochemical pathway in the pathophysiology and progression of HF. Specifically, oxidative stress appears to play important roles in deranged cardiac excitation–contraction coupling, myocyte hypertrophy and apoptosis, key underpinnings of the remodeling process. This chapter reviews the biochemical and (patho) physiologic evidence supporting a role for oxidative stress in HF and left ventricular (LV) remodeling.

BIOCHEMICAL MECHANISMS OF OXIDATIVE STRESS

Oxidative stress is defined as an imbalance between the production and degradation of endogenously produced, highly reactive free radicals, also known as reactive oxygen species (ROS). Free radicals are atoms or molecules with an unpaired electron. In this regard, superoxide (O_2^-) and nitric oxide (NO) (Fig. 1) represent primary examples of endogenously produced free radicals. The reactivity of free radicals derives from their propensity to abstract an electron from a second molecule, thereby oxidizing that molecule (5). Oxidative stress is implicated in HF pathophysiology from evidence that both myocardial free radical production (6) and circulating markers of oxidative stress are elevated in HF (7).

In healthy individuals, the physiologic generation of ROS is balanced by natural antioxidant mechanisms [e.g., enzymes such as catalase and superoxide dismutase (SOD)]. However, in pathological states, an increase in ROS production or a reduction in antioxidant reserves may interfere with normal physiological processes. Figure 1 illustrates some of the important free radical reactions in cardiovascular pathophysiology.

There are two general mechanisms by which free radical generation may lead to cardiac toxicity: (i) oxidative stress may contribute to tissue damage or cellular loss by

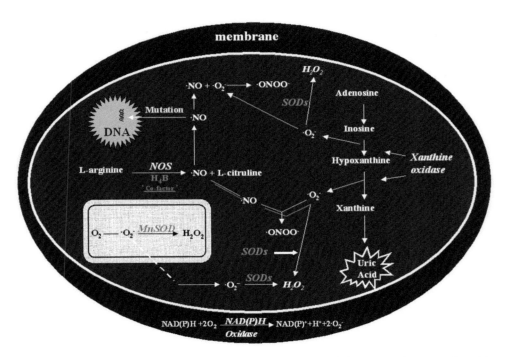

Figure 1 Intracellular production of ROS and antioxidant defenses. ROS are produced in the mitochondria, endoplasmic reticulum, plasma membrane, and cytosol. Superoxide is generated during mitochondrial respiration and is converted to H_2O_2 by SOD, while cytosolic O_2^- is converted to H_2O_2 by other SODs. Other sources of O_2^- include xanthine oxidase in the cytosol, NADPH oxidase in the membrane and NOS. nNOS, eNOS and iNOS are responsible for NO production, and are found in distinct subcellular locations—nNOS in the sarcoplasmic reticulum, eNOS in sarcolemmal caveolae, and iNOS throughout the cell after induction by cytokines. H_4B is a co-factor of NOS. NO can react with membrane lipids and can also cause mutations in DNA. In addition, reactions between NO and O_2^- lead to the formation of $ONOO^-$, a high toxic ROS, which can induce lipid peroxidation. *Abbreviations*: NADPH, nicotinamide adenine dinucleotide phosphate; NO, nitric oxide; NOS, nitric oxide synthase; O_2^-, superoxide; $ONOO^-$, peroxynitrite; SOD, superoxide dismutase. *Source*: From Ref. 51.

activating apoptotic or oncotic pathways (7,8) or (ii) oxidation of proteins may alter critical cellular functions, a phenomenon that may occur reversibly (9,10). For example, mitochondrial respiratory complexes (6), thiol-containing proteins involved in calcium cycling [e.g., the ryanodine receptor (11), sarcoplasmic reticulum (SR) Ca^{2+}-ATPase (12), and the sarcolemmal sodium–calcium exchanger (13) undergo oxidative modification, which may in turn lead to diminished myocardial performance].

EVIDENCE FOR INCREASED OXIDATIVE STRESS IN HF

Animal Model

One of the most convincing pieces of evidence that free radicals are implicated in the disruption of cardiac function is myocardial stunning (14)—the phenomenon of prolonged cardiac dysfunction after brief periods of ischemia in the absence of irreversible injury. There is strong evidence suggesting that such stunning is associated with generation of ROS. For example, Przyklenk et al. (15) administered free radical scavengers, SOD, and catalase, before and during coronary occlusion, and demonstrated attenuation of myocardium stunning (16).

ROS formation is also implicated in anthracycline-related cardiomyopathy, which results primarily from oxidative injury such that antioxidant therapies such as vitamin E or probucol attenuate cardiotoxicity (17,18). In addition, direct evidence of increased myocardial hydroxyl radical (OH^-) generation has been demonstrated in catecholamine-induced cardiomyopathy (19). Furthermore, in the rapid pacing model of cardiomyopathy, allopurinol, a xanthine oxidase (XO) inhibitor, led to acute improvements in myocardial function and energetics, further suggesting a role of ROS in this form of experimental cardiac dysfunction (20).

Human Studies

Many studies link oxidative stress and HF in humans. McMurray et al. (21) reported increased oxidative stress in patients with HF with or without coronary artery disease as evidenced by increased levels of thiobarbituric reactive substances (TBARS) and plasma thiol (PSH). Elevation in expired pentane, a product of lipid peroxidation and indirect marker of oxidative stress, has been observed in HF patients (22). More recently, 8-iso-prostaglandin F2alpha (8-iso-$PGF_{2\alpha}$), a specific and quantitative marker of oxidant stress in vivo, was found to be elevated in the pericardial fluid of HF patients and correlated with the degree of LV impairment (23). Unfortunately, the aforementioned studies are limited by small sample size and indirect measurements of oxidative stress. Furthermore, these studies fail to demonstrate any positive effect of controlling oxidative stress in patients with HF. To clarify the precise role of oxidative stress in HF, clinical studies designed to illustrate the positive effects of disrupting the oxidative stress pathway will ultimately determine whether this pathway represents a valuable therapeutic target.

SOURCES OF OXIDATIVE STRESS IN HF

It is important to note that ROS are produced physiologically. ROS consumption (or quenching) is mediated primarily by extra and intracellular SODs (8) and the glutathione peroxidase (GPx) reaction (24). In biological systems, there are many sources of free radical generation. In the heart, the main sources are the mitochondrial electron transport chain (a byproduct of mitochondrial respiration), XO (25), nitric oxide synthase (NOS), and NADH/NADPH oxidase. These systems are discussed in detail below (Fig. 1).

Mitochondria

Mitochondrial electron transport represents a major source of myocardial ROS production. Mitochondria generate adenosine triphosphate (ATP) by a series of oxidation–reduction reactions mediated by the respiratory complexes NADH-CoQ reductase (complex I), succinate-CoQ reductase (complex II), cytochrome c reductase (complex III), and cytochrome c oxidase (complex IV) (Fig. 2). Superoxide O_2^- generation in the mitochondria occurs as a result of the incomplete reduction of O_2 to H_2O_2. Under normal conditions, 1% to 2% of the O_2 used in the respiratory chain is converted to O_2^-, which occurs via ubiquinone at the level of complex I and II (26,27). Normally, mitochondrial respiration is physiologically modulated by NO, which reversibly inhibits complex IV (cytochrome c oxidase), thereby reducing O_2 consumption (28). The small amount of ROS production in the physiological state is used for signaling and is balanced by antioxidant enzymes, such as manganese SOD and GPx, which also reside in the mitochondria.

Under pathological conditions, oxidative stress is due to either excess production and/or insufficient buffering of ROS. Using a pacing-induced HF model, Ide et al. (6) reported a 2.8-fold increase in O_2^- production in failing versus control hearts. This finding was associated with approximately 50% reduction in complex I activity (6) but no change in SOD activity, supporting excess production of ROS

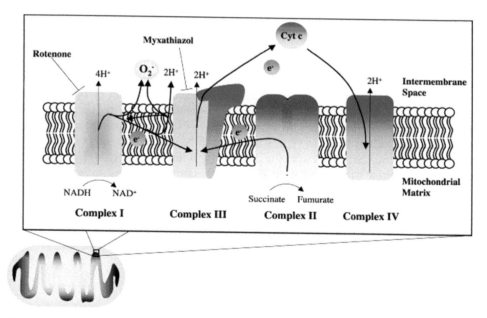

Figure 2 Major subunits of the electron transport chain and sites of O_2^- production. Complex I and complex III can both serve as site of O_2^- production possible through the formation of semiquinone. *Abbreviation*: O_2^-, superoxide. *Source*: From Ref. 58.

as a primary mechanism of disturbed mitochondrial respiration. Other studies, however, provide support for a reduced antioxidant capacity in the failing heart (29,30). Indeed, the finding that mice null for manganese SOD die of a dilated cardiomyopathy soon after birth underscores the fundamental importance of the antioxidant buffering system in maintaining cardiac structure (31).

Xanthine Oxidase

There is increasing evidence for the pathophysiological role of the xanthine oxidoreductase (XOR) signaling pathway in HF (25). XO is the product of the xanthine oxidoredutase gene that encodes xanthine dehydrogenase (XDH), a 150 Kda protein, which is converted to XO by proteolytic cleavage or sulfhydryl modification (32,33). XDH reduces NAD^+ to NADH, while XO produces O_2^- using O_2 as an electron acceptor as a byproduct of the terminal two steps of urate production: the conversion of hypoxanthine to xanthine and the subsequent conversion of xanthine into uric acid (32,33) (Fig. 1). Although both isozymes are capable of generating O_2^- (34), XO is thought to have greater physiological relevance.

Under normal conditions, XO accounts for a small proportion of ROS production. However, XO is upregulated within the heart in both experimental (~4-fold) (20,35) and human (36,37) HF and has been implicated in ischemia and reperfusion injury (33). Much has been written about the difficulty in identifying XO within the hearts of certain mammalian species, including humans (38). It is now clear that although XO is produced in greatest abundance in the liver and gut, it may also circulate in the blood and adhere to endothelium at distant sites (39). Moreover, XO is expressed in cardiac myocytes and may participate in intracrine signaling (40).

From a functional standpoint, XO participates in both vascular and cardiac dysfunction in HF. Several recent studies have demonstrated that inhibition of XO improves endothelial dysfunction in patients with congestive HF in association with a reduction in circulating markers of oxidative stress (41,42). With regard to cardiac dysfunction, XO has been linked to the process of mechanoenergetic uncoupling (43), whereby myocardial oxygen consumption remains the same or increases while cardiac stroke work falls dramatically. Inhibition of XO with allopurinol enhances the contractile response of failing myocardium to dobutamine and exercise (44), and improves the ratio of ventricular work to oxygen consumption in both experimental (20) and human (36) HF models. These benefits are due to both increases in myocyte calcium sensitivity and crosstalk between NOS and XO signaling pathways. Saavedra et al. (40), from our group, have shown that an intact NOS signaling pathway is necessary for the beneficial cardiac effects of XO inhibitors to occur. Using the pacing-induced HF model, they showed that NOS inhibition with L-NMMA (20 mg; i.v.) prevented allopurinol-derived benefits on myocardial mechanoenergetics (40). While the precise mechanism of this interaction remains to be fully elucidated, two general mechanisms are thought to be possible. First, NO itself may have a direct antioxidant effect, or, second, NO and O_2^- may compete to modify protein function. An example of the latter is that both NO and O_2^- modify the ryanodine receptor—NO reversibly thereby allowing dynamic regulation of SR Ca^{2+} release, and O_2^- irreversibly causing SR Ca^{2+} leak (12,45,46).

The body of work regarding XO inhibition not only implicates this enzyme as a source of oxidative stress in HF but also elucidates pathophysiologic consequences of oxidative stress. Clearly, ROS interfere with cardiac excitation–contraction coupling (20,36,40) to diminish the utilization of energy into cardiac work. Potential sites for

XO-derived ROS include mitochondria (40), SR, and myofilaments themselves (47). The acute responsiveness to XO inhibition clearly establishes the reversibility of the effects of oxidative stress on cardiac function, and suggests clinical applicability.

Nitric Oxide Synthase

NO is an endogenously synthesized free radical, first characterized as the noneicosanoid component of endothelial-derived relaxation factor (EDRF) (48). NO is produced by the oxidative deamination of L-arginine to L-citrulline catalyzed by the enzyme NOS (Fig. 1). NOS is expressed in the sarcolemma, SR, and mitochondria of myocytes. There are three members of the NOS family of enzymes: neuronal NOS (nNOS or type I), inducible NOS (iNOS or type II), and endothelial NOS (eNOS or type III) (49). Two isoforms, nNOS and eNOS, are constitutively expressed, and their activity is regulated by intracellular calcium concentration and calmodulin (50,51). The "calcium independent" isoform, iNOS, is usually induced in macrophages after stimulation by cytokines, lipopolysaccharides, and other immunologic agents (52). NO can also be generated nonenzymatically through the reduction of nitrite to NO under acidic conditions, such as those seen in ischemia (53). NOS participates in oxidative stress in several ways. First, NO may serve as both an endogenous antioxidant, quenching O_2^- or, when produced in excess, as a contributor to oxidative stress via direct chemical reaction with other ROS. Second, NOS may produce ROS directly as a result of electron uncoupling.

As a pro-oxidant, NO can be converted to various other reactive nitrogen species (RNS), such as nitrosonium cation (NO^+) or nitroxyl anion (NO^-). The exact mechanism(s) whereby NO becomes deleterious are under active investigation, but several potential pathways have been described. At the mitochondrial level, excess NO production has been shown to inactivate respiration by nitrosylating thiols in complex I (28). This process may be worsened by NO-mediated inactivation of several antioxidant enzymes, including catalase, GPx, and SOD (54), along with the generation of harmful peroxynitrite ($ONOO^-$), a highly reactive species resulting from a reaction between NO and O_2^-, capable of oxidizing cellular proteins (55–57). Peroxynitrite anion production is determined by fluctuations in the relative concentrations of O_2^- and NO and an imbalance between the two may affect mechanoenergetic uncoupling (43). Peroxynitrite has been linked to depressed cardiac efficiency via ATP consumption (55). In addition to affecting cellular energetic processes, excess NO may react directly with DNA, causing mutations and triggering apoptosis (58).

Our group has demonstrated the importance of NO as an antioxidant within the heart. In an animal model, the improvement of cardiac mechanical efficiency obtained with XO inhibition by allopurinol was eliminated in HF animals pretreated with L-NMMA (40), a NOS inhibitor. Similarly, acute administration of L-NMMA in normal animals caused mechanoenergetic uncoupling. These findings reinforce the protective role of NO in oxidative stress.

NOS, in addition to its participation in oxidant pathways through NO production, may also produce ROS directly as a result of electron uncoupling. In the process of generating NO from L-arginine and O_2, NOS can transfer an electron to an O_2 molecule instead of L-arginine, thus generating O_2^- (59,60) (Fig. 3). Tetrahydrobiopterin (H_4B), a co-factor of NOS, optimizes NO production, and states of H_4B deficiency favor O_2^- production from NOS (61). Furthermore, administration of H_4B increases NO production (62) and improves endothelial dysfunction in several conditions such as hypercholesterolemia and diabetes (63–65).

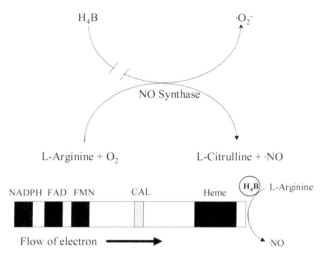

Figure 3 The nitric oxide synthase reaction. (*Top*) NOS utilizes the L-arginine and O_2 to produce L-citrulline and nitric oxide (NO). H_4B [(6R-5,6,7,8-tetrahydrobiopterin] is an essential cofactor, which facilitates electron transport in this reaction. Electrons donated by reduced nicotinamide-adenine dinucleotide phosphate (NADPH) are shuttled through the reduced flavins, FAD and FMN toward the oxidase domain. (*Bottom*) The system is used to oxidize L-arginine to NO. This reaction is dependent on the presence of H_4B. In states of H_4B low levels, NOS generates O_2^-. *Abbreviations*: FAD, flavin adenine dinucleotide; FMN, flavin mononucleotide; NADPH, nicotinamide adenine dinucleotide phosphate; NO, nitric oxide; NOS, nitric oxide synthase; O_2^-, superoxide. *Source*: From Ref. 61.

NADH/NADPH Oxidase

NAD(P)H oxidases are multi-subunit enzymes that catalyze single-electron reduction reactions using NAD(P)H as sources of electrons (66). NAD^+ is also an important electron acceptor in many oxidation–reduction reactions, notably within the mitochondria, and as an electron acceptor for XOR. NAD(P)H is utilized in extra-mitochondrial reactions (Fig. 4). Structurally, the difference between these co-factors is the phosphorylation at the C2 position of the adenosine ribosyl moeity of NAD^+ (Fig. 4). NAD(P)H oxidase has been classified as phagocytic and nonphagocytic (cardiovascular) oxidases. While phagocytic NAD(P)H oxidase structure and function is well documented, cardiovascular NAD(P)H oxidase is still under intense investigation. Both NAD(P)H oxidases are composed of a membrane-spanning cytochrome b558 complex (formed by subunits gp91[phox] and p22[phox]), in addition to two cytoplasmic polypeptide subunits, p47[phox] and p67[phox] (Fig. 5 and Table 1). A cytosolic guanine nucleotide-binding protein Rac2 (sometimes Rac1), which is a member of the Ras family, is necessary for oxidase activation (67) (Fig. 5 and Table 1). Recently, in vascular smooth muscle cells, a new subunit called *mox1*, a homolog of gp91[phox], has been demonstrated (68). Heymes et al. (69) have recently demonstrated upregulation of NAD(P)H subunits in human HF.

Despite the structural similarity between both NAD(P)H oxidases, the precise interaction and function of the subunits in the cardiovascular NAD(P)H oxidase is not completely elucidated. The cardiovascular NAD(P)H oxidase differs from that found in neutrophils, in that its estimated capacity is one third of the neutrophil NAD(P)H oxidase (70,71). Recent studies of cardiac NAD(P)H oxidase indicate that the p67[phox] and gp91[phox] subunits are involved in angiotensin II signaling, by

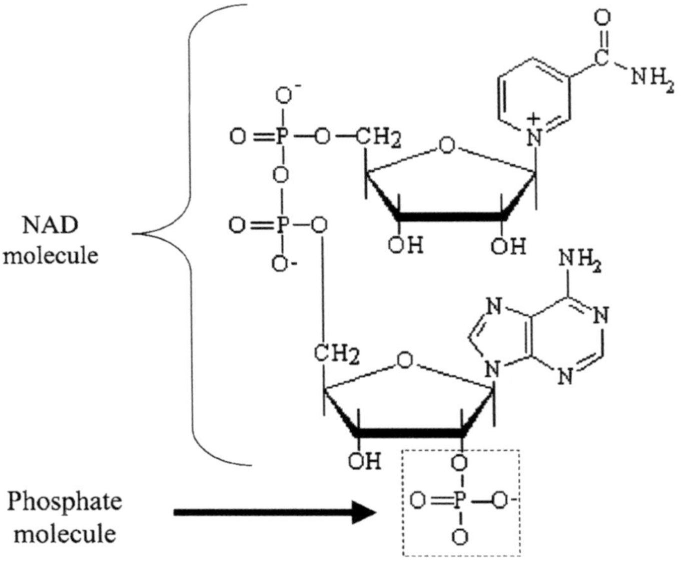

Figure 4 Molecular structure of NAD^+ and $NADP^+$. The presence of phosphate instead of a hydroxyl group at the boxed position differentiates NADP from NAD. *Abbreviations*: NAD, nicotinamide adenine dinuleotide; NADP, nicotinamide adenine dinucleotide phosphate.

stimulating O_2^- production in the vessel wall, which in turn contributes to regulating vascular tone and smooth muscle cell growth (66,72,73).

Experimental studies have linked granulocyte NAD(P)H oxidase to inflammatory reperfusion injury. Studies in dogs, following 60 to 90 minutes of coronary occlusion, have provided strong evidence that NAD(P)H oxidase present in neutrophils is one of the main sources of oxidative stress in reperfused hearts, playing a major role in microvascular obstruction ("no-reflow" phenomenon) (74). This finding is supported by the fact that anti-NAD(P)H oxidase interventions, such as

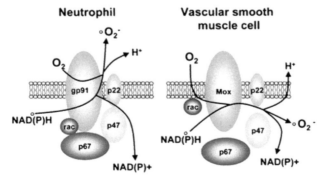

Figure 5 Structure of the NAD(P)H oxidase. (*Left*) Structure of the neutrophil NAD(P)H oxidase. gp91[phox] and p22[phox] are responsible for electron transfer while p47[phox], p67[phox], and the G protein rac modulate the enzymes activity. (*Right*) Components of the neutrophil oxidase that have been identified in VSMCs. *Abbreviations*: NADPH, nicotinamide adenine dinucleotide phosphate; VSMC, vascular smooth muscle cells. *Source*: From Ref. 66.

Table 1 Evidence for the Presence of Oxidase Subunits in Isolated Nonphagocytic Cells

	gp91phox		mox		p22phox		rac		p47phox		p67phox	
	RNA	Protein	RNA	Protein	RNA	Protein	RNA	Protein	RNA	Protein	RNA	Protein
Neutrophil	+++	+++	–	ND	+++	+++	+++	+++	+++	+++	+++	+++
VSMC	±	–	++	ND	+++	+++	+++	+++	+	+?	+	?
EC	+	±	ND	ND	+	+	+	+	±	±	+++	±
Fibroblast	±	±	–	ND+	+	+	+	+	+	+	+++	±
Mesangial cell	–	–?	ND	ND	+	+	+	+	+	+	+	+

Abbreviations: ND, not determined; –, not expressed; +, expressed; and +++, strongly expressed; EC, endothelial cells.

antibody against CD18, a neutrophil adhesion protein, confer tissue protection against postischemic reperfusion injury (75,76).

Another important feature of NAD(P)H oxidase is its capacity to respond to a variety of hormonal, hemodynamic, and local metabolic changes. For example, studies have shown that increases in lactate secondary to reoxygenation stimulate NAD(P)H oxidase activity in cardiac myocytes (77). In addition, as mentioned above, there is strong evidence that angiotensin II increases O_2^- production by NAD(P)H oxidase in smooth muscular cells and aortic adventitial fibroblasts in culture (66). Interestingly, in smooth muscle cells, angiotensin II appears to induce hypertrophy through intracellular generation of H_2O_2 (78). This finding is corroborated by the fact that attenuation of NAD(P)H oxidase activity inhibits hypertrophy in smooth muscle cells (79).

CONSEQUENCES OF OXIDATIVE STRESS ON CARDIAC REMODELING

Apoptosis

Cell death occurs by one of two distinct pathways, apoptosis or necrosis (58). Necrosis is the result of severe and overwhelming stress, leading to disorganized cellular function, subsequent ATP depletion, cellular swelling, and plasma membrane rupture (58). Apoptosis, or "programmed cell death," is a distinct mechanism of cell death. Apoptosis has been implicated in cell replacement, clearance, and tissue remodeling active in pathological states, including cardiomyopathy (80,81), ischemia (82), and transplant rejection (83). Morphologically, apoptosis is characterized by cell shrinkage, chromatin condensation, DNA fragmentation, and the formation of "apoptotic bodies" (58).

Oxidative stress is implicated in the pathophysiology of apoptosis. Direct addition of ROS (84) or inhibition of ROS scavengers induces myocardial apoptosis and administration of antioxidants such as N-acetylcysteine inhibits this process (58). In the pacing model of heart failure, there is a direct correlation between increased oxidative stress and apoptosis (85). Cesselli et al. (85) showed increased levels of apoptosis in proportion to oxidative stress as indicated by levels of the pro-apoptotic proto-oncogene p66[shc]. Although there is ample evidence that oxidative stress and apoptosis are associated phenomena, the causal relationship between the two remains an area of active investigation.

The mechanisms by which ROS instigate the apoptotic process are under intense investigation. Apoptosis appears to be triggered by a family of cysteine proteases known as caspases and by mitochondria (86). Caspases contain a thiol group, sensitive to redox modification, such as S-nitrosylation or oxidation (58). While the S-nitrosylation of caspase 3, a final step in the apoptotic chain-reaction, protects against apoptosis, activation of caspase 1 and/or caspase 8 stimulates it. Mitochondria are an important source of ROS, which may themselves trigger apoptosis. Mitochondria generate ATP via successive oxidation–reduction reactions that, under normal conditions, generate O_2^- as a product of incomplete reduction of O_2 to H_2O. In HF, this production of O_2^- is greatly augmented. This overproduction may exacerbate caspase activation through S-nitrosylation or oxidation of their thiol groups. In addition to O_2^--triggered apoptosis, mitochondria contribute directly to apoptosis through the release of cytochrome c, which stimulates apoptosis activating factor (Apaf-1) and caspases 9 and 3 (Fig. 6) (87). It has been postulated that

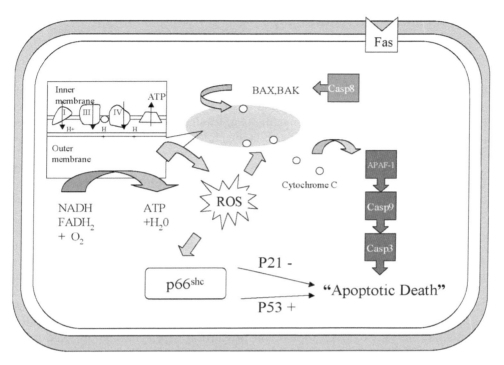

Figure 6 Central role for ROS in myocardial apoptosis. Cellular sources of O_2^- include mitochondria, XO, and NAD(P)H oxidase. Elevated O_2^- production in heart failure contributes to the release of cytochrome c from mitochondria and the activation of p66[shc], both of which participate in initiating or promoting apoptotic cascades. Thus, formation of ROS represents a central phenomenon for apoptosis in the failing heart. p53 and p21 are proapoptotic and antiapoptotic molecules, respectively, that are activated by p66[shc]. Bax and Bak are members of the BCL-2 protein family that participate in mitochondrial membrane pore-opening responsible for cytochrome c release into the cytoplasm. Fas is a cell surface receptor that activates apoptotic cascades when stimulated by its receptor. *Abbreviations*: NADPH, nicotinamide adenine dinucleotide phosphate; O_2^-, superoxide; ROS, reactive oxygen species; XO, xanthine oxidase. *Source*: From Ref. 87.

cytochrome c release may impair the flow of electrons through the respiratory cycle thereby contributing to further O_2^- production (86).

Redox status thus regulates apoptosis. ROS can act either as an instigator or as an inhibitor of apoptotic pathways. Similarly, NO exerts different effects on apoptosis in a concentration-dependent manner. At low concentrations (10 nm–1 μm), NO protects against cell death by inhibiting lipid peroxidation and caspase activity, scavenging peroxyl radicals and preventing cleavage of Bcl-2 (58,88), which plays an important role in the final steps of apoptosis (89). Cleavage of Bcl-2 results in activation or further amplification of downstream caspases on the apoptotic process. Another consequence of Bcl-2 cleavage is formation of the Bax protein, a protein involved in mitochondrial pore creation (88) and cytochrome c release (Fig. 6). At higher levels, NO induces cell death by inhibiting mitochondrial ATP synthesis, generating $ONOO^-$ and indirectly activating Fas receptors ("cell death receptors") (90). Future studies are needed to further elucidate this finely tuned and important regulatory process.

Oxidative Stress and Hypertrophy

Cardiac hypertrophy, defined as increased cardiomyocyte size, independent of gross heart weight, ventricular morphology, or any particular pattern of gene expression (91), may occur as an adaptive response to various stimuli, including excess hemodynamic load, or in response to circulating growth factors. Normally, this hypertrophic response is physiologic and adaptive in nature, but in situations of sustained excess workload, may be maladaptive, leading to pathologic cardiac decompensation and subsequent HF. Indeed, epidemiological studies have suggested that increased LV mass is an independent risk factor for cardiac morbidity and mortality as well as all-cause mortality (92). There is increasing evidence demonstrating that ROS participates in the development of cardiac hypertrophy through the activation of mitogen activated protein kinases (MAPKs) (93,94), including extracellular signal-regulated kinases (ERKs), and c-Jun NH_2-terminal kinase (JNK).

Chen and colleagues (95) demonstrated that pulse doses of H_2O_2 lead to significant increases in cell size, with concomitant increases in cell surface area, volume, and protein content. Furthermore, tumor necrosis factor, angiotensin, and adrenergic receptor stimulation have been shown to cause myocyte hypertrophy through ROS-dependent pathways (96,97), and α-1 adrenoreceptor and angiotensin stimulated hypertrophic signaling in ventricular myocytes may involve ROS generated by NAD(P)H oxidase (98). Interestingly, the HMG-CoA reductase inhibitors (statins) are known to have antioxidant properties, which may be involved in the ability of this class of drugs to inhibit cardiac hypertrophy (99,100).

Hypertrophied myocardium may exhibit alterations in function secondary to defective ROS handling. Endothelial-derived NO enhances myocardial relaxation. However, in experimental models of LV hypertrophy, ROS may impair myocardial relaxation. Using aortic banded swine hearts, MacCarthy et al. (101) found that the antioxidants, vitamin C and deferoxamine, restored LV relaxation responses to the NO agonists bradykinin and substance P, as evidence by a significant decrease in the time to onset of LV relaxation (tdP/dt_{min}). Furthermore, their study revealed an increase in expression of the NAD(P)H oxidase subunits gp91[phox] and p67[phox] and an increase in myocardial NAD(P)H oxidase activity, suggesting that this impaired relaxation response may be secondary to NAD(P)H oxidase produced ROS. Other studies have shown that hypertrophied cardiac myocytes exhibit an upregulation of NOS type II protein and activity, are more susceptible to NO inhibition of mitochondrial respiration, and that impaired contractility in hypertrophied myocardium is restored by NOS inhibition (102).

There are limited human data supporting the role of antioxidant therapies in the prevention and treatment of cardiac hypertrophy. As mentioned above, evidence from animal models indicates that statins inhibit cardiac hypertrophy and improve LV performance independent of their cholesterol lowering effect (101,103,104). These results have not yet been described in human models but, if present, could be of major benefit in the treatment of HF. The most compelling evidence for a potential role of antioxidants in human cardiomyopathy has been provided by Rustin and colleagues (105,106). They investigated the role of free radical scavengers in Friedreich's ataxia, a disease characterized by neurodegeneration and irreversible cardiac hypertrophy, thought to result from oxidative damage. Idebenone, a quinone with antioxidant properties, reduced LV mass after six months of therapy in patients with Friedreich's ataxia (105,107).

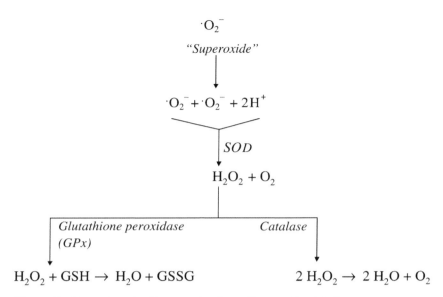

Figure 7 Natural antioxidant mechanisms. See text for details.

NATURAL ANTIOXIDANT DEFENSE MECHANISMS

In most tissues, antioxidant reserves consist of enzymes and nonenzymatic antioxidant substances that occur in both the intracellular and extracellular environment. A brief overview of these defenses is discussed below (Fig. 7).

Compartmentalization

ROS formation during cellular respiration is physically restricted within the cell, thereby limiting its toxicity. For example, large quantities of ROS can be produced by activated macrophages and neutrophils in response to microorganisms or neoplastic cells. This massive production of ROS is restricted to phagosomes, thus allowing the inflammatory reaction to be targeted against unrecognized cells, while minimizing host damage (108).

ANTIOXIDANT ENZYME SYSTEMS

Superoxide dismutase. SOD is an enzyme that catalyzes the dismutation of O_2^- to H_2O_2, thereby controlling the production of O_2^- anion. SOD activity maintains O_2^- at low intracellular levels (10^{-11} M) throughout normal metabolism (109). Impaired SOD activity in humans has been associated with several chronic diseases, including diabetes and HF (37,110,111). Animal models of SOD deficiency caused by administrating diethyldithiocarbamic acid (DDC), an inhibitor of cytosolic (Cu, Zn) and extracellular SOD, exhibit increased susceptibility to oxidative stress in cardiac myocytes, and consequent myocardial hypertrophy and apoptosis (8). In addition, SOD2 mutant mice, with complete absence of intramitochondrial SOD (Mn) activity, die within the first 10 days of life, manifesting dilated cardiomyopathy, accumulation of lipid in liver and skeletal muscle, and metabolic acidosis (31). These findings indicate the importance of SOD in neutralizing O_2^- for normal biological function.

Catalase. Superoxide dismutation leads to H_2O_2 formation, which requires a further metabolic step for elimination. There are many enzymatic pathways that metabolize H_2O_2, catalase being one of the most important. Upregulation of catalase gene expression has been shown in patients in end-stage HF (112).

Glutathione peroxidase. GPx controls H_2O_2 levels by catalyzing the reaction of H_2O_2 with reduced glutathione. GPx is an important endogenous antioxidant in the myocardium. Experimental studies, evaluating peroxidation activity, suggest a dominant role of GPx compared with catalase (113). In models of ischemic reperfusion injury, infusions of GPx analogs, such as ebselen, demonstrate an apoptotic protective effect (114). This fact is reinforced in mice overexpressing cellular glutathione (over 400% increase vs. controls), which experience decreased infarct size and improved recovery of myocardial contractility, relative to wild type (115). Interestingly, knockout animals exhibit the opposite effect, an increase in infarct size (116).

ANTIOXIDANT THERAPIES FOR HF

Antioxidant Effects of Existing Therapies

Many of the current treatment drugs that improve mortality in HF [e.g., angiotensin-converting enzyme (ACE) inhibitors and adrenergic nervous system blockers], exert beneficial effect on ventricular remodeling, decreasing chamber end-diastolic and systolic volumes, and improving ejection fraction (117). Most current HF treatments also possess antioxidant properties, which may contribute, at least in part, to reverse remodeling.

Angiotensin converting enzyme inhibitors. ACE inhibitors (118–120) exert antioxidant effects by inhibiting nuclear factor, (NF)-κB, a transcription factor, which regulates the expression of a variety of genes associated with inflammatory response, such as cytokines, chemokines, growth factors, and cell adhesion molecules (118–122). In addition, AT1 receptors have been linked to activation of NAD(P)H oxidase in vascular cells (123–125), and some groups have suggested a connection between angiotensin II and the generation of ROS by NOS uncoupling (125). Furthermore, ACE inhibitors exert a positive effect on cardiac remodeling, independent of blood pressure control (126). Lonn et al. (127,128), in studies in normotensive postinfarct populations, demonstrated that ACE inhibitors improve LV mass index and LV end-diastolic volume, as well as the extent of atherosclerosis (126–128). The precise contribution of ROS activity to this effect remains to be determined.

Hydralazine. Other vasodilators, such as hydralazine, when given concomitantly with isosorbide dinitrate, reduce mortality in HF patients (129,130). Hydralazine also acts by increasing sensitivity to nitroglycerin, which is implicated in improved remodeling following acute myocardial infarction (MI) (131), and attenuating nitrate tolerance (132,133), possibly by interfering with vascular O_2^- production. Münzel and colleagues (133) demonstrated that in isolated rabbit aortic segments, nitroglycerin led to a greater than two-fold increase in O_2^- production, mostly from an NADH-dependent membrane-associated oxidase. They found that hydralazine normalized this increase in O_2^- production and that acute addition of hydralazine to nitroglycerin-tolerant vessels immediately inhibited O_2^- production and NADH oxidase activity. These data suggest that the mechanism by which hydralazine prevents nitrate tolerance is via inhibition of a membrane-associated NADH oxidase, normally activated by chronic nitroglycerin treatment. It is interesting to note that, in this study, even aortic segments not exposed to nitroglycerin exhibited a significant decrease in O_2^- production after hydralazine treatment.

β-Blockers. Improvements in clinical outcomes due to the use of β-blockers in HF treatment may also be related to antioxidant effects. Carvedilol and metoprolol, both in clinical use for HF, also possess antioxidant activity (134). Carvedilol's antioxidant properties are implicated in improved cardiac function in HF (135). By ameliorating reperfusion injury, carvedilol decreases apoptosis (136) and infarct extension (137) in experimental models. Although carvedilol likely possesses stronger antioxidant properties than metoprolol (138) and is associated with profound beneficial effects on LV geometry and mass (139–141), one study has demonstrated that administration of either drugs is associated with similar decreases in markers of oxidative stress in correlation with clinical improvement (134).

Statins. Hydroxymethylglutaryl CoA reductase inhibitors (statins), even though not a component of the current HF treatment, have been demonstrated to significantly decrease cardiovascular morbidity and mortality in hypercholesterolemic patients (142,143). However, there is a growing body of evidence suggesting that these drugs have clinically relevant effects beyond lipid lowering (144). Indeed, statins have antioxidant properties, which may mediate the vasoprotective effects of statins. Atorvastatin significantly reduces angiotensin II induced ROS production, downregulates mRNA expression of critical NAD(P)H oxidase subunits, and upregulates catalase expression (145). In a recent study, Shishehbor and colleagues (146) administered atorvastatin to 35 hypercholesterolemic subjects, who had no known history of coronary artery disease. After 12 weeks of therapy, they reported significant reductions in the oxidative markers chlorotyrosine, NO2Tyr, and dityrosine, which were independent of decreases in lipids and lipoproteins. Further studies are needed to determine the clinical importance of the antioxidant properties of statins and their place in therapy.

ANTIOXIDANT THERAPIES IN DEVELOPMENT

Despite the fact that many drugs currently in use for HF have antioxidant properties, none were specifically designed to act by this mechanism. Novel therapeutic approaches aimed specifically at controlling ROS generation hold great potential as novel treatments for HF. At present, only a few antioxidant drugs have been tested, of which allopurinol, probucol, and certain vitamins (C, E and β-carotene) have shown the greatest potential.

Xanthine oxidase inhibitors. Allopurinol and its active metabolite oxypurinol represent a promising new therapy for HF. XO inhibition reduces the formation of O_2^- and oxidative stress. In the failing heart, allopurinol exhibits not only an anti-oxidant effect, but also a desirable inotropic response. This particular inotropic property seems to be the direct result of an increase in myocyte sensitivity to Ca^{+2} (47) as well as an interaction with NO pathway (40). Most of the inotropic drugs used in clinical practice improve myocardial contractility at the cost of increasing cardiac energy (oxygen) consumption. Therefore, in the failing and energy-depleted heart, these therapeutic approaches tend to be harmful, contributing to remodeling with worsening HF and subsequently, prognosis. Accordingly, allopurinol has unique features, improving mechanoenergetic coupling while stimulating myocardial contraction (20,36,40). In addition, XO inhibition has preliminarily been shown to reduce LV remodeling post-MI in experimental murine models (147).

Oxypurinol, the active metabolite of allopurinol, is currently undergoing clinical evaluation to test whether XO inhibition produces clinical benefits for patients with

HF. A phase II–III prospective, randomized, double-blind, placebo-controlled efficacy and safety study of oxypurinol added to standard therapy in patients with NYHA Class III–IV congestive heart failure (OPT–CHF), is a 400-patient study currently enrolling patients in the United States and Canada. If positive, the results of this trial have the potential to substantially influence HF management in the future.

Probucol. This lipid-lowering agent acts as a chain-breaking antioxidant, inhibiting lipid peroxidation and oxidative modification of LDL particles in vitro. Probucol has beneficial effect on LV dysfunction and remodeling in several HF models, including ventricular pacing (148), adriamycin-induced cardiomyopathy (17), and ischemic cardiomyopathy (149). These beneficial effects are thought to be mediated by its antioxidant activity and/or its anti-inflammatory properties. Administration of probucol is associated with reduced apoptosis (150), pro-inflammatory cytokines and cardiac fibrosis (149).

Ubiquinone (co-enzyme Q_{10}). Co-enzyme Q10 (CoQ10) or ubiquinone (2,3, dimethoxy-5-methyl-6-multiprenyl-1,4-benzoquinone) is a lipophilic molecule found in the phospholipid bilayer of virtually all cell membranes. Though it has multiple cellular functions, the best described functions are its activity as an electron carrier for ATP synthesis and its role as an antioxidant in the inner mitochondrial membrane (151).

Folkers and colleagues (152) described decreased levels of CoQ10 in the myocardium of patients with cardiac failure. Lower levels of CoQ10 correlated with worsening cardiomyopathy, as classified by the New York Heart Association guidelines, and these levels improved after six to eight months of CoQ10 supplementation. In contrast, Harker-Murray and colleagues (153) examined CoQ10 levels and supplementation in a canine tachycardia-induced CHF model. Serum and myocardial levels of CoQ10 did not change with the development of cardiomyopathy and CoQ10 supplementation increased serum but not myocardial levels of CoQ10. Interestingly, however, while CoQ10 supplementation had no effect on ventricular dilation, systolic, or diastolic dysfunction, CoQ10-treated dogs had less hypertrophy compared with untreated dogs.

The therapeutic use of CoQ10 in HF remains a topic of controversy. Some small studies have suggested a possible benefit from CoQ10 supplementation (154,155). However, randomized, double-blind, placebo-controlled trials in patients with LV dysfunction have shown no effect of CoQ10 supplementation on LV systolic function, peak oxygen consumption, exercise duration, or quality of life despite a doubling of blood levels of CoQ10 (156,157). These conflicting findings may be partially due to the fact that CoQ10 is extremely insoluble in water and therefore not taken up by cells to any significant degree, hindering its use as a therapeutic agent (158). One approach to circumvent this problem has been to synthesize ubiquinone analogs such as idebenone and decylubiquinone (159). However, there is concern that some CoQ10 analogs may actually have pro-oxidant effects (160). In summary, although CoQ10 may show promise as a potential adjuvant to HF therapy, more work must be performed to establish its role.

Vitamins. The most commonly used antioxidant vitamins in clinical trials are vitamin C (ascorbic acid), vitamin E (α-tocopherol), and β-carotene. Although these vitamins are not used clinically in the treatment of HF, there is growing evidence to suggest that they reduce remodeling in HF (161,162).

Vitamin C is an important hydrophilic antioxidant in extracellular fluid. In animal models of HF, it has been shown to attenuate remodeling (161) and enhance the mechanical efficiency of the heart (40). In humans with HF, vitamin C has been shown to reduce oxidative stress and improve endothelial function (163). Moreover, vitamin C was found to confer protection from myocardial apoptosis in HF patients (164). Treated patients had reduced plasma levels of circulating apoptotic microparticles (164), an effect thought to result from reductions in the release of cytochrome C from mitochondria and caspase inhibition (164).

Vitamin E is a lipophilic antioxidant with limited capacity for scavenging O_2^- (165). There is currently a great deal of controversy surrounding the effects of vitamin E on ischemic cardiovascular events. Some trials have shown a decrease in the incidence of cardiac ischemia with administration of vitamin E (166,167), whereas other authors have reported the opposite effect (168). In comparison, little has been done to evaluate the efficacy of α-tocopherol treatment in HF. Some experimental studies have shown that vitamin E reduces markers of oxidative stress and improves cardiac function (24,30,169,170). However, no significant improvements in the prognostic or functional indices of HF were observed in the only controlled clinical trial to date (171).

β-Carotene is more lipophilic than vitamin E and permeates the cell membrane more easily. β-Carotene acts principally by scavenging peroxyl radicals (172). Although the depletion of β-carotene has been linked to adriamycin-induced heart failure (173), there is no proof of any clear benefit of β-carotene supplementation in HF.

CONCLUSIONS

Heart failure is the final common pathway of a variety of ischemic, hemodynamic, and inflammatory insults to the myocardium. Despite dramatic improvements in the management of patients with HF, this condition remains one of the main causes of hospitalization and death. This chapter has provided an overview of the importance of ROS in the structure and function of the failing heart. Because ROS participate in several key features of HF, including mechanoenergetic uncoupling and remodeling of the failing heart, which result from enhanced myocyte apoptosis and hypertrophy, antagonism of the production of these molecules offers hope as a new therapeutic strategy. While much experimental data support the benefits of inhibitors of oxidative stress, trials designed to test the clinical efficacy of this strategy are currently underway. These trials will greatly clarify the importance of oxidative stress in the pathophysiology of HF.

REFERENCES

1. Jessup M, Brozena S. Heart failure. N Engl J Med 2003; 348:2007–2018.
2. Konstam MA. Progress in heart failure management? Lessons from the real world. Circulation 2000; 102:1076–1078.
3. The CONSENSUS Trial Study Group. Effects of enalapril on mortality in congestive heart failure. Results of the cooperative north scandinavian enalapril survival study. N Engl J Med 1987; 316:1429–1435.
4. Packer M, Coats AJ, Fowler MB, Katus HA, Krum H, Mohacsi P, Rouleau JL, Tendera M, Castaigne A, Roecker EB, Schultz MK, DeMets DL. Effect of carvedilol on survival in severe chronic heart failure. N Engl J Med 2001; 344:1651–1658.
5. Katz AM. Physiology of the heart. 3rd ed. Philadelphia: Lippincott Williams & Wilkins, 2001.
6. Ide T, Tsutsui H, Kinugawa S, Utsumi H, Kang D, Hattori N, Uchida K, Arimura K, Egashira K, Takeshita A. Mitochondrial electron transport complex I is a potential source of oxygen free radicals in the failing myocardium. Circ Res 1999; 85:357–363.

7. Givertz MM, Colucci WS. New targets for heart-failure therapy: endothelin, inflammatory cytokines, and oxidative stress. Lancet 1998; 352(suppl I):34–38.

8. Siwik DA, Tzortzis JD, Pimental DR, Chang DL, Pagano PJ, Singh K, Sawyer DB, Colucci WS. Inhibition of copper-zinc superoxide dismutase induces cell growth, hypertrophic phenotype, and apoptosis in neonatal rat cardiac myocytes in vitro. Circ Res 1999; 85:147–153.

9. Chiamvimonvat N, O'Rourke B, Kamp TJ, Kallen RG, Hofmann F, Flockerzi V, Marbán E. Functional consequences of sulfhydryl modification in the pore-forming subunits of cardiovascular Ca^{2+} and Na^{+} channels. Circ Res 1995; 76:325–334.

10. Thomas JA, Poland B, Honzatko R. Protein sulfhydryls and their role in the antioxidant function of protein S-thiolation. Arch Biochem Biophys 1995; 319:1–9.

11. Campbell DL, Stamler JS, Strauss HC. Redox modulation of L-type calcium channels in ferret ventricular myocytes. Dual mechanism regulation by nitric oxide and S-nitrosothiols. J Gen Physiol 1996; 108:277–293.

12. Xu L, Eu JP, Meissner G, Stamler JS. Activation of the cardiac calcium release channel (Ryanodine receptor) by poly-S-nitrosylation. Science 1998; 279:234–237.

13. Zeitz O, Maass AE, Van Nguyen P, Hensmann G, Kogler H, Moller K, Hasenfuss G, Janssen PM. Hydroxyl radical-induced acute diastolic dysfunction is due to calcium overload via reverse-mode Na(+)-Ca(2+) exchange. Circ Res 2002; 90:988–995.

14. Pomar F, Cosin J, Portoles M, Faura M, RenauPiqueras J, Hernandiz A, Andres F, Colomer JL, Graullera B. Functional and ultrastructural alterations of canine myocardium subjected to very brief coronary occlusions. Eur Heart J 1995; 16:1482–1490.

15. Przyklenk K, Kloner RA. Superoxide dismutase plus catalase improve contractile function in the canine model of the "stunned myocardium". Circ Res 1986; 58:148–156.

16. Sussman MS, Bulkley GB. Oxygen-derived free radicals in reperfusion injury. Methods Enzymol 1990:711–723.

17. Singal PK, Siveskiiliskovic N, Hill M, Thomas TP, Li TM. Combination therapy with probucol prevents adriamycin-induced cardiomyopathy. J Mol Cell Cardiol 1995; 27:1055–1063.

18. Siveskiiliskovic N, Hill M, Chow DA, Singal PK. Probucol protection against adriamycin cardiomyopathy without compromising antitumor action. Faseb J 1995; 9:A129.

19. Singal PK, Kapur N, Dhillon KS, Beamish RE, Dhalla NS. Role of free-radicals in catecholamine-induced cardiomyopathy. Canad J Physiol Pharmacol 1982; 60:1390–1397.

20. Ekelund UEG, Harrison RW, Shokek O, Thakkar RN, Tunin RS, Senzaki H, Kass DA, Marbán E, Hare JM. Intravenous allopurinol decreases myocardial oxygen consumption and increases mechanical efficiency in dogs with pacing-induced heart failure. Circ Res 1999; 85:437–445.

21. McMurray J, Chopra M, Abdullah I, Smith WE, Dargie HJ. Evidence of oxidative stress in chronic heart failure in humans. Eur Heart J 1993; 14:1493–1498.

22. Sobotka PA, Brottman MD, Weitz Z, Birnbaum AJ, Skosey JL, Zarling EJ. Elevated breath pentane in heart-failure reduced by free-radical scavenger. Free Rad Biol Med 1993; 14:643–647.

23. Mallat Z, Philip I, Lebret M, Chatel D, Maclouf J, Tedgui A. Elevated levels of 8-isoprotaglandin F2α in pericardial fluid of patients with heart failure. Circulation 1998; 97:1536–1539.

24. Dhalla AK, Hill MF, Singal PK. Role of oxidative stress in transition of hypertrophy to heart failure. J Am Coll Cardiol 1996; 28:506–514.

25. Berry C.E., Hare JM. Xanthine oxidoreductase and cardiovascular disease: molecular mechanisms and pathophysiologic implications. J Physiol 2004; 16:555 (pt3):589–606.

26. Boveris A, Cadenas E, Stoppani AOM. Role of ubiquinone in mitochondrial generation of hydrogen-peroxide. Biochem J 1976; 156:435–444.

27. Riobo NA, Clementi E, Melani M, Boveris A, Cadenas E, Moncada S, Poderoso JJ. Nitric oxide inhibits mitochondrial NADH : ubiquinone reductase activity through peroxynitrite formation. Biochem J 2001; 359:139–145.

28. Clementi E, Brown GC, Feelisch M, Moncada S. Persistent inhibition of cell respiration by nitric oxide: crucial role of S-nitrosylation of mitochondrial complex I and protective action of glutathione. Proc Natl Acad Sci USA 1998; 95:7631–7636.

29. Dhalla AK, Singal PK. Antioxidant changes in hypertrophied and failing guinea-pig hearts. Amer J Physiol 1994; 266:H1280–H1285.

30. Hill MF, Singal PK. Antioxidant and oxidative stress changes during heart failure subsequent to myocardial infarction in rats. Amer J Pathol 1996; 148:291–300.

31. Li Y, Huang TT, Carlson EJ, Melov S, Ursell PC, Olson JL, Noble LJ, Yoshimura MP, Berger C, Chan PH. Dilated cardiomyopathy and neonatal lethality in mutant mice lacking manganese superoxide dismutase. Nat Genet 1995; 11:376–381.

32. Saugstad OD. Role of xanthine oxidase and its inhibitor in hypoxia: reoxygenation injury. Pediatr 1996; 98:103–107.

33. McCord JM. Oxygen-derived free radicals in postischemic tissue injury. N Eng J Med 1985; 312:159–163.

34. Sanders SA, Eisenthal R, Harrison R. NADH oxidase activity of human xanthine oxidoreductase generation of superoxide anion. Eur J Biochem 1997; 245:541–548.

35. de Jong JW, Schoemaker RG, de Jonge R, Bernocchi P, Keijzer E, Harrison R, Sharma HS, Ceconi C. Enhanced expression and activity of xanthine oxidoreductase in the failing heart. J Mol Cell Cardiol 2000; 32:2083–2089.

36. Cappola TP, Kass DA, Nelson GS, Berger RD, Rosas GO, Kobeissi ZA, Marban E, Hare JM. Allopurinol improves myocardial efficiency in patients with idiopathic dilated cardiomyopathy. Circulation 2001; 104:2407–2411.

37. Landmesser U, Spiekermann S, Dikalov S, Tatge H, Wilke R, Kohler C, Harrison DG, Hornig B, Drexler H. Vascular oxidative stress and endothelial dysfunction in patients with chronic heart failure: role of xanthine-oxidase and extracellular superoxide dismutase. Circulation 2002; 106:3073–3078.

38. de Jong JW. Xanthine oxidoreductase activity in perfused hearts of various species, including humans. Circ Res 1990; 67:770–773.

39. White CR, Darley-Usmar V, Berrington WR, McAdams M, Gore JZ, Thompson JA, Parks DA, Tarpey MM, Freeman BA. Circulating plasma xanthine oxidase contributes to vascular dysfunction in hypercholesterolemic rabbits. Proc Natl Acad Sci USA 1996; 93:8745–8749.

40. Saavedra WF, Paolocci N, St John ME, Skaf MW, Stewart GC, Xie JS, Harrison RW, Zeichner J, Mudrick D, Marban E, Kass DA, Hare JM. Imbalance between xanthine oxidase and nitric oxide synthase signaling pathways underlies mechanoenergetic uncoupling in the failing heart. Circ Res 2002; 90:297–304.

41. Doehner W, Schoene N, Rauchhaus M, Leyva-Leon F, Pavitt DV, Reaveley DA, Schuler G, Coats AJ, Anker SD, Hambrecht R. Effects of xanthine oxidase inhibition with allopurinol on endothelial function and peripheral blood flow in hyperuricemic patients with chronic heart failure: results from 2 placebo-controlled studies. Circulation 2002; 105:2619–2624.

42. Farquharson CA, Butler R, Hill A, Belch JJ, Struthers AD. Allopurinol improves endothelial dysfunction in chronic heart failure. Circulation 2002; 106:221–226.

43. Suga H. Ventricular energetics. Physiol Rev 1990; 70:247–277.

44. Ukai T, Cheng CP, Tachibana H, Igawa A, Zhang ZS, Cheng HJ, Little WC. Allopurinol enhances the contractile response to dobutamine and exercise in dogs with pacing-induced heart failure. Circulation 2001; 103:750–755.

45. Khan SA, Skaf MW, Harrison RW, Lee K, Minhas KM, Kumar A, Fradley M, Shoukas AA, Berkowitz DE, Hare JM. Nitric oxide regulation of myocardial contractility and calcium cycling: independent impact of neuronal and endothelial nitric oxide synthases. Circ Res 2003; 92:1322–1329.

46. Eu JP, Sun J, Xu L, Stamler JS, Meissner G. The skeletal muscle calcium release channel: coupled O_2 sensor and NO signaling functions. Cell 2000; 102:499–509.

47. Perez NG, Gao WD, Marban E. Novel myofilament Ca2+-sensitizing property of xanthine oxidase inhibitors. Circ Res 1998; 83:423–430.

48. Rubbo H, DarleyUsmar V, Freeman BA. Nitric oxide regulation of tissue free radical injury. Chem Res Toxicol 1996; 9:809–820.

49. Hare JM. Nitric oxide and excitation-contraction coupling. J Mol Cell Cardiol 2003; 35:719–729.

50. Ruan J, Xie Q, Hutchinson N, Cho H, Wolfe GC, Nathan C. Inducible nitric oxide synthase requires both the canonical calmodulin-binding domain and additional sequences in order to bind calmodulin and produce nitric oxide in the absence of free Ca^{2+}. J Biol Chem 1996; 271:22,679–22,686.

51. Barouch LA, Harrison RW, Skaf MW, Rosas GO, Cappola TP, Kobeissi ZA, Hobai IA, Lemmon CA, Burnett AL, O'Rourke B, Rodriguez RE, Huang PL, Lima JAC, Berkowitz DE, Hare JM. Nitric oxide regulates the heart by spatial confinement of nitric oxide synthase isoforms. Nature 2002; 416:337–340.

52. Balligand J-L, Ungureanu-Longrois D, Simmons WW, Pimental D, Malinski TA, Kapturczak M, Taha Z, Lowenstein CJ, Davidoff AJ, Kelly RA, Smith TW, Michel T. Cytokine-inducible nitric-oxide synthase (iNOS) expression in cardiac myocytes: characterization and regulation of iNOS expression and detection of iNOS activity in single cardiac myocytes in vitro. J Biol Chem 1994; 269:27,580–27,588.

53. Zweier JL, Wang P, Samouilov A, Kuppusamy P. Enzyme-independent formation of nitric oxide in biological tissues. Nat Med 1995; 1:804–809.

54. Chandra J, Samali A, Orrenius S. Triggering and modulation of apoptosis by oxidative stress. Free Radic Biol Med 2000; 29:323–333.

55. Ferdinandy P, Panas D, Schulz R. Peroxynitrite contributes to spontaneous loss of cardiac efficiency in isolated working rat hearts. Am J Physiol 1999; 276:H1861–H1867.

56. Miyamoto Y, Akaike T, Yoshida M, Goto S, Horie H, Maeda H. Potentiation of nitric oxide-mediated vasorelaxation by xanthine oxidase inhibitors. Proc Soc Exp Biol Med 1996; 211:366–373.

57. Ferdinandy P, Danial H, Ambrus I, Rothery RA, Schulz R. Peroxynitrite is a major contributor to cytokine-induced myocardial contractile failure. Circ Res 2000; 87:241–247.

58. Curtin JF, Donovan M, Cotter TG. Regulation and measurement of oxidative stress in apoptosis. J Immunol Methods 2002; 265:49–72.

59. Wang W, Wang S, Yan L, Madara P, Del Pilar CA, Wesley RA, Danner RL. Super-oxide production and reactive oxygen species signaling by endothelial nitric-oxide synthase. J Biol Chem 2000; 275:16,899–16,903.

60. Stuehr D, Pou S, Rosen GM. Oxygen reduction by nitric-oxide synthases. J Biol Chem 2001; 276:14,533–14,536.

61. Cosentino F, Luscher TF. Tetrahydrobiopterin and endothelial nitric oxide synthase activity. Cardiovasc Res 1999; 43:274–278.

62. Wever RM, van Dam T, van Rijn HJ, de Groot F, Rabelink TJ. Tetrahydrobiopterin regulates superoxide and nitric oxide generation by recombinant endothelial nitric oxide synthase. Biochem Biophys Res Commun 1997; 237:340–344.

63. Stroes E, Kastelein J, Cosentino F, Erkelens W, Wever R, Koomans H, Luscher T, Rabelink T. Tetrahydrobiopterin restores endothelial function in hypercholesterolemia. J Clin Invest 1997; 99:41–46.

64. Pieper GM. Acute amelioration of diabetic endothelial dysfunction with a derivative of the nitric oxide synthase cofactor, tetrahydrobiopterin. J Cardiovasc Pharmacol 1997; 29:8–15.

65. Higman DJ, Strachan AM, Buttery L, Hicks RC, Springall DR, Greenhalgh RM, Powell JT. Smoking impairs the activity of endothelial nitric oxide synthase in saphe-nous vein. Arterioscler Thromb Vasc Biol 1996; 16:546–552.

66. Griendling KK, Sorescu D, Ushio-Fukai M. NAD(P)H oxidase: role in cardiovascular biology and disease. Circ Res 2000; 86:494–501.

67. Irani K, Xia Y, Zweier JL, Sollott SJ, Der CJ, Fearon ER, Sundaresan M, Finkel T, Goldschmidt-Clermont PJ. Mitogenic signaling mediated by oxidants in Ras-transformed fibroblasts. Science 1997; 275:1649–1652.

68. Suh YA, Arnold RS, Lassegue B, Shi J, Xu X, Sorescu D, Chung AB, Griendling KK, Lambeth JD. Cell transformation by the superoxide-generating oxidase Mox1. Nature 1999; 401:79–82.

69. Heymes C, Bendall JK, Ratajczak P, Cave AC, Samuel JL, Hasenfuss G, Shah AM. Increased myocardial NADPH oxidase activity in human heart failure. J Am Coll Cardiol 2003; 41:2164–2171.

70. Griendling KK, Ushio-Fukai M. Redox control of vascular smooth muscle proliferation. J Lab Clin Med 1998; 132:9–15.

71. Keisari Y, Braun L, Flescher E. The oxidative burst and related phenomena in mouse macrophages elicited by different sterile inflammatory stimuli. Immunobiol 1983; 165:78–89.

72. Lassegue B, Sorescu D, Szocs K, Yin Q, Akers M, Zhang Y, Grant SL, Lambeth JD, Griendling KK. Novel gp91(phox) homologues in vascular smooth muscle cells: nox1 mediates angiotensin II-induced superoxide formation and redox-sensitive signaling pathways. Circ Res 2001; 88:888–894.

73. Pagano PJ, Chanock SJ, Siwik DA, Colucci WS, Clark JK. Angiotensin II induces p67phox mRNA expression and NADPH oxidase superoxide generation in rabbit aortic adventitial fibroblasts. Hypertension 1998; 32:331–337.

74. Duilio C, Ambrosio G, Kuppusamy P, Dipaula A, Becker LC, Zweier JL. Neutrophils are primary source of O-2 radicals during reperfusion after prolonged myocardial ischemia. Am J Physiol-Heart Circ Physiol 2001; 280:H2649–H2657.

75. Arai M, Lefer DJ, So T, Dipaula A, Aversano T, Becker LC. An anti-CD18 antibody limits infarct size and preserves left ventricular function in dogs with ischemia and 48-hour reperfusion. J Am Coll Cardiol 1996; 27:1278–1285.

76. Perez RG, Arai M, Richardson C, Dipaula A, Siu C, Matsumoto N, Hildreth JEK, Mariscalco MM, Smith CW, Becker LC. Factors modifying protective effect of anti-CD18 antibodies on myocardial reperfusion injury in dogs. Am J Physiol-Heart Circulat Physiol 1996; 39:H53–H64.

77. Mohazzab H, Kaminski PM, Wolin MS. Lactate and PO2 modulate superoxide anion production in bovine cardiac myocytes: potential role of NADH oxidase. Circulation 1997; 96:614–620.

78. Zafari AM, Ushio-Fukai M, Akers M, Yin Q, Shah A, Harrison DG, Taylor WR, Griendling KK. Role of NADH/NADPH oxidase-derived H_2O_2 in angiotensin II-induced vascular hypertrophy. Hypertension 1998; 32:488–495.

79. Ushio-Fukai M, Zafari AM, Fukui T, Ishizaka N, Griendling KK. p22phox is a critical component of the superoxide-generating NADH/NADPH oxidase system and regulates angiotensin II-induced hypertrophy in vascular smooth muscle cells. J Biol Chem 1996; 271:23,317–23,321.

80. Narula J, Pandey P, Arbustini E, Haider N, Narula N, Kolodgie FD, Dal Bello B, Semigran MJ, Bielsa-Masdeu A, Dec GW, et al. Apoptosis in heart failure: release of cytochrome c from mitochondria and activation of caspase-3 in human cardiomyopathy. Proc Natl Acad Sci USA 1999; 96:8144–8149.

81. Wencker D, Chandra M, Nguyen K, Miao W, Garantziotis S, Factor SM, Shirani J, Armstrong RC, Kitsis RN. A mechanistic role for cardiac myocyte apoptosis in heart failure. J Clin Invest 2003; 111:1497–1504.

82. Palojoki E, Saraste A, Eriksson A, Pulkki K, Kallajoki M, Voipio-Pulkki LM, Tikkanen I. Cardiomyocyte apoptosis and ventricular remodeling after myocardial infarction in rats. Am J Physiol Heart Circ Physiol 2001; 280:H2726–H2731.

83. Xu B, Sakkas LI, Slachta CA, Goldman BI, Jeevanandam V, Oleszak EL, Platsoucas CD. Apoptosis in chronic rejection of human cardiac allografts. Transplantation 2001; 71:1137–1146.

84. Lennon SV, Martin SJ, Cotter TG. Dose-dependent induction of apoptosis in human tumour cell lines by widely diverging stimuli. Cell Prolif 1991; 24:203–214.

85. Cesselli D, Jakoniuk I, Barlucchi L, Beltrami AP, Hintze TH, Nadal-Ginard B, Kajstura J, Leri A, Anversa P. Oxidative stress-mediated cardiac cell death is a major determinant of ventricular dysfunction and failure in dog dilated cardiomyopathy. Circ Res 2001; 89:279–286.

86. Finkel E. The mitochondrion: is it central to apoptosis? Science 2001; 292:624–626.

87. Hare JM. Oxidative stress and apoptosis in heart failure progression. Circ Res 2001; 89:198–200.

88. Kim YM, Kim TH, Seol DW, Talanian RV, Billiar TR. Nitric oxide suppression of apoptosis occurs in association with an inhibition of Bcl-2 cleavage and cytochrome c release. J Biol Chem 1998; 273:31,437–31,441.

89. Cheng EH, Kirsch DG, Clem RJ, Ravi R, Kastan MB, Bedi A, Ueno K, Hardwick JM. Conversion of Bcl-2 to a Bax-like death effector by caspases. Science 1997; 278:1966–1968.

90. Stassi G, De Maria R, Trucco G, Rudert W, Testi R, Galluzzo A, Giordano C, Trucco M. Nitric oxide primes pancreatic beta cells for Fas-mediated destruction in insulin-dependent diabetes mellitus. J Exp Med 1997; 186:1193–1200.

91. Dorn GW, Robbins J, Sugden PH. Phenotyping hypertrophy: eschew obfuscation. Circ Res 2003; 92:1171–1175.

92. Levy D, Garrison RJ, Savage DD, Kannel WB, Castelli WP. Prognostic implications of echocardiographically determined left ventricular mass in the framingham heart study. N Engl J Med 1990; 322:1561–1566.

93. Aikawa R, Nagai T, Tanaka M, Zou Y, Ishihara T, Takano H, Hasegawa H, Akazawa H, Mizukami M, Nagai R, Komuro I. Reactive oxygen species in mechanical stress-induced cardiac hypertrophy. Biochem Biophys Res Commun 2001; 289:901–907.

94. Pimentel DR, Amin JK, Xiao L, Miller T, Viereck J, Oliver-Krasinski J, Baliga R, Wang J, Siwik DA, Singh K, Pagano P, Colucci WS, Sawyer DB. Reactive oxygen species mediate amplitude-dependent hypertrophic and apoptotic responses to mechanical stretch in cardiac myocytes. Circ Res 2001; 89:453–460.

95. Chen QM, Tu VC, Wu Y, Bahl JJ. Hydrogen peroxide dose dependent induction of cell death or hypertrophy in cardiomyocytes. Arch Biochem Biophys 2000; 373:242–248.

96. Nakamura K, Fushimi K, Kouchi H, Mihara K, Miyazaki M, Ohe T, Namba M. Inhibitory effects of antioxidants on neonatal rat cardiac myocyte hypertrophy induced by tumor necrosis factor-alpha and angiotensin II. Circulation 1998; 98:794–799.

97. Amin JK, Xiao L, Pimental DR, Pagano PJ, Singh K, Sawyer DB, Colucci WS. Reactive oxygen species mediate alpha-adrenergic receptor-stimulated hypertrophy in adult rat ventricular myocytes. J Mol Cell Cardiol 2001; 33:131–139.

98. Xiao L, Pimentel DR, Wang J, Singh K, Colucci WS, Sawyer DB. Role of reactive oxygen species and NAD(P)H oxidase in $\alpha(1)$-adrenoceptor signaling in adult rat cardiac myocytes. Am J Physiol Cell Physiol 2002; 282:C926–C934.

99. Nakagami H, Takemoto M, Liao JK. NADPH oxidase-derived superoxide anion mediates angiotensin II-induced cardiac hypertrophy. J Mol Cell Cardiol 2003; 35:851–859.

100. Takemoto M, Node K, Nakagami H, Liao Y, Grimm M, Takemoto Y, Kitakaze M, Liao JK. Statins as antioxidant therapy for preventing cardiac myocyte hypertrophy. J Clin Invest 2001; 108:1429–1437.

101. Luo JD, Xie F, Zhang WW, Ma XD, Guan JX, Chen X. Simvastatin inhibits noradrenaline-induced hypertrophy of cultured neonatal rat cardiomyocytes. Br J Pharmacol 2001; 132:159–164.

102. MacCarthy PA, Grieve DJ, Li JM, Dunster C, Kelly FJ, Shah AM. Impaired endothelial regulation of ventricular relaxation in cardiac hypertrophy: role of reactive oxygen species and NADPH oxidase. Circulation 2001; 104:2967–2974.

103. Luo JD, Zhang WW, Zhang GP, Guan JX, Chen X. Simvastatin inhibits cardiac hypertrophy and angiotensin-converting enzyme activity in rats with aortic stenosis. Clin Exp Pharmacol Physiol 1999; 26:903–908.

104. Patel R, Nagueh SF, Tsybouleva N, Abdellatif M, Lutucuta S, Kopelen HA, Quinones MA, Zoghbi WA, Entman ML, Roberts R, et al. Simvastatin induces regression of cardiac hypertrophy and fibrosis and improves cardiac function in a transgenic rabbit model of human hypertrophic cardiomyopathy. Circulation 2001; 104:317–324.

105. Hausse AO, Aggoun Y, Bonnet D, Sidi D, Munnich A, Rotig A, Rustin P. Idebenone and reduced cardiac hypertrophy in Friedreich's ataxia. Heart 2002; 87:346–349.

106. Rustin P, Rotig A, Munnich A, Sidi D. Heart hypertrophy and function are improved by idebenone in Friedreich's ataxia. Free Radic Res 2002; 36:467–469.

107. Dai L, Brookes PS, Darley-Usmar VM, Anderson PG. Bioenergetics in cardiac hypertrophy: mitochondrial respiration as a pathological target of NO*. Am J Physiol Heart Circ Physiol 2001; 281:H2261–H2269.

108. Droge W. Free radicals in the physiological control of cell function. Physiol Rev 2002; 82:47–95.

109. MacCarthy PA, Shah AM. Oxidative stress and heart failure. Coron Artery Dis 2003; 14:109–113.

110. Kang JH. Modification and inactivation of human Cu,Zn-superoxide dismutase by methylglyoxal. Mol Cells 2003; 15:194–199.

111. Seghrouchni I, Drai J, Bannier E, Riviere J, Calmard P, Garcia I, Orgiazzi J, Revol A. Oxidative stress parameters in type I, type II and insulin-treated type 2 diabetes mellitus; insulin treatment efficiency. Clin Chim Acta 2002; 321:89–96.

112. Dieterich S, Bieligk U, Beulich K, Hasenfuss G, Prestle J. Gene expression of antioxidative enzymes in the human heart: increased expression of catalase in the end-stage failing heart. Circulation 2000; 101:33–39.

113. Simmons TW, Jamall IS. Relative importance of intracellular glutathione peroxidase and catalase in vivo for prevention of peroxidation to the heart. Cardiovasc Res 1989; 23:774–779.

114. Maulik N, Yoshida T. Oxidative stress developed during open heart surgery induces apoptosis: reduction of apoptotic cell death by ebselen, a glutathione peroxidase mimic. J Cardiovasc Pharmacol 2000; 36:601–608.

115. Yoshida T, Watanabe M, Engelman DT, Engelman RM, Schley JA, Maulik N, Ho YS, Oberley TD, Das DK. Transgenic mice overexpressing glutathione peroxidase are resistant to myocardial ischemia reperfusion injury. J Mol Cell Cardiol 1996; 28:1759–1767.

116. Yoshida T, Maulik N, Engelman RM, Ho YS, Magnenat JL, Rousou JA, Flack JE III, Deaton D, Das DK. Glutathione peroxidase knockout mice are susceptible to myocardial ischemia reperfusion injury. Circulation 1997; 96:11–20.

117. Cohn JN, Ferrari R, Sharpe N. Cardiac remodeling—concepts and clinical implications: a consensus paper from an international forum on cardiac remodeling. Behalf of an international forum on cardiac remodeling. J Am Coll Cardiol 2000; 35:569–582.

118. Ruiz-Ortega M, Lorenzo O, Ruperez M, Konig S, Wittig B, Egido J. Angiotensin II activates nuclear transcription factor kappa B through AT(1) and AT(2) in vascular smooth muscle cells: molecular mechanisms. Circ Res 2000; 86:1266–1272.

119. Marui N, Offermann MK, Swerlick R, Kunsch C, Rosen CA, Ahmad M, Alexander RW, Medford RM. Vascular cell adhesion molecule-1 (VCAM-1) gene transcription and expression are regulated through an antioxidant-sensitive mechanism in human vascular endothelial cells. J Clin Invest 1993; 92:1866–1874.

120. Pi XJ, Chen X. Captopril and ramiprilat protect against free radical injury in isolated working rat hearts. J Mol Cell Cardiol 1989; 21:1261–1271.

121. Ruiz-Ortega M, Bustos C, Hernandez-Presa MA, Lorenzo O, Plaza JJ, Egido J. Angiotensin II participates in mononuclear cell recruitment in experimental immune complex nephritis through nuclear factor-kappa B activation and monocyte chemoattractant protein-1 synthesis. J Immunol 1998; 161:430–439.

122. Chen F, Castranova V, Shi X, Demers LM. New insights into the role of nuclear factor-kappa B, a ubiquitous transcription factor in the initiation of diseases. Clin Chem 1999; 45:7–17.

123. Griendling KK, Minieri CA, Ollerenshaw JD, Alexander RW. Angiotensin II stimulates NADH and NADPH oxidase activity in cultured vascular smooth muscle cells. Circ Res 1994; 74:1141–1148.

124. Nickenig G, Harrison DG. The AT(1)-type angiotensin receptor in oxidative stress and atherogenesis: part II: AT(1) receptor regulation. Circulation 2002; 105:530–536.

125. Nickenig G, Harrison DG. The AT(1)-type angiotensin receptor in oxidative stress and atherogenesis: part I: oxidative stress and atherogenesis. Circulation 2002; 105:393–396.

126. Sleight P. Angiotensin II and trials of cardiovascular outcomes. Am J Cardiol 2002; 89:11A–16A.

127. Lonn E, Yusuf S, Dzavik V, Doris C, Yi Q, Smith S, Moore-Cox A, Bosch J, Riley W, Teo K. Effects of ramipril and vitamin E on atherosclerosis: the study to evaluate carotid ultrasound changes in patients treated with ramipril and vitamin E (SECURE). Circulation 2001; 103:919–925.

128. Lonn EM, Shaishkoleslami R, Yi Q, Bosch J, Magi A, Yusuf S. Effects of ramipril on left ventricular mass and function in normotensive, high-risk patients with normal ejection fraction. A substudy of HOPE. J Am Coll Cardiol 2001; 37(suppl 2A):165A.

129. Cohn JN, Archibald DG, Ziesche S, Franciosa JA, Harston WE, Tristani FE, Dunkman WB, Jacobs W, Francis GS, Flohr KH. Effect of vasodilator therapy on mortality in chronic congestive heart failure. Results of a veterans administration cooperative study. N Engl J Med 1986; 314:1547–1552.

130. Cohn JN, Johnson G, Ziesche S, Cobb F, Francis G, Tristani F, Smith R, Dunkman WB, Loeb H, Wong M. A comparison of enalapril with hydralazine-isosorbide dinitrate in the treatment of chronic congestive heart failure. N Engl J Med 1991; 325:303–310.

131. McDonald KM, Francis GS, Matthews J, Hunter D, Cohn JN. Long-term oral nitrate therapy prevents chronic ventricular remodeling in the dog. J Am Coll Cardiol 1993; 21:514–522.

132. Bauer JA, Fung HL. Concurrent hydralazine administration prevents nitroglycerin-induced hemodynamic tolerance in experimental heart failure. Circulation 1991; 84:35–39.

133. Munzel T, Kurz S, Rajagopalan S, Thoenes M, Berrington WR, Thompson JA, Freeman BA, Harrison DG. Hydralazine prevents nitroglycerin tolerance by inhibiting activation of a membrane-bound NADH oxidase. A new action for an old drug. J Clin Invest 1996; 98:1465–1470.

134. Kukin ML, Kalman J, Charney RH, Levy DK, Buchholz-Varley C, Ocampo ON, Eng C. Prospective, randomized comparison of effect of long-term treatment with metoprolol or carvedilol on symptoms, exercise, ejection fraction, and oxidative stress in heart failure. Circulation 1999; 99:2645–2651.

135. Flesch M, Maack C, Cremers B, Baumer AT, Sudkamp M, Bohm M. Effect of beta-blockers on free radical-induced cardiac contractile dysfunction. Circulation 1999; 100:346–353.

136. Yue TL, Ma XL, Wang X, Romanic AM, Liu GL, Louden C, Gu JL, Kumar S, Poste G, Ruffolo RR Jr, Feuerstein GZ. Possible involvement of stress-activated protein kinase signaling pathway and Fas receptor expression in prevention of ischemia/reperfusion-induced cardiomyocyte apoptosis by carvedilol. Circ Res 1998; 82:166–174.

137. Brunvand H, Frlyland L, Hexeberg E, Rynning SE, Berge RK, Grong K. Carvedilol improves function and reduces infarct size in the feline myocardium by protecting against lethal reperfusion injury. Eur J Pharmacol 1996; 314:99–107.

138. Lysko PG, Webb CL, Gu JL, Ohlstein EH, Ruffolo RR, Yue TL. A comparison of carvedilol and metoprolol antioxidant activities in vitro. J Cardiovasc Pharmacol 2000; 36:277–281.

139. Lowes BD, Gill EA, Abraham WT, Larrain JR, Robertson AD, Bristow MR, Gilbert EM. Effects of carvedilol on left ventricular mass, chamber geometry, and mitral regurgitation in chronic heart failure. Am J Cardiol 1999; 83:1201–1205.

140. Hall SA, Cigarroa CG, Marcoux L, Risser RC, Grayburn PA, Eichhorn EJ. Time course of improvement in left ventricular function, mass and geometry in patients with congestive heart failure treated with beta-adrenergic blockade. J Am Coll Cardiol 1995; 25:1154–1161.

141. Bristow MR, Gilbert EM, Abraham WT, Adams KF, Fowler MB, Hershberger RE, Kubo SH, Narahara KA, Ingersoll H, Krueger S, et al. Carvedilol produces dose-related improvements in left ventricular function and survival in subjects with chronic heart failure. MOCHA investigators. Circulation 1996; 94:2807–2816.

142. Shepherd J, Cobbe SM, Ford I, Isles CG, Lorimer AR, Macfarlane PW, McKillop JH, Packard CJ. Prevention of coronary heart-disease with pravastatin in men with hypercholesterolemia. N Engl J Med 1995; 333:1301–1307.

143. Pedersen TR, Kjekshus J, Berg K, Haghfelt T, Faergeman O, Thorgeirsson G, Pyorala K, Miettinen T, Wilhelmsen L, Olsson AG et al. Randomized trial of cholesterol-lowering in 4444 patients with coronary-heart-disease—the scandinavian simvastatin survival study (4S). Lancet 1994; 344:1383–1389.

144. Bonetti PO, Lerman LO, Napoli C, Lerman A. Statin effects beyond lipid lowering-are they clinically relevant? Eur Heart J 2003; 24:225–248.

145. Wassmann S, Laufs U, Muller K, Konkol C, Ahlbory K, Baumer AT, Linz W, Bohm M, Nickenig G. Cellular antioxidant effects of atorvastatin in vitro and in vivo. Arterio Thromb Vasc Biol 2002; 22:300–305.

146. Shishehbor MH, Brennan ML, Aviles RJ, Fu XM, Penn MS, Sprecher DL, Hazen SL. Statins promote potent systemic antioxidant effects through specific inflammatory pathways. Circulation 2003; 108:426–431.

147. Engberding N, Heineke A, Wiencke A, Schaefer A, Spiekermann S, Fuchs M, Hornig B, Landmesser U. Allopurinol, a xanthine oxidase inhibitor, prevents left ventricular remodeling and attenuates lv dysfunction after myocardial infarction. A new action for an old drug? Circulation 2003; 108, No 17, IV-175.

148. Nakamura R, Egashira K, Machida Y, Hayashidani S, Takeya M, Utsumi H, Tsutsui H, Takeshita A. Probucol attenuates left ventricular dysfunction and remodeling in tachycardia-induced heart failure: roles of oxidative stress and inflammation. Circulation 2002; 106:362–367.

149. Sia YT, Parker TG, Liu P, Tsoporis JN, Adam A, Rouleau JL. Improved post-myocardial infarction survival with probucol in rats: effects on left ventricular function, morphology, cardiac oxidative stress and cytokine expression. J Am Coll Cardiol 2002; 39:148–156.

150. Kumar D, Kirshenbaum LA, Li T, Danelisen I, Singal PK. Apoptosis in adriamycin cardiomyopathy and its modulation by probucol. Antioxid Redox Signal 2001; 3:135–145.

151. Lass A, Sohal RS. Electron transport-linked ubiquinone-dependent recycling of α-tocopherol inhibits autooxidation of mitochondrial membranes. Arch Biochem Biophys 1998; 352:229–236.

152. Folkers K, Vadhanavikit S, Mortensen SA. Biochemical rationale and myocardial tissue data on the effective therapy of cardiomyopathy with coenzyme Q10. Proc Natl Acad Sci USA 1985; 82:901–904.

153. Harker-Murray AK, Tajik AJ, Ishikura F, Meyer D, Burnett JC, Redfield MM. The role of coenzyme Q10 in the pathophysiology and therapy of experimental congestive heart failure in the dog. J Card Fail 2000; 6:233–242.

154. Manzoli U, Rossi E, Littarru GP, Frustaci A, Lippa S, Oradei A, Aureli V. Coenzyme Q10 in dilated cardiomyopathy. Int J Tissue React 1990; 12:173–178.

155. Langsjoen PH, Vadhanavikit S, Folkers K. Response of patients in classes III and IV of cardiomyopathy to therapy in a blind and crossover trial with coenzyme Q10. Proc Natl Acad Sci USA 1985; 82:4240–4244.

156. Khatta M, Alexander BS, Krichten CM, Fisher ML, Freudenberger R, Robinson SW, Gottlieb SS. The effect of coenzyme Q(10) in patients with congestive heart failure. Ann Int Med 2000; 132:636–640.

157. Watson PS, Scalia GM, Galbraith A, Burstow DJ, Bett N, Aroney CN. Lack of effect of coenzyme Q on left ventricular function in patients with congestive heart failure. J Am Coll Cardiol 1999; 33:1549–1552.

158. Geromel V, Darin N, Chretien D, Benit P, DeLonlay P, Rotig A, Munnich A, Rustin P. Coenzyme Q(10) and idebenone in the therapy of respiratory chain diseases: rationale and comparative benefits. Mol Genet Metab 2002; 77:21–30.

159. Armstrong JS, Whiteman M, Rose P, Jones DP. The coenzyme Q10 analog decylubiquinone inhibits the redox-activated mitochondrial permeability transition: role of mitochondrial respiratory complex III. J Biol Chem 2003.

160. Genova ML, Pich MM, Biondi A, Bernacchia A, Falasca A, Bovina C, Formiggini G, Castelli GP, Lenaz G. Mitochondrial production of oxygen radical species and the role of coenzyme Q as an antioxidant. Exp Biol Med (Maywood) 2003; 228:506–513.

161. Liang C, Rounds NK, Dong E, Stevens SY, Shite J, Qin F. Alterations by norepinephrine of cardiac sympathetic nerve terminal function and myocardial beta-adrenergic receptor sensitivity in the ferret: normalization by antioxidant vitamins. Circulation 2000; 102:96–103.

162. Shite J, Qin F, Mao W, Kawai H, Stevens SY, Liang C. Antioxidant vitamins attenuate oxidative stress and cardiac dysfunction in tachycardia-induced cardiomyopathy. J Am Coll Cardiol 2001; 38:1734–1740.

163. Ellis GR, Anderson RA, Lang D, Blackman DJ, Morris RH, Morris-Thurgood J, McDowell IF, Jackson SK, Lewis MJ, Frenneaux MP. Neutrophil superoxide anion—generating capacity, endothelial function and oxidative stress in chronic heart failure: effects of short- and long-term vitamin C therapy. J Am Coll Cardiol 2000; 36:1474–1482.

164. Rossig L, Hoffmann J, Hugel B, Mallat Z, Haase A, Freyssinet JM, Tedgui A, Aicher A, Zeiher AM, Dimmeler S. Vitamin C inhibits endothelial cell apoptosis in congestive heart failure. Circulation 2001; 104:2182–2187.

165. Visioli F. Effects of vitamin E on the endothelium: equivocal? alpha-tocopherol and endothelial dysfunction. Cardiovasc Res 2001; 51:198–201.

166. Stephens NG, Parsons A, Schofield PM, Kelly F, Cheeseman K, Mitchinson MJ. Randomised controlled trial of vitamin E in patients with coronary disease: Cambridge heart antioxidant study (CHAOS). Lancet 1996; 347:781–786.

167. Blot WJ, Li JY, Taylor PR, Guo W, Dawsey S, Wang GQ, Yang CS, Zheng SF, Gail M, Li GY. Nutrition intervention trials in Linxian, China: supplementation with specific vitamin/mineral combinations, cancer incidence, and disease-specific mortality in the general population. J Natl Cancer Inst 1993; 85:1483–1492.

168. Jialal I, Devaraj S. Vitamin E supplementation and cardiovascular events in high-risk patients. N Engl J Med 2000; 342:1917–1918.

169. Li RK, Sole MJ, Mickle DA, Schimmer J, Goldstein D. Vitamin E and oxidative stress in the heart of the cardiomyopathic syrian hamster. Free Radic Biol Med 1998; 24:252–258.

170. Prasad K, Gupta JB, Kalra J, Lee P, Mantha SV, Bharadwaj B. Oxidative stress as a mechanism of cardiac failure in chronic volume overload in canine model. J Mol Cell Cardiol 1996; 28:375–385.

171. Keith ME, Jeejeebhoy KN, Langer A, Kurian R, Barr A, O'Kelly B, Sole MJ. A controlled clinical trial of vitamin E supplementation in patients with congestive heart failure. Am J Clin Nutr 2001; 73:219–224.

172. D'Aquino M, Dunster C, Willson RL. Vitamin A and glutathione-mediated free radical damage: competing reactions with polyunsaturated fatty acids and vitamin C. Biochem Biophys Res Commun 1989; 161:1199–1203.

173. Danelisen I, Palace V, Lou H, Singal PK. Maintenance of myocardial levels of vitamin A in heart failure due to adriamycin. J Mol Cell Cardiol 2002; 34:789–795.

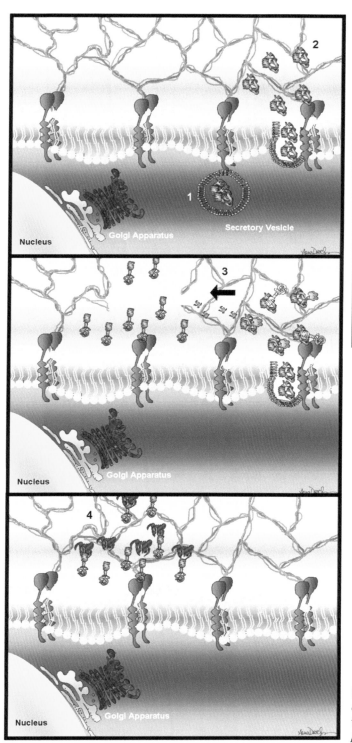

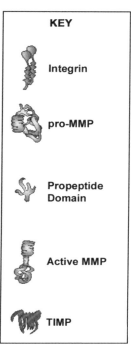

KEY

Integrin

pro-MMP

Propeptide
Domain

Active MMP

TIMP

Figure 3-2 Schematic of MMP secretion, activation, and inhibition. (*See p. 33 for full caption.*)

1

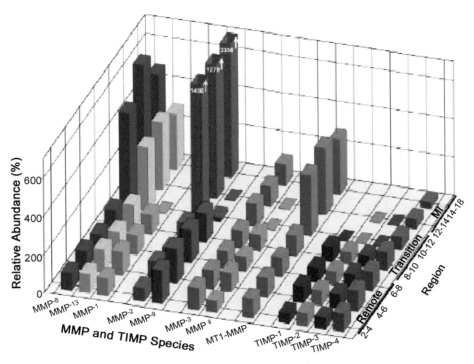

Figure 3-5 Three-dimensional histogram demonstrating regional distribution of MMPs and TIMPs after MI. *Source*: From Ref. 51. (*See p. 36 for full caption.*)

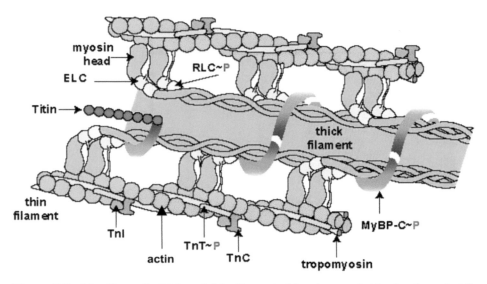

Figure 5-1 Myofilament's thick and thin filaments. Myosin cross-bridge heads project from the thick filament to interact with actin on the thin filament. (*See p. 58 for full caption.*)

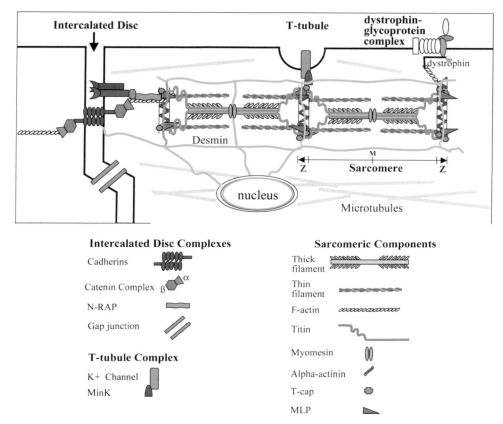

Figure 6-1 Cytoskeletal elements of the cardiac myocyte. Microtubules are found through-out the cytoplasm and are most dense near the nuclei. Desmin forms intermediate filaments that connect myofibrils laterally at the level of the Z-disk and M-line and that connect succes-sive Z-disks longitudinally. They also link myofibrils to nuclei and cell membranes. F-actin fila-ments are part of the sarcomeric thin filaments and constitute various linking systems that connect Z-disks to the plasma membrane and intercalated disk. They are linked to the cadherins of the intercalated disks by interacting with catenins and to the DGC via binding to dystrophin. Titin spans half-sarcomeres, with its terminal ends overlapping in the Z-disk and at the M-line. Titin is associated with the other parts of the cytoskeleton at various locations. T-cap interacts with the Z-disk end of titin and with MLP, forming part of an MLP-dependent stretch sensing complex. MLP interacts with N-RAP, a protein that may be involved in mediating interactions between myofibrils and the cell membrane at adherens junctions through its possible interactions with cadherin-based protein complexes and/or integrin-associated vinculin. T-cap also interacts with the potassium channel subunit, minK, at T-tubules forming a complex that may be involved in the stretch-dependent regulation of potassium flux. *Abbreviations*: DGC, dystrophin–glycoprotein complex; MLP, muscle LIM protein. *Source*: From Ref. 101.

3

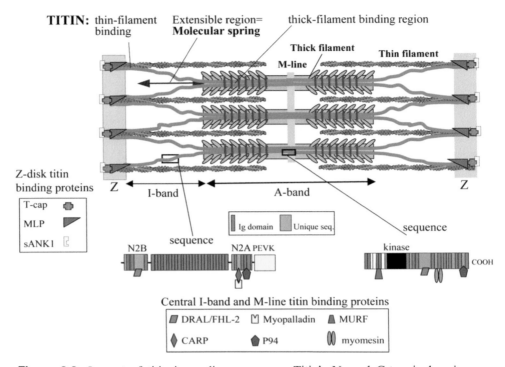

Figure 6-2 Layout of titin in cardiac sarcomere. Titin's N- and C-terminal regions are embedded in the Z-disk and M-line, respectively (73,102). Titin's I-band region is extensible and functions as a molecular spring that contains three distinct elements: tandem Ig segments (formed by serially linked Ig domains, shown as red rectangles), the PEVK segment (yellow), and the unique sequence (blue) that is part of the cardiac specific N2B element. Two main classes of cardiac titin isoforms exist: 1) The N2B isoform with a short spring that consists of tandem Ig segments with ~40 Ig domains, the N2B element and a short PEVK (~180 residues). 2) The N2BA isoform has a much longer spring, consisting of ~60 Ig domains, the N2B element, a ~800 residue proline (P), glutamate (E), valine (V) and lysine (K) rich domain (PEVK), and in addition the N2A element. (The shown sequence is part of the I-band sequence of N2BA cardiac titin.) Outside the molecular spring region the isoforms are largely identical in sequence. Titin-binding ligands located in Z-disk, I-band and M-line are also shown. The N-terminal region of titin interacts with T-cap and MLP, forming part of a MLP-dependent stretch sensing complex, as well as s-ANK1 that may have a structural role. Titin's N2B element (found in both N2B and N2BA isoforms) interacts with DRAL/FLH-2, a protein involved in recruiting multiple metabolic enzymes to the I-band. Titin's N2A element (only present in N2BA titin) interacts with CARP, a potential regulator of gene expression during stretch-induced myocyte hypertrophy, and the calpain protease p94. CARP also interacts with the muscle-specific nuclear and sarcomeric protein, myopalladin. The M-line region of titin contains a kinase domain of unknown function, and a MURF-interacting region both of which may be involved in signaling. Additionally, titin's M-line region interacts with myomesin, a structural protein. (Vertical gray lines in A-band region denote MyBP-C. Note that the figure is not to scale.) *Abbreviation*: s-ANK, small ankyrin-1. *Source*: From Ref. 103.

4

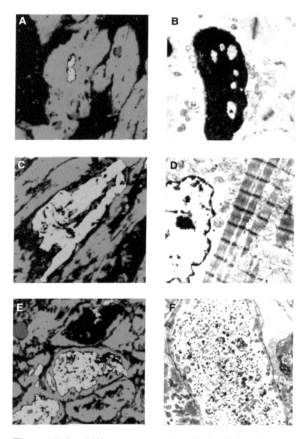

Figure 7-1 Different types of cell death in the myocardium. (A), (C), and (E): Confocal microscopic pictures (*dark grey is actin*); **B**, **D**, and **F**: Electron microscopic pictures (*all bars = 2 μm*). **A** and **B**: Apoptotic cell death demonstrated by TUNEL method. **A**, nuclei with DNA fragmentation are light grey. **B**, nuclei show condensed chromatin. **C** and **D**, necrotic cell death. **C**, single cell necrosis demonstrated by C9-labeling (*seen as light grey*). **D**, nuclei are electron-lucent with clumped chromatin, mitochondria are damaged with flocculent densities. **E** and **F**, autophagic cell death. **E**, Ubiquitin deposition (*seen as light grey*). **F**, ultra-structural appearance with numerous autophagic vacuoles. *Abbreviations*: TUNEL, terminal deoxynucleotidyl transferase biotin-dUTP nick end labeling. *Source*: From Ref. 24.

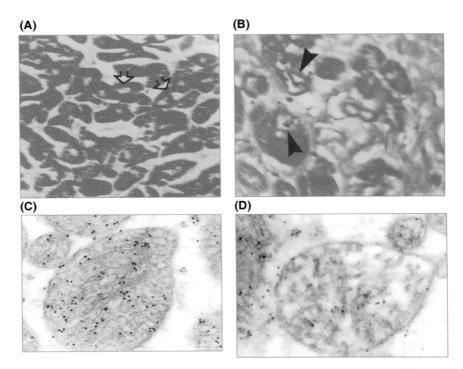

Figure 7-2 Histologic and ultrastructural evidence of apoptosis in heart failure. A histologic specimen from a victim of a motor vehicle accident shows absence of TUNEL staining (clear nuclear regions, open arrows) (**A**). Two myocytes in the cardiomyopathic heart demonstrate TUNEL-positive nuclei (blue-black color, arrowheads) (**B**). The ultrastructural examination with immunoelectron microscopy reveals cytochrome c (immunogold particles) localized in the mitochondrial compartment in the normal heart (**C**). On the other hand, cytochrome c is released from the mitochondrial to the cytoplasmic compartment in a cardiomyopathic heart, which is seen sprinkled on to the background of contractile proteins (**D**).

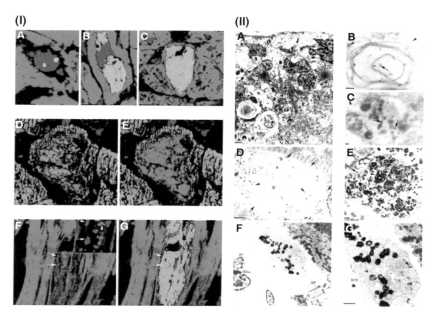

Figure 7-3 (**I**) Immunoconfocal patterns of ubiquitin-accumulations, sarcomeric proteins, and signs of autophagy in myocytes with ubiquitin-inclusions (*ubiquitin seen as light grey*). (**A**), Only two small dots of ubiquitin signal in a myocyte nucleus. (**B**), Punctate nuclear and massive cytoplasmic ubiquitin labeling in one myocyte. (**C**), Massive deposition of ubiquitin in a myocyte without a nucleus (*dark grey is actin*). (**D**) and (**E**), Double labeling for ubiquitin and myosin showing colocalization of these proteins and that ubiquitinated myosin lacks a typical cross-striated pattern (*dark grey is myosin*). F and G, Double labeling for ubiquitin (*light grey*) with monodansylcadaverine (*dark grey shown with arrows in F*). Shown with arrows in F are the autophagic vacuoles. G, Double labeling for ubiquitin and monodansylcadaverine showing colocalization of these two signals. (**II**) Ultrastructural features of autophagic myocytes in patients with DCM. A, Extensive cytoplasmic accumulations of autophagic vacuoles and myelin figures. Severe disintegration of the nuclei into 7 particles and only a narrow rim of the contractile material confined to the cell periphery is seen (arrows). B and C, Immunogold electron micrographs of autophagic myocytes show that ubiquitin is localized in the cytosol (*arrowhead*) and within a vacuolar-lamellar structure (*arrow*) corresponding to those shown in A with arrowheads. Similar distribution of ubiquitin was found in vacuolar-lysosomal structures (C) corresponding to those shown with arrowheads in D and E. D, Myocyte (asterisk) with numerous autophagic vacuoles in the absence of contractile filaments and a nucleus. E, Myocyte (*asterisk*) displaying electron dense autophagic vacuoles. Myocytes in D and E show still existing cell-to-cell contacts (*arrow*) at the intercalated disk (ID) region of the neighboring myocyte. F, Very small myocyte (*asterisk*) detached at the ID from an apparently normal myocyte. Macrophage (Mac) is seen in close vicinity with the degenerating myocyte. G, Enlargement of the central part of the degenerating cell (*asterisk*) shown in F displaying sarcoplasmic reticulum (SR) structures, remnants of Z-disks (*arrow*), and a basement membrane (BM). Scale bars = 5 μm in A and D, 2 μm in E through G, 250 nm in B and C. *Source*: From Ref. 24.

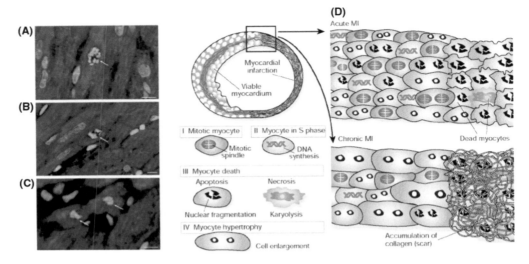

Figure 7-4 Evidence of mitotic activity in the cardiomyocytes following myocardial infarction in humans. Myocyte division a–c, Stages of mitosis in human myocytes: (**A**) metaphase chromosomes, (**B**) karyokinesis, and (**C**) cytokinesis are shown by fluorescence of propidium iodide (*seen here as light grey*). Dark grey reflects cardiac myosin antibody staining of myocyte cytoplasm. Arrows indicate mitosis. (**D**), Changes in myocyte growth in acute and chronic infarcts. MI, myocardial infarct. *Source*: From Ref. 37.

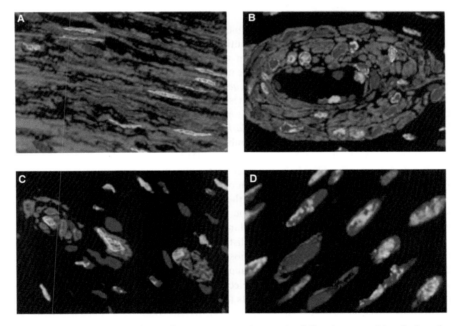

Figure 7-5 Regeneration of myocytes and vessels following cytokine-induced stem cell mobilization to the infarcted heart in mice. Ki-67 (**A** and **B**) and BrdUrd (**C** and **D**) labeling of myocytes (**A**), smooth muscle cells (**B** and **C**), and endothelial cells (EC); (**D**) in the forming myocardium. Myocytes are stained by cardiac myosin (**A**), SMC by a -smooth muscle actin (**B** and **C**), and EC by factor VIII (**D**). Bright fluorescence of nuclei (seen here as light grey) reflects the combination of propidium iodide (PI) and Ki-67 (**A** and **B**) or PI and BrdUrd (**C** and **D**). *Source*: From Ref. 47.

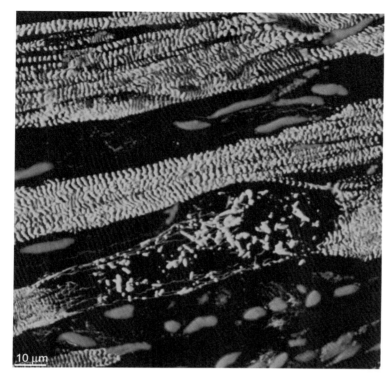

Figure 8-2 α-actinin (*green, nuclei are red*) for myocyte degeneration (note the severe disorganization of the sarcomeric structure in the myocyte in the lower part of the picture.

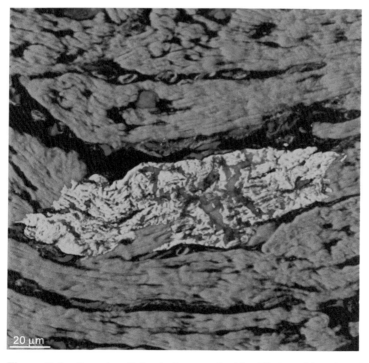

Figure 8-3 Oncotic cell death characterized by C9 deposition (*green*) in hypertrophied myocardium (*myocytes are red, nuclei blue*).

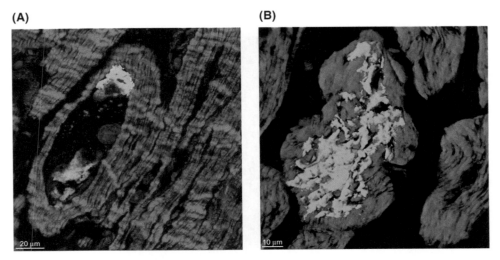

Figure 8-4 Autophagic cell death identified by ubiquitin labeling (*green*) in hypertrophied myocardium (*cardiomyocytes are red, nuclei blue*): (**A**) Moderate accumulation of ubiquitin/ protein complexes in a severely degenerated cell with loss of sarcomeres. This cell is still viable because it contains a nucleus. (**B**) Significant accumulation of ubiquitin/protein conjugates in a myocyte doomed to die because the nucleus is lacking.

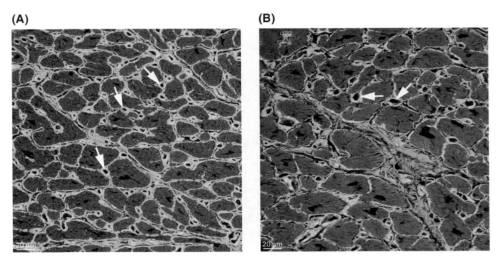

Figure 8-5 Fibronectin (*green*) for determination of fibrosis. Myocytes are red and black spots are nuclei: (**A**) Normal myocardium shows fine septa of ECM between cardiomyocytes. (**B**) In hypertrophied myocardium, the myocytes are enlarged and the amount of replacement fibrosis is increased. Note the distinct labeling of capillaries (*arrows*) and their reduced number in B. *Abbreviation*: ECM, extra cellular matrix.

10

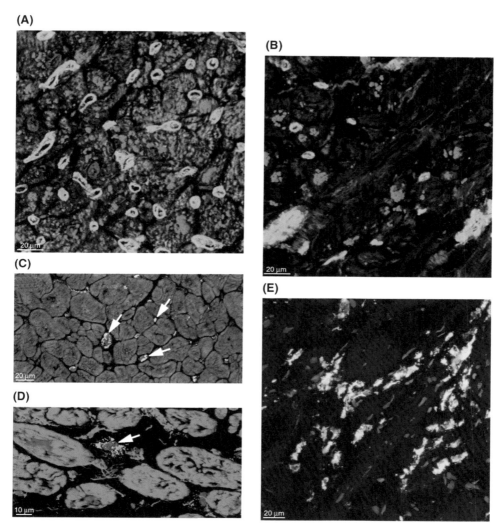

Figure 8-6 Changes in the microvasculature and the extracellular space. (A) and (B) Capillaries in human myocardium labeled by CD 31 (*green*). Lipofuscin (*red*) is abundant, myocytes are greenish: (**A**) Myocardium from a patient from group I shows a slightly reduced number of capillaries. (**B**) Tissue from group III exhibits only a few myocytes and a reduced number of capillaries. (**C**) Several capillaries are positive for ACE (*arrows, green, nuclei are blue*). (**D**) Occurrence of TGF-β_1 (*green, arrow*) in a macrophage within fibrotic tissue (*black area*). (**E**) An increased number of macrophages (*CD 68, green, nuclei are red*) is typical of fibrotic tissue in hypertrophied myocardium. *Abbreviation*: ACE, angiotensin-converting enzyme.

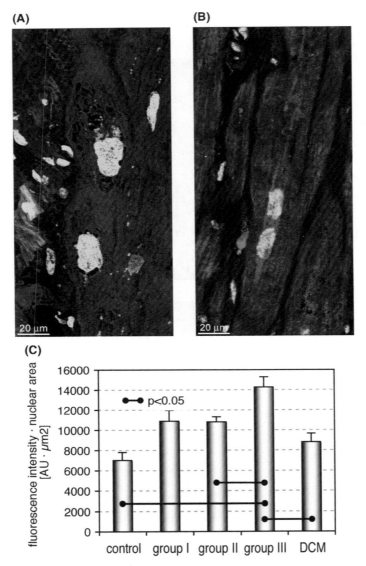

Figure 8-7 Nuclear staining: (**A**) Feulgen staining for DNA content (*yellow*), myocytes are reddish. (**B**) Sc-35 (*green*) in nearly black myocytes. Lipofuscin is red in (**A**) and (**B**). (**C**): Quantitative evaluation of the Sc-35 content in myocyte nuclei of controls and patients with aortic stenosis or DCM. Whereas a significant increase in Sc-35 is present in hypertrophied hearts, especially in group III, in patients with overt heart failure the Sc-35 content is signficantly reduced. *Abbreviation*: DCM, dilated cardiomyopathy.

12

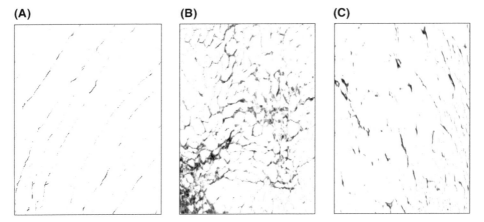

Figure 10-1 Representative staining for interstitial fibrosis with collagen-specific picosirius red. Interstitial fibrosis is indicated by dark purple staining. Magnification 200X. (**A**) Left-ventricular (LV) section from normal canine myocardium. (**B**) LV tissue section from micorembolization-induced failing canine myocardium. (**C**) LV tissue section from a canine treated with eprosartan, an angiotensin-1 receptor antagonist. Eprosartan ameliorates LV dysfunction associated with heart failure in the canine coronary microembolization model. *Source*: From Ref. 95.

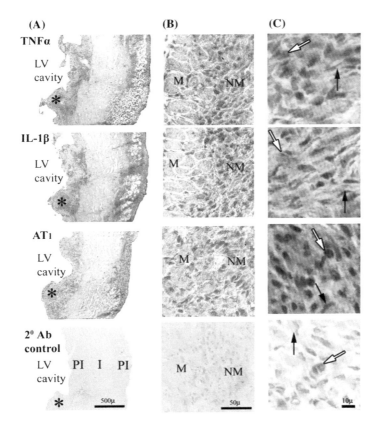

Figure 12-3 TNF-α, IL-1β and AT₁ receptor immunostaining is associated with non-myocytes in the post-MI heart. Immunostaining of the infarct region is shown in three magnifications as indicated by the scale bars. (*See p. 194 for full caption.*)

13

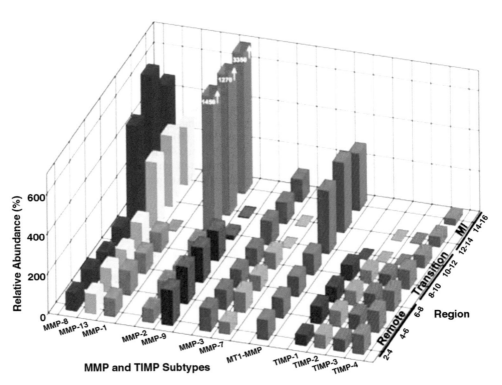

Figure 20-1 Regional variation in MMP and TIMP activity in remodeling myocardium. A significant increase in MT1-MMP and a reduction in TIMP-4 activity are noted in the transition and infarct territories. A marked increase in MMP-8, MMP-2, and MMP-13 activity were present in the transition and infarcted regions, while MMP-1 and MMP-9 levels were virtually undetectable in the infarcted tissue. TIMP levels were likewise markedly decreased in infarcted areas. *Abbreviations*: MMP, Matrix metalloproteinases; TIMP, tissue inhibitors of matrix metalloproteinases *Source*: Modified from Ref. 48.

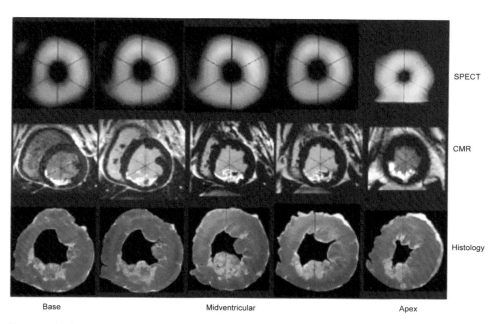

Figure 20-6 Comparison of SPECT and CMR images compared with histology (TTC-staining) in a canine infarct model. It was found that the reduced spatial resolution of SPECT led to underdetection of nontransmural infarctions, which was confirmed in patient examinations. *Abbreviations*: SPECT, single photon emission computed tomography; CMR, cardiac magnetic reasonance. *Source*: Modified from Ref. 121.

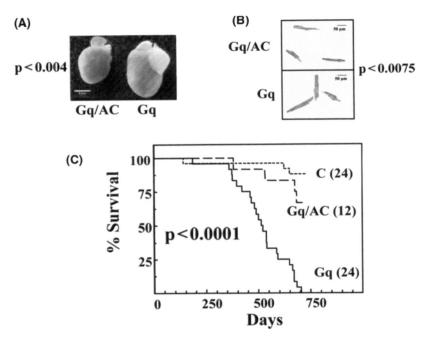

Figure 25-2 (**A**) Representative hearts from Gq/AC and Gq mice. (**B**) Representative cardiac myocytes isolated from Gq/AC and Gq mice. (**C**) Kaplan-Meier curve showing mortality rate in Gq ($n = 24$), Gq/AC ($n = 12$) and Control mice ($n = 25$). *Source*: From Ref. 52. (*See p. 517 for full caption.*)

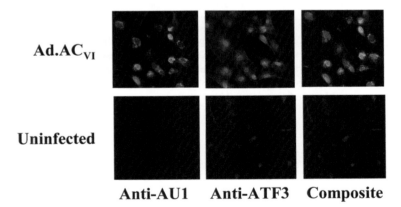

Figure 25-5 ATF3 protein localization after AC_{VI} gene transfer. Immunofluorescence staining with antiAU1 antibody for the expression of AC_{VI} (*green*) and anti ATF3 antibody (*red*) showed localization of AC_{VI} transgene in cell membrane and nuclear localization of ATF3 in cardiac myocytes overexpressing AC_{VI} (*top row*). Sparse amounts of endogenous AC_{VI} and ATF3 are seen (*bottom row*). *Source*: From Ref. 79.

16

16

Natural History of Cardiac Remodeling

Gary S. Francis and W. H. Wilson Tang
Department of Cardiovascular Medicine, Cleveland Clinic Foundation, Cleveland, Ohio, U.S.A.

INTRODUCTION

Left ventricular (LV) remodeling is the structural hallmark of heart failure. In keeping with a key principle of biology, abnormal organ structure leads to altered organ function. The structural changes observed in cardiac remodeling are directly related to impairment of cardiac performance. There is, therefore, a direct relation between remodeling and the progression of heart failure (1). It is now widely accepted that as heart disease progresses to chronic heart failure, the size of the heart increases, the heart changes shape and becomes more spheroidal, myocardial performance is altered, and symptoms of heart failure eventually ensue (Fig. 1). The syndrome of heart failure is not simply a problem of inadequate contractile function, but represents a complex interplay between response to injury, structural adaptation, impaired cardiac function, neurohormonal activation, major alterations in distribution of blood flow to peripheral organs, and blunted reflex mechanisms. At the crux of the problem of chronic heart failure is the change in the size and shape of the heart, the so-called "remodeling" process. In support of the remodeling hypothesis, the most effective therapies for heart failure are angiotensin converting enzyme (ACE) inhibitors and β-adrenergic blocking drugs, agents that knowingly slow the progression of heart failure by reducing or even reversing the remodeling process (2,3). LV remodeling has a major role in the pathophysiology of heart failure, and appears to have a relatively well-defined natural history, particularly following large anterior myocardial infarction (MI) (4).

THE INDEX EVENT

Any form of heart disease can lead to heart failure. At the root cause of remodeling of the heart is the so-called "index event," a process that initiates and sets into motion LV remodeling. However, the index event is not always an obvious clinical phenomenon or single point in time, such as a large anterior wall MI with sudden loss of contractile tissue. The index event can be the slow phenotypic expression of a clinically silent mutation, leading to structural change in the heart that unfolds

Parameter	3 weeks	1 year
End-diastolic volume	302 mL	377 mL
End-systolic volume	186 mL	271 mL
Circumference	59.5 cm	62.8 cm
Contractile segment	30.5 cm	33.8 cm
Non-contractile segment	23.7 cm	23.5 cm
Diastolic sphericity index	0.71	0.74
Systolic sphericity index	0.60	0.77

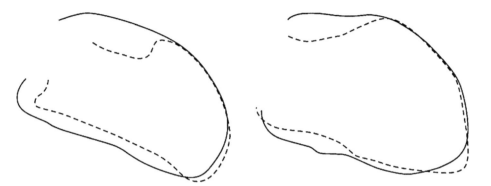

Figure 1 Late ventricular enlargement in a patient with anterior MI, indicating marked increase in volume resulting from increased circumference and sphericity. *Abbreviation*: MI, myocardial infarction. *Source*: From Ref. 91.

over many years, eventually leading to dilated cardiomyopathy and heart failure (5–10). In dilated cardiomyopathy, the mutated gene(s) may code for abnormal microtubules or calcium channels rather than abnormal contractile proteins. Any number of signaling proteins that interact between the cell surface, the contractile elements, and the nucleus can be absent or dysfunctional. These abnormal proteins, though not directly responsible for altered contraction, may eventually lead to secondary disruption of myocardial performance by altering cellular architecture.

In most cases during the index event, there is an important interaction with the environment. Loading conditions are often altered. The natural history of familial dilated cardiomyopathy may be different than the natural history of acute MI because the index event is quite different (i.e., mutation vs. acute myocardial injury). However, both may ultimately demonstrate similar large, rounded, dysfunctional hearts. Likewise, sarcomeric mutations in the tropomyosin, troponin T, titin, and β-myosin heavy chain gene may lead to either dilated cardiomyopathy or hypertrophic cardiomyopathy (11), each with a very different phenotype, cardiac shape, contractile function, and natural history. There is quite a pronounced plasticity in the heart failure phenotype. Therefore, the natural history of cardiac remodeling is to some extent determined by a highly variable index event and the changing environment. Chronic LV remodeling is characterized by an enormous variety of phenotypes, and can be in the form of an enlarged, dilated cavity or a small LV cavity with thickened LV walls. In some patients a hybrid of these two phenotypes exists (e.g., end-stage hypertensive heart disease). In the late, advanced stages, however, there is often a coalescence to a large, poorly functioning organ.

Examples of Varying Index Events: Hypertension and Diabetes Mellitus

Some patients with heart failure have an index event in the form of hypertension or onset of diabetes mellitus (12–15). In these disorders, LV remodeling manifests the phenotype of LV hypertrophy followed later by cavity dilatation and finally heart failure. The inciting event may be the insidious development of hypertension and/or diabetes mellitus. These are clinically silent index events that slowly manifest themselves over many years. However, given the same degree of hypertension or glucose impairment, there is substantial variance in the amount of LV remodeling that occurs in individual patients. This would suggest that other factors, such as genetic polymorphisms and the environment, are operating to drive the variance.

The occurrence of LV remodeling in the setting of poorly controlled hypertension is undoubted, and may initially take the form of LV hypertrophy followed later by cavity dilatation and reduced ejection fraction. There may be important gender differences in how the remodeling is expressed. For a similar degree of afterload stress, men are more likely to have more cavity enlargement, a lower ejection fraction, and increased diastolic myocardial stiffness (16), while women may have a smaller LV cavity and more hypertrophy than men for any given load. There has been an avid search for a signaling molecule that serves as a master switch for clinical hypertrophy, but redundant signaling pathways are more likely to modulate load-induced hypertrophy. Many signals and signaling cascades likely operate to modulate load-induced hypertrophy and include angiotensin II, calcineurin, and norepinephrine. We know that hypertensive LV hypertrophy is associated with increased sympathetic activity that is largely confined to the heart (17).

It is important to remember that an increase in LV mass by itself does not result in muscle dysfunction. Myocyte hypertrophy is an adaptive process that is considered physiologic in infancy and childhood, pregnancy and exercise. However, these physiologic states are intermittent or transient, unlike pathophysiologic loads, which are constant. In hypertension, there is a strong and independent relationship of LV mass to subsequent cardiovascular morbidity (18). LV hypertrophy in hypertension is also associated with both coronary vascular remodeling and attenuated endothelial and nonendothelial coronary blood flow reserve (19). The factors other than the high blood pressure per se that are important are highlighted by the observation that only half of the LV mass variability is explained by the blood pressure (20). Hereditary factors play a small but discernible proportion of the variance in LV mass (21). Eventually, there is a LV cavity dilatation and reduced systolic function, resulting in so-called "hypertensive cardiovascular disease." Diastolic impairment is common in patients with hypertensive LV hypertrophy and usually precedes systolic dysfunction by many years. The high prevalence and powerful risk of LV hypertrophy among black patients with heart disease carries a greater physiologic morbidity (22,23).

It has been 30 years since the work of Rubler and colleagues who pointed out that patients with diabetes mellitus and no evidence of atherosclerosis can develop a distinct cardiomyopathy (24). The Framingham studies showed that men with insulin-treated diabetes mellitus have a 2.4-fold increased incidence of heart failure, while women had a fivefold increase in incidence of heart failure (14). Diabetes mellitus is an independent contributor to LV mass and wall thickness, and is also associated with an increased LV end-diastolic dimension in women and a reduced fractional shortening in men (25). These abnormalities are undoubtedly amplified in many patients by concomitant hypertension. Diastolic dysfunction is also commonly found in diabetics, even in type II diabetic patients, despite adequate metabolic control and no clinical heart disease (26,27). Asymptomatic diastolic dysfunction is common in patients with type I diabetes mellitus and its severity is correlated with poor glycemic control (28). Even insulin resistance in the absence of overt diabetes mellitus can be associated with heart failure, and may be an important mechanism in the pathogenesis of myocardial contractile dysfunction in diabetic cardiomyopathy (15). Impaired glucose tolerance, perhaps in part owing to insulin resistance, is associated with increased LV wall thickness and mass (29).

Although the precise mechanism of diabetic cardiomyopathy and subsequent remodeling is not clear, the initial adaptation and subsequent maladaptation of the heart to a glycemic environment can be traced to a complex system of metabolic signals, increased fatty acid oxidation, and limited pyruvate oxidation (30,31). There is an accumulation of glycolytic intermediates in contractile muscle that may activate transcription factors, altering structure and function. Intramyocardial lipids may accumulate, and lipotoxicity may lead to the development of contractile dysfunction. Both myocardial and skeletal muscle energy metabolism can be impaired (32). In the end-stages of heart failure, the mechanism by which the heart arrived at this point is of less importance. In the early and intermediate stages, the index event is of great importance, as it determines the natural history to some extent.

MYOCARDIAL FIBROSIS AND DIASTOLIC DYSFUNCTION

The syndrome of "heart failure" usually incorporates some element of diastolic dysfunction, and in the end-stages often manifests clear evidence of both systolic and diastolic dysfunctions (33). Nearly all patients with heart failure manifest some excessive collagen deposition. The extracellular matrix is, in fact, an extraordinarily dynamic "organ," contrary to what was believed years ago (34). The major fibrillar collagens in the human heart are type I and type III.

During LV remodeling, several patterns of fibrosis have been recognized. First, there is perivascular fibrosis with type I collagen within the adventia of the myocardial coronary arteries and arterioles; this may extend into neighboring interstitial spaces to create reactive "interstitial fibrosis"; this pattern is often greatly increased in dilated (nonischemic) cardiomyopathy. Second, there is replacement or reparative fibrosis, that restores the structural integrity of tissues lost because of necrosis (e.g., MI). This pattern is greatly increased in so-called "ischemic cardiomyopathy." Both types of collagen deposition can frequently be seen together. Diffuse fibrosis within the myocardium is accompanied by abnormal chamber stiffness. Myocardial cell hypertrophy per se is not always associated with fibrosis (35). In the absence of fibrosis the hypertrophied ventricle retains normal stiffness.

The genes and environment stressors necessary for production of reactive and reparative fibrosis are just now beginning to be understood. Several neurohormones are involved, including norepinephrine, angiotensin II, aldosterone, and B-type natriuretic peptide (BNP). Chronic mineralocorticoid excess can enhance reactive fibrosis (36), and this observation has led to the "rediscovery" of aldosterone receptor blockers for the management of heart failure (37,38). These agents appear to have important antiremodeling effects (39,40), which may well account for their striking clinical efficacy. In addition, BNP may have important antigrowth and antifibrotic properties.

OTHER EXAMPLES OF LV REMODELING—VALVULAR HEART DISEASE AND OTHER CAUSES

The gradual onset of valvular heart disease can be the index event and lead to hypertrophy and dilatation of multiple cardiac chambers. Chamber remodeling is in part owing to pressure and/or volume overload created by the valvular abnormality. Pressure and volume overload produce a distinctly different phenotype at the cardiomyocyte level (41). A pure pressure overload leads to a thickening (increased transverse diameter) of cardiac myocytes with additional sarcomeres lining up in parallel, whereas a pure volume overload leads to elongation of cardiac myocytes with more sarcomeres assembled in series. Both types of overload can occur together, leading to a hybrid of concentric and eccentric hypertrophy. Sarcomere length itself does not change. Eventually the heart enlarges. In some patients the hypertrophy (i.e., new sarcomeres) fails to keep pace with the hemodynamic overload, leading to increased wall stress and diminished pump performance, setting the stage for the development of heart failure. The muscle LIM domain protein (MLP) gene appears to play a critical role as part of a two-disc telethonin–titin complex that serves as a component of the cardiac muscle stress sensor apparatus, thus transmitting increased wall stress signal to the sarcomere assembly regulatory proteins. However, dilated cardiomyopathy can occur from a missense mutation in the MLP gene in humans (42), even in the absence of a pressure or volume overload. The precise mechanism whereby external loading conditions are sensed and genetic expression in the cells of the heart is altered is a subject of intense research interest.

Systematic and locally active neurohormonal systems contribute importantly to LV remodeling (43–45). Patients with pheochromocytoma can develop either dilated or hypertrophic cardiomyopathy (46). Overproduction of angiotensin II and norepinephrine locally is well known to facilitate LV remodeling, no matter what the index event may be. These neurohormones in excessive quantities are also directly toxic to cardiac myocytes. Neurohormones act initially through membrane receptors, transmitting signals through various pathways to activate the necessary genes to drive the remodeling process. The remarkable role of the renin–angiotensin–aldosterone system and the adrenergic nervous system is highlighted by the fact that their pharmacologic blockade consistently attenuates the remodeling process. In a sense, any toxic or perverse milieu, whether chemical, genetic, or mechanical, has the potential to initiate cardiac remodeling. The index event may be obvious (i.e., acute MI) or clinically silent (a gene mutation). More unusual examples of remodeling inciting or index events include acute inflammatory myocarditis, various types of cancer chemotherapy, pheochromocytoma, and incessant tachycardia. The remodeling may take the form of primary hypertrophy (with a small LV cavity), primary dilatation (with a dilated chamber), or both. LV remodeling is a classic example of the environment interacting with genes to produce an early adaptive response that over time becomes noticeably more maladaptive, ultimately leading to heart failure.

THE ROLE OF LV GEOMETRY

The relatively normal balance between wall thickness and cavity dilatation is roughly 1:2 and is critical to the natural history of remodeling. For example, patients with severe, chronic aortic regurgitation may have a relatively preserved hypertrophy/dilation index (h/R ratio, normal $= 0.48 \pm 0.19$), thus allowing for a normal LV function and prolonged, asymptomatic natural history. On the contrary, patients with idiopathic dilated cardiomyopathy may have more dilatation relative to hypertrophy, leading in some cases to early symptoms and a foreshortened survival (47). Therefore the natural history of LV remodeling is highly dependent not only in the nature of the index event but also in the ability to adapt to the altered wall stress by the appropriate assembly of new sarcomeres.

The maintenance of adequate hypertrophy allows for adequate myocardial systolic performance. Impaired cardiac performance in some patients with pressure overload hypertrophy occurs because of inappropriately high wall stress rather than reduced myocardial contractility. Patients with h/R ratio > 0.36 have higher ejection fraction, velocity of circumferential fiber shortening, and stroke work index than those with lower values (48). Therefore, LV wall thickness, chamber size, and geometry are critical in determining overall cardiac systolic performance in patients with hypertrophy. In addition to the index event, the related mechanical LV stresses and local neurohormonal activation will also drive hypertrophy and the subsequent remodeling process to varying extents. Cells drop out (apoptosis) occurs and excess collagen is deposited. There is no better example for this process than post-MI remodeling, a very common clinical problem that has been extensively studied and is closely linked to the development of chronic heart failure.

POST-INFARCTION REMODELING

It has been widely recognized over the past 25 years that large transmural infarcts, especially anterior infarcts, lead to complex alterations that involve both the infarcted and uninfarcted zones of the heart (49). Sometimes profound distortion of the topographic architecture leads to extensive remodeling (Fig. 2). This has been the subject

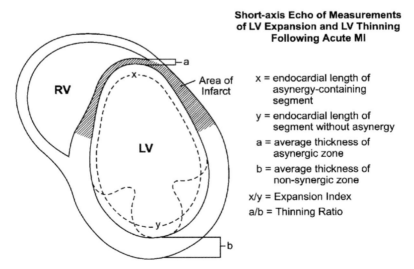

Figure 2 Short-axis two-dimensional echocardiographic measurement of LV expansion and thinning following acute MI. *Abbreviations*: LV, left ventricular; MI, myocardial infarction. *Source*: From Ref. 92.

of numerous experimental, pathological, and clinical studies. The natural history of post-MI remodeling, to some extent modified by modern reperfusion techniques, is now relatively well defined. Acute (expansion) and chronic components are worthy of discussion.

Myocardial Expansion

The concept of "infarct expansion" was first brought to our attention in 1978 by Hutchins and Bulkley who studied 76 consecutive patients by autopsy who died of acute MI within 30 days at the Johns Hopkins University Hospital (50). They found that the first few hours of necrosis, edema, and inflammation are localized to the infarct. This is followed by a long-term phase of fibroblast proliferation and collagen deposition. Before extensive scar formation but occurring at least five days after the acute infarction, there is acute dilatation and thinning of the area of infarction not accompanied by additional myocardial necrosis—referred to as infarct expansion. Later studies by the same research group indicated that a decrease in the number of cells across the wall occurs, accounting for most of the thinning (51). In the noninfarcted regions of the heart, wall thinning was believed to be due to slippage between muscle bundles (51). With increasing degrees of expansion, there was distortion of topography with greater separation of the relative positions of the papillary muscles, as well as the anterior and posterior intersections of left and right ventricular free wall and interventricular septum (Fig. 2). The very largest infarcts were associated with early patient death, as these hearts did not have time to develop infarct expansion. Expansion was limited to transmural infarcts, and was common; it was occurring in 72% of transmural infarcts between day 5 and day 7.

Early topographic changes as a consequence of acute MI were studied by sequential echo in patients by Eaton and his Hopkins' colleagues (52). The fundamental hypothesis was that the topographic changes observed by serial echocardiograms in 28 patients with acute transmural MI occur as a primary event and are not dependent on further myocardial necrosis. This hypothesis was also supported by the careful pathology studies performed earlier by the same investigative group. Regional infarct expansion and subsequent myocardial dilatation of the noninfarcted zone increase myocardial oxygen demand and likely accounts for ischemic consequences of late congestive heart failure (52). In "infarct expansion," the LV dilates, there is no new necrosis, and there is no associated new or discrete clinical event. LV aneurysm and myocardial rupture are likely extreme forms of myocardial expansion.

Infarct Extension

Infarct extension, unlike expansion, is caused by new myocardial necrosis, is a far less frequent event, and the amount of myocardium involved is usually insufficient to further compromise myocardial function (50). There are often clinical signs and symptoms accompanying infarct extension such as chest pain and electrocardiographic changes. Myocardial markers of necrosis such as troponin and creatine kinase are positive.

Changes in the Uninfarcted Zone

The uninfarcted zone is also of great interest. Changes here are more chronic, but also contribute to chamber enlargement. Here the cardiac myocytes elongate using eccentric hypertrophy, presumably in response to a relatively volume overloaded state, essentially transforming and maintaining stroke volume despite a reduced ejection fraction (53). The LV chamber progressively dilates and the heart becomes more spheroidal and eventually less efficient.

There are some links between early infarct expansion and later cavity dilatation because of changes in the uninfarcted zone of the heart. Erlebacher and colleagues (54) studied 13 patients by echo beginning 10 to 21 days after anterior transmural MI over 3 to 30 months and demonstrated that infarct expansion appears to be the main contributor to early LV dilatation, associated with impaired myocardial function, and predicts chronic progressive ventricular enlargement. McKay and colleagues (55) demonstrated by serial contrast ventriculograms that LV enlargement occurs gradually after transmural MI, not due to increased LV filling pressure (LV end-diastolic pressure is actually lower), and that the observed hemodynamic improvement is at the expense of a significant increase in LV volume. Increased LV chamber stiffness is detected 24 to 48 hours after acute MI, but returns to normal within several days (56). The chamber stiffness is higher in patients with large, anterior MIs, but does not necessarily impair LV systolic function. Of course, increased chamber stiffness could alter the pressure/volume relation to point of predisposing the patient to acute pulmonary edema. Using sequential gated radionuclide ventriculography in post-MI patients, Warren and colleagues (57) demonstrated that post-infarction LV dilatation is common (20 of 36 patients), is more severe in anterior MIs, and late thrombolysis (i.e., $5 + 1$ hours) does not prevent subsequent LV dilation.

Changes in the remaining noninfarcted segments in the LV are extensive. Gerdes and colleagues (58) noted an increase in myocyte length, which they considered as an intracellular event. In general, myocyte length in human hearts with ischemic cardiomyopathy explanted at the time of heart transplant is about 40% longer than normal, whereas cell width is unchanged. Sarcomere length also remains fixed. Gerdes and colleagues (58) concluded that chamber dilatation is, in large measure, a product of myocyte elongation. These observations were confirmed in related studies by Beltrami et al. (59). However, these investigators also described deposition of collagen, cell dropout, and mural slippage of cells. Together, these events contributed to what was essentially a 4.6-fold expansion of LV cavity volume and a 56% reduction in the expected mass-to-chamber volume ratio. The mechanism of cavity enlargement remains controversial, and the subject of "slippage" of muscle bundles is particularly contentious. There is an agreement that lengthening of myocytes owing to new sarcomeres is important in the process (60), but the quantitative contributions of mural slippage, apoptosis, and myocyte replication to the process are not uniformly agreed upon.

Progressive LV remodeling in the modern era of direct reperfusion, ACE inhibitors, β-adrenergic blockers, and statins still occurs, but is not prevalent, and may be attenuated or may occur later (61). It is important to recognize, however, that early infarct expansion is associated with continued (late) cavity enlargement. The late process of eccentric hypertrophy and cavity dilatation can continue long after the infarct has healed, and bears a strong relation to the development of progressive heart failure. The precise link between early infarct expansion and later, progressive chamber dilatation leading to heart failure is highly complex and involves multiple mechanisms, including altered loading conditions, neurohormones, apoptosis, possibly muscle bundle slippage, collagen deposition, and oxidative stress with activation of matrix metallic proteinases.

The remodeling process and its obligatory reduced LV ejection fraction are an excellent surrogate when gauging response to therapy. Drugs that improve or retard remodeling, especially after acute MI, are strongly linked to improve survival in most randomized, controlled trials where remodeling was measured (62–74). Failure to slow or improve remodeling with drug therapy is associated with little or no improvement in survival (3,75,76).

Experimental Models

A series of elegant experiments performed in the laboratory of Janice and Marc Pfeffer in the 1980s were very enlightening regarding the post-MI remodeling process. Using the rat model of MI, Pfeffer and colleagues (77,78) were able to make comparisons of chamber dilatation at a common distending LV pressure. The survival rate was noted to be related to infarct size, and chamber dilatation was a function of the extent of histological damage and the time after infarction. The pressure/volume relation gradually shifted to the right (Fig. 3) (79), indicating that the LV chamber enlargement was to some extent independent of distending pressure. As in humans, cavity enlargement continued long after the infarct was healed, usually by as much as 30% (80). Patients, following transmural MI, can also demonstrate marked remodeling unexplained by increased distending or filling pressures. This is also commonly seen in patients with idiopathic dilated cardiomyopathy, and may explain why some patients demonstrate no increase in plasma brain natriuretic peptide despite large, dilated hearts and low ejection fractions (81).

ADAPTATION

LV remodeling or enlargement can be viewed as a short-term adaptive response to reduced contractile shortening. The stroke volume can be maintained in the face of low ejection fraction if the heart dilates to contract from a larger end-diastole volume. This may be one reason why patients who develop rapid cardiogenic shock before the

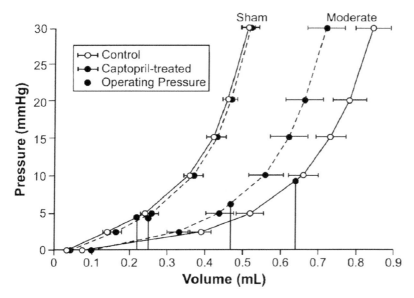

Figure 3 Pressure–volume relation following acute MI in rats. *Abbreviations*: MI, myocardial infarction. *Source*: From Ref. 37.

hearts have time to dilate exhibit a very high mortality—they are simply unable to increase their stroke volume. Over time, however, the extensive changes in the shape and the size of the LV cavity will make the heart less efficient. The wall tension will increase further raising myocardial oxygen demand and reducing pump performance. Reducing afterload stress by retarding progressive LV enlargement is one of the primary goals of therapy. Treatments that improve survival are generally associated with reduced progressive LV remodeling. In some cases it may be possible to even reverse the remodeling process with medical therapy or specialized devices (Fig. 4) (3,71–74).

Remodeling is the common disordered biology that underlies heart failure—reduced survival is the ultimate consequence of the process. Gaudron and colleagues (82) showed in the modern reperfusion era that about 25% of patients develop LV dilation within four weeks of a first transmural MI. This initial LV expansion is adaptive and helps to restore stroke volume and cardiac output, thus preserving exercise tolerance. A somewhat smaller group of patients (20%) develop a more progressive, chronic structural LV dilatation. These patients often go on to severe global ventricular dysfunction and heart failure. Infarct size, anterior infarct, and a permanently occluded left anterior descending coronary artery predict progressive dilatation and the development of heart failure (82). Late LV dilation beyond four weeks after acute MI is likely noncompensatory and portends the development of progressive dilation and heart failure (83). Spontaneous recovery of some LV function within four weeks of acute MI is common, and is usually due to improvement in transiently stunned or hibernating myocardium (84). For this reason measurement of LV size and function in a clinical setting is usually deferred for a highly variable period of time.

PROGNOSIS

We have known for 30 years that roentgenographic evidence of cardiac enlargement after an acute MI is associated with chronic heart failure and a poor prognosis (85–87). Quantitative left ventriculography has provided additional evidence that small increases in LV volume are potent predictors of a poor survival rate, especially an increased end-systolic volume (88). Heart volume is in fact a more powerful predictor of prognosis than the extent of coronary disease (88). In the MI model popularized by Pfeffers, the animals that exhibited the greatest reduction in volume with captopril derived the greatest survival benefit (80). The close link between LV remodeling, LV volume or chamber size, prognosis, and survival supports the hypothesis that progressive remodeling is the dominant phenotype of the heart failure syndrome. Along with plasma BNP, LV remodeling is the most powerful surrogate marker of heart failure.

In addition to progressive chamber enlargement leading to impairment of systolic performance and heart failure, altered LV architecture provides an important substrate for the initiation of high-grade ventricular arrhythmia (89). However, there are elusive additional factors that modulate this substrate that can lead to lethal arrhythmias and sudden cardiac death (90). To date, we have been unable to consistently identify those patients who benefit most from an implantable cardioverter-defibrillator before a lethal event occurs. The fact that LV remodeling is potentially reversible in some cases with medical or surgical therapy (3,75,76) and that specific antiremodeling therapies reduce sudden death in the post-infarction setting [such as

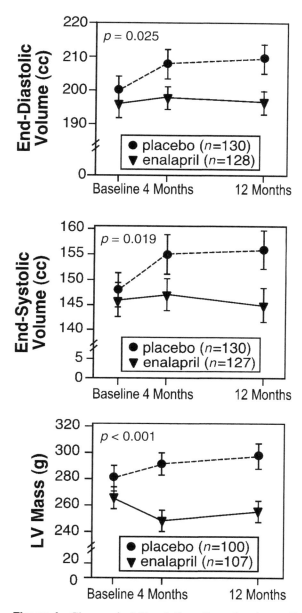

Figure 4 Changes in LV end-diastolic and end-systolic volumes and mass between enalapril and placebo: SOLVD echocardiography substudy data. *Abbreviation*: LV, left ventricular. *Source*: From Ref. 74.

eplerenone in EPHESUS (38)] provide hope that control of remodeling may to some extent inhibit the triggers of high-grade ventricular arrhythmias.

SUMMARY

The natural history of LV remodeling is reasonably well understood, thanks to seminal experimental studies by Pfeffers, Anversa, Braunwald, Gaudron, Jugdutt, Healy, and many others. The early clinical observations by the Hopkins group and careful sequential volumetric follow-up studies by Gaudron, Ertl, and many others in patients who had sustained an acute MI have provided much insight. Whether it is dilated cardiomyopathy or ischemic post-infarction cardiomyopathy, the fundamental sequence of events of LV remodeling seems to be the same: index event, adaptive response, progressive dilatation, and heart failure. It will now be important to recognize who is at greatest risk early on to develop progressive remodeling, and to take preventive steps to interdict the process at the earliest point possible. The goal is to intervene early with aggressive antiremodeling therapy before late cavity dilatation becomes evident.

REFERENCES

1. Cohn JN, Ferrari R, Sharpe N. Cardiac remodeling—concepts and clinical implications: a consensus paper from an international forum on cardiac remodeling. J Am Coll Cardiol 2000; 35:569–582.
2. Eichhorn EJ, Bristow MR. Medical therapy can improve the biological properties of the chronically failing heart. A new era in the treatment of heart failure. Circulation 1996; 94:2285–2296.
3. Kawai K, Takaoka H, Hata K, Yokota Y, Yokoyama M. Prevalence, predictors and prognosis of reversal of maladaptive remodeling with intensive medical therapy in idiopathic dilated cardiomyopathy. Am J Cardiol 1999; 84:671–676.
4. Tang WHW, Francis GS. Natural history of heart failure. In: Kukin ML, Fuster V, eds. Oxidative Stress and Cardiac Failure. Elmsford, New York: Blackwell Futura, 2002:3–47.
5. Vatta M, Stetson SJ, Perez-Verdia A, Entman ML, Noon GP, Torre-Amione G, Bowles NE, Towbin JA. Molecular remodelling of dystrophin in patients with end-stage cardiomyopathies and reversal in patients on assistance-device therapy. Lancet 2002; 359:936–941.
6. DiMauro S, Schon EA. Mitochondrial respiratory-chain diseases. N Engl J Med 2003; 348:2656–2668.
7. Kamisago M, Sharma SD, DePalma SR, Solomon S, Sharma P, McDonough B, Smoot L, Mullen MP, Woolf PK, Wigle D, et al. Mutations in sarcomere protein genes as a cause of dilated cardiomyopathy. N Engl J Med 2000; 343:1688–1696.
8. Baig MK, Goldman JH, Caforio ALP, Connar AS, Keeling PJ, McKenna WJ. Familial dilated cardiomyopathy: cardiac abnormalities are common in asymptomatic relatives and may represent early disease. J Am Coll Cardiol 1998; 31:195–201.
9. Taylor MRG, Fain PR, Sinagra G, Robinson ML, Robertson AD, Carniel E, Di Lenarda A, Bohlmeyer TJ, Ferguson DA, Brodsky GL, et al. Natural history of dilated cardiomyopathy due to lamin A/C gene mutations. J Am Coll Cardiol 2003; 41:771–780.
10. Schonberger J, Seidman CE. Many roads lead to a broken heart: the genetics of dilated cardiomyopathy. Am J Hum Genet 2001; 69:249–260.
11. Chien KR. Genotype, phenotype: upstairs, downstairs, in the family of cardiomyopathy. J Clin Invest 2003; 111:175–178.
12. Cohn JN. From hypertension to heart failure. Eur Heart J 2000; 2(suppl):A2–A5.
13. Levy D, Garrison RJ, Savage DD, Kannel WB, Castelli WP. Prognostic implications of echocardiographically determined left ventricular mass in the Framingham Heart Study. N Engl J Med 1990; 322:1561–1566.
14. Kannel WB, Hjortland M, Castelli WP. Role of diabetes in congestive heart failure: the Framingham study. Am J Cardiol 1974; 34:29–34.
15. Tang WHW, Young JB. Cardiomyopathy and heart failure in diabetes. Endocrin Metab Clin N Am 2001; 30:1031–1046.
16. Villari B, Campbell SE, Schneider J, Vassalli G, Chiariello M, Hess OM. Sex dependent differences in left ventricular function and structure in chronic pressure overload. Eur Heart J 1995; 16:1410–1419.

17. Schlaich MP, Kaye DM, Lambert E, Sommerville M, Socratous F, Esler MD. Relation between cardiac sympathetic activity and hypertensive left ventricular hypertrophy. Circulation 2003; 108:560–565.
18. Verdecchia P, Carini G, Circo A, Dovellini E, Giovannini E, Lombardo M, Solinas P, Gorini M, Maggioni AP. Left ventricular mass and cardiovascular morbidity in essential hypertension. The MAVI study. J Am Coll Cardiol 2001; 38:1829–1835.
19. Hamasaki S, Al Suwaidi J, Higano ST, Miyauchi K, Holmes DR, Lerman A. Attenuated coronary flow reserve and vascular remodeling in patients with hypertension and left ventricular hypertrophy. J Am Coll Cardiol 2000; 35:1654–1660.
20. Devereux RB, Roman MJ, de Simone G, O'Grady MJ, Paranicas M, Yeh JL, Fabsitz RR, Howard BV. Relations of left ventricular mass to demographic and hemodynamic variables in American Indians. The Strong Heart Study. Circulation 1997; 96:1416–1423.
21. Post WS, Larson MG, Myers RH, Galderisi M, Levy D. Heritability of left ventricular mass. The Framingham Heart Study. Hypertension 1997; 30:1025–1028.
22. Liao Y, Cooper RS, McGee DL, Mensah GA, Ghali JK. The relative effects of left ventricular hypertrophy, coronary artery disease, and ventricular dysfunction on survival among black adults. JAMA 1995; 273:1592–1597.
23. Houghton JL, Prisant LM, Carr AA, Flowers NC, Frank MJ. Racial differences in myocardial ischemia and coronary flow reserve in hypertension. J Am Coll Cardiol 1994; 23:1123–1129.
24. Rubler S, Dlugash J, Yuceoglu YZ, Kumral T, Branwood AW, Grishman A. New type of cardiomyopathy associated with diabetic glomerulosclerosis. Am J Cardiol 1972; 30:595–602.
25. Galderisi M, Anderson KM, Wilson PW, Levy D. Echocardiographic evidence for the existence of a distinct diabetic cardiomyopathy (the Framingham Heart Study). Am J Cardiol 1991; 68:85–89.
26. Schannwell CM, Schneppenheim M, Perings S, Plehn G, Strauer BE. Left ventricular diastolic dysfunction as an early manifestation of diabetic cardiomyopathy. Cardiology 2002; 98:33–39.
27. Zabalgoitia M, Ismaeil MF, Anderson L, Maklady FA. Prevalence of diastolic dysfunction in normotensive, asymptomatic patients with well-controlled type 2 diabetes mellitus. Am J Cardiol 2001; 87:320–323.
28. Shishehbor MH, Hoogwerf BJ, Schoenhagen P, Marso SP, Sun JP, Li J, Klein AL, Thomas JD, Garcia MJ. Relation of hemoglobin A1c to left ventricular relaxation in patients with type 1 diabetes mellitus and without overt heart disease. Am J Cardiol 2003; 91:1514–1517.
29. Ilercil A, Devereux RB, Roman MJ, Paranicas M, O'Grady MJ, Welty TK, Robbins DC, Fabsitz RR, Howard BV, Lee ET. Relationship of impaired glucose tolerance to left ventricular structure and function: The Strong Heart Study. Am Heart J 2001; 141:992–998.
30. Taegtmeyer H, McNulty P, Young ME. Adaptation and maladaptation of the heart in diabetes: Part I: General concepts. Circulation 2002; 105:1727–1733.
31. Young ME, McNulty P, Taegtmeyer H. Adaptation and maladaptation of the heart in diabetes: Part II: Potential mechanisms. Circulation 2002; 105:1861–1870.
32. Scheuermann-Freestone M, Madsen PL, Manners D, Blamire AM, Buckingham RE, Styles P, Radda GK, Neubauer S, Clarke K. Abnormal cardiac and skeletal muscle energy metabolism in patients with type 2 diabetes. Circulation 2003; 107:3040–3046.
33. Chen HH, Lainchbury JG, Senni M, Bailey KR, Redfield MM. Diastolic heart failure in the community: clinical profile, natural history, therapy, and impact of proposed diagnostic criteria. J Card Fail 2002; 8:279–287.
34. Mann DL, Taegtmeyer H. Dynamic regulation of the extracellular matrix after mechanical unloading of the failing human heart. Recovering the missing link in left ventricular remodeling. Circulation 2001; 104:1089–1091.
35. Weber KT, Brilla CG, Janicki JS. Myocardial fibrosis: functional significance and regulatory factors. Cardiovasc Res 1993; 27:341–348.

36. Weber KT, Brilla CG. Pathological hypertrophy and cardiac interstitium. Fibrosis and rennin–angiotensin–aldosterone system. Circ Res 1991; 83:1849–1865.
37. Pitt B, Zannad F, Remme WJ, Cody R, Castaigne A, Perez A, Palensky J, Wittes J. The effect of spironolactone on morbidity and mortality in patients with severe heart failure. Randomized Aldactone Evaluation Study Investigators. N Engl J Med 1999; 341: 709–717.
38. Pitt B, Remme W, Zannad F, Neaton J, Martinez F, Roniker B, Bittman R, Hurley S, Kleiman J, Gatlin M. Eplerenone, a selective aldosterone blocker, in patients with left ventricular dysfunction after myocardial infarction. N Engl J Med 2003; 348:1309–1321.
39. Zannad F, Dousset B. Treatment of congestive heart failure. Interfering the aldosterone–cardiac extracellular matrix relationship. Hypertension 2001; 38:1227–1232.
40. Hayashi M, Tsutamoto T, Wada A, Tsutsui T, Ishii C, Ohno K, Fujii M, Taniguchi A, Hamatani T, Nozato Y, et al. Immediate administration of mineralocorticoid receptor antagonist spironolactone prevents post-infarct left ventricular remodeling associated with suppression of a marker of myocardial collagen synthesis in patients with first anterior acute myocardial infarction. Circulation 2003; 107:2559–2565.
41. Grossman W. Cardiac hypertrophy: useful adapation or pathologic process? Am J Med 1980; 69:576–584.
42. Knoll R, Hoshijima M, Hoffman HM, Person V, Lorenzen-Schmidt I, Bang ML, Hayashi T, Shiga N, Yasukawa H, Schaper W, et al. The cardiac mechanical stretch sensor machinery involves a Z disc complex that is defective in a subset of human dilated cardiomyopathy. Cell 2002; 111:943–955.
43. Morgan HE, Baker KM. Cardiac hypertrophy: mechanical, neural, and endocrine dependence. Circulation 1991; 83:13–25.
44. Katz AM. Maladaptive growth in the failing heart: the cardiomyopathy of overload. Cardiovasc Drug Ther 2002; 16:245–249.
45. Francis GS, McDonald KM, Cohn JN. Neurohormonal activation in preclinical heart failure. Remodeling and the potential for intervention. Circulation 1993; 87:IV90–IV96.
46. Dalby MCD, Burke M, Radley-Smith R, Banner NR. Pheochromocytoma presenting after cardiac transplantation for dilated cardiomyopathy. J Heart Lung Transplant 2001; 20:773–775.
47. Benjamin IJ, Schuster EH, Bulkley BH. Cardiac hypertrophy in idiopathic dilated congestive cardiomyopathy: a clinicopathologic study. Circulation 1981; 64:442–447.
48. Gunther S, Grossman W. Determinants of ventricular function in pressure-overload hypertrophy in man. Circulation 1979; 59:679–688.
49. Pfeffer MA, Braunwald E. Ventricular remodeling after myocardial infarction. Experimental observations and clinical implications. Circulation 1990; 81:1161–1172.
50. Hutchins GM, Bulkley BH. Infarct expansion versus extension: two different complications of acute myocardial infarction. Am J Cardiol 1978; 41:1127–1132.
51. Weisman HF, Bush DE, Mannisi JA, Weisfeldt ML, Healy B. Cellular mechanisms of myocardial infarct expansion. Circulation 1988; 78:186–201.
52. Eaton LW, Weiss JL, Bulkley BH, Garrison JB, Weisfeldt ML. Regional cardiac dilatation after acute myocardial infarction. Recognition by two-dimensional echocardiography. N Engl J Med 1979; 300:57–62.
53. Ross JJ, McCullagh WH. Nature of enhanced performance of the dilated left ventricle in the dog during chronic volume overloading. Circ Res 1972; 30:549–556.
54. Erlebacher JA, Weiss JL, Eaton LW, Kallman C, Weisfeldt ML, Bulkley BH. Late effects of acute infarct dilation on heart size: a two dimensional echocardiographic study. Am J Cardiol 1982; 49:1120–1126.
55. McKay RG, Pfeffer MA, Pasternak RC, Markis JE, Come PC, Nakao S, Alderman JD, Ferguson JJ, Safian RD, Grossman W. Left ventricular remodeling after myocardial infarction: a corollary to infarct expansion. Circulation 1986; 74:693–702.

56. Popovic AD, Neskovic AN, Marinkovic J, Lee JC, Tan M, Thomas JD. Serial assessment of left ventricular chamber stiffness after acute myocardial infarction. Am J Cardiol 1996; 77:361–364.

57. Warren SE, Royal HD, Markis JE, Grossman W, McKay RG. Time course of left ventricular dilation after myocardial infarction: influence on infarct-related artery and success of coronary thrombolysis. J Am Coll Cardiol 1988; 11:12–19.

58. Gerdes AM, Kellerman SE, Moore JA, Muffly KE, Clark LC, Reaves PY, Malec KB, McKeown PP, Schocken DD. Structural remodeling of cardiac myocytes in patients with ischemic cardiomyopathy. Circulation 1992; 86:426–430.

59. Beltrami CA, Finato N, Rocco M, Feruglio GA, Puricelli C, Cigola E, Quaini F, Sonnenblick EH, Olivetti G, Anversa P. Structural basis of end-stage failure in ischemic cardiomyopathy in humans. Circulation 1994; 89:151–163.

60. Beltrami CA, Finato N, Rocco M, Feruglio GA, Puricelli C, Cligola E, Sonnenblick EH, Olivetti G, Anversa P. The cellular basis of dilated cardiomyopathy in humans. J Mol Cell Cardiol 1995; 27:291–305.

61. Bellenger NG, Swinburn JMA, Rajappan K, Lahiri A, Senior R, Pennell DJ. Cardiac remodelling in the era or aggressive medical therapy: does it still exist? Int J Cardiol 2002; 83:217–225.

62. The CAPRICORN Investigators. Effect of carvedilol on outcomes after myocardial infarction in patients with left-ventricular dysfunction: the CAPRICORN randomised trial. Lancet 2001; 357:1385–1390.

63. CIBIS-II Investigators and Committees. The Cardiac Insufficiency Bisoprolol Study II (CIBIS-II): a randomised trial. Lancet 1999; 353:9–13.

64. Effect of metoprolol CR/XL in chronic heart failure: Metoprolol CR/XL Randomized Intervention Trial in Congestive Heart Failure (MERIT-HF). Lancet 1999; 353:2001–2007.

65. Bristow MR, Gilbert EM, Abraham WT, Adams KF, Fowler MB, Hershberger RE, Kubo SH, Narahara KA, Ingersoll H, Krueger S, Young S, Shusterman N. Carvedilol produces dose-related improvements in left ventricular function and survival in subjects with chronic heart failure. Circulation 1996; 94:2807–2816.

66. Doughty RN, Whalley GA, Gamble G, MacMahon S, Sharpe N. Left ventricular remodeling with carvedilol in patients with congestive heart failure due to ischemic heart disease. J Am Coll Cardiol 1997; 29:1060–1066.

67. Sharpe N, Murphy J, Smith H, Hannan S. Treatment of patients with symptomless left ventricular dysfunction after myocardial infarction. Lancet 1988; 6:225–229.

68. Lamas GA, Vaughan DE, Parisi AF, Pfeffer MA. Effects of left ventricular shape and captopril therapy on exercise capacity after anterior wall acute myocardial infarction. Am J Cardiol 1989; 63:1167–1173.

69. St John Sutton M, Pfeffer MA, Moye L, Plappert T, Rouleau JL, Lamas GA, Rouleau J, Parker JO, Arnold MO, Sussex B, et al. Cardiovascular death and left ventricular remodeling two years after myocardial infarction: baseline predictors and impact of long-term use of captopril: information from the Survival and Ventricular Enlargement (SAVE) trial. Circulation 1997; 96:3294–3299.

70. St John Sutton M, Pfeffer MA, Plappert T, Rouleau JL, Moye LA, Dagenais GR, Lamas GA, Klein M, Sussex B, Goldman S, et al. Quantitative two-dimensional echocardiographic measurements are major predictors of adverse cardiovascular events after acute myocardial infarction. Circulation 1994; 89:68–75.

71. Pfeffer MA, Lamas GA, Vaughan DE, Parisi AF, Braunwald E. Effect of captopril on progressive ventricular dilatation after anterior myocardial infarction. N Engl J Med 1988; 319:80–86.

72. Packer M, Fowler MB, Roecker EB, Coats AJS, Katus HA, Krum H, Mohacsi P, Rouleau JL, Tendera M, Staiger C, et al. Effect of carvedilol on the morbidity of patients with severe chronic heart failure. Results of the Carvedilol Prospective Randomized Cumulative Survival (COPERNICUS) Study. Circulation 2002; 2002:2194–2199.

73. Konstam MA, Kronenberg MW, Rousseau MF, Udelson JE, Melin J, Stewart D, Dolan N, Edens TR, Ahn S, Kinan D, et al. Effects of angiotensin converting enzyme inhibitor enalapril on the long-term progression of left ventricular dilatation in patients with asymptomatic systolic dysfunction. Circulation 1993; 88:2277–2283.

74. Greenberg B, Quinones MA, Koilpillai C, Limacher M, Shindler D, Benedict C, Shelton B. Effects of long-term enalapril therapy on cardiac structure and function in patients with left ventricular dysfunction: results of the SOLVD echocardiography substudy. Circulation 1995; 91:2573–2581.

75. Tang WHW, Larson MS, Prikazsky L, Hu BS, Fowler MB. Reversal remodeling following long-term carvedilol therapy is associated with improvement in survival: The Stanford carvedilol echocardiographic registry [Abstract #1062MP-121]. J Am Coll Cardiol 2002; 39:141A.

76. Metra M, Nodari S, Parrinello G, Giubbini R, Manca C, Dei Cas L. Marked improvement in left ventricular ejection fraction during long-term beta-blockade in patients with chronic heart failure: clinical correlates and prognostic significance. Am Heart J 2003; 145:292–299.

77. Pfeffer MA, Pfeffer JM, Fishbein MC, Fletcher PJ, Spadaro J, Kloner RA, Braunwald E. Myocardial infarct size and ventricular function in rats. Circ Res 1979; 44:503–512.

78. Pfeffer MA, Pfeffer JM, Steinburg C, Finn P. Survival after an experimental myocardial infarction: beneficial effects of long-term therapy with captopril. Circulation 1985; 72:406–412.

79. Fletcher PJ, Pfeffer JM, Pfeffer MA, Braunwald E. Left ventricular diastolic pressure–volume relations in rats with healed myocardial infarction. Circ Res 1981; 49:618–626.

80. Pfeffer JM, Pfeffer MA, Braunwald E. Influence of chronic captopril therapy on the infarcted left ventricle of the rat. Circ Res 1985; 57:84–95.

81. Tang WHW, Girod JP, Lee MJ, Starling RC, Young JB, Francis GS. Prevalence and clinical characteristics of patients with chronic systolic heart failure and normal plasma levels of B-type natriuretic peptide [Abstract #888-3]. J Am Coll Cardiol 2003; 41:221A.

82. Gaudron P, Eilles C, Kugler I, Ertl G. Progressive left ventricular dysfunction and remodeling after myocardial infarction. Potential mechanisms and early predictors. Circulation 1993; 87:755–763.

83. Gaudron P, Eilles C, Ertl G, Kochsiek K. Compensatory and noncompensatory left ventricular dilatation after myocardial infarction: time course and hemodynamic consequences at rest and during exercise. Am Heart J 1992; 123:377–385.

84. Solomon SD, Glynn RJ, Greaves S, Ajani U, Rouleau JL, Menapace F, Arnold JMO, Hennekens C, Pfeffer MA. Recovery of ventricular function after myocardial infarction in the reperfusion era: the Healing and Early After Reducing Therapy study. Ann Intern Med 2001; 134:451–458.

85. Kostuk WJ, Kazamias TM, Gander MP, Simon AL, Ross JJ. Left ventricular size after acute myocardial infarction: serial changes and their prognostic significance. Circulation 1973; 47:1174–1179.

86. Shanoff HM, Little JA, Csima A, Yano R. Heart size and ten-year survival after uncomplicated myocardial infarction. Am Heart J 1969; 78:608–614.

87. Hammermeister KE, De Rouen TA, Dodge HT. Variables predictive of survival in patients with coronary disease: selection by univariate and multivariate analyses from the clinical electrocardiographic, exercise, arteriographic, and quantitative angiographic evaluations. Circulation 1979; 59:421–430.

88. White HD, Norris RM, Brown MA, Brandt PWT, Whitlock RML, Wild CJ. Left ventricular end-systolic volume as the major determinant of survival after recovery from myocardial infarction. Circulation 1987; 76:44–51.

89. St John Sutton M, Lee D, Rouleau JL, Goldman S, Plappert T, Braunwald E, Pfeffer MA. Left ventricular remodeling and ventricular arrhythmias after myocardial infarction. Circulation 2003; 107:2577–2582.

90. Zipes DP. Editorial: less heart is more. Circulation 2003; 107:2531–2532.

91. Mitchell GF, Lamas GA, Vaughan DE, Pfeffer MA. Left ventricular remodeling in the year after first anterior myocardial infarction: a quantitative analysis of contractile segment lengths and ventricular shape. J Am Coll Cardiol 1992; 19:1136–1144.

92. Jugdutt BI, Michorowski BL. Role of infarct expansion in rupture of the ventricular septum after acute myocardial infarction: a two-dimensional echocardiographic study. Clin Cardiol 1987; 10:641–652.

17

Assessment of Ventricular Remodeling in Heart Failure

Inder S. Anand

Division of Cardiology, University of Minnesota Medical School,
and VA Medical Center, Minneapolis, Minnesota, U.S.A.

INTRODUCTION

Heart failure is a complex clinical syndrome characterized by hemodynamic abnormalities, neurohumoral and cytokine activation, fluid retention, reduced exercise capacity, and poor prognosis. Critical to our understanding of this disorder are observations that the progression of this disease is related to progressive alterations in structure and function of the heart. Recent clinical studies have shown that progressive left ventricular (LV) hypertrophy, enlargement, and cavity distortion over time, termed "ventricular remodeling," are directly related to deterioration of LV performance and increase in mortality and morbidity (1–3). The importance of ventricular remodeling has increased with the observations that only drugs such as angiotensin-converting enzyme (ACE) inhibitors, beta-blockers, and aldosterone receptor blockers that attenuate or reverse ventricular remodeling improve survival (4–12). In contrast, drugs that do not improve ventricular remodeling have no beneficial effect and may even be deleterious (13–16). Ventricular remodeling is therefore emerging as an important therapeutic target in heart failure. Moreover, the appropriate use of ACE-I and beta-blockers in the management of heart failure has had a remarkable effect in reducing the annual mortality seen in clinical trials of patients with moderate to severe heart failure to around 6% to 10% (17,18). This has made it very difficult to assess the effects of newer approaches on mortality in heart failure. Accordingly, it has been suggested to use measurements of ventricular remodeling as an additional endpoint in the assessment of newer therapies for heart failure (19,20).

Accurate and reproducible assessment of ventricular dimensions, mass, and function is therefore important not only for establishing the presence of ventricular dysfunction, defining prognosis, guiding, and assessing therapy (1,21,22), but may also become a surrogate for assessing newer heart failure therapies (19,20). Most often, clinical decisions in heart failure are based on measurements of LV ejection fraction alone using one of several widely available imaging techniques. Evaluation of ventricular volumes and/or mass is seldom used in routine clinical practice, because the methods used are relatively cumbersome and not widely practiced. In the setting of clinical trials, however, analysis of serial changes in ventricular volumes and mass has provided substantial insight into the remodeling process. Study of the effect of drug and device interventions on the process of remodeling by noninvasive imaging techniques has also illuminated important mechanisms by which particular pharmacologic therapies improve morbidity and mortality in patients with heart failure and following MI.

This chapter will describe the most widely used noninvasive imaging techniques available for the accurate assessment of serial changes in ventricular structure and function and discuss their implications for use in heart failure clinical trials.

Assessment of Serial Ventricular Volume Changes over Time

The ideal imaging technique should be widely available, accurate, reproducible, and should not be unsafe to the patient. Numerous factors influence the ability of a technique to determine the true value of a particular measurement. The term *accuracy* refers to the ability of a test to measure a result relative to a "gold standard." The term "reproducibility" is used to describe the degree of agreement between two tests in which the same measurement is made. Reproducibility is inversely proportional to the *variability* of a measurement. Three components may contribute to total variability. Intraobserver variability refers to the deviation in repeat measurements by a single observer and reflects the ability of a single observer to make the same measurement consistently (e.g., LV volume). This type of variability can come from a number of sources, including variability in the selection of the images and the ability to trace or measure endocardial boundaries. Interobserver variability is the variability in the measurements made by different observers and reflects the ability of a second observer to select the same boundaries relative to the first observer. One of the strengths of analyzing images using a core laboratory in a clinical trial is that reproducibility is enhanced by having a single technologist performing image analysis, reducing the contribution to variability of multiple operators (23).

When measurements are performed to monitor changes over time in LV dimensions and function, the difference between two repeat measurements is the interstudy variability. The interstudy variability may be technical and of biologic origin (24) (Table 1). Technical factors include variability in data acquisition and more importantly in analysis. The largest contributor to variations in LV dimensions and function comes from a number of biological factors. These include changes in the loading conditions of the heart due to variations in intravascular volume, adrenergic drive, or whether measurements are made in the fasting or post-absorption state.

"Gold Standard" for Assessing the LV Mass and Dimensions

For decades, contrast angiography has been considered the "gold standard" for measuring LV volumes. There are, however, several theoretical and practical limitations of contrast angiography. The most obvious is that angiographic measurements are dependent on the assumption of a three-dimensional geometric model. Because the shape of the ventricle varies with disease states, especially in patients after myocardial infarction, this assumption is often invalid. Moreover, contrast angiographic techniques require substantial operator intervention to draw endocardial borders, and are often subject to significant influences of premature ventricular beats during the acquisition. Few catheterization laboratories at present routinely quantify LV volumes using angiography because of these limitations.

Table 1 Sources of Interstudy Variability

Technical factors
  Acquisition
    Use of internal landmarks
    Gain control
    Rigorous technique
  Analysis
    Same versus different observer
Biological intrinsic factors (not easily controlled)
  Inotropic "adrenergic" state
    Blood pressure
    Heart rate
Extrinsic factors (controllable)
  Time of day
  Medications
  Hours postprandial
  Respiratory phase

Source: Modified from Ref. 24.

Given the confounding issues in assessing the accuracy of noninvasive imaging techniques in comparison to contrast angiography, some studies in the literature have used comparison with postmortem assessments of mass and volumes as a gold standard. Such studies are difficult to perform, may not fully reflect the wide range of volumes or mass measurements, and do not appropriately define or correct for the loading conditions at the time the noninvasive imaging test is performed. MRI tomography provides in vivo measurements of LV mass and volumes that approximate most closely with those measured at postmortem examination (Table 2) (25,26).

Thus, there is difficulty in establishing absolute accuracy of noninvasive estimates of ventricular volumes or mass. However, most techniques used in clinical trials have shown reasonable correlation with some type of gold standard technique. Perhaps more important than the ability of a test to exactly measure ventricular volume is the ability to assess the changes in volumes or mass over time in a manner that is reproducible enough to be useful for clinical trial applications.

Determination of Sample Size for Clinical Trials

One of the most important determinations to be made when planning a clinical trial is the number of patients necessary to demonstrate biologically and clinically important changes. Calculation of an accurate sample size is dependent on the intrinsic accuracy of the method as well as the biologic variability in the population. The

Table 2 Magnetic Resonance Imaging Tomographic vs. Necropsy Assessment of LV Mass and Volume

Study	Method	n	r	SE	SD
Katz et al. (25)	Spin-echo (mass)	10	0.99	6.8 g	22 g
Rehr et al. (26)	Spin-echo (volume)	15	0.997	4.3 mL	17 mL

Abbreviations: LV, left ventricular; SE, standard error of estimation; SD, standard deviation; n, number of subjects (men) studied; r, correlation coefficient.

standard deviation of the measurement thus reflects both of these factors as taken into account for sample size calculation? For example, to determine the number of patients required for a study of post-MI ventricular remodeling in which patients were assigned to one of three treatment assignments, the standard deviation from the HEART trial was utilized and the sample size necessary to detect a clinically meaningful difference of 5 mL in ventricular volume between therapies was determined (sample size required = 200 patients per group).

For the purpose of measuring changes in LV volumes or mass in a clinical trial, the number of patients required assessing a given volume or mass change with a certain degree of confidence (i.e., the sample size) is in part dependent on the reproducibility of the test being used to assess the change. This variability should be assessed in serial tests over the period of time for which the trial is expected to last, rather than repeat analysis of a single image. This will incorporate the intraobserver, interobserver, and interstudy variability.

ECHOCARDIOGRAPHY

Technical Considerations

As a two-dimensional technique, echocardiography provides a cross-sectional image of an interrogated structure. Measurements of LV mass by this technique has been compared to necropsy assessment (Table 3). Unlike cardiac MRI, echocardiography is not a true tomographic technique; the image obtained and the measurements made from the image are dependent on the imaging plane that is chosen by the sonographer. Three-dimensional measures of ventricular size—such as ventricular volume— are never derived directly from echocardiographic images but are estimated based on formulas that make assumptions about ventricular shape. As ventricular volume has been the most commonly accepted measurement of ventricular size, volume calculations are typically derived from two-dimensional images. A number of approaches are commonly employed to perform these calculations (Table 4). Each of these methods relies on some assumptions about ventricular geometry and rarely provides as accurate a representation of ventricular volume as a true tomographic technique, in which multiple two-dimensional slices are used to reconstruct a volume. Approaches that make the fewest assumptions about ventricular geometry—such as the modified Simpson's rule method—offer the most accurate estimations of ventricular volume (27). Nevertheless, mistakes made on tracing a two-dimensional ventricular contour are amplified when volumes are calculated.

The biggest limitation for an accurate assessment of volumes with echocardiography is the digitization of the endocardial borders. There are currently no automated methods for tracing and digitization of the endocardial border that are sufficiently accurate to allow for reproducible measurement of ventricular size or function. Techniques that can enhance endocardial border definition such as contrast

Table 3 Two-Dimensional Echocardiographic vs. Necropsy Assessment of LV Mass

Study	Method	n	r	SE	SD
Reichek et al. (81)	Area-length (mass)	21	0.93	31 g	142 g
Devereux et al. (82)	Penn-cube (mass)	52	0.92	6.0 g	43 g

Abbreviations: LV, left ventricular; SE, standard error of estimation; SD, standard deviation; n, number of subjects (men) studied; r, correlation coefficient.

Table 4 Methods Utilized to Calculate LV Volumes

Method	Formula	Comments
Short axis "bullet" area length method	$V = \dfrac{5}{6} AL$	Dramatically reduced accuracy in patients with regional abnormalities involving the apex
Long axis area length method	$v = \dfrac{5}{6} \dfrac{A^2}{L}$	Similar accuracy to single-plane Simpson's rule
Modified Simpson's rule	Single plane: $$v = h \sum_{i=1}^{n} \pi \left(\dfrac{d}{2}\right)^2$$ Biplane: $$v = h \sum_{i=1}^{n} \dfrac{\pi d_1 d_2}{4}$$	Best two-dimensional measure of volume in patients with regional abnormalities

Abbreviations: LV, left ventricular; d, d_1, d_2, diameters; h, thickness; n, number of slices.
Source: From Ref. 80.

echocardiography can be of tremendous benefit. Second harmonic imaging improves endocardial border definition by improving the signal-to-noise ratio of the image (28). Ultrasonic contrast agents—albumin- or lipid-based microspheres filled with an inert gas—are small enough to pass through the pulmonary circulation and thus opacify the LV cavity, thereby improving endocardial border definition (29). These techniques can dramatically improve the accuracy of assessment of ventricular size and function.

To avoid the inherent limitations of calculating volumes from two-dimensional data, the SAVE investigators measured ventricular areas instead of ventricular volumes and have argued that this may be a more accurate approach (30). The limitation of this approach is that the ejection fraction, the primary reason to estimate volumes in clinical practice, cannot be calculated.

Echocardiography is also useful for assessing alterations in regional LV function. A number of clinical trials have utilized wall motion scoring to assess regional ventricular function. TRACE and VALIANT trials used wall motion scoring to identify patients at increased risk and related regional abnormalities to clinical outcomes (31,32). In addition, other parameters of ventricular morphology including measures of ventricular shape have been used to assess ventricular remodeling (33).

Reproducibility, Variability, and Sample Size Calculation

Measurement of ventricular volume by echocardiography utilizing the Simpson's rule method has been shown to compare well with known ventricular volumes in an in vitro canine model (27). Measurements of ventricular size made over time should not vary if there has been no biologic change apart from the variability discussed before. Nevertheless, changes in ventricular size and ejection fraction from beat to beat are common. While beat-to-beat variability is minimal in patients in sinus rhythm, ventricular volumes or ejection fraction can vary substantially owing to the respiratory cycle or in patients with rhythm disturbances, such as atrial fibrillation. In clinical practice, measurements are often made from one cardiac cycle, and thus, beat-to-beat variation can play a major role in reproducibility. In research setting, three or more beats are usually recorded, and the results are averaged.

For echocardiography, the accuracy of the measurement is largely dependent on image quality, as the blood–tissue interface can be difficult to identify clearly when image quality is less than optimal, particularly in patients with low cardiac output. The reproducibility of echocardiography for the measurement of LV dimensions, mass, and ejection fraction has been studied utilizing a variety of study designs and statistical methods (34–37). Although important information has been generated on interobserver variability and intraobserver variability, beat-by-beat variability, and interstudy variability in normal subjects, similar data are not widely available in a heart failure population, particularly repeated over a long term. Therefore, the value by which LV mass and volume must change to exceed methodologic variability on sequential examination in heart failure patients within a multicenter study is not well established.

M-mode echocardiography can be of some value in monitoring changes over time in LV mass and dimensions, in particular measurement of ventricular wall thickness (34,38). However, two-dimensional echocardiography has proven to be more reproducible than M-mode for measuring LV mass in normal subjects (34,36) and in patients with a distorted LV contour (39).

Outcome Studies in LV Dysfunction

Much of our current knowledge about LV remodeling is derived from clinical trials in which patients have undergone serial echocardiographic studies to assess changes in the size and shape of the heart in the context of specific drug therapies (4–11,30). These ECHO substudies taught us that ACE-I therapy in both the post-MI patients (SAVE trial) (30) and in patients with established LV dysfunction (SOLVD trial) (4) attenuates LV remodeling. In contrast, the ECHO substudy of the ANZ carvedilol trial (5) showed that beta-blockers work by reversing ventricular remodeling. Each of these trials has contributed to our understanding of the remodeling process and has also helped in answering specific questions regarding pharmacologic therapy of myocardial infarction or heart failure.

Hence, echocardiography has proven to be a clinically useful, reproducible, and accurate technique for assessment of LV remodeling and changes in LV function. As a clinical trial tool, echocardiography has a number of advantages over other techniques. The noninvasive echocardiographic examination is standardized throughout the world, can be performed virtually everywhere, and is relatively inexpensive. Finally, echocardiography can offer additional information about cardiac valvular function and diastolic function that are difficult to obtain with other modalities. Nevertheless, this technique suffers from marked variation in image quality that cannot be predicted, and approximately 10% to 15% of patients cannot be adequately imaged for quantitative purposes. For these reasons, the reproducibility and overall accuracy of echocardiography may not be as high as some of the other available techniques.

RADIONUCLIDE VENTRICULOGRAPHY

One of the main advantages of radionuclide ventriculography (RNV) is its ability to capture information on LV function on virtually any patient with a wide variety of cardiovascular disease syndromes. A wealth of data on the diagnostic and prognostic implications of changes in ejection fraction in response to exercise is available in a variety of patients. RNV has also been used extensively to examine serial changes in ejection fraction in various clinical settings, such as in patients treated with cardiotoxic or potentially cardiotoxic chemotherapy as well as in patients being followed serially with significant valvular heart disease.

Measurement of ejection fraction using RNV is straightforward as ejection fraction is a relative measure of stroke volume divided by end-diastolic volume. Background corrected radionuclide counts within a regional of interest could be used to determine ejection fraction without the need to determine absolute volumes. However, calculation of absolute LV volume is more complex, requiring several corrections and assumptions regarding the relation between count activity as measured in a two-dimensional picture of the left ventricle and absolute volume. Nonetheless, several laboratories have reported very good correlations with several different methods for RNV.

Technical Considerations

Common to all radionuclide techniques used to measure absolute ventricular volumes is the need to determine the relative activity of radionuclide counts to represent a known blood volume value as well as the need to estimate attenuation effects created by structures in between the LV blood pool and the imaging camera, which include the chest wall as well as the LV wall and lung fields. Several methods have been proposed for this purpose and will not be reviewed in detail.

The RNV technique is independent of left and right ventricular geometries and/or abnormalities in regional wall motion. All count changes (representing volume changes) within the chambers are captured and the technique does not rely on an operator-defined analysis of regional changes in wall motion. An important advantage of RNV over other techniques is that it requires less operator interaction on a frame-by-frame basis, particularly in the settings where multiple regional wall motion abnormalities are expected, such as in patients with ischemic cardiomyopathy.

The main disadvantage of the radionuclide technique is that it is not universally available and the analysis of left and right ventricular volumes is rarely undertaken in most clinical radionuclide laboratories. Moreover, the RNV techniques require meticulous attention to detail and to the measurements that are made at a clinical site in a clinical trial to estimate the activity/blood volume ratio as well as the depth correction for attenuation. In the setting of a clinical trial, all the RNV data as well as the raw imaging data are transferred to a core laboratory for analysis, so that the operator interactions, such as establishing the appropriate region of interest and background correction region, are performed in a uniform and high quality manner. Nevertheless, the image acquisition quality as performed at individual clinical sites can have an important impact on the final results. The same holds true for all imaging modalities in this regard, despite ultimate analysis by a core laboratory.

Reproducibility, Variability, and Sample Size Calculation

One of the strengths of the RNV technique is the substantial reproducibility and the low intra- and interobserver variability that have been reported in the literature. Several studies have assessed the reproducibility of RNV in both normal subjects and patients with heart diseases. Upton et al. (40) determined the intrinsic variability of radionuclide measurements of LV function at rest and during exercise in 10 normal subjects. The interobserver variability for ejection fraction ($2.1 \pm 1.0\%$) and for end-diastolic volume (7.5 ± 4.7 mL) at rest was considerably smaller than those reported in studies using contrast ventriculography (41,42). The interstudy variability in ejection fraction in this study was $4.0 \pm 3.8\%$ at rest and $3.2 \pm 2.5\%$ during exercise, suggesting that at 95% confidence level, repeat ejection fraction should not vary by more than 8% at rest and 5% during exercise. A difference in end-diastolic volume of at least 20 mL between rest and exercise studies was suggested to be required for the change to be considered meaningful (40). A similar variability in ejection fraction of $4.4 \pm 3.6\%$ was reported by Marshall et al. (43) in 20 patients with cardiac disease who had three resting radionuclide ventriculograms separated by an average of 4.3 days.

In an established core laboratory, extremely low coefficients of variation for left and right ventricular ejection fractions were 3.9% and 3.3%, respectively and of LV end-diastolic and end-systolic were 3.4% and 3.1%, respectively. Correlation coefficients between the duplicate measurements were 0.99. Thus, when the same measurement is obtained twice, variability is low.

Van Royen et al. (44) examined intra- and interobserver variability of LV ejection fraction measurements from equilibrium RNV in a series of over 70 patients across a wide range of ejection fraction values. Correlation between values determined by one operator on separate occasions was extremely high, as was correlation of values measured by two separate operators on the same raw data. These investigators also examined the frequency of absolute differences in repeat assessments of ejection fraction that may be of potential clinical relevance (i.e., a difference greater than or equal to 10 ejection fraction units) for RNV as well as visual estimation of LV ejection fraction by echocardiography. They found that potentially clinical relevant clinical differences of this degree did not occur on repeat processing of equilibrium RVNs, while potentially clinical relevant differences occurred in 8% to 26% of studies on repeat analysis of the echocardiograms. While these authors did not employ quantitative echocardiographic techniques, the data do suggest that RNV analysis of ejection fraction in this setting is highly reproducible, with very low intra- and interobserver variability.

As mentioned above, there are many potential sources of variability in the measurement of LV volume changes that can potentially impact serial measurements. One such variable is the inherent beat-to-beat biologic variability in LV function itself. Using the RNV technique, this aspect of variability in measurement is minimized, as the volumetric imaging data that are processed are displayed as a cine-looped beat representing the average of the several hundred beats that were acquired during the image acquisition. Thus, in patients with atrial fibrillation, the acquisition can be tailored to acquire sufficient beats for analysis. In patients with irregular heart rates, therefore, RNV remains the most accurate noninvasive method of measuring ventricular volumes.

Based on data from several studies, it is estimated that sample sizes for clinical trials of LV remodeling using RNV techniques will usually require approximately 50 patients per intervention group, to identify a difference in the change of LV end-diastolic volume of approximately $11 \, \text{mL/m}^2$ with 80% power or approximately $13 \, \text{mL/m}^2$ with 90% power. These changes are smaller than that observed in the SOLVD trial (11) between the placebo and the enalapril groups ($28 \, \text{mL/m}^2$ difference). In the contemporary era where most patients are receiving background therapy with ACE-I and beta-blockers, the changes in LV volumes over time are likely to be smaller than those outlined above. Hence, larger number of patients may be required to demonstrate significant changes with newer add-on therapies.

Table 5 The Long-Term Natural History Outcomes in Large Clinical Trials

Study	Drug	LV volume	Mortality
SOLVD, treatment arm	Enalapril	Decrease (4–11)	Decrease (83)
SOLVD, prevention arm	Enalapril	Mild decrease (11)	Mild decrease (84)
PRIME II	Ibopamine	Increase (13)	Increase (85)
ELITE I and II	Captopril versus losartan	Decrease with captopril (14)	Decrease with losartan (86) and trend to increase with captopril (87)
IMPRESS	Omapatrilat	Neutral (86)	OVERTURE— neutral results

Abbreviations: LV, left ventricular; SOLVD, studies of LV dysfunction; PRIME II, second prospective randomized study of ibopamine on mortality and efficacy; ELITE, evaluation of losartan in the elderly study; IMPRESS, inhibition of metaloproteinase by BMS186716 in a randomized exercise and symptoms study; OVERTURE, omapatrilat versus enalapril randomized trial of utility in reducing events.
Source: From Ref. 80.

Outcome Studies in LV Dysfunction

Numerous studies using radionuclide techniques in heart failure patients have documented that serial changes in LV volumes determined by the RNV are related to long-term prognosis. In the SOLVD trial (10,11), enalapril caused a favorable effect on LV remodeling compared to placebo. This was accompanied by a beneficial effect on mortality. However, other studies listed in Table 5 also used RNV to examine changes in LV volume during drug therapy but did not show favorable effects on LV volumes. In these studies also the outcomes reflect changes in LV volumes.

Thus, RNV allows a relatively accurate, reproducible assessment of serial changes in LV volumetrics suitable for analysis in clinical trials of new therapies in heart failure and other cardiovascular diseases. It is less costly than Doppler echocardiography or MRI, and it is widely available, at least in the United States. However, the exact volumetric techniques are no longer widely practiced for purely clinical care purposes. Nonetheless, laboratories with experience in these techniques as well as laboratories with physicians and technologists familiar with the techniques can be trained to acquire high quality data for the purposes of conducting a clinical trial of new therapeutics for examining changes in ventricular volumes, reflecting the process of LV remodeling. In contrast to echocardiography, the RNV technique provides technically adequate data virtually in all patients, and it is far less affected by ventricular geometry and wall motion abnormalities than either quantitative echocardiographic techniques or MRI techniques that rely on geometric assumptions of LV shape. The RNV methodologies have changed little over the last 20 years since their initial development, and electronic data transmission from a clinical site to a core laboratory is now fairly routine.

Like all such techniques however, meticulous attention to detail and assembling a trial group with expertise and ability to acquire high quality studies are important for the successful RNV assessment of the remodeling process. These attributes are of course important for all noninvasive or invasive imaging analyses of pathophysiologic phenomena.

MAGNETIC RESONANCE IMAGING

Technical Considerations

In recent years two tomographic techniques have been increasingly applied to the diagnosis of cardiovascular disease and the quantification of cardiac dimensions and function. These techniques, ultrafast computed tomography and MRI, can acquire tomographic images at multiple levels encompassing the entire heart and thereby yielding a three-dimensional data set. Computed tomography has, however, some major drawbacks; it requires rapid infusion of intravenous contrast material and exposes patients to radiation, both of which carry an element of risk (45,46). MRI does not have these limitations and can be performed in any imaging plane so that a data set parallel or perpendicular to the long axis of the ventricle is produced. Because a three-dimensional data set is acquired, accurate LV volume and mass measurements can be obtained. Several studies (47–51) have shown a correspondence between LV volumes and ejection fractions measured with MRI and contrast angiography or echocardiography or both. MRI is a rapidly evolving technique for the noninvasive assessment of cardiovascular disease. It is of particular interest because it is the one method most likely to offer a comprehensive evaluation of the heart, incorporating anatomical, structural, and functional data. Spin-echo techniques enable accurate anatomical assessment of the heart and great vessels. By use of gradient cine imaging, function of the left and right ventricles can be accurately measured by obtaining contiguous breath-hold gradient cine slices from the base to apex of the heart (26,47,52).

Imaging Technique

Briefly, before acquisition of the cine MRI, scout images are acquired for proper true short axis image slice plane orientation (Figs. 1 and 2). The short axis scout image localizer is used to define the horizontal long axis image plane. The slice for the horizontal long axis cine should intersect the center of the left ventricle and be

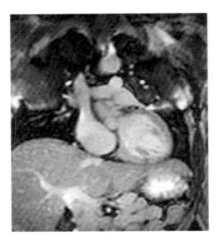 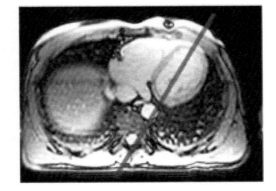

Figure 1 Coronal and transverse scout images acquired with single-show steady-state free precession imaging technique. The transverse scout image is used to position the slice for the pseudo-long axis localizer. The slice position and orientation is shown in the image on the right as a dark grey bar.

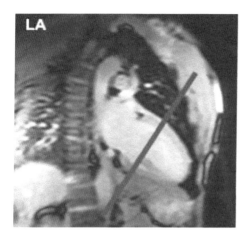
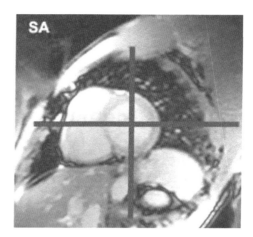

Figure 2 The image on the left (pseudo vertical long axis view) shows the slice positioning for the double-oblique short axis view of the heart (*dark bar*). Multiple short axis slices are acquired as shown on the right. The slice orientation for the cine acquisition of the horizontal, long axis view should pass through the center of the left ventricle, and intersect with the middle of the septal wall, as shown by the dark in the image on the right.

perpendicular to the mid-septum. For the vertical long axis, a slice parallel to the septum, centered in the LV cavity and positioned from the LV apex, through the mid portion of the mitral valve is prescribed. The image planes for the short axis studies start at the base of the heart at least 2 cm above the mitral and tricuspid valve planes to ensure complete coverage of the right ventricle. The slice position is then moved from base to apex of the heart in increments of 8 mm or less slice thickness (Fig. 3). For a slice thickness of 8 mm, no interslice gap is left, whereas for slice thickness of 5 to 6 mm a small (<3 mm) interslice gap is left. Thinner slices result in better definition of the endocardial border and better delineation of defects, if no gap is left. The slice thickness for the long axis slices can be as thin as 5 mm to optimize the contrast between blood pool and myocardium. For patients with low cardiac output thin slices improve contrast.

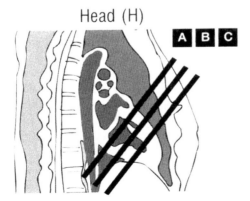

Figure 3 Schematic illustration of the position of short-axis image planes on long axis localizers.

Segmented Image Acquisition

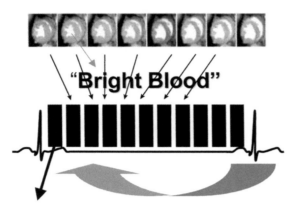

"Bright Blood"

"Window" for acquisition of image segment with 30–100 ms duration = Temporal resolution

Total Acquisition takes 15–25 R-to-R intervals, i.e. need stable HR

Figure 4 Cine MRI acquisition during a single RR interval. The temporal resolution for the cine frames is of the order of 30 to 60 milliseconds, depending on the heart rate. Typically, 15 to 20 frames are acquired per R-to-R interval.

Cine MRI studies are ECG gated and are acquired for a horizontal long axis section, a vertical long axis section of the heart, and adjacent short axis sections from base to apex during a breath-hold. The temporal resolution for the cine frames is of the order 30 to 60 milliseconds, depending on the heart rate. Typically, 15 to 20 frames are acquired per R-to-R interval (Fig. 4).

By use of computer software that can trace the endocardial and epicardial border in systole and diastole, ventricular volumes and mass can be calculated. MRI can also accurately assess cardiac motion and regional function. Similar to echo techniques, cine MRI allows qualitative assessment of wall thickening and motion with the additional freedom from acoustic window limitations and the ability to assess motion in any chosen plane (Fig. 5). New techniques such as MRI myocardial tagging offer a method for quantifying cardiac wall motion. The technique involves the application of a low-signal grid across the myocardium at end-diastole, and the subsequent pattern of grid deformation can be analyzed (53,54). It can examine changes in contractility and wall stress and assess myocardial viability (55–57).

More recently, the technique of delayed contrast-enhanced MRI using intravenous injection of gadolinium has allowed simultaneous and accurate quantitative estimates of global and regional (transmural) extent of myocardial infarction and global and regional LV dysfunction, using objective analysis techniques. Measurement of viability and function can be made during the same scanning session to eliminate misregistration errors, another advantage of MRI over other imaging modalities. These data can be used to assess the effect of revascularization on the return of regional and global LV function and to compare the relationships between infarct size and LV function in the acute and chronic setting.

Baseline Study (April 1998)

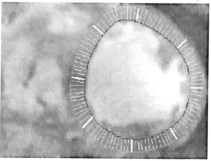

EDV	288 mL
ESV	230 mL
SV	58 mL
EF	20 %
LV mass	254 g
Mass/Vol	0.88

One year later (May 1999)

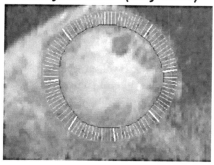

EDV	204 mL
ESV	130 mL
SV	74 mL
EF	36 %
LV mass	244 g
Mass/Vol	1.2

Figure 5 Cine MRI of a patient with severe LV dysfunction before (April 1998) and a year later (May 1999) after therapy with ACE-I and beta-blockers. Note considerable reverse remodeling with decrease in LV volumes, increase in EF, improvement SV, and mass/volume ratio. *Abbreviations*: EDV, endiastolic volume; ESV, endsystolic volume; EF, ejection fraction; LV, left ventricular; SV, stroke volume.

The assessment of LV structure and function by MRI has advantages, over echocardiography and RVN by its ability to provide accurate and reproducible tomographic static and dynamic images of high spatial and temporal resolution in any desired plane without limitation by acoustic access. MRI produces images with natural contrast between tissue and flowing blood thereby removing the need for a contrast agent in routine imaging. MRI sections eliminate overlapping of structures and provide three-dimensional imaging. No geometric assumptions are therefore necessary to determine ventricular dimensions. This is particularly important in evaluating ventricles of patients with heart failure that may have distorted morphology as a result of infarction and have regional differences in wall motion. However, in patients with severe LV dysfunction blood pool and tissue contrast may not be very different making edge detection more difficult.

Conventional ECG-gated, free-breathing, gradient-echo cine sequences require a total scanning time of approximately 30 minutes, but with more modern breath-hold fast low angle shot (FLASH) cine sequences, each one of the contiguous short axis slices may be acquired in a single breath-hold, reducing the total scanning time to less than 10 minutes. In patients with heart failure the required breath-hold of approximately 10 seconds may sometimes be problematic but the development of advanced respiratory gating techniques allows acquisition of short axis images of comparable or better quality during free breathing, within the same total time (58). The latest generation of ultrafast scanners may eradicate the problem of multiple breath-hold altogether as they can acquire the entire short axis three-dimensional data set in a single breath-hold of approximately 12 seconds (59), and real time solutions are also applicable during free breathing (60,61).

Several limitations are preventing the routine utilization of MRI for the evaluation of cardiovascular disease, including limited access to MRI, expense of the studies, and the duration of the procedure required to achieve images of the entire heart. Claustrophobia is a problem with some patients, but may be resolved with the newer open-sided scanners. Absolute contraindications to MRI are few, but patients with implantable devices such as pacemakers and defibrillators, cochlear implants, and cerebral aneurism clips usually cannot be scanned except under very controlled and clinically urgent circumstances (62). The time for quantitative analysis of imaging data may also be an issue with some of the manual techniques involved, but the development of new software and edge detection systems has reduced this considerably (63,64).

Reproducibility, Variability, and Sample Size Calculation

MRI is now considered the reference technique and the "gold standard" for the noninvasive assessment of LV dimensions, mass, and function. This method is accurate to about 2% (47) and, because it is highly reproducible, it offers an ideal means of serial assessment of disease progression and/or response to treatment in an individual patient (65) as well as allowing smaller sample sizes for studies of changes in ventricular function with pharmacological treatment (66). MRI has been shown to be superior to echocardiography, contrast ventriculography, and RVN with regard to the accuracy (67,68) and interobserver and interstudy reproducibility of LV measurements of cavity volumes, ejection fraction, and mass both in normal and in diseased hearts (69–73), including hearts deformed by acute infarctions (74). Pattynama et al. (69) compared spin-echo and gradient-echo MRI and found a similar reproducibility for measurements of LV ejection fraction and mass with standard deviations 4% and 6%, respectively. The high reproducibility of MRI has also been demonstrated in patients with heart failure (Table 6) (75). For research purpose, this results in a considerable reduction in the sample size required to show a given change, when compared with echocardiography (75). Bellenger et al. (75) have found that to demonstrate a 10 mL difference in end-diastolic volume and end-systolic volume and a 10 g difference in mass with power of 90% and p-value of 0.05, only 12, 10, and 9 patients would be needed for each of these parameters, respectively, by MRI versus 97, 53, and 190 patients by two-dimensional echocardiography. These values are based on MRI studies repeated over a short period of only one week. However, in the setting of large multicenter heart failure clinical trials, where the effect of interventions are measured over months and years, the variability is much

Table 6 Comparison of Percent Intraobserver, Interobserver, Interstudy, and Total Variability in ECHO and Two MRI studies

	Echo[a]	MRI[b]	MRI[c]
Intraobserver	4.50	2.6	0.9
Interobserver	3.19	4.2	0.6
Interstudy	7.54	3.1	8.8
Total	9.34	5.8	8.9

Source: Refs. 88[a], 75[b], 89[c].

Table 7 Comparison of Sample Size to Detect the Same Change by MRI Variables Using SD of Change from Reproducibility Studies Done at One Week and Six Months Interval (power 90%, $\alpha = 0.05$).

	Clinical change	MRI study one week apart		MRI study six months apart	
		SD	N	SD	N
EDV	10 mL	9.08	22	32.0	215
ESV	10 mL	10.37	28	31.4	207
EF	3%	3.93	14	5.2	63
Mass	10 g	7.1	13	16.2	56

Abbreviations: SD, standard deviation,

larger because of progressive remodeling over time, and these numbers may be much too conservative. Table 7 shows how the standard deviation of change in MRI variables increases when reproducibility is measured at one week and six months apart, resulting in a significant increase in the sample size. Indeed, we have recently reported similar reproducibility data in a large clinical trial that tested the effect of the endothelin blocker darusentan in heart failure (16).

Outcome Studies in LV Dysfunction

MRI is a relatively new technique that is getting increasing attention in designing new large-scale clinical trials. In the metoprolol randomized interventional trial in heart failure (MERIT-HF), 41 patients with chronic heart failure underwent serial MRIs for the assessment of LV dimensions and function (7). In this small study population, MRI permitted to demonstrate the antiremodeling effects of metoprolol CR/XL on the left ventricle (Fig. 6). The results from this MRI substudy provide an explanation for the highly significant decrease in mortality from worsening heart failure found in the metoprolol CR/XL randomized intervention trial in congestive heart failure trial (MERIT-HF) (76). Some other randomized clinical trials including the valsartan heart failure trial (18), the beta-blocker evaluation survival trial (77), and the randomized Etanercept North American strategy to study antagonism of cytokines trial have had MRI substudies with the results awaited. The endothelin$_A$ receptor antagonist trial in heart failure (16) was the first study to use changes in LV end-systolic volume over time with serial MRI as a primary endpoint of a study. The approach was to see whether the drug has favorable effects on remodeling before launching a large mortality trial. In this dose ranging study in 642 patients with moderate to severe failure, the endothelin ETA receptor antagonistic darusentan did not cause any change in the LVESV for up to two years. This was associated with lack of any clinical benefit of the drug (16).

Thus, MRI has shown that it has promise in substantially lowering the confidence limits of LV dimensions and mass measurements (25,78,79). Because of good accuracy and superior reproducibility, MRI may be considered the gold standard for quantification of LV mass, dimensions, and ejection fraction. Even though this method may be more expensive than two-dimensional echocardiography, its greater reproducibility may make it suitable for research studies where the greater cost could be offset by the savings from recruiting and studying fewer patients.

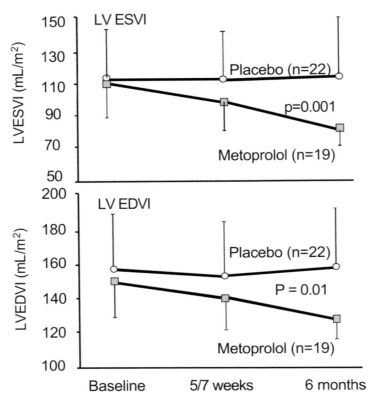

Figure 6 MERIT-HF MRI substudy showing significant reverse remodeling occuring within five to seven weeks after starting metoprolol therapy. No significant change in LV volumes was seen in the placebo group for up to six months.
Source: From Ref. 7.

CONCLUSIONS

There is a substantial literature to suggest that noninvasive imaging assessment of LV remodeling can provide considerable information regarding potential outcome in a population with LV dysfunction and heart failure as well as provide signals

Table 8 Comparison of Echocardiographic, Nuclear, and MRI Techniques

	Echocardiography	RVN	MRI
Accuracy	++	++	+++
Reproducibility	++	+++	+++
Cost	$	$	$$$
Availability	++++	+++	+
Ease of interpretation	++	+++	+
Patient discomfort	+++	+++	++
Technical adequacy	+++ (80%)	++++ (>98%)	++

Abbreviations: MRI, magnetic resonance imaging; RVN, radionuclide ventriculography.
Source: From Ref. 80.

for potentially favorable therapeutic drug effects. Echocardiography, RVN, and MRI have all been used in this setting. Each technique has its strengths and weaknesses regarding measurement variability and availability (Table 8) (80). Careful analysis of potential measurement variability and emphasis on the highest possible degree of quality control in image acquisition and analysis can theoretically overcome some of these differences and maximize the value of the data obtained by these techniques.

REFERENCES

1. White HD, Norris RM, Brown MA, Brandt PW, Whitlock RM, Wild CJ. Left ventricular end-systolic volume as the major determinant of survival after recovery from myocardial infarction. Circulation 1987; 76(1):44–51.
2. Hammermeister KE, DeRouen TA, Dodge HT. Variables predictive of survival in patients with coronary disease. Selection by univariate and multivariate analyses from the clinical, electrocardiographic, exercise, arteriographic, and quantitative angiographic evaluations. Circulation 1979; 59(3):421–430.
3. Vasan RS, Larson MG, Benjamin EJ, Evans JC, Levy D. Left ventricular dilatation and the risk of congestive heart failure in people without myocardial infarction. N Engl J Med 1997; 336(19):1350–1355.
4. Greenberg B, Quinones MA, Koilpillai C, Limacher M, Shindler D, Benedict C, Shelton B. Effects of long-term enalapril therapy on cardiac structure and function in patients with left ventricular dysfunction. Results of the SOLVD echocardiography substudy. Circulation 1995; 91(10):2573–2581.
5. Doughty RN, Whalley GA, Gamble G, MacMahon S, Sharpe N. Left ventricular remodeling with carvedilol in patients with congestive heart failure due to ischemic heart disease. Australia-New Zealand Heart Failure Research Collaborative Group. J Am Coll Cardiol 1997; 29(5):1060–1066.
6. St. John Stutton M, Pfeffer MA, Moye L, Plappert T, Rouleau JL, Lamas G, Rouleau J, Parker JO, Arnold MO, Sussex B, Braunwald E. Cardiovascular death and left ventricular remodeling two years after myocardial infarction: baseline predictors and impact of long-term use of captopril: information from the Survival and Ventricular Enlargement (SAVE) trial. Circulation 1997; 96(10):3294–3299.
7. Groenning BA, Nilsson JC, Sondergaard L, Fritz-Hansen T, Larsson HB, Hildebrandt PR. Antiremodeling effects on the left ventricle during beta-blockade with metoprolol in the treatment of chronic heart failure. J Am Coll Cardiol 2000; 36(7):2072–2080.
8. Doughty RN, Whalley GA, Walsh H, Gamble G, Sharpe N. Effects of carvedilol on left ventricular remodeling in patients following acute myocardial infarction: the CAPRICORN echo substudy [abstr]. Circulation 2001; 104(suppl)(17):II-517.
9. St. John Sutton M, Pfeffer MA, Plappert T, Rouleau JL, Moye LA, Dagenais GR, Lamas GA, Klein M, Sussex B, Goldman S, et al. Quantitative two-dimensional echocardiographic measurements are major predictors of adverse cardiovascular events after acute myocardial infarction. The protective effects of captopril. Circulation 1994; 89(1):68–75.
10. Konstam MA, Rousseau MF, Kronenberg MW, Udelson JE, Melin J, Stewart D, Dolan N, Edens TR, Ahn S, Kinan D, et al. Effects of the angiotensin converting enzyme inhibitor enalapril on the long-term progression of left ventricular dysfunction in patients with heart failure. Circulation 1992; 86(2):431–438.
11. Konstam MA, Kronenberg MW, Rousseau MF, Udelson JE, Melin J, Stewart D, Dolan N, Edens TR, Ahn S, Kinan D, et al. Effects of the angiotensin converting enzyme inhibitor enalapril on the long-term progression of left ventricular dilatation in patients with asymptomatic systolic dysfunction. SOLVD (Studies of Left Ventricular Dysfunction) Investigators. Circulation 1993; 88(5 Pt 1):2277–2283.

12. Modena MG, Aveta P, Menozzi A, Rossi R. Aldosterone inhibition limits collagen synthesis and progressive left ventricular enlargement after anterior myocardial infarction. Am Heart J 2001; 141(1):41–46.

13. Rousseau MF, Konstam MA, Benedict CR, Donckier J, Galanti L, Melin J, Kinan D, Ahn S, Ketelslegers JM, Pouleur H. Progression of left ventricular dysfunction secondary to coronary artery disease, sustained neurohormonal activation and effects of ibopamine therapy during long-term therapy with angiotensin-converting enzyme inhibitor. Am J Cardiol 1994; 73(7):488–493.

14. Konstam MA, Patten RD, Thomas I, et al. Effects of losartan and captopril on left ventricular volumes in elderly patients with heart failure: results of the ELITE ventricular function substudy. Am Heart J 2000; 139(6):1081–1087.

15. Udelson JE, Antonopoulos G, Proulx G, Susses BA, Arnold JM, Kouz S, Mehta SS, Konstam MA. Comparison of long-term dual vasopeptidase inhibition with omapatrilat to ACE inhibition with lisinopril on ventricular volumes in patients with heart failure. Circulation 2000; 102:II-536.

16. Anand IS, McMurray J, Cohn JN, Konstam MA, Notter T, Quitzau K, Ruschitzka F, Luscher TF, investigator obotE. Long-term effects of darusentan on left-ventricular remodelling and clinical outcomes in the EndothelinA Receptor Antagonist Trial in Heart Failure (EARTH): randomized, double-blind, placebo-controlled trial. Lancet 2004; 364(9431):347–354.

17. Metoprolol CR/XL Randomized Intervention Trial in Congestive Heart Failure Group. Effect of metoprolol CR/XL in chronic heart failure: Metoprolol CR/XL Randomized Intervention Trial in Congestive Heart Failure (MERIT-HF). Lancet 1999; 353(9169): 2001–2007.

18. Cohn JN, Tognoni G. A randomized trial of the angiotensin-receptor blocker valsartan in chronic heart failure. N Engl J Med 2001; 345(23):1667–1675.

19. Anand IS, Florea VG, Fisher L. Surrogate end points in heart failure. J Am Coll Cardiol 2002; 39(9):1414–1421.

20. Konstam MA, Udelson JE, Anand IS, Cohn JN. Ventricular remodeling in heart failure: A credible surrogate endpoint. J Card Fail 2003; 9(5):350–353.

21. Dunn FG, Pringle SD. Sudden cardiac death, ventricular arrhythmias and hypertensive left ventricular hypertrophy. J Hypertens 1993; 11(10):1003–1010.

22. Levy D, Garrison RJ, Savage DD, Kannel WB, Castelli WP. Prognostic implications of echocardiographically determined left ventricular mass in the Framingham Heart Study. N Engl J Med 1990; 322(22):1561–1566.

23. Pollick C, Fitzgerald PJ, Popp RL. Variability of digitized echocardiography: size, source, and means of reduction. Am J Cardiol 1983; 51(3):576–582.

24. Gordon EP, Schnittger I, Fitzgerald PJ, Williams P, Popp RL. Reproducibility of left ventricular volumes by two-dimensional echocardiography. J Am Coll Cardiol 1983; 2(3):506–513.

25. Katz J, Milliken MC, Stray-Gundersen J, Buja LM, Parkey RW, Mitchell JH, Peshock RM. Estimation of human myocardial mass with MR imaging. Radiology 1988; 169(2):495–498.

26. Rehr RB, Malloy CR, Filipchuk NG, Peshock RM. Left ventricular volumes measured by MR imaging. Radiology 1985; 156(3):717–719.

27. Schapira JN, Kohn MS, Beaver WL, Popp RL. In vitro quantitation of canine left ventricular volume by phased-array sector scan. Cardiology 1981; 67(1):1–11.

28. Kasprzak JD, Paelinck B, Ten Cate FJ, Vletter WB, de Jong N, Poldermans D, Elhendy A, Bouakaz A, Roelandt JR. Comparison of native and contrast-enhanced harmonic echocardiography for visualization of left ventricular endocardial border. Am J Cardiol 1999; 83(2):211–217.

29. Mulvagh SL, DeMaria AN, Feinstein SB, Burns PN, Kaul S, Miller JG, Monaghan M, Porter TR, Shaw LJ, Villanueva FS. Contrast echocardiography: current and future applications. J Am Soc Echocardiogr 2000; 13(4):331–342.

30. St John Sutton M, Pfeffer MA, Plappert T, et al. Quantitative two-dimensional echo-cardiographic measurements are major predictors of adverse cardiovascular events after acute myocardial infarction. The protective effects of captopril. Circulation 1994; 89(1):68–75.

31. Kober L, Torp-Pedersen C, Carlsen J, Videbaek R, Egeblad H. An echocardiographic method for selecting high risk patients shortly after acute myocardial infarction, for inclusion in multi-centre studies (as used in the TRACE study). TRAndolapril Cardiac Evaluation. Eur Heart J 1994; 15(12):1616–1620.

32. Pfeffer MA, McMurray JJ, Velazquez EJ, Roulean JL, Kober L, Maggioni AP, Solomon SD, Swedberg K, Van de Werf F, White H, Leimberger JD, Henis M, Edwards S, Zelen-kofske S, Sellers MA, Califf RM. Valsartan, captopril, or both in myocardial infarction complicated by heart failure, left ventricular dysfunction, or both. N Engl J Med 2003; 349(20):1893–1906.

33. Mitchell GF, Lamas GA, Vaughan DE, Pfeffer MA. Left ventricular remodeling in the year after first anterior myocardial infarction: a quantitative analysis of contractile segment lengths and ventricular shape. J Am Coll Cardiol 1992; 19(6):1136–1144.

34. Collins HW, Kronenberg MW, Byrd BF. Reproducibility of left ventricular mass measurements by two-dimensional and M-mode echocardiography. J Am Coll Cardiol 1989; 14(3):672–676.

35. Crawford MH, Grant D, O'Rourke RA, Starling MR, Groves BM. Accuracy and repro-ducibility of new M-mode echocardiographic recommendations for measuring left ventri-cular dimensions. Circulation 1980; 61(1):137–143.

36. Himelman RB, Cassidy MM, Landzberg JS, Schiller NB. Reproducibility of quantitative two-dimensional echocardiography. Am Heart J 1988; 115(2):425–431.

37. Pietro DA, Voelkel AG, Ray BJ, Parisi AF. Reproducibility of echocardiography. A study evaluating the variability of serial echocardiographic measurements. Chest 1981; 79(1):29–32.

38. Devereux RB, Hammond IW, Lutas EM, Spitzer MC, Alderman MH, Laragh JH. Year-to-year variability of echocardiographic measurements in normal subjects. J Am Coll Cardiol 1984; 3:516A.

39. Reichek N. Standardization in the measurement of left ventricular wall mass. Two-dimensional echocardiography. Hypertension 1987; 9(2 Pt 2):II30–2.

40. Upton MT, Rerych SK, Newman GE, Bounous EP Jr, Jones RH. The reproducibility of radionuclide angiographic measurements of left ventricular function in normal subjects at rest and during exercise. Circulation 1980; 62(1):126–132.

41. Cohn PF, Levine JA, Bergeron GA, Gorlin R. Reproducibility of the angiographic left ventricular ejection fraction in patients with coronary artery disease. Am Heart J 1974; 88(6):713–720.

42. Chaitman BR, DeMots H, Bristow JD, Rosch J, Rahimtoola SH. Objective and subjec-tive analysis of left ventricular angiograms. Circulation 1975; 52(3):420–425.

43. Marshall RC, Berger HJ, Reduto LA, Gottschalk A, Zaret BL. Variability in sequential measures of left ventricular performance assessed with radionuclide angiocardiography. Am J Cardiol 1978; 41(3):531–536.

44. van Royen N, Jaffe CC, Krumholz HM, Johnson KM, Lynch PJ, Natale D, Atkinson P, Deman P, Wackers FJ. Comparison and reproducibility of visual echocardiographic and quantitative radionuclide left ventricular ejection fractions. Am J Cardiol 1996; 77: 843–850.

45. vanSonnenberg E, Neff CC, Pfister RC. Life-threatening hypotensive reactions to con-trast media administration: comparision of pharmacologic and fluid therapy. Radiology 1987; 162(1 Pt 1):15–19.

46. Katzberg RW. Renal effects of contrast media. Invest Radiol 1988; 23(suppl 1):S157–160.

47. Longmore DB, Klipstein RH, Underwood SR, Firmin DN, Hounsfield GN, Watanabe M, Bland C, Fox K, Poole-Wilson PA, Rees RS, et al. Dimensional accuracy of magnetic resonance in studies of the heart. Lancet 1985; 1(8442):1360–1362.

48. Caputo GR, Tscholakoff D, Sechtem U, Higgins CB. Measurement of canine left ventri-cular mass by using MR imaging. Am J Roentgenol 1987; 148(1):33–38.

49. Utz JA, Herfkens RJ, Heinsimer JA, Bashore T, Califf R, Glover G, Pelc N, Shimakawa A. Cine MR determination of left ventricular ejection fraction. Am J Roentgenol 1987; 148(5):839–843.

50. Van Rossum AC, Visser FC, Sprenger M, Van Eenige MJ, Valk J, Roos JP. Evaluation of magnetic resonance imaging for determination of left ventricular ejection fraction and comparison with angiography. Am J Cardiol 1988; 62(9):628–633.

51. Cranney GB, Lotan CS, Dean L, Baxley W, Bouchard A, Pohost GM. Left ventricular volume measurement using cardiac axis nuclear magnetic resonance imaging. Validation by calibrated ventricular angiography. Circulation 1990; 82(1):154–163.

52. Sechtem U, Pflugfelder PW, Gould RG, Cassidy MM, Higgins CB. Measurement of right and left ventricular volumes in healthy individuals with cine MR imaging. Radiology 1987; 163(3):697–702.

53. Rogers WJ Jr, Shapiro EP, Weiss JL, Buchalter MB, Rademakers FE, Weisfeldt ML, Zerhouni EA. Quantification of and correction for left ventricular systolic long-axis shortening by magnetic resonance tissue tagging and slice isolation. Circulation 1991; 84(2):721–731.

54. Axel L, Dougherty L. MR imaging of motion with spatial modulation of magnetization. Radiology 1989; 171(3):841–845.

55. Power TP, Kramer CM, Shaffer AL, Theobald TM, Petruolo S, Reichek N, Rogers WJ, Jr. Breath-hold dobutamine magnetic resonance myocardial tagging: normal left ventricular response. Am J Cardiol 1997; 80(9): 1203–1207.

56. Geskin G, Kramer CM, Rogers WJ, Theobald TM, Pakstis D, Hu YL, Reichek N. Quantitative assessment of myocardial viability after infarction by dobutamine magnetic resonance tagging. Circulation 1998; 98(3):217–223.

57. Kramer CM, Lima JA, Reichek N, et al. Regional differences in function within noninfarcted myocardium during left ventricular remodeling. Circulation 1993; 88(3): 1279–1288.

58. Bellenger NG, Gatehouse PD, Rajappan K, Keegan J, Firmin DN, Pennell DJ. Left ventricular quantification in heart failure by cardiovascular MR using prospective respiratory navigator gating: comparison with breath-hold acquisition. J Magn Reson Imaging 2000; 11(4):411–417.

59. Epstein FH, Wolff SD, Arai AE. Segmented k-space fast cardiac imaging using an echo-train readout. Magn Reson Med 1999; 41:609–613.

60. Nagel E, Schneider Y, Schalla S, et al. Magnetic resonance real-time imaging for the evaluation of left ventricular function. J Cardiovasc Magn Reson 2000; 2:7–14.

61. Yang PC, Kerr AB, Liu AC, Liang DH, Hardy C, Meyer CH, Macovski A, Pauly JM, Hu BS. New real-time interactive cardiac magnetic resonance imaging system complements echocardiography. J Am Coll Cardiol 1998; 32(7):2049–2056.

62. Gimbel JR, Johnson D, Levine PA, Wilkoff BL. Safe performance of magnetic resonance imaging on five patients with permanent cardiac pacemakers. Pacing Clin Electrophysiol 1996; 19(6):913–919.

63. Balzer P, Furber A, Cavaro-Menard C, Croue A, Tadei A, Geslin P, Jallet P, Le Jenne JJ. Simultaneous and correlated detection of endocardial and epicardial borders on short-axis MR images for the measurement of left ventricular mass. Radiographics 1998; 18(4):1009–1018.

64. Nachtomy E, Cooperstein R, Vaturi M, Bosak E, Vered Z, Akselrod S. Automatic assessment of cardiac function from short-axis MRI: procedure and clinical evaluation. Magn Reson Imaging 1998; 16(4):365–376.

65. Eichstadt HW, Felix R, Langer M, Gutmann ML, Dougherty FC, Huben HJ, Schmutzler H. Use of nuclear magnetic resonance imaging to show regression of hypertrophy with ramipril treatment. Am J Cardiol 1987; 59(10):98D–103D.

66. Bellenger NG, Burgess MI, Ray SG, Lahiri A, Coats AJ, Cleland JG, Pennell DJ, Pennell DJ. Comparison of left ventricular ejection fraction and volumes in heart failure by echo-

cardiography, radionuclide ventriculography and cardiovascular magnetic resonance; are they interchangeable?. Eur Heart J 1992; 13(12):1677–1683.

67. Mogelvang J, Stokholm KH, Saunamaki K, Reimer A, Stubgaard M, Thomsen C, Fritz-Hansen P, Heuriksen O. Assessment of left ventricular volumes by magnetic resonance in comparison with radionuclide angiography, contrast angiography and echocardiographyEur Heart J 1992; 13(12):1677–1683.

68. Higgins CB. Which standard has the gold? J Am Coll Cardiol 1992; 19(7):1608–1609.

69. Pattynama PM, Lamb HJ, van der Velde EA, van der Wall EE, de Roos A. Left ventricular measurements with cine and spin-echo MR imaging: a study of reproducibility with variance component analysis. Radiology 1993; 187(1):261–268.

70. Semelka RC, Tomei E, Wagner S, Mayo J, Kondo C, Suzuki J, Caputo GR, Higgins CB. Normal left ventricular dimensions and function: interstudy reproducibility of measurements with cine MR imaging. Radiology 1990; 174(3 Pt 1):763–768.

71. Semelka RC, Tomei E, Wagner S, Mayo J, Caputo G, O'Sullivan M, Parmley WW, Chatterjee K, Wolfe C, Higgins CB. Interstudy reproducibility of dimensional and functional measurements between cine magnetic resonance studies in the morphologically abnormal left ventricle. Am Heart J 1990; 119(6):1367–1376.

72. Debatin JF, Nadel SN, Paolini JF, Sostman HD, Coleman RE, Evans AJ, Beam C, Spritzer CE, Bashore TM. Cardiac ejection fraction: phantom study comparing cine MR imaging, radionuclide blood pool imaging, and ventriculography. J Magn Reson Imaging 1992; 2(2):135–142.

73. Germain P, Roul G, Kastler B, Mossard JM, Bareiss P, Sacrez A. Inter-study variability in left ventricular mass measurement. Comparison between M-mode echography and MRI. Eur Heart J 1992; 13(8):1011–1019.

74. Shapiro EP, Rogers WJ, Beyar R, Soulen RL, Zerhoani EA, Lima JA, Weiss JL. Determination of left ventricular mass by magnetic resonance imaging in hearts deformed by acute infarction. Circulation 1989; 79(3):706–711.

75. Bellenger NG, Davies LC, Francis JM, Marcus NJ, Pennell DJ. Reduction in sample size for studies of remodelling in heart failure by the use of cardiovascular magnetic resonance. J Cardiovasc Magn Reson 2000; 2:271–278.

76. The MERIT-HF Investigators. Effect of metoprolol CR/XL in chronic heart failure: Metoprolol CR/XL Randomized Intervention Trial in Congestive Heart Failure (MERIT-HF). Lancet 1999; 353:2001–2007.

77. The BEST Steering Committee. Design of the Beta-Blocker Evaluation Survival Trial (BEST). Am J Cardiol 1995; 75(17):1220–1223.

78. Schulman SP, Weiss JL, Becker LC, Gottlieb SO, Woodruff KM, Weisfeldt ML, Gerstenblith G. The effects of antihypertensive therapy on left ventricular mass in elderly patients [see comments]. N Engl J Med 1990; 322(19):1350–1356.

79. Aurigemma G, Davidoff A, Silver K, Boehmer J. Left ventricular mass quantitation using single-phase cardiac magnetic resonance imaging. Am J Cardiol 1992; 70(2): 259–262.

80. Anand IS, Florea VG, Solomon SD, Konstam MA, Udelson JE. Noninvasive assessment of left ventricular remodeling: concepts, techniques, and implications for clinical trials. J Card Fail 2002; 8(6 suppl):S452–464.

81. Reichek N, Helak J, Plappert T, Sutton MS, Weber KT. Anatomic validation of left ventricular mass estimates from clinical two-dimensional echocardiography: initial results. Circulation 1983; 67(2):348–352.

82. Devereux RB, Alonso DR, Lutas EM, Gottlieb GJ, Campo E, Sachs I, Reichek N. Echocardiographic assessment of left ventricular hypertrophy: comparison to necropsy findings. Am J Cardiol 1986; 57(6):450–458.

83. The SOLVD Investigators. Effect of enalapril on survival in patients with reduced left ventricular ejection fractions and congestive heart failure. N Engl J Med 1991; 325(5): 293–302.

84. The SOLVD Investigators. Effect of enalapril on mortality and the development of heart failure in asymptomatic patients with reduced left ventricular ejection fractions. N Engl J Med 1992; 327(10):685–691.

85. Hampton JR, van Veldhuisen DJ, Kleber FX, Cowley AJ, Ardia A, Block P, Cortina A, Cserhalmi L, FollathF, Jensen G, Kayanakis J, Lie KI, Mancia G, Skene AM. Randomized study of effect of ibopamine on survival in patients with advanced severe heart failure. Second Prospective Randomized Study of Ibopamine on Mortality and Efficacy (PRIME II) Investigators. Lancet 1997; 349(9057):971–977.

86. Pitt B, Segal R, Martinez FA, Meurers G, Cowley AJ, Thomas I, Deedwania Pc, Ney De, Snavely DB, Chang PI. Randomized trial of losartan versus captopril in patients over 65 with heart failure (Evaluation of Losartan in the Elderly Study, ELITE). Lancet 1997; 349(9054):747–752.

87. Pitt B, Poole-Wilson PA, Segal R, Martinez FA, Dickstein K, Camm AJ, Konstam MA, Riegger G, Klinger GH, Neaton J, Sharma D, Thiyagarajan B. Effect of losartan compared with captopril on mortality in patients with symptomatic heart failure: randomized trial–the Losartan Heart Failure Survival Study ELITE II. Lancet 2000; 355(9215):1582–1587.

88. Otterstad JE, Froeland G, St John Sutton M, Holme I. Accuracy and reproducibility of biplane two-dimensional echocardiographic measurements of left ventricular dimensions and function. Eur Heart J 1997; 18(3):507–513.

89. Panse PM, Muehling O, Zenovich A, Panse N, Seethamraju R, Huang Y, Swingen C, Herold MJ, Wike N, Anand I. Inter-study reproducibility of left ventricular structure and function in heart failure patients with cine MR imaging. J Am Coll Cardiol 2001; 37(suppl-A):1267.

18

Treatment of Remodeling: Inhibition of the Renin–Angiotensin–Aldosterone System

Richard D. Patten and Marvin A. Konstam
New England Medical Center, Boston, Massachusetts, U.S.A.

Left ventricular (LV) remodeling refers to alterations in ventricular mass, chamber size, and geometry that result from myocardial injury, pressure, or volume overload. The ultrastructural changes of the remodeled ventricle are the direct result of myocyte hypertrophy, fibroblast proliferation, and abnormal accumulation of the extracellular matrix. A substantial amount of experimental and clinical data now exists that supports the pivotal role of the renin–angiotensin–aldosterone system (RAAS) in these cellular processes. This chapter reviews experimental data supporting the pathophysiologic roles of both angiotensin II (Ang II) and aldosterone in ventricular remodeling. In addition, experimental data exploring the effects and mechanisms of pharmacologic agents that block components of RAAS signaling on ventricular remodeling are reviewed. Furthermore, this chapter presents pertinent clinical evidence supporting the utility of these agents to prevent remodeling and their benefits in the management of heart failure.

MOLECULAR PATHWAYS OF THE RAS

The renin–angiotensin system (RAS) consists of both circulating and local tissue compartments, the activation of which leads to the formation of Ang II, the primary hormonal mediator of the RAS (Fig. 1). In the circulating RAS, decreased renal blood flow results in the release of renin from the juxtaglomerular apparatus. Angiotensinogen released by the liver is cleaved by circulating renin resulting in the decapeptide, angiotensin I (Ang I). Ang I is then cleaved into the octapeptide, Ang II, by angiotensin converting enzyme (ACE) present in the endothelial bed. Ang II then activates its receptors causing vasoconstriction, fluid retention, and heightened sympathetic outflow (Fig. 2). The type 1 Ang II (AT_1) receptor mediates these cardiovascular effects of Ang II. Activation of the AT_1 receptor also accounts for the cell growth-promoting effects of Ang II. Other known Ang II receptors include the type 2 (AT_2), and type 4 (AT_4) receptors. The latter has been identified on endothelial cells and may promote the release of procoagulant substances such as plasminogen activator inhibitor-1 (1). Although the precise role of the AT_2 receptor remains

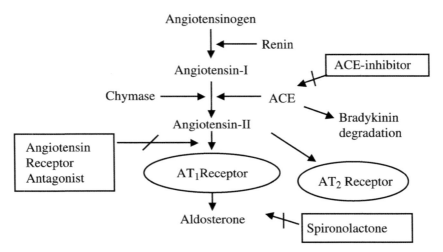

Figure 1 RAAS activation pathways: flow diagram depicting points at which pharmacologic agents have their effects. *Abbreviations*: ACE, angiotensin converting enzyme; AT_1, angiotensin II type I; AT_2, angiotensin II type 2; RAAS, renin–angiotensin–aldosterone system.

controversial (2), several studies suggest that under some circumstances, AT_2 receptor activation may counter the effects of AT_1 receptor activation (reviewed below).

Although circulating Ang II contributes importantly to the adverse cardiovascular effects of RAS activation, further exploration of RAS components elucidated the presence of an active RAS at the local tissue level. For example, Hirsch et al. (3) examined rat hearts 85 days following myocardial infarction (MI), at which time the animals exhibited evidence of heart failure. They observed a significant up regulation of ACE mRNA along with enzyme activity within the noninfarcted myocardium.

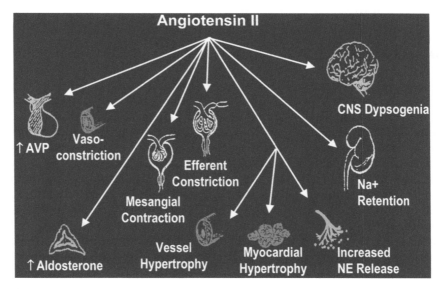

Figure 2 Cardiovascular effects of Ang II: schematic representation of the adverse effects of Ang II within the cardiovascular system. *Abbreviations*: AVP, arginine vasopressin; NE, norepinephrine; Ang II, angiotensin II.

Thus, in addition to the role of Ang II as a potent vasoconstrictor, its local formation within the myocardium may act as a growth factor in an autocrine and/or paracrine manner. In support of this notion, Sadoshima and Izumo (4,5) demonstrated the localization of Ang II within myocyte secretory granules 30 minutes following mechanical stretch. These investigators showed further that an AT_1 receptor antagonist largely blocks the increase in myocyte protein synthesis following mechanical stretch. Furthermore, Ang II stimulates collagen production and proliferation of cardiac fibroblasts in vitro via activation of the AT_1 receptor (5). To examine the effects of Ang II on cell growth in vivo, Schunkert et al. (6) directly infused Ang II into isolated rat heart preparations and observed a marked rise in protein synthesis that was blocked by an AT_1 receptor antagonist. These experimental observations support that Ang II can stimulate the myocyte and fibroblast growth responses in vitro and in vivo independent of effects on load suggesting that Ang II acts as a growth factor contributing to the cellular events that are central to the pathophysiology of ventricular remodeling.

EXPERIMENTAL MODELS AND INHIBITION OF THE RAAS

ACE-Inhibition in Animal Models of Ventricular Remodeling

Numerous experimental studies have confirmed that ACE inhibitors mitigate progressive LV remodeling in animal models of heart failure. Pfeffer et al. (7) developed a rat MI model to study ventricular remodeling and found that captopril therapy not only improved survival, but also reduced the extent of LV chamber enlargement (8,9). Figure 3 displays passive pressure–volume (P–V) relationships obtained immediately post-mortem in rats following MI. Hearts from untreated infarcted animals demonstrate significant increases in LV chamber size for a given pressure, manifest by a

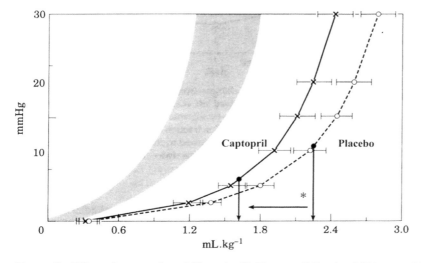

Figure 3 Effect of captopril on LV passive P–V curves following MI in rats. LV passive P–V relationships (mean ± sem) of untreated (o curve, $n = 8$) and captopril treated (× curve; $n = 8$) rats with large MIs three months following coronary ligation. Shaded area represents the volume ± 2SD of noninfarcted controls. Vertical arrows indicate operating volumes of treated and untreated rats. $^*p < 0.05$ untreated versus treated for operating volume index. *Abbreviations*: MI, myocardial infarction; LV, left ventricular; P–V, pressure–volume. *Source*: From Ref. 9.

marked rightward shift of the P–V relation. Captopril reduced the extent of LV dilatation resulting in a leftward shift of the P–V relation compared to untreated infarcts. Moreover, authors (10–12) have shown in rat MI models that myocardial hypertrophy and increased collagen deposition within the noninfarct zone are reduced by ACE-inhibition. Similarly, Michel et al. (13) showed in a rat MI model that ACE-inhibition ameliorated increases in LV mass and collagen deposition, while reducing the expression of the atrial natriuretic peptide (ANP) gene, a marker for the conversion of cardiomyocytes to the hypertrophic phenotype (14). In a mouse MI model, we demonstrated that ACE-inhibition markedly reduced myocardial hypertrophy along with ANP and collagen type I gene expression in the noninfarct zone (15) (examples are shown in Fig. 4) consistent with inhibition of myocyte growth and the production

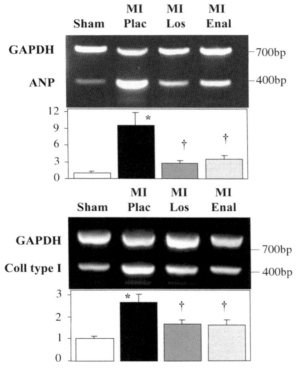

Figure 4 Gene expression within the noninfarct zone following MI in mice. Upper panel: Ethidium bromide stained agarose gel in which the levels of ANP and GAPDH mRNA were measured via multiplex rtPCR. Within the ventricular myocardium ANP expression is specific to cardiac myocytes and indicates a conversion to the hypertrophic phenotype. The bar graph below the gel image displays the mean $\pm$ SEM for the respective groups. The Plac MI group demonstrates increased ANP gene expression compared to shams, which is significantly inhibited in the MI mice, treated MI with either the AT_1 receptor antagonist, Los or the ACE-inhibitor, Enal. $^*p < 0.01$ versus shams; $^\dagger p < 0.01$ versus Plac. *Abbreviations*: Coll, collagen; MI, myocardial infarction; bp, base pairs; Plac, placebo; ANP, atrial natriuretic peptide; GAPDH, glyceraldehyde-3-phosphate dehydrogenase; Los, losartan; Enal, enalapril. Lower panel: Examples of collagen type 1 and GAPDH gene expression measured by rtPCR with bar graphs below corresponding to the mean $\pm$ sem for all four groups. $^*p < 0.01$ versus shams; $^\dagger p < 0.05$ versus Plac. *Abbreviations*: Coll, collagen; MI, myocardial infarction; bp, base pairs; Plac, placebo; ANP, atrial natriuretic peptide; GAPDH, glyceraldehyde-3-phosphate dehydrogenase; Los, losartan; Enal, enalapril. *Source*: From Ref. 15.

of type I collagen by cardiac fibroblasts. Finally, McDonald et al. (16) observed reductions in LV mass with ACE-inhibitors in a direct myocardial injury model in the dog. Similar findings have been observed in pressure overload models of ventricular hypertrophy, in which ACE-inhibitors reduce both the increases in LV mass (17–19) and interstitial collagen deposition (18,19).

Although it is apparent from the above studies that ACE-inhibitors consistently limit the extent of ventricular remodeling among a wide range of species, whether these effects are independent of load reduction is not clear. To directly explore this question, Linz et al. (18) administered the ACE-inhibitor, ramipril, at a low dose (10 µg/kg/day) that did not affect blood pressure, and high dose (1 mg/kg/day) that did lower mean arterial pressure in a rat pressure overload model. They observed equivalent decreases in LV hypertrophy and myocardial fibrosis with both doses supporting that the anti-remodeling effects of ramipril were independent of its hemodynamic effects. A similar study by Grimm et al. (20) using a rat pressure overload model demonstrated that ramipril reduced LV mass and fibrosis despite similar LV systolic pressures among the ramipril-treated and control group. Furthermore, hydralazine had no effect on LV hypertrophy or fibrosis. These data support that the effects of ACE-inhibitors in mitigating LV hypertrophy and myocardial fibrosis are, in part, independent of their load reducing effects.

Role of Tissue ACE in Ventricular Remodeling

Several lines of evidence support that tissue ACE contributes significantly to the cellular responses of the remodeling ventricle, and inhibition of tissue ACE may be important to the anti-remodeling effects of ACE-inhibitors. Following MI in the rat, myocardial tissue ACE activity and ACE mRNA levels have been shown to increase twofold. The level of both correlated well with infarct size in this study (3). Because ACE-inhibitors possess variable capacity to inhibit local, tissue ACE (21,22), some agents may not adequately suppress local increases in Ang II and, therefore, may have reduced ability to inhibit remodeling. In support of this notion, Ruzicka and colleagues (23) reported that prevention of LV hypertrophy in a rat volume overload model was dependent on inhibition of local (myocardial) ACE. In a rat MI model, Wollert (24) found that more potent inhibition of tissue ACE activity was associated with improved survival and greater reduction in both LV mass and ventricular ANP gene expression. These studies suggest therefore that the degree of tissue ACE-inhibition may be important to the prevention of remodeling in some animal models.

Contribution of Bradykinin to Antigrowth Effects of ACE-Inhibition

ACE also acts as a kininase and contributes significantly to the degradation of bradykinin at the local, tissue level. Thus, ACE-inhibition augments local levels of endogenous bradykinin. Several experimental studies have suggested that bradykinin contributes to the anti-remodeling effects of ACE-inhibitors. For example, Linz (25) demonstrated in a pressure overload model that the bradykinin type 2 (B2 kinin) receptor antagonist, HOE140, eliminated the effect of ACE-inhibitors in reducing myocardial hypertrophy. Furthermore, in their model of transmural myocardial injury in the dog, McDonald (26) showed that administration of the same B2 kinin receptor antagonist abolished the effects of ACE-inhibition in preventing hypertrophy of the noninjured myocardium. In the rat MI model, coadministration of a B2 kinin receptor antagonist blocked the decrease in interstitial fibrosis seen in animals treated with an ACE-inhibitor alone compared to placebo (27). Moreover, Yang et al. (28) demonstrated in B2 kinin receptor knockout mice that the ACE-inhibitor, ramipril, had no effect on interstitial collagen within the noninfarct zone, whereas ramipril prevented collagen accumulation in wild type mice following MI. These experimental data therefore support that augmentation of local bradykinin levels, in part, mediates the antigrowth effects of ACE-inhibitors in the remodeled ventricle (29) at the myocyte and fibroblast levels.

Non-ACE–Mediated Ang II Forming Pathways

In addition to ACE, other proteases have been identified that may contribute to the conversion of Ang I to Ang II [reviewed in (30)]. For example, chymase is a serine protease released by cardiac mast cells which converts Ang I to Ang II. In landmark studies by Urata et al., chymase was shown to contribute to cardiac Ang II formation in human cardiac membrane preparations (31,32) and in the ischemic dog heart (33). Based on these data, it is clear that ACE-inhibitors may lack the ability to completely suppress Ang II formation. As a result, great interest has been generated in agents that directly block the AT_1 receptor.

AT_1 Receptor Antagonists in Animal Models of Remodeling

It has been demonstrated in a mouse MI model that AT_1 receptor blockade limits the degree of myocardial hypertrophy and prevents the rise in ANP and collagen type I gene expression and interstitial collagen content in the noninfarct zone (15). These effects were equivalent to those obtained with an ACE-inhibitor in the same study (15) (also Fig. 4) and suggest that both agents equally inhibit the myocyte and fibroblast growth responses in this model. Many investigators utilizing the rat MI model have demonstrated that AT_1 receptor blockade also limits LV hypertrophy and increased interstitial collagen content (27,34,35). Blockade of the AT_1 receptor inhibits its activation by Ang II and leads to the unopposed stimulation of the AT_2 receptor subtype, an effect that may account for some of the observed benefits of AT_1 blockade in the rat heart following MI (27,36). In contrast to ACE-inhibitors, however, AT_1 receptor antagonists do not reduce LV remodeling in some animal models for example, in the rat MI model, we found that AT_1 receptor blockade limited the degree of LV hypertrophy, yet did not significantly alter the rise in interstitial collagen or nonmyocyte proliferation within the noninfarct zone when compared to an ACE-inhibitor (10). Van Krimpen's group demonstrated similar findings in a rat MI model (11,34). Furthermore, in dog models of direct myocardial injury (16), and mitral regurgitation (37), and in the pig model of chronic tachycardia-induced dilated cardiomyopathy (38), AT_1 receptor blockade had no effect in reducing LV hypertrophy or dilatation. The reasons for the disparate effects of ACE-inhibition versus AT_1 receptor antagonism are not clear but these data support that the non-Ang II related effects of ACE-inhibitors may be critical to their anti-remodeling effects. Moreover, the differences observed between these classes of drugs in the prevention of remodeling in animal models may be due to observed species variations in myocardial levels of Ang II, its receptors, or other components important to Ang II signaling (39,40).

The above summarized comparative studies suggest that the mechanisms by which ACE-inhibitors inhibit remodeling may be different from AT_1 receptor antagonists and support the idea that combining these agents may provide synergistic anti-remodeling effects. For example, in Dahl salt-sensitive rats, the combination of the ACE inhibitor, benazepril, with the AT_1 receptor antagonist, valsartan, improved survival, reduced heart failure markers, and inhibited LV hypertrophy to a greater degree than either agent alone (41). In a sheep model of MI, Mankad et al. (42) examined end-diastolic and end-systolic volumes using gated magnetic resonance imaging, comparing the effects of the ACE inhibitor, ramipril, alone or in combination with losartan. After eight weeks of therapy, combined ACE-inhibition and AT_1 blockade affected a greater reduction in end-diastolic and end-systolic volumes. These experimental data support, therefore, that combined ACE-inhibition and AT_1 receptor blockade may be more effective in the prevention of LV remodeling than either agent alone.

Role of the AT_2 Receptor

The precise role for AT_2 receptor activation in ventricular remodeling remains controversial (2). Earlier studies in the rat MI model demonstrated that reductions in LV end-diastolic volume, LV end-systolic volume, and myocyte cross-sectional area that resulted from treatment with an AT_1 receptor antagonist were completely reversed by simultaneous administration of an AT_2 receptor antagonist (27). This important study by Liu et al. suggested that unopposed AT_2 receptor activation might contribute to the anti-remodeling effects of AT_1 blockade in this model. The advent of transgenic mice that overexpress the AT_2 receptor and of mice harboring a targeted deletion of the AT_2 receptor gene have allowed the careful evaluation of its role in cardiac function and remodeling. Accordingly, Ichihara demonstrated that mice lacking the AT_2 receptor had greater than twice the rate of cardiac rupture one week following coronary ligation as compared to wild type littermates. This heightened risk of rupture post-MI may, in part, have been related to reduced infarct healing secondary to the reduced expression of genes encoding the extracellular matrix components, collagen type I, type III, and fibronectin (43). Furthermore, Adachi et al. (44) showed that AT_2 deficient mice had a greater incidence of heart failure within two weeks of MI that was associated with a greater reduction in fractional shortening in the early period following MI (two days) compared to wild types. Further evidence in support of a protective role for the AT_2 receptor comes from a similar study performed by Oishi et al. (45) in which AT_2 receptor knockout mice developed greater LV hypertrophy and dilatation coupled with an increase in mortality following MI compared to wild types. These data, therefore, suggest that AT_2 receptor activation plays a cardioprotective role in the post-MI heart. However, in a pressure overload model, AT_2 receptor knockout mice did not develop LV hypertrophy or systolic dysfunction that was observed in wild types (46) favoring a growth-promoting effect of AT_2 receptor activation in these models. The exact contribution of AT_2 receptor activation in pathophysiology of ventricular remodeling may be species/strain-specific and may depend on the stimulus for cardiac overload (47).

Role of Aldosterone in Ventricular Remodeling

The steroid hormone aldosterone is yet another component of the RAAS that contributes to the development of adverse ventricular remodeling in patients with LV systolic dysfunction, independent of Ang II-mediated effects. Aldosterone secretion

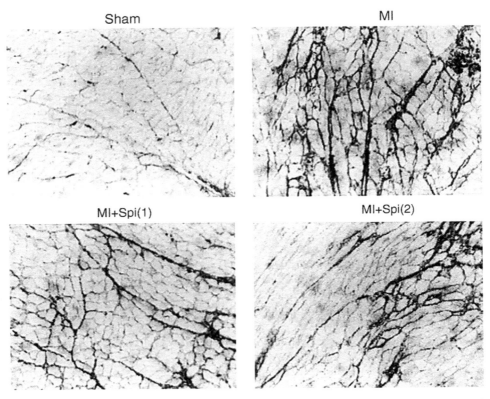

Figure 5 Effect of spironolactone on interstitial fibrosis following MI in rats. Histologic images of myocardial sections stained with the collagen specific dye, Sirius red. In this photomicrograph, collagen appears black and myocytes white. Shown are noninfarct zone sections from rats 25 days post-MI. Untreated rats (MI) demonstrate a significant increase in noninfarct zone collagen content compared to shams. Treatment with the aldosterone antagonist, spironolactone, at low [20 mg/kg/day-MI + Spi(1)] and high doses [80 mg/kg/day-MI + Spi(2)] decreased interstitial collagen content as shown. *Abbreviations*: MI, myocardial infarction. *Source*: From Ref. 54.

is partially under the control of Ang II via activation of the AT_1 receptor. However, many other factors influence the secretion of aldosterone including serum sodium and potassium concentrations, adrenocorticotropic hormone, ANP, and endothelin (48). Thus, inhibition of Ang II formation by ACE-inhibitors or blockade of AT_1 receptors may not be sufficient to inhibit the secretion of aldosterone. Similar to components of the RAS, mineralocorticoid receptors are present in the heart (49), and studies have shown that aldosterone is produced within the myocardium both in experimental heart failure models and in patients with hypertension or heart failure (50,51).

Aldosterone has been primarily implicated in the fibroblast responses within the myocardium that accompany remodeling of the left ventricle. In vitro, aldosterone induces an increase in collagen synthesis by cardiac fibroblasts (52). Young et al. demonstrated in rats that aldosterone-induced systolic hypertension is accompanied by LV hypertrophy and a 2.5-fold increase in interstitial collagen content (52a). In a rat model of renovascular hypertension, Brilla (53) reported that the aldosterone antagonist, spironolactone, prevented the increase in interstitial collagen within the myocardium. In a rat MI model, Silvestre et al. (54) demonstrated that spironolactone partially inhibits the increase in interstitial collagen content within the noninfarcted myocardium (Fig. 5). In addition, the novel mineralocorticoid receptor antagonist, eplerenone, prevented the increase in interstitial fibrosis within the noninfarct zone four weeks following MI in rats (55). Furthermore, in the dog microembolization model of LV dysfunction and heart failure, eplerenone prevented the increase in LV end-diastolic and end-systolic volumes and decrease in fractional shortening observed in placebo-treated dogs (56). Surprisingly, eplerenone reduced cardiomyocyte hypertrophy as assessed by myocyte cross-sectional area measurements, coupled with a reduction in extracellular matrix accumulation. These experimental studies therefore support a role for aldosterone in the pathophysiology of ventricular remodeling following myocardial injury or pressure overload. Given the recent clinical evidence supporting a role for aldosterone antagonism in the treatment of heart failure, the next decade will likely yield further studies examining the molecular pathways involved in the activation of mineralocorticoid receptors in the heart.

CLINICAL STUDIES

ACE Inhibitors

The effects of ACE-inhibitors on ventricular remodeling have been investigated extensively in humans (57) (Table 1). Indeed, based on their work in a rat MI model (8,9,58), Pfeffer tested the hypothesis that the ACE-inhibitor, captopril, would inhibit remodeling and improve survival after a large MI in humans. Their study in a group of patients following anterior MI showed a favorable trend for reduced LV enlargement (59) in those patients treated with captopril. In a subsequent study, Sharpe et al. (60) demonstrated that captopril administered 24 to 48 hours following a transmural MI resulted in significantly less increase in LV end-diastolic volume over three months as compared to the placebo group. The beneficial impact of captopril was later confirmed in the survival and ventricular enlargement (SAVE) trial in which captopril was administered 3 to 16 days following MI in patients with an LVEF of less than 40%. In this multicenter trial, captopril reduced the cardiovascular event rate by approximately 30% (61). In the cardiac imaging substudy of SAVE, captopril treatment limited the increase in LV chamber size observed in the placebo group (Fig. 6) (62).

The beneficial impact of ACE-inhibitors on ventricular remodeling following MI was also shown in the Gruppo Italiano per lo Studio della Sopravvivenza nell'infarto Miocardico 3 (GISSI 3) trial using lisinopril (63). In this study, patients assigned to lisinopril with LV dysfunction determined by an echocardiographic wall motion score displayed significantly less LV enlargement after six weeks of therapy (64). A withdrawal study at the end of the six-week follow-up period confirmed a sustained reduction in volumes in the lisinopril treated patients. This observation supports that ACE-inhibitors exert beneficial effects on ventricular structure, beyond the ongoing influence of load reduction.

The early administration (within 24 hours) of ACE-inhibitors in the post-MI period may inhibit ventricular remodeling in high-risk patients such as those with an anterior MI, those with a prior history of MI, or patients presenting in heart failure. Both GISSI 3 (63) and the fourth international study of infarct survival (65)

Table 1 Selected Human Studies of RAAS Antagonists in Remodeling

Reference	Clinical trial (if applicable) and agents used	Patient population	Imaging modality	Clinical outcomes (if applicable)	Remodeling effects (time period)
ACE-inhibitor trials					
Pfeffer et al. (59)	Captopril versus placebo	Acute anterior MI	Ventriculography	Improved exercise time in captopril treated patients	Trend for reduced LV dilation with captopril; significant reduction in LV dilatation in patients with persistent occlusion of the infarct related vessel (one year).
Sharpe et al. (60)	Captopril versus placebo	24 to 48 hours following Q-wave MI	Echo		Prevention of increased LVEDVI and reduced LVESVI with improved EF (three months)
St. John Sutton et al. (62)	SAVE, captopril versus placebo	Day 3 to 16 post-MI, LVEF < 40%	Echo	19% reduction in death, 25% reduction in HF hospitalization	Prevention of increases in LVED area and ES area with captopril (one year)
Bonarjee et al. (68,69)	CONSENSUS II enalapril versus placebo	<24 hours after presenting with an acute MI	Ventriculography	No difference in survival; trend favored placebo	Reduction in LV dilatation only in patients with anterior MI (six months)
Nicolosi et al. (64)	GISSI 3, lisinopril versus placebo	<24 hours after presenting with an acute MI	Echo	11% reduction in mortality, 10% reduction in mortality, and severe LV dysfunction at six weeks	Reduced LV dilatation in patients with wall motion asynergy >27% by Echo (Treatment for six weeks, Echo eval at six weeks and six months)

Schulman et al. (66)	Enalapril versus placebo	<24 hours after presenting with an acute MI	RVG and MRI		Reduction in LVEDV at one month primarily driven by reduced infarct expansion (one month)
Konstam et al. (73)	SOLVD prevention trial, enalapril versus placebo	LVEF ≤35%, asymptomatic	RVG	20% reduction in death and HF hospitalizations	Prevention of increases in LVEDVI and ESVI with enalapril (one year)
Konstam et al. (74)	SOLVD treatment trial, enalapril versus placebo	LVEF ≤35%, Class II–IV HF	RVG	26% reduction in death and HF hospitalizations	Prevention of increases in LVEDVI and ESVI with enalapril (one year)
Greenberg et al. (75)	SOLVD prevention and treatment trials, enalapril versus placebo	LVEF ≤35%, asymptomatic and Class II–IV HF	Echo	See above for both trials	Prevention of increases in LVEDV and ESV with enalapril. Reduced LV mass with enalapril (one year)
AT1 receptor antagonists					
Konstam et al. (81)	ELITE 1, captopril versus losartan	Age > 65, LVEF < 40%, Class II–IV heart failure	RVG	Reduction in mortality with losartan versus captopril (no difference in mortality in ELITE 2)	Both agents reduced LVEDVI and ESVI. Trends favored captopril. Reduction in LVEDVI was sustained after drug withdrawal with captopril only (48 weeks)
McKelvie et al. (83)	RESOLVD (candesartan/enalapril combination vs. candesartan and enalapril alone)	LVEF < 40%, Class II–IV HF, six minutes walk < 500 m	RVG	No differences in outcomes (QOL, NYHA, six minutes walk)	Reduced LVEDVI and LVESVI in combination of candesartan/enalapril versus enalapril alone (43 weeks)

(Continued)

Table 1 Selected Human Studies of RAAS Antagonists in Remodeling (*Continued*)

Reference	Clinical trial (if applicable) and agents used	Patient population	Imaging modality	Clinical outcomes (if applicable)	Remodeling effects (time period)
Wong et al. (86)	Val-HEFT (valsartan vs. placebo with background ACE-inhibitor therapy)	LVEF < 40%, Class II–IV HF	Echo	13% reduction in death and HF hospitalization	Greater reduction in LVIDd and improvement in LVEF versus placebo (24 months)
Aldosterone antagonists					
Tsutamoto et al. (92)	Spironolactone versus placebo	Non-ischemic CMP, LVEF < 45%, NYHA Class II–III	Echo		Reduced LVEDVI, LVESVI, and LV mass index compared to placebo (four months)
Kasama et al. (91)	Spironolactone versus placebo	Non-ischemic CMP, NYHA Class II–III	Echo		Reduced LVEDVI, improved LVEF compared to placebo (six months)
Hayashi et al. (90)	Spironolactone versus placebo	First anterior wall MI with successful PCI	Ventriculography		Reduced LV dilation (less increase in LVEDVI), greater improvement in LVEF compared to placebo (six months)

Abbreviations: Echo, echocardiography; "Class," New York Heart Association symptom class; CMP, cardiomyopathy; LVEF, left ventricular ejection fraction; LVEDVI, left ventricular end-diastolic volume index; LVESVI, left ventricular end-systolic volume index; LVIDd, left ventricular end-diastolic diameter; MI, myocardial infarction; MRI, magnetic resonance imaging; PCI, percutaneous coronary intervention; QOL, quality of life; RAAS, renin–angiotensin–aldosterone system; RVG, radionuclide ventriculography; SAVE, survival and ventricular enlargement; CONSENSUS II, cooperative new Scandinavian enalapril survival study II; GISSI 3, Gruppo Italiano per lo Studio della Sopravvivenza nell'infarto Miocardico 3; SOLVD, studies of left ventricular dysfunction; RESOLVD, randomized evaluation strategies for left ventricular dysfunction.

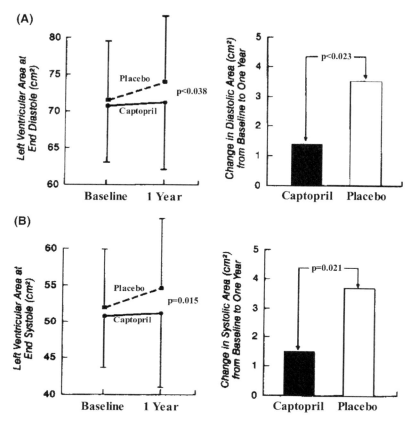

Figure 6 LV remodeling data from the SAVE trial. Graphs of LV end-diastolic (**A**) and end-systolic (**B**) areas at baseline and at one year in the two treatment groups demonstrate that captopril attenuates LV enlargement in the first year following an MI in patients with a post-MI LVEF of 40%. *Abbreviations*: LV, left ventricular; MI, myocardial infarction; SAVE, survival and ventricular enlargement; LVEF, left ventricular ejection fraction. *Source*: From Ref. 62.

showed that the administration of ACE-inhibitors early in acute MI was safe and reduced mortality at one month. In fact, in GISSI 3, the survival curves between the placebo and lisinopril group separated early (day 1). Schulman et al. (66) examined the effect of early intravenous enalaprilat followed by titrated oral enalapril compared to placebo in patients within hours of presenting with an acute MI. They demonstrated that enalapril given within hours of presenting with an acute MI caused a reduction in both LV end-diastolic and end-systolic volumes, whereas the placebo group displayed increases in these parameters at one month. The differences between the enalapril and placebo groups were even more pronounced in the group of patients with an anterior wall MI. Using magnetic resonance imaging to examine the infarct and noninfarct zones, they observed less infarct expansion in the anterior MI patients treated with enalapril. These studies support that early administration of ACE-inhibitors in acute MI reduces stretching and expansion of the infarct zone, thus preventing acute changes in LV geometry and increases in LV volumes.

Not all studies, however, have demonstrated a benefit from early administration of ACE-inhibitors in acute MI. The cooperative new Scandinavian enalapril survival study II (CONSENSUS II) was one of the first trials to explore the effects of early ACE-inhibition in patients with an acute MI (67). In this trial, patients presenting within 24 hours of symptom onset with definite or highly probable evidence of an acute MI were randomized to enalapril or placebo. Enalapril was administered intravenously typically within hours following randomization. This study was terminated early when the survival curves demonstrated no favorable effect of enalapril. In fact, there was an increase in number of deaths in the treatment arm that did not reach statistical significance. Although the reasons for this are not entirely clear, an excess of hypotensive episodes in the treatment arm, particularly among elderly patients, may have contributed to the trend for greater events in the enalapril group. However, it is important to note that in a substudy of CONSENSUS II, high risk patients (those with an anterior wall MI) treated with enalapril benefited with evidence of reduced LV dilation at one and six months post-MI (68,69).

To further address the question as to whether ACE-inhibition is beneficial in the early period (<24 hours) of an acute MI, a meta-analysis of all acute MI, ACE-inhibitor trials (including CONSENSUS II) demonstrated a modest but statistically significant benefit of ACE-inhibition within the first day of treatment leading to an absolute reduction in deaths up to 0.2% (70). Thus, available evidence supports the use of early initiation of oral ACE-inhibitors in acute MI in patients who are not hypotensive (SBP < 100) or in cardiogenic shock.

The studies of left ventricular dysfunction (SOLVD) have also demonstrated a favorable impact of ACE-inhibitors on both morbidity and mortality in asymptomatic and symptomatic patients with LV systolic dysfunction (71,72). These studies included patients with an LVEF <35% from any etiology; however, 70% to 80% of patients had LV dysfunction secondary to coronary artery disease. SOLVD consisted of two trials: patients without symptoms of heart failure were included in the prevention trial, and those with symptoms were entered into the SOLVD treatment trial. As part of a radionuclide ventriculographic substudy of the SOLVD trials, we demonstrated that the ACE-inhibitor, enalapril, inhibits progressive LV enlargement in patients with (73) and without (74) symptoms of heart failure. In both of these studies, enalapril prevented the increase in LV end-diastolic volume observed in the placebo-treated patients. Figure 7 shows examples of P–V loops obtained from patients participating in the remodeling substudy of the SOLVD prevention trial. While placebo-treated patients developed increased LV chamber size during this study (indicated by a rightward shift of the P–V loops), ACE-inhibition resulted in reduced LV volumes. This effect was sustained even after withdrawal of enalapril indicating that the observed reduction in LV volumes compared with placebo was not merely due to a decrease in load. These findings were later substantiated in the echocardiographic substudy of the combined SOLVD prevention and treatment trials (75) that included more than 300 patients. Greenberg et al. demonstrated that enalapril prevented the increase in LVEDV and LVESV observed in the placebo-treated patients. In addition, LV mass was significantly lower in the enalapril group compared to placebo after one year of treatment.

While the exact mechanism of benefit of ACE-inhibitors in the clinical setting has not been elucidated, inhibition of Ang II formation likely contributes at least partially to their anti-remodeling effects. As reviewed above, recent experimental studies suggest that augmentation of local bradykinin levels may contribute to the observed clinical benefits (36,76). Witherow and colleagues, in their study performed in humans, demonstrated that endogenous bradykinin contributes to ACE-inhibitor effects on vascular relaxation in patients with heart failure and LV systolic dysfunction (77). Thus,

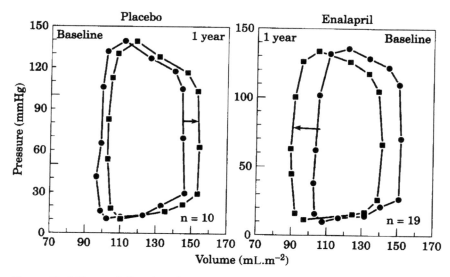

Figure 7 LV remodeling data from the SOLVD trial. Mean LVP–V relationships at baseline and one year in patients enrolled in the studies of LV dysfunction prevention trial randomized to chronic treatment with placebo or enalapril. Reproduced with permission (74). *Abbreviations*: LV, left ventricular; P–V, pressure–volume; SOLVD, studies of left ventricular dysfunction. *Source*: From Ref. 74.

augmented local bradykinin levels may also contribute partially to the clinical benefits of ACE-inhibitors in humans. Although the exact mechanisms by which ACE-inhibitors produce their clinical benefits are not entirely clear, the ability of these agents to attenuate LV remodeling most certainly contributes to their well-documented survival benefits in the treatment of heart failure. Thus, given the overwhelming evidence supporting the beneficial impact of ACE-inhibitors on ventricular remodeling in addition to morbidity and mortality, ACE-inhibitors remain the cornerstone for the treatment of heart failure associated with LV dilation.

Despite their therapeutic effectiveness, it is clear in clinical studies that ACE-inhibitors do not block completely the formation of Ang II. Petrie (78) recently demonstrated that the vasoconstrictor response to Ang I infusions in heart failure patients was only blocked completely by a combination of ACE-inhibitor and chymase inhibitor, whereas neither agent alone had any significant effect. This study suggests that chymase contributes to Ang II formation in the vasculature of heart failure patients. Moreover, Rousseau et al. (79) demonstrated that heart failure patients with LV systolic dysfunction had persistently elevated plasma Ang II levels despite therapeutic doses of ACE-inhibitors. Based on these observations, the development of Ang II receptor antagonists has allowed a means to more completely block the effects of Ang II in the cardiovascular system.

AT$_1$ Receptor Antagonists

The effect of AT$_1$ receptor antagonists on ventricular remodeling in the clinical setting is less well studied than ACE-inhibitors. The evaluation of losartan in the elderly (ELITE) trial compared the effect of losartan, an AT$_1$ antagonist, with the ACE inhibitor, captopril, on renal function in elderly patients with heart failure and systolic dysfunction (EF < 40%). While no significant difference was noted in the primary endpoint of renal function, losartan treatment was associated with better survival compared with captopril that was of borderline statistical significance (80). The radionuclide substudy of the ELITE trial evaluated LV function and volumes following 48 weeks of therapy. In this study, captopril and losartan demonstrated statistically equivalent effects on reducing LV end-diastolic and end-systolic volumes although trends favored a greater benefit by captopril compared with losartan (81) (Table 1 and Fig. 8). Because mortality was not a primary endpoint of the ELITE trial, the "ELITE 2" trial was undertaken in which the primary endpoint was total mortality. Keeping in mind the effects of both agents on remodeling in ELITE 1, it is particularly noteworthy that ELITE 2 did not show a survival benefit from losartan compared to captopril (82).

The study by McKelvie et al. (83) compared the effects of the AT_1 antagonist, candesartan, with the ACE inhibitor, enalapril, and their combination on LV remodeling as part of the randomized evaluation strategies for left ventricular dysfunction

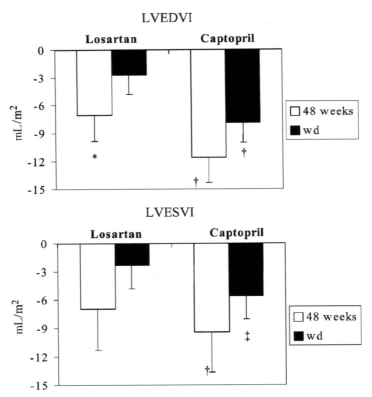

Figure 8 LV remodeling in the ELITE trial. (*Upper panel*): Change from baseline in LV end-diastolic volume (LVEDVI) after 48 weeks of therapy (white bars) and five days following drug withdrawal (solid bars) in losartan and captopril groups. *$p < 0.05$ versus baseline value. †$p < 0.01$ versus baseline value. (*Lower panel*) change from baseline in LV end-systolic volume (LVESVI) after 48 weeks of therapy (white bars) and five days following drug withdrawal (solid bars) in both treatment groups. †$p < 0.01$ versus baseline. ‡$p = 0.074$ versus baseline. *Abbreviations*: wd, withdrawal; LV, left ventricular; ELITE, evaluation of losartan in the elderly. *Source*: From Ref. 81.

(RESOLVD) pilot study. The combination of agents in this study resulted in a significant reduction in LV end-systolic volume at 43 weeks compared with enalapril treatment alone, suggesting that combination of an ACE inhibitor and AT_1 receptor antagonist in the clinical setting may be additive. Tonkon (84) reported that heart failure patients treated with irbesartan added to background ACE inhibitor therapy demonstrated reduction in LV volumes and greater LVEF compared to those treated with an ACE inhibitor alone. The valsartan heart failure trial (Val-HEFT) compared the AT_1 antagonist, valsartan, versus placebo added to standard medical therapy (including an ACE-inhibitor) in patients with LV systolic dysfunction and NYHA Class II–IV heart failure (Fig. 9). This study demonstrated that treatment with valsartan caused a significant 13% reduction in the combined endpoint of death and heart failure hospitalizations (85). In this trial, valsartan caused a modest but significant decrease in LV end-diastolic diameter and an increase in LV ejection fraction compared to placebo-treated patients (86). These studies support that AT_1 receptor antagonists favorably affect LV remodeling in patients with heart failure. However, given that withdrawal data were lacking in the RESOLVD and Val-HEFT trials, the question as to whether these effects were secondary to beneficial changes in myocardial structure or purely related to afterload reduction remains unanswered.

Aldosterone Antagonists

The results of the randomized aldactone evaluation study (RALES) (87) recently reported that treatment with low dose spironolactone resulted in a 30% reduction

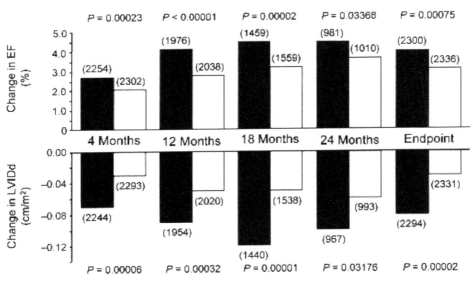

Figure 9 LV remodeling in the Val-Heft study. Shown are effects of valsartan on LV end-diastolic diameter corrected for body surface area (LVIDd/BSA) and EF compared to placebo: change from baseline to endpoint. Groups are separated by the time point from the initiation of treatment, and include those randomized to valsartan (*black bars*) and placebo (*white bars*). Bar graphs represent change expressed in absolute units for EF (%) and LVIDd/BSA (cm/m^2). *Abbreviations*: LV, left ventricular; EF, ejection fraction. *Source*: From Ref. 86.

in mortality and 35% reduction in heart failure admissions in patients with severe heart failure due to LV systolic dysfunction. This landmark clinical trial has resulted in a resurgence of interest in the potential adverse effects of aldosterone on the myocardium. A subsequent study using the novel and specific aldosterone antagonist, eplerenone, demonstrated a 17% reduction in cardiovascular deaths and a 13% reduction in cardiovascular deaths and hospitalizations after a mean follow-up of 16 months in patients with depressed LV function (EF < 40%) and heart failure complicating an acute MI (88).

In a substudy of the RALES trial, Zannad reported that patients with elevation above the median of procollagen type III N-terminal peptide (PCIIINP), a serum marker of myocardial collagen synthesis, had a greater mortality benefit from spironolactone compared with those with levels below the median (RR 0.44 vs. 1.11). Additionally, treatment with spironolactone resulted in normalization of PCIIINP levels consistent with diminished collagen synthesis within the myocardium (89). Hayashi et al. (90) performed a randomized trial of spironolactone in patients immediately following a first anterior wall MI. They found that spironolactone-treated patients developed less LV dilatation and systolic dysfunction post-MI compared to placebo-treated patients in association with markedly diminished trans-cardiac extraction of aldosterone and lower PCIIINP. In this study, a strong correlation existed between change in PCIIINP and change in LV end-diastolic volume index ($r = 0.685$). In two recent studies, spironolactone given to patients with nonischemic cardiomyopathy in a randomized fashion was associated with marked reductions in LV volumes after four and six months of therapy, respectively (91,92). These studies were performed with approximately 50% of patients taking beta-blockers and nearly all patients taking ACE-inhibitors. Whether aldosterone antagonists have important anti-remodeling effects in patients on combined ACE-inhibitor and β-blocker therapy warrants further investigation. Thus, although further studies are needed, the limited data available to date suggest that aldosterone antagonists favorably influence ventricular remodeling in the post-MI heart and in nonischemic dilated cardiomyopathy.

SUMMARY

Ventricular remodeling in the setting of heart failure and LV dilation is a progressive, maladaptive response that contributes to heightened morbidity and mortality in this syndrome. A wealth of experimental data supports that the RAAS plays a pivotal role in the cardiomyocyte and fibroblast growth responses that are central to the pathophysiology of ventricular remodeling. Given the vast amount of supportive experimental and clinical evidence, ACE-inhibitors have become the mainstay of therapy to mitigate remodeling in the overloaded heart. In the case of AT_1 receptor antagonists, experimental data support that these agents can ameliorate LV remodeling in a species-specific fashion. Clinical data, while limited, generally support that AT_1 receptor antagonists limit LV enlargement in patients with systolic dysfunction, but further studies are needed given the paucity of data compared with the large body of evidence supporting the use of ACE-inhibitors. Similarly, in the case of aldosterone antagonists, experimental studies support that they limit the fibroblast responses in the remodeling ventricle. However, only a limited number of small clinical studies support that these agents reduce remodeling in patients with heart failure, and thus, further exploration is needed.

REFERENCES

1. Kerins DM, Hao Q, Vaughan DE. Angiotensin induction of PAI-1 expression in endothelial cells is mediated by the hexapeptide angiotensin IV. J Clin Invest 1995; 96:2515–2520.
2. Opie LH, Sack MN. Enhanced angiotensin II activity in heart failure: reevaluation of the counterregulatory hypothesis of receptor subtypes. Circ Res 2001; 88:654–658.
3. Hirsch AT, Talsness CE, Schunkert H, Paul M, Dzau VJ. Tissue-specific activation of cardiac angiotensin converting enzyme in experimental heart failure. Circ Res 1991; 69:475–482.
4. Sadoshima J, Xu Y, Slayter HS, Izumo S. Autocrine release of angiotensin II mediates stretch-induced hypertrophy of cardiac myocytes. Cell 1993; 75:977–984.
5. Sadoshima J, Izumo S. Molecular characterization of angiotensin II-induced hypertrophy of cardiac myocytes and hyperplasia of cardiac fibroblasts: critical role of the AT_1 receptor subtype. Circ Res 1993; 73:413–423.
6. Schunkert H, Sadoshima J-I, Cornelius T, Kagaya Y, Weinberg EO, Izumo S, Riegger G, Lorell BH. Angiotensin II–induced growth responses in isolated adult rat hearts: evidence for load-independent induction of cardiac protein synthesis by angiotensin II. Circ Res 1995; 76:489–497.
7. Pfeffer MA, Pfeffer JM, Braunwald E. Myocardial infarct size and ventricular function in rats. Circ Res, 1979; 44:503–512.
8. Pfeffer MA, Pfeffer JM, Steinberg C, Finn P. Survival after an experimental myocardial infarction: beneficial effects of long-term therapy with captopril. Circulation 1985; 72: 406–412.
9. Pfeffer JM, Pfeffer MA, Braunwald E. Influence of chronic captopril on the infarcted left ventricle of the rat. Circ Res 1985; 57:84–95.
10. Taylor K, Patten RD, Smith JJ, Aronovitz MJ, Wight J, Salomon RN, Konstam MA. Divergent Effects of angiotensin converting enzyme inhibition and angiotensin II receptor antagonism on myocardial cellular proliferation and collagen deposition after myocardial infarction in rats. J Cardiovasc Pharmacol 1998; 31:654–660.
11. van Krimpen C, Schoemaker RG, Cleutjens JPM, Smits JFM, Struyker-Boudier HAJ, Bosman FT, Daemen MJAP. Angiotensin I converting enzyme inhibitors and cardiac remodeling. Basic Res Cardiol 1991; 86:149–157.
12. van Krimpen C, Smits JFM, Cleutjens JPM, Debets JJM, Schoemaker RG, Struyker Boudier HAJ, Bosman FT, Daemen MJAP. DNA synthesis in the non-infarcted cardiac interstitium after left coronary artery ligation in the rat: effects of captopril. J Mol Cell Cardiol 1991; 23:1245–1253.
13. Michel JB, Lattio AL, Salzmann JL, Cerol M, Philippe M, Camilleri JP, Corvol P. Hormonal and cardiac effects of converting enzyme inhibition in rat myocardial infarction. Circ Res 1988; 62:641–650.
14. Chien KR, Knowlton KU, Zhu H, Chien S. Regulation of cardiac gene expression during myocardial growth and hypertrophy: molecular studies of an adaptive physiologic response. FASEB J 1991; 5:3037–3046.
15. Patten RD, Aronovitz MJ, Einstein M, Lambert M, Pandian NG, Mendelsohn ME, Konstam MA. Effects of angiotensin II receptor blockade versus angiotensin-converting-enzyme inhibition on ventricular remodelling following myocardial infarction in the mouse. Clin Sci (Lond) 2003; 104:109–118.
16. McDonald KM, Garr MD, Carlyle PF, Francis GS, Hauer K, Hunter DW, Parish T, Stillman A, Cohn JN. Relative effects of α_1-adrenoceptor blockade, converting enzyme inhibitor therapy, and angiotensin II subtype 1 receptor blockade on ventricular remodeling in the dog. Circulation 1994; 90:3034–3046.
17. Baker KM, Chernin MI, Wixson SK, Aceto JF. Renin-angiotensin system involvement in pressure-overload cardiac hypertrophy in rats. Am J Physiol 1990; 259: H324–H332.

18. Linz W, Schaper J, Wiemer G, Albus U, Schölkens BA. Ramipril prevents left ventricular hypertrophy with myocardial fibrosis without blood pressure reduction: a one year study in rats. Br J Pharmacol 1992; 107:970–975.

19. Brilla CG, Janicki JS, Weber KT. Cardioreparative effects of lisinopril in rats with genetic hypertension and left ventricular hypertrophy. Circulation 1991; 83:1771–1779.

20. Grimm D, Kromer EP, Bocker W, Bruckschlegel G, Holmer SR, Riegger GA, Schunkert H. Regulation of extracellular matrix proteins in pressure-overload cardiac hypertrophy: effects of angiotensin converting enzyme inhibition. J Hypertens 1998; 16:1345–1355.

21. Kinoshita A, Urata H, Bumpus Fm, Husain A. Measurement of angiotensin I converting enzyme inhibition in the heart. Circ Res 1993; 73:51–60.

22. Cushman DW, Wang F, Fung WC, Harvey CM, DeForrest JM. Differentiation of angiotensin converting enzyme (ACE) inhibitors by their selective inhibition of ACE in physiologically important target organs. Am J Hypertension 1989; 2:294–306.

23. Ruzicka M, Skarda V, Leenen FHH. Effects of ACE inhibitors on circulating versus cardiac angiotensin II in volume overload-induced cardiac hypertrophy in rats. Circulation 1995; 92:3568–3573.

24. Wollert KC, Struder R, von Bülow B, Drexler H. Survival after myocardial infarction in the rat: role of tissue angiotensin-converting enzyme inhibition. Circulation 1994; 90:2457–2467.

25. Linz W, Schölkens BA. A specific B_2-bradykinin receptor antagonist HOE 140 abolishes the antihypertrophic effect of ramipril. Br J Pharmacol 1992; 105:771–772.

26. McDonald KM, Mock J, D'Aloia A, Parrish T, Hauer K, Francis G, Stillman A, Cohn JN. Bradykinin antagonism inhibits the antigrowth effect of converting enzyme inhibition in the dog myocardium after discrete transmural myocardial necrosis. Circulation 1995; 91:2043–2348.

27. Liu Y-H, Yang Y-P, Sharov VG, Nass O, Sabbah HN, Peterson E, Carretero OA. Effects of angiotensin-converting enzyme inhibitors and angiotensin II type 1 receptor antagonists in rats with heart failure: role of kinins and angiotensin II type 2 receptors. J Clin Invest 1997; 99:1926–1935.

28. Yang X-P, Liu Y-H, Mehta D, Cavasin MA, Shesely E, Xu J, Liu F, Carretero OA. Diminished cardioprotective response to inhibition of angiotensin-converting enzyme and angiotensin II type 1 receptor in B2 kinin receptor gene knockout mice. Circ Res 2001; 88:1072–1079.

29. Linz W, Wiemer G, Schölkens BA. Contribution of bradykinin to the cardiovascular effects of ramipril. J Cardiovasc Pharmacol 1993; 22:S1–S8.

30. Dell'Italia LJ, Husain A. Dissecting the role of chymase in angiotensin II formation and heart and blood vessel diseases. Curr Opin Cardiol 2002; 17:374–379.

31. Urata H, Healy B, Stewart RW, Bumpus FM, Husain A. Angiotensin II-forming pathways in normal and failing human hearts. Circ Res 1990; 66:883–890.

32. Urata H, Boehm KD, Philip A, Kinoshita A, Gabrovsek J, Bumpus FM, Husain A. Cellular localization and regional distribution of an angiotensin II-forming chymase in the human heart. J Clin Invest 1993; 91:1269–1281.

33. Noda K, Sasaguri M, Ideishi M, Ikeda M, Arakawa K. Role of locally formed angiotensin II and bradykinin in the reduction of myocardial infarct size in dogs. Cardiovasc Res 1993; 27:334–340.

34. Smits JFM, van Krimpen C, Schoemaker RG, Cleutjens JPM, Daemen MJAP. Angiotensin II receptor blockade after myocardial infarction in rats: effects on hemodynamics, myocardial DNA synthesis, and interstitial collagen content. J Cardiovasc Pharm 1992; 20:772–778.

35. Schieffer B, Wirger A, Meybrunn M, Seitz S, Holtz J, Riede UN, Drexler H. Comparative effects of chronic angiotensin-converting enzyme inhibition and angiotensin II type 1 receptor blockade on cardiac remodeling after myocardial infarction in the rat. Circulation 1994; 89:2273–2282.

36. Wollert KC, Studer R, Doerfer K, Schieffer E, Holubarsch C, Just H, Drexler H. Differential effects of kinins on cardiomyocyte hypertrophy and interstitial collagen matrix in the surviving myocardium after myocardial infarction in the rat. Circulation 1997; 95: 1910–1917.

37. Perry GJ, Wei C-C, Hankes GH, Dillon SR, Rynders P, Mukherjee R, Spinale FG, Dell'Italia LJ. Angiotensin II receptor blockade does not improve left ventricular function and remodeling in subacute mitral regurgitation in the dog. J Am Coll Cardiol 2002; 39:1374–1379.

38. Spinale FG, de Gasparo M, Whitebread S, Hebbar L, Clair MJ, Melton DM, Krombach RS, Mukherjee R, Iannini JP, O SJ. Modulation of the renin-angiotensin pathway through enzyme inhibition and specific receptor blockade in pacing-induced heart failure: I. Effects on left ventricular performance and neurohormonal systems. Circulation 1997; 96:2385–2396.

39. Dostal DE, Baker KM. The cardiac renin-angiotensin system: conceptual, or a regulator of cardiac function? Circ Res 1999; 85:643–650.

40. Gallagher AM, Bahnson TD, Yu H, Kim NN, Printz MP. Species variability in angiotensin receptor expression by cultured cardiac fibroblasts and the infarcted heart. Am J Physiol 1998; 274:H801–H809.

41. Kim S, Yoshiyama M, Izumi Y, Kawano H, Kimoto M, Zhan Y, Iwao H. Effects of combination of ACE inhibitor and angiotensin receptor blocker on cardiac remodeling, cardiac function, and survival in rat heart failure. Circulation 2001; 103:148–154.

42. Mankad S, d'Amato TA, Reichek N, McGregor WE, Lin J, Singh D, Rogers WJ, Kramer CM. Combined angiotensin II receptor antagonism and angiotensin-converting enzyme inhibition further attenuates postinfarction left ventricular remodeling. Circulation 2001; 103:2845–2850.

43. Ichihara S, Senbonmatsu T, Price E Jr, Ichiki T, Gaffney FA, Inagami T. Targeted deletion of angiotensin II type 2 receptor caused cardiac rupture after acute myocardial infarction. Circulation 2002; 106:2244–2249.

44. Adachi Y, Saito Y, Kishimoto I, Harada M, Kuwahara K, Takahashi N, Kawakami R, Nakanishi M, Nakagawa Y, Tanimoto K, Saitoh Y, Yasuno S, Usami S, Iwai M, Horiuchi M, Nakao K. Angiotensin II type 2 receptor deficiency exacerbates heart failure and reduces survival after acute myocardial infarction in mice. Circulation 2003; 107:2406–2408.

45. Oishi Y, Ozono R, Yano Y, Teranishi Y, Akishita M, Horiuchi M, Oshima T, Kambe M. Cardioprotective role of AT2 receptor in postinfarction left ventricular remodeling. Hypertension 2003; 41:814–818.

46. Senbonmatsu T, Ichihara S, Price E Jr, Gaffney FA, Inagami T. Evidence for angiotensin II type 2 receptor-mediated cardiac myocyte enlargement during in vivo pressure overload. J Clin Invest 2000; 106:R25–R29.

47. Inagami T, Senbonmatsu T. Dual effects of angiotensin II type 2 receptor on cardiovascular hypertrophy. Trends Cardiovas Med 2001; 11:324–328.

48. Mulrow PJ. Angiotensin II and aldosterone regulation. Regul Pept 1999; 80:27–32.

49. Lombes M, Oblin ME, Gasc JM, Baulieu EE, Farman N, Bonvalet JP. Immunohistochemical and biochemical evidence for a cardiovascular mineralocorticoid receptor. Circ Res 1992; 71:503–510.

50. Mizuno Y, Yoshimura M, Yasue H, Sakamoto T, Ogawa H, Kugiyama K, Harada E, Nakayama M, Nakamura S, Ito T, Shimasaki Y, Saito Y, Nakao K. Aldosterone production is activated in failing ventricle in humans. Circulation 2001; 103:72–77.

51. Yamamoto N, Yasue H, Mizuno Y, Yoshimura M, Fujii H, Nakayama M, Harada E, Nakamura S, Ito T, Ogawa H. Aldosterone is produced from ventricles in patients with essential hypertension. Hypertension 2002; 39:958–962.

52. Brilla CG, Zhou G, Matsubara L, Weber KT. Collagen metabolism in cultured adult rat cardiac fibroblasts: response to angiotensin II and aldosterone. J Mol Cell Cardiol 1994; 26:809–820.

52a. Young M, Fullerton M, Dilley R, Funder J. Mineral corticoids, hypertension, and cardiac fibrosis. J Clin Invest 1994; 93:2578–2583.

53. Brilla CG, Pick R, Tan LB, Janicki JS, Weber KT. Remodeling of the rat right and left ventricles in experimental hypertension. Circ Res 1990; 67:1355–1364.

54. Silvestre JS, Heymes C, Oubenaissa A, Robert V, Aupetit-Faisant B, Carayon A, Swynghedauw B, Delcayre C. Activation of cardiac aldosterone production in rat myocardial infarction: effect of angiotensin II receptor blockade and role in cardiac fibrosis. Circulation 1999; 99:2694–2701.

55. Delyani JA, Robinson EL, Rudolph AE. Effect of a selective aldosterone receptor antagonist in myocardial infarction. Am J Physiol Heart Circ Physiol 2001; 281:H647–654.

56. Suzuki G, Morita H, Mishima T, Sharov VG, Todor A, Tanhehco EJ, Rudolph AE, McMahon EG, Goldstein S, Sabbah HN. Effects of long-term monotherapy with eplerenone, a novel aldosterone blocker, on progression of left ventricular dysfunction and remodeling in dogs with heart failure. Circulation 2002; 106:2967–72.

57. Konstam MA. Role of angiotensin converting enzyme inhibitors in preventing left ventricular remodeling following myocardial infarction. Eur Heart J 1995; 16 (Suppl K):42–48.

58. Pfeffer MA, Pfeffer JM. Ventricular enlargement and reduced survival after myocardial infarction. Circulation 1987; 75:IV-93–IV-97.

59. Pfeffer MA, Lamas GA, Vaughan DE, Parisi AF, Braunwald E. Effect of captopril on progressive ventricular dilatation after anterior myocardial infarction. N Engl J Med 1988; 319:80–86.

60. Sharpe N, Smith H, Murphy J, Greaves S, Hart H, Gamble G. Early prevention of left ventricular dysfunction after myocardial infarction with angiotensin-converting enzyme inhibition. Lancet 1991; 337:872–876.

61. Pfeffer MA, Braunwald E, Moye' LA, Basta L, Brown EJ, Cuddy TE, Davis BR, Geltman EM, Goldman S, Flaker GC, et al., for the SAVE investigators. Effect of captopril on mortality and morbidity in patients with left ventricular dysfunction after myocardial infarction. N Eng J Med 1992; 327:669–677.

62. St. John Sutton M, Pfeffer MA, Plappert T, Rouleau JL, Moye' LA, Dagenais GR, Lamas GA, Klein M, Sussex B, Goldman S, et al., for the SAVE investigators. Quantitative two dimensional echocardiographic measurements are major predictors of adverse cardiovascular events after acute myocardial infarction, the protective effect of captopril. Circulation 1994; 89:68–75.

63. GISSI-3 Investigators. Effects of lisinopril and transdermal glyseryl trinitrate singly and together on
6-week mortality and ventricular function after acute myocardial infarction. Lancet 1994; 343:1115–1122.

64. Nicolosi GL, Latini R, Marino P, Maggioni AP, Barlera S, Franzosi MG, Geraci E, Santoro L, Tavazzi L, Tognoni G, et al. The prognostic value of predischarge quantitative two-dimensional echocardiograpic measurements and the effects of early lisinopril treatment on left ventricular structure and function after acute myocardial infarction in the GISSI-3 trial. Eur Heart J 1996; 17:1646–1656.

65. ISIS-4. A randomized factorial trial assessing early oral captopril, oral mononitrate, and intravenous magnesium sulfate in 58,050 patients with suspected acute myocardial infarction. Lancet 1995; 345:669–685.

66. Schulman SP, Weiss JL, Becker LC, Guerci AD, Shapiro EP, Chandra NC, Siu C, Flaherty JT, Coombs V, Taube JC, et al. Effect of early enalapril therapy on left ventricular function and structure in acute myocardial infarction. Am J Cardiol 1995; 76: 764–770.

67. Swedberg K, Held P, Kjekshus J, Rasmussen K, Ryden L, Wedel H. Effects of the early administration of enalapril on mortality in patients with acute myocardial infarction. Results of the cooperative new Scandinavian enalapril survival study II (CONSENSUS II). N Engl J Med 1992; 327:678–684.

68. Bonarjee VV, Omland T, Nilsen DW, Carstensen S, Berning J, Edner M, Caidahl K. Left ventricular volumes, ejection fraction, and plasma proatrial natriuretic factor (1-98) after withdrawal of enalapril treatment initiated early after myocardial infarction. CONSENSUS II multi-echo study group. Br Heart J 1995; 73:506–510.

69. Bonarjee VV, Carstensen S, Caidahl K, Nilsen DW, Edner M, Berning J. Attenuation of left ventricular dilatation after acute myocardial infarction by early initiation of enalapril therapy. CONSENSUS II multi-echo study group. Am J Cardiol 1993; 72: 1004–1049.

70. ACE Inhibitor Myocardial Infarction Collabrative Group. Indications for ACE inhibitors in the early treatment of acute myocardial infarction: systematic overview of individual data from 100,000 patients in randomized trials. Circulation 1998; 97:2202–2212.

71. The SOLVD Investigators. Effect of angiotensin converting enzyme inhibition with enalapril on survival in patients with reduced left ventricular ejection fraction and congestive heart failure: results of the treatment trial of the studies of left ventricular dysfunction (SOLVD); a randomized double blind trial. N Engl J Med 1991; 325:293–302.

72. The SOLVD Investigators. Effect of enalapril on mortality and the development of heart failure in asymptomatic patients with reduced left ventricular ejection fractions. N Engl J Med 1992; 327:685–691.

73. Konstam MA, Rousseau MF, Kronenberg MW, Udelson JE, Melin J, Stewart D, Dolan N, Edens TR, Ahn S, Kinan D, et al., the SOLVD investigators. Effects of the angiotensin converting enzyme inhibitor, enalapril, on the long-term progression of left ventricular dysfunction in patients with heart failure. Circulation 1992; 86:431–438.

74. Konstam MA, Kronenberg MW, Rousseau MF, Udelson JE, Melin J, Stewart D, Dolan N, Edens TR, Ahn S, Kinan D, et al., for the SOLVD investigators. Effects of the angiotensin converting enzyme inhibitor, enalapril, on the long-term progression of left ventricular dilatation in patients with asymptomatic systolic dysfunction. Circulation 1993; 88:2277–2283.

75. Greenberg B, Quinones MA, Koilpillai C, Limacher M, Shindler D, Benedict C, Shelton B. Effects of long-term enalapril therapy on cardiac structure and function in patients with left ventricular dysfunction: results of the SOLVD echocardiography substudy. Circulation 1995; 91:2573–2581.

76. Wollert KC, Drexler H. The kallikrein-kinin system in post-myocardial infarction cardiac remodeling. Am J Cardiol 1997; 80(suppl 3):158A–161A.

77. Witherow FN, Helmy A, Webb DJ, Fox KAA, Newby DE. Bradykinin contributes to the vasodilator effects of chronic angiotensin-converting enzyme inhibition in patients with heart failure. Circulation 2001; 104:2177–2181.

78. Petrie MC, Padmanabhan N, McDonald JE, Hillier C, Connell JMC, McMurray JJV. Angiotensin converting enzyme (ACE) and non-ACE dependent angiotensin II generation in resistance arteries from patients with heart failure and coronary heart disease. J Am Coll Cardiol 2001; 37:1056–1061.

79. Rousseau MF, Konstam MA, Benedict CR, Donckier J, Galanti L, Melin J, Kinan D, Ahn S, Ketelslegers J, Pouleur H. Progression of left ventricular dysfunction secondary to coronary artery disease, sustained neurohormonal activation and effects of ibopamine therapy during long-term therapy with angiotensin-converting enzyme inhibitor. Am J Cardiol 1994; 73:488–493.

80. Pitt B, Segal R, Martinez FA, Meurers G, Cowley AJ, Thomas I, Deedwania PC, Ney DE, Snavely DB, Chang PI. Randomized trial of losartan versus captopril in patents over 65 with heart failure (Evaluation of Losartan in the Elderly Study, ELITE). Lancet 1997; 349:747–752.

81. Konstam MA, Patten RD, Thomas I, Ramahi T, La Bresh K, Goldman S, Lewis W, Gradman A, Self KS, Bittner V, et al. Effects of losartan and captopril on left ventricular volumes in elderly patients with heart failure: results of the ELITE ventricular function substudy. Am Heart J 2000; 139:1081–1087.

82. Pitt B, Poole-Wilson PA, Segal R, Martinez FA, Dickstein K, Camm AJ, Konstam MA, Riegger G, Klinger GH, Neaton J, et al. Effect of losartan compared with captopril on

mortality in patients with symptomatic heart failure: randomised trial–the Losartan heart failure survival study ELITE II. Lancet 2000; 355:1582–1587.

83. McKelvie RS, Yusuf S, Pericak D, Avezum A, Burns RJ, Probstfield J, Tsuyuki RT, White M, Rouleau J, Latini R, et al. Comparison of candesartan, enalapril, and their combination in congestive heart failure: randomized evaluation of strategies for left ventricular dysfunction (RESOLVD) pilot study. The RESOLVD pilot study investigators. Circulation 1999; 100:1056–1064.

84. Tonkon M, Awan N, Niazi I, Hanley P, Baruch L, Wolf RA, Block AJ. A study of the efficacy and safety of irbesartan in combination with conventional therapy, including ACE inhibitors, in heart failure. Irbesartan heart failure group. Int J Clin Pract 2000; 54:11–14, 16–18.

85. Cohn JN, Tognoni G. A randomized trial of the angiotensin-receptor blocker valsartan in chronic heart failure. N Engl J Med 2001; 345:1667–1675.

86. Wong M, Staszewsky L, Latini R, Barlera S, Volpi A, Chiang YT, Benza RL, Gottlieb SO, Kleemann TD, Rosconi F, et al. Valsartan benefits left ventricular structure and function in heart failure: Val-HeFT echocardiographic study. J Am Coll Cardiol 2002; 40:970–975.

87. Pitt B, Zannad F, Remme WJ, Cody R, Castaigne A, Perez A, Palensky J, Wittes J. The effect of spironolactone on morbidity and mortality in patients with severe heart failure. Randomized aldactone evaluation study investigators [see comments]. N Engl J Med 1999; 341:709–17.

88. Pitt B, Remme W, Zannad F, Neaton J, Martinez F, Roniker B, Bittman R, Hurley S, Kleiman J, Gatlin M, the eplerenone post-acute myocardial infarction heart failure efficacy and survival study Investigators. Eplerenone, a selective aldosterone blocker, in patients with left ventricular dysfunction after myocardial infarction. N Engl J Med 2003; 348:1309–1321.

89. Zannad F, Alla F, Dousset B, Perez A, Pitt B. Limitation of excessive extracellular matrix turnover may contribute to survival benefit of spironolactone therapy in patients with congestive heart failure: insights from the randomized aldactone evaluation study (RALES). Circulation 2000; 102:2700–2706.

90. Hayashi M, Tsutamoto T, Wada A, Tsutsui T, Ishii C, Ohno K, Fujii M, Taniguchi A, Hamatani T, Nozato Y, Kataoka K, et al. Immediate administration of mineralocorticoid receptor antagonist spironolactone prevents post-infarct left ventricular remodeling associated with suppression of a marker of myocardial collagen synthesis in patients with first anterior acute myocardial infarction. Circulation 2003; 107:2559–2565.

91. Kasama S, Toyama T, Kumakura H, Takayama Y, Ichikawa S, Suzuki T, Kurabayashi M. Effect of spironolactone on cardiac sympathetic nerve activity and left ventricular remodeling in patients with dilated cardiomyopathy. J Am Coll Cardiol 2003; 41:574–581.

92. Tsutamoto T, Wada A, Maeda K, Mabuchi N, Hayashi M, Tsutsui T, Ohnishi M, Sawaki M, Fujii M, Matsumoto T. Effect of spironolactone on plasma brain natriuretic peptide and left ventricular remodeling in patients with congestive heart failure. J Am Coll Cardiol 2001; 37:1228–1233.

19

Effects of Adrenergic Blockade on Cardiac Remodeling

Henry Ooi and Wilson S. Colucci
Boston University School of Medicine, Boston Medical Center, Boston, Massachusetts, U.S.A.

INTRODUCTION

Activation of the adrenergic nervous system acts in the short term to compensate for reduced cardiac performance in heart failure and attempts at treating heart failure centered initially on augmenting adrenergic support to the failing heart. However, several lines of evidence in the early 1980s suggested that sustained activation of the sympathetic nervous system may actually play a role in the pathophysiology of disease progression: (i) plasma norepinephrine levels, a surrogate marker of sympathetic adrenergic activation, are elevated in heart failure and independently associated with a worse outcome (1–4), (ii) β-adrenergic receptor density in the failing myocardium is reduced and their pathways are partially uncoupled, findings typically found following exposure to excessive adrenergic drive (5–7), (iii) therapies increasing inotropic support to the heart were associated with increased mortality (8–12), (iv) small studies demonstrating a favorable clinical response to β-adrenergic receptor blockers (β-blockers) in patients with cardiomyopathy (13–15). These findings prompted large clinical trials which have established the efficacy of β-adrenergic receptor blockade in heart failure and have been arguably the most significant advance made in the treatment of heart failure in the past century.

Despite the success of the β-blocker trials, the complexity, long duration, and cost of these studies have rekindled interest in surrogate end points as an index of clinical efficacy. Historically, the use of surrogate end points like evidence of neurohormonal modulation and acute hemodynamic response to drug therapy has been disappointing, with little or no relation to hard health outcomes like mortality and hospitalizations (16–18). However, there is now a large body of evidence that the reversal or blunting of myocardial remodeling not only has a profound effect on cardiac function, but is closely coupled to improved clinical outcomes (see Chapter 20). Prevention of remodeling is not only a valid surrogate for health outcomes in heart failure, but appears central to the clinical efficacy of β-adrenergic receptor blockade. Furthermore, despite the lack of prospective study data, inhibiting remodeling in patients with asymptomatic or mild heart failure seems a logical target, with the aim of preventing progression of disease and hence morbidity and mortality.

The adrenergic pathways that are implicated in the pathophysiology of myocardial remodeling and dysfunction are discussed in Chapter 15 and will not be covered in detail here. We will discuss the pharmacology of β-blockers currently available and their effects on cardiac remodeling and survival in chronic heart failure and following myocardial infarction (MI). In contrast to the consistent benefit observed with β-adrenergic receptor blockade, there is little evidence for clinical benefit or a favorable impact on remodeling following α-adrenergic receptor blockade in heart failure. These studies will also be reviewed, as will the results of work involving central inhibition of adrenergic drive.

PHARMACOLOGY OF β-BLOCKERS

β-blockers are a heterogeneous group of agents which differ substantially in their properties, their biological effects, and perhaps their impact on clinical outcomes (19). β-blockers can be divided into three classes based on their β-adrenergic receptor blocking profiles, vasodilating activity, and ancillary properties. Propanolol is the classic first generation agent, a group with nonselective β_1- and β_2-receptor antagonistic properties. Propanolol is a pure competitive antagonist, with no capacity to activate β-adrenergic receptors (i.e., no intrinsic sympathomimetic activity or ISA). Some β-blockers also have the property of inverse agonism or the capacity to deactivate spontaneously active receptors (20). The second generation or "cardioselective" agents such as metoprolol, atenolol, and bisoprolol provide more selective β_1-receptor antagonism (Table 1). Metoprolol is approximately 75-fold more selective for the β_1-receptor than for the β_2-receptor, and bisoprolol approximately 120-fold more selective (19). Third generation β-blockers are distinguished by their ancillary properties which include blockade of α-adrenergic receptors, vasodilation, and antioxidant effects, which may play a role in their biological effects. Most drugs in this class are nonselective β-blockers and examples include labetalol, carvedilol, and bucindolol. Labetalol is a blocker of α_1- but not α_2-adrenergic receptors and nonselectively blocks both β_1- and β_2-adrenergic receptors. Estimates as to the selectivity of labetalol for α_1- and β-adrenergic receptors vary; the ratio of β- to α_1-adrenergic receptor selectivity has been reported to be approximately three- to 10-fold (21–23). However, labetalol has not been well studied in heart failure; in contrast, carvedilol and bucindolol have been extensively evaluated (24–31). Carvedilol is a β-blocker with mildly selective (approximately sevenfold) β_1-receptor blocking properties at lower doses, with less selectivity at higher doses (19). Carvedilol is also a potent α-blocking agent, with two- to threefold selectivity for the β- compared to the α-receptor, and powerful antioxidant (19). Bucindolol is a nonselective β-blocker with mild direct vasodilating properties (19). Bucindolol may possess ISA, or the ability for partial agonist activity at the receptor, and has significantly less inverse agonism than most of the major β-blockers in use today.

The β-blockers may differ in their antiadrenergic potential (Table 2) based on their receptor selectivity, clearance of norepinephrine, and effects on β-adrenergic receptor density. Carvedilol and bucindolol have theoretically greater antiadrenergic potential than metoprolol—they are nonselective β-blockers, do not upregulate

Table 1 Adrenergic, Receptor, Blocking Affinities of β-Blocking Agents in Human Receptors

Generation/class	Compound	$K(\beta_1)$, nmol/L	$K(\beta_2)$, nmol/L	β_1/β_2 selectivity	$K(\alpha_1)$, nmol/L	β_1/α_1 selectivity
First/nonselective	Propranolol	4.1	8.5	2.1	...	...
Second/selective β_1						
	Metoprolol	45	3345	74	...	...
	Bisoprolol	121	14390	119	...	...
Third/β-blocker–vasodilator						
	Carvedilol	4.0	29	7.3	9.4	2.4
	Bucindolol	3.6	5.0	1.4	238	66(19)[a]
	Nebivolol	5.8	1700	293	...	...

β-Receptors are the average of data from radioligand binding data in myocardial membranes and recombinant receptors, and inhibition in functional essays; α_1-receptors are from myocardial membranes. Metoprolol and bisoprolol data are from radioligand binding data in myocardial membranes. $K(\beta_1)$ is the average high-affinity dissociation constant determined from ^{125}ICYP competition curves in human ventricular myocardial membranes, dissociation constant determined from competition curves in transfected cells expressing recombinant human β_1-receptors, and dissociation constant determined from inhibition of isoproterenol-mediated stimulation of muscle contraction in preparations of nonfailing human heart. $K(\beta_2)$ is the average low-affinity dissociation constant determined from ^{125}ICYP competition curves, dissociation constant determined from simple curve fitting in transfected cells expressing recombinant human β_2-receptors, and dissociation constant determined from inhibition of isoproterenol-mediated stimulation of adenyl cyclase in membrane preparations of human heart. $K(\alpha_1)$ is the dissociation constant determined from ^{125}IBE2254 competition curves in human ventricular myocardial membranes.
[a]Based on an α_1 K_1 of 69 nmol/L in human saphenous vein ring segments (RI Tackett, personal communication, 1999).
Source: From Ref. 19.

β-receptors, lower cardiac norepinephrine levels, and in the case of carvedilol, it possesses significant additional α-receptor blocking properties. In contrast, metoprolol spares β_2-receptors, upregulates β_1-receptors, and has no effect on cardiac norepinephrine. It remains unclear what importance, if any, these additional antiadrenergic, vasodilating, and ancillary properties (e.g., antioxidant capacity) play in the clinical effects of β-blockers.

Table 2 Comparison of Antiadrenergic Properties of β-Blockers

Property	Metoprolol	Bucindolol	Carvedilol
β_1-Blockade	++	++	++
β_2-Blockade	0	+	+
α_1-Blockade	0	0	+
Downregulation of β_1-receptors	−	+	+
Lowering of cardiac norepinephrine	0	+	+
Lowering of systemic norepinephrine	0	+	0
Lowering of AT II	+	+	+

Abbreviations: AECI, anigotensin converting enzyme inhibitor; AT II, angiotensin II.
Source: From Ref. 174.

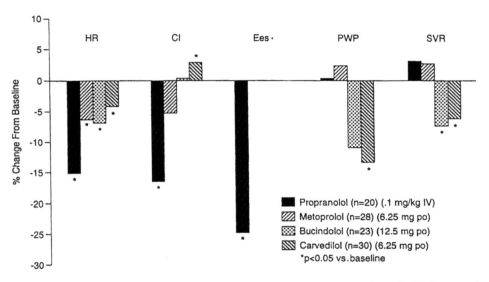

Figure 1 Comparative acute hemodynamic effects of first-, second-, and third-generation
β-blockers. Response to oral β-blockers was taken from peak effect data obtained at two or four
hours (metoprolol, carvedilol) or at four hours after administration (bucindolol). *Abbreviations*:
CI, cardiac index; Ees, end-systolic elastance, a measure of load independent of systolic function
(no data generated for orally administered β-blockers); HR, heart rate; PWP, mean pulmonary
artery wedge pressure; SVR, systemic vascular resistance. *Source*: From Ref. 37.

β-ADRENERGIC RECEPTOR BLOCKADE IN HEART FAILURE

Acute Administration

Acute administration of β-blockers slows the heart rate and decreases myocardial
contractility as a result of reducing adrenergic support for the failing heart. This
may result in a clinical worsening of heart failure when introducing β-blockers in
patients with heart failure (Fig. 1). This initial worsening of cardiac function tends
to be worse with first generation β-blockers like propanolol that cause vasoconstriction via peripheral β_2-receptor blockade, increase systemic vascular resistance, and a
fall in cardiac output (32). For these reasons, propanolol tends to be less well tolerated in patients with heart failure (33). In contrast, β_1-receptor selective blockers like
metoprolol are better tolerated (34,35); the failing heart continues to receive adrenergic support via cardiac β_2-receptors, and peripheral vascular β_2-receptors are left
relatively intact. Third generation β-blockers with vasodilating properties directly
reduce afterload, which helps to preserve cardiac output. However, there may also
be an increased risk of orthostatic hypotension with the vasodilating β-blockers.
Overall, over 90% of patients are able to tolerate introduction of second or third
generation β-blockers like metoprolol and carvedilol (28,35,36), whereas the intolerance
rate for propanolol is in the order of 20% (33).

Long-Term Effects

The long-term administration of β-blockers produces a diverse range of effects on
myocardial structure and function which become evident after several weeks of
therapy, and is usually manifest by an improvement in measures of clinical outcome.

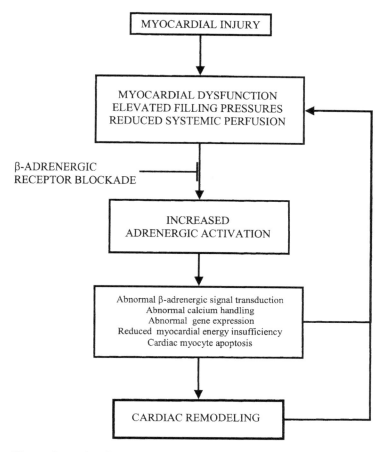

Figure 2 Role of cardiac adrenergic activation in pathophysiology of myocardial dysfunction and remodeling.

However, the precise mechanisms by which β-blockers exert their beneficial effect remain uncertain (37). There is evidence that β-blockade improves myocardial function and remodeling through a number of mechanisms (Fig. 2).

Improving Myocardial Contractility

Myocytes isolated from failing hearts demonstrate impaired contractile function which can be prevented by chronic β-blockade (38). Chronic administration of β-blockers results in an improvement in myocardial contractility through the following means:

(1) *Reversal of abnormal β-adrenergic signal transduction.* Numerous abnormalities of β-adrenergic signal transduction in heart failure have been described (39,40), including downregulation of β_1- and uncoupling of β_1- and β_2-adrenergic receptors (5–7,41–44), increase in the activity of the inhibitory G protein G_I (41,43,45), downregulation of adenyl cyclase (43,44), and an increase in expression of β-adrenergic receptor kinase (which phosphorylates and desensitizes β-adrenergic receptors) (46–48). β_1-selective blockers upregulate β_1-receptor density (49), an effect which may contribute to the improved left ventricular (LV) function seen in clinical trials.

β-blockers may also recouple uncoupled β-adrenergic receptors (50–53) and increase signaling activity in the β-adrenergic pathway downstream from the receptor (50,52,53). However, correction of β-adrenergic signaling cannot be the sole mechanism responsible for the improved LV systolic function. The nonselective β-blocker carvedilol, which does not upregulate β_1-receptor density (54), produces at least an equivalent if not greater improvement in LV systolic function and clinical outcomes compared to metoprolol (see later).

(2) *Improvement in myocardial calcium handling.* Abnormalities of calcium handling have been implicated in the pathophysiology of impaired systolic function. Downregulation of sarcoplasmic reticulum Ca^{2+} ATPase (SERCA) (55–57) and abnormal function of the cardiac calcium release channel or ryanodine receptor 2 have both been reported in failing myocardium (58). β-blockers upregulate expression of SERCA (59) toward normal and restore ryanodine receptor function and β-agonist response in failing hearts from patients undergoing transplantation (60).

(3) *Correction of abnormalities of gene expression.* The expression of several categories of genes that may modify contractility or hypertrophy is changed in heart failure. Some of these genes are described above; other changes that have been reported include alterations in atrial natriuretic peptide and α- and β-myosin heavy chain mRNA (61,62). Lowes et al. have recently shown that patients with a favorable response to β-blockade had an increase in SERCA mRNA and α-myosin heavy chain mRNA as well as a decrease in β-myosin heavy chain mRNA (59).

(4) *Improving myocardial energy efficiency.* Myocardial substrate utilization shifts to a greater dependence on free fatty acid (FFA) metabolism and a decrease in glucose use in heart failure (63,64). FFA metabolism, though providing a higher energy yield, is less energy efficient than glucose oxidation because of a greater oxygen consumption per mole of substrate. β-adrenoceptor blockade in patients with heart failure decreases myocardial oxygen consumption and improves myocardial efficiency, an improvement linked to decreased FFA metabolism and perhaps increased glucose oxidation (63–67).

(5) *Induction of a relative bradycardia and afterload reduction.* In normal hearts, there is a correlation between increased heart rate and myocyte contractility (the Treppe phenomenon). In contrast, failing hearts may have an abnormal inverse relationship between heart rate and contractility (41,68–70). Slowing the heart rate may preserve contractility; furthermore, at slower heart rates, the action potential is prolonged allowing more Ca^{2+} influx through the Na^+–Ca^{2+} exchanger and more complete relaxation and reloading of the sarcoplasmic reticulum (71).

Afterload reduction unloads the left ventricle and enhances ejection fraction (72–75). This property is theoretically a means by which the vasodilating β-blockers carvedilol and bucindolol might achieve greater improvements in LV systolic function and/or remodeling compared to nonvasodilating β-blockers. However, there is as yet no definitive evidence that this is true, and in the case of bucindolol, less clinical benefit was actually found when compared to the improvements reported in studies of carvedilol. Additionally, it remains unclear if other ancillary properties of these drugs, for example, their antioxidant properties, play a significant role in reverse remodeling.

Prevention of Cardiac Myocyte Apoptosis

An increased rate of apoptosis, or programmed cell death, has been reported in the failing heart by several groups and has been associated with worsening cardiac function and remodeling (68–70,76–78). β-adrenergic stimulation has been demonstrated to promote apoptosis in cardiac myocytes (68–70,79–81), and conversely, β-blockade prevents apoptosis and myocardial remodeling in transgenic mice overexpressing cardiac $G_{s\alpha}$, a central component of the β-adrenergic signaling pathway (82).

Reversal of Cardiac Remodeling

Clinical trials of β-blockade in heart failure have almost universally documented reversal of cardiac remodeling and improvements in mortality and hospitalizations. However, the precise relationship between changes in intrinsic myocardial contractile function or cardiac myocyte biology induced by β-blockade and reverse remodeling remains elusive (37). The primary stimulus for myocardial remodeling appears to be intrinsic myocardial dysfunction and its accompanying neurohormonal milieu, with cardiac dilatation and remodeling an attempt by the heart to normalize stroke volume and cardiac output through recruitment of preload by the Frank–Starling relationship. As myocardial contractile function improves and neurohormonal activation recedes following β-blockade, expanded cardiac volumes are no longer required to maintain stroke volume, and the heart adapts through the process of reverse remodeling.

STUDIES OF THE EFFECTS OF β-BLOCKADE ON REMODELING IN HEART FAILURE

Early studies suggesting a clinical benefit from blockade of the β-adrenergic system in patients with idiopathic dilated cardiomyopathy were performed in Sweden in the 1970s (13–15). Administration of practolol or alprenolol to small groups of patients with congestive cardiomyopathy resulted in an improvement in LV ejection fraction (LVEF), hemodynamics, clinical status, and a lower mortality when compared to historical controls. Despite the promise of these early studies, it was not until 1990 that a placebo-controlled trial demonstrated the superiority of β-blockade in improving indices of cardiac function and clinical status (31). This was followed by the first large multicenter randomized trial in 1993 examining the effects of metoprolol in patients with idiopathic dilated cardiomyopathy (31). There have now been a succession of studies establishing the beneficial effects of different β-blockers on remodeling and on clinical end points in heart failure (Table 3). It is worth noting that unlike angiotensin converting enzyme (ACE) inhibitors, the pharmacological differences within the family of β-blockers make this a heterogeneous group of drugs, a factor which may play a significant role in their clinical efficacy (19,37). Accordingly, the clinical experience using the different agents will be described individually; possible explanations for differing clinical effects are discussed.

Metoprolol

The first large multicenter randomized trial of β-blockade in heart failure was published in 1993 (34). The metoprolol in dilated cardiomyopathy (MDC) trialists administered metoprolol tartrate or placebo to 383 patients with idiopathic dilated cardiomyopathy, an ejection fraction <40% and predominantly NYHA functional

Table 3 Large Clinical Trials of β-Blockade in Heart Failure

Study	Drug	No. of patients	NYHA class	Effect on mortality	Effect on LVEF (compared to placebo, absolute EF change)
MDC	Metoprolol	383	II–III	↓ 34%[a] ($p = 0.058$)	↑ + 7% ($p < 0.0001$)
CIBIS-I	Bisoprolol	641	III–IV	↓ 20% ($p = 0.22$)	↑ + 11% ($p < 0.05$)[b]
ANZ	Carvedilol	415	II–III	↓ 23% (NS)	↑ + 5.3% ($p < 0.0001$)
United States-carvedilol	Carvedilol	1094	II–IV	↓ 65% ($p < 0.001$)	Dose related ↑
MERIT	Metoprolol	3991	II–IV	↓ 34% ($p = 0.00009$)	↑ + 10% ($p = 0.03$)[b]
CIBIS-II	Bisoprolol	2647	III–IV	↓ 34% ($p < 0.0001$)	–
BEST	Bucindolol	2708	III–IV	↓ 10% ($p = 0.13$)	–
COPERNICUS	Carvedilol	2289	IV	↓ 35% ($p = 0.0014$)	

[a]Risk of death or hospitalization.
[b]Substudy analysis.
Abbreviations: LVEF, left ventricular ejection fraction; MDC, metoprolol in dilated cardiomyopathy trial (34); CIBIS-I–II, cardiac insufficiency bisoprolol studies (94,95); ANZ, Australia–New Zealand heart failure trial (86); MERIT-heart failure, metoprolol CR/XL randomized intervention trial in congestive heart failure (35); BEST, beta blocker evaluation of survival trial (99); COPERNICUS, Carvedilol Prospective Randomized Cumulative Survival Study Group (36).

class II–III heart failure. Follow-up was for a period of 12 to 18 months. Patients receiving metoprolol had a greater increase in ejection fraction compared to placebo (13% compared to 6%), less frequently had progression of heart failure requiring cardiac transplantation, lower pulmonary capillary wedge pressure, and longer exercise time. No survival benefit was seen with metoprolol, though this may have been confounded by the greater need for transplantation in the placebo group.

Fisher et al. reported the results of a small study using metoprolol tartrate in patients with heart failure and known coronary artery disease (83). Use of a β-blocker was associated with increased ejection fraction, improved clinical status, and reduction in the number of hospitalizations. The largest study yet performed of metoprolol in heart failure has been the metoprolol CR/XL randomized intervention trial in congestive heart failure (MERIT-HF) (35). Three thousand nine hundred and ninety one patients with NYHA functional class II–IV heart failure (65% with an ischemic etiology) received a long-acting metoprolol succinate preparation or placebo over a mean follow-up time of one year. A 34% reduction in mortality was observed following treatment with metoprolol succinate. Treatment with metoprolol prevented sudden cardiac death as well as death from progressive heart failure and was equally effective across different tertiles of ejection fraction. Although remodeling end points were not examined in the main study, a magnetic resonance imaging substudy was performed in 41 patients (84). This showed decreases in LV end-diastolic and end-systolic volume indices and a significant increase in LVEF from 29% to 37% in metoprolol-treated patients compared to placebo.

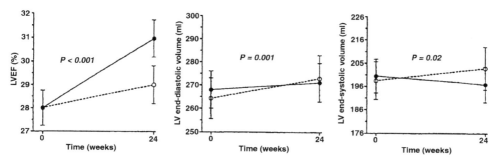

Figure 3 Changes in LVEF and LV volumes in response to metoprolol (●) versus placebo (○). Data are mean ± SEM. *Abbreviations*: LV, left ventricular; LVEF, left ventricular ejection fraction. *Source*: From Ref. 85.

The randomized evaluation of strategies for left ventricular dysfunction (RESOLVD) pilot study evaluated the effects of metoprolol CR in addition to the angiotensin receptor blocker candesartan, the ACE inhibitor enalapril, or the combination of candesartan and enalapril (85). Effects on ventricular volumes and function as well as a range of clinical and neurohumoral parameters were studied in 426 patients with NYHA functional class II–III heart failure. Following an initial 17 week period in which patients received candesartan, enalapril, or the combination, they were randomized to receive metoprolol CR or placebo and followed for an additional 24 weeks. Patients receiving placebo had no change in ejection fraction and had a significant increase in ventricular volumes (Fig. 3). In contrast, metoprolol increased LVEF and prevented the increase in end-systolic and end-diastolic volumes. Although the study was not powered to detect a difference in clinical end points, a trend was seen toward lower mortality with metoprolol, and angiotensin II and renin levels decreased in the metoprolol-treated group.

The multicenter, dose–response study to assess the effect of extended-release metoprolol succinate on cardiac remodeling in symptomatic chronic heart failure study will provide additional information on the optimal dose of metoprolol succinate (Toprol-XL) to reverse remodeling. Enrollment of up to 240 patients with NYHA class II–III heart failure has begun and will examine changes in LVEF (assessed by radionuclide ventriculography) over a range of target doses of Toprol-XL from 25 to 200 mg daily.

Carvedilol

Several large placebo-controlled trials have established the beneficial effects of carvedilol in patients with mild, moderate, and severe heart failure. The Australia/New Zealand heart failure trial recruited 415 patients with NYHA functional class II–III heart failure due to ischemic heart disease (24,86,87). After 12 months, carvedilol reduced the rate of death or hospitalization by 26%. LVEF increased from 28.4% to 33.5%, LV end-diastolic and end-systolic diameters (LVEDD and LVESD) were 1.7 and 3.2 mm smaller in the carvedilol group than in the placebo group (Fig. 4). These changes were apparent by six months of treatment and maintained to at least 12 months.

The United States Carvedilol Heart Failure Program consisted of four stratified treatment protocols in which patients with mild, moderate, or severe heart failure were

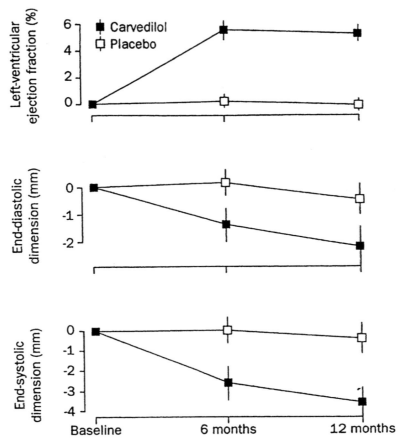

Figure 4 Changes in LVEF and LV dimensions in carvedilol and placebo groups during 12 months of follow-up. Values at six and 12 months represent mean change from baseline (SEM). *Abbreviations*: LV, left ventricular; LVEF, left ventricular ejection fraction. *Source*: From Ref. 86.

assigned to one of the four protocols on the basis of their exercise capacity (25–28,88). One thousand and ninety four patients were randomized to receive carvedilol or placebo; the study was terminated prematurely by the data and safety committee because of a highly significant 65% decrease in mortality in the group receiving carvedilol. In addition, carvedilol therapy reduced the risk of hospitalization for cardiovascular causes by 27%, and the combined risk of hospitalization or death by 38% (28). Carvedilol reduced clinical progression in patients with all grades of heart failure and dose-dependently increased LVEF and survival (Fig. 5). In the small group of patients with severe heart failure, cardiac performance improved (LVEF increased), hemodynamics and submaximal exercise tolerance improved, and of risk of hospitalization, ventricular tachyarrhythmias and death was reduced (88).

Despite the evidence of safety and benefit provided by the carvedilol and metoprolol studies, concern remained over the safety of β-blockade in patients with advanced heart failure, in whom the initial cardiodepressant effects could significantly worsen symptoms. This question was addressed in the carvedilol prospective randomized cumulative survival study (COPERNICUS) which examined the safety and efficacy of carvedilol in patients with severe heart failure (36); 2289 patients with symptoms of heart failure at rest for at least two months and an LVEF <25% were randomized to receive carvedilol or placebo. Carvedilol was well tolerated, with a lower withdrawal rate in the β-blocker group compared to placebo. A reduction in mortality of 35% and death or hospitalization of 24% was achieved, results similar to that seen in patients with lesser degrees of heart failure (Fig. 6). This benefit was apparent in all subgroups studied, even in those at highest risk.

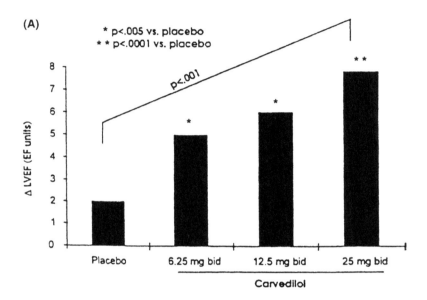

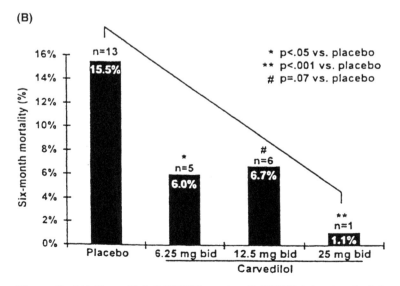

Figure 5 (**A**) Carvedilol (MOCHA protocol): LVEF data at end of six-month maintenance period as change (Δ) from baseline values. (**B**) Carvedilol (MOCHA protocol): six-month crude mortality as deaths per randomized patients × 100. *Abbreviations*: LVEF, left ventricular ejection fraction. *Source*: From Ref. 25.

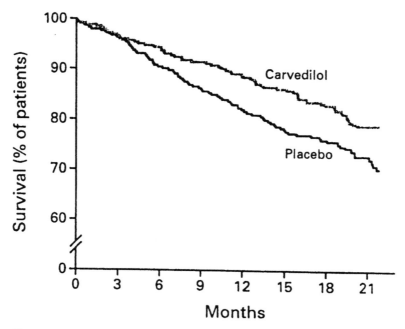

Figure 6 Kaplan–Meier analysis of time to death in the placebo and carvedilol group. The Carvedilol Prospective Randomized Cumulative Survival Study Group. *Source*: From Ref. 36.

Other small studies have demonstrated that carvedilol improves remodeling in heart failure (89–93). Quaife et al. demonstrated that after four months of therapy, carvedilol significantly increased LVEF, prevented an increase in LV end-diastolic volume, and significantly decreased LV end-systolic volume (Fig. 7) (89). Lowes et al. showed a decrease in LV wall thickness, LV mass, and mitral regurgitation as well as an increase in LVEF and an improvement in LV sphericity index (long axis length/short axis length), a measure of LV geometry, with carvedilol treatment (Fig. 8) (90).

Bisoprolol

The first cardiac insufficiency bisoprolol study (CIBIS-I) was a placebo-controlled trial of the effects of bisoprolol on mortality in 641 patients with predominantly NYHA functional class III heart failure of various etiologies (94). A 20% reduction in mortality by β-blockade was noted which was not statistically significant, mainly because of a lower than expected event rate in the control group. This was followed by the much larger CIBIS-II study which recruited 2647 patients from a similar population of patients with heart failure (95). CIBIS-II was stopped early after a mean 1.3 years of follow-up owing to a significant 34% mortality benefit from bisoprolol. Treatment effects were independent of the severity or cause of heart failure.

It is worth noting however that the annualized mortality rate in the placebo group of CIBIS-I and -II was 10.9% and 13.2%, respectively, suggesting that the patient population of these studies had only moderately severe heart failure despite their NYHA functional class III–IV classification [annual mortality rate in the predominantly NYHA class II–III MERIT-HF population and in the NYHA class IV patients of COPERNICUS was 11% and 19%, respectively (35,36)]. An

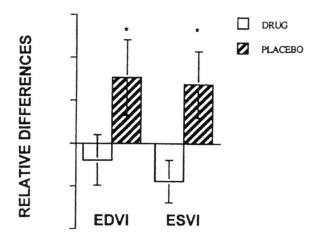

Figure 7 Mean ± SEM relative difference values [four months – baseline value (mL)] for EDVI and ESVI following carvedilol therapy. The EDVI and ESVI decreased after carvedilol therapy (*open bars*), whereas they tended to increase in the placebo group (*hatched bars*). *Abbreviations*: EDVI, end-diastolic volume index; ESVI, end-systolic volume index. *Source*: From Ref. 89.

explanation for this discordancy may be that CIBIS-I and -II recruited heavily from Eastern European countries, where interpretation of symptoms may be different compared to western Europe and the United States.

Remodeling end points have been less extensively studied for bisoprolol than for carvedilol and metoprolol (96). Dubach et al. used magnetic resonance imaging to examine the effects of bisoprolol on exercise capacity, ventricular volumes, and

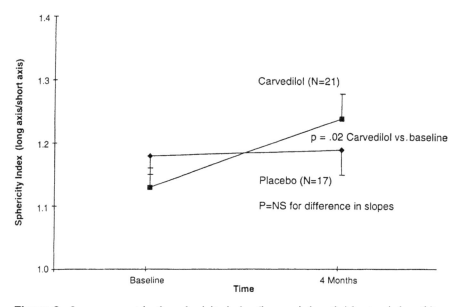

Figure 8 Improvement in the sphericity index (long axis length/short axis length), a measure of LV geometry, in carvedilol-treated patients between baseline and four months. *Abbreviation*: LV, left ventricular. *Source*: From Ref. 90.

function in 28 patients (97). A trend toward a fall in LV end-diastolic and LV end-systolic volumes was seen in patients receiving bisoprolol which became apparent between six and 12 months after initiating treatment (Table 4). Although the changes in volume were substantial (22% and 32% reduction in end-diastolic and end-systolic volumes, respectively), the changes did not reach statistical significance, most likely because of the study being insufficiently powered. An echocardiographic substudy from CIBIS-I examined remodeling and its interaction with survival (70). In 557 patients studied, LV fractional shortening (LVFS) significantly increased after five months of treatment in the bisoprolol group (+4% compared to −0.1% in the placebo group) as did LV end-systolic index. LV end-diastolic dimensions did not change significantly. An improvement in LVFS over time was significantly correlated with survival, and the authors concluded that preservation of LV function appears to play a key role in the bisoprolol-induced beneficial effects on prognosis in chronic heart failure. Despite the results of the CIBIS trials, approval for the use of bisoprolol by the Food and Drug Administration in the United States remains limited to the treatment of hypertension.

Bucindolol

Gilbert et al. published the results of the first placebo-controlled study in 1990 showing an improvement in ventricular function from β-blockade (31). Bucindolol or placebo was administered to 28 patients with idiopathic cardiomyopathy, and effects on clinical end points, hemodynamics, and LVEF were examined. LVEF increased from 26% to 35% with bucindolol, and this improvement was associated with a fall in plasma norepinephrine levels and improvements in functional class and hemodynamics. A phase 2 trial confirmed a dose-dependent improvement in LV function and a strong trend toward reduction in end-systolic and end-diastolic volumes (98). Based on the dose ranging and results of the phase 2 trial, a large phase 3 trial (the beta-blocker evaluation of survival trial—BEST) was performed on 2708 patients with advanced (NYHA functional class III–IV) heart failure (99). The study was terminated by the study's sponsors following a mean two years of follow-up when it became clear that a significant reduction in total mortality would not be achieved. Although the 10% reduction in mortality with bucindolol was not statistically significant ($p = 0.10$), the rate of death from cardiovascular causes, the combined end point of death or heart transplantation, and the rate of heart failure hospitalizations were all significantly reduced as were the plasma norepinephrine levels. LVEF showed greater improvement with bucindolol at three and at 12 months (7.3% vs. 3.3% in the placebo group). Surprisingly, a planned subgroup analysis showed an interaction effect for race, suggesting a significant survival benefit only in nonblacks.

The reason for the disappointing lack of survival benefit with bucindolol therapy despite the favorable effects on remodeling seen in prior studies remains hotly debated. The interaction of race and treatment raised the possibility of race-specific responses to pharmacological therapy. However, an analysis from the U.S. Carvedilol Program and a later study suggested an equivalent beneficial effect in blacks and nonblacks, albeit in small patient groups with a low number of events (100,101). Unique pharmacological differences of bucindolol may have played a role; bucindolol possesses the lowest amount of inverse agonism or the ability of an antagonist to inactivate active-state receptors, among the major β-blockers (19). Bucindolol is a more balanced β_1–β_2-adrenergic receptor antagonist than carvedilol and is a potent sympatholytic agent; excessive sympatholysis has been reported to be

Table 4 Magnetic Resonance Imaging of Ventricular Function Following Treatment with Bisoprolol

	Bisoprolol fumarate			Placebo			P value between groups
	Baseline	6 Months	1 Year	Baseline	6 Months	1 Year	
LVEDV (mL)	252.1 ± 78	231.4 ± 85	197.8 ± 105	200.9 ± 55	202.6 ± 64	202.8 ± 55	.14
LVESV (mL)	190.9 ± 68	191.2 ± 94	129.2 ± 85	147.7 ± 51	149.7 ± 59	152.0 ± 54	.25
EF (%)	25.0 ± 7	29.2 ± 8	36.2 ± 9[a]	27.0 ± 13	27.8 ± 10	26.2 ± 11	.33
LVSV (mL/beat)	61.1 ± 17	62.8 ± 6	68.7 ± 26	53.4 ± 27	52.9 ± 18	50.9 ± 19	.74
CO (L/min)	4.82 ± 1.8	3.66 ± 0.44	4.11 ± 1.4	3.97 ± 2.1	3.82 ± 1.7	3.36 ± 1.0	.13
PR	1772 ± 547	2127 ± 464	2289 ± 794	2728 ± 1462	2415 ± 653	2563 ± 581	.20

[a] $p < .05$ versus baseline within group.

Abbreviations: LVEDV, left ventricular end-diastolic volume; LVESV, left ventricular end-systolic volume; EF, ejection fraction; LVSV, left ventricualr stroke volume; CO, cardiac output; PR, peripheral resistance.

Source: From Ref. 97.

potentially harmful through withdrawal of adrenergic drive, especially in patients with more advanced heart failure (102–105). Of note, substantial falls in plasma catecholamine levels were achieved in BEST, suggesting a significant degree of sympatholysis. Although the patient population of BEST generally had more severe heart failure than MERIT-HF, CIBIS-II, or the U.S. Carvedilol Program (annual mortality of 17% in the placebo group), severity of heart failure in itself is unlikely to be a negative determinant of response to β-blockade given the positive results of COPERNICUS (36). Finally, bucindolol is believed by some (106–108), but not all (19,109–113), authorities to possess ISA. This property may also contribute to the neutral effects of bucindolol on heart failure through stimulation of β-adrenergic signaling with all its attendant adverse effects described previously.

Taken together, these studies confirm a beneficial effect of β-adrenergic receptor blockade on cardiac remodeling, mortality, and other clinical end points in chronic heart failure. These studies have shown a consistent improvement in LVEF, which appears to be driven by a relatively more pronounced effect of β-blockade on reducing end-systolic dimensions and volume. This is in keeping with declining LV contractile function secondary to chronic hyperadrenergic stimulation being the main impetus in progressive cardiac remodeling. Long-term β-adrenergic receptor blockade suppresses cardiac adrenergic stimulation, improves LV contractility, reduces end-systolic volumes, and secondarily prevents or reverses further cardiac dilatation. Clear clinical benefit of chronic β-blockade in heart failure has been demonstrated with carvedilol, metoprolol, and bisoprolol; despite beneficial effects of bucindolol on remodeling, a mortality benefit has yet to be established. This emphasizes the potentially significant pharmacological differences between agents and the desirability of implementing therapies with a proven track record in heart failure. While a clear relationship between dose of β-blocker and benefit has been established in several studies (25,98), concern remains that excessive adrenergic suppression may also be detrimental.

TIME COURSE OF IMPROVEMENT

The majority of studies have suggested that beneficial effects of β-blockade on remodeling and mortality are evident from three to six months of initiation of therapy (26,31,34,85,88,91,114). Krum et al. (115) in an analysis of data from COPERNICUS reported that survival curves began to diverge a mere 21 days following initiation of treatment with carvedilol in patients with severe heart failure. Hall et al. examined the time course of ventricular function improvement in patients with dilated cardiomyopathy and the long-term effects on ventricular mass and geometry (116). Patients received metoprolol or standard therapy and underwent serial echocardiography; those receiving metoprolol had an initial decline (day 1 vs. day 0) in ventricular function (increase in end-systolic volume and decrease in ejection fraction), but improved between months one and three (Fig. 9). The improvement in LVEF at three months was primarily because of a decrease in end-systolic volume. LV mass regressed at 18 months (333 ± 85 to 275 ± 53 g), and LV shape became less spherical (sphericity index 1.5 ± 0.2 to 1.7 ± 0.2).

WHO BENEFITS THE MOST FROM β-BLOCKADE?

The benefit of β-blockade in heart failure has been established virtually across every subgroup of patient studied including age, sex, concomitant therapy, blood pressure,

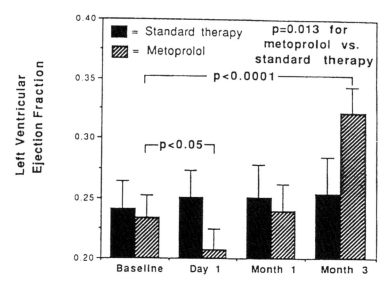

Figure 9 Changes in LVEF from baseline to day 1, month 1, and month 3 in the metoprolol and standard therapy groups. LVEF did not change in the standard therapy group. *Abbreviation*: LVEF, left ventricular ejection fraction. *Source*: From Ref. 116.

presence of diabetes mellitus, LVEF, and etiology of heart failure (19,21,28,34–36,83,86). Questions have been raised over the effect of race on response to treatment as described above; however, the available evidence and existing heart failure guidelines support the ongoing use of β-blockade across all racial groups (100,101,117).

Underlying Etiology of Heart Failure

While the clinical benefit from β-blockade is maintained in patients with idiopathic dilated or ischemic cardiomyopathy, the improvement in LVEF has been reported to be more heterogeneous in patients with ischemic heart disease (118,119). In the Australia–New Zealand Carvedilol Program which recruited only patients with chronic stable heart failure due to ischemic heart disease, carvedilol improved indices of ventricular remodeling and rate of death or hospital admission when compared to placebo, but a significant affect on functional status or episodes of worsening heart failure could not be demonstrated (24,86). A possible explanation for this finding is that myocardium from patients with coronary artery disease has a more varied composition; it may range from infarcted, nonviable scar which would not be expected to respond to pharmacological treatment, to hibernating myocardium (viable ischemic myocardium with contractile failure), or remodeled myocardium with preserved systolic function. The carvedilol hibernating reversible ischemia: marker of success trial sought to establish if improved LVEF from β-blockade with carvedilol was associated with the volume of hibernating myocardium in patients with ischemic cardiomyopathy (120). Myocardial hibernation was defined as a severe wall motion abnormality with preserved myocardial perfusion as assessed by technetium sestamibi scintigraphy. Treatment with carvedilol increased LVEF as expected; however, the role of hibernating myocardium in determining LVEF response to carvedilol was less clear cut; while hibernation status (hibernator or nonhibernator) was not associated with an increased LVEF, there was a linear relationship between volume of hibernating myocardium and the increase in LVEF with carvedilol. In addition, carvedilol slowed the decline in myocardial viability seen over the course of the study. A potentially smaller effect on remodeling in patients with reduced LV function and coronary artery disease should not be a reason for withholding β-blockers; β-blockers may have additional beneficial effects in patients with myocardial ischemia, for example, amelioration of ischemia and prevention of sudden cardiac death, which may improve symptoms and survival.

Despite the more unpredictable effect of β-blockade in patients with ischemic heart failure, the pleiotropic effects of β-blockade on the myocardium including reducing acute coronary events and malignant arrhythmias make it likely that the majority of patients with ischemic heart failure will derive a clinical benefit. The overwhelming weight of evidence from large clinical studies favors the use of β-blockade irrespective of etiology of heart failure (28,35,36,95).

Asymptomatic LV Dysfunction

Less data exist on the efficacy of β-blockade in patients with asymptomatic LV dysfunction. A retrospective analysis of the studies of left ventricular dysfunction database identified 1015 (24% of the study population) Prevention trial patients and 197 (8% of the study population) Treatment trial patients receiving β-blockers (121). β-blocker use was associated with fewer symptoms, higher ejection fractions, and a synergistic reduction in mortality with enalapril in the asymptomatic patients of the Prevention trial. Colucci et al. showed that in 366 patients with minimally symptomatic heart failure, carvedilol reduced clinical progression of heart failure and mortality; LVEF, NYHA class, and symptom scores all improved (26). The reversal of ventricular remodeling with Toprol XL study is examining the effects of metoprolol succinate on ventricular remodeling in 164 patients with asymptomatic LV dysfunction (class B, NYHA class I). Study enrollment has now closed, but results of the study are not anticipated until 2005.

While the data are relatively scarce, intuitively one would expect prevention of remodeling and heart failure progression if treatment were continued for a sufficient time period in this population. Current recommendations for treatment of heart failure suggest considering initiation of a β-blocker in patients with Stage A heart failure (patients with coronary artery disease, hypertension, diabetes, and/or valvular heart disease at high risk of developing heart failure, without known LV dysfunction) and to initiate treatment in all patients with Stage B heart failure (decreased systolic function without clinical heart failure symptoms) (117,122).

COMPARATIVE EFFECTS OF β-ADRENERGIC RECEPTOR BLOCKADE AND ACE INHIBITION ON REMODELING

The role of renin–angiotensin system (RAS) inhibition using ACE inhibitors in preventing or reversing remodeling and reducing mortality has been established in earlier studies (123–131). There is considerable cross-talk and cross-regulation between the RAS and the sympathetic nervous system (132). The release of renin from the kidney is partially under adrenergic control, and an additional mechanism by which β-blockers may prevent remodeling is through additional inhibition of the RAS. It is also conceivable that inhibition of either the RAS or adrenergic pathways alone may be sufficient to prevent remodeling, or alternatively, there may be a synergistic or an additive effect in preventing remodeling through dual RAS/adrenergic blockade.

The studies of ACE inhibition in heart failure and post-MI were predominantly performed in the 1980s and early 1990s when β-blockade in heart failure was thought to be contraindicated, and therefore had a very low percentage of patients taking β-blockers. The results of significant studies of ventricular remodeling with ACE inhibition or with β-blockade are presented in Table 5. In general, these results suggest that the magnitude of the effect of ACE inhibition on remodeling is less than that seen with β-blockade in the heart failure. However, it is worth nothing that virtually all the trials of β-blockade showing beneficial effects on remodeling and mortality were performed in patients already on standard therapy including ACE inhibitors, suggesting additive effects of β-blockade on remodeling when administered in association with an ACE inhibitor.

There are only a few data directly comparing the effects of ACE inhibition and β-blockade on cardiac remodeling in heart failure. Khattar et al. (133) randomized 57 patients with heart failure to three months of treatment with captopril or carvedilol, followed by three months of combined treatment. Carvedilol monotherapy produced significant reductions in end-systolic volume and a greater increase in ejection fraction compared to captopril monotherapy (Fig. 10). Combination therapy resulted in additional improvements in LVEF and end-systolic volume. End-diastolic volume did not change significantly with either carvedilol or captopril monotherapy. Each drug reduced LV mass, chamber sphericity, and pulmonary artery wedge pressure to a similar degree during monotherapy and combined therapy.

Preliminary results of the carvedilol ACE inhibitor remodeling in mild heart failure evaluation have recently been presented at the American College of Cardiology Scientific Sessions (2003). Five hundred and seventy two patients with predominantly NYHA class II–III heart failure were assigned to carvedilol, enalapril, or combination therapy; following an 18 month follow-up period, LV systolic volume decreased significantly in the groups receiving combination therapy or carvedilol monotherapy (by $6.0 \, mL/m^2$ and $4.7 \, mL/m^2$, respectively) but not in the group receiving enalapril monotherapy (increased by $0.6 \, mL/m^2$).

Though experience remains limited, on the basis of these studies it appears safe to initiate treatment with a β-blocker rather than an ACE inhibitor if the clinical situation dictates. Available evidence suggests that the anti-remodeling effects of β-adrenergic system suppression appear to be greater in comparison with RAS inhibition by ACE inhibitors, with an additive or synergistic effect of combined β-adrenergic/RAS suppression on remodeling. Despite this, the wealth of clinical experience, clinical trial data, safety, and ease of initiation of ACE inhibitors in heart failure continue to mandate that they be the drug of first choice. Current guidelines and standard practice remain to initiate ACE inhibitors in association with diuretics if there is evidence of elevated cardiac filling pressures, followed by initiation of β-blockers when the patient is clinically stable.

β-ADRENERGIC RECEPTOR BLOCKADE FOLLOWING MI

LV dysfunction following MI is an important indicator of prognosis. Loss of myocardial tissue and scar formation increases ventricular load and wall stress which initiates a process of progressive remodeling involving not only the scar but also remote myocardium. Infarct expansion occurs rapidly and is accompanied by

Table 5 Effects of ACE Inhibition and β-Blockade on Indices of Myocardial Remodeling

| | | | | Change compared to baseline | | | | | |
| | | | | EDV (mLs) | | ESV (mLs) | | LVEF (EF%) | |
Study	No. of patients	Patient population	Follow-up period	Placebo	Drug	Placebo	Drug	Placebo	Drug
SOLVD (enalapril)	301	Chronic heart failure	One year	+10	+1	+8	−1	0	+1
SAVE (captopril)	512	↓LVEF post-MI	One year	+3.5	+1.4	+3.7	+1.5	−4.4	0
ANZ (carvedilol)	123	Chronic heart failure	One year	+19	−9	+13	−15	−1.4	+5.5
RESOLVD (metoprolol)	426	Chronic heart failure	24 weeks	+9	+3	+6	−4	+1	+3
Dubach et al. (bisoprolol)	28	Chronic heart failure	One year	+1.9	−54.3	+4.3	−61.7	−0.8	+11.2

Abbreviations: EDV, end-diastolic volume; ESV, end-systolic volume; LVEF, left ventricular ejection fraction; SOLVD, studies of left ventricular dysfunction (131); SAVE, survival and ventricular enlargement trial (175); ANZ, Australia–New Zealand heart failure trial (86); RESOLVD, randomized evaluation of strategies for left ventricular dysfunction (85).

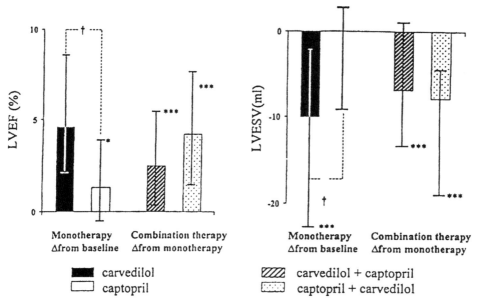

Figure 10 Median changes in LVEF and LVESV with carvedilol and captopril, from baseline to monotherapy, and monotherapy to combination therapy. *Abbreviations*: LVEF, left ventricular ejection fraction; LVESV, left ventricular end-systolic volume. *Source*: From Ref. 133.

changes in remote myocardium similar to that seen in nonischemic cardiomyopathies. The efficacy of ACE inhibitors in attenuating remodeling and improving survival post-MI has been well established (124,125,134,135). A considerable literature also exists on the role of β-blockade in reducing mortality and ischemic events post-MI (136), and β-blockers are routinely prescribed post-MI in the absence of contraindications. However, relatively little is known on the effects of β-blockade on remodeling post-MI.

In an animal model of MI, carvedilol was administered to rats following ligation of the left coronary artery (137). Although carvedilol attenuated the fall in LV function and preserved intrinsic myocardial contractility, LV dilatation was not prevented and early mortality was actually increased with carvedilol. The authors speculated that the early mortality may have been related to excessive hypotension. Basu et al. (138) administered carvedilol intravenously to patients post-MI and continued oral treatment for a further six months. Carvedilol was well tolerated, and in a subgroup with LVEF <45%, remodeling was attenuated. In a follow-up study by the same group, 49 patients with an LVEF <45% post-MI received carvedilol for a total of six months (139). Approximately 50% of the patients had congestive heart failure at presentation, traditionally considered a contraindication to β-blockade. Carvedilol reduced wall thickness and LV mass compared to placebo and preserved LV geometry, suggesting that early administration of a β-blocker in the post-MI period was effective in attenuating LV remodeling in patients with LV dysfunction, even in the presence of symptomatic congestion. The carvedilol postinfarct survival control in LV dysfunction (CAPRICORN) study randomized 1959 patients with an LVEF ≤40% post-MI and symptoms or signs of heart failure to carvedilol or placebo in addition to standard therapy including ACE inhibitors (140). All-cause mortality was reduced by 23% in the carvedilol group compared to placebo; in a substudy analysis involving 129 patients, Doughty et al. (141) reported that improvements in stroke volume and LVEF were evident within one month in patients receiving carvedilol. At six months, there were significant reductions in LV end-diastolic and end-systolic volumes with carvedilol. These results suggest an early functional improvement from carvedilol possibly related to its vasodilating properties, with reverse remodeling effects evident by six months. The results of the study emphasized the high-risk nature of the patient with LV dysfunction post-MI and the additive benefit of β-blockade and ACE inhibition. Given the results of CAPRICORN and the smaller studies, β-blockade post-MI appears to be safe and effective in attenuating remodeling and improving outcomes, even in the presence of symptomatic LV dysfunction.

WHICH β-ADRENERGIC RECEPTOR BLOCKER TO USE?

The success of β-blockers in the treatment of heart failure and the heterogeneity of the class of drugs as a whole has generated considerable interest in identifying the "ideal" agent for treating heart failure (109,142,143). Most debate has centered on the relative merits of carvedilol, a third generation, nonselective, β/α_1-adrenergic receptor blocker with antioxidant properties and the second generation, β_1-"cardioselective" receptor blocker metoprolol. Is combined β_1/β_2-receptor blockade more effective than β_1-receptor blockade alone? Are the α_1-receptor blocking/vasodilating and antioxidant properties of carvedilol clinically important?

There is experimental and theoretical reason to suggest that all three types of adrenergic receptors present in the cardiac myocyte—β_1, β_2, and α_1—might contribute to the pathophysiology of myocardial remodeling and failure (144,145). While the effects of the β_1-receptor and its associated G_s-coupled signaling pathway appear to be primarily deleterious (144,146–148), the effects of β_2- and α_1-receptors are more complex, exerting effects which might be either detrimental or beneficial. Overexpression of the β_1-receptor in mouse myocardium is detrimental, leading to myocardial failure even at low levels of expression (147). In contrast, when β_2-receptors are expressed at similar low levels there is improved myocardial function (149). However, when expressed at higher levels, β_2-receptors become detrimental, with effects not dissimilar to the β_1-receptor (149). Likewise, overexpressing the α_1-receptor in the mouse myocardium may yield diverse phenotypes ranging from hypertrophy without failure (150) to dilated failure (151). Thus, based solely on in vitro and animal studies it is difficult to predict whether it would be more or less advantageous to block all three receptor types in the myocardium.

As carvedilol not only blocks β_1-, β_2-, and α_1-receptors but also has significant antioxidant properties, it has been suggested that it might provide more beneficial clinical effects than β_1-selective blockers such as metoprolol or bisoprolol (19,142). A number of relatively small studies in animals and in humans have attempted to address this question (Table 6) (54,152–156). Results from these studies have been conflicting. None have been powered to demonstrate a difference in mortality, and some (54,152,154,156) but not all (153,155) have shown an advantage of carvedilol over metoprolol in terms of remodeling or other end points (Figs. 11 and 12). It is worth noting that of the four studies directly comparing the effects of metoprolol and carvedilol on remodeling (152–155), the two studies (152,154) showing a superiority of carvedilol over metoprolol in improving LVEF had the longest period of follow-up (~12 months); this suggests that a longer duration of treatment may be

Table 6 Trials Comparing Metoprolol and Carvedilol in Heart Failure

Study	Study design	Duration (months)	No. of patients	Mean dose metoprolol/ carvedilol reached (mg per day)	Effects on remodeling
Di Lenarda et al. (152)	Open label, parallel[a]	12	30	142/74	Carvedilol ↓ EDV, ESV, ↑ LVEF compared to metoprolol
Kukin et al. (153)	Randomized	6	67	83%/81% reached target dose (50 mg daily of each drug)	No difference between groups (both groups ↑ LVEF)
Sanderson et al. (155)	Randomized, double-blind	3	51	N/A (target dose 100 mg/50 mg daily)	Similar ↑ LVEF between groups; carvedilol ↓ EDV compared to metoprolol.
Metra et al. (154)	Randomized, double-blind	12	150	125/49	Carvedilol ↑ LVEF compared to metoprolol, no difference in EDV/ESV between groups

[a]Poor responders to metoprolol treatment were randomized to continue metoprolol or cross over to carvedilol.
Abbreviations: LVEF, left ventricular ejection fraction; EDV, left ventricular end-diastolic volume; ESV, left ventricular end-systolic volume.

required to document the relatively small differential in effects on LVEF. It is also worth noting that in the study by Kukin et al. (153), parallel declines in plasma thiobarbituric acid-reactive substances, a marker of oxidative stress, were seen in patients receiving carvedilol or metoprolol; this implies that carvedilol exerts no additional clinical antioxidant effects compared to therapy with metoprolol. Packer et al. (157) performed a meta-analysis of 19 studies comparing the effects of carvedilol and metoprolol; this analysis showed a greater increase in LVEF with carvedilol compared to metoprolol. Furthermore, this analysis again suggested that the magnitude of the effect of β-blockade on ejection fraction appeared to be greater in patients with nonischemic cardiomyopathy compared to those with coronary artery disease.

The carvedilol or metoprolol European trial (COMET) was designed to directly compare the effects of carvedilol and metoprolol on mortality and morbidity in 3029 patients with mild to severe chronic heart failure (158). While COMET had by far the longest follow-up period (mean duration of 58 months) of all the major heart failure trials and showed a 17% mortality reduction in favor of carvedilol, this study has been criticized for several reasons: firstly, the preparation of metoprolol used, the short-acting tartrate preparation, was not that used in the landmark MERIT-HF trial (long-acting metoprolol succinate); secondly, the average dose of

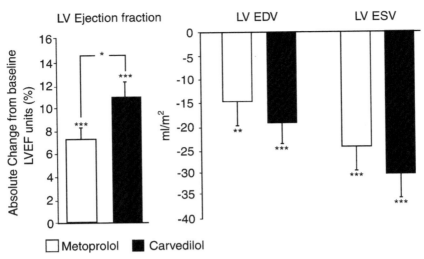

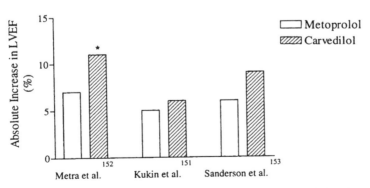

Figure 11 Greater improvement in LV remodeling following treatment with carvedilol compared to metoprolol. Absolute changes (mean ± SEM) from baseline in LVEF, EDV, and ESV after treatment with metoprolol or carvedilol for 13 to 15 months. Symbols immediately above or below the columns designate significance of differences from baseline; symbols between columns designate significance of differences between groups. $^*p < 0.05$; $^{**}p < 0.01$; $^{***}p < 0.001$. *Abbreviations*: LVEF, left ventricular ejection fraction; EDV, end-diastolic volume; ESV, end-systolic volume. *Source*: From Ref. 154.

metoprolol tartrate used (85 mg/day) was considerably less than the dose of metoprolol succinate achieved in MERIT-HF (159 mg/day, 100 mg tartrate preparation biologically equivalent to approximately 150 mg succinate preparation); finally, the doses of carvedilol and metoprolol used (25 mg twice daily and 50 mg twice daily, respectively) may not have provided equivalent degrees of β-blockade, judging by

Figure 12 Studies directly comparing the effects of metoprolol and carvedilol on remodeling in heart failure. Several prospective, randomized studies have directly compared the relative effects of the β-adrenergic receptor blockers metoprolol and carvedilol on ventricular remodeling in heart failure (153–155). While all three studies showed a significant improvement in LVEF with both agents compared to baseline, only the largest study involving 150 patients, performed by Metra et al. (154), showed a significant advantage of carvedilol over metoprolol in terms of improvement in LVEF. $^*p < 0.05$, change in LVEF from baseline, carvedilol compared to metoprolol.

relative reductions in resting heart rate in this study and in maximal exercise rate achieved with similar doses of β-blocker in a prior study (154). Although the balance of current evidence appears to favor carvedilol as the β-blocker of choice in heart failure, it would be unwise to make a definitive statement at this point as to the superiority of carvedilol over metoprolol for the treatment of heart failure. What can be concluded is that both agents are effective in preventing or reversing remodeling and improving survival in heart failure. Clinically significant differences between the agents, while probably real, are likely to be relatively minor.

Initiating and Titrating β-Blockers in Heart Failure

As discussed above, β-blockers are now recommended for the treatment of patients with any severity of heart failure, including asymptomatic patients (class B, NYHA functional class I) and patients with severe symptoms (class D, NYHA functional class IV) (New Heart Failure Society of America practice guidelines, in press) (117,122,159). Treatment with β-blockers is initiated at a low dose in patients with stable, compensated symptoms of heart failure, and the dosage gradually uptitrated over a number of weeks until a target dose or the highest tolerated dosage is reached. Hypotension may be minimized by avoidance of hypovolemia or by temporarily reducing the dose of ACE enzyme inhibitor. Intolerance of a vasodilating β-blocker like carvedilol may sometimes be circumvented by substituting a nonvasodilating β-blocker. Mild symptoms of worsening heart failure can usually be managed by increasing the dosage of diuretics or by further afterload reduction using intensified ACE inhibition. Past practice had been to postpone initiation of β-blocker therapy to subsequent clinic visits following an admission for decompensated heart failure owing to concerns over exacerbation of symptoms; these concerns have been alleviated by the results of the initiation management predischarge process for assessment of carvedilol therapy for heart failure study which were recently presented by Gheorghiade et al. (160). This study randomized patients admitted for treatment of worsening heart failure to initiation of carvedilol prior to discharge, or to initiation during clinic follow-up as per usual practice. The primary end point of the study was the number of patients treated with carvedilol at 60 days. Initiation of β-blockers was well tolerated; not only was a larger proportion of patients receiving carvedilol at 60 days in the predischarge initiation group, but the mean dose reached was also greater in this group compared to the usual care group. A trend was seen toward a lower cardiovascular event rate in the patients receiving carvedilol, but the study was not powered to show a reduction in clinical events. This approach, if widely adopted, may lead to more rapid uptitration of dose compared to routine initiation postdischarge and potentially better outcomes.

Bradycardia or bronchospasm has also been considered contraindications to β-blockade. It is, however, difficult to withhold a treatment that will potentially have more of an impact on morbidity and survival than any other form of pharmacological intervention the patient will receive. It is reasonable to initiate a trial of low dose β-blocker under careful observation in patients with all but the most severe bradycardia or bronchospasm and to cautiously uptitrate the dose as tolerated. As described previously, even a small dose of β-blocker will provide a significant degree of clinical benefit. Alternatively, the magnitude of the benefit derived from β-blockade is such that strong consideration should be given to implantation of a permanent cardiac pacemaker to facilitate β-blockade in patients with significant bradycardia.

α₁-ADRENERGIC RECEPTOR BLOCKADE

Data from animal models suggest that overexpression of the α_{1B}-adrenergic receptor may result in different phenotypic patterns including myocyte hypertrophy (150), reduced agonist-stimulated LV contractility (161), or a phenotype resembling dilated cardiomyopathy depending on the level of overexpression and type of receptor expressed (constitutively active mutant or wild type) (151). Despite these data and in contrast to the substantial literature on β-adrenergic system blockade in preventing remodeling and improving survival in heart failure, there is no evidence that α_1-adrenergic receptor blockade exerts a favorable effect on ventricular remodeling in vivo. The first veterans affairs cooperative vasodilator-heart failure trial showed no effect on ventricular function or survival following α_1-adrenergic receptor inhibition with prazosin, unlike the improvements seen with hydralazine-isosorbide dinitrate (73). Similarly, McDonald et al. (162) found that terazosin failed to prevent ventricular remodeling in a canine model of heart failure. Furthermore, the doxazosin arm of the largest study of antihypertensive treatment ever performed, the antihypertensive and lipid-lowering treatment to prevent heart attack trial (ALLHAT), was terminated prematurely because of an excess incidence of congestive heart failure (163,164).

Taken together, these data would suggest that isolated α_1-adrenergic blockade is insufficient to prevent cardiac remodeling once the process is initiated. β-adrenergic activation appears central and necessary for ventricular remodeling in heart failure, and β-adrenergic inhibition is required as an integral part of any attempt to interrupt remodeling through inhibition of the adrenergic nervous system. This does not exclude a synergistic effect of combined α_1/β-adrenergic blockade on ventricular remodeling; however, this will be difficult to prove in a clinical setting given the results of ALLHAT and the low likelihood of any significant future study involving α-adrenergic blockade in heart failure.

CENTRAL SYMPATHETIC NERVOUS SYSTEM INHIBITION

The success of β-adrenergic inhibition in treating heart failure has led to interest in means of suppressing central sympathetic outflow to protect the heart. Attempts have been made to inhibit central sympathetic outflow using the central α_2- and imidazoline I_1-receptor agonist clonidine in a rat model of hypertension and heart failure as well as in small human studies of heart failure (165–167). While improvement in LVEF has been shown, it is unclear if this relates to the action of clonidine in causing arterial dilation and reducing cardiac afterload or to a specific effect on the myocardium. The unpleasant side effects associated with clonidine (dizziness, dry mouth, sedation, rebound hypertension) have also precluded more widespread enthusiasm in its use.

An alternative approach has been through the use of the selective imidazoline receptor agonist moxonidine. The drug stimulates medullary imidazoline I_1-receptors, inhibiting central sympathetic activity and hence reducing blood pressure with fewer α_2-mediated side effects (168). The moxonidine congestive heart failure (MOXCON) and moxonidine safety and efficacy (MOXSE) trials evaluated the use of moxonidine in patients with heart failure (104,169). While use of moxonidine in MOXSE was associated with marked reduction in plasma norepinephrine levels and a modest increase in LVEF, significant dose-related adverse effects and an increase in mortality related to moxonidine were seen in MOXCON and prompted early termination of the study. The serious adverse events noted have been widely interpreted as being because of excessive sympatholysis leading to cardiac depression and an adverse outcome in patients dependent on adrenergic drive. This occurred despite favorable effects on reverse remodeling, and has parallels to BEST, in which bucindolol was associated with reverse remodeling and significant falls in plasma norepinephrine, but a neutral effect on survival (99). The valsartan in heart failure trial provided additional circumstantial evidence that excessive sympatholysis may be associated with adverse effects on outcome (102,105). Addition of the angiotensin receptor antagonist valsartan to standard therapy including ACE inhibitor was associated with reverse remodeling and fewer heart failure hospitalizations; however, in a post hoc analysis stratified by baseline treatment with ACE inhibitor or β-blocker, benefit derived from valsartan was confined to patients receiving one or neither of these drugs, with an adverse effect in patients already receiving combined therapy.

The recently published results of the candesartan in heart failure assessment of reduction in mortality and morbidity program (CHARM) have quelled concerns over excessive neurohormonal blockade (170–173). The program consisted of three parallel, independent, randomized, double-blind, clinical trials comparing candesartan with placebo in distinct but complementary populations of patients with symptomatic heart failure. The patient populations studied were (1) intolerant of ACE inhibitors, (2) currently receiving ACE inhibitors and standard therapy, (3) had LVEF >40% (presumed diastolic dysfunction). A reduction in the primary end point of cardiovascular deaths and hospital admissions for heart failure was observed in the overall program. Significantly, in the CHARM-added trial comparing the addition of candesartan or placebo to standard therapy including ACE inhibitors in 2548 patients with NYHA class II–IV heart failure and LVEF <40%, candesartan reduced cardiovascular death and hospital admission for heart failure by 15% ($p = 0.01$). Of note, 56% of patients were receiving β-blockers at baseline, and candesartan reduced risk to an equivalent degree in this group. These findings suggest further clinical benefit from additional neurohormonal blockade using an angiotensin receptor antagonist in combination with ACE inhibition and β-adrenergic receptor blockade; furthermore, there is no need to withhold add-on treatment with candesartan if clinically indicated to patients already receiving ACE inhibitors and β-blockers.

The results with bucindolol and moxonidine sound a cautionary note amid the success of recent trials of β-adrenergic receptor blockade. Despite the consistent advantageous effects of β-blockade on myocardial remodeling, there appears to remain a dissociation between favorable effects on remodeling and "hard" clinical outcomes like mortality and morbidity with some agents. However, it seems likely that this dissociation may be at least partially explained by the ancillary properties of these agents that may act to counteract the beneficial effects of β-blockade, rather than a lack of clinical benefit from reverse ventricular remodeling.

CONCLUSION

The success of β-adrenergic receptor blockade in chronic heart failure has undoubtedly been one of the great advances in the treatment of heart failure. In association with this, there is substantial evidence that β-adrenergic receptor blockade favorably influences the features of cardiac remodeling that are characteristic of progressive heart failure. Furthermore, the magnitude of reverse remodeling with β-blockers has typically been greater than that seen with ACE inhibition alone. It remains to be seen if combined β-blockade and ACE inhibition will exert additional beneficial effects on cardiac remodeling and clinical end points, or if β-blockade alone will be sufficient to attain all the documented benefits. Metoprolol, bucindolol, carvedilol, and bisoprolol have all been shown to reverse remodeling in chronic heart failure; however, unlike the other agents, no improvement in mortality or morbidity has been demonstrated for bucindolol. Treatment is generally well tolerated, and reverse remodeling is evident by as early as three months after initiation of treatment. There is typically a more pronounced effect on LV end systolic compared to end-diastolic volumes and therefore a concomitant increase in LVEF. It however remains unclear if β_2-receptor blockade in addition to β_1-receptor blockade provides incremental improvement, or if auxiliary properties like the antioxidant effects of carvedilol play any role in their clinical efficacy.

In contrast to the beneficial effects of β-blockade, there is little evidence of a positive effect from α-adrenergic receptor blockade, and central sympatholysis has been associated with adverse outcomes. Additional efforts at reversing remodeling may focus on "fine-tuning" of antiadrenergic suppression or the customization of treatment protocols for individual patients.

REFERENCES

1. Cohn JN, Levine TB, Olivari MT, Garberg V, Lura D, Francis GS, Simon AB, Rector T. Plasma norepinephrine as a guide to prognosis in patients with chronic congestive heart failure. N Engl J Med 1984; 311:819–823.
2. Francis GS, Cohn JN, Johnson G, Rector TS, Goldman S, Simon A. Plasma norepinephrine, plasma renin activity, and congestive heart failure. Relations to survival and the effects of therapy in V-HeFT II. The V-HeFT VA Cooperative Studies Group. Circulation 1993; 87:VI40–VI48.
3. Francis GS, McDonald KM, Cohn JN. Neurohumoral activation in preclinical heart failure. Remodeling and the potential for intervention. Circulation 1993; 87:IV90–IV96.
4. Swedberg K, Eneroth P, Kjekshus J, Wilhelmsen L. Hormones regulating cardiovascular function in patients with severe congestive heart failure and their relation to mortality. CONSENSUS Trial Study Group. Circulation 1990; 82:1730–1736.
5. Bristow MR, Ginsburg R, Umans V, Fowler M, Minobe W, Rasmussen R, Zera P, Menlove R, Shah P, Jamieson S, et al. Beta 1- and beta 2-adrenergic-receptor subpopulations in nonfailing and failing human ventricular myocardium: coupling of both receptor subtypes to muscle contraction and selective β_1-receptor down-regulation in heart failure. Circ Res 1986; 59:297–309.
6. Bristow MR, Hershberger RE, Port JD, Minobe W, Rasmussen R. Beta 1- and beta 2-adrenergic receptor-mediated adenylate cyclase stimulation in nonfailing and failing human ventricular myocardium. Mol Pharmacol 1989; 35:295–303.
7. Bristow MR. Changes in myocardial and vascular receptors in heart failure. J Am Coll Cardiol 1993; 22:61A–71A.
8. Colucci WS. Do positive inotropic agents adversely affect the survival of patients with chronic congestive heart failure? III. Antagonist's viewpoint. J Am Coll Cardiol 1988; 12:566–569.
9. Uretsky BF, Jessup M, Konstam MA, Dec GW, Leier CV, Benotti J, Murali S, Herrmann HC, Sandberg JA. Multicenter trial of oral enoximone in patients with moderate to moderately severe congestive heart failure. Lack of benefit compared with placebo. Enoximone Multicenter Trial Group. Circulation 1990; 82:774–780.

10. Packer M, Carver JR, Rodeheffer RJ, Ivanhoe RJ, DiBianco R, Zeldis SM, Hendrix GH, Bommer WJ, Elkayam U, Kukin ML, et al. Effect of oral milrinone on mortality in severe chronic heart failure. The PROMISE Study Research Group. N Engl J Med 1991; 325:1468–1475.

11. Cohn JN, Goldstein SO, Greenberg BH, Lorell BH, Bourge RC, Jaski BE, Gottlieb SO, McGrew F 3rd, DeMets DL, White BG. A dose-dependent increase in mortality with vesnarinone among patients with severe heart failure. Vesnarinone Trial Investigators. N Engl J Med 1998; 339:1810–1816.

12. Feldman AM, Bristow MR, Parmley WW, Carson PE, Pepine CJ, Gilbert EM, Strobeck JE, Hendrix GH, Powers ER, Bain RP, et al. Effects of vesnarinone on morbidity and mortality in patients with heart failure. Vesnarinone Study Group. N Engl J Med 1993; 329:149–155.

13. Swedberg K, Hjalmarson A, Waagstein F, Wallentin I. Prolongation of survival in congestive cardiomyopathy by beta-receptor blockade. Lancet 1979; 1:1374–1376.

14. Swedberg K, Hjalmarson A, Waagstein F, Wallentin I. Beneficial effects of long-term beta-blockade in congestive cardiomyopathy. Br Heart J 1980; 44:117–133.

15. Waagstein F, Hjalmarson A, Varnauskas E, Wallentin I. Effect of chronic beta-adrenergic receptor blockade in congestive cardiomyopathy. Br Heart J 1975; 37:1022–1036.

16. Anand IS, Florea VG, Fisher L. Surrogate end points in heart failure. J Am Coll Cardiol 2002; 39:1414–1421.

17. Fleming TR, DeMets DL. Surrogate end points in clinical trials: are we being misled? Ann Intern Med 1996; 125:605–613.

18. Gheorghiade M, Adams KF Jr, Gattis WA, Teerlink JR, Orlandi C, O'Connor CM. Surrogate end points in heart failure trials. Am Heart J 2003; 145:S67–S70.

19. Bristow MR. Beta-adrenergic receptor blockade in chronic heart failure. Circulation 2000; 101:558–569.

20. Chidiac P, Hebert TE, Valiquette M, Dennis M, Bouvier M. Inverse agonist activity of beta-adrenergic antagonists. Mol Pharmacol 1994; 45:490–499.

21. Frishman WH. Carvedilol. N Engl J Med 1998; 339:1759–1765.

22. Goa KL, Benfield P, Sorkin EM. Labetalol – A reappraisal of its pharmacology, pharmacokinetics and therapeutic use in hypertension and ischemic heart-disease. Drugs 1989; 37:583–627.

23. Hoffman BB. Catecholamines, sympathomimetic drugs, and adrenergic receptor antagonists. In: Hardman JG, Limbird LE, Goodman AG, eds. Goodman and Gilman's The Pharmacological Basis of Therapeutics. New York: McGraw-Hill, 2001:215–268.

24. Australia-New Zealand Heart Failure Research Collaborative Group. Effects of carvedilol, a vasodilator-beta-blocker, in patients with congestive heart failure due to ischemic heart disease. Circulation 1995; 92:212–218.

25. Bristow MR, Gilbert EM, Abraham WT, Adams KF, Fowler MB, Hershberger RE, Kubo SH, Narahara KA, Ingersoll H, Krueger S, et al. Carvedilol produces dose-related improvements in left ventricular function and survival in subjects with chronic heart failure. Circulation 1996; 94:2807–2816.

26. Colucci WS, Packer M, Bristow MR, Gilbert EM, Cohn JN, Fowler MB, Krueger SK, Hershberger R, Uretsky BF, Bowers JA, et al. Carvedilol inhibits clinical progression in patients with mild symptoms of heart failure. US Carvedilol Heart Failure Study Group. Circulation 1996; 94:2800–2806.

27. Packer M, Colucci WS, Sackner-Bernstein JD, Liang CS, Goldscher DA, Freeman I, Kukin ML, Kinhal V, Udelson JE, Klapholz M, et al. Double-blind, placebo-controlled study of the effects of carvedilol in patients with moderate to severe heart failure. The PRECISE Trial. Prospective randomized evaluation of carvedilol on symptoms and exercise. Circulation 1996; 94:2793–2799.

28. Packer M, Bristow MR, Cohn JN, Colucci WS, Fowler MB, Gilbert EM, Shusterman NH. The effect of carvedilol on morbidity and mortality in patients with chronic heart failure. U.S. Carvedilol Heart Failure Study Group. N Engl J Med 1996; 334:1349–1355.

29. Cohn JN, Fowler MB, Bristow MR, Colucci WS, Gilbert EM, Kinhal V, Krueger SK, Lejemtel T, Narahara KA, Packer M, et al. Safety and efficacy of carvedilol in severe heart failure. The U.S. Carvedilol Heart Failure Study Group. J Card Fail 1997; 3:173–179.

30. Eichhorn E, Domanski M, Krause-Steinrauf H et al. A trial of the beta-blocker bucindolol in patients with advanced chronic heart failure. N Engl J Med 2001; 344:1659–1667.

31. Gilbert EM, Anderson JL, Deitchman D, Yanowitz FG, O'Connell JB, Renlund DG, Bartholomew M, Mealey PC, Larrabee P, Bristow MR. Long-term beta-blocker vasodilator therapy improves cardiac function in idiopathic dilated cardiomyopathy: a double-blind, randomized study of bucindolol versus placebo. Am J Med 1990; 88: 223–229.

32. Haber HL, Simek CL, Gimple LW, Bergin JD, Subbiah K, Jayaweera AR, Powers ER, Feldman MD. Why do patients with congestive-heart-failure tolerate the initiation of beta-blocker therapy? Circulation 1993; 88:1610–1619.

33. Talwar KK, Bhargava B, Upasani PT, Verma S, Kamlakar T, Chopra P. Hemodynamic predictors of early intolerance and long-term effects of propranolol in dilated cardiomyopathy. J Card Fail 1996; 2:273–277.

34. Waagstein F, Bristow MR, Swedberg K, Camerini F, Fowler MB, Silver MA, Gilbert EM, Johnson MR, Goss FG, Hjalmarson A. Beneficial effects of metoprolol in idiopathic dilated cardiomyopathy. Metoprolol in Dilated Cardiomyopathy (MDC) Trial Study Group. Lancet 1993; 342:1441–1446.

35. MERIT-HF Study Group. Effect of metoprolol CR/XL in chronic heart failure: Metoprolol CR/XL randomized intervention trial in congestive heart failure (MERIT-HF). Lancet 1999; 353:2001–2007.

36. Packer M, Coats AJ, Fowler MB, Katus HA, Krum H, Mohacsi P, Rouleau JL, Tendera M, Castaigne A, Roecker EB, et al. Carvedilol Prospective Randomized Cumulative Survival Study Group. Effect of carvedilol on survival in severe chronic heart failure. N Engl J Med 2001; 344:1651–1658.

37. Bristow MR. Mechanism of action of beta-blocking agents in heart failure. Am J Cardiol 1997; 80:L26–L40.

38. Tsutsui H, Spinale FG, Nagatsu M, Schmid PG, Ishihara K, DeFreyte G, Cooper G 4th, Carabello BA. Effects of chronic beta-adrenergic-blockade on the left-ventricular and cardiocyte abnormalities of chronic canine mitral regurgitation. J Clin Invest 1994; 93:2639–2648.

39. Bristow MR, Gilbert EM. Improvement in cardiac myocyte function by biological effects of medical therapy – a new concept in the treatment of heart-failure. Eur Heart J 1995; 16:20–31.

40. Eichhorn EJ, Bristow MR. Medical therapy can improve the biological properties of the chronically failing heart – a new era in the treatment of heart failure. Circulation 1996; 94:2285–2296.

41. Alpert C, Ramdev N, George D, Loscalzo J. Detection of S-nitrosothiols and other nitric oxide derivatives by photolysis-chemiluminescence spectrometry. Anal Biochem 1997; 245:1–7.

42. Bristow MR, Ginsburg R, Minobe W, Cubicciotti RS, Sageman WS, Lurie K, Billingham ME, Harrison DC, Stinson EB. Decreased catecholamine sensitivity and beta-adrenergic-receptor density in failing human hearts. N Engl J Med 1982; 307:205–211.

43. Bristow MR, Anderson FL, Port JD, Skerl L, Hershberger RE, Larrabee P, O'Connell JB, Renlund DG, Volkman K, Murray J, et al. Differences in beta-adrenergic neuroeffector mechanisms in ischemic versus idiopathic dilated cardiomyopathy. Circulation 1991; 84:1024–1039.

44. Bristow MR, Minobe W, Rasmussen R, Larrabee P, Skerl L, Klein JW, Anderson FL, Murray J, Mestroni L, Karwande SV, et al. Beta-adrenergic neuroeffector abnormalities

in the failing human heart are produced by local rather than systemic mechanisms. J Clin Invest 1992; 89:803–815.

45. Feldman AM, Cates AE, Veazey WB, Hershberger RE, Bristow MR, Baughman KL, Baumgartner WA, Van Dop C. Increase of the 40,000-mol wt pertussis toxin substrate (G-Protein) in the failing human-heart. J Clin Invest 1988; 82:189–197.

46. Inglese J, Freedman NJ, Koch WJ, Lefkowitz RJ. Structure and mechanism of the G protein-coupled receptor kinases. J Biol Chem 1993; 268:23,735–23,738.

47. Ungerer M, Bohm M, Elce JS, Erdmann E, Lohse MJ. Altered expression of beta-adrenergic receptor kinase and beta 1-adrenergic receptors in the failing human heart. Circulation 1993; 87:454–463.

48. Ungerer M, Parruti G, Bohm M, Puzicha M, DeBlasi A, Erdmann E, Lohse MJ. Expression of beta-arrestins and beta-adrenergic receptor kinases in the failing human heart. Circ Res 1994; 74:206–213.

49. Heilbrunn SM, Shah P, Bristow MR, Valantine HA, Ginsburg R, Fowler MB. Increased β-receptor density and improved hemodynamic response to catecholamine stimulation during long-term metoprolol therapy in heart failure from dilated cardiomyopathy. Circulation 1989; 79:483–490.

50. Bohm M, Deutsch HJ, Hartmann D, Rosee KL, Stablein A. Improvement of postreceptor events by metoprolol treatment in patients with chronic heart failure. J Am Coll Cardiol 1997; 30:992–996.

51. Hall JA, Kaumann AJ, Brown MJ. Selective beta 1-adrenoceptor blockade enhances positive inotropic responses to endogenous catecholamines mediated through beta 2-adrenoceptors in human atrial myocardium. Circ Res 1990; 66:1610–1623.

52. Iaccarino G, Tomhave ED, Lefkowitz RJ, Koch WJ. Reciprocal in vivo regulation of myocardial G protein-coupled receptor kinase expression by beta-adrenergic receptor stimulation and blockade. Circulation 1998; 98:1783–1789.

53. Yoshikawa T, Port JD, Asano K, Chidiak P, Bouvier M, Dutcher D, Roden RL, Minobe W, Tremmel KD, Bristow MR. Cardiac adrenergic receptor effects of carvedilol. Eur Heart J 1996; 17(suppl B):8–16.

54. Gilbert EM, Abraham WT, Olsen S, Hattler B, White M, Mealy P, Larrabee P, Bristow MR. Comparative hemodynamic, left ventricular functional, and antiadrenergic effects of chronic treatment with metoprolol versus carvedilol in the failing heart. Circulation 1996; 94: 2817–2825.

55. Arai M, Matsui H, Periasamy M. Sarcoplasmic reticulum gene expression in cardiac hypertrophy and heart failure. Circ Res 1994; 74:555–564.

56. Hasenfuss G, Reinecke H, Studer R, Meyer M, Pieske B, Holtz J, Holubarsch C, Posival H, Just H, Drexler H. Relation between myocardial function and expression of sarcoplasmic reticulum Ca(2+)-ATPase in failing and nonfailing human myocardium. Circ Res 1994; 75:434–442.

57. Mercadier JJ, Lompre AM, Duc P, Boheler KR, Fraysse JB, Wisnewsky C, Allen PD, Komajda M, Schwartz K. Altered sarcoplasmic reticulum Ca2(+)-ATPase gene expression in the human ventricle during end-stage heart failure. J Clin Invest 1990; 85: 305–309.

58. Brillantes AM, Allen P, Takahashi T, Izumo S, Marks AR. Differences in cardiac calcium release channel (ryanodine receptor) expression in myocardium from patients with end-stage heart failure caused by ischemic versus dilated cardiomyopathy. Circ Res 1992; 71:18–26.

59. Lowes BD, Gilbert EM, Abraham WT, Minobe WA, Larrabee P, Ferguson D, Wolfel EE, Lindenfeld J, Tsvetkova T, Robertson AD, et al. Myocardial gene expression in dilated cardiomyopathy treated with beta-blocking agents. N Engl J Med 2002,; 346:1357–1365.

60. Reiken S, Wehrens XH, Vest JA, Barbone A, Klotz S, Mancini D, Burkhoff D, Marks AR. Beta-blockers restore calcium release channel function and improve cardiac muscle performance in human heart failure. Circulation 2003; 107:2459–2466.

61. Lowes BD, Minobe W, Abraham WT, Rizeq MN, Bohlmeyer TJ, Quaife RA, Roden RL, Dutcher DL, Robertson AD, Voelkel NF, et al. Changes in gene expression in the intact human heart. Downregulation of alpha-myosin heavy chain in hypertrophied, failing ventricular myocardium. J Clin Invest 1997; 100(9):2315–2324.

62. Miyata S, Minobe W, Bristow MR, Leinwand LA. Myosin heavy chain isoform expression in the failing and nonfailing human heart. Circ Res 2000; 86:386–390.

63. Paolisso G, Gambardella A, Galzerano D, D'Amore A, Rubino P, Verza M, Teasuro P, Varricchio M, D'Onofrio F. Total-body and myocardial substrate oxidation in congestive heart failure. Metabolism 1994; 43:174–179.

64. Taylor M, Wallhaus TR, Degrado TR, Russell DC, Stanko P, Nickles RJ, Stone CK. An evaluation of myocardial fatty acid and glucose uptake using PET with [18F]fluoro-6-thia-heptadecanoic acid and [18F]FDG. J Nucl Med 2001; 42:55–62.

65. Beanlands RS, Nahmias C, Gordon E, Coates G, deKemp R, Firnau G, Fallen E. The effects of beta(1)-blockade on oxidative metabolism and the metabolic cost of ventricular work in patients with left ventricular dysfunction: A double-blind, placebo-controlled, positron-emission tomography study. Circulation 2000; 102:2070–2075.

66. Eichhorn EJ, Heesch CM, Barnett JH, Alvarez LG, Fass SM, Grayburn PA, Hatfield BA, Marcoux LG, Malloy CR. Effect of metoprolol on myocardial function and energetics in patients with nonischemic dilated cardiomyopathy: a randomized, double-blind, placebo-controlled study. J Am Coll Cardiol 1994; 24:1310–1320.

67. Wallhaus TR, Taylor M, DeGrado TR, Russell DC, Stanko P, Nickles RJ, Stone CK. Myocardial free fatty acid and glucose use after carvedilol treatment in patients with congestive heart failure. Circulation 2001; 103:2441–2446.

68. Kjekshus JK. Importance of heart rate in determining beta-blocker efficacy in acute and long-term acute myocardial infarction intervention trials. Am J Cardiol 1986; 57:43F–49F.

69. Alpert NR, Mulieri LA. Human heart failure: determinants of ventricular dysfunction. Adv Exp Med Biol 1997; 430:97–108.

70. Lechat P, Escolano S, Golmard JL, Lardoux H, Witchitz S, Henneman JA, Maisch B, Hetzel M, Jaillon P, Boissel JP, Mallet A. Prognostic value of bisoprolol-induced hemodynamic effects in heart failure during the Cardiac Insufficiency BIsoprolol Study (CIBIS). Circulation 1997; 96:2197–2205.

71. Barry WH. Na(+)-Ca(2+) exchange in failing myocardium: friend or foe? Circ Res 2000; 87:529–531.

72. Cintron G, Johnson G, Francis G, Cobb F, Cohn JN. Prognostic significance of serial changes in left ventricular ejection fraction in patients with congestive heart failure. The V-HeFT VA Cooperative Studies Group. Circulation 1993; 87:VI17–VI23.

73. Cohn JN, Archibald DG, Ziesche S, Franciosa JA, Harston WE, Tristani FE, Dunkman WB, Jacobs W, Francis GS, Flohr KH, et al. Effect of vasodilator therapy on mortality in chronic congestive heart failure. Results of a Veterans Administration Cooperative Study. N Engl J Med 1986; 314:1547–1552.

74. Wong M, Johnson G, Shabetai R, Hughes V, Bhat G, Lopez B, Cohn JN. Echocardiographic variables as prognostic indicators and therapeutic monitors in chronic congestive heart failure. Veterans Affairs cooperative studies V-HeFT I and II. V-HeFT VA Cooperative Studies Group. Circulation 1993; 87:VI65–VI70.

75. Wong M, Germanson T, Taylor WR, Cohen IS, Perry G, Baruch L, Deedwania P, Lopez B, Cohn JN. Felodipine improves left ventricular emptying in patients with chronic heart failure: V-HeFT III echocardiographic substudy of multicenter reproducibility and detecting functional change. J Card Fail 2000; 6:19–28.

76. Mani K, Kitsis RN. Myocyte apoptosis: Programming ventricular remodeling. J Am Coll Cardiol 2003; 41:761–764.

77. Wencker D, Chandra M, Nguyen K, Miao W, Garantziotis S, Factor SM, Shirani J, Armstrong RC, Kitsis RN. A mechanistic role for cardiac myocyte apoptosis in heart failure. J Clin Invest 2003; 111:1497–1504.

78. Abbate A, Biondi-Zoccai GG, Bussani R, Dobrina A, Camilot D, Feroce F, Rossiello R, Baldi F, Silvestri F, Biasucci LM, et al. Increased myocardial apoptosis in patients with unfavorable left ventricular remodeling and early symptomatic post-infarction heart failure. J Am Coll Cardiol 2003; 41:753–760.

79. Communal C, Singh K, Pimentel DR, Colucci WS. Norepinephrine stimulates apoptosis in adult rat ventricular myocytes by activation of the beta-adrenergic pathway. Circulation 1998; 98:1329–1334.

80. Communal C, Singh K, Sawyer DB, Colucci WS. Opposing effects of beta(1)- and beta(2)-adrenergic receptors on cardiac myocyte apoptosis: role of a pertussis toxin-sensitive G protein. Circulation 1999; 100:2210–2212.

81. Communal C, Colucci WS, Singh K. p38 mitogen-activated protein kinase pathway protects adult rat ventricular myocytes against beta -adrenergic receptor-stimulated apoptosis. Evidence for G_i-dependent activation. J Biol Chem 2000; 275:19395–19400.

82. Asai K, Yang GP, Geng YJ, Takagi G, Bishop S, Ishikawa Y, Shannon RP, Wagner TE, Vatner DE, Homcy CJ, Vatner SF. Beta-adrenergic receptor blockade arrests myocyte damage and preserves cardiac function in the transgenic G(salpha) mouse. J Clin Invest 1999; 104:551–558.

83. Fisher ML, Gottlieb SS, Plotnick GD, Greenberg NL, Patten RD, Bennett SK, Hamilton BP. Beneficial effects of metoprolol in heart failure associated with coronary artery disease: a randomized trial. J Am Coll Cardiol 1994; 23:943–950.

84. Groenning BA, Nilsson JC, Sondergaard L, Fritz-Hansen T, Larsson HB, Hildebrandt PR. Antiremodeling effects on the left ventricle during beta-blockade with metoprolol in the treatment of chronic heart failure. J Am Coll Cardiol 2000; 36:2072–2080.

85. Effects of metoprolol CR in patients with ischemic and dilated cardiomyopathy: the randomized evaluation of strategies for left ventricular dysfunction pilot study. Circulation 2000; 101:378–384.

86. Australia/New Zealand Heart Failure Research Collaborative Group. Randomised, placebo-controlled trial of carvedilol in patients with congestive heart failure due to ischaemic heart disease. Lancet 1997; 349:375–380.

87. Doughty RN, Whalley GA, Gamble G, MacMahon S, Sharpe N. Left ventricular remodeling with carvedilol in patients with congestive heart failure due to ischemic heart disease. Australia-New Zealand Heart Failure Research Collaborative Group. J Am Coll Cardiol 1997; 29:1060–1066.

88. Krum H, Sackner-Bernstein JD, Goldsmith RL, Kukin ML, Schwartz B, Penn J, Medina N, Yushak M, Horn E, Katz SD, et al. Double-blind, placebo-controlled study of the long-term efficacy of carvedilol in patients with severe chronic heart failure. Circulation 1995; 92:1499–1506.

89. Quaife RA, Gilbert EM, Christian PE, Datz FL, Mealey PC, Volkman K, Olsen SL, Bristow MR. Effects of carvedilol on systolic and diastolic left ventricular performance in idiopathic dilated cardiomyopathy or ischemic cardiomyopathy. Am J Cardiol 1996; 78:779–784.

90. Lowes BD, Gill EA, Abraham WT, Larrain JR, Robertson AD, Bristow MR, Gilbert EM. Effects of carvedilol on left ventricular mass, chamber geometry, and mitral regurgitation in chronic heart failure. Am J Cardiol 1999; 83:1201–1205.

91. Olsen SL, Gilbert EM, Renlund DG, Taylor DO, Yanowitz FD, Bristow MR. Carvedilol improves left ventricular function and symptoms in chronic heart failure: a double-blind randomized study. J Am Coll Cardiol 1995; 25:1225–1231.

92. Metra M, Nardi M, Giubbini R, Dei Cas L. Effects of short- and long-term carvedilol administration on rest and exercise hemodynamic variables, exercise capacity and clinical conditions in patients with idiopathic dilated cardiomyopathy. J Am Coll Cardiol 1994; 24:1678–1687.

93. Metra M, Nodari S, Parrinello G, Giubbini R, Manca C, Dei Cas L. Marked improvement in left ventricular ejection fraction during long-term beta-blockade in patients with

chronic heart failure: clinical correlates and prognostic significance. Am Heart J 2003; 145:292–299.

94. CIBIS Investigators and Committees. A randomized trial of beta-blockade in heart failure. The Cardiac Insufficiency Bisoprolol Study (CIBIS). Circulation 1994; 90: 1765–1773.

95. CIBIS II Investigators and Committees. The Cardiac Insufficiency Bisoprolol Study II (CIBIS-II): a randomised trial. Lancet 1999; 353:9–13.

96. Cohn JN, Ferrari R, Sharpe N. Cardiac remodeling—concepts and clinical implications: a consensus paper from an international forum on cardiac remodeling. On behalf of an International Forum on Cardiac Remodeling. J Am Coll Cardiol 2000; 35:569–582.

97. Dubach P, Myers J, Bonetti P, Schertler T, Froelicher V, Wagner D, Scheidegger M, Stuber M, Luchinger R, Schwitter J, et al. Effects of bisoprolol fumarate on left ventricular size, function, and exercise capacity in patients with heart failure: analysis with magnetic resonance myocardial tagging. Am Heart J 2002; 143:676–683.

98. Bristow MR, O'Connell JB, Gilbert EM, French WJ, Leatherman G, Kantrowitz NE, Orie J, Smucker ML, Marshall G, Kelly P, et al. Dose-response of chronic beta-blocker treatment in heart failure from either idiopathic dilated or ischemic cardiomyopathy. Bucindolol Investigators. Circulation 1994; 89:1632–1642.

99. A trial of the beta-blocker bucindolol in patients with advanced chronic heart failure. N Engl J Med 2001; 344:1659–1667.

100. Freudenberger R, Kalman J, Mannino M, Buchholz-Varley C, Ocampo O, Kukin M. Effect of race in the response to metoprolol in patients with congestive heart failure secondary to idiopathic dilated or ischemic cardiomyopathy. Am J Cardiol 1997; 80: 1372–1374.

101. Yancy CW, Fowler MB, Colucci WS, Gilbert EM, Bristow MR, Cohn JN, Lukas MA, Young ST, Packer M; U.S. Carvedilol Heart Failure Study Group. Race and the response to adrenergic blockade with carvedilol in patients with chronic heart failure. N Engl J Med 2001; 344:1358–1365.

102. Cohn JN, Tognoni G. A randomized trial of the angiotensin-receptor blocker valsartan in chronic heart failure. N Engl J Med 2001; 345:1667–1675.

103. Floras JS. The "unsympathetic" nervous system of heart failure. Circulation 2002; 105:1753–1755.

104. Swedberg K, Bristow MR, Cohn JN, Dargie H, Straub M, Wiltse C, Wright TJ; Moxonidine Safety and Efficacy (MOXSE) Investigators. Effects of sustained-release moxonidine, an imidazoline agonist, on plasma norepinephrine in patients with chronic heart failure. Circulation 2002; 105:1797–1803.

105. Wong M, Staszewsky L, Latini R, Barlera S, Volpi A, Chiang YT, Benza RL, Gottlieb SO, Kleemann TD, Rosconi F, et al. Val-HeFT Heart Failure Trial Investigators. Valsartan benefits left ventricular structure and function in heart failure: Val-HeFT echocardiographic study. J Am Coll Cardiol 2002; 40:970–975.

106. Andreka P, Aiyar N, Olson LC, Wei JQ, Turner MS, Webster KA, Ohlstein EH, Bishopric NH. Bucindolol displays intrinsic sympathomimetic activity in cultured human and rat myocardial strips. Eur Heart J 2000; 21:439.

107. Willette RN, Mitchell MP, Ohlstein EH, Lukas MA, Ruffolo RR Jr. Evaluation of intrinsic sympathomimetic activity of bucindolol and carvedilol in rat heart. Pharmacology 1998; 56:30–36.

108. Willette RN, Aiyar N, Yue TL, Mitchell MP, Disa J, Storer BL, Naselsky DP, Stadel JM, Ohlstein EH, Ruffolo RR Jr. In vitro and in vivo characterization of intrinsic sympathomimetic activity in normal and heart failure rats. J Pharmacol Exp Ther 1999; 289:48–53.

109. Bristow M. Antiadrenergic therapy of chronic heart failure: surprises and new opportunities. Circulation 2003; 107:1100–1102.

110. Bristow MR, Larrabee P, Minobe W, Roden R, Skerl L, Klein J, Handwerger D, Port JD, Muller-Beckmann B. Receptor Pharmacology of Carvedilol in the Human Heart. J Cardiovasc Pharmacol 1992; 19:S68–S80.

111. Bristow MR, Roden RL, Lowes BD, Gilbert EM, Eichhorn EJ. The role of third-generation beta-blocking agents in chronic heart failure. Clin Cardiol 1998; 21:3–13.

112. Brixius K, Bundkirchen A, Bolck B, Mehlhorn U, Schwinger RH. Nebivolol, bucindolol, metoprolol and carvedilol are devoid of intrinsic sympathomimetic activity in human myocardium. Br J Pharmacol 2001; 133:1330–1338.

113. Sederberg J, Wichman SE, Lindenfeld J et al. Bucindolol has no intrinsic sympathomimetic activity (ISA) in nonfailing human ventricular preparations. J Am Coll Cardiol 2000; 35:207A.

114. Metra M, Nodari S, D'Aloia A, Bontempi L, Boldi E, Cas LD. A rationale for the use of beta-blockers as standard treatment for heart failure. Am Heart J 2000; 139:511–521.

115. Krum H, Roecker EB, Mohacsi P, Rouleau JL, Tendera M, Coats AJ, Katus HA, Fowler MB, Packer M; Carvedilol Prospective Randomized Cumulative Survival (COPERNICUS) Study Group. Effects of initiating carvedilol in patients with severe chronic heart failure: results from the COPERNICUS Study. JAMA 2003; 289:712–718.

116. Hall SA, Cigarroa CG, Marcoux L, Risser RC, Grayburn PA, Eichhorn EJ. Time course of improvement in left ventricular function, mass and geometry in patients with congestive heart failure treated with beta- adrenergic blockade. J Am Coll Cardiol 1995; 25:1154–1161.

117. Hunt SA, Baker DW, Chin MH, Cinquegrani MP, Feldman AM, Francis GS, Ganiats TG, Goldstein S, Gregoratos G, Jessup ML, et al.; American College of Cardiology/American Heart Association Task Force on Practice Guidelines (Committee to Revise the 1995 Guidelines for the Evaluation and Management of Heart Failure); International Society for Heart and Lung Transplantation; Heart Failure Society of America. ACC/AHA Guidelines for the Evaluation and Management of Chronic Heart Failure in the Adult: Executive Summary A Report of the American College of Cardiology/American Heart Association Task Force on Practice Guidelines (Committee to Revise the 1995 Guidelines for the Evaluation and Management of Heart Failure): Developed in Collaboration With the International Society for Heart and Lung Transplantation; Endorsed by the Heart Failure Society of America. Circulation 2001; 104:2996–3007.

118. Woodley SL, Gilbert EM, Anderson JL, O'Connell JB, Deitchman D, Yanowitz FG, Mealey PC, Volkman K, Renlund DG, Menlove R, et al. Beta-blockade with bucindolol in heart failure caused by ischemic versus idiopathic dilated cardiomyopathy. Circulation 1991; 84:2426–2441.

119. O'Keefe JH Jr, Magalski A, Stevens TL, Bresnahan DR Jr, Alaswad K, Krueger SK, Bateman TM. Predictors of improvement in left ventricular ejection fraction with carvedilol for congestive heart failure. J Nucl Cardiol 2000; 7:3–7.

120. Cleland JG, Pennell DJ, Ray SG, Coats AJ, Macfarlane PW, Murray GD, Mule JD, Vered Z, Lahiri A. Carvedilol hibernating reversible ischaemia trial: marker of success investigators. Myocardial viability as a determinant of the ejection fraction response to carvedilol in patients with heart failure (CHRISTMAS trial): randomised controlled trial. Lancet 2003; 362:14–21.

121. Exner DV, Dries DL, Waclawiw MA, Shelton B, Domanski MJ. Beta-adrenergic blocking agent use and mortality in patients with asymptomatic and symptomatic left ventricular systolic dysfunction: a post hoc analysis of the Studies of Left Ventricular Dysfunction. J Am Coll Cardiol 1999; 33:916–923.

122. Gheorghiade M, Colucci WS, Swedberg K. Beta-blockers in chronic heart failure. Circulation 2003; 107:1570–1575.

123. Pfeffer MA, Pfeffer JM, Steinberg C, Finn P. Survival after an experimental myocardial infarction: beneficial effects of long-term therapy with captopril. Circulation 1985; 72:406–412.

124. Pfeffer MA, Lamas GA, Vaughan DE, Parisi AF, Braunwald E. Effect of captopril on progressive ventricular dilatation after anterior myocardial infarction. N Engl J Med 1988; 319:80–86.

125. Pfeffer MA, Braunwald E, Moye LA, Basta L, Brown EJ Jr, Cuddy TE, Davis BR, Geltman EM, Goldman S, Flaker GC, et al. Effect of captopril on mortality and morbidity in patients with left ventricular dysfunction after myocardial infarction. Results of the survival and ventricular enlargement trial. The SAVE Investigators. N Engl J Med 1992; 327:669–677.

126. Pfeffer MA. ACE inhibitors in acute myocardial infarction: patient selection and timing. Circulation 1998; 97:2192–2194.

127. Cohn JN, Johnson G, Ziesche S, Cobb F, Francis G, Tristani F, Smith R, Dunkman WB, Loeb H, Wong M, et al. A comparison of enalapril with hydralazine-isosorbide dinitrate in the treatment of chronic congestive heart failure. N Engl J Med 1991; 325: 303–310.

128. Dyadyk AI, Bagriy AE, Lebed IA, Yarovaya NF, Schukina EV, Taradin GG. ACE inhibitors captopril and enalapril induce regression of left ventricular hypertrophy in hypertensive patients with chronic renal failure. Nephrol Dial Transplant 1997; 12: 945–951.

129. The SOLVD Investigators. Effect of enalapril on survival in patients with reduced left ventricular ejection fractions and congestive heart failure. N Engl J Med 1991; 325: 293–302.

130. The SOLVD Investigators. Effect of enalapril on mortality and the development of heart failure in asymptomatic patients with reduced left ventricular ejection fractions. The SOLVD Investigators. N Engl J Med 1992; 327:685–691.

131. Greenberg B, Quinones MA, Koilpillai C, Limacher M, Shindler D, Benedict C, Shelton B. Effects of long-term enalapril therapy on cardiac structure and function in patients with left ventricular dysfunction. Results of the SOLVD echocardiography substudy. Circulation 1995; 91:2573–2581.

132. Saxena PR. Interaction between the renin-angiotensin-aldosterone and sympathetic nervous systems. J Cardiovasc Pharmacol 1992; 19(suppl 6):S80–S88.

133. Khattar RS, Senior R, Soman P, van der Does R, Lahiri A. Regression of left ventricular remodeling in chronic heart failure: Comparative and combined effects of captopril and carvedilol. Am Heart J 2001; 142:704–713.

134. GISSI-3: effects of lisinopril and transdermal glyceryl trinitrate singly and together on 6-week mortality and ventricular function after acute myocardial infarction. Gruppo Italiano per lo Studio della Sopravvivenza nell'infarto Miocardico. Lancet 1994; 343:1115–1122.

135. Sharpe N, Smith H, Murphy J, Greaves S, Hart H, Gamble G. Early prevention of left ventricular dysfunction after myocardial infarction with angiotensin-converting-enzyme inhibition. Lancet 1991; 337:872–876.

136. Yusuf S, Peto R, Lewis J, Collins R, Sleight P. Beta blockade during and after myocardial infarction: an overview of the randomized trials. Prog Cardiovasc Dis 1985; 27:335–371.

137. Sia YT, Parker TG, Tsoporis JN, Liu P, Adam A, Rouleau JL. Long-term effects of carvedilol on left ventricular function, remodeling, and expression of cardiac cytokines after large myocardial infarction in the rat. J Cardiovasc Pharmacol 2002; 39: 73–87.

138. Basu S, Senior R, Raval U, van der Does R, Bruckner T, Lahiri A. Beneficial effects of intravenous and oral carvedilol treatment in acute myocardial infarction. A placebo-controlled, randomized trial. Circulation 1997; 96:183–191.

139. Senior R, Basu S, Kinsey C, Schaeffer S, Lahiri A. Carvedilol prevents remodeling in patients with left ventricular dysfunction after acute myocardial infarction. Am Heart J 1999; 137:646–652.

140. Dargie HJ. Effect of carvedilol on outcome after myocardial infarction in patients with left-ventricular dysfunction: the CAPRICORN randomised trial. Lancet 2001; 357:1385–1390.

141. Doughty RN, Whalley GA, Gamble GD, Walsh H, Sharpe N. Carvedilol and left ventricular remodeling following acute myocardial infarction: variable effects over time and possible mechanisms. The CAPRICORN echo substudy. Circulation 2004; 109:201–206.

142. Bristow MR. What type of beta-blocker should be used to treat chronic heart failure? Circulation 2000; 102:484–486.

143. Stevenson LW. Beta-blockers for stable heart failure. N Engl J Med 2002; 346: 1346–1347.

144. Dorn GW. Adrenergic pathways and left ventricular remodeling. J Card Fail 2002; 8:S370–S373.

145. Singh K, Xiao L, Remondino A, Sawyer DB, Colucci WS. Adrenergic regulation of cardiac myocyte apoptosis. J Cell Physiol 2001; 189:257–265.

146. Bisognano JD, Weinberger HD, Bohlmeyer TJ, Pende A, Raynolds MV, Sastravaha A, Roden R, Asano K, Blaxall BC, Wu SC, et al. Myocardial-directed overexpression of the human beta(1)-adrenergic receptor in transgenic mice. J Mol Cell Cardiol 2000; 32:817–830.

147. Engelhardt S, Hein L, Wiesmann F, Lohse MJ. Progressive hypertrophy and heart failure in beta1-adrenergic receptor transgenic mice. Proc Natl Acad Sci USA 1999; 96:7059–7064.

148. Vatner SF, Vatner DE, Homcy CJ. beta-adrenergic receptor signaling: an acute compensatory adjustment-inappropriate for the chronic stress of heart failure? Insights from $G_{s\alpha}$ overexpression and other genetically engineered animal models. Circ Res 2000; 86:502–506.

149. Liggett SB, Tepe NM, Lorenz JN, Canning AM, Jantz TD, Mitarai S, Yatani A, Dorn GW. Early and delayed consequences of beta(2)-adrenergic receptor overexpression in mouse hearts: critical role for expression level. Circulation 2000; 101:1707–1714.

150. Milano CA, Dolber PC, Rockman HA, Bond RA, Venable ME, Allen LF, Lefkowitz RJ. Myocardial expression of a constitutively active alpha 1B- adrenergic receptor in transgenic mice induces cardiac hypertrophy. Proc Natl Acad Sci USA 1994; 91(21): 10,109–10,113.

151. Lemire I, Ducharme A, Tardif JC, Poulin F, Jones LR, Allen BG, Hebert TE, Rindt H. Cardiac-directed overexpression of wild-type alpha1B-adrenergic receptor induces dilated cardiomyopathy. Am J Physiol Heart Circ Physiol 2001; 281:H931–H938.

152. Di Lenarda A, Sabbadini G, Salvatore L, Sinagra G, Mestroni L, Pinamonti B, Gregori DCiani F, Muzzi A, Klugmann S, et al. Long-term effects of carvedilol in idiopathic dilated cardiomyopathy with persistent left ventricular dysfunction despite chronic metoprolol. The Heart-Muscle Disease Study Group. J Am Coll Cardiol 1999; 33: 1926–1934.

153. Kukin ML, Kalman J, Charney RH, Levy DK, Buchholz-Varley C, Ocampo ON, Eng C. Prospective, randomized comparison of effect of long-term treatment with metoprolol or carvedilol on symptoms, exercise, ejection fraction, and oxidative stress in heart failure. Circulation 1999; 99:2645–2651.

154. Metra M, Giubbini R, Nodari S, Boldi E, Modena MG, Dei Cas L. Differential effects of beta-blockers in patients with heart failure: A prospective, randomized, double-blind comparison of the long-term effects of metoprolol versus carvedilol. Circulation 2000; 102:546–551.

155. Sanderson JE, Chan SK, Yip G, Yeung LY, Chan KW, Raymond K, Woo KS. Beta-blockade in heart failure: a comparison of carvedilol with metoprolol. J Am Coll Cardiol 1999; 34:1522–1528.

156. Yaoita H, Sakabe A, Maehara K, Maruyama Y. Different effects of carvedilol, metoprolol, and propranolol on left ventricular remodeling after coronary stenosis or after permanent coronary occlusion in rats. Circulation 2002; 105:975–980.

157. Packer M, Antonopoulos GV, Berlin JA, Chittams J, Konstam MA, Udelson JE. Comparative effects of carvedilol and metoprolol on left ventricular ejection fraction in heart failure: results of a meta-analysis. Am Heart J 2001; 141:899–907.

158. Poole-Wilson PA, Swedberg K, Cleland JG, Di Lenarda A, Hanrath P, Komajda M, Lubsen J, Lutiger B, Metra M, Remme WJ, et al. Carvedilol Or Metoprolol European Trial Investigators. Comparison of carvedilol and metoprolol on clinical outcomes in patients with chronic heart failure in the Carvedilol Or Metoprolol European Trial (COMET): randomised controlled trial. Lancet 2003; 362:7–13.

159. Heart Failure Society of America (HFSA) practice guidelines. HFSA guidelines for management of patients with heart failure caused by left ventricular systolic dysfunction–pharmacological approaches. J Card Fail 1999; 5:357–382.

160. Gheorghiade M, Gattis WA, Gallup DS, Chandler AB, Chu AA, Hasselblad V, O'Connor CM. Pre-discharge initiation of carvedilol in patients stabilized during an admission for decompensated heart failure is a effective strategy to improve the use of beta-blocker therapy: results of the IMPACT-HF trial. Circulation 2003; 108:668.

161. Akhter SA, Milano CA, Shotwell KF, Cho MC, Rockman HA, Lefkowitz RJ, Koch WJ. Transgenic mice with cardiac overexpression of alpha1B-adrenergic receptors. In vivo alpha1-adrenergic receptor-mediated regulation of beta-adrenergic signaling. J Biol Chem 1997; 272:21,253–21,259.

162. McDonald KM, Garr M, Carlyle PF, Francis GS, Hauer K, Hunter DW, Parish T, Stillman A, Cohn JN. Relative effects of alpha 1-adrenoceptor blockade, converting enzyme inhibitor therapy, and angiotensin II subtype 1 receptor blockade on ventricular remodeling in the dog. Circulation 1994; 90:3034–3046.

163. The ALLHAT Officers and Coordinators for the ALLHAT Collaborative Research Group. Major Cardiovascular Events in Hypertensive Patients Randomized to Doxazosin vs. Chlorthalidone: The Antihypertensive and Lipid-Lowering Treatment to Prevent Heart Attack Trial (ALLHAT). JAMA 2000; 283:1967–1975.

164. The ALLHAT Officers and Coordinators for the ALLHAT Collaborative Research Group. Major outcomes in high-risk hypertensive patients randomized to angiotensin-converting enzyme inhibitor or calcium channel blocker vs. diuretic: the antihypertensive and lipid-lowering treatment to prevent heart attack trial (ALLHAT). JAMA 2002; 288: 2981–2997.

165. Thomas L, Gasser B, Bousquet P, Monassier L. Hemodynamic and cardiac anti-hypertrophic actions of clonidine in Goldblatt one-kidney, one-clip rats. J Cardiovasc Pharmacol 2003; 41:203–209.

166. Giles TD, Thomas MG, Quiroz A, Rice JC, Plauche W, Sander GE. Acute and short-term effects of clonidine in heart failure. Angiology 1987; 38:537–548.

167. Manolis AJ, Olympios C, Sifaki M, Handanis S, Bresnahan M, Gavras I, Gavras H. Suppressing sympathetic activation in congestive heart failure. A new therapeutic strategy. Hypertension 1995; 26:719–724.

168. Messerli F. Moxonidine: a new and versatile antihypertensive. J Cardiovasc Pharmacol 2000; 35:S53–S56.

169. Coats AJ. Heart Failure 99—the MOXCON story. Int J Cardiol 1999; 71:109–111.

170. Pfeffer MA, Swedberg K, Granger CB, Held P, McMurray JJ, Michelson EL, Olofsson B, Ostergren J, Yusuf S, Pocock S. CHARM Investigators and Committees. Effects of candesartan on mortality and morbidity in patients with chronic heart failure: the CHARM-Overall programme. Lancet 2003; 362:759–766.

171. McMurray JJ, Ostergren J, Swedberg K, Granger CB, Held P, Michelson EL, Olofsson B, Yusuf S, Pfeffer MA. CHARM Investigators and Committees. Effects of candesartan in patients with chronic heart failure and reduced left-ventricular systolic function taking angiotensin-converting-enzyme inhibitors: the CHARM-Added trial. Lancet 2003; 362:767–771.

172. Granger CB, McMurray JJ, Yusuf S, Held P, Michelson EL, Olofsson B, Ostergren J, Pfeffer MA, Swedberg K. CHARM Investigators and Committees. Effects of candesar-

tan in patients with chronic heart failure and reduced left-ventricular systolic function intolerant to angiotensin-converting-enzyme inhibitors: the CHARM-Alternative trial. Lancet 2003; 362:772–776.

173. Yusuf S, Pfeffer MA, Swedberg K, Granger CB, Held P, McMurray JJ, Michelson EL, Olofsson B, Ostergren J. CHARM Investigators and Committees. Effects of candesartan in patients with chronic heart failure and preserved left-ventricular ejection fraction: the CHARM-Preserved Trial. Lancet 2003; 362:777–781.

20

The Effects of Cardiac Revascularization on the Remodeling Process

Victor A. Ferrari, Craig H. Scott, and Martin St. John Sutton

Noninvasive Imaging Laboratory, Cardiac Imaging Program,
Cardiovascular Medicine Division, University of Pennsylvania
Medical Center, Philadelphia, Pennsylvania, U.S.A.

INTRODUCTION

Myocardial infarction (MI) results from the interaction of multiple cardiac risk factors and a genetic predisposition for atherosclerosis that triggers intramural plaque instability, rupture of the fibrous cap and acute thrombotic occlusion. The resulting interuption in coronary flow causes loss of contracting myocytes. The clinical outcome of acute myocardial infarction depends upon a number of important factors including initial infarct size, infarct thickness (i.e., partial vs. transmural) infarct location, and local tissue factors. History of prior infarction, diabetes and/or hypertension and the presence of early ventricular arrhythmias, also affects the outcome.

Early reperfusion of the infarct [either spontaneously or by opening the infarct-related artery with thrombolytic therapy or percutaneous coronary intervention (PCI)], suppression of life-threatening ventricular arrhythmias, blunting of neuro-hormonal activation with angiotensin converting enzyme (ACE) inhibitors or angiotensin receptor blockers, beta adrenergic receptor biockers, and aldosterone antagonists, and blunting of platelet actions with aspirin have all improved mortality and morbidity following acute myocardial infarction. However, in spite of these interventions, a proportion of patients experience progressive cardiac dilatation caused by a dynamic process known as ventricular remodeling that culminates in congestive heart failure. The importance of this process is emphasized by its strong association with the risk of future cardiovascular events. Progressive left ventricular (LV) remodeling occurs in between one-third and one-half of survivors of acute myocardial infarction with ejection fractions <40% (1). It is the cause of the majority of cases of heart failure in patients with systolic dysfunction in the United States and it will account for around 2/3 of new cases, diagnosed each year.

The purpose of this article is to describe the effects of early and late reperfusion on postinfarction remodeling of (LV) myocardium and to outline potential mechanisms by which the beneficial impact on clinical outcome may be mediated. This requires a brief outline of recent insights into the current understanding of the patho-physiology, biochemistry, and mechanics of postinfarction remodeling.

VENTRICULAR REMODELING

LV remodeling is a well-characterized process through which the heart responds to cellular injury and to changes in loading conditions. Remodeling may be physiologic and reversible, or pathologic and irreversible—or only partially reversible. The simplest form of physiologic remodeling occurs with normal growth during which the heart increases in size many fold from birth to adulthood but maintains similar chamber architecture by an almost constant relationship between cavity volume and myocardial mass. Similarly, during pregnancy the heart increases its output between 30% and 40% and then returns back to its normal baseline within three months of parturition. Pathologic irreversible remodeling is triggered by an increase in ventricular loading conditions. This may be due to a chronic condition as in hypertension or aortic stenosis or as a consequence of an acute increase in loading conditions as occurs in myocardial infarction where there is a loss of contracting myocytes. Postinfarction LV remodeling is a dynamic process. The major component is progressive ventricular dilatation that is associated with deterioration in contractile function, distortion of cavity geometry, and mitral regurgitation. LV dilatation begets mitral regurgitation by disrupting the architecture of the mitral annulus and subvalve apparatus and in a vicious cycle mitral regurgitation begets further dilatation (2). The extent of remodeling reflects a change in the dynamic equilibrium between the distending forces and the restraint imposed by the extracellular matrix (ECM), which provides a visco-elastic collagen scaffold.

POSTINFARCTION REMODELING

Immediately after acute myocardial infarction, the ischemic region stops contracting (3) and within one hour, the infarct zone expands (stretches) and the LV begins to diiate. Stroke volume and ejection fraction acutely decrease as both ischemic and nonischemic myocardium adapt to the mechanical overload from early LV dilatation and myocardial stretching (4,5). Initially the Frank–Starling mechanism in nonischemic myocardium partially compensates for the reduction in stroke volume (4,6,7). However, an increase in regional wall stress and loss of myocytes causes progressive LV dilatation and decreased ejection fraction (2). Postinfarction remodeling is arbitrarily divided into an early phase (<72 hours), confined to the infarct zone (infarct stretching/expansion), and a late phase (>72 hours) that involves stretching of the entire LV myocardium including the adjacent border zone and normally perfused remote myocardium. Early infarct expansion due to wall thinning increases ventricular volumes resulting in increased wall stress (2,6,8). Further infarct expansion and LV dilatation occur over the first few days, but then cease in compensated hearts with small infarcts (9–11). Late LV dilatation involves both infarcted and remote ventricular myocardium and continues in large anterior transmural infarcts that are not reperfused (9–20). This continued remodeling often progresses to congestive heart failure.

LV remodeling is initially characterized by progressive dilatation, distortion of chamber architecture, mural hypertrophy, and deterioration in contractile function. It involves structural changes in the myocytes and the extracellular collagen framework, which if unchecked result in heart failure. An increase in LV chamber size leads to a change in the dynamic equilibrium between collagen synthesis and degradation, favoring the extracellular collagen matrix degradation and realignment of the myofilaments to obviate disruption of the sarcomeric proteins. In addition, left ventricular hypertrophy results in an increase in wall thickness that redistributes wall force homogeneously to the left ventricular walls. Remodeling following the abrupt loss of contracting myocytes begins within one hour after acute myocardial infarction and is initiated by the abrupt increase in ventricular loading conditions (21). Left ventricular dilatation increases the mitral valve annular diameter, which in turn disturbs the normal geometry of the mitral valve apparatus causing mitral regurgitation. The volume overload imposed by mitral regurgitation further increases left ventricular loading conditions and may escalate the deterioration in contractile function that leads to the development of heart failure. Thus, left ventricular dilatation begets mitral regurgitation, heart failure, and sudden death from ventricular arrhythmias (1,22). Development of heart failure imposes limitations on exercise capacity impairs quality of life. Leads to recurrent hospital admission and has grave clinical prognostic implications for survival.

Wall Stress

Regional wall stress is the major driving force for LV remodeling after acute myocardial infarction (2,6,7,10,23,24). Current therapeutic interventions that beneficially affect LV remodeling and survival after infarction are based on strategies designed to reduce wall stress by limiting LV dilatation: (1) by reducing LV loading conditions with angiotensin converting enzyme (ACE) inhibitors, angiotensin receptor blocking (ARB) agents (25–36), nitrates (37), β-blockers (38–41), and mechanical assist devices (42,43), (2) by salvaging myocytes with early myocardial reperfusion with primary angioplasty, thrombolytic therapy or spontaneous thrombolysis (44,45) or (3) by altering the material properties of the myocardium by late reperfusion well after the time window for myocyte salvage. Late reperfusion increases myocardial stiffness, thereby attenuating stretch and subsequent ventricular dilatation (46,47).

Remodeling has both mechanical and biochemical origins. Wall stress is the most important mechanical parameter that precipitates remodeling. It is related to left ventricular cavity pressure, wall thickness, and cavity dimensions/geometry. Wall stress increases with cavity pressure and radius of curvature of the LV by the law of LaPlace. The physiologic response of myocardium to a prolonged increase in wall stress is hypertrophy since an increase in wall thinckness will normalize the strain. This is the reason why diseases that increase afterload result in concentric LV hypertrophy. Infarcted tissue, however, cannot hypertrophy. Wall stress increases in the infarcted myocardium soon after infarction. Neurohormonal activation caused by anxiety, stimulation of baroreceptors due to relative hypotension, and other hemodynamic factors contribute to the increase in wall stress. Plasma norepinephrine, atrial and brain natriuretic peptides, angiotensin, and aldosterone rise soon after infarction and can remain elevated for several days until myocardial repair begins. High sympathetic tone increases afterload, rate of oxygen consumption, and heart rate, all of which contribute to the increase in wall stress. An increase in concentration of angiotensin II leads to constriction of arterial resistance vessels and fluid retention thereby resulting in an increase in both preload and afterload. Increased aldosterone levels postinfarction help to maintain blood pressure by the retention of sodium, thereby causing plasma volume expansion. However, chronically elevated levels of aldosterone cause pathological myocardial and vascular fibrosis.

Increased wall stress will result in hypertrophy of noninfarcted regions of the heart in order to normalize wall stress. When there is insufficient viable myocardium and/or if the hypertrophic response is insufficient, two detrimental processes that affect the structure of the extra-cellular matrix (ECM) can occur: (1) increased wall stress can cause lengthening of collagen fibers by mechanical means and (2) surface mechanoreceptors transduce intracellular signaling for hypertrophy and activate matrix metalloproteinases (MMP) which then soften the collagen scaffold or interstitial matrix of the infarcted region, thereby worsening collagen fiber lengthening. There are many different MMPs classified into general categories that include the collagenases, the gelatinases, the stromelysins/matrilysins, and the membrane-type MMPs. The MMPs are regulated by tissue inhibitors of matrix metalloproteinases (TIMPs), which blunt its activity. In the living cell, the MMPs and TIMPs assist in the repair of collagen by providing a baseline level of collagen turnover. An excess of MMP over TIMP may initiate the changes seen in remodeling, favoring destruction of collagen. In areas of infarction, certain species of MMP rise while TIMP levels fall precipitously (48) (Fig. 1).

Most research has concentrated on systolic wall stress, which is significantly higher in the infarct region. However, some researchers have noted an increase in remote

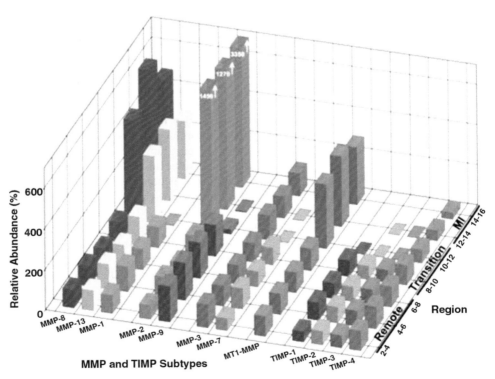

Figure 1 Regional variation in MMP and TIMP activity in remodeling myocardium. A significant increase in MT1-MMP and a reduction in TIMP-4 activity are noted in the transition and infarct territories. A marked increase in MMP-8, MMP-2, and MMP-13 activity were present in the transition and infarcted regions, while MMP-1 and MMP-9 levels were virtually undetectable in the infarcted tissue. TIMP levels were likewise markedly decreased in infarcted areas. *Abbreviations*: MMP, Matrix metalloproteinases; TIMP, tissue inhibitors of matrix metalloproteinases *Source*: Modified from Ref. 48. (*See color insert.*)

zone diastolic wall stress due to increased filling pressures. This pressure increase results in the increased expression of endothelin-1 and atrial naturetic peptide (49) in the remote zone, but not the infarct zone or the border zone of the infarction. At this time, it is unclear whether these findings are important clinically, considering the fact that most therapies that reduce systolic strain will eventually reduce diastolic strain.

Extracellular Matrix

The importance of the ECM in regulating postinfarction remodeling has only recently been appreciated. The ECM comprises a stress-tolerant scaffold composed of collagen types I and III (>90% of total myocardial collagen). The ECM has a number of functions including harnessing myocytes with inter-myofilament collagen struts that insure parallel fiber alignment for optimal force development (50–53). It also maintains critical spatial relations between myofilaments and the capillary microcirculation and it prevents ventricular dilatation (54). Thus, the ECM forms a mechanical–chemical interface that enables the myocardium to respond to the environmental signals such as changes in pressure and volume (stretch) (55). This bidirectional interaction between environmental and intracellular signaling is mediated via integrins which serve as mechano-transducers and play a fundamental role in the interaction between the myocytes and fibroblasts (56–59). Integrins do not possess enzymatic activity, but trigger growth factors and cytokines by mechanical stretch-induced changes in the integrin-cytoskeleton complex. Thus, mechanical stretch plays a pivotal role by activating a number of signaling pathways postinfarction.

Intracellular Signaling

Myocyte necrosis following acute myocardial infarction releases a number of growth factors and signaling proteins involved in the regulation and integration of two important parallel processes that take place during myocardial repair: the expression of early response genes for synthesis of fibrillar collagen type I and type III, and collagen degradation (60). Specific signaling proteins each play a role in myocardial repair (61–67), important among which is TNFα-1, a proinflammatory cytokine that activates MMPs causing loss of fibrillar collagen, progressive degradation of the ECM, and ventricular dilatation (5,8,62,64,65,68,92,95). ET-1 is a signaling protein and a stretch-induced arterial vasoconstrictor that plays a role in neurohumoral activation and reducing myocardiai contractile function (67). TGFβ-1 is a profibrogenic growth factor (66,69–72) that stimulates macrophage and fibroblast chemotaxis, fibroblast proliferation, and phenotypic transformation of fibroblasts to myofibroblasts (60). Myofibroblasts express genes encoding for fibrillar procollagen types I and III (60) that are modulated by complex costimulatory regulatory factors including angiotensin II, fibrobiastic growth factor (FGF), platelet derived growth factor (PDGF), bradykinin (BK), and aldosterone.

Matrix Metalloproteinases

Collagen degradation is resistant to proteolytic enzymes except a family of matrix metalloproteinases (MMPs) (55,60,73,74). MMPs are controlled at 3 levels: transcription, activation of proenzymes, and inhibition by specific endogenous tissue inhibitor metalloproteinases (TIMPS) (63,74,75). TNFα-1 activates MMPs and TGFβ-1 upregulates expression of MMPs 2 and 9, while decreasing the expression of MMP-1 and MMP-3. Plasmin acts on the propeptide domain of MMPs, that cause conformational changes that allow cleavage of the activation site by other MMPs, and thus, initiating the cascade of collagen degradation (61,75) and simultaneous inactivation of TIMPS (76). Collagen degradation begins within three hours of AMI in the infarct zone. Restraining acute infarct expansion with Marlex mesh decreased MMP-1 and MMP-2 activities, altered the ratio of type I/type III fibrillar collagen in the remodeling border zone myocardium, and increased collagen accumulation in remote myocardium (77). While studies in different animal models have postulated that myocardial stretch (strain) modulates LV remodeling, we demonstrated a direct relationship between the regional myocardial strain and the MT1-MMP activity and the TIMP-4 activity in a sheep infarct model (48).

Apoptosis and Remodeling

Preclinical work in rodent models indicates that the majority of cell death following coronary occlusion occurs via apoptosis in the first two to four hours, followed by necrosis from 6 to 24 hours (78,79). In addition, reperfusion itself in acute myocardial infarction was initially felt to stimulate additional apoptotic mechanisms, resulting in further cell death (80). However, subsequent works proved that reperfusion facilitated apoptosis in nonviable zones, and in fact, prevented this process in viable tissue (78). Apoptosis has been implicated in the chronic postinfarction remodeling process, particularly in the remote (noninfarcted) zone (81). In a recent study, apoptotic rates were found to be increased in the remote myocardium of patients who died subsequent to infarction. These investigators also found increased apoptotic rates in patients with occluded infarct-related arteries, late after infarction (82), and a much greater negative impact of an occluded infarct-related artery in high-risk patients (ischemic cardiomyopathy with low EF), suggesting a syngeneic, and potentially causal relationship between ischemia, neurohumoral activation, and apoptosis (83). Thus, future strategies designed to limit remodeling could potentially be targeted to antiapoptotic pathways, and may include broader recommendations for revascularization (84).

CLINICAL STUDIES WITH NONINVASIVE IMAGING

The acute changes in LV size and the immediate onset of regional contractile dysfunction/loss of wall thickening were appreciated after acute infarction in experimental animal models several decades ago. However, the concept of progressive LV remodeling did not emerge until noninvasive imaging enabled serial assessment of LV size and function in the late 1970s. Two seminal studies by Erlebacher and colleagues using quantitative echocardiography demonstrated that LV dilatation occurred postinfarction in almost 50% of patients with anterior infarction (14). In addition, early LV enlargement occurring within 72 hours was confined to the infarct zone (infarct expansion), whereas late LV dilatation (beyond 72 hours) involved the entire myocardium, even the normally perfused tissue remote from the infarct zone.

Almost a decade later, the temporal sequence and magnitude of postinfarction LV remodeling were clarified in the placebo controlled treatment arms of large multicenter clinical trials in which longitudinal quantitative echocardiography was used prospectively to assess the effects of ACE inhibitor therapy on adverse cardiovascular events long-term (6,20,29). These studies unequivocally showed that LV dilatation was a powerful predictor of adverse cardiovascular events including death, reinfarction, and heart failure, and that LV end-systolic size was the strongest predictor of clinical outcome postinfarction (19,20). A plethora of subsequent clinical trials corroborated the important clinical consequences of unchecked LV remodeling after myocardial infarction.

REVASCULARIZATION AND INFARCT ARTERY PATENCY

Two important but unrelated studies were published, implicating the impact of coronary artery patency on postinfarction LV remodeling. The first was a small observational study (13) in which patients, one month after their index infarction, with an open infarct-related coronary artery or a closed vessel but collateral flow to the infarct zone, sustained significantly less LV dilatation than those patients with an occluded vessel and no collateral flow to the infarct. The second study demonstrated that patients in the GISSI trial who received thrombolytic therapy had less LV dilatation and improved mortality due to arrested LV remodeling (85). In addition, in patients matched for initial infarct size and in whom the status of infarct-related vessel patency was known, those patients with perfused infarcts had significantly smaller end-systolic volumes than patients with nonperfused infarcts—suggesting that reperfusion exerts a restraining effect on infarct expansion beyond its primary effect of infarct size reduction (86). The mortality benefits of thrombolysis early postinfarction has been confirmed in numerous thrombolytic trials (87–89). These and other studies espoused the "open artery hypothesis" as being critically important in limiting infarction size, and thereby the clinical outcome. Subgroup analysis of larger trials and subsequent focused studies suggested that some patients derived benefit from revascularization late (>12 hours) after symptom onset, and that perhaps an open infarct-related artery was advantageous beyond the period of myocardial salvage (90–92).

Early Reperfusion

Early reperfusion postinfarction salvaged myocytes, and the earlier the administration of thrombolytic therapy, the less LV dilatation, the smaller the infarction, the better LV function, and the greater the impact on morbidity and mortality. The beneficial effect of thrombolysis on LV size and function were demonstrable in the earliest phases of acute infarction but subsequent remodeling was primarily mediated through patency of the infarct-related artery (93). Thus, thrombolytic therapy and infarct artery patency have additive and complementary effects on reducing LV dilatation postinfarction. Coronary reocclusion of the infarct-related artery within three months after successful thrombolysis is associated with progressive LV dilatation (94) and deterioration in LV function five years after the index myocardial infarction (Fig. 2) (95). Similarly, patients who develop the "no reflow" phenomenon after initial coronary reflow experience more extensive remodeling, deterioration in pump function, and greater incidence of congestive heart failure and cardiovascular death (96). Early reperfusion with primary angioplasty confers additional benefits to thrombolytic therapy which are believed to result from establishing superior early flow that translates into greater myocardial salvage and improved survival compared with thrombolysis as shown by Bolognese et al. (87,97,98). However, early reperfusion and a patent infarct-related artery does not altogether preclude subsequent LV remodeling, nor does it alter the predictive power of age, initial infarct size, or LV size (99). These investigators found that patients with the most severe myocardial damage had the greatest continuous stimulus for remodeling. They also found that the pattern of dilatation (occurring at 24 hours, one month, or

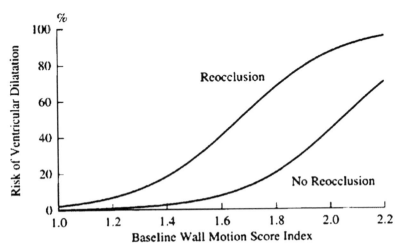

Figure 2 Relationship between baseline WMSI and risk of ventricular dilatation, relative to presence of reocclusion. WMSI predicts dilatation, which is always augmented in the setting of reocclusion. For example, for a given WMSI (1.8), the risk of dilatation in patients with no reocclusion is 21% vs. 68% in patients with reocclusion. *Abbreviation*: WMSI, wall motion score index. *Source*: Modified from Ref. 95.

six months) was independent of the type of reperfusion used and the clinical outcome. The simple presence of a >20% increase in end-diastolic volume from baseline to six months postinfarction was the most powerful predictor of long-term risk, even in patients with small initial end-diastolic volumes (Fig. 3).

Angiographic parameters for assessment of the microcirculation have been proven to have prognostic importance as well, and aided the interpretation of a "lytic-plus-rescue percutaneous coronary intervention (PCI) at 90 minutes" substudy of the TIMI 10B trial. While rescue PCI decreased overall long-term mortality, improved microvascular perfusion (using TIMI myocardial perfusion grades) prior to PCI also resulted in improved mortality—regardless of the infarct artery patency or performance of PCI (100). These data suggest that early microvascular perfusion is vital to a favorable remodeling and prognosis. Whether improved microvascular integrity can be achieved by the earliest possible reperfusion (including prehospital lytic therapy) and/or adjunctive PCI therapies (combination platelet inhibition/ anticoagulants, etc.) remains to be proven.

A recent study expanded on these findings, and demonstrated that after primary coronary angioplasty for acute infarction, the loss of the integrity of the microvascular circulation as evaluated by intracoronary myocardial contrast echo-cardiography was an important determinant both of LV remodeling and adverse long-term outcome (Fig. 4) (101). This study confirms the principle that despite infarct-reiated artery patency, dysfunctional microcirculation due to embolization of plaque-related platelet thrombus, platelet aggregation, or reperfusion injury to the vascular bed remains a major factor in adverse LV remodeling. Until recently, few studies have accounted for the integrity of the microcirculation, which has been a major shortcoming in comparing the studies of outcome and prognosis, particularly in the early post-MI period. Ideally, a truly noninvasive tool for evaluation of the microcirculation would be preferred for serial measurements and risk stratification. In a preclinical study using intravenous contrast-enhanced MRI methods,

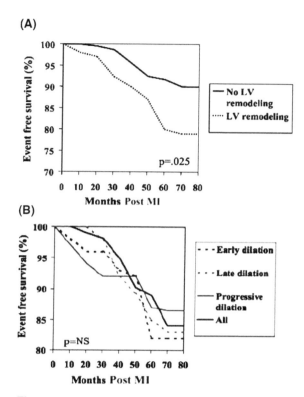

Figure 3   (A) EVS in patients with and without LV remodeling. Note greater EVS in non-remodeling group at each time point. (B) The pattern of LV dilatation did not significantly impact event-free survival. *Abbreviations*: EVS, event-free survival; LV, left ventricular. *Source*: Modified from Ref. 99.

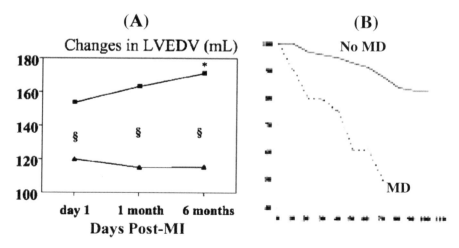

Figure 4   (A) Change in LV EDV over time with (■) and without (▲) MD. Note increased EDV at all time points in MD group. (B) Event-free survival vs. time in groups with and without MD. Event-free survival was significantly greater at all time points in the non-MD group. *Abbreviations*: EDV, end-diastolic volume; MD, microvascular dysfunction; *Source*: Modified from Ref. 10.

Lima et al. (102) demonstrated that infarct size and microvascular obstruction could be quantified and followed serially.

In a subsequent study, it was found that during the early (<48 hours) reperfusion period, the extent of microvascular obstruction (EMO) using MRI techniques predicted LV dilatation independent of infarct size, and that the EMO correlated with the degree of myocardial dysfunction in the infarct and adjacent zones (103). When applied to patients after first MI, Wu et al. (104) observed an increased risk of cardiac complications two years post-MI, and found that EMO was an independent predictor of the outcome (Fig. 5).

Late Reperfusion

Establishing flow in the infarct-related coronary artery early after acute myocardial infarct, either with primary angioplasty and stenting or thrombolysis salvages myocytes, minimizes infarct size and preserves LV function. For this reason, primary coronary intervention early after the onset of acute chest pain syndrome with elevated cardiac enzymes has become the optimal standard therapeutic strategy. In the mid 1980s through late 1990s, a body of evidence was developed using several experimental animal models. It was shown that late reperfusion after acute myocar-

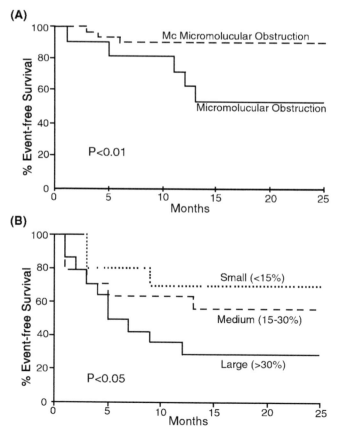

Figure 5 Panel A, Event-free survival in patients with and without MO, and Panel B, by size of MO region. *Abbreviation*: MO, microvascular obstruction. *Source*: Modified from Ref. 104.

dial infarction—beyond the time for myocyte salvage and infarct reduction—had beneficial effects on attenuating LV dilatation and contractile dysfunction, compared with controls without reperfusion (105,106). Although no clearly defined mechanism accounts for the beneficial effects of late reperfusion, accelerated rate of infarct repair, and replacement of necrotic myocytes with scar, without changes in the infarct size has been postulated (46,107,108). A further explanation worthy of consideration is that late reperfusion alters the material properties of the infarct and border zone myocardium, thus limiting ventricular dilatation, and thereby preserving contractile function. Similar salutary effects of late reperfusion were described in man in small observational studies (109) which were independent of the limitation of infarct size (110). These initial findings have been corroborated by multiple subsequent studies (90,91,95) and beg the question, "Is myocardial revascularization ever too late in patients who have severe residual stenoses in the infarct-related artery?"

DIAGNOSIS/PROGNOSIS

We can predict with some certainty which infarctions will remodel and which will not. In general, indicators of infarct size and transmural extent, state of the coronary arteries supplying the infarcted area, and markers of increased wall strain are all useful in predicting remodeling. Peak and mean values for CPK correlate well with infarct size. Echocardiographic findings of dyskinesis early after infarction indicate transmural infarction. If more than one-third of the myocardium is infarcted, remodeling is likely. The presence of Q waves early after infarction also suggests a more transmural event. The number of leads with Q waves and ST segment elevation correlates with infarct size. Coronary artery distribution suggested by echo or ECG or confirmed by selective coronary angiography is also very helpful. The left anterior descending (LAD) typically supplies the greatest percentage of myocardium, including the LV apex. Therefore, LAD infarction is more likely to remodel and the more proximal the infarct lesion, the more the myocardium involved. Percutaneous revascularization failure, either by the inability to cross the culprit lesion or by the "no reflow" phenomenon, where the anatomic blockage is relieved but flow is not reestablished, increase the likelihood of remodeling. In fact, after controlling for infarct size and LV function, the extent of microvascular dysfunction as estimated by intracoronary myocardial contrast echo was the most powerful predictor of LV dilatation (101). The authors felt that microvascular obstruction (MO), despite restoration of epicardial flow, could represent the "pathophysiological missing link" between reperfusion, remodeling, and the outcome in acute myocardial infarction, mediated primarily through altering the mechanical properties of the infarcted (and adjacent) myocardium. A newer noninvasive method using intravenous contrast-enhanced MRI techniques can accurately estimate infarction size, degree of MO, and provide prognostic information in humans independent of infarction size. The presence of MO in patients after a first anterior MI was associated with a 17% greater risk of cardiovascular events (death, stroke, congestive heart failure) at two years post-MI, while those with no MO had an excellent prognosis (104).

Echocardiography

Transthoracic echocardiography (TTE) is an excellent noninvasive diagnostic modality which can be used to guide therapy during infarction, can determine the risk of complications, including remodeling, and allows for the longitudinal evaluation of infarcted myocardium useful for both clinical and research information. Specifically, TTE can assess global and regional LV function, the extent of an acute cardiac injury, evaluate complications such as acute mitral regurgitation, ventricular septal defect, abnormal diastolic filling, etc. With time, infarcted myocardium thins, the chamber dilates, and in the worst cases, aneurysms of the left ventricle can cause thrombus formation in areas of hypokinesis, or akinesis may occur and these can be readily assessed by TTE.

Early risk stratification with TTE is possible. Patients with persistent wall motion abnormalities have more severe ischemia and adverse events (111). Patients without signs of cardiogenic shock can still have significant regional wall motion abnormalities (111). In addition, increased end-systolic volume, decreased ejection fraction, and the presence of even small degrees of mitral regurgitation predict more negative outcomes. Doppler findings of diastolic dysfunction are present with large infarctions and evidence of restrictive mitral filling also predicts a poorer outcome (112). Repeated TTE measurements have been shown to be an excellent way to follow and study remodeling (113).

Three-dimensional TTE is a new technique for the assessment of LV function and allows for improved assessment of LV size, shape, and volumes. It has become the gold standard for echocardiography-based research. Other echocardiographic techniques such as tissue Doppler imaging, strain rate imaging, and tissue characterization can also be used to assess remodeling. Techniques such as harmonic imaging or the use of contrast agents that can traverse the lungs improve the ability of TTE to assess regional wall motion in patients with technical limitations. Finally, myocardial contrast echocardiography (MCE) is a type of perfusion imaging allowing rapid assessment of the area of myocardium at risk after acute coronary occlusion (114). MCE has been used to assess blood flow after reperfusion. Up to 30% of arteries with TIMI III (Thrombolysis in Myocardial Infarction trial) flow after revascularization exhibit no-reflow phenomenon and necrosis. MCE has the potential to demonstrate a lack of capillary perfusion in those areas, thereby providing additional prognostic information beyond what the TIMI grading provides.

Cardiac Magnetic Resonance

Recent studies have demonstrated the ability of cardiac magnetic resonance (CMR) to accurately detect the extent of transmural fibrosis after infarction and to distinguish viable from nonviable myocardium, even within the first hour after infarction (115,116). This technique is also useful in determining the potential for recovery of function after revascularization (117–119). One very important advantage of CMR techniques is that they permit detailed assessment of remodeling myocardium, with respect to infarct resorption, remote zone hypertrophy, and adjacent zone structure and function (120). These methods permit early and late risk stratification in patients with CAD, and compare favorably (and are frequently superior) to existing studies, particularly regarding detection of nontransmural infarctions (Fig. 6) (121). As discussed earlier, the ability of CMR techniques to depict the extent of microvascular obstruction and the degree of transmural involvement using a straightforward noninvasive imaging method permits serial assessment of remodeling and evaluation of prognosis. Their noninvasive nature and relative ease of use provide valuable data on a broad range of patients, and should contribute important data for decisions regarding revascularization of individual patients.

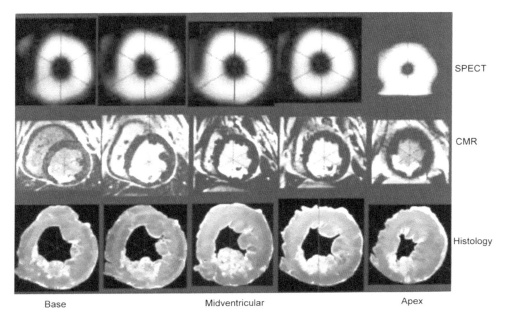

SPECT

CMR

Histology

Base   Midventricular   Apex

Figure 6 Comparison of SPECT and CMR images compared with histology (TTC-staining) in a canine infarct model. It was found that the reduced spatial resolution of SPECT led to underdetection of nontransmural infarctions, which was confirmed in patient examinations. *Abbreviations*: SPECT, single photon emission computed tomography; CMR, cardiac magnetic reasonance. *Source*: Modified from Ref. 121. (*See color insert.*)

Cardiac Catheterization

Diagnostic catheterization performed during an acute infarction can quickly assess the coronary anatomy, in order to determine the infarct artery patency and interventional methods, such as angioplasty with or without stenting, which can alleviate the obstruction and restore flow to the infarct artery. Catheterization can be used as a primary technique if quickly available, or in conjunction with thrombolytic therapy as a salvage technique if thrombolysis fails to open the artery, or as a diagnostic tool prior to coronary artery bypass grafting. As discussed above, TIMI grading of flow can risk-stratify patients. Catheterization ill be discussed more fully in the Therapy section.

Prediction Algorithms

Quantitative echocardiography, creatinine phosphokinase, hemodynamics, infarction location, patient history, and other factors can be combined to improve our ability to predict remodeling post-infarction, deKam et al. (122). have published an algorithm which can predict, with 80% accuracy, whether an infarction will remodel using data acquired at the time of infarction; specifically, the gender, peak CPK, echocardiographically determined LV volume indices, and the interaction between peak CPK and time. The most important component of this algorithm is the echocardiographic measurement of volume which should be acquired as soon as possible, since significant dilatation can occur even in the first 72 hours after the initial event.

THERAPY

We now understand that reduction of wall stress and restoring blood flow to the infarct area, as early as possible, can minimize myocyte damage and improve the tensile strength of formed scar, thereby limiting infarct size and remodeling (123). The best results are achieved when these therapies are begun as soon as practical after infarction is diagnosed (124). In fact, despite successful reperfusion, longer than two to three hours after symptoms begin, myocardial salvage is decreased, and recovery of LV function is only modest (125,126) (Fig. 7). Much of the inflammation and scar formation occurs within the first few days to a week after infarction and is usually complete by one to two months. Any recovery of function due to stunning will occur in that time frame, and strategies aimed at providing the greatest potential improvement in remodeling and reducing the loss of ventricular function should be instituted as early as possible during this period.

Selection of Patients for Revascularization

Urgent Revascularization

The best way to minimize myocardial damage after infarction is to restore blood flow to the infarct artery as soon as possible. While reperfusion may not change the fate of irreversibly injured myocytes at the infarct core (due to death via apoptotic or necrotic mechanisms), maintaining blood flow to stunned or border zone myocardium will help prevent loss of additional myocytes and potentially stabilize these territories. Reperfusion injury is a well-known consequence of restitution of blood flow to acutely infarcted myocardium, which occurs via free radical injury and other

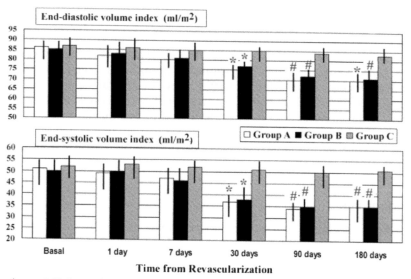

$* = p < 0.05, \# = p < 0.01$

Figure 7: Effect of primary PTCA on LV post-infarction remodeling. Greatest benefit is seen in patients who undergo PTCA less than 4 hours after infarction. Time after PTCA - Group A: < 2 h, Group B: 2 to 4 h, Group C: > 4 h. *Abbreviations*: PTCA, percutaneous transluminal coronary angioplasty; LV, left ventricular. *Source*: Modified from Ref. 126.

mechanisms, and may even induce additional apoptotic mechanisms (80). However, there is an obvious benefit in limiting the ongoing ischemic injury and facilitating the delivery of nutrients, oxygen, and blood-borne reparative agents [endogenous and exogenous (i.e., drug therapies)].

Three revascularization strategies are available: thrombolytic therapy, primary angioplasty/stenting (PCI) or emergency CABG. The choice of technique depends on the time to presentation of the patient, the status of the patient, contraindications to the procedures, and availability of the therapy. If primary angioplasty is available, there is a small additional benefit of this approach over thrombolytics (127,128). However, this benefit is greatest if time-to-treatment for catheterization is less than 90 minutes from symptom onset (126,129). Recent data suggest that for patients presenting within two to three hours of symptom onset, if PCI is unavailable or would be delayed by greater than 90 minutes, fibrinolytic therapy is the best option (provided that contraindications are absent). Even paramedic-administered lytic therapy in the field, prior to hospital arrival, is under investigation as an alternative strategy to reduce time to reperfusion, but studies are small at present, and the ultimate risk/benefit is uncertain.

Patients presenting after two to three hours from symptom onset with continued pain and/or ST-segment elevations, but with a PCI delay of greater than two hours, should receive lytic therapy (absent contraindications) and be considered for early PCI. Those patients presenting more than six hours from symptom onset will achieve only modest benefit from lytic therapy, but angiography will provide useful information regarding suitability for PCI versus CABG. This period of benefit may be greater (perhaps up to 12 hours) in patients with established coronary collateral circulation, continued pain, and persistent ST-segment elevation (130). The potential benefits of opening occluded vessels after even greater delays (>48 hours) are the subject of ongoing clinical investigations (Occluded Artery Trial, TOSCA) (131,132).

Studies evaluating multiple-vessel angioplasty show it to be safe and at least as effective as CABG (133). In the acute MI patient, however, it is important to remember that the infarct artery patency is of primary concern. Stenting has improved infarct-related and the overall artery patency in general. If the patient has contraindications to thrombolytic use, then primary angioplasty is indicated. If the patient has contraindications to both thrombolytic therapy and PCI, emergent CABG should be considered. In addition, emergent CABG is indicated for acute MI under the following circumstances: cardiogenic shock, left main or proximal left anterior descending coronary artery disease, evidence of extensive stunned or hibernating myocardium, infarction complicated by acute mitral regurgitation or ventricular septal defect, or after a failed angioplasty. Likewise, in patients with life-threatening ventricular arrhythmias and $\geq 50\%$ left main stenosis and/or triple vessel disease, emergent CABG is warranted (134).

Delayed Revascularization

Revascularization may have some benefit at 12 to 24 hours or even weeks after the initial event (135), though definitive proof via randomized clinical trial data is currently lacking. While infarct size may not be affected directly, border zone and/or stunned myocardium at risk for infarct extension may be salvaged, thereby reducing the risk of remodeling. Under less urgent conditions, the choice of revascularization technique in multivessel disease (multivessel stenting vs. CABG) remains uncertain. In the Arterial Revascularization Therapy Study, stenting was favored at one year in a cost analysis with comparable outcomes otherwise. At three years, however, the cumulative cost of treatment of restenosis favored CABG (136). The effect of drug-eluting stents on outcome and cost/benefit remain to be determined, but could favor stenting (137). Adjunctive therapy to promote greater success with PCI (improved patency, with fewer adverse effects) is evolving, and larger trials utilizing new techniques which include cost/benefit analyses are being evaluated (138,139).

Viable/Hibernating Myocardium

Myocardial stunning refers to the temporary mechanical dysfunction occurring after a transient severe ischemic episode, which is reversible over time without PCI or other intervention (140). Hibernating myocardium, however, represents the chronically dysfunctional tissue distal to a severe stenosis (or stenoses), which has been subjected to persistent or repeated ischemia. This tissue has markedly downregulated metabolic function, and may be capable of improved mechanical function after revascularization by PCI or CABG (141,142). Whether hibernation precedes apoptosis is uncertain; however, large number of apoptotic cells have been detected in areas of hibernating myocardium (143). There appears to be a careful balance between the severity of chronic hypoperfusion and the signaling mechanisms for apoptosis (144,145). Some investigators have suggested that the incomplete functional recovery of hibernating myocardium post-revascularization may be partly due to apoptotic activity (146).

Viability may be defined in a number of ways. Functional recovery after revascularization is a commonly used method to describe and test for viable myocardium. However, nontransmural infarction may complicate assessment of wall thickening by its effect on subendocardial mechanical function. Specifically, resting hypokinesis at follow-up evaluation post-revascularization [due to prior subendocardial infarction (SEMI) or tethering] may lead to underestimation of the impact of revascularization on midwall and subepicardial function (147). Likewise, the response to inotropic stimulation is another method used to detect viable myocardium in the presence of baseline hypokinesis. However, passive wall thickening and circumferential shortening can mimic active contractile behavior in the setting of SEMI (148); and this could exaggerate the potential for recovery. Recovery of function is also affected by the completeness of revascularization, intervening ischemic events, and adjunctive medical therapy. There are several studies documenting the benefits of revascularization regardless of the lack of recovery of global LV function (149) (Fig. 8). Thus, viable myocardium should not be defined by recovery of mechanical function alone.

In fact, the presence of viable myocardium is actually a double-edged sword. When revascularized, improved outcomes and global function may result. However, the presence of viable/hibernating tissue, which is not revascularized, is associated with unfavorable outcomes in long-term follow-up of patients with ischemic cardiomyopathy. Although it would seem preferable to have viable tissue present but, if it remains vulnerable to ischemia, the risk of unfavorable cardiac outcomes is greater than if the tissue was nonviable (150–152).

Given the potential inaccuracies of viability testing, is there sufficient justification for testing patients for viable myocardium prior to revascularization? In the group of patients with moderate or greater risk for surgery and suboptimal target vessels, testing seems reasonable. However, in lower risk patients with ischemic cardiomyopathy and adequate target vessels, current data appear to favor revascularization, even in the observe of testing for viability.

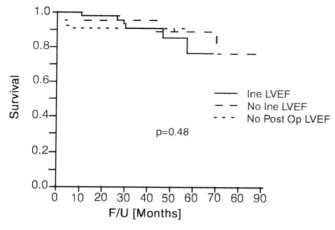

Figure 8 Actuarial survival free of cardiac death between study groups A (increased [Inc] in LVEF) and B (No increase in LVEF) and group C, patients in whom postoperative assessment of LVEF was not available. There were no statistically significant differences in survival free of cardiac death between the three groups. *Abbreviation*: LVEF, left ventricular ejection fraction. *Source*: Modified from Ref. 149.

Mechanical Support

The intra-aortic balloon pump (IABP) is a useful adjunct which decreases afterload and increases coronary perfusion pressure. It is commonly used in patients with cardiogenic shock. Although there have been no clinical studies to date evaluating the balloon pump in the prevention of remodeling, one preclinical study demonstrated a reduction of microvascular obstruction and limitation of the "no-reflow" phenomenon in IABP-supported dogs post-MI, as assessed with cardiac MRI techniques (153). However, another mechanical support device, the left ventricular assist device (LVAD) has been shown to affect numerous cellular and tissue adaptive responses in the direction of normal (154), including regression in fibrosis, reduction in myocardial apoptosis and myocytolysis, and improved myocyte function (155). Decreased neurohumoral activation, increased β-receptor density, and decreased concentration of serum cytokines have also been demonstrated with the LVAD. The trigger for these changes seems to be unloading of the myocardium with resultant beneficial changes in the cardiac and systemic neurohormonal pathways. However, these changes are typically insufficient to restore cardiac function in a clinically meaningful way.

Ventricular Surgery

There are multiple surgical procedures designed to decrease left ventricular size and therefore reduce wall strain by the law of LaPlace (155). The Dor procedure isolates the apex of the LV with a baffle. The Batista procedure resects a wedge of the free wall of the left ventricle from base to apex, while sparing the function of the papillary muscles. Surgical mortality is high and the procedures tend to work best if the surgery is done to resect a ventricular aneurysm. A new experimental device, the ACORN device, is a fiber sock that fits over the epicardial surface of the left ventricle. The properties of the device prevent excess diastolic expansion without adversely causing the mechanical equivalent of constrictive pericarditis. The resulting reduction in diastolic wall stress may prevent or even reverse remodeling. This device has shown great promise in animal trials (156) and is presently being evaluated in patients.

Medical Management

Medical therapy is utilized in all patients with acute MI and also in the chronic treatment of infarction survivors. Even if an invasive strategy is attempted, medical therapy helps to stabilize patients destined for the cardiac catheterization lab or operating room. Chronic use of medications can prevent or limit remodeling. In patients who are not eligible for intervention, due to comorbidities, failure to meet criteria for an interventional approach, delay in presentation, or other concerns, medications are usually the primary therapy. At present, a combination of ACE inhibitors or angiotensin receptor blocking (ARB) agents, beta blockers, and aldosterone receptor antagonists are given. Diuretic agents also administration when these is evidence of volume overload. The net effect of these agents is the alleviation of wall stress and a reduction in the neurohormonal stimulus for myocyte growth and fibrosis. The ACE inbitors and ARBs have been well studied and they show benefit in preventing remodeling and improving survival after infraction. The ACE inhibitors were first noted to improve survival in patients with cardiomyopathy (29,157), many of whom had ischemic cardiomyopathy. Patients with normal (158) and reduced (26) ejection fraction had improved survival, after the administration of the ACE inhibitors captopril and enalapril following infarction. The greatest benefit was seen when the ACE inhibition was begun within 24 hours of the onset of symptoms. Chronic ACE inhibition in patients with normal or depressed ejection fraction due to coronary artery disease diminishes the prevalence of heart failure, recurrent myocardial infarction, stroke, and death from cardiovascular causes, in part by attenuating progressive remodeling (35).

Beta blockade was not initially widely accepted due to the perception that the negative inotropy induced by this medication class would be detrimental in patients with left ventricular dysfunction. Studies have unequivocally demonstrated that beta blockers improve survival in all but the most unstable patients by multiple mechanisms: (1) Decreases in heart rate and contractility reduce oxygen consumption and may better preserve border space zone myocardium, (2) reductions in heart rate and contractility (and therefore blood pressure) also decrease wall stress acutely, thereby preventing infarct expansion; and (3) reductions in the incidence of malignant ventricular arrhythmias. Recent trials evaluated the use of carvedilol in patients with ischemic cardiomyopathy, demonstrating a survival benefit and an improvement in symptoms in more severe cases (EF < 25%, NYHA class III or IV) (159). Carvedilol has also been shown to prevent adverse ventricular remodeling after infarction. In patients with persistently reduced ejection fraction post-MI, carvedilol prevented chamber dilatation and remote zone hypertrophy when combined with conventional therapy (160).

Studies using intravenous nitroglycerine during the peri-infarction period (0–24 hours) limited infarct size and expansion (30,31,161); but long-term use of nitroglycerine did not confer additional benefit. Calcium channel blockers have no role in the treatment of infarction or the prevention of remodeling, except for a controversial role in treatment of the "no-reflow" phenomenon.

Acute activation of the renin/angiotensin/aldosterone system is detrimental during infarction due to an increased blood volume and therefore an increased preload. Chronically elevated levels can cause myocardial and vascular fibrosis. ACE inhibition will initially suppress aldosterone, but the levels eventually rise despite a high dose ACE inhibition (162). Spironolactone used with ACE inhibition reduces the mortality of heart failure patients (163), most of whom had significant CAD. In addition, the addition of eplerenone ACE inhibitor and beta blocked therapy improves survival in post-MI patients with LV dysfunction and evidence of heart failure (163).

It is unclear as to what agents should be administred preferentially in the patient with marginal blood pressure, in whom the combination of nitrates, ACE inhibitors, beta blockers, and diuresis would result in unacceptably low perfusion pressures. Some of the decision can be based on the clinical state of the patient. Massive volume overload with pending pulmonary edema will likely benefit from diuresis, while patients with frequent ectopy may benefit from the arrhythmia-suppressing effect of beta blockers. In general, however, the ACE inhibitor class holds the most benefit for a single agent given its ability to reduce preload and afterload. A comparison of captopril and the ARB losartan (164,165) shows a similar survival benefit in patients with cardiomyopathy, but better patient tolerance of losartan. Similar results were demonstrated when the ARB valsartun was compared to the ACE inhibitor captopril in post-MI patients (165,166). A rational approach would therefore be to use ACE inhibitors as first line therapy and ARBs if the ACE inhibitor is not tolerated.

CONCLUSION

The negative long-term effects of remodeling after myocardial infarction result in significant morbidity and mortality. Patients at higher risk for remodeling have larger, infarcts which are transmural in their extent. The concomitant effects of increased wall stress and abnormal neurohormonal and MMP activity cause early and continued infarct expansion. These processes can be aborted or minimized by an early intervention to restore blood flow after infarction, the use of afterload reduction, and perhaps, the direct blocking of MMP activity. The impact of microvascular obstruction on myocardial salvage and remodeling, despite epicardial vessel patency, is a topic of great interest and may hold important answers to limiting unfavorable remodeling early post-MI. Late reperfusion, myocardial apoptosis, and their mechanical and functional consequences remain topics of active investigation, which should yeid important therapeutic insights. Until this time, current recommendations indicate that late revascularization should be undertaken only in patients with symptomatic myocardial ischemia, ischemic LV dysfunction, or hemodynamic/electrical instability (134,166). Since remodeling increases risk of progression to heart failure and long-term mortality the early diagnosis and treatment of MI with more aggressive therapy aimed at preventing or attenuating this process will hopefully become more widespread. As it does, we can anticipate beneficial effects on the clinical course of patients at risk.

REFERENCES

1. St John Sutton M, Pfeffer M, Moye L, et al. Cardiovascular death and left ventricular remodeling two years after myocardiai infarction. Baseline predictors and impact of long term use of captopril: information from the survival and ventricular enlargement (SAVE) trial. Circulation 1997; 96:3294–3299.
2. Pfeffer M, Braunwald E. Ventricular remodeling after myocardial infarction: Experimental observations and clinical implications. Circulation 1990; 81:1161–1172.

3. Heyndrickx G, Amano J, Patrick T, et al. Effects of coronary arter reperfusion on regional myocardial blood flow and function in conscious baboons. Circulation 1985; 71:1029–1037.

4. Lew W, Chen Z, Guth B, et al. Mechanisms of augmented segment shortening in non-ischemic areas during acute ischemia of the canine left ventricle. Circ Res 1985; 56: 351–358.

5. Bogen D, Rabinowitz S, Needleman A, et al. An analysis of the mechanical disadvantage of myocardial infarction in the canine left ventricle. Circ Res 1980; 47:728–741.

6. Gaudron P, Eilles C, Kugler I, et al. Progressive left ventricular dysfunction and remodeling after myocardial function. Circulation 1993; 87:755–763.

7. Grossman W. Cardiac hypertrophy: Useful adaptation or pathologic process. Am J Med 1980; 69:576–584.

8. Erlebacher J, Weiss J, Eaton L, et al. Late effects of acute infarct dilation on heart size: A two dimensional echocardiographic study. Am J Cardiol 1982; 49:1120.

9. Stone P, Raabe D, Jaffe A, et al. Prognostic significance of location and type of myocardial infarction: Independent adverse outcome associated with anterior location. Journal American College of Cardiology 1988; ll:453–463.

10. Pouleur H, Rousseau M, van Eyll C, et al. Cardiac mechanics during development of heart failure. Circulation 1993; (suppl IV):IV14–IV20.

11. Hutchins G, Bulkley B. Infarct expansion versus extension: Two different complications of acute myocardial infarction. American Journal of Cardiology 1978; 41:1127–1132.

12. Visser C, David G, Meltzer R, et al. Two-dimensional echocardiography during PTCA. American Heart Journal 1986; 111:1035–1041.

13. Jeremy R, Hackwarthy R, Bautovich G, et al. Infarct artery perfusion and changes in left ventricular volume in the month after acute myocardial infarction. J Am Coll Cardiol 1987; 9:989–995.

14. Erlebacher J, Weiss J, Weisfeldt M, et al. Early dilation of the infarcted segment in acute transmural myocardial infarction: Role of infarct expansion in acute left ventricular enlargement. J Am Coll Cardiol 1984; 4:201–208.

15. Abernethy M, Sharpe N, Smith H, et al. Echocardiographic prediction of left ventricular volume after myocardial infarction. J Am Coll Cardiol 1991; 17:1527–1532.

16. Mitchell G, Lamas G, Vaughan D, et al. Left ventricular remodeling in the year after anterior myocardial infarction: A quantitative analysis of contractile segment lengths and ventricular shape. J Am Coll Cardiol 1992; 19:1136–1144.

17. Lamas G, Mitchell G, Flaker G, et al. Clinical significance of mitral regurgitation after acute myocardial infarction. Circulation 1997; 96:827–833.

18. Eaton L, Weiss J, Bulkley B, et al. Regional cardiac dilation after acute myocardial infarction. N Engl J Med 1979; 300:57.

19. White H, Norris R, Brown M, et al. Left ventricular end-systolic volume as the major determinant of survival after recovery from myocardial infarction. Circulation 1987; 76:44–51.

20. St John Surton M, Pfeffer M, Plappert T, et al. Quantitative two- dimensional echocardiographic measurements are major predictors of adverse cardiovascular events after acute myocardial infarction: the protective effects of captopril. Circulation 1994; 89:68–75.

21. Gheorghiade M, Bonow R. Chronic heart failure in the United States. A manifestation of coronary artery disease.. Circulation 1998; 98:282–289.

22. St John Sutton M, Lee D, Rouleau J-L, et al. Left ventricular remodeling and ventricular arrhythmias after myocardial infarction. Circulation 2003; 107:2577–2582.

23. McKay R; Pfeffer M, Pasternak R, et al. Left ventricular remodeling after myocardial infarction: A corollary to infarct expansion. Circulation 1986; 74:693–702.

24. Jackson B, Gorman JI, Salgo I, et al. Increased wall stress due to altered borderzone geometry as assessed by perfusion echocardiography. Am J Physiology 2003; 284: H475–H479.

25. Kober L, Torp-Pedersen C, Carlsen J, et al. A clinical trial of the angiotensin-converting-enzyme inhibitor trandolapril in patients with left ventricular dysfunction after myocardial infarction. Trandolapril Cardiac Evaluation (TRACE) Study Group. N Engl J Med 1995; 333:1670–1676.

26. Pfeffer M, Braunwald E, Moye L, et al. Effects of captopril on mortality and morbidity in patients with left ventricular dysfunction after myocardial infarction: Results of the Survival and Ventricular Enlargement Trial. N Engl J Med 1992; 327:669–667.

27. The Acute Infarction Rarmipril Efficacy (AIRE) Study Investigators. Effect of ramipril on mortality and morbidity of survivors of acute myocardial infarction with clinical evidence of heart failure. Lancet 1993; 342:821–828.

28. Ambrosioni E, Borghi C, Magnani B, et al. The effect of the angiotensin converting-enzyme inhibitor zofenopril on mortality and morbidity after anterior infarction. N Engl J Med 1995; 332:80–85.

29. The SOLVD Investigators. Effects of enalapril on survival in patients with reduced ejection fraction and congestive heart failure. N Engl J Med 1991; 325:293–302.

30. Gruppo Italiano per lo Studio della Sopravivenza nell'Infarto Miocardico (GISSI-3). Effects of lisinopril and transdermal glyceryl trinitrate singly and together on 6-week mortality and ventricular function after acute myocardial infarction. Lancet 1994; 343:1115–1122.

31. ISIS-4 Collaborative Group. Fourth International Study of Infarct Survival: a randomized factorial trial assessing early captopril, oral mononitrate, and intravenous magnesium sulphate in 58,050 patients with suspected acute myocardial infarction. Lancet 1995; 345:669–685.

32. Ertl G, Gaudron P, Hu K. Efficacy of early angiotensin-converting enzyme inhibitor therapy in attenuation of left ventricular remodelling and clinical outcome. In: St John Sutton M, ed. Left Ventricular Remodelling After Acute Myocardial Infarction. London: Science Press Ltd, 1996:60–76.

33. Pfeffer M, Lamas G, Vaughan D, et al. Effect of captopril on progressive ventricular dilatation after anterior myocardial infarction. N Engl J Med 1988; 319:80–86.

34. Sharpe N, Murphy J, Smith H, et al. Treatment of patients with symptomless left ventricular dysfunction after myocardial infarction. Lancet 1988; 1:255–259.

35. The Heart Outcomes Prevention Evaluation Study Investigators. Effects of an angiotensin-converting enzyme inhibitor, ramipril, on cardiovascular events in high risk patients. N Engl J Med 2000; 342:145–153.

36. Sun Y, Cleutjens J, Dias-Arias A, et al. Cardiac angiotensin enzyme and myocardial fibrosis in the rat. Cardiovasc Res 1994; 28:1423–1432.

37. Jugdutt B. Does early unloading of the left ventricle with intravenous nitrates after myocardial infarction prevent infarct stretching and progressive remodeling? In: St John Sutton M, ed. Left Ventricular Remodeling After Acute Myocardial Infarction. London: Science Press Ltd, 1996:11–29.

38. The MIAMI Trial Research Group. Metoprolol in acute myocardial infarction: Enzymatic estimation of infarct size. Am J Cardiol 1985; 56:27–33.

39. The International Collaborative Study Group. Reduction of infarct size with the early use of timolol in acute myocardial infarction. N Engl J Med 1984; 310:9–16.

40. ISIS-1 FISoIS. Collaborative Group: Randomized trial of intravenous atenolol among 16,027 cases of suspected acute myocardial infarction. Lancet 1986; 2:57–66.

41. Yusuf S, PetoR, Lewis J, et al. Beta blockade during and after myocardial infarction: An overview of the randomized trials. Prog Cardiovasc Dis 1985; 27:335–371.

42. McCarthy P, Golding L. Temporary mechanical circulatory support. In: Edmunds LJ, ed. Cardiac Surgery in the Adult. New York: McGraw-Hill, 1997:319–338.

43. Ratcliffe M, Bavaria J, Wenger R, et al. Left ventricular mechanics of ejecting, postischemic hearts during left ventricular circulatory assistance. J Thorac Cardiovasc Surg 1991; 101:345–355.

44. Braunwald E. The open-artery theory is alive and well again. N Engl J Med 1993; 329:1650–1652.

45. Lavie C, O'Keefe J, Chesebro J, et al. Prevention of the ventricular dilatation after acute myocardial infarction by successful thrombolytic reperfusion. Am J Cardiol 1990; 66:31–36.

46. Boyle M, Weisman H. Limitation of infarct expansion and ventricular remodeling by late reperfusion. Study of time course and mecahnism in a rat model. Circulation 1993; 88:2872–2883.

47. Libby P, Lee R. Matrix matters. Circulation 2000; 102:1874–1876.

48. Wilson E, Moainie S, Baskin J, et al. Region-and type-specific induction of matrix metalloproteinases in post-myocardial infarction remodeling. Circulation 2003; 107:2857–2863.

49. Loennechen J, Stoylen A, Beisvag V, et al. Regional expression of endothelian-1, ANP, IGF-1, and LV wall stress in the infarcted rat heart. Am J Phys Heart Circ Physiol 2001; 280:H2902–H2910.

50. Zhao M, Zhang H, Robinson T, et al. Profound structural alterations of the extracellular collagen matrix in post-ischemic dysfunctional ("stunned") but viable myocardium. Am J Cardiol 1987; 10:1322–1334.

51. Whitaker P. Unravelling the mysteries of collagen and cicatrix after myocardial infarction. Cardiovasc Res. 1996; 31:19–27.

52. Whitaker P, Boughner D, Kloner R. Analysis of healing after myocardial infarction using polarized light microscopy. Am J Pathol 1989; 134:879–893.

53. Franzen L, Ghassemifar M, Salerud G, et al. Actin fiber orientation in connective tissue contraction: a quantitative study with the perforated rat mesentery model. Wound Rep Reg 1996; 4:454–460.

54. Cheung P-Y, Sawicki GI, Wozniak M, et al. Matrix metalloproteinase-2 contributes to ischemia-reperfusion injury in the heart. Circulation 2000; 101:1833–1839.

55. Ross R, Borg T. Integrins and the myocardium. Circ Res 2001; 88:1112–1119.

56. MacKenna D, Dolfi F, Vuori K, et al. Extracellular signal-regulated kinase and c-jun NH2-terminal kinase activation by stretch is integren dependent and matrix specific in rat cardiac fibroblasts. J Clin Invest 1998; 101:301–310.

57. Urbich C, Walter D, Zeiher A, et al. Laminar shear stress upregulates integrin expression: Role in endothelial cell adhesion and apoptosis. Circ Res 2000; 87:683–689.

58. Ingber D. Mechanical signaling and cellular response to extracellular matrix in angiogenesis and cardiovascular physiology. Circ Res 2002; 91:877–887.

59. Kuppuswamy D. Importance of integrin signaling in myocyte growth and survival. Circ Res 2002; 90:1240–1242.

60. Weber K. Extracellular matrix remodeling in heart failure. A role for de novo angiotensin II generation. Circulation 1997; 96:4065–4082.

61. Spinale F. Bioactive peptide signaling within the myocardial interstitium and the matrix metalloproteinases. Circ Res 2002; 91:1082–1084.

62. Bozurt B, Kribbs S, Clubb F, et al. Pathophysiologically relevant concentrations of tumor necrosis factor-α promote progressive left ventricular dysfunction and remodeling in rats. Circulation 1998; 1382–1391.

63. Spinale F. Matrix metalloproteinases: regulation and dysregulation in the failing heart. Circ Res 2002; 520–530.

64. Mann D. Inflammatory mediators and the failing heart: past, present, and the foreseeable future. Circ Res 2002; 988–998.

65. Sivasubrumanian N, Coker M, Kurrelmeyer K, et al. Left ventricular remodeling in transgenic mice with cardiac restricted overexpression of tumor necrosis factor. Circulation 2001; 104:826–831.

66. Kuwahara F, Kai H, Tokuda K, et al. Transforming growth factor-β function blocking prevents myocardial fibrosis and diastolic dysfunction in pressure-overloaded rats. Circulation 2002; 106:130–135.

67. Ergul A, Walker C, Goldberg A, et al. ET-1 in the myocardial interstitium: relation to myocyte ECE activity and expression. Am J Physiol 2000; 278:H2050–H2056.
68. Li Y, Feng Y, Kadokami T, et al. Myocardial extracellular matrix remodeling in transgenic mice overexpressing tumor necrosis factor α can be modulated by anti-tumor necrosis factor α therapy. Proc Nat Acad Sci 2000; 97:12,746–12,751.
69. Casscells W, Basenbeny F, Speir E, et al. Transforming growth factor beta-1 in normal heart and in myocardial infarction. Ann NY Acad Sci 1990; 593:148–160.
70. Massague J. The TGF-(β) family of growth and differentation factors. Cell 1987; 49:437–438.
71. Border W, Ruoslahti E. Transforming growth factor-(β) in disease: The dark side of tissue repair. J Clin Invest 1992; 90:1–7.
72. Weber K, Brilla C, Janicki J. Myocardial fibrosis: Functional significance and regulatory factors. Cardiovasc Res. 1993; 27:341–348.
73. Spinale F, Zellner J, Johnson W, et al. Cellular and extracellular remodeling with the development and recovery from tachycardia- induced cardiomyopathy: changes in fibrillar collagen, myocyte adhesion capacity and proteoglycans. J Mol Cell Cardiol 1996; 28:1591–1608.
74. Dollery C, McEwan J, Henney A. Matrix metalloproteinases and cardiovascular diseases. Circ Res 1995; 77:864–868.
75. Mann D, Spinale F. Activation of matrix metalloproteinases in the failing human heart. Breaking the tie that binds. Circulation 1998; 98:1699–1702.
76. Tyagi S, Lewis K, Pikes D, et al. Stretch-induced membrane type matrix metalloproteinase and tissue plasminogen activator in cardiac fibroblast cells. J Cell Physiol 1998; 176:374–382.
77. Bowen F, Jones S, Narula N, et al. Restraining acute infarct expansion decreases collagenase activity in borderzone myocardium. Ann Thorac Surg 2001; 72:1950–1956.
78. Fliss H, Gattinger D. Apoptosis in ischemic and reperfused rat myocardium. Circ Res 1996; 79:949–956.
79. Kajstura J, Cheng W, Reiss K, et al. Apoptotic and necrotic myocyte cell deaths are independent contributing variables of infarct size in rats. Lab Invest 1996; 74:86–107.
80. Gottlieb R, Burleson K, Kloner R, et al. Reperfusion injury induces apoptosis in rabbit cardiomyocytes. J Clin Invest 1994; 94:1621–1628.
81. Abbate A, Biondi-Zoccai G, Bussani R, et al. Increased myocardial apoptosis in patients with unfavorable left ventricular remodeling and early symptomatic post-infarction heart failure. J Am Coll Cardiol 2003; 41:753–760.
82. Abbate A, Bussani R, Biondi-Zoccai G, et al. Persistent infarct-related artery occlusion is associated with an increased myocardial apoptosis at postmortem examination in humans late after an acute myocardial infarction. Circulation 2002; 106:1051–1054.
83. Abbate A, Biondi-Zoccai G, Bussani R, et al. High-risk clinical features predict increased post-infarction myocardial apoptosis and the benefits as a result of an open infarct-related artery. Eur J Clin Invest 2003; 33:662–668.
84. Mani K, Kitsis R. Myocyte apoptosis: Programming ventricular remodeling. J Am Coll Cardiol 2003; 41:761–764.
85. Marino P, Zanolla L, Zardini P. Effect of streptokinase on left ventricular modeling and function after myocardial infarction: The GISSI (Gruppo Italiano per lo Studio della Streptochinasi nell'Infarto Miocardico) Trial. JACC 1989; 14:1149–1158.
86. Marino P, Destro G, Barbieri E, et al. Reperfusion of the infarct-related coronary artery limits left ventricular expansion beyond myocardial salvage. Am Heart J 1992; 123:1157–1165.
87. The GUSTO Angiographic Investigators. The effects of tissue plasminogen activator, streptokinase, or both on coronary-artery patency, ventricular function, and survival after acute myocardial infarction. N Engl J Med 1993; 329:1615–1622.

88. Gruppo Italiano per lo Studio della Sopravivenza nell'Infarto Miocardico (GISSI). Effectiveness of intravenous thrombolytic therapy in acute myocardial infarction. Lancet 1986; 1:397–402.

89. Dalen J, Gore J, Braunwald E, et al. Six and twelve-month follow-up of the phase I Thrombolysis in Myocardial Infarction (TIMI) trial. Am J Cardiol 1988; 62:179–185.

90. LATE Study Group. Late Assessment of Thrombolytic Effect study with alteplase 6-24 hours after onset of acute myocardial infarction. Lancet 1993; 342:759–766.

91. Nidorf S, Siu S, Galambos G, et al. Benefit of late coronary perfusion on ventricular morphology and function after myocardial infarction. J Am Coll Cardiol 1993; 21:683–691.

92. Kim C, Braunwald E, et al. Potential benefits of late reperfusion of infarcted myocardium. The open artery hypothesis. Circulation 1993; 88:2426–2436.

93. Popovic A, Neskovic A, Babic R, et al. Independent impact of thrombolytic therapy and vessel patency on left ventricular dilation after myocardial infarction. Serial echocardiographic follow-up. Circulation 1994; 90:800–807.

94. Meijer A, Verheugt F, van Eenige M, et al. Left ventricular function at three months after successful thrombolysis: impact of reocclusion without reinfarction on ejection fraction, regional function, and remodeling. Circulation 1994; 90:1706–1714.

95. Nijland F, Kamp 0, Verheugt F, et al. Long-term implication of reocclusion on left ventricular size and function after successful thrombolysis for first anterior myocardial infarction. Circulation 1997; 95:111–117.

96. Ito H, Maruyama A, Katsuomi I, et al. Clinical implications of the 'no reflow' phenomenon: A predictor of complication and left ventricular remodeling in reperfused anterior wall myocardial infarction. Circulation 1996; 93:223–228.

97. Zijlstra F, Hoomtje J, de Boer M, et al. Long-term benefit of primary angioplasty as compared with thrombolytic therapy for acute myocardial infarction. NEJM 1999; 341:1413–1419.

98. Grines C, Browne K, Marco J, et al. A comparison of immediate angioplasty with thrombolytic therapy for acute myocardial infarction. N Engl J Med 1993; 328:673–679.

99. Bolognese L, Neskovic A, Parodi G, et al. Left ventricular remodeling after primary coronary angioplasty: Patterns of Left Ventricular Dilation and Long-Term Prognostic Implications. Circulation 2002; 106:2351–2357.

100. Gibson C, Cannon C, Murphy S, et al. Relationship of the TIMI myocardial perfusion grades, flow grades, frame count, and percutaneous coronary intervention to long-term outcomes after thrombolytic administration in acute myocardial infarction. Circulation 2002; 105.

101. Bolognese L, CarrabbaN, Parodi G, et al. Impact of microvascular dysfunction on left ventricular remodeling and long-term clinical outcome after primary coronary angioplasty for acute myocardial infarction. Circulation 2004; 109:l121–1126.

102. Lima J, Judd R, Bazille A, et al. Regioanl heterogeneity of human myocardial infarcts demonstrated by contrast-enhanced MRI: potential mechanisms. Circulation 1995; 92:1117–1125.

103. Gerber B, Rochitte C, Melin J, et al. Microvascular obstruction and left ventricular remodeling early after myocardial infarction. Circulation 2000; 101:2734–2741.

104. Wu K, Kim R, Bluemke D, et al. Quantification and time course of microvascular obstruction by contrast-enhanced echocardiography and magnetic resonance imaging following acute myocardial infarction and reperfusion. J Am Coll Cardiol 1998; 32:1756–1764.

105. Hochman J, Choo H. Limitation of myocardial infarct expansion by reperfusion independent of myocardial salvage. Circulation 1987; 75:299–306.

106. Michael L, Ballantyne C, Zachariah J, et al. Myocardial infarction and remodeling in mice: Effect of reperfusion. Am J Physiol 1999; 277:H660–H668.

107. Vincent R, Murry C, Reimer K. Healing of myocardial infarcts in dogs: Effects of late reperfusion. Circulation 1995; 92:1891–1901.

108. Force T, Kemper A, Leavitt M, et al. Acute reduction in functional infarct expansion with late coronary reperfusion: assessment with quantitative two-dimensional echocardiography. J Am Coll Cardiol 1988; 11:192–200.

109. Lamas G, Pfeffer M, Braunwald E. Patency of the infarct-related coronary artery and ventricular geometry. Am J Cardiol 1991; 68:41–51.

110. Hirayama A, Adachi T, Asada S, et al. Late reperfusion for acute myocardial infarction limits the dilation of left ventriular without the reduction of infarct size. Circulation 1993; 88:2565–2574.

111. Romano S, Dagianti A, Penco M. Usefulness of echocardiography in the prognostic evaluation of non-Q wave myocardial infarction. Am J Cardiol 2000; 86:43G–45G.

112. Cerisano G, Bolognese L, Carrabba N. Doppler-derived mitral deceleration time: A early strong predictor of LV remodeling after reperfused anterior acute myocardial infarction. Circulation 1999; 99:230–236.

113. Korup E, Kober L, Torp-Pedersen C. Prognostic usefulness of repeated echocardiographic evaluation after acute myocardial infarction (TRACE study group). Am J Cardiol 1999; 83:1559–1562.

114. Lepper W, Hoffmann R, Kamp O. Assessment of myocardial reperfusion by intravenous myocardial contrast echocardiography and coronary flow reserve after primary percutaneous transluminal coronary angioplasty in patients with acute myocardial infarction. Circulation 2000; 101:2368–2374.

115. Kim R, Fieno D, Parrish T, et al. Relationship of MRI delayed contrast enhancement to irreversible injury, infarct age, and contractile function. Circulation 1999; 100:1992–2002.

116. Fieno D, Kim R, Chen E, et al. Contrast-enhanced magnetic resonance imaging of myocardium at risk: distinction between reversible and irreversible injury throughout infarct healing. J Am Coll Cardiol 2000; 36:1985–1991.

117. Kim R, Wu E, Rafael A, et al. The use of contrast-enhanced magnetic resonance imaging to identify reversible myocardial dysfunction. N Engl J Med 2000; 343:1445–1453.

118. Beek A, Kuhl H, Bondarenko O, et al. Delayed contrast-enhanced magnetic resonance imaging for prediction of regional functional improvement after acute myocardial infarction. J Am Coll Cardiol 2003; 42:895–901.

119. Rogers W, Kramer C, Geskin G, et al. Early contrast-enahnced MRI predicts late functional recovery after reperfused myocardial infarction. Circulation 1999; 99:744–750.

120. Fieno D, Hillenbrand H, Rehwald W, et al. Infarct resorption, compensatory hypertrophy, and differing patterns of ventricular remodeling following infarctions of varying size. J Am Coll Cardiol 2004; 43:2124–2131.

121. Wagner A, Mahrholdt H, Holly T, et al. Contrast-enhanced MRI and routine single photon emission computed tomography (SPECT) perfusion imaging for detection of subendocardial infarcts: an imaging study. Lancet 2003; 361:374–379.

122. deKam P, Nicolosi G, Voors A. Prediction of six months left ventricular dilatation after myocardial infarction in relation to cardiac morbidity and mortality. Application of a new dilatation model to GISSI-3 data.. Eur Heart J 2002; 23:536–542.

123. Milavetz J, Giebel D, Christian T, et al. Time to therapy and salvage in myocardial infarction. J Am Coll Cardiol 1998:31.

124. ACE Inhibitor MI Collaborative Group. Indications for ACE inhibitors in the early treatment of acute myocardial infarction: systematic overview of individual data from 100,000 patients in randomized trials. Circulation 1998; 97:2202–2212.

125. Brodie B, Stuckey T, Wall T, et al. Importance of time to reperfusion for 30-day and late survival and recovery of left ventricular function after primary angioplasty for acute myocardial infarction. J Am Coll Cardiol 1998; 32:1312–1319.

126. Sheiban I, Fragasso G, Rosano G, et al. Time course and determinants of left ventricular function recovery after primary angioplasty in patients with acute myocardial infarction. J Am Coll Cardiol 2001; 38:464–471.

127. Weaver W, Simes R, Betriu A. Comparison of primary coronary angioplasty and intravenous thrombolytic therapy for acute myocardial infarction: a quantitative review. JAMA 1997; 278:2093–2098.
128. Keely E, et al. Primary angioplasty versus intravenous thrombolytic therapy for acute myocardial infarction: A quantitative review of 23 randomised trials. Lancet 2003; 361:13–20.
129. Nallamothu B, Bates E. Percutaneous coronary intervention versus fibrinolytic therapy in acute myocardial infarction: is timing (almost) everything? Am J Cardiol 2003; 42:824–826.
130. Giugliano R, Braunwald E. Selecting the best reperfusion strategy in ST-elevation myocardial infarction. It's all a matter of time. Circulation 2003:108.
131. Sadanandan S, Buller C, NMenon V, et al. The late open artery hypothesis: a decade later. Am Heart J 2001; 142:411–421.
132. Dzavik V, Carere R, Mancini G, et al. Predictors of improvement in left ventricular function after percutaneous revascularization of occluded coronary arteries: A report from the Total Occlusion Study of Canada (TOSCA). American Heart Journal 2001; 142:301–308.
133. BARI Investigators. Comparison of coronary bypass surgery with angioplasty in patients with multivessel disease. The bypass angioplasty revascularisation investigation (BARI) investigators. N Engl J Med 1996; 335:217–225.
134. Antman E, Smith S, Alpert J, et al. ACC/AHA Practice Guidelines for the Management of Patients with ST-Elevation Myocardial Infarction - Executive Summary. Circulation 2004; 110:588–636.
135. Pizzetti G, Belotti G, Margonato A, et al. Coronary recanalization by elective angioplasty prevents ventricular dilatation after anterior myocardial infarction. J Am Coll Cardiol 1996; 28:837–845.
136. Legrand V, Serruys P, Unger F, et al. Three-year outcome after coronary stenting versus bypass surgery for the treatment of multivessal disease. Circulation 2004; 109.
137. Saia F, Lemos P, Lee C, et al. Sirolimus-eluting stent implantation in ST-elevation acute myocardial infarction: a clinical and angiographic study. Circulation 2003; 108: 1927–1929.
138. Stone G, Grines C, Cox D, et al. Comparison of angioplasty with stenting, with or without abciximab, in acute myocardial infarction. N Engl J Med 2002; 346:957–966.
139. Lincoff A, et al., for the REPLACE-2 Investigators. Bivalirudin and provisional glycoprotein IIb/HIa blockade compared with heparin and planned glycoprotein IIb/IHa blockade during percutaneous coronary intervention: REPLACE-2 randomized trial. JAMA 2003; 289:853–863.
140. Braunwald E, Kloner R. The stunned myocardium: Prolonged, postischemic ventricular dysfunction. Circulation 1982; 66:1146–1149.
141. Braunwald E, Rutherford J. Reversible ischemic left ventricular dysfunction: Evidence for the 'hibernating' myocardium. J Am Coll Cardiol 1986; 8:1467–1470.
142. Wijns W, Vatner S, Camici P. Hibernating myocardium. N Engl J Med 1998; 339: 173–181.
143. Lim H, Fallavollita J, Hard R, et al. Profound apoptosis-mediated regional myocyte loss and compensatory hypertrophy in pigs with hibernating myocardium. Circulation 1999; 100:2380–2386.
144. Chen C, Ma L, Linfert D, et al. Myocardial cell death and apoptosis in hibernating myocardium. J Am Coll Cardiol 1997; 30:1407–1412.
145. Elsasser A, Müller K, Slkcwara W, et al. Severe energy deprivation of human hibernating myocardium as possible common pathomechanism of contractile dysfunction, structural degeneration and cell death. J Am Coll Cardiol 2002; 39:1189–1198.
146. Elsasser A, Schlepper M, Klövekorn W, et al. Hibernating myocardium - an incomplete adaptation to ischemia. Circulation 1997; 96:2920–2931.

147. Armstrong W. "Hibernating" myocardium: asleep or part dead. J Am Coll Cardiol 1996; 28:530–535.
148. Garot J, Bluemlce D, Osman N, et al. Transmural contractile reserve after reperfused myocardial infarction in dogs. J Am Coll Cardiol 2000; 36:2339–2346.
149. Samady H, Elefteriades J, Abbott B; et al. Failure to improve left ventricular function after coronary revascularization for ischemic cardiomyopathy Is not associated with worse outcome. Circulation 1999; 100:1298–1304.
150. Lee K, Marwick T, Cook S, et al. Prognosis of patients with left ventricular dysfunction, with and without viable myocardium after myocardial infarction: Relative efficacy of medical therapy and revascularization. Circulation 1994; 90:2687–2694.
151. Gioia G, Powers J, Heo J, et al. Prognostic value of rest-redistribution tomographic thallium-201 imaging in ischemic cardiomyopathy. Am J Cardiol 1995; 75:759–762.
152. Williams M, Odabashian J, Lauer M, et al. Prognostic value of dobutamine echocardiography in patients with left ventricular dysfunction. J Am Coll Cardiol 1996; 27: 132–139.
153. Amado L, Kraitchman D, Gerber B, et al. Reduction of "no-reflow" phenomenon by intra-aortic balloon counterpulsation in a randomized magnetic resonance imaging experimental study. J Am Coll Cardiol 2004; 43:1291–1298.
154. Margulies K. Reversal mechanisms of left ventricular remodeling: lessons from left ventricular assist device experiments. J Card Fail 2002; 8:S500–S505.
155. Alfieri O, Maisano F, Schreuder J. Surgical methods to reverse left ventricular remodeling in congestive heart failure. Am J Cardiol 2003; 91:81F–87F.
156. Pilla J, Bloom A, Brockman D, et al. Ventricular constraint using the ACORN cardiac support device reduces myocardial akinetic area in an ovine model of acute infarction. Circulation 2002; 106:1207–1211.
157. The CONSENSUS Trial Study Group. Effects of enalapril on mortality in severe congestive heart failure. Results of the Cooperative North Scandinavian Enalapril Survival Study (CONSENSUS). N Engl J Med 1987; 316:1429–1435.
158. The SOLVD Investigators. Effect of enalapril on mortality and the development of heart failure in asymptomatic patients with reduced left ventricular ejection fractions. N Engl J Med 1992; 327:685–691.
159. Packer M, Coats A, Fowler M, et al. Effect of carvedilol on survival in severe chronic heart failure. N Engl J Med 2001; 344:1651–1658.
160. Senior R, Basu B, Kinsey C, et al. Carvedilol prevents remodeling in patients with left ventricular dysfunction after acute myocardial infarction. Am Heart J 1999; 137: 646–652.
161. Jugdutt B, Warnica J. Intravenous nitroglyerin therapy to limit myocardial infarct size, expansion, and complications; effect of timing, dosage, and infarct location. Circulation 1988; 78:906–919.
162. Struthers A. Aldosterone escape during angiotensin-converting enzyme inhibitor therapy in chronic heart failure. J Card Fail 1996; 2:47–54.
163. Pitt B, Zannad F, Remme W, et al. The effect of spironolactone on morbidity and mortality in patients with severe heart failure. N Engl J Med 1999; 341:709–717.
164. Pitt B, Remme W, Zannad F, et al. (2003) Eplerenone, a selective aldosterone blocket, in patients with left ventricular dysfunction after myocardial infraction. N Engl J Med 348:1309–1321.
165. Pitt B, Segal R, Martinez F, et al. Randomized trial of losartan versus I captopril in patients over 65 with heart failure (Evaluation of Losartan in the Elderly, ELITE). Lancet 1997; 349:747–752.
166. Pitt B, Poole-Wilson P, Segal R, et al. Effect of losartan compared with captopril on mortality in patients with symptomatic heart failure: randomised trial-the Losartan Heart Failure Survival Study ELITE II. Lancet 2000; 355:1582–1587.

167. Pfeffer MA, McMurray JJ, Velazquez et al. Valsartan, captapril or both in myocardial infraction complicated by heart failure left ventricular dysfunction or both. N Engl J Med (2003); 349:1893–1906.

168. Smith SJ, Dove J, Jacobs A, et al. ACC/AHA guidelines for percutaneous coronary interventions (revision of the 1993 PTCA guidelines). J Am Coll Cardiol 2001; 37:2215–2239.

21

Reverse Remodeling After Heart Valve Replacement and Repair

Lynne Hung and Shahbudin H. Rahimtoola
Griffith Center, Division of Cardiovascular Medicine, Department of Medicine, LAC-USC Medical Center, Keck School of Medicine, University of Southern California, Los Angeles, California, U.S.A.

INTRODUCTION

Ventricular remodeling may be defined as the alteration of the left ventricular (LV) size, shape, mass, and function. The altered loading conditions of various valve lesions results in changes in LV size, shape, and mass. Superimposed changes in LV function leads to additional changes in these parameters. Following correction of the valve lesion(s), these changes in cardiac structure often return to or toward normal ("reverse remodeling"). This chapter examines reported changes in LV size (volumes and/or dimension), function, and mass after valve replacement/repair, a process that has been termed LV "reverse remodeling".

REVERSE VENTRICULAR REMODELING

Difficulties are encountered in evaluating LV reverse remodeling: (i) there is no standardized method to evaluate LV size, function, and mass. Several modalities, such as contrast ventriculography, echocardiography, CT, or MRI may be utilized to assess the LV after valve replacement but they may yield measurements that may be very diverse from each other. (ii) different techniques may have been utilized pre and postoperatively; and (iii) the follow-up periods may also differ from study to study. Thus, it is difficult to assess the LV changes with time, and particularly, when comparing data from different studies.

Moreover, all techniques that measure LV mass include myocardium, blood vessels, connective tissue, and fibrous tissue. Thus, although data are frequently labeled as left ventricular hypertrophy (LVH), the measurement is actually of LV mass and the contribution of each tissue component is not specified, which may be of considerable importance in assessing remodeling and its reversal. For instance, it is difficult to envisage that extensive fibrosis would regress to the same degree as muscle after valve surgery; thus, the preoperative constituents of LV mass influence the observed changes in LV mass. In addition to the above factors, the patients that were excluded from studies prior to postoperative evaluation may have an important influence on the assessment of reverse remodeling. If patients who died prior to evaluation experienced less remodeling, their exclusion in postoperative studies might lead to an exaggeration of the observed beneficial changes in LV remodeling.

Factors Affecting Reverse Remodeling

Despite these technique related problems, other factors that affect reverse ventricular remodeling after valve surgery can be divided into preoperative, operative, and postoperative (Table 1). The preoperative factors include the amount of hypertrophy, LV dilatation, and associated coronary artery disease. The operative factors include the extent of myocardial revascularization, techniques of myocardial protection, and amount of myocardial damage. Finally, the postoperative factors that influence the reverse remodeling process include valve prosthesis–patient mismatch, complications from valve surgery, and the duration of time after surgery that the LV is evaluated (1–3).

Mechanisms of Reverse Remodeling

The mechanisms of reverse remodeling (Table 2) can be divided into four categories: mechanical, molecular, neurohormonal, and metabolic. The tissue changes that develop in response to the altered loading conditions, imposed by a valve lesion, strongly influence reverse remodeling after valve surgery. Biopsies taken of LV myocardium from patients with aortic stenosis (AS) revealed a higher percentage of fibrosis in the subendocardium (19%) than in the subepicardium (13%). The ultrastructural myocardial changes associated with LV dysfunction in severe AS include loss of myofibrils and accumulation of Z band material, unusual large nuclei, large cytoplasmic areas that were devoid of contractile material but filled with glycogen, ribosomes, and increased amounts of collagen fibers in the interstitial space (4). The correlation between percent fibrosis and ejection fraction (EF) is not significant statistically (5). Although the above described ultrastructural changes do not prevent improvement in EF, they may account for the difference in the rate of recovery as well as time of recovery of LV function in patients with AS after aortic valve

Table 1 Factors Affecting Left Ventricular "Reverse Remodeling" After Valve Surgery

---

Preoperative
  Excessive hypertrophy
  Amount of fibrosis
  Associated coronary artery disease
  Myocardial infarct
Operative
  Amount of myocardial damage
  Inadequate myocardial revascularization
Postoperative
  Valve prosthetic patient mismatch
  Complications of prosthetic heart valve
  Time of evaluation after surgery

Table 2 Factors That Influence Reverse Ventricular Remodeling

Mechanical
  Loss of myofibrils
  Increase fibrosis
  Cardiomyocyte deformity with changes in intracellular contents (i.e., ribosome,
  mitochondria)
Molecular
  Elevated levels of matrix metalloproteinases
  Increased cytokines (i.e., tumor necrosis factor-α)
  Elevated endothelial-selectin levels
Neurohormonal
  Elevated levels of atrial natriuretic peptides
  Activation of renin–angiotensin–aldosterone system
Metabolic
  Lower ratio of phosphocreatine to ATP

Abbreviation: ATP, adenine triphosphate.
Source: From Refs. 4, 8, 9, 11, 12, 13, 15.

replacement (AVR). In general, the extent of preoperative LVH has been found to be a poor prognostic factor postoperatively. In cases which reverse remodeling occurs, the decrease in muscle mass is related to the regression of myocardial cellular hypertrophy (the volume of myofibril) and the percent of interstitial fibrosis persisting after AVR (6). The volume of myofibrils may decrease "early" but it may take several years before any significant improvements in LV mass are seen after surgery because fibrous tissue may not regress or may regress slowly (5).

Some studies suggest that the regression of "LVH" is associated with changes in extracellular matrix gene expression. In AS, gene expression for matrix metalloproteinases (MMP) have been shown to increased significantly. After surgical correction, there is a return of the MMP gene expression to baseline (7). Similarly, in patients with aortic valve disease endothelial selectin (E-selectin) levels have been reported to be increased compared to a control population and then to significantly decrease after AVR (8). Of note is the finding that after AVR, the E-selectin levels were unchanged at six months postoperatively, but were significantly lower at 18 months postoperatively. These results raise the possibility that the expression of genes involved in the regulation of extracellular matrix proteins may be reliable "markers" of reverse remodeling.

In patients with valvular heart disease, circulating neurohormonal levels may be elevated above normal. Atrial natriuretic factor, which is usually elevated in patients with heart failure and in patients with elevated left atrial (LA) pressure, is significantly higher in patients with mitral regurgitation (MR) or AS than in patients without these lesions (9). Levels of atrial natriuretic factor are higher in patients with MR than in patients with AR or in normal subjects (60.3 ± 47 vs. 19.0 ± 11 and 12.4 ± 5.2 fmol/mL, $P < 0.0005$) (10). Similarly, angiotens in II levels are also elevated in patients with chronic MR and there is evidence that the adrenergic nervous system is activated in patients with heart failure due to mitral stenosis (11). Neurohormones levels have been shown to decrease significantly after mitral valve surgery: plasma renin activity (3.1 ng/mL/hr to 1.2 ng/mL/hr, $P < 0.05$), aldosterone (160 pg/mL to 93 pg/mL, $P < 0.05$), and atrial natriuretic peptide (75 pmol/L to 62 pmol/L, $P < 0.05$) (12). Since many of those neurohormones are associated with cell growth and cardiac remodeling the reduction that occurs following correction of the valve lesion would be expected to be associated with a return to normal pressure and volume.

Proinflammatory cytokines, such as tumor necrosis factor-α (TNF-α), TNF-α receptors 1 and 2, are also increased in heart failure patients and in patients with chronic MR (13). After mitral valve surgery, plasma levels of TNF-α and of TNF-α receptors 1 and 2 significantly decreased to levels similar to those of control patients. As the plasma cytokine levels normalized, LV dimensions also normalized (13). After mitral valve replacement (MVR), there is an inverse relationship between myocardial TNF-α expression and regression in left ventricular end-diastolic volume (LVEDV) and left ventricular end-systolic volume (LVESV). Since increased TNF-α levels have been shown to promote LV dysfunction (at least in some experimental models) (14), the postoperative reduction of TNF-α may contribute to and be a marker for an improvement in LV function.

Recent studies indicate that the ratio of phosphocreatine (PCr) to adenosine triphosphate (ATP) is a helpful indicator of MR severity. Reduction in PCr/ATP has been observed in cardiomyopathies, ischemia, and other disorders involving the LV. Severe MR has a lower PCr/ATP (1.29 ± 0.32, $P < 0.01$) than moderate MR (1.49 ± 0.18, $P < 0.05$), which is lower than mild MR (1.73 ± 0.17, $P < 0.05$) (15). Thus, the more severe the MR, the lower the PCr/ATP ratio. This suggests that the impaired LV function associated with the severe MR results in reduction of high-energy phosphate metabolism. With surgical correction of the lesion, the PCr/ATP is increased, suggesting improved "contractility" of the LV.

The type of prosthetic heart valve (PHV) and its size do not appear to affect reverse ventricular remodeling. The Veteran Administration trial (16) showed that mechanical and bioprosthetic valves yield similar hemodynamic results and that the effect of valve replacement on patient functional status and changes in cardiac hemodynamics are principally determined by patient-related factors. Therefore, it can be assumed that LV remodeling may be similar with use of either mechanical or bioprosthetic PHV. In addition to the type of PHV, the size of the PHV does not appear to affect the reverse ventricular remodeling process provided that the PHV is of reasonable size and is appropriate for the patient's body size; in other words, if valve prosthesis–patient mismatch is only mild, or at most moderate in severity, reverse remodeling should not be adversely affected. Tasca and colleagues evaluated 19-mm, 21-mm, and 23-mm Carpentier-Edwards Perimount pericardial bioprosthesis in a total of 88 patients (Fig. 1) (17). They found that the LV mass was significantly reduced without any PHV size-related differences. This is probably because patients who received the smaller prosthesis were largely women who had smaller body size, and thus, had only mild to moderate valve prosthesis–patient mismatch. This probably applies as well to patients with AR and with MR, but so far, this has not been documented.

Results after homograft valve implantation are comparable to those seen after implantation of heterograft valves. A few studies have suggested that within the biological valves (18), the stentless valves may be associated with greater improvement of LV function and earlier LV mass regression than the conventional stented valves (19–22). However, current data suggest that regardless of the valve type, surgical correction of the valve lesion leads to reverse remodeling (Table 3). Both the Edinburgh (23) and Veterans Administration Cooperative Study (16) showed that there is no difference between bioprosthetic valve and mechanical valve in survival, reoperation, and valve-related complications up to 10 years after surgery.

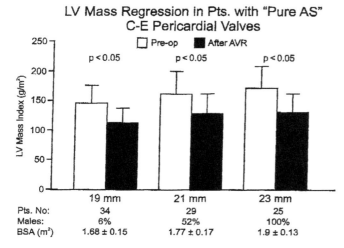

Figure 1 LV mass regression occurred after AVR in patients who received the 19, 21, and, 23 mm size Carpentier-Edwards pericardial valve. Those who received the smallest valve were almost all women who were smaller than the men who were bigger and received the largest valve. Thus, valve prosthesis–patient mismatch was mild to moderate in severity. *Abbreviations:* AVR, aortic valve replacement; LV, left ventricle. *Source:* Developed from reference 17 and reproduced from Rahimtoola SH: The Year in Valvular Heart Disease. J Am Coll Cardiol 2005; 45(1):111–112.

In summary, the choice of PHV or valve size does not appear to be a major factor in the reverse remodeling process (24–28), provided that the size of PHV is appropriate for the patient's body size. LV reverse remodeling seems to be related to: (i) the patient's preoperative status, including the severity and duration of the loading condition(s), extent of LV dysfunction, type of overload (pressure or volume), and associated changes in the myocardium and connective tissue. (ii) operative factors, e.g., myocardial damage, if any, and completeness of myocardial revascularization; and (iii) postoperative factors, e.g., severity of valve prosthesis–patient mismatch, complications of PHV, time of evaluation after surgery, and the percent of patients who died prior to postoperative evaluation. As noted previously the percent of patients who died prior to postoperative evaluation may influence the

Table 3 LV MASS (g/m^2)

Valve	Pre-op	Post-op	LV mass	Reduction %
Starr-Edwards	229 ± 43	123 ± 25	106	46
Homograft	210 ± 105	147 ± 10	63	30
Toronto SPV	225 ± 105	144 ± 8	81	36
Carpentier-Edwards Pericardial	265 ± 83[a]	208 ± 70[a]	57[a]	22[a]
19 mm	212 ± 77	145 ± 21	67	32
21 mm	135 ± 25	114 ± 35	21	16
23 mm	170 ± 101	143 ± 83	27	16

[a]Grams.
Source: From Refs. 21, 22, 34.

observed beneficial changes in LV remodeling since it is possible that postoperative death and failure to reverse remodel could be associated.

AORTIC STENOSIS

LVEDV/LVEDVI

Data from a composite of studies that included pre and postoperative measurements show that the average preoperative left ventricular end-diastolic volume index (LVEDVI) decreased significantly after AVR from $98 \pm 23 \, \text{mL/m}^2$ preoperatively to $81 \pm 22 \, \text{mL/m}^2$ postoperatively ($P = 0.05$) (Fig. 2; Table 4). Although LVEDVI decreased postoperatively, the change is not as great as that seen in LV mass (Table 4). This is due to the fact that in the absence of LV systolic dysfunction, LVEDV is normal preoperatively. In contrast, LV mass is increased proportionately to the severity of afterload increase with AS. Evidence that LVEDVI still declined significantly after AVR is probably related to the fact that, in some studies, LVEF was reduced and also that some patients had afterload mismatch both of which may increase LVEDVI. Much of the reduction in LV chamber size appears to occur early after AVR. In one study, the LV end-diastolic dimension (LVEDD) decreased from $5.9 \pm 0.7 \, \text{cm}$ preoperatively to $5.5 \pm 0.6 \, \text{cm}$ at one week postoperatively ($P = 0.02$) (29). Most studies did not evaluate changes in LV end-systolic volume index (LVESVI). However, since LVEDVI decreased and LVEF increased (see below), LVESVI must have decreased.

LVEF

From a composite of data that were analyzed in patients with AS, the LVEF was found to increase postoperatively (Fig. 3). The improvement in LVEF may not occur

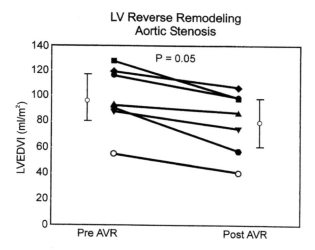

Figure 2 Aortic stenosis. LVEDVI decreases after AVR. This change is statistically significant ($P = 0.05$). The mean LVEDVI preoperatively is $98 \pm 23 \, \text{mL/m}^2$ and decreases to a postoperative value of $81 \pm 22 \, \text{mL/m}^2$. *Abbreviations*: AVR, aortic valve replacement; LVEDVI, left, ventricular end-diastolic volume index; LV, left ventricle.

Table 4 Aortic Stenosis

References: first author and citation number	# of pts	Mean age (yrs)	Mean LVEDV (mL)		Mean LVEDVI (mL/m²)		Mean LVESV (mL)		Mean LVESVI (mL/m²)		Mean LV mass (g/m²)		Mean LVEF	
			Pre	Post	Pre	Post	Pre	Post	Pre	Post	Pre	Post	Pre	Post
Pantely et al. (34)	10	55			93	87					229	123	0.65	0.76
Smith et al. (33)	10	65			119	107							0.34	0.63
Schwarz et al. (43)	5	48			90	58					181	128	0.59	0.73
Krayenbuehl et al. (42)	27	52			127	98					186	94	0.59	0.57
Kurnik et al. (38)	17	63			56	41	36	31	36	31	159	105	0.64	0.69
Villari et al. (40)	22	44			116	99			51	44	202	116	0.55	0.60
Harpole et al. (35)	10	65					55	37			216	139	0.41	0.62
De Paulis et al. (27)	10	65	135	110									0.63	0.61
Lund et al. (36)	46	61			88						200	148	0.59	0.64
Murakami et al. (41)	29	59				75					215	159		
Ikonomidis et al. (26)	41										187	179		
Legarra et al. (18)	30	30									179	96		
Kuhl et al. (37)	26	65	97	88			37	26			148	109	0.65	0.70
Mean	22	56	116	99	98	81	43	31	44	38	191	127	0.56	0.66
SD	12	12	19	11	23	22	9	4	7	6	23	26	0.10	0.06
P-value					0.05						0.02		0.19	

Mean changes in LVEDV, LVEDVI, LVESV, LVESVI, LV mass, and LVEF before and after AVR.

Abbreviations: LVEDV, left ventricular end-diastolic volume; LVEDVI, left ventricular end-diastolic volume index; LVEF, left ventricular ejection fraction; LVESV, left ventricular end-systolic volume; LVESVI, left ventricular end-systolic volume index.

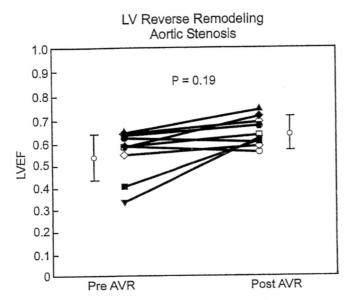

Figure 3 Aortic stenosis. LVEF increases after AVR. The mean EF before AVR is 0.56 ± 0.10 and increases to 0.66 ± 0.6 after AVR. This change is most evident in patients with LV dysfunction. *Abbreviations*: AVR, aortic valve replacement; EF, ejection fraction; LV, left ventricle; LVEF, left ventricular ejection fraction.

immediately after surgery, but may take months before any change is evident (29). Patients with severe AS and normal or reduced LVEF who underwent AVR were found to have no improvement by one week postoperatively. Nevertheless, over a course of several months, there was a significant increase in postoperative LVEF, which was most evident in patients with LV dysfunction.

There are, however, at least three conditions when LVEF may improve rapidly or even dramatically following AVR. These are: (i) when LVEF is reduced preoperatively largely because of afterload mismatch; (ii) in patients with very severe AS (i.e., aortic valve area (AVA) is $\leq 0.5 \, cm^2$), in whom severe afterload mismatch is rapidly reduced by surgery; and (iii) in patients with preoperative severe LV dysfunction. Patients with LV dysfunction may have greater reverse remodeling and thus, there is a potential for greater benefit (30–32). In addition to reversal of remodeling, AVR for AS also improves long-term survival, and the New York Heart Association (NYHA) Functional Class (33).

LV Mass

Data from a composite of studies show that LV mass undergoes the most significant change postoperatively (Table 4). The changes in LVEDVI and LVEF are less marked because, preoperatively, when LV remodeling is occurring, these parameters have smaller changes than are seen for LV mass. On the other hand, LV mass is increased early and often markedly in order to normalize stress.

In an analysis of a series of studies (Table 4), the preoperative average LV mass was $191 \, g/m^2$, and it decreased significantly to $127 \, g/m^2$ postoperatively ($P = 0.02$). In all studies, there was a significant decline in LV mass after AVR (Fig. 4). The fact

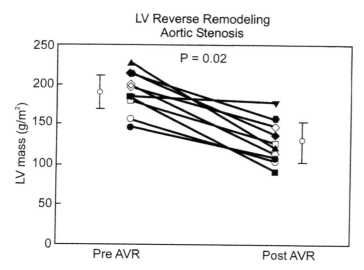

Figure 4 Aortic stenosis. LV mass regress after AVR. The mean LV mass is $191 \pm 23\,\text{g/m}^2$ before surgery and decreases to $127 \pm 26\,\text{g/m}^2$ after surgery. This is a significant change from the preoperative value ($P = 0.02$). *Abbreviations*: AVR, aortic valve replacement; LV, left ventricle.

that these findings were confirmed in studies using different imaging modalities, including angiography, echocardiography, and ultrafast computed tomography increases the level of confidence in their validity (6,34–37). Information regarding the time course for the regression of LV mass is also available.

Results indicate that LV mass decreases by an average of 25% at four months and by 34% at eight months after AVR (38). The regression of LV mass may begin as early as several hours after AVR and it continues for years after the correction of the primary hemodynamic abnormality (24,34,39). Several studies indicate that diastolic dysfunction, which is associated with LVH, improves after AVR, but that it may take years to normalize owing to the very slow regression of muscular and nonmuscular tissue (25,40–43). This suggests that the process of reverse remodeling can extend years into the postoperative period.

AORTIC REGURGITATION

LVEDV/LVEDVI

Patients with chronic aortic regurgitation (AR) have a significant decline in LVEDVI after AVR. A composite of studies (Table 5) showed that the mean preoperative LVEDVI was $189\,\text{mL/m}^2$ and was reduced postoperatively to $105\,\text{mL/m}^2$ ($P = 0.03$). The postoperative change, that is, the reduction in LVEDD and left ventricular end-systolic dimension (LVESD), is greater in the patients with LVEDD $> 80\,\text{mm}$ than in those with $< 80\,\text{mm}$. Most of the LV dimensional changes occur ≥ 6 to 8 months postoperatively. In a sequential study, the LVEDD declined significantly by six to eight months, with most of the changes occurring in the first several weeks after AVR, but no further decline or changes in subsequent years (45). In patients with normal LVEF, about 30% of the patients will normalize their LVEDD late after surgery

Table 5 Aortic Regurgitation

References: first author and citation number	Mean # of pts	Mean age (yrs)	Mean LVEDVI (mL/m²)		Mean LVESVI (mL/m²)		Mean LV mass (g/m²)		Mean LVEF	
			Pre	Post	Pre	Post	Pre	Post	Pre	Post
Pantely et al. (34)	8	42	205	140			222	128	0.50	0.49
Schwarz et al. (43)	8	35	164	97			193	146	0.53	0.63
Clark et al. (33)	10	44	133	82	133	82	234	170	0.43	0.49
Taniguchi et al. (47)	23	42	216	98	110	37	243	158	0.50	0.62
Bonow et al. (45)	61	43							0.43	0.56
Krayenbuehl et al. (42)	17	43	226	109			201	93	0.60	0.64
Murakami et al. (41)	41	54					238	145		
Mean	24	43	189	105	121	59	222	140	0.49	0.57
SD	6		39	22	11	22	21	27	0.06	0.07
P-value				0.03				0.01		0.11

Mean changes in LVEDVI, LVESVI, LV mass, and LVEF preoperatively and postoperatively.
Abbreviations: LVEDVI, left ventricular end-diastolic volume index; LVEF, left ventricular ejection fraction; LVESVI, left ventricular end-systolic volume index.

(3–7 years). The ratio of ventricular end-diastolic radius to wall thickness may be of prognostic value in patients with AR. If the preoperative end-diastolic radius to wall thickness ratio (R/Th) is greater than 3.8 or a product of R/Th and left ventricular systolic pressure (PR/Th) exceeds 600, then patients will likely have persistent or slow regression of LV dilatation postoperatively (46).

LVEF

Although LVEF can increase after AVR in chronic AR (Fig. 5), the improvement in LVEF may not be evident immediately. In fact, patients with AR and with chronic MR, will usually have decreased LVEF immediately after surgery. The reason for this early reduction in postoperative LVEF is related to the abrupt absence of regurgitant volume (RgV) due to the surgical procedure. For example, if prior to AVR the LVEDV is 200 mL, RgV fraction is 0.50, and LVEF is 0.60, then the LVESV is 80 mL, and LVSV is 120 mL. Total LV stroke volume (SV) is partition so that RgV is 60 mL, and the forward SV is 60 ml. At a heart rate of 70 bpm, the cardiac output (CO) would be 4.2 L/min before surgery. Immediately after surgery, the LVEDV does not decrease and is still 200 mL, the forward SV still has to be only 60 mL to maintain the same CO of 4.2 L/min. However, with a RgV that is now is 0, LVEF will be 0.30 (60 mL/200 mL) (Table 6).

In the late postoperative period, change in LVEF are highly variable with some patients improving and others, showing no change. LVEF can continue to improve for years after AVR. This has substantial prognostic significance since patients who show an improvement in LVEF and normalization of the LV dimensions have an increased survival benefit (44,47). The extent of remodeling that develops in response to AR is an important predictor of the long-term postoperative clinical course. In one study, patients with preoperative end-systolic volume index $< 200 \, \text{mL/m}^2$ were shown to have better survival and were more likely to have reversal of LV dysfunction postoperatively (47). In another study, however, preoperative LVEF and

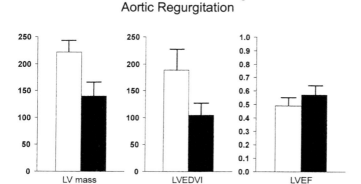

Figure 5 Aortic regurgitation. Comparison of LV mass, LVEDVI, and LVEF in preoperative and postoperative patients with aortic regurgitation. As the LV mass and the LVEDVI decreases significantly, the LVEF remains statistically unchanged. The greatest change occurred in the LV mass after surgery. *Abbreviations*: LV, left ventricle; LVEDVI, left ventricular, end-diastolic volume index; LVEF, left ventricular ejection fraction.

Table 6 Chronic Valvular Regurgitation

	Normal	Pre-op	Immediately post-op[b]	Late[a] post-op scenario1	Late[a] post-op scenario2
LVEDV (mL)	100	200	200	200	120
LVESV (mL)	40	80	140	140	60
LVEF	0.60	0.60	0.30	0.30	0.50
LVSV (mL)	60	120	60	60	60
Rg fraction	0	50%	0	0	0
RgV (mL)	0	60	0	0	0
FSV (mL)	60	60	60	60	60
Heart rate (beats/min)	70	70	70	70	70
CO (L/min)	4.2	4.2	4.2	4.2	4.2

[a]Late: months to years after surgery.
[b]The late Dwight McGoon of the Mayo Clinic pointed out in 1975 at a national meeting that immediately after valve replacement for severe AR or MR, LVEF is reduced because of the elimination of valvular regurgitation. However, the LVSV is now the same as FSV and thus, the same CO can be maintained at a lower total LVSV, which means LVEF is reduced because LVEDV does not decrease immediately.
Abbreviations: LVSV, left ventricular stroke volume; Rg fraction, regurgitation fraction; RgV, regurgitant volume; FSV, forward stroke volume to the body; CO, cardiac output.
Source: From Ref. 51.

cardiac index were better predictors of late survival after AVR than LV diastolic parameters and clinical status (48).

In chronic AR, the duration of LV dysfunction and poor exercise tolerance are important predictors of reverse remodeling. Patients with LV dysfunction of brief duration (<12 months) have a high likelihood of normalizing their LVEF and experiencing a substantial reduction of LVEDV, such that these parameters are not significantly different from patients who had normal LVEF prior to AVR. Patients with prolonged duration of LV dysfunction with good exercise tolerance and those with LV dysfunction with poor exercise tolerance may have no significant change in LVEF (Fig. 6A) and the reduction in LVEDV (Fig. 6B) is less than that seen in patients with normal LVEF. Furthermore, patients with brief duration of LV dysfunction have early (6–8 months) and late (3–7 years) improvement in LVEF after surgery, while patients with prolonged duration of LV dysfunction show minimal to no changes in LVEF after surgery (45).

Using a "contractility" score, one group (49) showed that LV function and myocardial contractility improves and in some cases, normalizes after AVR. Postoperatively, the LV is still dilated (increased LVEDVI is still present), but the RgV is significantly reduced, therefore the CO is maintained or increased. The LVEF increases after a reduction of the LVEDVI (50–52). Even patients with low LVEF (0.25 to 0.49) still show clinical benefits of AVR (53). In addition, the absolute rest and exercise LVEF also improves after AVR (54,55).

LV Mass

A composite of data shows that LV mass decreased after AVR (from $222 \, g/m^2$ to $140 \, g/m^2$). In patients with AR, the left ventriculr mass index (LVMI) decreases less rapidly than LVEDVI leading to an increase in the LVMI/LVEDVI ratio (56). The

LVEDVI may normalize within two weeks after AVR, while LV mass takes at least six months to show significant changes (57). The amount of LV mass regression correlates with attainment of NYHA Functional Class I status (38). Even though the LV size is normalized, the improvement in LVEF from 49% to 57% was not statistically significant ($P = 0.11$) in these series of patients (Table 5). One reason for this may be that most of the studies evaluated patients with normal to only mild reductions in LVEF.

MITRAL STENOSIS

LVEDV/LVEDVI

Compared to patients without mitral valve disease, patients with mitral stenosis (MS) have LVEDVI and stroke volume index (SVI) that were significantly lower, but the LVESVI was similar (63). In patients who underwent mitral valve repair or commissurotomy, LVEDV is often increased after intervention; the mean preoperative LVEDV increased from 98 mL to 122 mL postoperatively (Table 7). However, this increase is not significant ($P = 0.39$). Although there were a limited number of studies to evaluate, the LVEDVI and LVESVI do not change significantly postoperatively from the preoperative values mainly because LVEDVI is normal preoperatively in most patients. In some studies (58–60), the LVEDV remained similar before and after valvotomy. This may be related to the time after surgery that LV changes were evaluated. LVEDVI changes by only a small amount immediately after valvotomy, but may continue to increase over time (61). SVI increased from $34 \pm 10 \, \text{mL/m}^2$ to $41 \pm 12 \, \text{mL/m}^2$ ($P < 0.05$) immediately after valvotomy and increased further to $50 \pm 11 \, \text{mL/m}^2$ ($P < 0.001$). The increase in LV volume and mass has been suggested as a mechanism through which the increase in preload stimulates myocardial growth for the recovery of LV function postoperatively (62).

LVEF

Data from a composite of studies showed that the preoperative LVEF, EDVI, and ESVI change insignificantly postoperatively (Table 7). However, cardiac arrhythmias can alter the preoperative LVEF, LVEDVI, and ESVI in patient with MS (64). Patients with MS and chronic atrial fibrillation had a lower EF than patients with MS in normal sinus rhythm (0.56 ± 0.10 vs. 0.49 ± 0.10, $P < 0.05$). Compared to patients with MS and in normal sinus rhythm, patients with MS and in atrial fibrillation have a higher LV wall stress and LVEDVI. Furthermore, the cardiac index is also lower in patients with atrial fibrillation compared to patients in normal sinus rhythm ($2.9 \pm 0.7 \, \text{L/min/m}^2$ vs. $2.1 \pm 0.6 \, \text{L/min/m}^2$). The LV long axis is $7.1 \pm 0.8 \, \text{cm}$ in those in normal sinus rhythm versus $6.6 \pm 0.7 \, \text{cm}$ in those in atrial fibrillation in whom the increased ventricular rate is probably one reason for the smaller LV size.

LV Mass

Changes in LV mass in MS are uncertain because there are few studies that assessed LV mass preoperatively and postoperatively. The available data show that the mean LV mass decreases, but that this change is not statistically significant (62,65). Since

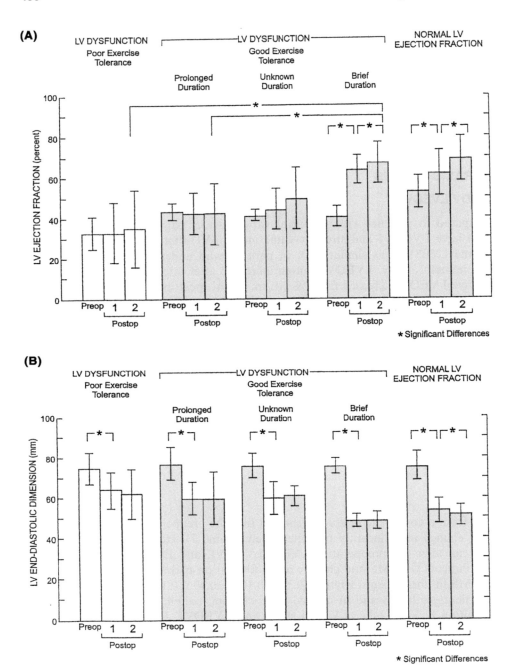

Figure 6 (*Caption on facing page*)

LV volumes and LV systolic pressure are normal in most patients with MS, LV mass should also be normal; an increase in LV mass suggests the presence of associated systemic hypertension, aortic valve disease, MR, or reduced LVEF.

MITRAL REGURGITATION

The difficulties in evaluating data that assess LV reverse remodeling in MR include: (i) there are not enough studies with valve replacement and valve repair to make a meaningful comparison; (ii) there are an insufficient number of studies before and after mitral valve repair; (iii) the etiology of MR is frequently not provided and even when it is indicated, changes in parameters of LV remodeling are not separated by etiology; and (iv) some data are presented as actual values and other values are corrected for body size.

LVEDV/ LVEDVI

The data in Table 8 reflect both mitral valve replacement and repair. Unfortunately there are not enough published studies to adequately assess mitral valve replacement versus repair in reverse LV remodeling. Regardless of replacement or repair, however, the LVEDVI decreases postoperatively. A summary of composite data indicates that the LVESVI did not decrease consistently and the change did not reach statistical significance. However, patients with preoperative end-systolic diameter >50 mm have a poorer outcome (death and heart failure) after MVR, despite chordal preservation, when compared to patients with normal end-systolic diameter (66). Table 8 shows that the LVEDVI decreased significantly from a mean preoperative value of $126 \, \text{mL/m}^2$ to a postoperative value of $85 \, \text{mL/m}^2$, $P < 0.05$. The reason for the decrease in LVEDVI is that the RgV, which had been part of the LVEDVI preoperatively, is now eliminated postoperatively (see AR and Table 6). The data from the VA trial show the LVEDVI decreases according to the amount of reduction of RgV. The fact that LVESVI remains unchanged (Table 9) indicates that LVEF is reduced (60). Once the mitral valve lesion is corrected, LV reverse remodeling is initiated and LV dimensions and function may normalize with time (65–68). In both mitral valve replacement and repair the LVEDV may decrease significantly almost immediately after surgery (69). If the LVED dimensions and peak systolic stress

---

Figure 6 (*Facing page*) (**A**) Patients with brief duration of LV dysfunction and good exercise tolerance have similar improvement in postoperative LVEF as patients with normal preoperative LVEF. These changes are seen at six to eight months (1) and three to seven years (2) after surgery. Patients with prolonged LV dysfunction and good exercise tolerance have no changes in postoperative LVEF, which was also seen in patients with LV dysfunction and poor exercise tolerance. (**B**) Patients with brief duration of LV dysfunction and good exercise tolerance have similar significant reduction in LVED dimension as patients with normal LVEF. These changes are seen at six to eight months (1) and continue into three to seven years (2) postoperatively. Patients with prolonged LV dysfunction and good exercise tolerance have significant decline in LVED dimension early postoperatively (1), but the LVED dimension does not continue to decline in long-term follow-up (2). This pattern of changes is similar to patients with LV dysfunction and poor exercise tolerance. *Abbreviations*: LV, left ventricle; LVED, left ventricular end diastole; LVEF, left ventricular ejection fraction. *Source*: From Ref. 45.

Table 7 Mitral Stenosis

References: first author and citation number	Mean # of pts	Mean age (yrs)	Mean LVEDV (mL)		Mean LVEDVI (mL/m²)		Mean LVESV (mL)		Mean LVESVI (mL/m²)		Mean LV mass (g/m²)		Mean LVEF	
			Pre	Post	Pre	Post	Pre	Post	Pre	Post	Pre	Post	Pre	Post
Kennedy et al. (65)	6	49			86	72			43	37	106	120	0.51	0.49
Crawford et al. (60)	33	55			79	72			41	39			0.48	0.47
Harpole et al. (74)	4	65											0.51	0.52
Tischler et al. (62)	15	59	69	82	40	46	35	37	20	20	106	114	0.50	0.55
Harrison et al. (59)	27	44	124	125										
Grover-Mckay et al. (58)	11	48	102	158			29	27					0.73	0.78
Fawzy et al. (61)	17	27			60	72			25	24			0.57	0.71
Mean	16	50	98	122	66	66			32	30			0.55	0.59
SD	10	12	28	38	21	13			11	9			0.09	0.13
P-value				0.39		0.22				0.16				0.16

Mean changes in LVEDV, LVEDVI, LVESV, LVESVI, LV mass, and LVEF before after intervention.
Abbreviations: LVEDV, left ventricular end-diastolic volume; LVEDVI, left ventricular end-diastolic volume index; LVESV, left ventricular end-systolic volume; LVESVI, left ventricular end-systolic volume index; LVEF, left ventricular ejection fraction; LVESV, left ventricular end-systolic volume index.

Table 8 Mitral Regurgitation

References: first author and citation number	Mean # of pts	Mean age (yrs)	Mean LVEDV (mL)		Mean LVEDVI (mL/m²)		Mean LVESV (mL)		Mean LVESVI (mL/m²)		Mean LV mass (g/m²)		Mean LVEF	
			Pre	Post	Pre	Post	Pre	Post	Pre	Post	Pre	Post	Pre	Post
Kennedy et al.[a] (65)	7	48			124	96			58	57			0.55	0.43
Boucher et al.[a] (73)	20	53			131	78			45	42			0.66	0.48
Crawford et al.[a] (60)	48	57			117	89			54	50			0.56	0.45
Shyu et al.[b] (79)	55	41	204	120	133	79	67	46	44	30	211	134	0.69	0.64
Ren et al.[b] (69)	29	66	127	107			56	45					0.51	0.56
Mean	32	53			126	85			50	45			0.59	0.51
SD	20	9			7	9			7	12			0.08	0.09
P-value						0.04				0.13				0.42

Mean changes in LVEDVI, LVESVI, LV mass, and LVEF preoperatively and postoperatively.
[a]Mitral valve replacement.
[b]Mitral valve repair.
Abbreviations: LVEDV, left ventricular end-diastolic volume; LVEDVI, left ventricular end-diastolic volume index; LVEF, left ventricular ejection fraction; LVESV, left ventricular end-systolic volume; LVESVI, left ventricular end-systolic volume index.

Table 9 Effect of MV Replacement on LV Function in MR

	Before MVR	After MVR
LVESVI (mL/m^2)	54 ± 12	50 ± 25
RgV (mL)	59 ± 45	11 ± 17
LVSVI (mL/m^2)		
LVEDVI (mL/m^2)	117 ± 51	89 ± 27
LVEF	0.56 ± 0.15	0.45 ± 0.13

Abbreviations: LVEDVI, left ventricular end-diastolic volume index; LVEF, left ventricular ejection fraction; LVESVI, left ventricular end-systolic volume index; LVSVI, left ventricular stroke volume index; RgV, regurgitant volume.
Source: From Ref. 60.

are elevated preoperatively, then these values will decline to normal after surgery (70). Echocardiography has shown a significant decrease in short-axis dimension postoperatively versus preoperatively (2.8 ± 0.6 cm vs. 3.1 ± 0. 7 cm; $P < 0.009$), an increase in long-axis dimension (5.8 ± 0.9 cm vs. 5.3 ± 0.8 cm; $P < 0.005$), and an increase in eccentricity index (2.1 ± 0.4 vs. 1.7 ± 0.4; $P < 0.0001$) at end-systole (69). This represents a "reversal" toward a more normal configuration. The ratio of end-systolic wall stress to end-systolic volume index (ESWS/ESVI) may be of predictive value for patients undergoing mitral valve surgery (71). Patients with a lower ESWS/ESVI ratio had a higher mortality rate postoperatively (71).

LVEF

The LVEF may get worse postoperatively; as occurs in most regurgitant lesions immediately after valve replacement (Table 6) (72,73) This is most evident in AR because of the larger LVEDVI, but it is also seen in MR. The data from the VA study, performed six months after MVR, confirms that the theoretical considerations relating to correction of a regurgitant valve (describe in the section about AR) actually do occur in patients. (Table 9)(60). Furthermore, for LVEF to improve after surgery, the LVEDV needs to decrease. LV dysfunction, however, is often a complication of the LV dilatation and hypertrophy that develops in patients with severe MR. If the LV myocardium does not hypertrophy adequately for the dilatation, then LV dysfunction can develop in patients with severe MR. The LV dysfunction may reverse with mitral valve surgery. Studies have shown that LVEDVI, LV mass index, and LVESVI decrease after mitral valve surgery (60,65,68,74). Additionally, in patients with mitral valve disease, the changes in the LV function after surgery are dependent on whether the valve was repaired or replaced and whether the chordae was preserved or resected because the preserved chordae favorably influence LV torsional dynamics (69,75,76). If the chordal apparatus was retained in MVR, then the LV dimensions decrease and LV function improves postoperatively to a greater extent than if the chordal apparatus was resected (77). Furthermore, complete retention of the mitral subvalvular apparatus results in a greater decrease in LV dimensions when compared to partial chordal preservation (78). Overall, changes in LVEF after MVR are related to several factors: (i) loss of low impedance leak; (ii) perioperative myocardial infarction; (iii) resection of chordae tendinae; and (iv) suboptimal reduction of LVEDV.

LV Mass

There is no sufficient data to evaluate the effect of mitral valve repair or replacement on LV mass. However, it can be assumed that if the LVEDVI decreases after surgical correction, then LV mass should decrease accordingly. In one study, the LV mass index decreased significantly from a preoperative value of $211 \pm 82 \, \text{g/m}^2$ to a post-operative value of $134 \pm 52 \, \text{g/m}^2$ $(P < 0.001)$ (79).

CONCLUSIONS

- Moderate and severe valvular disease results in LV remodeling.
- After surgical correction of the valve disease, LV reverse remodeling occurs. The completeness of the reverse remodeling is dependent on many factors which include:
  - Preoperative condition of the LV
  - Myocardial damage, if any, at surgery
  - Type of corrective surgery
  - Time of assessment of LV change after surgery
- There is no standard method to assess LV reverse remodeling.
- Time after surgery to evaluate LV reverse remodeling is not standardized. In prospective studies, we suggest that the evaluation be done at the time of hospital discharge at 6 and/or 12 months and at 5 and 10 years after valve surgery.
- In AS, there is a significant decrease in LVEDVI and LV mass, and an increase in LVEF after AVR.
- In AR, the LVEDVI and LV mass decrease significantly after AVR, but the LVEF may remain unchanged or may increase to a less degree as compared to AS.
- In MS, most patients have a LV that is not remodeled, so no reverse remodeling occurs.
- In MR, LVESVI is usually unchanged. If LVEDVI declines after surgery, LVEF usually increases. Frequently LVEF is decreased.
- In AR and MR the LVEF is decreased immediately after valve surgery, because of elimination of RgV.

APPENDIX

AR	Aortic regurgitation
AS	Aortic stenosis
AVA	Aortic valve area
AVR	Aortic valve replacement
CAD	Coronary artery disease
CO	Cardiac output
EDV	End-diastolic volume
EDVI	End-diastolic volume index
EF	Ejection fraction
ESV	End-systolic volume
ESVI	End-systolic volume index

LV	Left ventricle
LVH	Left ventricular hypertrophy
LVM	Left ventricular mass
LVMI	Left ventricular mass index
MR	Mitral regurgitation
MS	Mitral stenosis
PHV	Prosthetic heart valve
RgV	Regurgitant volume
SV	Stroke volume
SVI	Stroke volume index

REFERENCES

1. Duran CG, Revuelta JM, Gaite L, Alonso C, Fleitas MG. Stability of mitral reconstructive surgery at 10–12 years for predominantly rheumatic valvular disease. Circulation 1988; 78(suppl I):I-91–I-96

2. Munoz S, Gallardo J, Diaz-Gorrin JR, Medina O. Influence of surgery on the natural history of rheumatic mitral and aortic valve disease. Am J Cardio 1975; 35:234–242.

3. Cohn LH, Allred EN, Cohn LA, Austin JC, Sabik J, Disesa VJ, Shemin RJ, Collins JJ Jr. Early and late risk of mitral valve replacement. J Thorac Cardiov Surg 1985; 90:872–881.

4. Ferrans VJ. Human cardiac hypertrophy: Structural aspects. Eur Heart J 1982; 3(suppl A): 15–27.

5. Hess OM, Villari B, Krayenbuehl HP. Diastolic dysfunction in aortic stenosis. Circulation 1993; 87(suppl IV):73–76.

6. Orsinelli DA, Aurigemma GP, Battista S, Krendel S, Gaash WH. Left ventricular hypertrophy and mortality after aortic valve replacement for aortic stenosis. A high risk subgroup identified by preoperative relavtive wall thickness. J of Am Coll of Card 1993; 22(6):1679–1683.

7. Walther T, Schubert A, Falk V, Binner C, Kanev A, Bleiziffer S, Walther C, Doll N, Autschbach R, Mohr FW. Regression of left ventricular hypertrophy after surgical therapy for aortic stenosis is associated with changes in extracelllular matrix gene expression. Circulation 2001; 104(suppl I):I-54-58.

8. Ghaisas NK, Foley JB, O'Briain S, Crean P, Kelleher D, Walsh M. Adhesion Molecules in Nonrheumatic aortic valve disease: endothelial expression, serum levels and effects of valve replacement. JACC 2000; 36(7):2257–2262.

9. Qi W, Mathisen P, Kjekshus J, Simonsen S, Bjornerheim R, Endressen K, Hall C. Natriuretic peptides in patients with aortic stenosis. Am Heart J 2001; 142(4):725–732.

10. Roman MJ, Devereux RB, Atlas SA, Pini R, Ganau A, Hochreiter C, Niles NW, Borer JS, Laragh JH. Relationship of atrial natriuretic factor to left ventricular volume and mass. Am Heart J 1989; 118:1236–1242.

11. LeTourneau T, de Groote PD, Millaire A, Foucher C, Savoye C, Pigny P, Prat A, Warembourg H, Lablanche JM. Effect of mitral valve surgery on exercise capacity, ventricular ejection fraction and neurohormonal activation in patients with severe mitral regurgitation. JACC 2000; 36:2263–2269.

12. Imamura Y, Ando H, Ashihara T, Fukuyama T. Myocardial adrenergic nervous activity is intensified in patients with heart failure without left ventricular volume or pressure overload. JACC 1996; 28(2):371–375.

13. Oral H, Sivasubramanian N, Dyke DB, Mehta RH, Grossman PM, Briesmiester K, Fay WP, Pagdni ED, Bolling SF, Mann DL, Starling MR. Myocardial proinflammatory cytokine expression and left ventricular remodeling in patients with chronic mitral regurgitation. Circulation 2003; 107:831–837.

14. Bozkurt B, Kribbs SB, Clubb FJ Jr, Michael LH, Didenko VV, Hornsby PJ, Seta Y, Oral H, Spinale FG, Mann DL. Pathophysiologically relevant concentrations of tumor necrosis factor-α promote progressive left ventricular dysfunction and remodeling in rats. Circulation 1998; 97:1382–1391.

15. Conway MA, Bottomley PA, Ouwerkerk R, Radda GK, Rajagopalan B. Mitral regurgitation. Impaired systolic function, eccentric hypertrophy, and increased severity are linked to lower phosphocreatine/ATP ratios in humans. Circulation 1998; 97:1716–1723.

16. Hammermeister K, Sethi GK, Henderson WG, Grover Fl, Oprian C, Rahimtoola SH. Outcomes 15 years replacement with a mechanical vs. bioprosthesis: Final report of the VA randomized trial. J Am Coll Cardiol 2000; 36:1152–1158.

17. Tasca G, Brunelli F, Cirillo M. Mass regression in aortic stenosis after valve replacement with small size pericardial bioprosthesis. Ann Thorac Surg 2003; 76:1107–1113.

18. Legarra JJ, Concha M, Casares J, Merino C, Munoz I, Alados P. Left ventricular remodeling after pulmonary autograft replacement of the aortic valve (Ross operation). J of Heart Valve Dis 2001; 10:43–48.

19. Collinson J, Henein M, Flather M, Pepper JR, Gibson DG. Valve replacement for aortic stenosis in patients with poor left ventricular function. Comparison of early changes with stented and stentless valves. Circulation 1999; 100(suppl II):II-1–II-5.

20. Walther T, Falk V, Langebartels G, Kruger M, Berhardt U, Diegeler A, Gummert J, Autschbacher R, Mohr FW. Prospectively randomized evaluation of stentless versus conventional biological valves: impact on early regression of left ventricular hypertrophy. Circulation 1999; 100(suppl II):II-6–II-10.

21. Khan SS, Siegel RJ, DeRobertis MA, Blanche LE, Kass RM, Cheng W, Fontana GP, Trento A. Regression of hypertrophy after Carpentier-Edwards pericardial aortic valve replacement. Annals of Thoracic Surgery 2000; 69(2):531–535.

22. Jin XY, Pillai R, Westaby S. Medium-term determinants of left ventricular mass index after stentless aortic valve replacement. Ann of Thor Surgery 1999; 67(2):411–416.

23. Oxenham H, bloomfield P, Wheatley DJ, Lee RJ, Cunningham J, Prescott RJ, Mitter HC. Twenty year comparison of a Bjork-Shiley mechanical heart valve with porcine bioprostheses. Heart 2003; 89:715–721.

24. Gonzalez-Juanatey JR, Vega FM, Gude FM, Duran Munoz DD, Iglesias C. Influence of prosthesis size and left ventricular mass on left ventricular diastolic reserve in patients with aortic valve prostheses. J Heart Valve Dis 2001; 10(5):611–618.

25. Ikonomidis I, Tsoukas A, Parthenakis F, Gournizakis A, Kassimatis A, Rallidir L, Nihopamopaochs P. Four year follow up of aortic valve replacement for isolated aortic stenosis:a link between reduction in pressure overload, regression of left ventricular hypertrophy, and diastolic function. Heart 2001; 86(3):309–316.

26. Gelsomino S, Frassani R, Morocutti G, Nucifora R, Da Col P, Minen G, Morelli A, Liui U. Time course of left ventricular remodeling after stentless aortic valve replacement. Am Heart J 2001; 142:556–562.

27. De Paulis R, Sommariva L, De Matteis GM, Caprara E, Tomai F, Penta de Peppo A, Polisca P, Cassan C, Chiariello L. Extent and pattern of regression of left ventricular hypertrophy in patients with small size CarboMedics aortic valves. J Thorac Cardiovasc Surg 1997; 113:901–909.

28. Bech-Hanssen O, Caidahl K, Wall B, Myken D, Larsson S, Wallentin I. Influence of aortic valve replacement, prosthesis type, and size on functional outcome and ventricular mass in patients with aortic stenosis. J Thorac Cardiovasc Surg 1999; 118(1):57–65.

29. Robiolio PA, Rigolin VH, Hearne SE, Baker NA, Kisslo KB, Pierce CH, Baslore TM, Harrison JK. Left ventricular performance improves late after aortic valve replacement in patients with aortic stenosis and reduced ejection fraction. Am J Cardiol 1995; 76:612–615.

30. Connolly HM, Oh JK, Orzulak TA, Osborn SL, Roger VL, Hodge DO, Failey RR, Seward JB, Tajik AJ. Aortic valve replacement for aortic stenosis with severe left ventricular dysfunction. Circulation 1997; 95(10):2395–2400.

31. Pela G, La Canna G, Metra M, Ceconi M, Berra Centurmi P, Alfieri O, Visioli O. Long-term changes in left ventricular mass, chamber size and function after valve replacement in patients with severe aortic stenosis and depressed ejection fraction. Cardiology 1997; 88:315–322.

32. Croke RP, Pifarre R, SullivanH, Gunnar R, Loeb H. Reversal of advanced left ventricular dysfunction following aortic valve replacement for aortic stenosis. The Annals of Thoracic Surgery 1977; 24(1):38–43.

33. Smith N, McAnulty JH, Rahimtoola SH. Severe aortic stenosis with impaired left ventricular function and clinical heart failure: results of valve replacement. Circulation 1978; 58(2):255–264.

34. Pantely G, Morton M, Rahimtoola SH. Effects of successful, uncomplicated valve replacement on ventricular hypertrophy, volume, and performance in aortic stenosis, and in aortic incompetence. J Thoracic and Cardiovasc Surg 1978; 75:383–391.

35. Harpole DH, Gall SA, Wolfe WG, Ramkin JS, Jones RH. Effects of valve replacement on ventricular mechanics in mitral regurgitation and aortic stenosis. Ann Thorac Surg 1996; 62:756–761.

36. Lund O, Erlandsen M. Changes in left ventricular function and mass during serial investigations after valve replacement for aortic stenosis. J Heart Valve Dis 2000; 9(4): 583–593.

37. Kuhl HP, Franke A, Puschmann D, Schondube FA, Hoffmann R, Hanrath P. Regression of left ventricular mass one year after aortic valve replacement for pure severe aortic stenosis. Am J Cardiol 2002; 89(4):408–413.

38. Kurnik PB, Innerfield M, Wachspress JD, Eldredge WJ, Waxman H. Left ventricular mass regression after aortic valve replacement measured by ultrafast computed tomography. Am Heart J 1990; 120:919–927.

39. Monrad ES, Hess OM, Murakami T, Nonogi H, Corin WJ, Krayenbuehl HP. Time course of regression of left ventricular hypertrophy after aortic valve replacement. Circulation 1988; 77(6):1345–1355.

40. Villari B, Vassalli G, Monrad ES, Chiariello M, Turina M, Hess OM. Normalization of diastolic dysfunction in aortic stenosis late after valve replacement. Circulation 1995; 91:2353–2358.

41. Murakami T, Kikugawa D, Endou K, FukuhiroY, Ishida A, Monta I, Masaki H, Inada H, Fugiwara T. Changes in patterns of left ventricular hypertrophy after aortic valve replacement for aortic stenosis and regurgitation with St. Jude medical cardiac valves. Artificial organs 2000; 24(12):953–958.

42. Krayenbuehl HP, Hess OM, Monrad ES, Schneider J, Nall G, Turina M. Left ventricular myocardial structure in aortic valve disease before, intermediate, and late after aortic valve replacement. Circulation 1989; 79:744–755.

43. Schwarz F, Flemeng W, Schaper J, Langebartels F, Sesto M, Hehrlein, Schlepper M. Myocardial structure and function in patients with aortic valve disease and their relation to postoperative results. Am J of Cardiol 1978; 41:661–670.

44. Klodas E, Enriquez-Sarano M, Tajik AJ, Mullany CJ, Bailey KR, Seward JB. Aortic regurgitation complicated by extreme left ventricular dilation: long-term outcome after surgical correction. JACC 1996; 27:670–677.

45. Bonow RO, Dodd JT, Maron BJ, O'Gara PT, White GG, McIntosch Cl, Clark RE, Epstein BE. Long-term serial changes in left ventricular function and reversal of ventricular dilatation after valve replacement for chronic aortic regurgitation. Circulation 1988; 78:1108–1120.

46. Gaasch WH, Carroll JD, Levine HJ, Criscitiello MG. Chronic aortic regurgitation: prognostic valve of left ventricular end-systolic dimension and end-diastolic radius/thickness ratio. JACC 1983; 1(3):775–782.

47. Taniguchi K, Nakano S, Hirose H, Matsuda H, Shirakura R, Cakai K, Kawamoto T, Sakaki S, Kawashima Y. Preoperative left ventricular function: minimal requirement

for successful late results of valve replacement for aortic regurgitation. JACC 1987; 10:510–518.

48. Greves J, Rahimtoola SH, McAnulty JH, DeMot SH, Clark DG, Greenberg B, Starr A. Preoperative criteria predictive of late survival following valve replacement for severe aortic regurgitation. Am Heart J 1981; 101:300–308.

49. Di Biasi P, Paje A, Salati M, Bozzi G, Viecca ON, Gialfi A, DiBiasi M, Guzzetti S, Santoli C. Surgical timing in aortic regurgitation: left ventricular function analysis by contractility score. Ann Thorac Surg 1994; 58:509–515.

50. Levine HJ, Gaasch WH. Ratio of regurgitant volume to end-diastolic volume: a major determinant of ventricular response to surgical correction of chronic volume overload. Am J of Cardiol 1983; 52:406–410.

51. Rahimtoola SH. Drug-related valvular heart disease: here we go again: will we do better this time? Mayo clinic proceeding 2002; 77(12):1275–1277.

52. Rumberger JA, Reed JE. Quantitative dynamics of left ventricular emptying and filling in pure aortic regurgitation and in normal subjects. Am J Cardiol 1992; 70:1045–1050.

53. Clark DG, McAnulty JH, Rahimtoola SH. Valve replacement in aortic insufficiency with left ventricular dysfunction. Circulation 1980; 61:411–421.

54. Boucher CA, Wilson RA, Kanarek DJ, Hutter AM Jr, Okada RD, Liberthson RR, Strauss HW, Pohost GM. Exercise testing in asymptomatic or minimally symptomatic aortic regurgitation: relationship of left ventricular ejection fraction to left ventricular filling pressure during exercise. Circulation 1983; 67:1091–1100.

55. Borer JS, Herrold EM, Hochreiter C, Roman M, Supino P, Devereux RB, Kligfield P, Nawaz H. Natural history of left ventricular performance at rest and during exercise after aortic valve replacement for aortic regurgitation. Circulation 1991; 84(suppl III):III-133–III-139.

56. Lamb HJ, Beyerbacht HP, de Roos A, Van der Laarse A, Vliegen HW, Leujes F, Bax JJ, Van der Wall EE. Left ventricular remodeling early after aortic valve replacement: differential effects on diastolic function in aortic valve stenosis and aortic regurgitation. JACC 2002; 40:2182–2188.

57. Carroll JD, Gaasch WH, Zile MR, Levine HJ. Serial changes in left ventricular function after correction of chronic aortic regurgitation: dependence on early changes in preload and subsequent regression of hypertrophy. Am J Cardiol 1983; 51:476–482.

58. Grover-McKay M, Weiss RM, Vandenberg BF, Burns TL, Weidner GJ, Winniford MD, Stanford W, McKay CR. Assessment of cardiac volumes and left ventricular mass by cine computed tomography before and after mitral balloon commissurotomy. Am Heart J 1994; 128:553–559.

59. Harrison JK, Davidson CJ, Hermiller JB, Harding MB, Hanemann JD, Curma JT, Kisslo KB, Bashore TM. Left ventricular filling and ventricular diastolic performance after percutaneous balloon mitral valvotomy. Am J Cardiol 1992; 69:108–112.

60. Crawford MH, Souchek, Oprian CA, Miller DC, Rahimtoola S, Giacomini JC, Hammermeister KE. Determinants of survival and left ventricular performance after mitral valve replacement. Circulation 1990; 81:1173–1181.

61. Fawzy ME, Choi WB, Mimish L, Sivanandam V, Lingamanaicker I, Khan A, Patel A, Klam B. Immediate and long-term effect of mitral balloon valvotomy on left ventricular volume and systolic function in severe mitral stenosis. Am Heart J 1996; 132:356–360.

62. Tischler MD, Sutton MS, Bittl JA, Parker JD. Effects of percutaneous mitral valvuloplasty on left ventricular mass and volume. Am J Cardiol 1991; 68:940–944.

63. Gash AK, Carabello BA, Cepin D, Spann JF. Left ventricular ejection performance and systolic muscle function in patients with mitral stenosis. Circulation 1983; 67(1):148–154.

64. Mohan JC, Arora R. Effects of atrial fibrillation on left ventricular function and geometry in mitral stenosis. Am J Cardiol 1997; 80:1618–1620.

65. Kennedy JW, Doces JG, Stewart DK. Left ventricular function before and following surgical treatment of mitral valve disease. Am Heart J 1979; 97(5):592–598.

66. Wisenbaugh T, Skudicky D, Sareli P. Myocardial function/valvular disease/hypertensive heart disease: prediction of outcome after valve replacement for rheumatic mitral regurgitation in the era of chordal preservation. Circulation 1994; 89(1):191–197.

67. Corin WJ, Sutsch G, Murakami T, Krogmann ON, Turina M, Hess OM. Left ventricular function in chronic mitral regurgitation: preoperative and postoperative comparison. JACC 1995; 25:113–121.

68. Sousa Uva MS, Dreyfus G, Rescigno G, Aile N, Mascagni R, La Marra M, Pouillart F, Pagaonkar S, Palsky E, Raffoul R, Scorsin M, Moera G, Lesstana A. Surgical treatment of asymptomatic and mildly symptomatic mitral regurgitation. J Thorac Cardiovasc Surg 1996; 112:1240–1249.

69. Ren JF, Aksut S, Lighty GW, Vigilante GI, Sink JD, Segal BC, Hargrove WC. Mitral valve repair is superior to valve replacement for the early preservation of cardiac function: relation of ventricular geometry to function. Am Heart J 1996; 131:974–981.

70. Goldfine H, Aurigemma GP, Zile MR, Gaasch WH. Left ventricular length-force shortening relations before and after surgical correction of chronic mitral regurgitation. JACC 1998; 31:180–185.

71. Carabello BA, Nolan SP, McGuire LB. Assessment of preoperative left ventricular function in patients with mitral regurgitation: value of the end-systolic wall stress-end-systolic volume ratio. Circulation 1981; 64(6):1212–1218.

72. Wisenbaugh T, Spann JF, Carabello BA. Differences in myocardial performance and load between patients with similar amounts of chronic aortic versus chronic mitral regurgitation. JACC 1984; 4(4):916–923.

73. Boucher CA, Bingham JB, Osbakken MD, Okada RD, Strauss HW, Block PC, Levine FH, Phillips HR, Pohost GM. Early changes in left ventricular size and function after correction of left ventricular volume overload. Am J of Cardiol 1981; 47(5):991–1004.

74. Harpole DH, Rankin JS, Wolfe WG, Clements FM, Trigt P, Young WG, Jones RH. Effects of standard mitral valve replacement on left ventricular function. Ann Thorac Surg 1990; 49:866–874.

75. Enriquez-Sarano M, Schaff HV, Orszulak TA, Tajik AJ, Bailey KR, Frye RL. Valve repair improves the outcome of surgery for mitral regurgitation. Circulation 1995; 91:1022–1028.

76. DeAnda A, Moon MR, Kwok LY, Yun KL, Daughter GT, Ingels NB Jr, Miller DC. Left ventricular torsional dynamics immediately after valve replacement. Circulation; 1994; 90[part 2]:II-339–II-346.

77. Hennein HA, Sawain JA, McIntosh CL, Bonow RO, Stone CD, Clark RE. Comparative assessment of chordal preservation versus chordal transection during mitral valve replacement. J Thorac Cardiovasc Surg 1990; 99:228–237.

78. Yun KL, Sintek CF, Miller DC, Pfeffer TA, Kochanba GS, Khonsan S, Zile MR. Randomized trial comparing partial versus complete chordal-sparing mitral valve replacement: effects on left ventricular volume and function. J Thorac Cardiovasc Surg 2002; 123(4):707–714.

79. Shyu KG, Chen JJ, Lin FY, Tsai CH, Lin TL, Tseng YZ, Lien WP. Regression of left ventricular mass after mitral valve repair of pure mitral regurgitation. Ann Thorac Surg 1994; 58:1670–1673.

22

Cardiac Remodeling After LVAD Placement

Kenneth B. Margulies
Cardiovascular Institute, University of Pennsylvania, Philadelphia, Pennsylvania, U.S.A.

INTRODUCTION

It is clear from other chapters in this text that pathological cardiac remodeling, including myocardial hypertrophy and chamber dilation, is a process that promotes progression of myocardial dysfunction and adverse clinical sequelae among patients with dilated cardiomyopathies, regardless of the etiology. Accordingly, the goal of delaying progression or inducing regression of pathological cardiac hypertrophy and dilation, so-called "reverse remodeling," has emerged as an important therapeutic target in the treatment of dilated cardiomyopathies. Other chapters in this section highlight the actions of a variety of medical and surgical therapies that have demonstrated the capacity to induce reverse remodeling of the hearts of at least some treated patients. To some degree, distinct mechanisms that promote or sustain pathological hypertrophy are implicated by the distinct therapies, which have been associated with clinically evident reverse remodeling. These mechanisms and their associated therapies include: neurohormonal activation (RAAS and/or adrenergic inhibitors), increased ventricular wall stress (valve surgery, surgical remodeling and/or passive restraint devices), ongoing ischemia (revascularization), and contractile dyssynchrony (multisite pacing).

In recent years, some of the most dramatic examples of reverse remodeling have been observed following placement of left ventricular assist devices (LVADs) in patients with medically refractory heart failure awaiting heart transplantation. Mechanical (LVADs), as illustrated in Figure 1, have become a reliable means of sustaining medically refractory patients with heart failure awaiting cardiac transplantation. Immediately after their implantation, LVADs induce profound cardiac unloading, including reductions in both preload and afterload. The profound decreases in cardiac loading conditions are well illustrated by the immediate decreases in left ventricular end-diastolic volume and increases in relative wall thickness observed immediately following LVAD implantation (1). Measurements of cardiac hemodynamics before and after initiation of LVAD support and confirm the striking and sustained reductions in myocardial preload and afterload and also

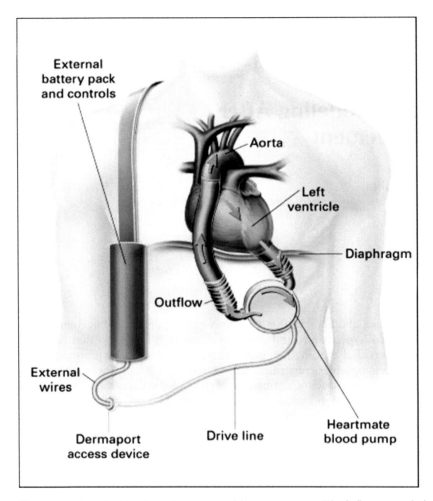

Figure 1   A typical implanted LVAD and its components. The inflow cannula is inserted into the apex of the left ventricle, and the outflow cannula is anastomosed to the ascending aorta. Blood returns from the lungs to the left side of the heart and exits through the left ventricular apex and across an inflow valve into the pumping chamber. Blood is then actively pumped through an outflow valve into the ascending aorta. The pumping chamber is placed within the abdominal wall. LVAD placement induces immediate and sustained decreases in left ventricular preload and afterload along with increases in effective cardiac output. *Abbreviations*: LVAD, left ventricular assist device. *Source*: From Ref. 1a.

demonstrate marked increases in effective cardiac output (2,3). Over time, LVAD support is associated with progressive decreases in the neurohormonal activation associated with advanced heart failure. Specifically, sustained LVAD support induces deactivation of the renin–angiotensin–aldosterone system, the sympathetic nervous system, arginine vasopressin, and natriuretic peptide systems (2–4). These combined effects of LVAD support seem ideally suited for inducing shifts in the pathological myocardial phenotype, though it may be difficult to determine whether hemodynamic unloading or neurohumoral deactivation are the most important determinants of the changes observed.

Beyond the strong stimuli for reverse remodeling engendered by sustained LVAD support, another key feature of the bridge to transplant use of LVAD support is the ability to document the myocardial adaptations to this therapeutic intervention at the tissue level. This opportunity derives from the typical removal of a core of transmural left ventricular tissue at the time of LVAD implantation and the removal of the remainder of the heart at the time of transplantation. These tissue specimens provide the opportunity for unique in vitro morphological, physiological, and molecular characterizations of the failing and recovering human heart using methodologies that are impractical in other clinical settings. Using both paired and unpaired study designs, studies from multiple laboratories have demonstrated that LVAD support induces regression of many of the typical cellular and molecular abnormalities of the failing myocardium and their in vivo correlates. The composite insights gained from such studies will be reviewed in this chapter. In general, these studies have supported the hypothesis that the progression and persistence of many of the myocardial abnormalities observed in dilated failing hearts, regardless of etiology, are a direct or indirect consequence of factors ameliorated by LVAD support, namely increased myocardial wall stress and sustained neurohumoral activation. Because the bridge to transplant use of LVADs is only applied to patients with the most severe degrees of myocardial failure and pathological remodeling, studies of LVAD-supported hearts also support the hypothesis that the potential for myocardial reverse remodeling is retained across a wide spectrum of disease severity. Nevertheless, it is likely that LVAD-associated myocardial adaptations, reflect a plasticity and potential for reverse remodeling that exists within most myopathic hearts.

LVAD-ASSOCIATED REGRESSION OF PATHOLOGICAL STRUCTURE

Myocyte Morphology

Given that both hemodynamic overload and neurohumoral activation have been implicated as primary triggers of cardiac myocyte hypertrophy, it is not surprising that the mechanical unloading and neurohumoral deactivation associated with LVAD support induces regression of myocyte hypertrophy. Indeed, several different studies have reported reductions in myocyte size following LVAD support (2,5–9). In the most comprehensive study to date, Zafeiridis and coworkers utilized isolated cardiac myocytes to evaluate changes in myocyte size, shape, and heterogeneity following sustained LVAD support. As shown in Table 1, these studies demonstrated that the average volume of failing human myocytes is nearly twice as great as that observed in nonfailing myocytes, and that LVAD support is associated with a 60% regression of this pathological hypertrophy. Similarly, LVAD support induces a 62% regression of the increase in average myocyte length and reduces the marked heterogeneity of myocyte lengths associated with advanced heart failure. Extending these observations, Rivello et al. (9) reported that sustained LVAD support is associated with decreases in cardiac myocyte nuclear size and DNA content. Finally, Barbone et al. (6) demonstrated that decrements in left ventricular myocyte size during LVAD support are not observed in the minimally unloaded right ventricle. These findings indicate a dominant role for mechanical unloading, rather than neurohomoral deactivation, in mediating the regression of hypertrophy observed in LVAD-supported hearts.

Table 1 Morphometric Data Obtained from Isolated Human Cardiac Myocytes from HF, HF/LVAD, and Hearts Obtained from Non-failing Donors

Variable	HF	HF/LVAD	Nonfailing
Rods, %	30 ± 3	24 ± 5	27 ± 8
Volume, μm³	51,888 ± 2067	37,443 ± 3307[a]	27,947 ± 1980[a]
Length, μm	201 ± 6	161 ± 7[a]	136 ± 41[a]
Width, μm	31.5 ± 0.9	25.1 ± 1.5[a]	26.2 ± 1.3[a]
Thickness, μm	10.9 ± 0.7	11.8 ± 0.7	10.1 ± 1.0
Length-to-thickness ratio	21.0 ± 1.7	14.2 ± 1.3[a]	14.0 ± 1.3
Mononucleated cells, %	48 ± 3	50 ± 2	75 ± 2
Binucleated cells, %	51 ± 3	50 ± 2	25 ± 2

All data are mean ± SEM.
[a]$P < 0.05$ versus HF.
Abbreviations: HF, failing hearts without LVAD support; HF/LVAD failing hearts with prior LVAD support; LVAD, left ventricular assist device; SEM, standard error of mean.
Source: From Ref. 5.

Extracellular Matrix

Studies conflict regarding the effects of LVAD support on the remodeling of the extracellular matrix (ECM). Some studies indicate no change in the ratio of matrix to myocyte (10,11), some suggest increased fibrosis (6,12–15), and some studies have reported decreased fibrosis (7,16). These disparities likely reflect differences in measurement techniques and heterogeneity among the samples included in the relatively small series that have been reported. One particularly important technical issue is whether reports of collagen or other ECM constituents are reported as an unadjusted concentration or whether measurements are normalized to account for the large changes in myocyte volume that accompany LVAD support, as underscored by recent reports in animal models (17). Sample heterogeneity arising from differences in etiology, sampling locations, the duration of mechanical support, and/or adjuvant therapies may also have a profound impact on measurements of ECM changes following LVAD support.

As with the importance of qualitative ECM changes during the progression of cardiac remodeling, recent studies suggest that qualitative ECM adaptations during LVAD support may be of primary importance. For example, Li et al. (10) observed that total myocardial collagen content was unchanged following two months of mechanical unloading, yet the relative concentration of un-denatured collagen was increased, consistent with an increase in collagen cross-linking. Associated findings of decreased matrix metalloproteinases (MMPs) and increases in endogenous tissue inhibitors of metalloproteinases (TIMPs) suggest a possible molecular basis for the apparent increases in collagen cross-linking. Moreover, other studies demonstrating striking increases in the slope of the end-diastolic pressure–volume relationship are consistent with increased chamber stiffness as might be expected with increased collagen cross-linking (18). Overall, the findings suggest that the changes in the extracellular fraction of the myocardium induced by mechanical unloading are subtle and time-dependent. However, the shift in myocardial remodeling enzyme activity and abundance suggests that the ECM has a highly regulated plasticity similar to that in myocytes. As with myocytes, changes in mechanical loading conditions appear to be a fundamental trigger for ECM remodeling. Indeed, the ECM combined with integrin-mediated signaling processes has been implicated in the mechanotransduction that triggers myocyte adaptations to alterations in mechanical loading conditions (19). However, along with mechanical factors, changes in myocardial cytokines and neurohormonal activation may also be contribute to MMP and TIMP regulation during LVAD support (20).

Organ Level Structure

In general, the reverse remodeling of cell dimensions is associated with marked reverse remodeling at the organ level. Previous studies have demonstrated reductions in LV mass that parallel reductions in myocyte volume (2,5). Chamber geometry is also significantly altered following sustained LVAD support, and previous reports have documented near normalization of LV end-diastolic diameter (2,13). Moreover, studies examining end-diastolic pressure–volume relationships have confirmed that decreases in echocardiographically defined LV dimensions reflect a true change in cardiac geometry rather than simply cardiac decompression (18,21). As noted above, these same studies suggested that LVAD support increases the chamber stiffness of the failing heart. In addition, it appears that decreases in chamber volume are limited during the first 40 days of LVAD support, but become much more significant thereafter in association with increases in relative collagen content within the LV myocardium (15).

LVAD-ASSOCIATED IMPROVEMENT IN ABNORMAL FUNCTION

Basal Contractile Function

Because LVAD support results in dramatic and sustained decreases in left ventricular preload and afterload, in vivo measures of contractile performance before and during support are difficult to compare. Accordingly, using in vitro techniques that avoid the confounding effects of differences in loading conditions upon measures of contractility, several investigators have demonstrated improvements in myocardial contractile function. In isolated human cardiac myocytes, Dipla et al. (22) observed that one to six months of LVAD support was associated with significant improvements in the fractional shortening of isolated myocytes with even more marked improvements in the rates of shortening and relaxation. These findings were confirmed by Terracciano et al. (23). In isolated cardiac muscle strips, Heerdt et al. (21) demonstrated 38% percent improvements in developed force following LVAD support, using a paired tissue analysis in which the pre-LVAD contractility of muscle strips from LV apical core were compared with the post-LVAD contractility of muscle strips from the LV free wall. In contrast, at slow stimulation frequencies, Ogletree-Hughes did not observe alterations in developed tension or other contractile parameters when isolated trabeculae from LVAD-supported hearts were compared with strips from other failing hearts (24). However, like others these investigators did observe shorter relaxation times in LVAD-supported failing hearts.

Prolongation of the cardiac myocyte action potential is another functional hallmark of the failing myocyte. In medically refractory failing hearts requiring LVAD placement, Harding et al. (25) reported reductions in action potential duration following sustained LVAD support. These decreases in action potential duration appear to reflect a reversal of the electrophysiological abnormalities characteristic of the failing human heart. Through modulation of voltage-dependent calcium fluxes, these changes in action potential duration may also contribute to changes in contractile function, including the faster myocardial relaxation observed following LVAD support.

Functional Reserve

Under normal circumstances, human myocardium demonstrates striking improvements in contractile performance during increases in stimulation frequency and/or increases in adrenergic stimulation, while defects in these sources of contractile reserve are characteristic of the failing heart (26–28). Employing in vitro techniques, several studies have demonstrated that LVAD support is associated with improved in vitro contractile performance during increases in stimulation frequency. For example, Dipla et al. (22) reported that failing human myocytes without LVAD-supported hearts exhibit a stepwise decrease in shortening magnitude when stimulation frequency is incremented from 0.2 to 0.5 to 1.0 Hz, but that failing myocytes with antecedent LVAD support exhibit better preserved shortening during increased stimulation frequency. Using isolated left ventricular trabeculae from human hearts and pacing rates from 1.0 to 2.5 Hz, Heerdt et al. (21) likewise demonstrated a negative force–frequency relationship in failing myocardium without prior LVAD support and a positive force–frequency response in muscle strips from both nonfailing and LVAD-supported human hearts, as illustrated in Figure 2. This study also observed

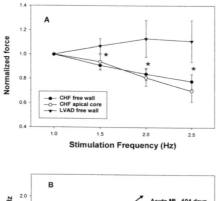

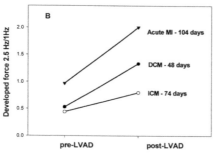

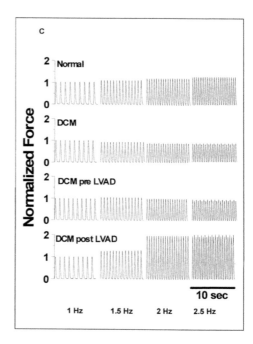

Figure 2 (**A**) Normalized force (mean ± SEM) developed at increasing stimulation frequency by isometric LV trabeculae from free wall of failing hearts (CHF free wall, n = 15), from apex before LVAD placement (CHF apical core, n = 5), and from free wall after LVAD support (LVAD free wall, n = 7). $^{a}P \leq 0.05$ for CHF free wall versus. LVAD free wall. (**B**) Ratio between force developed at 2.5 and 1 Hz before and after LVAD support in three individual hearts. MI indicates myocardial infarction. (**C**) Force tracings from isometrically contracting trabeculae isolated from one nonfailing (normal) and two DCM hearts. In one DCM heart, a negative FFR before LVAD support became markedly positive after LVAD support. *Abbreviations*: CHF, congestive heart failure; DCM, dilated cardiomyopathy; FFR force-frequency relationships; LV, left ventricle; LVAD, left ventricular asisst device; SEM, standard error of mean. *Source*: From Ref. 21.

improved frequency-dependent contractile performance during experiments in which cardiac trabeculae from the same patient were studied before and after LVAD support. While affirming that LVAD support induces improved frequency-dependent contractions of LV trabeculae, Barbone and coworkers demonstrated that a normalized force–frequency response is not consistently observed in right ventricular trabeculae (6). Given that the magnitude of hemodynamic unloading and remodeling induced by LVAD support is far less in the right ventricle compared with that of the left ventricle, these findings suggest that the reduction of hemodynamic load is a primary factor underlying improved contractile function and contractile reserve observed following LVAD support.

Complementing the results observed with frequency-dependent contractile reserve, several studies have elucidated beneficial effects of LVAD support upon myocardial β-adrenergic responses. In isolated myocytes, Dipla et al. (22) reported that isoproterenol induced much greater increments in shortening magnitude following LVAD support than in myocytes from failing hearts without prior LVAD support. Similarly, Ogletree-Hughes and coworkers observed that LVAD-supported failing muscle strips exhibited marked improvements in peak developed tension and positive and negative dP/dt responses to isoproterenol compared to failing human cardiac trabeculae without prior LVAD support (24). These investigators further demonstrated that these improved β-adrenergic responses did not differ significantly from responses observed in human cardiac trabeculae from nonfailing hearts. In both myocyte and muscle strip studies, improvements in adrenergic responses have been observed even when there is no difference in basal contractility between failing muscle strips and failing muscle strip preparations with LVAD support (22,24).

Mechanisms of Functional Improvements

Just as many separate pathogenic processes may contribute to contractile dysfunction in failing hearts, multiple mechanisms may contribute to improved contractile performance following LVAD support. Nevertheless, several lines of evidence implicate improvements in defective cardiac myocyte calcium handling as a cellular mechanism contributing to improved basal contractile function following LVAD support. First, studies in isolated myocytes have identified changes in the shape of the calcium transient that mirror changes in cellular shortening following LVAD support (22). Moreover, two separate studies have demonstrated LVAD-associated improvements in calcium uptake rates and binding in isolated sarcoplasmic reticulum membranes (21,29), while recent studies by Terracciano et al. (23) indicate that there are increases in sarcoplasmic reticulum calcium load following LVAD support. Consistent with these findings, myocardial expression of calcium handling proteins suggest that sarcoplasmic reticulum calcium ATPase (SERCA) mRNA and protein abundance may be increased following LVAD support (21). Alternatively, LVAD-associated decreases in sodium–calcium exchanger protein without alterations in SERCA were correlated with faster rates of Ca^{2+} transient decay in recent studies by Chaudhary et al. (30). These changes in Ca^{2+} transient decay are illustrated in Figure 3. Finally, because cellular relaxation is in part regulated by cellular repolarization, LVAD-associated reductions in action potential duration (25) may also contribute to faster rates of relaxation and improved rate-dependent contractile reserve observed following LVAD support.

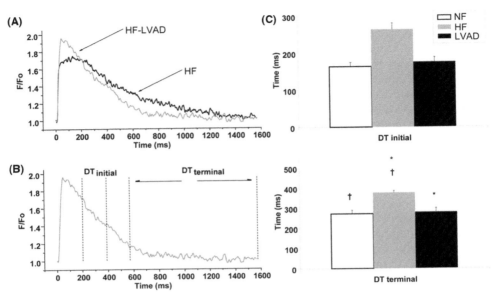

Figure 3 Effects of LVAD support on $[Ca^{2+}]_i$ transients in cardiac myocytes from failing hearts with and without prior LVAD support. Data are collected during field stimulation at 0.5 Hz and $[Ca^{2+}]_i$ was measured by Fluo-3 fluorescence. (**A**) Raw fluorescence intensity traces from representative myocytes. (**B**) Schematic illustrating curve-fitting technique for initial and delayed portions of the $[Ca^{2+}]_i$ transient decay. (**C**) Average initial and delayed time constants for the $[Ca^{2+}]_i$ transients in each of the three experimental groups. Data are expressed as mean ± SEM. *$P < 0.0001$ HF-only versus. HF-LVAD, †$P < 0.001$. *Abbreviations*: HF, failing heart; HF-LVAD, failing heart with LVAD support; LVAD, left ventricular assist device; NF, nonfailing heart. *Source*: From Ref. 30.

Because β-adrenergic–mediated improvements in contractility and relaxation rates depend, in part, on phosphorylation of the key calcium regulatory proteins (L-type calcium channel, ryanodine receptor, and phospholamban), some of the same factors affecting basal contractility and rate-dependent contractile reserve are likely contributing to improvements in adrenergic responsiveness following LVAD support. In addition, Ogletree-Hughes et al. (24) correlated improved adrenergic responses with increases in β-adrenergic receptor density following LVAD support. Improved adrenergic signaling with LVAD support likely extends beyond the receptor level to factors regulating the phosphorylation of key calcium handling proteins following adrenergic stimulation. For example, as illustrated in Figure 4, Chen et al. (31) demonstrated that increases in L-type calcium current density following isoproterenol were greater in LVAD-supported or nonfailing myocytes compared with nonsupported failing myocytes. These studies further indicated that the defects in the failing myocytes are related to hyperphosphorylation of the L-type calcium channel that appears to be ameliorated by LVAD support. Several reports provide possible triggers for these multilevel improvements in β-adrenergic signaling following LVAD support. First, at least two previous studies, have demonstrated gradual decreases in circulating epinephrine and norepinephrine over one to two months following initiation of LVAD support (3,29) and most patients are weaned off intravenous inotropes, including β-adrenergic agonists, soon after LVAD placement. Moreover, Muller and coworkers have reported decreases in circulating

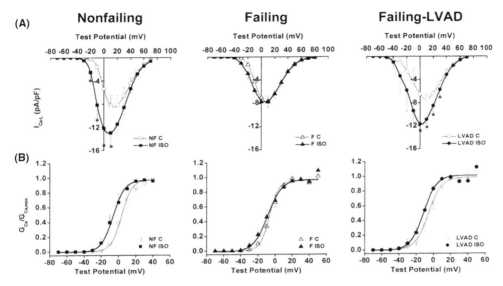

Figure 4 Effect of 1 μmol/L isoproterenol on the L-type calcium channel ($I_{Ca,L}$) current-voltage relationship in NF ($n = 6$, $N = 3$), F ($n = 14$, $N = 7$), and F-LVAD ($n = 10$, $N = 5$) human ventricular myocytes. (**A**) Isoproterenol increased $I_{Ca,L}$ significantly in NF and LVAD but had little effect in F ($I_{Ca,L,\ ISO}/I_{Ca,L,\ C}$ in NF vs. LVAD vs. F). (**B**) Isoproterenol caused a significant leftward shift of the voltage dependence of $I_{Ca,L}$ activation in NF and F-LVAD myocytes, but no significant shift in F myocytes. C indicates control, open symbols; ISO, 1 μmol/L isoproterenol, filled symbols. *Abbreviations*: F, failing heart; NF, nonfailing heart; LVAD, left ventricular assist device. *Source*: From Ref. 31.

anti–β1-adrenoreceptor autoantibodies during sustained LVAD support (16,32). Overall, the findings to date suggest that improved contractility, relaxation rates, and contractile reserve reflect LVAD-induced changes in cellular calcium handling. In turn, altered calcium handling dynamics and adrenergic responsiveness are likely a result of altered interactions between protein abundance, electrophysiological factors, and post-translational regulation (e.g. receptor internalization and phosphorylation) of key calcium regulatory proteins following LVAD support.

In Vivo Function

Several studies have examined in vivo LV contractile function following LVAD support. With the LVAD temporarily turned off, Frazier and coworkers observed a mean change in LV ejection fraction from 11% ± 5% to 22% ± 17% following LVAD support (29). Studies from the Berlin Heart Institute likewise documented increases in LV ejection fraction following LVAD support, yet these studies clearly demonstrate that improved ejection fraction is not a universal finding in LVAD-supported patients (33). In vivo inquiries have also extended the observations of changes in myocyte electrophysiology following LVAD support. Specifically, Harding et al. reported that decreases in action potential duration were paralleled by decrease in QTc duration on the surface ECG (25). These in vivo observations were not associated with any decreases in the QRS duration further supporting the implication that decreases in the QTc duration were a result of decreases in the duration and/or heterogeneity of myocyte repolarization.

LVAD-ASSOCIATED CHANGES IN MOLECULAR PHENOTYPE

Stretch-Induced Signaling and Apoptosis

While many of the studies highlighted above implicate decreases in mechanical loading conditions as a primary trigger of structural and functional "reverse remodeling", very few studies to date have elucidated the signal transduction pathways, which may mediate changes in myocardial phenotype during LVAD support of the failing human myocardium. In one study, Flesch et al. examined the family of mitogen activated protein kinases (MAPKs), including p44/42 extracellularly regulated kinase (ERK), p38 kinase, ad c-Jun N-terminal kinase (JNK), in LVAD-supported human hearts (34). Compared with failing hearts without LVAD support, these investigators observed decreased p44/42 phosphorylation and activity, an increase in p38 phosphorylation and activity, and a decrease in JNK1/2 abundance and activity among LVAD-supported hearts. These distinct changes in MAPK signaling molecules were associated with reductions in myocyte size and reduced rates of myocyte apoptosis based on an in situ DNA ligation assay. In another study by Baba et al. (35), several different signaling pathways implicated in stretch-related myocardial adaptations were altered following LVAD support in human myocardium. In this study, ERK and Akt activity were markedly decreased after LVAD support while the activity of GSK-3β was increased. Changes in the expression of these molecules were associated with a loss of their usual transmural gradient of expression in failing hearts, a finding that is consistent with their involvement in stretch-associated signaling. As in the studies by Flesch et al. (34), changes in MAPK activity were associated with decreases in myocardial apoptosis (36). In turn, findings of reduced apoptosis rates following LVAD support are in accord with other studies in which reduced apoptosis has been associated with LVAD-related decreases in the proapoptotic molecule Bcl-1 and increases in the anti-apoptotic molecules Bcl-xL, Mcl-1, and FasExo6Del/Fas (37–39).

Cytokine Pathways

Recognizing that increased myocardial cytokines have been observed in advanced cardiac failure and implicated as mediators of cardiac dysfunction, several studies have examined the expression of myocardial cytokines before and after LVAD support. For example, Torre-Amione et al. (40) observed consistent decreases in myocardial tumor necrosis factor alpha (TNF-α) immunostaining in eight consecutive LVAD-supported patients and reported that the magnitude of the TNF-α decrease was correlated with the clinical improvement score, although a subsequent study failed to confirm the association between TNF-α mRNA and clinical improvement (41). In addition to TNF-α, Barton et al. (42) reported that the cytokines interleukin-1β and interleukin-6 are significantly increased among patients requiring LVAD support, and that these cytokines are directly correlated with increases in myocardial MMP and TIMP expression. Employing microarray technology, Chen et al. (43) also observed reductions in the myocardial expression of several cytokine-associated molecules including TNF superfamily 10, interferon-induced protein, and small inducible cytokines A2 and A21.

Other Molecular Changes of Possible Functional Significance

In addition to changes in stretch responsive molecules and cytokine-associated pathways, other molecular changes reported following LVAD support of failing human hearts may provide insights into the mechanisms mediating the reverse remodeling process. For example, Razeghi et al. (44) reported alterations in metabolic genes, including the glucose transporters GLUT1 and GLUT4, muscle carnitine palmitoyl transferase-1, and uncoupling protein 3, and speculated that reductions in uncoupling protein might contribute to reduced oxygen-derived free radicals and oxidative stress after LVAD support. Complementary studies by Mital et al. (45) also demonstrated that LVAD support is associated with more efficient myocardial oxygen utilization via augmented nitric oxide–mediated regulation of mitochondrial respiration. Other studies demonstrating increases in caveolin expression following LVAD support provide a possible mechanism for improved β-adrenergic signaling following LVAD support (46), because previous studies have linked increased caveolin-dependent β$_2$-adrenergic receptor sequestration to improved coupling between the β$_2$-adrenergic receptor and adenylate cyclase (47). In addition, reports of decreases in nuclear factor-κB expression and binding activity (8) and decreases in myocardial dystrophin expression (48) suggest still other molecular changes that may contribute to the reverse remodeling observed following LVAD support.

Broad Trends in Gene Expression

Exploiting the paired human myocardial specimens available at the time of LVAD implantation and removal and the analytical power of cDNA microarrays capable of simultaneously exploring the mRNA abundance for thousands of genes, several investigators have examined broad trends in myocardial gene expression following LVAD support. In general, these are hypothesis-generating inquiries designed to identify key molecular signaling pathways driving reverse remodeling during LVAD support. In one study, Blaxall et al. demonstrated that LVAD support induces significant changes in the gene expression and that a distinct separation between the pre- and post-LVAD groups was consistent with a unique gene signature associated with reverse remodeling (49). These investigators also detected disease-related differences in the response to LVAD support such that the divergence of molecular phenotype between pre-LVAD and post-LVAD was greater for patients with nonischemic cardiomyopathy than it was for patients with heart failure due to coronary artery disease. In a second study, Chen et al. (43) observed that genes related to the process of transcription, cell growth/apoptosis/DNA repair, structural proteins, metabolism, and cell signaling were, in general, upregulated following LVAD support; and genes related to cytokines were downregulated. These findings are illustrated in Figure 5. In general, these observations are consistent with several previous observations concerning LVAD-associated changes in cytokine expression, signaling, apoptosis, and metabolism. (34,39,40,44). Interestingly, despite the tendency to focus on LVAD-associated changes in gene expression, a large-scale analysis from our laboratory has indicated that dysregulated genes in failing hearts are far more likely to remain abnormal than they are to recover toward a normal phenotype (50). In these studies, the general discordance between the microarray data indicating persistent pathological abnormalities and the structural and functional data suggesting significant improvements following LVAD support, suggest that many key pathological processes are regulated at a post-transcriptional level. Overall, it seems likely that the application of microarray technology to failing myocardium, including tissues subjected to LVAD support, will continue to generate new insights and hypotheses concerning the biology of reverse remodeling.

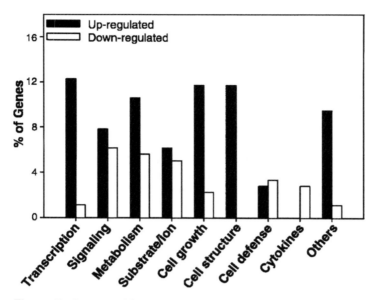

Figure 5 Percent of known genes in each functional category that were upregulated (*solid bars*) or downregulated (*open bars*) following sustained LVAD support of failing human hearts. Transcription, transcriptional factors; signaling, cell signaling/communication; substrate/ion, substrate/ion transport; cell growth, cell growth/apoptosis/DNA repair; cell structure, cell structure/extracellular matrix; cell defense, cell/organism defense. *Abbreviations*: LVAD, left ventricular assist device. *Source*: From Ref. 43.

FROM REVERSE REMODELING TO RECOVERY WITH LVADs

General Considerations

As reviewed above, the composite studies to date examining the impact of LVAD support on the failing myocardium clearly demonstrate that the combination of myocardial unloading and neurohormonal deactivation induced by these devices results in multilevel adaptations within the failing heart. However, while many of these adaptations suggest a substantial potential for structural, functional, and molecular plasticity within even the most diseased hearts, LVAD support has resulted in sustained myocardial recovery after device removal in only a small fraction of cases in which devices were placed for the purpose of stabilizing patients awaiting cardiac transplantation (33,51). It is likely that multiple factors contribute to the relative infrequency of sustained myocardial recovery following LVAD support their use as a last resort rather than earlier in the course of disease progression, the loss of excessive myocardium to necrosis or apoptosis, and/or the fact that it is often difficult and damaging to remove the devices typically used for bridge-to-transplant support. Beyond these realities, there are also many important aspects of the biology and therapeutics of reverse remodeling that remain incompletely understood. Some of the key questions are: How does one identify which hearts are intrinsically more recoverable? What is the time-dependence of the various facets of the remodeling/recovery process? How does the degree of unloading shape the recovery process? Which myocardial adaptations during LVAD support are beneficial and which are detrimental? What adjuvant therapies or interventions potentiate the recovery process? and Are there readily accessible biomarkers that can reliably identify necessary and/or sufficient degrees of myocardial recovery to allow weaning from LVAD support? At this point, there are no definitive answers to any of these questions. However, as highlighted below, existing data support the necessity for addressing such questions as part of a strategy for promoting sustained myocardial recovery.

Time-Dependence of Reverse Remodeling

Concerning the time-dependence of myocardial recovery, studies by Madigan et al. (15) indicate that myocyte hypertrophy regresses by about 40 days following initiation of LVAD support, but changes in SERCA expression occur faster, while changes in the absolute quantity of ECM constituents takes a longer time than the regression of myocyte hypertrophy. These studies, which exploited the clinical variability in the duration of LVAD support to derive insights into the tissue biology of reverse remodeling, are illustrated in Figure 6. Alternatively, for electrophysiological changes, in vivo observations by Harding et al. indicate a biphasic temporal response, such that LVAD-mediated decompression of the overloaded myocardium is associated with an immediate increase in the heart rate adjusted QT interval (QTc), but with sustained LVAD support, there was a secondary decrease in QTc duration that paralleled decreases in the cellular action potential duration measured in isolated human ventricular myocytes (30). On yet another level, Uray et al. (45) reported that increases in caveolin-1 transcription was greatest in patients with a six to 24 week duration of LVAD support but that it declined with longer support intervals. Clearly, the more discordant the time-dependence of the various beneficial aspects of reverse remodeling, the more challenging it is to define the optimal time for device removal, even without considering the potential for adverse myocardial adaptations following restoration of normal physiological loading conditions (52).

Adjuvant Interventions

Some investigators have begun exploring the utility of adjunctive therapeutic interventions that might improve the likelihood of durable myocardial recovery following

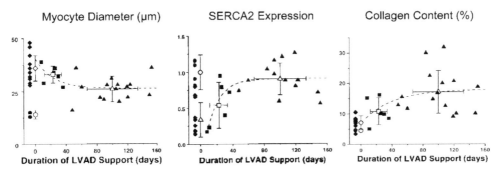

Figure 6 Effect of duration of LVAD support on myocyte diameter (*left*), SERCA2 mRNA abundance (*center*), and relative myocardial collagen content (*right*). Filled symbols represent individual patients and open symbols represent mean values. Squares indicate patients with <40 days of LVAD support, triangles indicate patients with >40 days of LVAD support, diamonds indicate non-LVAD failing hearts, and circles indicate nonfailing hearts. The dashed line indicates a nonlinear fit to LVAD data. The time constants for change are 24 days for myocyte diameter, 15 days for SERCA expression, and 39 days for collagen content. *Abbreviations*: LVAD, left ventricular assist device; SERCA, sarcoplasmic reticulum calcium ATPase.

sustained LVAD support. In general, the goal of these efforts is to address challenges to recovery posed by the underlying etiology of heart failure (such as ischemia, infarction, and valve disease) and/or potentially unfavorable adaptations to LVAD support (such as excessive loss of myocardial mass). In one study, investigators injected previously harvested, autologous skeletal muscle myoblasts into previously infracted regions of myocardium at the time of LVAD insertion in five patients with medically refractory heart failure (53). These studies demonstrated skeletal muscle cell engraftment and differentiation into mature myofibers without any significant adverse events. Although rates of myoblast engraftment were extremely low ($< 1\%$) and there were no effects of myoblast transplantation on structural or functional adaptations to LVAD support, these studies do provide an important proof of principle for the strategy of adjuvant cellular transplantation in the setting of LVAD support. Likewise, these studies demonstrate that the clinical stabilization provided by LVAD support may facilitate the deployment of novel interventions with delayed therapeutic benefits in critically ill patients. Other studies have begun to explore the efficacy of adjuvant pharmacological therapy designed to enhance rates of recovery in LVAD-supported hearts. For example, in addition to conventional heart failure therapy (beta blockers, ACE-inhibitors, angiotensin receptor antagonists, and spironolactone), Terracciano et al. used the β_2 receptor agonist clenbuterol in LVAD-supported patients to promote physiological hypertrophy and functional recovery (23). However, despite adjuvant pharmacological therapy, cell size regressed to normal levels during LVAD support and previously observed improvements in the shortening-frequency relationship (21,22) were not apparent in this study. Overall, there is still considerable uncertainty about the proper role of adjuvant therapies in augmenting LVAD-associated recovery phenomena with the goal of device weaning. Beyond autologous myoblasts and existing pharmacological therapies, potentially useful adjuvants might include multisite pacemakers, defibrillators, cardiac restraint devices, stem cells, growth factors, cytokine antagonists, and therapies that regulate extracellular matrix constituents. Though unproven, it seems logical that the choice of therapeutic adjuvants should be matched to the defects and risks specific to the host myocardium and the type and duration of mechanical support being employed. Clearly, this type of targeted recovery strategy will require far more research into the biology of reverse remodeling and the host responses to therapeutic adjuvants.

Biomarkers

Recognizing the close association between myocardial hypertrophy and the expression of natriuretic peptides within the ventricular myocardium, one of the earliest and most consistent changes reported with LVAD support is a reduction in expression of ANP and or BNP in the supported myocardium (2). Though changes in natriuretic peptide precursors within the ventricular myocardium are generally felt to be markers, of reverse-remodeling, rather than mediators, one study has shown that plasma levels of these secreted proteins may provide readily available noninvasive insights into the degree and tempo of reverse remodeling of the supported heart (4). Perhaps more importantly, Khan et al. (54) examined whether in vivo assessment of contractile reserve with graded doses of dobutamine might predict favorable responses following withdrawal of mechanical support without transplantation. In these studies, six of nine patients with favorable responses to dobutamine, including an increase in cardiac index, dP/dt, ejection fraction and a decrease in left ventricular end-diastolic diameter, survived for 12 months after LVAD removal without the need for transplantation or restoration of mechanical support. Although virtually all patients with LVAD support demonstrate decreases in myocardial mass and myocyte hypertrophy, these studies suggest that the restoration of appropriate contractile reserve is a prerequisite for device weaning, because it may help distinguish clinically meaningful recovery from persistent dysfunction.

CONCLUSION

In summary, a number of important lessons about cardiac remodeling can be derived from studies of LVAD-supported human hearts. Most obviously, LVAD support induces multilevel regression of the pathologic phenotype of the failing human heart. In this regard, changes observed during LVAD support demonstrate that reverse remodeling can occur even in the most advanced cases of dilated cardiomyopathy. Although local and systemic neurohormonal factors may play a role, it is increasingly apparent that mechanical loading conditions are as important in sustaining the pathologic features of the severely failing heart as they are in driving the progressive remodeling that occurs during transitions from normal to hypertrophy to failure. It is also likely that diverse signaling cascades contribute to the adaptations to myocardial unloading and the phenotypic transitions in the LVAD-supported heart. Mechanistically, it remains uncertain whether reverse remodeling simply involves a deactivation of those signaling pathways involved in forward remodeling or rather a unique activation of novel recovery pathways. At the same time, insights into broad molecular dynamics provided by early studies using microarray technology suggest that many of the structural adaptations typical of LVAD-supported failing hearts are likely regulated via post-transcriptional control mechanisms. Regardless of the specific signaling pathways involved, it is also likely that some LVAD-induced myocardial adaptations might help promote a sustained myocardial recovery while others might actually be maladaptive for a nonsupported heart (e.g. atrophy). Indeed, this concern and unaddressed host myocardial defects likely account for the relatively low rates of sustained clinical recovery in patients weaned off LVAD support. Therefore, further studies to help distinguish adaptive from maladaptive features of reverse remodeling will provide the foundation for future therapeutic strategies, including therapeutic adjuvants that address host myocardial defects and antagonize maladaptive processes, while still promoting favorable adaptations to LVAD support.

REFERENCES

1. Nakatani S, Thomas JD, Savage RM, Vargo RL, Smedira NG, McCarthy PM. Prediction of right ventricular dysfunction after left ventricular assist device implantation. Circulation 1996; 94:II216–21.
1a. Goldstein DJ, et al. N Engl J Med 1998; 339:1522–1533.
2. Altemose GT, Gritsus V, Jeevanandam V, Goldman B, Margulies KB. Altered myocardial phenotype after mechanical support in human beings with advanced cardiomyopathy. J Heart Lung Transplant 1997; 16:765–73.
3. James KB, McCarthy PM, Thomas JD, Vargo R, Hobbs RE, Sapp S, Bravo E. Effect of implantable levt ventricular assist device on neuroendocrine activation in heart failure. Circulation 1995; 92:II-191–II-195.

4. Milting H, A ELB, Kassner A, Fey O, Sarnowski P, Arusoglu L, Thieleczek R, Brinkmann T, Kleesiek K, Korfer R. The time course of natriuretic hormones as plasma markers of myocardial recovery in heart transplant candidates during ventricular assist device support reveals differences among device types. J Heart Lung Transplant 2001; 20:949–55.

5. Zafeiridis A, Jeevanandam V, Houser SR, Margulies KB. Regression of cellular hypertrophy after left ventricular assist device support. Circulation 1998; 98:656–62.

6. Barbone A, Holmes JW, Heerdt PM, The AH, Naka Y, Joshi N, Daines M, Marks AR, Oz MC, Burkhoff D. Comparison of right and left ventricular responses to left ventricular assist device support in patients with severe heart failure: a primary role of mechanical unloading underlying reverse remodeling. Circulation 2001; 104:670–5.

7. Bruckner BA, Stetson SJ, Perez-Verdia A, Youker KA, Radovancevic B, Connelly JH, Koerner MM, Entman ME, Frazier OH, Noon GP, et al. Regression of fibrosis and hypertrophy in failing myocardium following mechanical circulatory support. J Heart Lung Transplant 2001; 20:457–64.

8. Grabellus F, Levkau B, Sokoll A, Welp H, Schmid C, Deng MC, Takeda A, Breithardt G, Baba HA. Reversible activation of nuclear factor-κB in human end-stage heart failure after left ventricular mechanical support. Cardiovasc Res 2002; 53:124–30.

9. Rivello HG, Meckert PC, Vigliano C, Favaloro R, Laguens RP. Cardiac myocyte nuclear size and ploidy status decrease after mechanical support. Cardiovasc Pathol 2001; 10:53–7.

10. Li YY, Feng Y, McTiernan CF, Pei W, Moravec CS, Wang P, Rosenblum W, Kormos RL, Feldman AM. Downregulation of matrix metalloproteinases and reduction in collagen damage in the failing human heart after support with left ventricular assist devices. Circulation 2001; 104:1147–52.

11. Kinoshita M, Takano H, Takaichi S, Taenaka T, Nakatani T. Influence of prolonged ventricular assistance on myocardial histopathology in intact heart. Ann Thorac Surg 1996; 61:640–645.

12. Nakatani S, McCarthy PM, Kottke-Marchant K, Harasaki H, James KB, Savage RM, Thomas JD. Left ventricular echocardiographic and histologic changes: impact of chronic unloading by an implantable ventricular assist device. J Am Coll Cardiol 1996; 27:894–901.

13. McCarthy PM, Nakatani S, Vargo R, Kottke-Marchant K, Harasaki H, James KB, Savage RM, Thomas JD. Structural and left ventricular histologic changes after implantable LVAD insertion. Ann Thorac Surg 1995; 59:609–13.

14. Scheinin SA, Capek P, Radovancevic B, Duncan JM, McAllister HA, Jr., Frazier OH. The effect of prolonged left ventricular support on myocardial histopathology in patients with end-stage cardiomyopathy. Asaio J 1992; 38:M271–4.

15. Madigan JD, Barbone A, Choudhri AF, Morales DL, Cai B, Oz MC, Burkhoff D. Time course of reverse remodeling of the left ventricle during support with a left ventricular assist device. J Thorac Cardiovasc Surg 2001; 121:902–8.

16. Loebe M, Hennig E, Muller J, Spiegelsberger S, Weng Y, Hetzer R. Long-term mechanical circulatory support as a bridge to transplantation, for recovery from cardiomyopathy, and for permanent replacement. Eur J Cardiothorac Surg 1997; 11 Suppl: S18–24.

17. McGowan BS, Scott CB, Mu A, McCormick RJ, Thomas DP, Margulies KB. Unloading-induced remodeling in the normal and hypertrophic left ventricle. Am J Physiol Heart Circ Physiol 2003; 284:H2061–8.

18. Levin HR, Oz MC, Chen JM, Packer M, Rose EA, Burkhoff D. Reversal of chronic ventricular dilation in patients with end-stage cardiomyopathy by prolonged mechanical unloading. Circulation 1995; 91:2717–20.

19. Ross RS, Borg TK. Integrins and the myocardium. Circ Res 2001; 88:1112–9.

20. Spinale FG. Bioactive peptide signaling within the myocardial interstitium and the matrix metalloproteinases. Circ Res 2002; 91:1082–4.

21. Heerdt PM, Holmes JW, Cai B, Barbone A, Madigan JD, Reiken S, Lee DL, Oz MC, Marks AR, Burkhoff D. Chronic unloading by left ventricular assist device reverses contractile dysfunction and alters gene expression in end-stage heart failure. Circulation 2000; 102:2713–9.

22. Dipla K, Mattiello JA, Jeevanandam V, Houser SR, Margulies KB. Myocyte recovery after mechanical circulatory support in humans with end-stage heart failure. Circulation 1998; 97:2316.

23. Terracciano CM, Harding SE, Adamson D, Koban M, Tansley P, Birks EJ, Barton PJ, Yacoub MH. Changes in sarcolemmal Ca entry and sarcoplasmic reticulum Ca content in ventricular myocytes from patients with end-stage heart failure following myocardial recovery after combined pharmacological and ventricular assist device therapy. Eur Heart J 2003; 24:1329–39.

24. Ogletree-Hughes ML, Stull LB, Sweet WE, Smedira NG, McCarthy PM, Moravec CS. Mechanical unloading restores beta-adrenergic responsiveness and reverses receptor downregulation in the failing human heart. Circulation 2001; 104:881–6.

25. Harding JD, Piacentino V, 3rd, Gaughan JP, Houser SR, Margulies KB. Electrophysiological alterations after mechanical circulatory support in patients with advanced cardiac failure. Circulation 2001; 104:1241–7.

26. Rossman EI, Petre RE, Chaudhary KW, Piacentino V, Janssen PM, Gaughan JP, Houser SR, Margulies KB. Abnormal frequency-dependent responses represent the pathophysiologic signature of contractile failure in human myocardium. J Mol Cell Cardiol 2004; 36:33–42.

27. Holubarsch C, Ruf T, Goldstein DJ, Ashton RC, Nickl W, Pieske B, Pioch K, Ludemann J, Wiesner S, Hasenfuss G, et al. Existence of the Frank-Starling mechanism in the failing human heart. Investigations on the organ, tissue, and sarcomere levels. Circulation 1996; 94:683–9.

28. Houser SR, Margulies KB. Is depressed myocyte contractility centrally involved in heart failure? Circ Res 2003; 92:350–8.

29. Frazier OH, Benedict CR, Radovancevic B, Bick RJ, Capek P, Springer WE, Macris MP, Delgado R, Buja LM. Improved left ventricular function after chronic left ventricular unloading. Ann Thorac Surg 1996; 62:675–81; discussion 681–2.

30. Chaudhary KW, Rossman EI, Piacentino V III, Kenessey A, Weber C, Gaughan JP, Ojamaa K, Klein I, Bers DM, Houser SR, et al. Altered myocyte Ca^{2+} cycling after left ventricular assist device support in the failing human heart. J Am Coll Cardiol 2004; 44:837–45.

31. Chen X, Piacentino V, 3rd, Furukawa S, Goldman B, Margulies KB, Houser SR. L-type Ca^{2+} channel density and regulation are altered in failing human ventricular myocytes and recover after support with mechanical assist devices. Circ Res 2002; 91: 517–24.

32. Muller J, Wallukat G, Weng YG, Dandel M, Spiegelsberger S, Semrau S, Brandes K, Theodoridis V, Loebe M, Meyer, et. al R. Weaning from mechanical cardiac support in patients with idiopathic dilated cardiomyopathy. Circulation 1997; 96:542–9.

33. Hetzer R, Muller JH, Weng Y, Meyer R, Dandel M. Bridging-to-recovery. Ann Thorac Surg 2001; 71:S109–13; discussion S114–5.

34. Flesch M, Margulies KB, Mochmann HC, Engel D, Sivasubramanian N, Mann DL. Differential regulation of mitogen-activated protein kinases in the failing human heart in response to mechanical unloading. Circulation 2001; 104:2273–6.

35. Baba HA, Stypmann J, Grabellus F, Kirchhof P, Sokoll A, Schafers M, Takeda A, Wilhelm MJ, Scheld HH, Takeda N, et. al. Dynamic regulation of MEK/Erks and Akt/GSK-3beta in human end-stage heart failure after left ventricular mechanical support: myocardial mechanotransduction-sensitivity as a possible molecular mechanism. Cardiovasc Res 2003; 59:390–9.

36. Baba HA, Grabellus F, August C, Plenz G, Takeda A, Tjan TD, Schmid C, Deng MC. Reversal of metallothionein expression is different throughout the human myocardium

after prolonged left-ventricular mechanical support. J Heart Lung Transplant 2000; 19:668–74.

37. Bartling B, Milting H, Schumann H, Darmer D, Arusoglu L, Koerner MM, El-Banayosy A, Koerfer R, Holtz J, Zerkowski H. Myocardial Gene Expression of Regulators of Myocyte Apoptosis and Myocyte Calcium Homeostasis During Hemodynamic Unloading by Ventricular Assist Devices in Patients with End-Stage Heart Failure. Circulation 1999; 100: II–216–II–223.

38. Francis GS, Anwar F, Bank AJ, Kubo SH, Jessurun J. Apoptosis, Bcl-2, and proliferating cell nuclear antigen in the failing human heart: observations made after implantation of left ventricular assist device. J Card Fail 1999; 5:308–15.

39. de Jonge N, van Wichen DF, van Kuik J, Kirkels H, Lahpor JR, Gmelig-Meyling FH, van den Tweel JG, de Weger RA. Cardiomyocyte death in patients with end-stage heart failure before and after support with a left ventricular assist device: low incidence of apoptosis despite ubiquitous mediators. J Heart Lung Transplant 2003; 22:1028–36.

40. Torre-Amione G, Stetson SJ, Youker KA, Durand JB, Radovancevic B, Delgado RM, Frazier OH, Entman ML, Noon GP. Decreased expression of tumor necrosis factor-alpha in failing human myocardium after mechanical circulatory support : A potential mechanism for cardiac recovery. Circulation 1999; 100:1189–93.

41. Razeghi P, Mukhopadhyay M, Myers TJ, Williams JN, Moravec CS, Frazier OH, Taegtmeyer H. Myocardial tumor necrosis factor-alpha expression does not correlate with clinical indices of heart failure in patients on left ventricular assist device support. Ann Thorac Surg 2001; 72:2044–50.

42. Barton PJ, Birks EJ, Felkin LE, Cullen ME, Koban MU, Yacoub MH. Increased expression of extracellular matrix regulators TIMP1 and MMP1 in deteriorating heart failure. J Heart Lung Transplant 2003; 22:738–44.

43. Chen Y, Park S, Li Y, Missov E, Hou M, Han X, Hall JL, Miller LW, Bache RJ. Alterations of gene expression in failing myocardium following left ventricular assist device support. Physiol Genomics 2003; 14:251–60.

44. Razeghi P, Young ME, Ying J, Depre C, Uray IP, Kolesar J, Shipley GL, Moravec CS, Davies PJ, Frazier OH, et al. Downregulation of metabolic gene expression in failing human heart before and after mechanical unloading. Cardiology 2002; 97:203–9.

45. Mital S, Loke KE, Addonizio LJ, Oz MC, Hintze TH. Left ventricular assist device implantation augments nitric oxide dependent control of mitochondrial respiration in failing human hearts. J Am Coll Cardiol 2000; 36:1897–902.

46. Uray IP, Connelly JH, Frazier OH, Taegtmeyer H, Davies PJ. Mechanical unloading increases caveolin expression in the failing human heart. Cardiovasc Res 2003; 59:57–66.

47. Ostrom RS, Gregorian C, Drenan RM, Xiang Y, Regan JW, Insel PA. Receptor number and caveolar co-localization determine receptor coupling efficiency to adenylyl cyclase. J Biol Chem 2001; 276:42063–9.

48. Vatta M, Stetson SJ, Perez-Verdia A, Entman ML, Noon GP, Torre-Amione G, Bowles NE, Towbin JA. Molecular remodelling of dystrophin in patients with end-stage cardiomyopathies and reversal in patients on assistance-device therapy. Lancet 2002; 359: 936–941.

49. Blaxall BC, Tschannen-Moran BM, Milano CA, Koch WJ. Differential gene expression and genomic patient stratification following left ventricular assist device support. J Am Coll Cardiol 2003; 41:1096–106.

50. Margulies KB, Matiwala S, Cornejo C, Olsen H, Craven WA, Bednarik D. Mixed messages: transcription patterns in failing and recovering human myocardium. Circ Res 2005; 96:592–599.

51. Mancini DM, Beniaminovitz A, Levin H, Catanese K, Flannery M, DiTullio M, Savin S, Cordisco ME, Rose E, Oz M. Low incidence of myocardial recovery after left ventricular assist device implantation in patients with chronic heart failure. Circulation 1998; 98:2383–9.

52. Helman DN, Maybaum SW, Morales DL, Williams MR, Beniaminovitz A, Edwards NM, Mancini DM, Oz MC. Recurrent remodeling after ventricular assistance: is long-term myocardial recovery attainable? Ann Thorac Surg 2000; 70:1255–8.

53. Pagani FD, Dersimonian H, Zawadzka A, Wetzel K, Edge AS, Jacoby DB, Dinsmore JH, Wright S, Aretz TH, Eisen HJ, et al. Autologous skeletal myoblasts transplanted to ischemia-damaged myocardium in humans. Histological analysis of cell survival and differentiation. J Am Coll Cardiol 2003; 41:879–88.

54. Khan T, Delgado RM, Radovancevic B, Torre-Amione G, Abrams J, Miller K, Myers T, Okerberg K, Stetson SJ, Gregoric I, et al. Dobutamine stress echocardiography predicts myocardial improvement in patients supported by left ventricular assist devices (LVADs): hemodynamic and histologic evidence of improvement before LVAD explantation. J Heart Lung Transplant 2003; 22:137–46.

23

Cardiac Resynchronization Therapy in Heart Failure: A Powerful Tool for Reverse Ventricular Remodeling?

Chu-Pak Lau
Cardiology Division, University of Hong Kong, Queen Mary Hospital, Hong Kong

INTRODUCTION

Even with optimal medical therapy, patients with advanced heart failure HF have significant mortality and morbidity. In the Metoprolol CR/XL Randomized Interventional Trial in Congestive Heart Failure (MERIT-HF) study (1), patients in class III/IV HF on a combination of angiotensin converting enzyme inhibitors (ACEI) and metoprolol had a mortality rate of 7.2% per patient-year. More advanced HF patients were recruited in the selective aldosterone blocker eplerenone trial. Patients receiving this agent in addition to other effective neurohormonal blocking medications (86% on ACEI and 75% on beta-blockers) still had a mortality rate of 11.8% at one year (2). Thus, additional approaches are needed to improve survival in the heart failure population

Cardiac remodeling worsens heart function and increases morbidity (3). Best exemplified by the use of ACEI, inhibition of cardiac remodeling is directly associated with a favorable impact on the clinical course (4). To either slow or reverse the remodeling process is now increasingly recognized as a therapeutic target in HF of all etiologies. Cardiac resynchronization therapy (CRT) using biventricular (BV) pacing has recently emerged as a nonpharmacological adjunct to medications in the treatment of HF by correcting cardiac mechanical dyssynchrony secondary to electrical disorders. While demonstration of a survival advantage of using CRT alone on top of medical therapy is not yet available, preliminary evidence suggests a beneficial effect of CRT on the ventricular remodeling process.

ELECTROMECHANICAL CHANGES

Patients with HF frequently develop underlying alterations in electrical conduction pathways in the heart in addition to susceptibility to either atrial and/or ventricular arrhythmias. These conduction abnormalities include sinus node dysfunction, interatrial conduction delay, first degree atrio-ventricular (AV) block, and prolonged

Table 1 Adverse Cardiac Mechanical Consequences of Electrical Conduction Problems in Systolic Heart Failure

	Mechanical effects	Electrical causes
Inappropriate AV timing	Inadequate LV filling due to improper AV interval	First-degree AV block LV activation delay due to BBB
Interventricular dyssynchrony	RV–LV competition in filling and ejection	BBB IVCD
Intraventricular dyssynchrony	Lack of coordinated LV contraction and relaxation	BBB IVCD

Abbreviations: AV, atrio-ventricular; BBB, bundle branch block; IVCD, intraventricular conduction delay; LV, left ventricular; RV, right ventricular.

ventricular conduction (commonly called left bundle branch block (LBBB) or intraventricular conduction delay) (5). The adverse mechanical effects of these electrical abnormalities lead to inadequate left ventricular (LV) filling, interventricular dyssynchrony, and intraventricular dyssynchrony (Table 1). While the relative importance of these factors is likely to vary between patients, the net hemodynamic effects are a reduction in stroke volume, decrease in LV contractility, and worsening of mitral regurgitation. The accumulative effects of these changes are responsible for cardiac remodeling. A widened QRS complex is significantly associated with an impaired long-term survival (6,7).

AV Dyssynchrony

An effective and appropriately timed atrial systole maximizes LV filling at the end of aortic ejection. This is represented by an A wave on transmitral valvular Doppler that follows the passive LV filling (E wave) (Fig. 1A). In the presence of a prolonged PR interval, onset of the A wave will be brought forward while E wave timing remains unchanged, thus shortening the total duration of diastolic filling (8–10). Worse still, ventricular systole can occur during the period when the mitral valve (MV) is not adequately opposed, resulting in diastolic mitral regurgitation (8,10). AV dyssynchrony not only occurs with PR prolongation, but can result from LBBB and delayed mechanical LV contraction, which delays the onset of the E wave (Fig. 1B) (9). The effective filling time is inversely related to the heart rate, with a reduction of 80 ms for each 100 ms decrease in RR interval (10). Thus, in patients with LBBB and prolonged PR interval, LV filling may be so shortened as to significantly reduce the stroke volume, particularly at a fast heart rate.

Early studies using DDD and right ventricle (RV) apical pacing have shown a beneficial effect on LV function and cardiac output in these type of patients due to lateral cardiomyopathy, primarily by appropriately timing the AV interval such that diastolic filling time is improved and mitral regurgitation reduced (8,11). Using acute epicardial pacing at different AV intervals, maximum increase in LV dp/dt and pulse pressure using either LV or BV pacing occurs at a small range of AV intervals (30 ms) (12). It should be noted, however, that while an optimally timed AV interval contributes to cardiac hemodynamics, patients in chronic atrial fibrillation (AF) who developed HF after AV nodal ablation and RV pacing can benefit from upgrade to BV pacing (13), suggesting that CRT has an effect in addition to AV interval modulation.

(A)

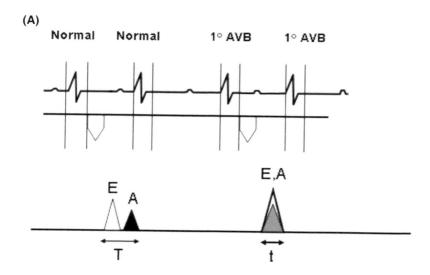

(B)

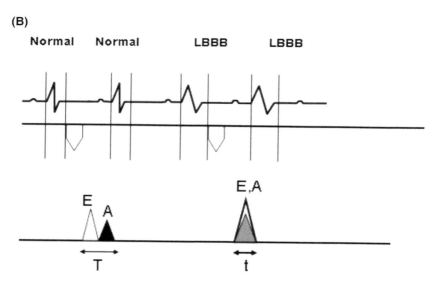

Figure 1 Diastolic filling time due to conduction abnormalities. (**A**) In the presence of a prolonged PR interval, the onset of transmitral Doppler A wave will be brought forward in relation to E wave, thus shortening the filling time from T to t. (**B**) In the presence of LBBB, the E wave, which begins after ventricular ejection, will be delayed while the A wave onset time remains unchanged, again shortening the filling time. Only one set transmitral Doppler is shown in each conduction state. *Abbreviation*: LBBB, left bundle branch block. *Source*: From Ref. 9.

Interventricular Dyssynchrony

A dyssynchronous effect of RV and LV filling and ejection can have deleterious effect on LV mechanical function. An early study (14) has shown that in the presence of LBBB, the total ejection time remains unchanged, but the duration of isovolumetric contraction of the ventricles is prolonged. This not only shortens the LV filling time for a given cardiac cycle length, but LV filling can occur during RV ejection,

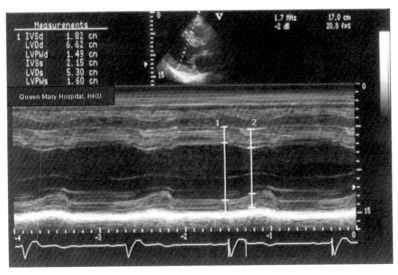

Figure 2 Parasternal long axis M mode echocardiogram in a patient with LBBB. Septal activation (RV contraction) occurred at the time of continued LV filling (*line 1*). Peak septal contraction (*line 2*) occurred prior to the onset of peak posterior wall contraction. *Abbreviations*: LBBB, left bundle branch block; LV, left ventricular; RV, right ventricular.

resulting in a reduction of the LV filling fraction (Fig. 2). The interventricular septum plays a crucial role, in determining LV mechanical function. The presence of paradoxical septum motion reduces the effective LV ejection. On echo-Doppler, this dyssynchrony is represented by a widened difference between the RV and LV pre-ejection period (normally <40 ms) (9). Using radionuclide imaging in 13 patients (15), the phase angle between RV and LV ejection is used to quantify interventricular dyssynchrony. The phase angle correlates inversely with the ejection fraction (EF), and the improvement in EF after CRT is dependent on the degree of asynchrony. When simultaneous RV and LV pressures are measured and plotted in a Lissajous phase plot, the pressure difference between the two ventricles can be represented by a RV–LV pressure loop (Fig. 3A) (16). In the absence of dyssynchrony, the loop size will be minimal. In LBBB, delayed LV contraction occurs, the loop size increases and the loop shifts rightward. This was found to occur in 22/30 (73%) of patients with LBBB in one study.

After CRT, interventricular synchrony was significantly reduced on radionuclide imaging (from 27.5° to 14.1°) (15). The magnitude of interventricular dyssynchrony present in sinus rhythm correlated with the magnitude of improvement in dyssynchrony during BV pacing. There was a significant reduction in the pre-ejection period difference between the RV and LV after BV pacing (9). CRT significantly reduced the simultaneous pressure loop area, and resulted in a rightward shift of the RV/LV loop (Fig. 3B). This effect is more marked in patients with significant pre-existing RV/LV dyssynchrony.

Interventricular Dyssynchrony

Paradoxical or dyssynchronous motion of the septum that results from LBBB as discussed above can be considered as a form of intraventricular dyssynchrony, since

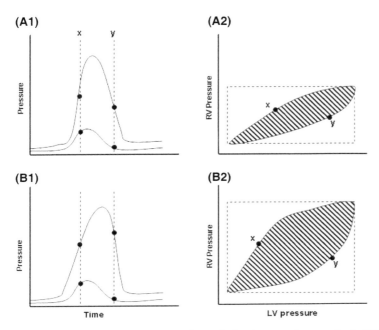

Figure 3 Schematic representation of simultaneously acquired RV and LV pressures in normal subjects (**A1** and **A2**) and in a patient with LBBB (**B1–B2**). The simultaneous pressures at *x* and *y* are depicted with RV as ordinates and LV as abscissca. Simultaneous onset of contraction between the RV and LV (**A1**) results in minimum pressure difference between the RV and LV at all times (**A2**). This resulted in a minimal loop size. In patient with LBBB with delayed LV ejection, the onset of peak LV ejection is delayed. As a result, RV and LV pressure difference is greater, the loop size increased and rightward rotated. *Abbreviations*: LV, left ventricular; RV, right ventricular. *Source*: From Ref. 16.

the septum is shared between the two ventricles. Using MRI tagging (17), there was a marked phase delay between early septal LV and late lateral LV time of contraction, with reciprocal stretch of the septal territory during the contraction of the lateral wall. In general, the early activated region (e.g., the septum) occurs at a time of inadequate filling, such that the contraction is ineffectual. On the other hand, the late activated region (e.g., lateral/posterior wall) would be stretched and excessively loaded by the septal contraction. This contributes to LV remodeling (17–19). Using tissue Doppler imaging, the peak systolic timing of different regions of the LV can be mapped to assess dyssynchrony (17,18,20–25). Figure 4 shows this timing from 12 LV segments and two RV sites in 25 patients with QRS >140 ms (25). The basal anteroseptal segment was the earliest peak sustained systolic movement and occurs at (148 ± 25 ms) after the R wave, whereas the basal lateral segment was the latest 216 ± 52 ms. There was a marked variation in the timing of different segments such that the SD of the timing of these contractions were >30 ms. After CRT with BV pacing, there was a more uniform contraction time in all segments, with reduction in the SD of contraction times. Radial LV dyssynchrony has also been studied with two-dimensional echocardiography, using phase analysis of the septal and lateral LV wall. In 34 patients with LBBB and HF (26), 50% of patients showed septal wall delay in contraction, 38% showed septal wall delay and paradoxical septal movement, and 12% showed no dyssynchrony. The best hemodynamic response was

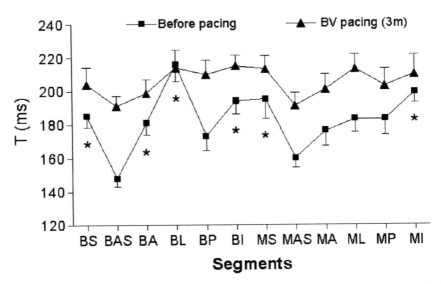

Figure 4 Changes in the time to peak regional sustained systolic contraction (T) before and after BV pacing. At baseline, there was marked regional variation in T among the LV segments. T was earliest in the basal anteroseptal segment and latest in the basal lateral segment. After BV pacing therapy, T of different segments were synchronized to that of the basal lateral segment so that regional variation in T was abolished. $^*P < 0.05$ versus basal anteroseptal segment at baseline. *Abbreviations*: B, basal; M, mid; A, anterior; AS, anteroseptal; I, inferior; L, lateral; P, posterior; S, septal. *Source*: From Ref. 25.

observed after CRT in patients with dyssynchrony but not in those without. Intraventricular dyssynchrony and its reversal by CRT have also been measured with M-mode echocardiography (16), three-dimensional echocardiography (27), myocardial contrast echocardiography (28), and myocardial strain (29).

ACUTE EFFECTS OF CRT

The acute hemodynamic effects of CRT in patients with HF and LBBB have been studied in 18 patients using atrial sensing, RV, BV, or LV free wall pacing mode (30). Figure 5 shows the effect of CRT on LV pressure and volume of the LV. RV apical pacing results in little changes in the pressure–volume loop, whereas either LV or BV pacing significantly increases the stroke volume and shifts the loop downward and to the left. Basal QRS duration correlated with the improvement, but the pacing efficacy was not associated with the degree of QRS narrowing. Using epicardial pacing in 27 patients with dilated cardiomyopathy (12), increases in LV dp/dt max (reflects contractility) and pulse pressure (reflects stroke volume) occurred only when LV or both ventricles were paced. Furthermore, maximum benefit at any site occurred at a patient-specific AV interval.

Myocardial Oxygen Consumption

The effect of CRT versus dopamine on myocardial oxygen demand to achieve similar increase in contractile function was compared in 10 patients (17). At the same

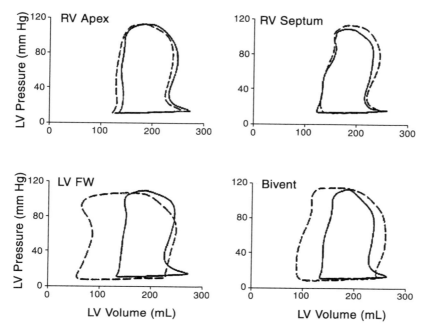

Figure 5 Pressure volume loops from a patient with baseline LBBB as a function of varying pacing site. Data are shown for the optimal AV interval at each site. Solid line indicates non-pacing control; dashed line, VDD pacing. There was negligible effect from RV apical or RV septal pacing. However, LV pacing produced loops with greater area (stroke work) and width (stroke volume) and a reduced systolic volume. The latter is consistent with increased contractile function and thus with elevation of dP/dtmax. *Abbreviations*: LV, left ventricular; RV, right ventricular. *Source*: From Ref. 30.

increase in dp/dt, myocardial oxygen consumption was increased from the baseline by 22% ± 11% during dopamine infusion, but was reduced by 8.0% + 6.5% after CRT. PET imaging was used in eight patients to assess myocardial oxidative metabolism (31). Global LV and RV metabolism were unchanged by CRT, but septal wall metabolism improved after CRT, probably because of an adequate preload. Work metabolic index of the LV was increased by 13% and stroke volume index by 10%, without an increase in the overall metabolism of the LV. Furthermore, the individual segmental contractility using tissue Doppler imaging was not increased after CRT (25). Thus, CRT appears to be energy inexpensive, most likely owing to improved synchronicity to enhance global LV contractility, without an overall increase in myocardial oxygen consumption.

LV REVERSE REMODELING

Early studies showed a significant reduction of left ventricular end diastolic volume (LVEDV) and systolic volume (LVESV) after prolonged CRT, a possible reverse remodeling effect (32,33). In a series of 25 patients (25), the change in LV volumes has been followed before implant, and at one week, one month, and three months after implant (Fig. 6). There was an initial reduction in LV volumes that occurred within one week of implant, which represent an acute CRT effect. There was

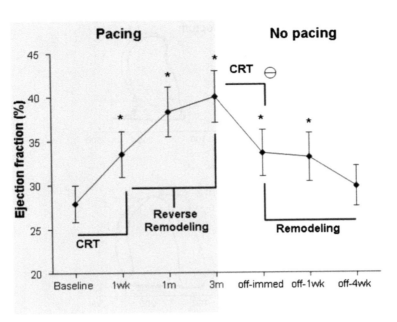

Figure 6 Changes in ejection fraction % (closed squares, end-diastolic, closed triangles, end-systolic) from baseline, one week to three months, and when pacing was withheld for up to four weeks. An early cardiac resynchronization effect (CRT ⊕) followed by a delayed cardiac reverse remodeling effect is observed. There is an initial loss of synchronization effect (CRT5) when pacing is withheld before remodeling finally occurs at four weeks. $^*P < 0.05$ versus baseline. *Source*: From Ref. 25.

progressive reduction in LV volumes that reached a peak at three months of the study period, representing a reverse remodeling effect. Importantly, when pacing was withheld for four weeks, mitral regurgitation recurred immediately owing to a loss of CRT effect. However, LV volumes only began to increase at the end of four weeks, representing a remodeling effect. This study provides strong evidence that CRT has both an acute synchronization effect and a reverse remodeling effect that may take months to occur.

A reverse remodeling effect has also been reported in several series, which do not involve a cross-over design (Table 2). Saxon et al. (34) reported the 12 weeks effect of CRT on 53 patients after BV pacing. At six weeks, no change in cardiac volumes was observed. However, by three months, there was significant reduction in LVEDV and LVESV, with a 10% decrease occurring in 18% and 39% of patients, respectively. There was no change in sphericity index or in LV mass, an effect that the authors attributed to the short follow-up period of the study. Interestingly, the left atrial volume index was reduced (34.4 ± 9 vs. 30.6 ± 10 mL/m^3) during the study, which may have important implications for AF incidence after CRT. In another study (24), the effect on reverse remodeling appears to be maintained up to 12 months.

The multicenter InSync Randomized Clinical Evaluation (MIRACLE) trial (35), randomized 453 patients to CRT or no CRT in a parallel design. In the echo substudy of MIRACLE (36), 323 patients had echo measured at baseline, three months and six months. LV volume changes were reported for both the control and the CRT arm. There was significant reduction in LV volumes, and improvement in EF at three months, which was maintained at six months (Fig. 7). This reverse

Table 2 Reverse LV Remodeling Effects of CRT

Authors	No. of patients	Duration (m)	ΔLVEDV (%)	ΔLVESV (%)	ΔEF (%)
Stellbrink (2001) (33)	25	6	−10.3	−13.9	18.2
Yu (2002) (25)	25	3	−18	−24	43
Saxon (2002) (34)	53	3	−6.3	−8.2	8.7
Sogaard (2002) (24)	20	12	−9.6	−16.5	21.7
Gras (2002) (55)	46	12	−1.5	—	20.3
St. John Sutton (2003) (36)	323	6	−9.2	−11.2	14.7
Mean		7	−9	−12	21
	Total = 492				

Abbreviations: ΔEF, mean change in left ventricular end diastolic volume compared to baseline; ΔLVEDV, mean change in left ventricular end systolic volume compared to baseline; ΔLVESV, mean change in ejection fraction compared to baseline.

remodeling was accompanied by a reduction in mitral regurgitation and a reduction in LV mass (−12.0 vs. −10.6 g, $P < 0.01$ at 6 months). Furthermore, both ischemic and nonischemic etiologies of their cardiomyopathy improved, although the nonischemic patients benefit more. Of note is the finding that the effects of CRT were not influenced by concurrent use of beta-blockers. This study also showed an improved myocardial performance index, prolonged LV filling time and reduced interventricular dyssynchrony, shortening of isovolumetric contraction time and diastolic function after CRT, both when compared to baseline and to the control group.

In the above studies that included a total of 492 patients, and with a duration of follow-up ranging from 3 to 12 months, a mean reduction of 9% and 21%, respectively was observed in both LVEDV and LVESV, along with a 21% increase in EF. In all these trials, a clinical improvement was also achieved.

Increase in plasma norepinephrine level has a negative impact on survival in HF. In an acute study of 13 patients with EF of 28% ± 7% (37), either LV or BV pacing significantly reduced sympathetic nervous activity in efferent, postganglionic peroneal muscle nerve recording when compared to atrio-synchronous RV pacing. Pacing was performed only for three minutes at an AV interval of 100 ms, and a baseline, no-pacing control was not available. It is arguable whether RV pacing actually increased the sympathetic neural discharges rather than BV or LV pacing reducing them. In a longer term study, Saxon et al. (34) measured serum norepinephrine and found that there was no significant change after three months of BV pacing. The lack of any increase is reassuring, as many drug therapies that augment systolic function may have a deleterious effect on survival, when these are associated with norepinephrine increase. Brain natriuretic peptide (BNP) release was used to follow a group of 17 patients after CRT (38). CRT was temporarily suspended for 10 days, which was associated with an elevation of BNP (382 ± 381 pg/mL), and the level of this circulating biomarker was significantly reduced when CRT was re-initiated. Interestingly, in the subgroup of patients found to have no ventricular remodeling, BNP levels did not change on re-initiation of therapy.

Presystolic mitral regurgitation is reduced by appropriate timing of the AV interval. Functional mitral regurgitation secondary to dilated mitral valve ring worsens LV function, increases LV work, and reduces forward cardiac output. In dilated cardiomyopathy, LV dilatation and increases in sphericity index increase the

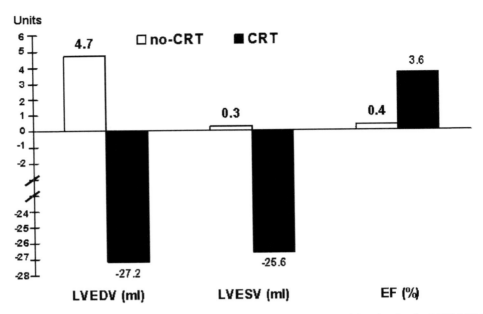

Figure 7 Changes in LVEDV, LVESV, and EF in patients participating in the MIRACLE trial. The control patients have an increase in LV volumes at six months, whereas reduced LV volumes were observed in the CRT group. Only the CRT group had a significant increase in EF. *Abbreviations*: EF, ejection fraction; LVEDV, left ventricular end-diastolic volume; LVESV, left ventricular and end-sybtolic volume. *Source*: From Ref. 36.

distance between the papillary muscles to the enlarged MV annulus as well as to each other, restricting leaflet motion and increasing the force needed for effective mitral valvular closure. Breithardt et al. (39) measured the transmitral valvular gradient as a marker of the force for MV closure. LV dp/dt increased after CRT, and this increase was associated with a decrease in mitral valve orifice from 25 ± 19 to $13 \pm 8\,mm^2$. This change in mitral valve orifice was directly correlated with the change in dp/dt. Furthermore, transmitral valvular gradient increased more rapidly and peaked at a higher value after CRT. Thus, CRT, which increased LV dp/dt, has a favorable effect on MV closure through increase in transmitral valvular gradient.

The antiarrhythmic effects of CRT are of interest and may be related to by reverse LV remodeling. In a study on Ventak CRT-ICD trial, patients were randomized to receive three months of ICD and three months of ICD with CRT in a crossover study (40). There was a significant reduction in the frequency of ICD therapies in the ICD with CRT arm (0.6 ± 2.1 vs. 1.4 ± 3.5 episodes) compared with ICD only arm. Anti-tachycardia pacing was more effective when delivered biventricularly rather than from the RV alone (41). This study also showed reduction in LVESV and LVEDV, and improvement in walking distance. A reduction in ventricular vulnerability occured after reduction in LV volumes in the CRT arm, but not with the RV pacing arm in another study (42). AF occurs frequently in patients with HF, and has implications on short- and long-term mortality and morbidity. Preliminary observational reports have suggested that CRT may reduce the incidence of AF and facilitate the maintenance of sinus rhythm.

Table 3 summarizes the proposed mechanisms for reverse remodeling of CRT in HF. It is likely that CRT has a multiple impact on the ventricular remodeling

Table 3 Proposed Mechanisms for Reverse Remodeling of CRT in Heart Failure

Redistribution of loading in different regions of the LV
Reduction of MR
Reduction of sympathetic activities
Reduction of LV mass
Impact on arrhythmias
Allow further optimization of medical therapy

process. In addition, it is likely that further optimization of medical therapy (e.g., beta blockers) made possible by an implanted device that limits bradycardia may confer additional benefit (see below).

CLINICAL TRIALS

Several registries and controlled trials have reported on the long-term functional benefits of CRT (Fig. 8) (12,35,43,44). The PATH-CHF and multisite stimulation in cardiomyopathy (MUSTIC) studies are cross-over studies, whereas the MIRACLE and CONTAK CD trials are essentially parallel arm studies with controls (device implanted but inactivated). While significant differences were not always achieved for all the parameters measured between trials, there is a uniform trend to beneficial effects with CRT compared with the no pacing arm. These include improvement in quality of life, six-minute walking distance, NYHA functional class, peak oxygen consumption and treadmill exercise time, a favorable change in cardiac structure and function, and fewer hospitalizations. In particular, while all-cause patient survival rate was not changed over a six-month period, there was a reduction in hospitalization and the need for intravenous diuretic therapy in the CRT group compared to controls in the MIRACLE study (35). In this study, there was an increase in

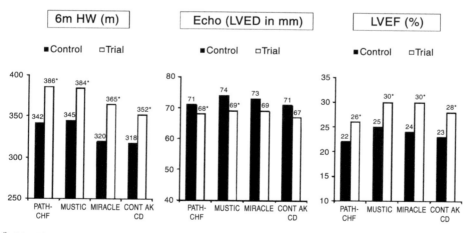

Significant Improvement

Figure 8 Change in six-minute walking distance (6 mWD), LVED, and LVEF in the CRT and control group in the randomized trials. *Abbreviations*: LVED, left ventricular end-diastalic diameter; LVEF, left ventricular ejection fraction. *Source*: From Refs. 12, 29, 35, 43.

Progressive Heart Failure Mortality

Trials	CRT No.	CRT No. (%) of Heart Failure Deaths	No CRT No.	No CRT No. (%) of Heart Failure Deaths	Odds Ratio (95% CI)
CONTAK CD	245	4 (1.5)	245	9 (3.7)	0.44 (0.13-1.43)
InSync ICD	272	6 (2.2)	282	10 (3.5)	0.61 (0.22-1.71)
MIRACLE	263	4 (1.5)	269	10 (3.7)	0.40 (0.12-1.29)
MUSTIC	29	0	29	0	
Overall	809	14 (1.7)	825	29 (3.5)	0.49 (0.25-0.93)

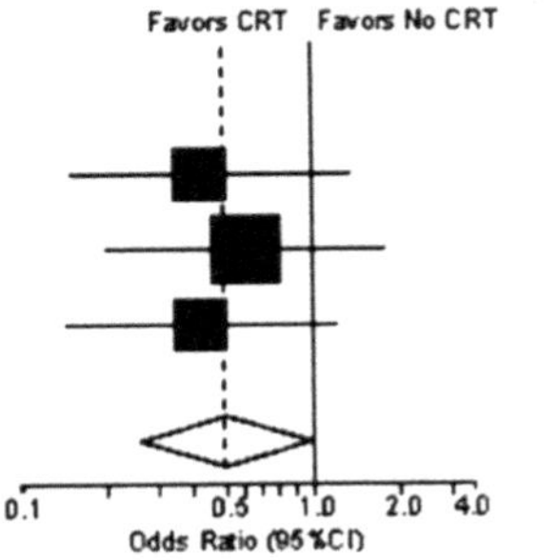

Hospitalisation

Trials	CRT No.	CRT No. (%) Hospitalized	No CRT No.	No CRT No. (%) Hospitalized	Odds Ratio (95% CI)
CONTAK CD	245	11 (4.5)	245	16 (6.5)	0.67 (0.31-1.48)
InSync ICD	272	14 (5.1)	282	17 (6.0)	0.85 (0.41-1.75)
MIRACLE	263	14 (5.3)	269	19 (7.1)	0.74 (0.36-1.51)
Overall	809	40 (4.9)	825	52 (6.3)	0.77 (0.51-1.18)

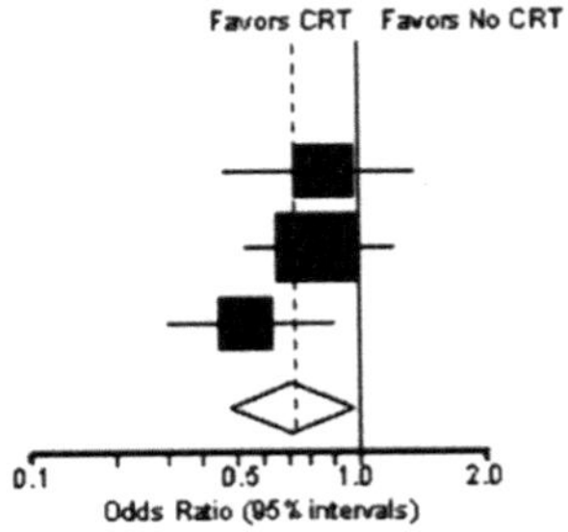

Figure 9 A meta-analysis of the odds ratio of progressive heart failure death and hospitalization in published trials. *Source*: From Ref. 45.

six-minute walking distance (+39 vs. +10 m, $P = 0.005$), functional class ($P < 0.001$), quality of life (−18.0 vs. −9.0 points, $P = 0.001$), time on the treadmill during exercise testing (+81 vs. +19 s, $P = 0.001$), and EF (+4.6% vs. −0.2%, $P < 0.001$).

Mortality and HF requiring hospitalization are of great interest. In a meta analysis, Bradley et al. (45) examined four trials of CRT for these parameters (Fig. 9). With up to 809 patients randomized, there was nearly 50% reduction of progressive HF mortality but without a significant reduction in overall mortality (OR 0.77). On the other hand, HF requiring hospitalization was significantly reduced. This meta analysis included two trials on ICD's, and the potential difference between CRT alone versus CRT with ICD was not analyzed.

Preliminary result of the COMPANION trial has been presented (46,47). This trial randomized 1520 patients with Class III/IV HF to receive optimal medical therapy, optimal medical therapy and CRT, and optimal medical therapy with CRT with ICD, in a 1:2:2 randomization. The composite primary endpoint of death or any hospitalization and the need of intravenous diuretic therapy was reduced at 12 months by CRT or CRT with ICD by about 18% to 19% compared to optimal medical therapy. However mortality difference was only observed in the CRT with ICD arm. This study suggests that in these patients, treatment of concomitant ventricular tachyarrhythmias was important to prolong survival. The use of CRT alone versus optimal medical therapy is the subject of an ongoing trial ("CARE"-HF).

A comparison of the efficacy of CRT with pharmacological therapy is summarized in Figure 10 (48). Overall, the data from the COMPANION study suggest high cost-effectiveness of CRT, with the number of patients needed to treat to save one

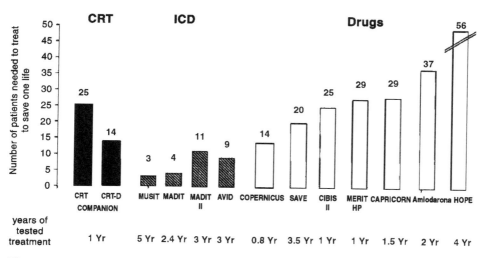

Figure 10 Comparison of different pharmacological and device therapies expressed as the number of patients needed to treat to save one's life. *Source*: From Ref. 48.

life very similar to that of using beta-blockers. This was achieved on top of optimal medical therapy.

IMPLANTATION

The best hemodynamic response of LV pacing is achieved through pacing in the mid lateral or posterior LV based on studies using epicardial electrodes (49). While mini-thoracotomy or thoroscopic lead placement has been used, LV pacing is usually achieved with leads placed in the posterior or lateral branches of the coronary sinus (CS) (endovascular placement but epicardial pacing). To achieve this, special guiding catheters are needed to cannulate the CS, and leads are designed to reach and stay in branches of the CS that give the best compromise between pacing and sensing parameters, and hemodynamic effects (Fig. 11).

There are only registry data on the success rate of LV lead implantation. A success rate of implantation of 90% to 93% was reported with the Medtronic Stylet-driven leads, with a dislodgement rate of 4% to 12%. The procedure time for the entire implant varied from 0.9 to 7.3 (mean 2.7 hours), which is patient-dependent and also depends on the physicians' experience. A 92% successful implant rate after cannulation of the CS with the guidant lead was reported, with 2% dislodgement and a very similar range of implant time (range 0.9–8, mean 2.8 hours) (50). In a small series, an 85% successful implant rate was reported for the stylet-driven Aescular leads (St. Jude), without any late dislodgement, with stable lead performance after implantation (51).

Despite these advances, cannulation of the CS in a dilated heart is a common cause of implant failure, and development like steerable sheaths, atraumatic tips and easier sheath removal techniques while maintaining lead positioning are desirable. Lead development to facilitate negotiation of sharp angles associated with the branching points of the CS or tortuosity of these branches are helpful, and investigation into the best method of lead extraction is forthcoming. As the hemodynamic benefit of CRT is likely to be LV pacing site dependent, lead positioning with on-line hemodynamic assessment will be very useful. Lead development for multiple LV

BEFORE   **AFTER**

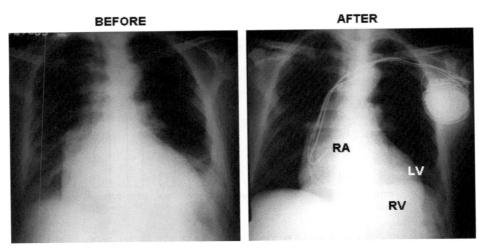

Figure 11   Chest radiograph before and after implantation of a biventricular ICD. Note the position of the CRT leads (RA, RV, and the CS). A substantial reduction in the size of the heart due to reverse remodeling was observed after CRT. *Abbreviations*: CS, coronary sinus; RA, rightatrium; RV, right ventricle.

site pacing or for pacing sites other than the CS may further promote optimal hemodynamic benefit to CRT.

INTERFACE OF DEVICE AND HF MANAGEMENT

While device implantation and follow-up are often handled by the electrophysiologists or device specialists, HF specialists play a pivotal role in the long-term care of these patients.

Indications of CRT

Table 4 summarizes the traditional indications of CRT in CRT trials. Overall, patients have been recruited with QRS width ≥ 120 ms, NYHC III/IV, EF $<30\%$ to 35%, and a dilated LV. CRT is now a Class IIa indication for pacing.

Optimization of Medical Therapy

The availability of backup pacing allows control of the rate and AV interval such that the dose of beta-blocker can be optimized. Potentially negative inotropic anti-arrhythmic medications can be withdrawn, as the patient is protected with an ICD. It is of interest to see if AF incidence can be reduced through the use of overdrive pacing, improvement of heart function and possible AF.

Monitoring of HF

CRT therapy opens the possibility of using implantable sensors for HF monitoring. In addition, it is possible to tailor device therapy based on the HF condition of the patients.

From a historical perspective, sensors have been incorporated into pacing leads to monitor the effect of exercise so that rate adaptation can be achieved. Amongst these, central venous oxygen saturation and RV dp/dt had been implanted in HF

Table 4 Classic Indications of CRT

	PATH-CHF (12)	MUSTIC (35)	Miracle (43)	CONTAK CD (44)	INSYN ICD (55)	Companion (46)
NYHC	≥ II–III	≥ III	≥ III	≥ II	≥ II	III–IV
QRS width (ms)	≥120	≥150	≥130	>120	≥130	>120
EF (%)	< 30	< 35	< 35	< 35	< 35	< 35
LVED (cm)	—	≥ 60	≥ 55	—	> 55	> 60

Abbreviations: EF, ejection fraction; LVED, left ventricular end diastolic diameter; NYHC, New York heart class for heart failure.

patients for monitoring. In 21 patients with HF, these sensors had high reproducibility for detecting RV systolic and diastolic pressures, and central venous oxygen saturation (≤0.91, 0.79, and 0.78, respectively) (52). Acute hemodynamic measurements showed an underestimation of the pressure by 4.5 mmHg and oxygen saturation by 1.6%, but these differences remained stable over one year. While these results were promising, there was a high failure rate of these sensors (12/21 oxygen and 2/21 pressure sensors over 1 year), raising concerns on the long-term stability of these devices.

More recently, an accelerometer has been incorporated in the tip of a lead to measure the so called "peak endocardial acceleration" (PEA), which may reflect cardiac contractility. This PEA sensor has been used to compare BV versus LV pacing (53). The PEA sensor was reported to be able to detect the advantage of LV or BV pacing over RV pacing alone. Further data is needed to see if PEA can be used to monitor HF and/or to optimize AV interval in a CRT device.

A gross measurement of the clinical well-being of a HF patient is the amount of activity of daily living that he/she can perform. This is conveniently measured by an activity sensor (usually a piezoelectric crystal or an accelerometer) in the pacemaker casing, which detects body movement. The quantity of daily movement reflects the activity of the patient, which correlated well with the condition of HF (Fig. 12).

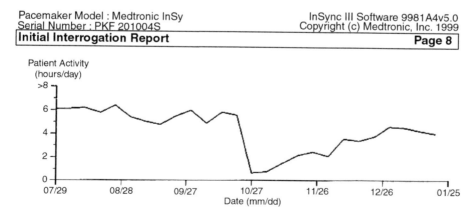

Figure 12 Increase in daily activity in a patient with InSyn III biventricular pacer (Model 5042, Medtronic Inc.) after implantation as recorded by the patient activity log. During an intercurrent heart failure (10/27), there was reduction in activity log. With intravenous diuretic therapy, the patient condition improved with increase in activity.

RENEWAL CRT+ICD
Post implant

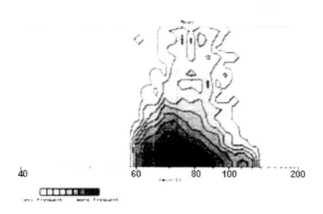

3 months after

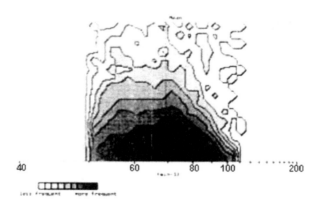

Figure 13 Change in "footprint": a measure of heart variability in the first and at third month after implantation of a renewal biventricular ICD (Model H135, Guidant Inc.). There is an increase in the spectrum of the heart rate (x-axis), and a higher variation at each rate range (y-axis).

ECG markers such as the paced QT interval and evoked QRS have been suggested to relate to intraventricular conduction and heart function. Heart rate variability is a convenient marker of HF, and this is easily incorporated in devices. An increase in this parameter may reflect better cardiac function (Fig. 13).

Intrathoracic impedance can measure change in heart volume by the change in resistivity across the heart. The "ventricular inotropic index" using unipolar impedance at the tip of the lead has been proposed to measure cardiac contractility. More recently, transthoracic impedance has been used to reflect lung edema; this parameter decreases when the lungs are congested. Early experience with such an

Table 5 Nonresponders to CRT

Pathological reasons
  No significant dyssynchrony despite a wide QRS complex
  Insufficient residual functioning myocardium
  Disease progression
  Atrial fibrillation
  RBBB/narrow QRS complex without mechanical dyssynchrony
Technical reasons
  Inadequate lead positioning (suboptimal LV and/or lead site)
  Improper programmed timing between AV, RV to LV and interatrial conduction
  Intolerance (e.g., diaphragmatic pacing)

Abbreviations: AV, atrio-ventricular; LV, left ventricular; RBBB, right bundle branch block; RV, right ventricular.

implantable device showed that impedance was reduced two weeks prior to onset of acute HF, thus making this a valuable marker for either the physician or the patients to take early corrective measures (54).

Development is underway to make HF monitoring data to be available to the HF specialists through telemetry, or through home or office-based internet data transmission.

Nonresponders to CRT

Up to 20% of patients with CRT devices do not respond to CRT. HF is often a progressive disease, and it is likely that disease progression may have limited CRT benefit. Table 5 lists possible reasons why CRT may not work. Proper selection of patients with CRT and proper lead positioning appear to be important for a response. Surface ECG is a poor marker for intraventricular dyssynchrony and direct measurement of ventricular dyssynchrony with various means appear to be a useful marker for response. The use of mechanical dyssynchrony may also extend the indication for CRT to patients with narrow QRS complex.

CONCLUSION

CRT using LV stimulation is an important breakthrough of pacing for a nonbradycardia indication. While published data on survival advantage of CRT over optimal medical therapy awaits a large randomized trial, the functional advantages and favorable cardiac structural changes with CRT have prompted major development in device technology. It is not unreasonable to expect that device therapy for HF will become an established and easy to achieve therapeutic option in selected HF patients. A multidisciplinary approach between the electrophysiologists, device experts, HF specialists and others is required to optimize the future of HF patients with ventricular conduction abnormalities.

REFERENCES

1. MERIT-HF Study Group. Effect of metoprolol CR/XL in chronic heart failure: Metoprolol CR/XL Randomized Intervention Trial in Congestive Heart Failure (MERIT-HF). Lancet 1999; 353:2001–2007.

2. Pitt B, Remme W, Zannad F, Neaton J, Martinez F, Roniker B, Bittman R, Hurley S, Kleiman J, Gatlin M. Eplerenone Post-Acute Myocardial Infarction Heart Failure Efficacy and Survival Study Investigators. Eplerenone, a selective aldosterone blocker, in patients with left ventricular dysfunction after myocardial infarction. N Engl J Med 2003; 348:1309–1321.

3. Cohn JN, Ferrari R, Sharpe N. On behalf of an International Forum on Cardiac Remodeling. Cardiac remodeling - concepts and clinical implications: A consensus paper from an international forum on cardiac remodeling. J Am Coll Cardiol 2000; 35:569–682.

4. Greenberg BH. Effects of angiotensin converting enzyme inhibitors on remodeling in clinical trials. J Card Fail 2002; 8:S486–S490.

5. Cleland JGF, Chattopadhyay S, Khand A, Houhton T, Kaye GC. Prevalence and incidence of arrhythmias and sudden death in heart failure. Heart Failure Reviews 2002; 7:229–242.

6. Aaronson KD, Schwartz JS, Chen TM, Wong KL, Goin JE, Mancini DM. Development and prospective validation of a clinical index to predict survival in ambulatory patients referred for cardiac transplant evaluation. Circulation 1997; 95:2660–2667.

7. Gottipaty VK, Krelis SP, Lu F, et al. The resting electrocardiogram provides a sensitive and inexpensive marker of prognosis in patients with chronic congestive heart failure (abstract). J Am Coll Cardiol 1999; 2:145A.

8. Brecker SJD, Xiao HB, Sparrow J, Gibson DG. Effects of dual-chamber pacing with short atrioventricular delay in dilated cardiomyopathy. Lancet 19920; 340:1308–1312.

9. Cazeau S, Bordachar P, Jauvert G, Lazarus A, Alonso C, Vandrell MC, Mugica J, Ritter P. Echocardiographic modeling of cardiac dyssynchrony before and during multisite stimulation: a prospective study. Pacing Clin Electrophysiol 2003; 26:137–143.

10. Ng KSK, Gibson DG. Impairment of diastolic function by shortened filling period in severe left ventricular disease. Br Heart J 1989; 62:246–252.

11. Hochleitner M, Hortnagl H, Ng CK, Gschnitzer F, Zechmann W. Usefulness of physiologic dual-chamber pacing in drug-resistant idiopathic dilated cardiomyopathy. Am J Cardiol 1990; 66:198–202.

12. Auricchio A, Stellbrink C, Block M, Sack S, Vogt J, Bakker P, Klein H, Kramer A, Ding J, Salo R, et al. Effect of pacing chamber and atrioventricular delay on acute systolic function of paced patients with congestive heart failure. The Pacing Therapies for Congestive Heart Failure Study Group. The Guidant Congestive Heart Failure Research Group. Circulation 1999; 99:2993–3001.

13. Leon AR, Greenberg JM, Kanuru N, Baker CM, Mera FV, Smith AL, Langberg JJ, Delurgio DB. Cardiac resynchronization in patients with congestive heart failure and chronic atrial fibrillation: effect of upgrading to biventricular pacing after chronic right ventricular pacing. J Am Coll Cardiol 2002; 39:1258–1263.

14. Grines CL, Bashore TM, Boudoulas H, Olson S, Shafer P, Wooley CF. Functional abnormalities in isolated left bundle branch block: the effect of interventricular asynchrony. Circulation 1989; 79:845–853.

15. Kerwin WF, Botvinick EH, O'Connell JW, Merrick SH, DeMarco T, Chatterjee K, Scheibly K, Saxon LA. Ventricular contraction abnormalities in dilated cardiomyopathy: effect of biventricular pacing to correct interventricular dyssynchrony. J Am Coll Cardiol 2000; 35:1221–1227.

16. Yu Y, Kramer A, Spinelli J, Ding J, Hoersch W, Auricchio A. Biventricular mechanical asynchrony predicts hemodynamic effect of uni- and biventricular pacing. Am J Physiol Heart Circ Physiol 2003; 285:H2788–H2796.

17. Nelson GS, Curry CW, Wyman BT, Kramer A, Declerck J, Talbot M, Douglas MR, Berger RD, McVeigh ER, Kass DA. Predictors of systolic augmentation from left ventricular preexcitation in patients with dilated cardiomyopathy and intraventricular conduction delay. Circulation 2000; 101:2703–2709.

18. Prinzen FW, Augustijn CH, Arts T, Allessie MA, Reneman RS. Redistribution of myocardial fiber strain and blood flow by asynchronous activation. Am J Physiol 1990; 259:H300–H308.

19. Verbeek XAAM, Vernooy K, Peschar M, Cornelussen RNM, Prinzen FW. Intra-ventricular resynchronization for optimal left ventricular function during pacing in experimental left bundle branch block. J Am Coll Cardiol 2003; 42:558–567.

20. Ansalone G, Giannantoni P, Ricci R, Trambaiolo P, Laurenti A, Fedele F, Santini M. Doppler myocardial imaging in patients with heart failure receiving biventricular pacing treatment. Am Heart J 2001; 142:881–896.

21. Bax JJ, Marwick TH, Molhoek SG, Bleeker, GB, Erven L, Boersma E, Steendijk P, Wall EE, Schalij MJ. Left ventricular dyssynchrony predicts benefit of cardiac resynchronization therapy in patients with end-stage heart failure before pacemaker implantation. Am J Cardiol 2003; 92:1238–1240.

22. Kanzaki H, Jacques D, Sade LE, Severyn DA, Schwartzman D, Gorcsan J III. Regional correlation by color-coded tissue Doppler to quantify improvements in mechanical left ventricular synchrony after biventricular pacing therapy. Am J Cardiol 2003; 92:752–755.

23. Schuster P, Faerestrand S, Ohm OJ. Colour tissue velocity imaging can show resynchronization of longitudinal left ventricular contraction pattern by biventricular pacing in patients with severe heart failure. Heart 2003; 89:859–864.

24. Soggard P, Egeblad H, Kim WY, Jensen HK, Pedersen AK, Kristensen BO, Mortensen PT. Tissue Doppler imaging predicts improved systolic performance and reversed left ventricular remodeling during long-term cardiac resynchronization therapy. J Am Coll Cardiol 2002; 40:723–730.

25. Yu CM, Chau E, Sanderson JE, Fan K, Tang MO, Fung WH, Lin H, Kong SL, Lam YM, Hill MRS, et al. Tissue Doppler echocardiographic evidence of reverse remodeling and improved synchronicity by simultaneously delaying regional contraction after biventricular pacing therapy in heart failure. Circulation 2002; 105:438–445.

26. Breithardt OA, Stellbrink C, Kramer AP, Sinha AM, Franke A, Salo R, Schiffgens B, Huvelle E, Auricchio A, for the PATH-CHF Study Group. Echocardiographic quantification of left ventricular asynchrony predicts an acute hemodynamic benefit of cardiac resynchronization therapy. J Am Coll Cardiol 2002; 40:536–545.

27. Kim WY, Soggard P, Mortensen PT, Jensen HK, Pedersen AK, Kristensen BO, Egeblad H. Three dimensional echocardiography documents haemodynamic improvement by biventricular pacing in patients with severe heart failure. Heart 2001; 85:514–520.

28. Kawaguchi M, Murabayashi T, Fetics BJ, Nelson GS, Samejima H, Nevo E, Kass DA. Quantitation of basal dyssynchrony and acute resynchronization from left or biventricular pacing by novel echo-contrast variability imaging. J Am Coll Cardiol 2002; 39:2052–2058.

29. Popovic ZB, Grimm RA, Perlic G, Chinchoy E, Geraci M, Sun JP, Donal E, Xu XF, Greenberg NL, Wilkoff BL, Thomas JD. Non invasive assessment of cardiac resynchronization therapy for congestive heart failure using myocardial strain and left ventricular peak power as parameters of myocardial synchrony and function. J Cardovasc Electrophysiol 2002; 13:1203–1208.

30. Kass DA, Chen CH, Curry C, Talbot M, Berger R, Fetics B, Nevo E. Improved left ventricular mechanics from acute VDD pacing in patients with dilated cardiomyopathy and ventricular conduction delay. Circulation 1999; 99:1567–1573.

31. Ukkonen H, Beanlands RSB, Burwash IG, de Kemp RA, Nahmias C, Fallen E, Hill MR, Tang ASL. Effect of cardiac resynchronization on myocardial efficiency and regional oxidative metabolism. Circulation 2003; 107:28–31.

32. Lau CP, Yu CM, Chau E, Fan K, Tse HF, Lee K, Tang MO, Wan SH, Law TC, Lee PY, et al. Reversal of left ventricular remodeling by synchronous biventricular pacing in heart failure. Pacing Clin Electrophysiol 2000; 23:1722–1725.

33. Stellbrink C, Breithardt OA, Franke A, Sack S, Bakker P, Auricchio A, Pochet T, Salo R, Kramer A, Spinelli J. Path-CHF Investigators, CPI Guidant Congestive Heart Failure

Research Group. Impact of cardiac resynchronization therapy using hemodynamically optimized pacing on left ventricular remodeling in patients with congestive heart failure and ventricular conduction disturbances. J Am Coll Cardiol 2001; 38:1957–1965.

34. Saxon LA, Marco T, Schafer J, Chatterjee K, Kumar UN, Foster E. VIGOR Congestive Heart Failure Investigators. Effects of long-term biventricular stimulation for resynchronization on echocardiograhic measures of remodeling. Circulation 2002; 105:1304–1310.

35. Abraham WT, Fisher WG, Smith AL, Delurgio DB, Leon AR, Loh E, Kocovic DZ, Packer M, Clavell AL, Hayes DL, et al. Cardiac resynchronization in chronic heart failure. N Engl J Med 2002; 346:1845–1853.

36. St John Sutton MG, Plappert T, Abraham WT, Smith AL, Delurgio DB, Leon AR, Loh E, Kocovic DZ, Fisher WG, Ellestad M, et al. Multicentre InSync Randomized Clinical Evaluation (MIRACLE) Study Group. Effect of cardiac resynchronization therapy on left ventricular size and function in chronic heart failure. Circulation 2003; 107:1985–1990.

37. Hamdan MH, Zagrodzky JD, Joglar JA, Sheehan CJ, Ramaswamy K, Erdner JF, Page RL, Smith ML. Biventricular pacing decreases sympathetic activity compared with right ventricular pacing in patients with depressed ejection fraction. Circulation 2000; 102:1927–1032.

38. Sinha AM, Filzmaier K, Breithardt OA, Kunz D, Graf J, Markus KU, Hanrath P, Stellbrink C. Usefulness of brain natriuretic peptide release as a surrogate marker of the efficacy of long-term cardiac resynchronization therapy in pateints with heart failure. Am J Cardiol 2003; 91:755–758.

39. Breithardt OA, Sinha AM, Schwammenthal E, Bidaoui N, Markus KU, Franke A, Stellbrink C. Acute effects of cardiac resynchronization therapy on functional mitral regurgitation in advanced systolic heart failure. J Am Coll Cardiol 2003; 41:765–770.

40. Higgins SL, Yong P, Sheck D, McDaniel M, Bollinger F, Vadecha M, Desai S, Meyer DB. Biventricular pacing diminishes the need for implantable cardioverter defibrillator therapy. Ventak CHF Investigators. J Am Coll Cardiol 2000; 36:824–827.

41. Kuhlkamp V, for the InSyn 7272 ICS world wide investigators. Initial experience with an implantable cardioverter-defibrillator incorporating cardiac resynchronization therapy. J Am Coll Cardiol 2002; 39:790–797.

42. Lee KLF, Fan K, Tse HF, Chau E, Yu CM, Lau CP. Chronic biventricular pacing is associated with a decrease in ventricular vulnerability to develop ventricular fibrillation in heart failure patients. Pacing Clin Electrophysiol 24(Part II), 666. 2001 [abstract].

43. Cazeau S, Leclercq C, Lavergne T, Walker S, Varma C, Linde C, Garrigue S, Kappenberger L, Haywood GA, Santini M, et al. Effects of multisite biventricular pacing in patients with heart failure and intraventricular conduction delay. N Engl J Med 2001; 344:873–880.

44. Higgins SL, Hummel JD, Niazi JK, Giudici MC, Worley SJ, Saxon LA, Boehmer JP, HIgginbotham MB, De Marco T, Foster E, et al. Cardiac resynchronization therapy for the treatment of heart failure in patients with intraventricular conduction delay and malignant ventricular tachyarrhythmias. J Am Coll Cardiol 2003; 42:1454–1459.

45. Bradley DJ, Bradley EA, Baughman KL, Berger RD, Calkins H, Goodman SN, Kass DA, Powe NR. Cardiac resynchronization and death from progressive heart failure: a meta-analysis of randomized controlled trials. JAMA 2003; 289:730–740.

46. Bristow MR, Feldman AM, Saxon LA. Heart failure management using implantable devices for ventricular resynchronization: Comparison of Medical therapy, Pacing, and defibrillation in chronic heart failure (COMPANION) trial. COMPANION Steering Committee and COMPANION Clinical Investigators. J Card Fail 2000; 6:276–285.

47. Bristow MR, Saxon LA, Boehmer J, et al., for the Comparison of Medical Therapy, Pacing, and Defibrillation in Heart Failure (COMPANION) Investigators. Cardiac-resynchronization therapy with or without an implantable defibrillator in advanced chronic heart failure. N Engl J Med 2004; 350:2140–2150.

48. Auricchio A, Abraham WT. Cardiac resynchronization therapy: current state of the art: cost versus benefit. Circulation 2004; 109:300–307.

49. Auricchio A, Stellbrink C, Sack S, Block M, Vogt J, Bakker P, Mortensen PT, Klein H. The pacing therapies for congestive heart failure (PATH-CHF) study: rationale, design, and endpoints of a prospective randomized multicenter study. Am J Cardiol 1999; 83:130D–135D.

50. Purerfellner H, Nesser HJ, Winter S, Schwierz T, Hornell H, Maertens S. Transvenous left ventricular lead implantation with the EASYTRAK lead system: the European experience. Am J Cardiol 2000; 86:K157–K164.

51. Tse HF, Yu C, Lee KL, Yu CM, Tsang V, Leung SK, Lau CP. Initial clinical experience with a new self-retaining left ventricular lead for permanent left ventricular pacing. Pacing Clin Electrophysiol 2000; 23:1738–1740.

52. Ohlsson A, Kubo SH, Steinhaus D, Connelly DT, Adler S, Bitkover C, Nordlander R, Ryden L, Bennett T. Continuous ambulatory monitoring of absolute right ventricular pressure and mixed venous oxygen saturation in patients with heart failure using an implantable haemodynamic monitor: results of a 1 year multicentre feasibility study. Eur Heart J 2001; 22:942–954.

53. Garrigue S, Bordachar P, Reuter S, Jais P, Kobeissi A, Gaggini G, Haissaguerre M, Clementy J. Comparison of permanent left ventricular and biventricular pacing in patients with heart failure and chronic atrial fibrillation: prospective haemodynamic study. Heart 2002; 87:529–534.

54. Yu CM, Lau CP, Tang MO, et al. Changes in device-based thoracic impedance in decompensating congestive heart failure (abstract). Circulation 2001; 104:II–419.

55. Gras D, Leclercq C, Tang AS, Bucknall C, Luttikhuis HO, Kirstein-Pedersen A. Cardiac resynchronization therapy in advanced heart failure the multicenter InSync clinical study. Eur J Heart Fail 2002; 4:311–320.

24

Other Surgical or Device-Based Approaches to Treating Cardiac Remodeling

V. Dor
Thoracic and Cardio-Vascular Surgery, Centre Cardio Thoracique, Monaco

M. Di Donato and M. Sabatier
Hemodynamic Department, Centre Cardio Thoracique, Monaco

F. Civaia
Cardiac Imaging, Centre Cardio Thoracique, Monaco

Some aspects of the surgical approaches to treating left ventricular (LV) remodeling have already been analyzed in Chapters 21 (revascularization), 22 (valve replacement or repair), and 23 (left ventricular assist device). None of these techniques "tackle" the diseased LV wall itself, which is the foundation of the remodeling.

THE LEFT VENTRICLE AFTER MYOCARDIAL INFARCTION

If coronary artery occlusion is the mechanism, or the cause, of myocardial infarction (MI), the ventricular wall is the final victim of acute or prolonged ischemia owing to diminished blood flow. When this occurs, the consequence both on the diseased and on the undamaged segments of the LV must be analyzed.

Diseased Area

The Lesion Itself

It has been well known for decades that the infarcted myocardium is first necrotic (within days), then fibrotic (within weeks), and finally calcified (within years). Necrosis, however, can affect full or partial thickness of the ventricular wall, so that an infarct may be either transmural or nontransmural. With modern treatment approaches that achieve early reperfusion, transmural infarcts are less frequent than in the past. With successful reperfusion, necrosis often affects essentially the subendocardial layer, with the subepicardial area of the ventricular wall protected by the recanalization of the occluded coronary artery (1). The external aspect of such an LV wall seems normal. During surgery, the wall collapses when the LV cavity is vented.

Upon opening this wall, the first appearance is 1 or 2 mm of normal muscle surrounding a fibrotic, white, subendocardial layer. This explains the classical limit of surgery, which fails to detect one-third of the dyskinetic areas evidenced by angiography. Cardiac magnetic resonance imaging (CMRI) can now easily demonstrate the presence of subendocardial scar by the use of gadolinium late enhancement (GLE), which shows the necrotic region (white scar) surrounded by normal muscle (Fig. 1). When the infarct is transmural, the whole wall is scarred. With progressive transmural fibrosis the wall thins and becomes dyskinetic, and there is evidence of bulging during systole. With subendocardial scar, the LV wall becomes akinetic. In this case the normal muscle surrounding the fibrous, and finally the calcified, endocardial scar becomes progressively useless. The potential recovery of a segment of the LV wall that has undergone this process is limited and may not be accurately represented by Thallium scintigraphy or positron emission tomography (PET) scan, which can be misleading. Echo stress or MRI assessment is more valuable in predicting the low likelihood of functional recovery in these areas.

Localization

Even though disease in the right coronary artery is the leading cause of infarction, the antero-apico–septal region is the most common location for left ventricular scar. This is probably because of the detrimental effect of occlusion of the left anterior descending (LAD) artery and its branches compared with the consequences of occlusion of the right or circumflex arteries with their balanced anatomy. The interventricular septum is not adequately analyzed by the right oblique projection of an angiogram. Echocardiography and, more precisely, MRI, with four projections [two chambers, four chambers, left ventricular outflow tract (LVOT), and short axis] illustrate in the majority of cases, the involvement of the septum. In our experience

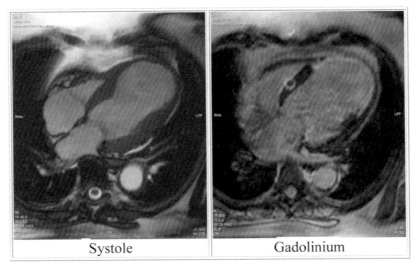

| Systole | Gadolinium |

Figure 1 CMR image of antero-septo–apical aneurysm: Four-chambers view in systole (*left*). The bulging affects the total scarred area. The mitral valve is closed. At late gadolinium enhancement (*right*), the transmural lesion is limited to the septum and in the apical and lateral walls. Scar localization is subendocardial and living myocardium is surrounding the necrosis. *Abbreviation*: CMR, cardiac magnetic resonance.

with ischemic LV wall repair, antero-septo–apical localization was found in nearly 90% of cases, posterior or postero-lateral (with frequent mitral insufficiency and ventricular arrhytmias) in 7% of cases, and purely lateral in only 2% of cases.

Type of Asynergy

Transmural infarct with an antero- or postero-apical scar extending to the exterior surface of the LV can be dyskinetic and bulge during systole. When this occurs it is a true aneurysm. However, the progressive accumulation of clots inside the ventricle and calcification of the wall often induce akinesia so that bulging is not seen. A partially scarred wall, or involvement of nonexteriorized areas like septum, is more often akinetic. As the anatomy and consequences of these two types of lesions are similar, as already mentioned by Gorlin (2) in 1967, the theoretical controversy between dyskinetic or akinetic lesions should be abandoned.

Extension of Asynergy

Measurement of the extension of the area of affected myocardium that fails to contract during the post-MI evolution helps to establish the prognosis. Gorlin mentioned that progressive dilatation occurs when 20% of the LV area is infarcted. Gaudron et al. (3) have shown that 20% of all infarct have a progressive evolution toward a depressed ejection fraction (EF) below 30% and LV dilatation after 18 months. In clinical practice, when more than 50% of LV circumference is asynergic, the progressive evolution to congestive heart failure (CHF) seems inescapable. This extension can be assessed by angiography with the centerline method in the right oblique projection (4), echocardiography, or radionuclide ventriculography. It can be accomplished even more easily and precisely providing there is no contraindication (implantable pacer or defibrillator) by MRI. By this latter method the LV wall and the presence of a necrotic scar (white line traced by GLE) are analyzed in four projections (two chambers, four chambers, LVOT, and short axis) (Fig. 2). The extension of the asynergic wall can be expressed as the ratio between the length of the diseased wall and the total length of the LV circumference. This measurement is not based on the same principle as the centerline method, which analyzes the chord shortening around the LV perimeter and relates it to the standard deviation of normal mean value (zero line). However, when comparing the two techniques in the same patients, the final percentages of asynergic wall are roughly similar.

The Undamaged, Remote Area

This area ndergoes "adaptive" changes that follow the healing of the infarcted zone. The undamaged myocardium is first normal, then hypertrophied to compensate for the lack of contractility of the necrotic wall, and finally dilated by physical mechanical forces. The dilatation increases LV stroke volume and temporarily improves the cardiac index (Starling law). However, the increased wall tension has a detrimental effect on myocardial contractility (Laplace law). This physical and mechanical explanation of the progressive dilatation of the heart is too simple and, in reality, the remodeling is a more complex process: Cell necrosis initiates an inflammatory reaction with granulocyte infiltration and release of proteolytic enzymes. Plasma concentration of several neurohormones are also increased after infarction. The neurohormonal hypothesis (5,6) postulates that neurohormones initially serve an adaptive role by maintaining cardiac output and tissue perfusion, but in the later stages the

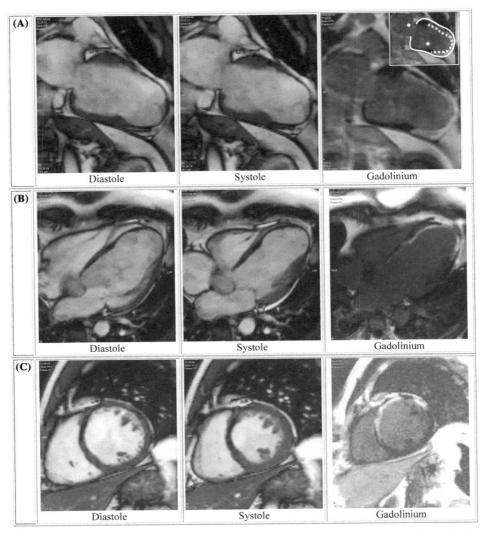

Figure 2 Extent of the asynergy: The asynergic scar is analyzed in two-chambers (**A**), four-chambers (**B**), and short-axis views (**C**) in diastole and systole. Gadolinium late enhancement shows that necrosis affects more than 60% of the LV circumference (long-axis view). The same measurement can be done on four-chambers or short-axis view. α is the left ventricle circumference length, β is the asynergic extension (GLH) length. *Abbreviations*: LV, left ventricle; GLE, gadolinium late enhancement.

responses become pathological and contribute to adverse remodeling and progressive ventricular failure. There is also an increase in sympathetic drive, which has both positive chronotropic and inotropic actions. While cardiac output is maintained, left ventricular loading increases, thereby aggravating infarct expansion. In addition, myocardial oxygen demand rises, precipitating further ischemia. Vasoconstriction and fluid retention are both features of activation of the renin–angiotensin axis, which serve to maintain blood pressure. The end of the process is a dilatation of the noninfarcted area. The ventricle assumes a more spherical shape and it becomes hypo- or akinetic. It must be pointed that this complex neurohormonal reaction is the consequence of the adaptation of the "organism" fighting against a loss of myocardial contractility. The cause of the remodeling is the asynergic scarred LV wall.

Left ventricular volume is a sensitive marker of postinfarction ventricular dysfunction, and left ventricular end-systolic volume is an important predictor of prognosis after MI (7,8). If 25/30 ml for end systolic volume index (ESVI) and 50/60 ml for end diastolic volume index (EDVI) are considered to be normal values, doubling of these indexes can be considered severe dilatation.

Two remarks summarize this survey of anatomic physiology of the infarcted LV wall:

1. The "primum movens" of this complex adaptation being the presence of a scarred asynergic wall, assessing precisely its extension is mandatory to determine by prospective and retrospective studies the percentage of scared LV circumference that can be considered at the trigger of the deleterious LV remodeling.
2. The progressive loss of performance of the non-damaged myocardium is an evolving process. Remodeling must be followed carefully by objective assessment by EKG or MRI, both of which are more reliable than the appearance of clinical symptoms, which tend to be inconsistent. Early recognition of remodeling provides a trigger to act before the global akinesia occurs.

SURGERY OF THE ISCHEMIC WALL

With current techniques one can now answer "yes" to the question, "Is it possible to act on this asynergic wall?" The goal of reconstructive surgery is to treat the disease (scarred LV wall) to suppress the cause of the LV remodeling and to reorganize the whole LV.

Classical Linear Suture

Since the 1950s, even before the heart-lung machine era, resection of bulging exteriorized aneurysms was performed (10,11). The technique for this procedure is simple: The dyskinetic external wall is resected and a long linear or in coat suture reinforced by a Teflon® strip closes the LV cavity. These spectacular techniques have progressively acquired a poor reputation in the cardiological field (12,13) as a large amount of the scarred area was not resected (i.e., the septum). In addition, the LV geometry was not generally improved, with the linear suture destroying the shape of the ventricle (an exception being the limited distal apical lesion). The LV EF was not much improved and the overall results of this type of surgery were particularly disappointing (with high mortality and lack of improvement) in cases in which there were large areas of akinesia and CHF was present.

Circular Ventricular Reconstruction

Since the 1980s, procedures that employed more physiological reconstruction essentially with circular reorganization of the LV wall were described: Jatene in 1985 (14) presented a large experience using a technique of external circular reconstruction of the left ventricular wall associated with plication of the akinetic septum let inside the cavity. A patch was sometimes used (10%) to close the reorganized wall, and coronary revascularization was accomplished in 20% of cases. In 1984 (15), we used a patch inserted inside the ventricle on contractile muscle to exclude all akinetic, non-resectable areas and to rebuild the ventricular cavity as it was before the infarct. This technique was called endoventricular circular patch plasty (EVCPP). Other similar techniques that were described include endoaneurysmorrhaphy by Cooley in 1989 (16) or tailored scar incision by Mickleborough et al. (17) in 1994.

The LV wall reorganization is combined, when needed, with coronary revascularization, mitral repair, and endocardectomy for ventricular arrhythmia, and is now known as left ventricular reconstruction (LVR) or restoration or reshaping, which aims at improving the LV by opposing the spontaneous, deleterious evolution of the remodeling of the infarcted ventricle.

Standard Techniques of EVCPP (Fig. 3)

As part of this procedure transesophageal echocardiography is used to assess the mitral valve. Surgery is conducted with a totally arrested heart. Coronary revascular-

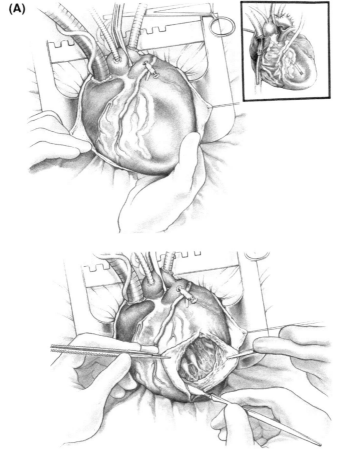

Figure 3 (**A–D**) The technique of standard EVCPP. (**A**) Opening of the LV at the center of the depressed wall after completion on an arrested heart. Coronary revascularization of the LAD is accomplished first.

(B)

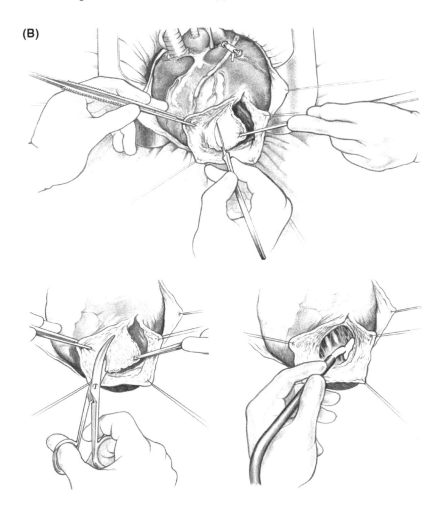

Figure 3 (**B**) Mobilization of the endocardial scar.

ization is accomplished first, then the LV wall is opened at the center of the depressed area (Fig. 3A), the clots are removed, and the endocardial scar is dissected, mobilized, and resected (Fig. 3B). In cases where the scar is calcified or if there is spontaneous or inducible ventricular tachycardia it is resected. In such circumstances, cryotherapy is applied in addition at the edge of the resection. For mitral insufficiency, the valve is inspected through the atrial and, eventually, the ventricular approach. If required, a reconstructive procedure is performed using posterior annuloplasty, Gore-tex® neochordae, Alfieri E-to-E suture, or mitral valve replacement if the posterior papillary muscle is totally diseased. The rebuilding of the left ventricular cavity is initiated by continuous purse string suture of 2/0 monofilament passed at the limit between fibrous and normal muscle (Fig. 3C) and tied over a soft rubber balloon inflated inside the left ventricular cavity at the theoretical diastolic capacity of the patient (40–50 ml/m^2 body surface area) to maintain a physiological volume for the rebuilt LV cavity. The endoventricular circular suture, in addition to restoring the curvature of myocardium to what it was before the infarct, also helps in the selection of the shape, size, and orientation of the patch. When the infarct scar is

(C)

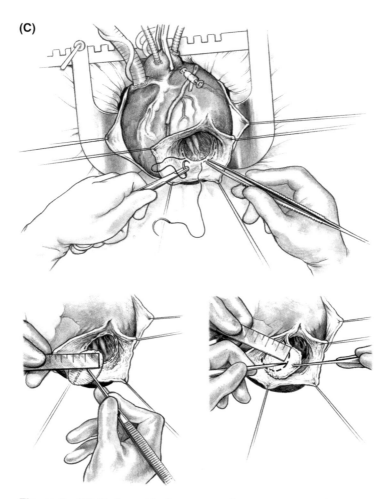

Figure 3 (**C**) Endoventricular purse string suture at the limit between scarred and normal myocardium, tied on a rubber balloon.

located in the antero-apical region, the septum and the apex are more involved than the lateral wall. The suture is placed far back in the septum, totally excluding the apex and the posterior wall below the posterior papillary muscle base and only a small portion of lateral wall above the antero-lateral papillary muscle base, so that the orientation of this new neck (and of the patch) is roughly aligned in the direction of the septum. A circular or ovalar patch of Dacron® is fashioned to the size of this neck (mean diameter between 1.5 and 2.5 cm, area 3.5 and 7 cm^2) and fixed to the clothesline (the endoventricular circular suture) with the same suture. Excluded areas are either resected or, more often, sutured above the patch.

Concomitant Procedures

Coronary Revascularization

Of all significantly stenosed coronary arteries supplying the contractile area is mandatory. Revascularization of the infarcted area is almost always possible, even with a thrombosed LAD artery that does not receive blood flow from homologous or

(D)

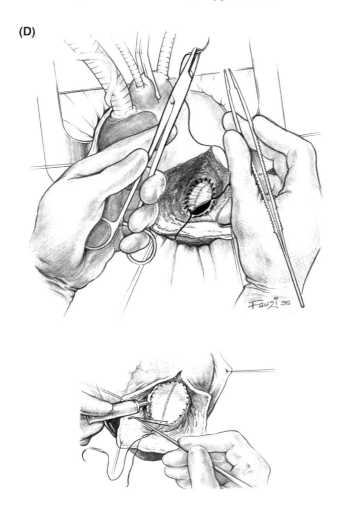

Figure 3 (**D**) Patch anchored on the closed line of the endoventricular suture. *Abbreviations*: EVCPP, endoventriclular circular patch plasty; LAD, left anterior descending; LV, left ventricle.

heterologous collaterals on the preoperative coronary angiogram. With the introduction of early revascularization over the past 10 years, the problem is now whether or not to bypass a culprit artery that was successfully recanalized by percutaneous coronary intervention (PCI) but which left an asynergic wall in its corresponding territory. In our recent series we found 25% of such cases.

Mitral Insufficiency

Mitral insufficiency is commonly associated with left ventricular scar, and the mitral valve must be assessed carefully before and during surgery using transesophageal echocardiography (TEE). When mitral regurgitation is quantified as grade II or if the regurgitation volume is above 30 ml or the mitral annulus is sized above 40 mm, mitral annuloplasty is advisable. This annuloplasty must be done when the annulus reaches 40 mm, even in the absence of regurgitation, to prevent the potential mitral insufficiency.

Ventricular Arrhythmias

Spontaneous ventricular tachycardias (which affected 13% of the patients in our series) and inducible ventricular tachycardias (which affected 25% of the patients in our series) are frequent and, in such circumstances, subtotal nonguided endocardectomy (18) is conducted on all the endoventricular scar. Cryotherapy at the limit of this resection completes the surgical excision.

RESULTS

LVR by EVCPP is no longer a new experimental technique as it has now been applied for almost 20 years. The technique has been presented by many authors in numerous publications. An increasing number of surgical teams in Europe, North and South America, Australia, Japan, etc., have adopted the concept of this technique (with variation in the details) as it gives obviously better result than classical simple resection. Over the past decade, only a few presentations (19–21) have continued to report results using the linear suture technique, which is efficient only for limited, exteriorized distal antero-apical lesions.

Based on our experience of more than 1000 LVR and on the recent literature, results must be analyzed regarding the operative risk, objective hemodynamic results, and late evolution at 1, 2, 5, and 10 years.

Immediate Results

LVR is feasible with an acceptable risk; the hospital mortality varies: 0% [Jakob et al. (22)], 3% [Mickleborough et al. (16)], 4.6% [Salati et al. (23)], 9.1% [Grossi et al. (24)]. In series between 20 and 90 cases. It can be above 10% to 15% when emergency cases are included (25). In our global series extending from April 1984 to December 2002, 77 patients among 1050 patients operated upon died during the first postoperative month (7.7%) [similar to the saver series where operative mortality rate was 6.6% (26)]. However, with increasing experience and improvement in the management of the severely ill patients, and recognition of the importance of the residual diastolic volume by balloon sizing, the operative risk of our recent series (1998–2002) was 4.8%. In the group of patients with severely depressed LV (EF < 30%) and extensive akinesia, it was 7.7%.

The postoperative LV angiogram (Fig. 4) following LVR shows return to a normalization of the LV shape, particularly in relation to the septal exclusion. Since the last two years, MRI allows a more precise approach to analyzing both morphology and performance of the heart (Fig. 5).

Both systolic and diastolic functions are improved. The mean increase of LVEF is between 10% and 15% (27), confirmed by Jakob (9%), Grossi (12%), Salati (19%). This improvement is noted for dyskinetic as well as akinetic lesions (28).

Diastolic function is also improved: analyzing by echo, peak filling rate rose from 1.79/s EDV/s preoperatively to 3.07 EDV/s at one month and remained at 2.73 after one year. In addition the time-to-peak filling rate decreased from 190 m/s to 110 m/s and 90 m/s, respectively, at the one month and one year time points. The tagging of the LV wall thickness is also a promising technique with MRI to precisely check diastolic function.

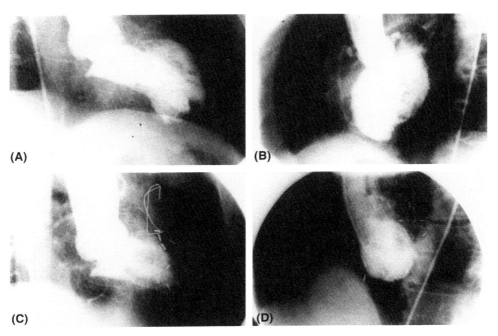

Figure 4 Large antero-septo–apical aneurysm: (**A** and **B**) Preoperative LV angiogram (right and left oblique projection or an antero-septo–apical dyskinesia). (**C** and **D**) Same patient postoperatively. Note the total disappearance of the septal akinesia. *Abbreviation*: LV, left ventricle.

Ventricular arrhythmias, chiefly ventricular tachycardia (spontaneous or inducible), are controlled in 90% of cases by LVR that includes extended endocardectomy and cryotherapy (29). Only 10% of these patients required the implantation of an intraventricular defibrillator owing to inducible ventricular arrhythmia.

The Whys and Wherefores of EVCPP

LVR enhances the beneficial, but partial, effects of coronary revascularization and of mitral valve repair (if necessary) by improving left ventricular function for the following reasons:

1. Septal scar is excluded.
2. The LV wall is reorganized. This suppresses the increase in wall tension in remote myocardial areas and improves contraction of these areas. This is clearly shown by the analysis of pressure–volume curves. The morphology of pressure–length regional loops (Fig. 6) of regions remote from the scarred area is normalized after LVR, whereas it was totally disorganized before, and this is true for both dyskinesia and akinesia (30).
3. The patch sized for the theoretical diastolic volume of the patient avoids excessive reduction of volume and maintains a reasonable physiological cavity.

After Months and Years

More than 90% of operated-upon patients are surviving.

(A) PRE-OP

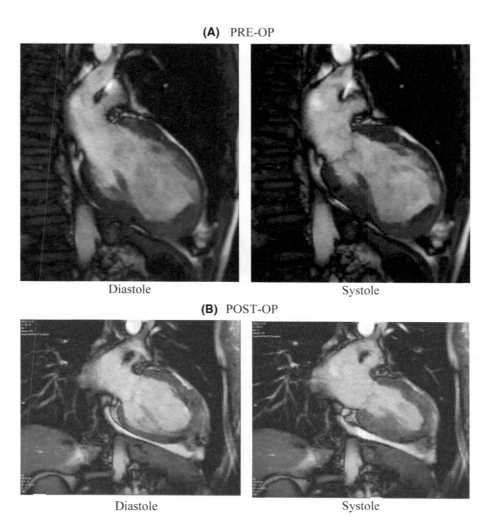

Diastole   Systole

(B) POST-OP

Diastole   Systole

Figure 5 CMR control of left ventricular reconstruction for large antero-septo–apical aneurysm. **(A)** Preoperative diastole (*left*) and systole (*right*): EF = 22%; ESVI = 102 ml/m^2; EDVI = 132 ml/m^2. **(B)** Postoperative: diastole (*left*) and systole (*right*): EF = 54%; ESVI = 28 ml/m^2; EDVI = 78 ml/m^2. *Abbreviations*: CMR, cardiac magnetic resonance; ESVI, end systolic volume index; EDVI, end diastolic volume index; EF, ejection fraction.

Disappointing Results

Although operative and hospital mortalities are not prohibitive for extensive LV akinesia and global dilatation, some disappointing late results have been noted. In our global experience approximately 25% of such patients had a tendency after one year for an increase in pulmonary pressure, and secondary mitral insufficiency could occur. The mechanism of the late failure is not clear. When the interval between infarction and LV reconstruction is above 40 months, ventricular remodeling appears to continue in spite of surgery (31). Moreover, diastolic volume or compliance and mitral competence must be carefully assessed. Since 1998, when the balloon sizing of the remaining ventricle was applied and mitral annuloplasty was more

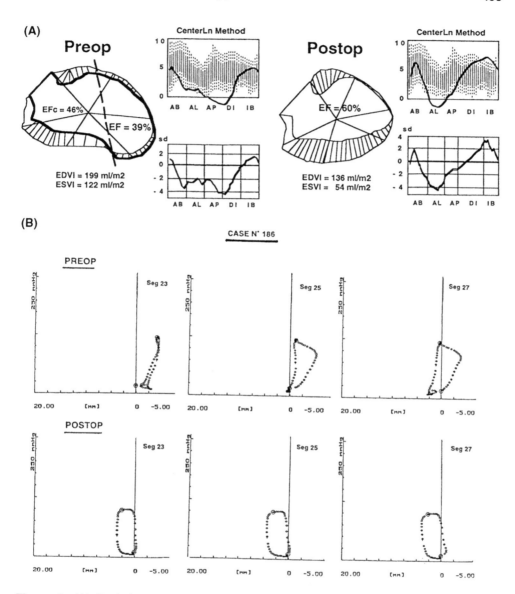

Figure 6 (A) Evolution of EF and volumes. (B) (Same patient) Modification of pressure motion loop with restoration of normal morphology and orientation in segments 23, 25, and 27, remote from the diseased area. *Abbreviations*: EDVI, end diastolic volume index; ESVI, end systolic volume index; EF, ejection fraction.

widely used to maintain the annulus below 30 mm diameter, the tendency for late impairment fell from 25% to 10% in such patients.

More Than 80% of Survivors are Stabilized or Improved

The remodeling evolution during the first postoperative year was analyzed with reliable values for EF and LV volumes in our institution by MRI: Since January 2002,

Table 1 Evolution of Hemodynamic Parameters During First Post-op Year: the Remodeling Regression

	Preoperative	1st month	1st Year
EF(%)	25 ± 5	41 ± 8^a	45 ± 9^b
EDVI (ml/ml^2)	126 ± 34	82 ± 17^a	81 ± 22
ESVI (ml/m^2)	90 ± 29	50 ± 14	46 ± 17^c

[a] $p = 0.0001$ vs. preoperative.
[b] $p = 0.0001$.
[c] $p = 0.004$ vs. one month postoperative.

125 patients had a ventricular reconstruction with exclusion of the asynergic zone (dyskinetic or akinetic) and MRI assessment was obtained after surgery during the first month and the first year. In 55 out of the 125 cases the main indication for surgery was residual ischemia associated with CHF II/III. This cohort, suitable for randomization with medical therapy and/or revascularization alone, was excluded. In 70 patients the main indication was permanent CHF III/IV (mean EF $\leq 25\%$ mean ESVI >80 ml, asynergy of circumference range from 35% to 69%, mean 49.5%) in spite of full medical therapy and previous coronary revascularization by PCI (35 cases) or surgery (3 cases). Follow-up was 100% completed. Six in-hospital deaths (8.6%) occurred: seven patients did non improve. Fifty-seven patients (80%) had clinical and hemodynamic progressive improvement (Table 1).

These data confirmed after other data (48) that surgical exclusion of the asynergic LV wall reverts the deleterious process of remodeling.

Postoperative Treatment—Life Expectancies and Indications

Postoperative Therapy

Optimal medical treatment is mandatory to maintain immediate improvement as the continuation of LV remodeling is influenced by mechanical, neurohormonal, or autoimmune reactions. In addition, Tanoue et al. (32) and Fantini et al. (33) demonstrated that after EVCPP, the LV contractility (end systolic elastance) and efficiency (ventricular coupling) are not improved when afterload (arterial elastance) does not change. Some patients need to be protected with medical therapy, including beta-blockers, diuretics, vasodilatators, and angiotensin-converting enzyme (ACE) inhibitors, to avoid this deleterious evolution during the first postoperative year.

Global Life Expectancy at Five Years

In a series of 207 surviving patients analyzed from 1991 to 1996 (34), is 82%. This percentage is above 90% for patients with preoperative ESVI below $120 \, ml/m^2$ and at 70% for those with ESVI above $120 \, ml/m^2$ (Fig. 7). At 10 years, in this last category of patients with very large dilated ventricles, the percentage of survival is 50%, whereas it is 80% for patients with ESVI below $90 \, ml/m^2$.

Indications for Reconstructive Surgery Post-MI or "When Is the Ideal Moment to Repair Scarred Ventricle?"

1. Too soon after MI, when tissues are necrotic, ventricular wall surgery has to be avoided, except for mandatory indications, as mechanical complications (mitral or free wall or septal ruptures).

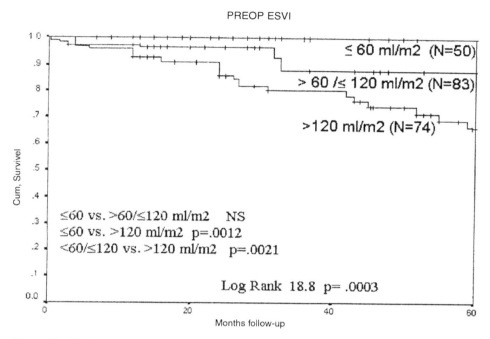

Figure 7 Kaplan Meyer five-year survival in 207 consecutive surviving patients according to preoperative ESVI. *Abbreviation*: ESVI, end systolic volume index.

2. Too late, four, five, or more years after MI in patients with asynergy above 50% of LV circumference, an evolution from dyskinesia to global akinesia, as in the case illustrated in Figure 8, is frequently seen. Surgery done at the first stage of such patients can avoid this type of fatal evolution. In the same way that indications for surgery in valvular diseases have changed (repair or replacement of the valve according to gradient or degree of regurgitation, before permanent cardiac enlargement and failure have developed), indications for LVR of ischemic wall motion abnormalities are also likely to emerge.

3. When a patient is stabilized under medical therapy, after the acute phase of a large MI, whatever the potency of the culprit artery, is measurement of LV volumes and performances every six months is recommended. If there is a tendency for LV volume to increase and/or for EF to decrease, LVR should be considered during the first or second year after infarction.

4. Very severe MI with immediate congestive heart failure, after days or weeks of intensive care, with patients under full medical therapy and period of intraaortic balloon. The LV wall should be assessed weeks after infarct to precise the extension of necrotic scar. Unstable patients can be operated on after six to eight weeks, when tissues are more fibrotic than necrotic.

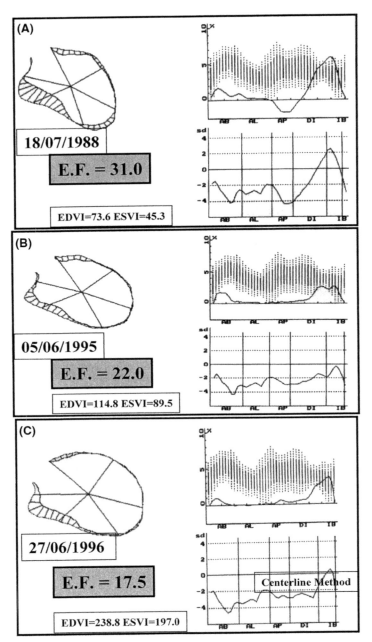

Figure 8 Evolution of a dyskinetic large anterior scar. (**A**) Forty-eight-year old patient, six months after anterior infarct. Surgery was not proposed. (**B**) Same patient, age 55. Surgery was refused by patient. (**C**) Same patient, age 56. The patient died before surgery. *Abbreviations*: EDVI, end diastolic volume index; ESVI, end systolic volume index; EF, ejection fraction.

SUMMARY

In patients at the late stage of remodeling, with class IV CHF, very large failing ventricle, LVR is possible with an acceptable risk (below 10%) and life expectancy at 2, 5, and 10 years, which is better than the natural evolution of the disease.

In patients with a large scarred ventricle after MI, EF below 40%, ESVI above $40\,ml/m^2$, and further deterioration of these parameters during followup evaluation, there is evidence that LVR can restore a more normal LV in terms of shape and performances. In these cases there is hope to avoid, suppress, or limit the remodeling process.

The results of the STICH (Surgical Treatment of Ischemic Heart failure) trial (35), a multicenter, international, randomized study of ischemic cardiomyopathy (cm) that will compare medical therapy versus medical therapy + surgical revascularization versus medical therapy + revascularization + LVR that is currently under way, are expected to confirm our observational results.

However, in the growing field of ischemic CM there remain cases of severe failing ventricle beyond all possibilities of randomization between LVR, already failed medical therapy, or a useless revascularization in the absence of residual ischemia. The only option is heart replacement or reconstruction: if a large asynergic LV scar can be excluded it is fair to give the rest of the contractile myocardium a chance to recover.

OTHER SURGICAL TECHNIQUES IN PROCESS

To treat dilated ischemic cardiomyopathy, many techniques have been or are still being used. Their goals can be classified into three categories.

Enhancement of the Myocardial Contraction

1. Coronary revascularization (analyzed in Chapter 26) is obviously the first logical indication if some part of the diseased myocardium remains ischemic. The revascularization can be conducted by endoluminal or surgical approach according to the type, size, and number of coronary stenosis or occlusions.
2. Gene transfer therapy (discussed in Chapter 21).
3. Cellular (stem cell or myoblast) transplantation (detailed in Chapter 22). This technique of injection of cells into a non-revascularizable area was the subject of numerous experimental studies during the last decade, as there is a tremendous hope to regenerate the myocardium (36–38).
   Some initial clinical trials such as the MAGIC (39) study, which combines cellular transplantation with coronary artery bypass grafting, are in process. The results will provide important information concerning the efficiency of these new myocytes and their synchronization with the patients rhythm.
4. Transmyocardial laser revascularization (TMLR): In 1996 there were numerous publications about the possibilities to create neomyocardial revascularization using laser techniques. The mechanism of this partial improvement, which was more subjective than objective, was never elucidated. Over the last two years, publications describing these techniques disappeared almost completely. The conclusion of Rhaman at The American

Heart Association (AHA) conference 2001 (40) about 256 patients treated by clinical angiogenesis and TMLR was "specific outcome measures need to be better defined … blinding is crucial to avoid the placebo effect."

Containment of Dilatation

Cardiomyoplasty

Cardiomyoplasty was a great hope 10 years ago, but the list of contraindications published by its promotor (41) makes this technique almost never suitable in the field of ischemic CHF. These contraindications include grade IV congestive heart failure, very large LV dilatation, mitral insufficiency, ventricular arrhythmias, and very low EF, which all of, of course, are characteristic of severe ischemic cardiomyopathy. In addition, combining complete revascularization and cardiomyoplasty seems rather difficult.

The Batista Procedure

The resection of a triangle of the lateral LV wall to reduce the volume of dilated cardiomyopathy (42) with mandatory mitral repair was proposed in the 1990s with some questionable results when applied in idiopathic cardiomyopathy (43). In the field of ischemic cardiomyopathy where, in the majority of cases, only the lateral wall is contracting, this technique is not applicable. Only rare cases of lateral scar with good anterior and septal wall contraction appears amenable to lateral resection with or without mitral repair.

Mitral Valve Repair

An ischemic cardiomyopathy is associated with mitral regurgitation, even of small volume (30 ml) or when there is a dilatation of the mitral annulus above 35 mm. This technique, which is one of the main parts of the LVR, is analyzed in Chapter 27.

"Sticks," "Baskets," "Bags," and "Nets"

Recently some original devices were proposed to treat dilated cardiomyopathy.

"Myosplint." McCarthy et al. (44) experimentally analyzed the effects of a "myosplint" device (Mayocor, Inc., Maple Grove, Minnesota, U.S.A.) based on the interest in reducing wall stress. It consists of an implantable transventricular splint and two epicardial pads that are adjusted to draw the walls of the LV together and thereby reduce the LV radius. It was assessed in dogs and some results from humans are available, but only for idiopathic cardiomyopathy. There are no follow-up data of this technique for ischemic patients treated by this method. In the same field, Kashem (45) presented the CardioClasp device (CardioClasp, Inc., Cincinnati, Ohio, U.S.A.) in which two bares or sticks are implanted in the LV to reshape it. In this experimental study on dogs there was an obvious positive effect on wall tension. To date there has been no clinical application of this approach, and no information on its effect on ischemic hearts.

"Bags," "Baskets," and "Nets." These devices consist of bags or mesh folded around the heart to reduce the diastolic dilatation. Experimental work has been performed using a variety of different materials. The most advanced of these clinical trials is the ACORN support device. However, concerning the use of restraint devices in ischemic cardiomyopathy, there are always two problems that have not been solved yet: The risk of inducing a restrictive cardiomyopathy by fibrosis of the epicardium and the competition between the material and the coronary bypasses.

In the series of Raman et al. (45), five patients were operated on with concomitant bypasses, and in four cases, in addition, endoventricular reconstruction. Therefore, it is difficult at present to reach any conclusion about the future benefits of this procedure.

In some experimental models (cardiac failure induced by rapid pacing in dogs) of idiopathic cardiomyopathies, dilatation per se can be hypothesized as the mechanism of failure, and volume reduction can temporarily or definitively improve the situation. But in ischemic cardiomyopathy, the mechanism of the disease is the lack of contractility of a part of the cavity. Dilatation is an adaptation to fight against lost contractility; therefore, volume reduction per se should not be the primary focus since it is not a cause of impairment.

MECHANICAL ASSISTANCE OF THE FAILING HEART

Left Ventricular Assist Devices (LVADs)

Partial assistance (LVADs) are discussed and analyzed in Chapter 28. These devices are partially or totally implantable. Progress is made each year in this type of device, having been used in clinical practice for more than 20 years. Recently, in "the REMATCH trial" (46) for patients needing a heart transplantation (HT), life expectancy was 25% at two years and a recent conclusion in the New England Journal of Medicine (47) has to be taken in consideration: "We now know that ventricular assist devices prolong life. We do not yet know for how long at what cost."

Artificial Heart

The biventricular total implantable artificial heart has been the "Holy Grail" of surgeons, engineers, and research labs in all developed countries for almost 50 years. It remains a reasonable hope for the future to fight against the increase of severe CHF, with lack of possibility for HT.

Heart Transplantation

HT remains in 2004 the only surgical curative radical technique of end-stage ischemic cardiomyopathy with permanent intractable congestive heart failure. However, this procedure has limited applicability since limitation in the availability of donor hearts mandate that

- only 10% of patients needing HT can obtain it;
- the long-term survival rate at 10 years is 50%.

The goal of conservative reconstruction surgery in ischemic cardiomyopathy is to avoid the necessity of HT, or delay it, to increase by one or two decades the life expectancy of the patient.

CONCLUSION

With well-recognized techniques like coronary revascularization, mitral repair, subtotal nonguided endocardectomy, LVR, and, more recently, attempt to modify LV dilatation and to improve the capital of cardiomyocyte, the natural history of deleterious LV remodeling can be favorably altered. In all cases surgical approaches should be combined with optimal treatment, which includes blockade of neurohormonal systems using agents such as ACE inhibitors or receptor blockers, aldosterone blockers, and beta blockers. The net effect of such combined therapy is a substantial improvement in the clinical course.

REFERENCES

1. Bogaert J, Maes A, et al. Functional recovery of subepicardial myocardial tissue in transmural myocardial infarction after successful reperfusion. Circulation 1999; 36–43.
2. Klein MD, Herman MV, et al. A hemodynamic study of left ventricular aneurysm. Circulation 1967; 35:614–630.
3. Gaudron P, Eilles C, et al. Progressive left ventricular dysfunction after myocardial infarction. Circulation 1993; 87:755–762.
4. Sheehan FH, Bolson EL, et al. Advantages and applications of the centerline method for characterizing regional ventricular function. Circulation 1986; 74:293–305.
5. McAlpine HM, Morton JJ, et al. Neuroendocrine activation after acute myocardial infarction. Br Heart J 1988; 60:117–124.
6. Packer M. The neurohormonal hypothesis: a theory to explain the mechanism of disease progression in heart failure. J Am Coll Cardiol 1992; 20:248–254.
7. Braunwald E, Pfeffer MA. Ventricular enlargement and remodeling following acute myocardial infarction: mechanisms and management. Am J Cardiol 1991; 68(suppl D): 1–6D.
8. Yamaguchi A, Ino T, et al. Left ventricular value predicts postoperative course in patients with ischemic cardiomyopathy. Ann Thorac Surg 1998; 65:434–438.
9. Likoff W, Bailey CP. Ventriculoplasty: excision of myocardial aneurysm. JAMA 1955; 158:915.
10. Cooley DA, Collins HA, et al. Ventricular aneurysm after myocardial infarction: surgical excision with use of temporary cardiopulmonary bypass. JAMA 1958; 167:557.
11. Froehlich RT, Falsetti HL, et al. Prospective study of surgery for left ventricular aneurysm. Am J Cardiol 1980; 45:923.
12. Cohen M, Packer M, et al. Indications for left ventricular aneurysmectomy. Circulation 1983; 67:717.
13. Jatene AD. Left ventricular aneurysmectomy resection or reconstruction. J Thorac Cardiovasc Surg 1985; 89:321–331.
14. Dor V, Kreitmann P, et al. Interest of "physiological closure (circunferential plasty on contractile areas) of left ventricle after resection and endocardectomy for eneurysm of akinetic zone comparison with classical technique about a series of 209 left ventricular resection (abstract). J Cardiovasc Surg 1985; 26:73.
15. Cooley D. Ventricular endoaneurysmorrhaphy: a simplified repair for extensive postinfarction aneurysm. J Cardiac Surg 1989; 4:200–205.
16. Mickleborough L, Maruyama H, et al. Results of left ventricular aneurysmectomy with a tailored scar excision and primary closure technique. J Thorac Cardiovasc Surg 1994; 107:690–698.
17. Josephson ME, Harken AH, et al. Endocardial excision: a new surgical technique for the treatment of recurrent ventricular tachycardia. Circulation 1979; 60:1430–1439.
18. Eleftriades JA, Solomon LW, et al. Linear left ventricular aneurysmectomy: modern imaging studies reveal improved morphology and function. Ann Thorac Surg 1993; 56:242–252.
19. Kesler KA, Fiore AC, et al. Anterior wall left ventricular aneurysm repair. A comparison of linear versus circular closure. J Thorac Cardiovasc Surg 1992; 103:841–848.
20. Tavakoli R, Bettex A, et al. Repair of post infarction dyskinetic LV aneurysm with either linear or patch technique. Eur J Cardiothorac Surg 2002; 22:129–134.

21. Jakob H, Zölch B, et al. Endoventricular patch plasty improves results of LV aneurysmectomy. Eur J Cardiothorac Surg 1993; 7:428–436.

22. Salati M, Di Biasi P, et al. Left ventricular geometry after endoventriculoplasty. Eur J Cardiothorac Surg 1993; 7:574–579.

23. Grossi E, Chimitz L, et al. Endoventricular remodeling of left ventricular aneurysm: functional, clinical and electrophysiological results. Circulation 1995; 92(suppl II): 98–100.

24. Di Donato M, Sabatier M, et al. Outcome of left ventricular aneurysmectomy with patch repair in patients with severely depressed pump function. Am J Cardiol 1995; 76:557–561.

25. Athanasuleas C, Stanley A, et al. Surgical anterior ventricular endocardial restoration (SAVER) in the dilated remodeled ventricle after anterior myocardial infarction. J Am Coll Cardiol 2001; 37:1199–1209.

26. Dor V, Sabatier M, et al. Late hemodynamic results after left ventricular patch repair associated with coronary grafting in patients with postinfarction akinetic or dyskinetic aneurysm of the left ventricle. J Thorac Cardiovasc Surg 1995; 110:1291–1301.

27. Dor V, Sabatier M, et al. Efficacy of endoventricular patch plasty in large postinfarction akinetic scar and severe left ventricular dysfunction: comparison with a series of large dyskinetic scars. J Thorac Cardiovasc Surg 1998; 116:50–59.

28. Dor V, Sabatier M, et al. Results of nonguided subtotal endocardiectomy associated with left ventricular reconstruction in patients with ischemic ventricular arrhythmias. J Thorac Cardiovasc Surg 1994; 107(5):1301–1308.

29. Di Donato M, Sabatier M, et al. Regional myocardial performance of non-ischaemic zones remote form anterior wall left ventricular aneurysm. Effects of aneurysmectomoy. Eur Heart J 1995; 16:1285–1292.

30. Louagie Y, Alouini T, et al. Left ventricular aneurysm complicated by congestive heart failure: analysis of long term results and risk factors of surgical treatment. J Cardiovasc Surg 1989; 30:648–655.

31. Tanoue Y, Ando H, et al. Ventricular energetics in endoventricular circular patch plasty for dykinetic anterior left ventricular aneurysm. Ann Thorac Surg 2003; 75:1205–1209.

32. Fantini F, Barletta G, et al. Effects of reconstructive surgery for left ventricular anteriore aneurysm on ventriculoarterial coupling. Heart 1999; 81:171–176.

33. Di Donato M, Toso A, et al. Intermediate survival and predictors of death after surgical ventricular restoration. Seminars in Thoracic and Cardiovascular Surgery 2002; 13(4), 2001:468–475.

34. Jones R. Is it time for a randomized trial of surgical treatment of ischemic heart failure? J Am Coll Cardiol 2001; 37:1211–1213.

35. Scorsin M, et al. Comparison of the effects of fetal cardiomyocytes and skeletal myoblast transplantation on postinfarct left ventricular function. J Thorac Cardiovasc Surg 2000; 119:1169–1175.

36. Pouzet B, et al. Is skeletal myoblast transplantation clinically relevant in the era of angiotensin-converting enzyme inhibition? Circulation 2001; 104:1223–1288.

37. Stamm C, et al. Autologous bone-marrow stem-cell transplantation for myocardial regeneration. Lancet 2003; 361:45–46.

38. Menasché P, et al. Autologous skeletal myoblast transplantation for severe postinfarction left ventricular dysfunction. J Am Coll Cardiol 2003; 41:1078–1083.

39. Laham R, et al. Clinical angiogenesis and laser myocardial revascularization studies: the need for better outcome measures. Abstract 2104. Circulation 2001; 104:444.

40. Carpentier A, Chachques JC, et al. Dynamic cardiomyoplasty at seven years. J Thorac Cardiovasc Surg 1993; 106;42–52.

41. Batista RJV, Santos JVL, et al. Partial left ventriculectomy to improve left ventricular function in end-stage heart disease. J Card Surg 1996; 11:96–97.

42. Franco-Cerecceda A, McCarthy P, et al. Partial left ventriculectomy for dilated cardiomyopathy: is this an alternative to transplantation? JTCVS 2001:879–893.

43. McCarthy P, Takagaki M, et al. Device-based change in left ventricular shape: a new concept for the treatment of dilated cardiomyopathy. J Thorac Cardiovasc Surg 2001; 122:482–490.

44. Kashem A, Hassan S, et al. Left ventricular reshaping: Effects on the pressure volume relationship. J Thorac Cardiovasc Surg 2003; 125:391–399.

45. Raman J, Power JM, et al. Ventricular containment as an adjunctive procedure in ischemic cardiomyopathy: early results. Ann Thorac Surg 2000; 70:1124–1126.

46. Rose EA, Moskowitz AJ, et al. The REMATCH trial: rationale, design, and end points-randomized evaluation of mechanical assistance for the treatment of congestive heart failure. Ann Thorac Surg 1999; 67:723–730.

47. Jessup M. Mechanical cardiac-support devices–dreams and devilish details. N Engl J Med 2001; 345:1490–1492.

48. Shenk S, McCarthy P, et al. Neurohormonal response to left ventricular reconstruction surgery in ischemic cardiomyopathy. J Thorac Cardiovasc Surg 2004; 128:38–43.

25

Gene Transfer and Left Ventricular Remodeling

H. Kirk Hammond
Department of Medicine, University of California, San Diego, California, U.S.A.

David M. Roth
Department of Anesthesia, University of California, San Diego, California, U.S.A.

INTRODUCTION

Early revascularization and pharmacological therapy have been used successfully to attenuate the extent of adverse remodeling associated with cardiovascular diseases such as myocardial infarction and hypertension. In some instances, revascularization and pharmacological treatment can reverse deleterious left ventricular remodeling and improve cardiac function. In the process of learning how left ventricular remodeling occurs and discovering ways to attenuate the effects, the expression of specific genes have become a focus for new therapeutic approaches that are not attainable through traditional pharmacotherapy. Gene transfer of these specific genes, it is reasoned, would provide a new approach in the treatment of heart failure and adverse remodeling. It is our goal to provide a general overview of exogenous gene transfer for the treatment of heart disease, with a focus on those strategies that may be best suited to attenuate or reverse adverse left ventricular remodeling. We will then review approaches that have been associated with alterations in left ventricular structure in preclinical studies that are likely to be tested in clinical gene transfer trials.

OVERVIEW OF CARDIAC GENE TRANSFER

We will not provide an extensive review of cardiac gene transfer. Instead, we will outline basic concepts simply to facilitate an understanding of translational research studies in remodeling. Three elements are required for cardiac gene transfer: (i) a gene of interest, (ii) a vector to carry and express the gene, and (iii) a suitable delivery method (Tables 1–3).

Table 1 Genes Tested in Preclinical Studies of Remodeling

Genes tested	Rationale	References
Growth hormone; IGF-I	Angiogenic, antiapoptotic, stimulates protein synthesis, inotrope	31,32
Hepatocyte growth factor	Angiogenic, antiapoptotic	33
Fibroblast growth factor-4	Angiogenic	10
Fibroblast growth factor-5	Angiogenic	14
VEGF-121	Angiogenic	36
Angiopoietin-1	Angiogenic	37
Bcl-2	Antiapoptotic	38
Mutant phospholamban peptide SERCA2a, sorcin	Inhibits PLB function; increases Ca^{2+} handling	22,41,42
Mutant β-adrenergic receptor kinase-1	Inhibits βARK; increases βAR signaling	43
Adenylyl cyclase type VI	Increases βAR signaling and Ca^{2+} handling; reduces PLB expression	12

Abbreviations: IGF, insulin-like growth factor; PLB, phospholamban; SERCA2a, sarcoplasmic reticulum Ca^{2+}-ATPase.

Gene of Interest

In general, selecting a gene for gene transfer to treat a specific abnormality of cardiac function is based on data indicating that the endogenous levels of the selected gene are insufficient, or that the elaboration of the gene—even if foreign to the heart—will have a salutary effect (Table 1). Because of difficulties in obtaining high efficiency gene transfer and expression, it is an advantage if the protein expressed has a paracrine effect so that even a limited degree of gene transfer and subsequent expression will ensure that the effect is amplified by the transgene protein having effects on multiple cells. An additional concern is that the transgene protein must have a means of being secreted from the target cell to manifest a paracrine effect.

Angiogenic genes serve as a suitable model to illustrate these points. It is presumed that a proper vector and delivery technique have been selected to achieve gene transfer to the cardiac myocyte. If the transgene was synthesized in the cardiac myocyte but could not exit the cell, then it would not serve as an angiogenic stimulus.

Table 2 Vectors for Cardiovascular Gene Transfer

Vector	Advantages	Disadvantages
Adenovirus	Direct intracoronary delivery; easy to produce	Expression not permanent; inflammation (dose, delivery method and tissue-dependent)
AAV	Long term expression; minimal inflammation	Limited insert size; direct intracoronary delivery inefficient; difficult to produce
Lentivirus	Long-term expression; large insert	Insertional mutagenesis; vascular delivery to heart doubtful; difficult to produce

Abbreviation: AAV, adeno-associated virus.

Table 3 Delivery Methods for Cardiac Gene Transfer

Method	Advantages	Disadvantages
Direct intracoronary	Simple; reduced morbidity vs. other methods	Less efficient than indirect intracoronary
Indirect intracoronary	Efficient gene transfer	Requires PA and aortic cross clamping and CP bypass and thoracotomy
Myocardial injection	Can select region of interest	Requires thoracotomy or sophisticated catheter methods; inefficient gene transfer

Abbreviations: CP, cardiopulmonary; PA, pulmonary artery.

This can be accomplished simply by selecting an angiogenic gene with an endogenous signal sequence (a short peptide sequence that earmarks a protein for extracellular secretion via the Golgi apparatus) or, alternatively, engineering a suitable signal sequence linked to the coding sequence of the transgene when one constructs the vector. Even if the transgene is expressed in a minority of cardiac myocytes, such a treatment may be effective because of the paracrine nature of the transgene.

In contrast, if a transgene has no likelihood for a paracrine-type function—a structural membrane protein, for example, then the extent of gene transfer may be a more important determinant of the effect. The level of cardiac gene transfer attainable with current vectors and delivery methods makes a nonparacrine transgene strategy less likely to be successful. However, just as is the case with traditional pharmacological therapy, the effects of exogenous gene expression are complex. The precise mechanisms for the effects may include altered expression of other genes that have salutary (or adverse) effects not anticipated based on the biological effects of the transgene per se. Furthermore, since cardiac myocytes are organized in a syncytium, an effect in a single cell may affect contraction of adjacent cardiac myocytes—with greater impact than anticipated from the extent of gene transfer.

Vectors

The choice of vector will depend on the cell type in the heart that is targeted for gene transfer. If the target is the cardiac myocyte, then most retrovirus vectors will not be suitable, because gene transfer with such vectors requires the target cell to enter a specific phase of cell division. An exception is lentivirus vectors, which will transfect terminally differentiated cells such as cardiac myocytes. Adenovirus and adeno-associated virus (AAV) are also suitable vectors, which can be used to obtain gene transfer in cardiac myocytes and other nondividing cells. Both AAV and lentivirus provide persistent transgene expression, as the transgene becomes integrated into the host chromosome. Adenovirus, in contrast, achieves transient transgene expression and is not generally viewed as a long-term expression vector. Although there are other virus and nonvirus vectors used in some preclinical studies, the most likely vectors for clinical cardiac gene transfer are adenovirus, AAV, and lentivirus (Table 2).

Of paramount importance in considering vectors is the degree to which the vector will incite an inflammatory response in the tissue and the host into which it is delivered. All these virus vectors have been modified so that they are unlikely to cause a clinical infection. However, there are virus proteins that are encoded by adenovirus vectors, for example, which are recognized as foreign antigens by the host with resultant cell-mediatved and humoral immune responses. In some applications this has been an insurmountable problem in the use of adenovirus vectors. Recent modifications, including deletions of the virus DNA such that virus protein expression is reduced, have provided vectors that may be associated with less immune response. The routes of delivery, dose, and targeted tissue are also important while considering the likelihood of inflammatory response. For example, with intracoronary delivery of first and second generation adenovirus vectors, there is very little evidence of inflammatory responses until a very high amount of virus (normalized per gram of heart perfused) is used Table 4 (1). The relative absence of inflammatory response following intracoronary delivery of adenovirus is in contrast to direct intramyocardial injection, in which the inflammatory response is dose-dependent but seen at smaller viral doses (2). This may have to do with high local concentrations of adenovirus and a direct cytopathic effect with subsequent inflammation.

An additional concern is the possibility of insertional mutagenesis, which could be associated with the use of retrovirus vectors. Indeed, a retrovirus vector used to treat patients with X-linked severe combined immunodeficiency resulted in the occurrence of T-cell acute lymphoblastic leukemia in two of the nine otherwise successfully treated patients. The complication is believed to have resulted from insertional mutagenesis (3). The possibility of this complication, combined with the unknown effects of persistent expression of exogenous genes, provides a rationale for the use of shorter term expression vectors, such as adenovirus, as a compromise for safety in these early days of clinical gene transfer trials.

Delivery Methods

In the process of selection from among adenovirus, AAV or lentivirus, the delivery method must be considered and will dictate the selection of vector. For example, lentivirus vectors are unsuitable for intracoronary delivery because they do not readily cross endothelial cells and therefore cannot gain access to the cardiac interstitium (4). In our studies, we achieve transgene expression in up to 50% of the left ventricle using indirect intracoronary delivery (see below) of adenovirus vectors in mice (5). When we apply the same methods but use lentivirus, we are unable to detect transgene expression in the heart (unpublished data).

Lentivirus vectors are effective when delivered via direct intramuscular injection into the myocardium (4). Intracoronary routes are possible with adenovirus and AAV (Table 3). However, the efficiency of gene transfer with these vectors after intracoronary delivery is quite variable, and generally not very efficient. Adenovirus is more efficient than AAV in achieving cardiac gene transfer after direct intracoronary delivery. Other routes of delivery have been used in both animal studies and in clinical trials. For example, direct intramyocardial injection has been used extensively with AAV and with adenovirus. However, gene transfer using this delivery method is limited to a small area adjacent to the needle tract, and the technique requires thoracotomy or sophisticated catheter devices and imaging techniques, which limits its utility in treating patients with severely compromised cardiac function.

Direct intracoronary delivery of adenovirus has been used with some success (Table 4) (6–21). These studies have occasionally used pharmacological agents that increase the extent of gene transfer, such as histamine (11), nitroprusside (12) or

Table 4 Direct Intracoronary Delivery of Adenovirus

Ref.	Species	Vector	vp/g	Days after Rx	Efficacy	Myocarditis
6	Human	Ad.FGF4 − E1	2.2×10^6–2.2×10^8	Up to 12 months	Yes[a]	No
7	Human	Ad.FGF4 − E1	7.1×10^7	56	Yes[a]	No
8	Pig	Ad.FGF5 − E1	1.7×10^9	14	Yes	No
9	Pig	Ad.FGF4 − E1	8.3×10^7–8.3×10^9	14	Yes	No
10	Pig	Ad.FGF4 − E1	8.3×10^9	21	Yes	No
11	Pig	Ad.AC$_{VI}$ − E1	1.2×10^{10}	10–60	Yes	No
12	Pig	Ad.AC$_{VI}$ − E1	1.2×10^{10}	14	Yes	No
14	Pig	Ad.FGF5 − E1	3.7×10^{10}	14	Yes	No
13	Rabbit	Ad.VEGF − E1/−E3	6×10^{10}	17	Yes	No
15	Minipig	Ad.lacZ − E1	9×10^{10}	31	Not relevant	No
16	Rabbit	Ad.lacZ − E1	10^{10}–10^{11}	5–60	Not relevant	No
17	Rabbit	Ad.ß2AR − E1/−E4	2×10^{11}	21 2d steroid	Yes	Not stated
18	Rabbit	Ad.ßARKct − E1/−E4	2×10^{11}	3–6 2d steroid	Yes	Not stated
19	Rabbit	Ad.ßARKct − E1/−E4	2×10^{11}	5	Yes	Not stated
20	Rabbit	Ad.V2 Rec − E1	4×10^{11}	Unclear vs. IM injection	Yes	Unclear vs. IM injection?
21	Pig	Ad.Gi -E1/-E3	7.5×10^{12}	7	Yes	Monocyte inflt, mild

Published studies using direct intracoronary delivery of adenovirus vectors—those procedures using thoracotomy and aortic cross clamping (indirect intracoronary delivery) are not reviewed. Studies are listed in increasing order of amount of adenovirus delivered per gram of myocardium served by the vessel(s) injected. Unless specifically stated in the paper, certain assumptions were made regarding left ventricular (LV) weight: rabbits weighing 3.0 kg have left ventricular (LV) weights of 10 g; pig LV weight is 3 g/kg body weight for normal pigs and 3.5 g/kg for minipigs; average human LV weight 140 g; AV nodal artery perfuses 1 g myocardium; 1 pfu = 100 vp.

[a]Subsequent larger scale clinical trials were ended—they were reportedly safe but data did not meet the efficacy end point required for continuing the trial.

Abbreviations: Ad, adenovirus; IM, intramuscular; Rec, receptor; d, day; -E1, E-1 deleted vector; inflt, infiltration vp, virus particles; g, gram; Gi, inhibitory GTP-binding protein; Rx, treatment.

Source: From Ref. 1.

sildenafil (21)—agents that share the ability to increase nitric oxide, which may promote transvascular movement of the vector (5,11,12).

Mechanical procedures to increase the time of contact between vector and heart (dwell time) have included cross clamping of the pulmonary artery and aorta in combination with hypothermia—to decrease injury to the brain resulting from sustained reduction in cerebral perfusion during aortic cross clamping. The vector is then delivered proximal to the aortic clamp, where it gains access to the heart via the coronary circulation. We have called this delivery method indirect intracoronary delivery (5). This technique is particularly suited for animals with coronary arteries too small for direct intracoronary injection. Although the procedure is invasive, it provides an ability to increase dwell time and thereby promote gene transfer to a greater degree than that achieved by the direct intracoronary route. Indirect intracoronary gene transfer is also increased by pharmacological agents that increase transvascular movement of the vector, including nitroprusside, histamine, serotonin, acetylcholine, and substance P (5,22).

The degree of cardiac gene transfer achieved by these methods is questionable. Claims of extensive transgene expression in the heart do not provide adequate methodological detail for rigorous evaluation. Using the indirect intracoronary delivery method in rats and mice, we expressed gene transfer as a percentage of the entire left ventricle (rather than the percent of a highly selected microscopic field). In rat heart using AAV, we obtained 25% to 30% gene transfer using aortic and pulmonary artery cross clamping and histamine (23); in mouse heart using adenovirus vectors we obtained gene expression in up to 50% of the entire left ventricle, using aortic and pulmonary artery cross clamping in conjunction with hypothermia 25°C and either serotonin or acetylcholine (5).

We have not systematically evaluated the extent of gene transfer obtained with direct intracoronary delivery in large animals such as the pig, but we would expect the extent of gene transfer to be less than that achieved by the more invasive indirect intracoronary delivery techniques just described. However, using enzyme assays as a reporter system, we detected transgene expression and activity in left ventricular samples even months after direct intracoronary delivery of the vector (9,11), and the extent of transgene expression is increased three-to four-fold by simultaneous delivery of nitroprusside (24). Finally, of the published papers using direct intracoronary direct delivery of adenovirus vectors (6–21), most report a significant biological effect of the transgene protein, suggesting that even small amounts of detectable gene expression can have powerful effects on physiological function (Table 4).

If gene transfer is to be applied successfully in clinical settings, the technique that is effective but also simplest and safest to apply will prevail as the method of choice. Emphasis on simplicity and safety will be particularly germane for application of such procedures to patients with severely compromised hearts in whom invasive procedures such as thoracotomy, cross clamping of major vessels, and cardiopulmonary bypass would be prohibitive.

GENE TRANSFER AND LEFT VENTRICULAR REMODELING

Currently, there are no published clinical studies that have used cardiac gene transfer to alter left ventricular remodeling. We will therefore focus on a review of preclinical studies. We have selected studies that are illustrative of promising strategies, which provide evidence of attenuation of adverse left ventricular remodeling.

Growth Hormone/Insulin-like Growth Factor-I (IGF-I)

Growth hormone exerts many of its actions through IGF-I. Growth hormone and IGF-I are angiogenic, antiapoptotic, positive inotropes, and stimulate protein synthesis (25–29). All these features seem desirable in the treatment of adverse remodeling associated with wall thinning following myocardial infarction. In addition, there have been a few published papers suggesting that growth hormone has favorable effects on left ventricular chamber dimensions and function (25). Unfortunately, this report, like many other clinical papers that have used growth hormone in patients with heart failure, was based on data that were not collected in a blinded manner. Animal studies that have used either growth hormone or IGF-I protein has substantiated, in general, that positive effects on remodeling and cardiac function after myocardial infarction can be achieved (26,27). However, systemic delivery of growth hormone is associated with hypertension and a diabetogenic state (28); high levels of plasma IGF-I have been associated with hypoglycemia, hypotension, and the development of cancer (29,30). It would be more desirable if the delivery of these agents could be isolated to the heart, to circumvent these dire consequences. A gene transfer strategy could, theoretically, accomplish this.

To determine if growth hormone gene transfer would prevent left ventricular remodeling, a study was conducted using the rat myocardial infarction model (31). At the time of coronary occlusion, simultaneous needle injection of adenovirus encoding growth hormone (Ad.GH) or an adenovirus vector without an insert (null vector; Ad.Null) was performed—adenovirus vector was injected in the border region of infarction and viable myocardium. Five injections were made and the total virus dose was 5×10^9 plaque forming units [pfu; a common method of estimating the infectivity of adenovirus, generally 100-fold less than the total virus particles (vp) of the same vector, although this ratio varies with a specific clone]. Six weeks after myocardial infarction and gene transfer, favorable effects were seen in the group that had received growth hormone gene transfer: end-diastolic dimension was reduced ($P < 0.001$) and left ventricular dP/dt was increased ($P < 0.001$). In addition, wall thickness in the infarct region was increased ($P < 0.001$). Increased expression of growth hormone was confirmed in rats that had received Ad.GH treatment. This study showed that gene transfer of growth hormone, when delivered at the time of myocardial infarction had salutary effects on remodeling. The limitation of the study is that to replicate this study in a clinical setting would necessitate gene transfer during the acute phase of myocardial infarction. Whether these salutary effects would be observed weeks or months after myocardial infarction were not addressed. This would be a likely window for clinical gene transfer, in contrast to immediately before or during the first minutes of coronary occlusion. Furthermore, since growth hormone exerts its actions through IGF-I and ultimately through Akt, which may play the major role in cardiac myocyte survival, one wonders if targeting specific distal events helps to delineate the precise mechanisms for the effects as well as to add specificity.

Indeed, when adenovirus encoding IGF-I was injected into the jeopardized perfusion bed just before coronary occlusion in rats, the extent of infarction was reduced by 50%, an effect thought primarily to be the result of reduced apoptosis (32). An important element in this study was that gene expression was detected in only 15% of the area at risk, but the effects on reducing infarct size were greater, suggesting a paracrine benefit of IGF-I. As mentioned previously, transgenes that operate in a paracrine manner are likely to be more effective than those that do not, owing to the relative paucity of gene transfer attainable. Finally, there was no detectable plasma IGF-I, a desirable outcome, given the deleterious consequences that have been associated with plasma levels of IGF-I (29,30). However, this study did not demonstrate an IGF-I effect on reversal of remodeling, and cannot be directly translated to clinical settings, since the timing of gene transfer was prior to the onset of myocardial infarction. In addition, it is surprising that a gene transfer strategy could provide an adequate amount of IGF-I fast enough to reduce apoptosis associated with acute myocardial infarction. Beneficial effects may be related to effects other than apoptosis.

Hepatocyte Growth Factor

Hepatocyte growth factor (HGF) has both angiogenic and antiapoptotic effects, and therefore may be well suited for the treatment of ischemic hearts (33). This was explored using the rat infarction model—adenovirus encoding HGF or null vector was injected into the myocardium surrounding the occluded perfusion bed [5 injections totaling 5×10^9 pfu]. Measurements were then taken six weeks after myocardial infarction, showing increased left ventricular dP/dt, increased blood flow (microspheres), reduced apoptosis, and evidence of regional myocardial hypertrophy in the region that received gene transfer. The limitation of this study was that, again, the intervention was made before acute myocardial infarction, so that while the data show a potential beneficial effect of HGF, they are not directly applicable to clinical settings. A more relevant study for clinical application of gene transfer would have been by treating rats after the acute phase of myocardial injury was complete.

FGF-4, FGF-5, VEGF-121, Angiopoeitin-1

To determine whether fibroblast growth factor–4 (FGF-4) could have favorable effects on heart failure, the rapid pacing model of heart failure in pigs was used. The pacing model of heart failure is associated with myocardial ischemia (34). Patients with dilated failing hearts, even when unassociated with coronary artery disease, have measurable myocardial ischemia (35), hence improved myocardial blood flow might improve left ventricular function in heart failure, despite the type of etiology. Therefore, it was proposed that angiogenic gene transfer may improve heart function in pacing-induced heart failure. Intracoronary injection of adenovirus encoding fibroblast growth factor-4 (Ad.FGF4) at a dose of 10^{12} vp per pig was delivered in a randomized blinded study (vs. intracoronary saline) (10). After three continuous weeks of pacing, animals that had received intracoronary Ad.FGF4 showed reduced left ventricular end-diastolic diameter ($P = 0.05$), increased regional wall thickening ($P = 0.002$), and reduced wall stress ($P = 0.004$). A shortcoming of the study was that gene transfer was performed immediately before pacing was initiated rather than inducing heart failure first and then treating. A subsequent study, using direct intramyocardial injection of adenovirus encoding vascular endothelial growth factor-121 (Ad.VEGF-121), also showed favorable effects on regional function in the same model (36), but no data were provided regarding left ventricular chamber geometry, so whether or not attenuation of adverse remodeling occurred is not known.

In a porcine model of left ventricular dysfunction associated with hibernating myocardium, an adenovirus encoding fibroblast growth factor-5 (Ad.FGF-5) was infused in each of the three proximal coronary arteries [10^{11} pfu, total per animal] (14). It was speculated that an angiogenic gene increases blood flow and regional myocardial function in hibernating myocardium. Two weeks after gene transfer, regional improvement in function was observed ($P < 0.05$). Although small increases in basal perfusion were seen, maximal perfusion was unchanged. However, Ad.FGF-5 treatment was associated with increased numbers of cardiac myocytes in mitosis and a 29% hypertrophy. These data indicate that effects of FGF-5 other than angiogenesis may be important in altering cardiac function in this setting.

Adenovirus encoding angiopoietin-1—an angiogenic factor associated with vascular maturation in synergy with vascular endothelial growth factor—was delivered (intramyocardial injection) at the time of coronary occlusion in rats (37). Four weeks later, treated animals showed increased left ventricular ejection fractions, reduced infarct size, and increased capillary density in treated regions compared to control animals. It is difficult, however, to reconcile that an angiogenic effect of the transgene could have any bearing on reduced infarct size, since the vector was delivered at the time of coronary occlusion. It seems unlikely that the transgene had sufficient time to be transcribed, translated, and that the transgene protein could then elaborate a functional vasculature, still within a time that would permit a reduction in infarct size. On the other hand, progressive chamber remodeling due to relative ischemia in the border region—perhaps, a later effect—may have been favorably affected by gene transfer, although this study did not find changes in wall thickness, function, or end-diastolic dimensions to support such an effect. Thus, the favorable effects observed do not have a ready mechanism to explain them.

Reduction of Apoptosis

It is difficult to see how gene transfer fits the paradigm so well-served by percutaneous coronary interventions vis-à-vis reducing acute myocardial infarct size. An effective antiapoptotic protein would outperform a gene transfer strategy in the acute setting—too much time would be lost in waiting for transgene expression. Finally, the concept of performing gene transfer in advance of a threatened infarction, to reduce apoptosis associated with acute MI—prophylactic gene transfer—is a concept ahead of its time. On the other hand, if apoptosis is an important element in progressive cardiac myocyte loss associated with some forms of experimental and perhaps clinical CHF, then inhibiting this process may attenuate cell loss and thereby preserve myocardial mass.

To begin to explore the effects of inhibiting apoptosis through gene transfer, left circumflex coronary artery occlusion in rabbits, followed by reperfusion was used (38). Indirect intracoronary delivery of adenovirus encoding Bcl-2, an inhibitor of apoptosis (39), versus empty vector at doses of 5×10^{10} pfu was performed at the same surgical procedure. Six weeks later, a second thoracotomy was performed and sonomicrometers were used to evaluate fractional shortening in the region bordering the infarction and in the viable region remote from the infarct. Echocardiography was used to assess global left ventricular function sequentially during the study.

Transgene expression was documented in rabbits that had received Ad.Bcl-2, and infarct size was not different between the two groups. However, gene transfer of Bcl-2, as compared to that of the control vector, was associated with higher left ventricular ejection fractions at two, four, and six weeks after infarction, and there was less left ventricular dilation and wall thinning. These favorable influences on left ventricular function and geometry were associated with reduced apoptosis. Their finding that infarct size was not reduced is expected, because performing gene transfer at the time of injury would not be expected to provide sufficient expression of the transgene to inhibit apoptosis acutely. However, the extent of beneficial effects over such a short period of time is striking, which calls into question whether the beneficial effects could solely relate to inhibition of apoptosis alone. Whatever the mechanism, this study does demonstrate that gene transfer of Bcl-2 has favorable effects on remodeling.

Gene Transfer to Increase Calcium Handling

Ablation of phospholamban, an endogenous muscle-specific inhibitor of sarcoplasmic reticulum Ca^{2+}-ATPase 2 (SERCA2a), has been shown, in transgenic mice, to prevent abnormalities associated with a genetic model of heart failure (40). A recent study used myocardial infarction in rats to determine the effects of phospholamban inhibition (22). Five weeks after myocardial infarction, rats with large infarctions (30% to 40% of the left ventricle, assessed by echocardiography) were randomized to receive saline or AAV encoding a pseudophosphorylated mutant phospholamban peptide, which inhibits the endogenous function of phospholamban. The AAV vector was delivered by an indirect intracoronary method in anesthetized, mechanically ventilated animals using occlusion of pulmonary artery and aorta via a thoracotomy, with injection into the proximal aorta, distal to the occlusion via an indwelling catheter. Hypothermia (26°C) was achieved during vector delivery, which permitted at least a sixty second occlusion, and substance P was used as an adjunct to increase gene transfer efficiency.

Animals were then studied two and six months after having received gene transfer or saline. The results of phospholamban inhibition included increased ejection fraction measured by echocardiography (vs. saline-treated rats), and a smaller rise over the time interval in left ventricular end-diastolic dimension. A terminal study, performed six months after treatment, showed that rats that had received gene transfer had higher peak positive and peak negative left ventricular dP/dt, and reductions in heart weight, cardiac myocyte size, and fibrosis.

This study has direct applicability to clinical settings, because treatment was delivered not at the time of infarct, but only after left ventricular remodeling had occurred, and provides evidence that remodeling can at least be attenuated following gene transfer. The use of a vector that provides prolonged transgene expression—a laudable goal for this specific treatment of a failing heart—was also commendable.

It is unlikely that the delivery approach that was required to obtain high efficiency AAV gene transfer—thoracotomy, hypothermia, and occlusion of the pulmonary artery and aorta—could be safely applied to patients with compromised left ventricular function. The generally superior efficiency of direct intracoronary adenovirus versus AAV (8,14,23) would, potentially, circumvent the need for these potentially morbid procedures. On the other hand, adenovirus is not expected to provide the prolonged gene expression conferred by AAV. Ideas and targets for intervention are developing at a faster pace than are new vectors and delivery strategies. This is a conundrum in the advancement of cardiovascular gene transfer.

There have been several attempts aimed at modification of myocardial calcium handling that merit discussion. For example, aortic banding of rats induced decreased systolic function with left ventricular chamber dilation and hypertrophy six months later (41). Indirect intracoronary delivery of adenovirus encoding SERCA2a (vs. adenovirus encoding lacZ, a reporter gene) was associated with improved survival four weeks later and left ventricular volumes were reduced.

Sorcin, a 21.6 kDa Ca^{2+} binding protein, which is a member of the penta EF-hand family, is another calcium modulating protein that was recently examined in a study using gene transfer (42). This study used streptozotocin-treated mice as a model of cardiomyopathy, with subsequent direct intramyocardial injection of adenovirus encoding sorcin (Ad. Sorcin) (vs. empty adenovirus vector) three weeks later. Five injections each containing 10^9 pfu Ad.Sorcin were performed per heart. Five days after gene transfer, the hearts were isolated and perfused, and contractile function assessed. Increased calcium transients in cardiac myocytes from hearts that had received sorcin gene transfer were associated with increased left ventricular dP/dt in studies of isolated hearts. Echocardiography showed increased fractional shortening. Unfortunately, no data were provided regarding left ventricular size; hence, whether the treatment was associated with attenuation or reversal of remodeling is not known.

β-Adrenergic Receptor Kinase Inhibition

A truncated form of the β-adrenergic receptor kinase-1 (βARK-1), referred to as βARKct, acts as a peptide inhibitor of endogenous βARK-1 and other protein kinases. Inhibition of such kinases can facilitate β-adrenergic receptor signaling and increase left ventricular function. An adenovirus encoding βARKct [5×10^{11} vp] was delivered by direct intracoronary infusion to rabbit hearts three weeks after experimentally induced myocardial infarction (18). Seven days after gene transfer, systolic shortening (sonomicrometry) was improved ($P<0.05$) in treated animals compared to that in control animals (empty adenovirus vector). Unfortunately, no data were reported to document an alteration in left ventricular chamber size, so whether this treatment actually had favorable effects on remodeling is not known.

Adenylyl Cyclase Type VI

In this final section pertaining to specific transgenes that may be effective in attenuating or reversing left ventricular remodeling, we provide an in-depth analysis of adenylyl cyclase type VI (AC_{VI}). The reasons to do this are twofold. First, this strategy is derived from papers published by the authors of the present chapter. Second, the progression of studies performed provides a suitable model of translational research starting from basic observations and ending with preparations for clinical gene transfer trials using AC_{VI}.

   Adenylyl cyclase (AC) has long been recognized as a pivotal effector molecule in cardiac myocytes and other cells. Molecular cloning studies indicate that there are at least ten different isoforms of AC, each with different structure, tissue distribution, and regulation (43,44). Cardiac myocytes appear to express AC_V and AC_{VI} predominantly (45,46). AC_{VI} plays a pivotal role in contractile responsiveness, is tightly linked to left ventricular function, and is functionally inadequate and shows reduced expression in the failing heart (46).

AC_{VI} Gene Transfer in Cultured Cardiac Myocytes

In studies designed to explore the effects of AC_{VI} gene transfer on cardiac myocyte biology, we used recombinant adenovirus to increase AC_{VI} expression in neonatal cardiac myocytes (47). Cells that overexpressed AC_{VI} responded to agonist stimulation with marked increases in cAMP production in proportion to protein expressed. Of particular note was the absence of increased cAMP in the unstimulated state (48), distinct from gene transfer of βAR or Gαs (the GTP-binding protein, which transduces βAR occupation with stimulation of AC), which results in sustained activation and steady increases in intracellular cAMP (49,50).

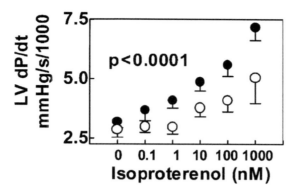

Figure 1 Hearts from transgenic mice were isolated and perfused, and peak positive left ventricular pressure development (LV dP/dt) in response to isoproterenol was measured. Open circles, data from six transgene negative hearts; closed circles, data from 11 AC_{VI} positive hearts. Heart rate was unchanged (*not shown*). *Source*: From Ref. 48.

Cardiac-Directed AC_{VI} Expression—Transgenic Mice

To explore the consequences of long-term and high level expression of AC_{VI} in the heart, we generated transgenic mice with cardiac-directed expression of AC_{VI} (48). A particular concern was that long-term expression of other elements of the βAR signaling pathway (βAR, Gαs) was associated with short-term improvements in cardiac function, but, eventually, with heart failure (49,50).

Mice with cardiac-directed AC_{VI} expression showed increased cardiac function (Fig. 1) and cardiac myocytes showed increased cAMP production. In contrast, basal cAMP and cardiac function were normal, and long-term transgene expression was not associated with abnormal histological findings or deleterious changes in cardiac structure and function (48).

Cardiomyopathy Treated by AC_{VI}

Cardiac-directed expression of Gαq is associated with left ventricular dilation, reduced heart function and impaired cAMP production, mimicking aspects of clinical heart failure. To determine if increased cardiac myocyte AC_{VI} content increases cardiac function in cardiomyopathy, we crossbred AC_{VI} transgenic mice and mice with cardiomyopathy induced by cardiac-directed Gαq expression (51). Cardiac-directed expression of AC_{VI} in this model of heart failure resulted in improved basal cardiac function in vivo. Left ventricular peak positive ($P<0.005$) and negative ($P<0.04$) dP/dt were increased, and cardiac myocytes showed increased cAMP production. Increased myocardial AC_{VI} content therefore improved cardiac function and responsiveness. Increased left ventricular function was facilitated by restoration of cAMP generating capacity in cardiac myocytes, and left ventricular end-diastolic dimensions were decreased ($P<0.0001$). In subsequent studies, we also found that cardiac hypertrophy was abrogated ($P<0.004$) and marked improvement in survival ($P<0.0001$) was observed (52) (Fig. 2).

Abnormalities in calcium handling are dominant in the failing heart, and cardiac AC_{VI} was found to restore SERCA2a affinity for calcium and maximum velocity of cardiac calcium uptake by sarcoplasmic reticulum ($P=0.012$) in murine dilated cardiomyopathy associated with Gαq (53). The return to normal of SERCA2a affinity for calcium was associated with decreased phospholamban protein expression

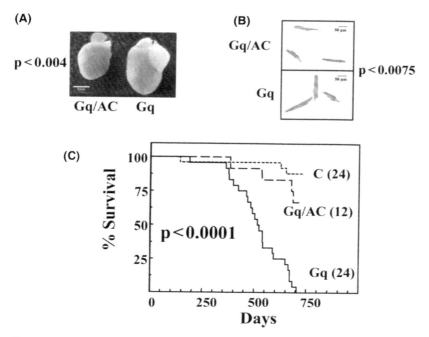

Figure 2 (A) Representative hearts from Gq/AC and Gq mice are shown. These hearts were obtained at 11 months of age—the time when mortality increases in Gq mice. These mice were killed electively for this photograph. Western blots of ventricular homogenates from these hearts confirmed increased expression of AC_{VI} and Gq proteins in the Gq/AC mouse and Gq protein in the Gq mouse (*data not shown*). (B) Representative cardiac myocytes isolated from Gq/AC and Gq mice are shown. These myocytes were obtained at 15 months of age—mice were killed electively for this photograph. Myocyte size was enlarged hearts from Gq animals, but was reduced to normal size by concurrent expression of cardiac AC_{VI}. 160×. (C) Kaplan-Meier curve showing mortality rate in Gq ($n=24$), Gq/AC ($n=12$) and Control mice ($n=25$). Increased survival was associated with expression of cardiac AC_{VI} in cardiomyopathy ($P<0.0001$). There was no difference in mortality between Gq/AC and Control mice. These data indicate a pronounced favorable effect on survival associated with cardiac-directed AC_{VI} expression in Gq cardiomyopathy. *Abbreviation*: AC, adenyl cyclase. *Source*: From Ref. 52. (*See color insert.*)

and increased phospholamban phosphorylation by PKA activation. These data suggest a mechanism by which AC_{VI}—unlike other signaling elements associated with increased cAMP generation—has a beneficial effect in the failing heart.

Abrogation of Myocardial Hypertrophy: Mechanistic Insights in Remodeling

The data from experiments in transgenic mice indicated that AC_{VI} expression prevents myocardial hypertrophy in the Gαq cardiomyopathy model. The precise molecular pathway involved in this phenomenon is unknown. G-protein-coupled receptors link via heterotrimeric G-proteins to either stimulate (Gαs) or inhibit (Gαi) AC activity and subsequent formation of cAMP. AC_V and AC_{VI} are inhibited by PKA, low concentrations of free Ca^{2+} (0.1–1 μM), Gαi and βγ dimer and stimulated by Gαs. However, differences exist between AC_V and AC_{VI} in regulation by protein kinase C and regulation by some hormone receptors (54–57). Benefits of isoform-specific differences in experimental or clinical cardiac hypertrophy, remodeling, and heart failure are unknown.

The role of AC and other βAR signaling elements in cardiac hypertrophy, fibrosis, remodeling, and the development of heart failure has been the focus of extensive basic and clinical investigation (58,59). Isolated neonatal (60) and adult rat cardiac myocytes (61) show hypertrophy in response to β_1AR stimulation. Transgenic mice with cardiac-directed β_1AR or β_2AR expression develop marked cardiac myocyte hypertrophy, fibrosis, and progressive heart failure with functional and histological defects similar to patients with heart failure (62–64). Mice with cardiac-directed Gαs expression also develop progressive cardiac hypertrophy, fibrosis, and heart failure (49,65). In contrast, cardiac-directed expression of AC_{VI} in mice produces increased cardiac βAR responsiveness, normal cardiac function, and no signs of cardiac hypertrophy, fibrosis, or cardiomyopathy when animals are studied at four and 21 months of age (48). This indicates that, that expression of different β-adrenergic signaling proteins is associated with different effects on hypertrophy, fibrosis, and remodeling.

Does hypertrophy and fibrosis produced by cardiac-directed expression of β-adrenergic receptors result from increased intracellular cAMP and downstream activation of PKA, or are other cell signaling pathways involved? βAR subtypes have distinct actions in cardiac myocytes (66). Hypertrophy in isolated rat cardiac myocytes in response to β_1AR stimulation may be independent of cAMP and cannot be reproduced by dibutyryl cAMP (61) or direct AC stimulation with forskolin (60). Other cell signaling pathways may play a role in βAR-promoted cardiac hypertrophy, including the extracellular signal-related kinase (ERK) subfamily of mitogen activated protein kinases (MAPK) (66) and cell signaling pathways involving activation of the serine threonine kinase Akt (67,68). Differences regarding hypertrophy and fibrosis in mice with cardiac-directed AC_{VI} expression versus cardiac-directed expression of other βAR signaling elements have prompted us to explore the effects of cardiac-directed AC expression in models of cardiac hypertrophy.

Gαq-linked signaling pathways are one of the most extensively studied pathways for cardiac hypertrophy and pathological cardiac growth (69,70). Mice with cardiac-directed expression of Gαq have provided a useful model to study cell signaling involved in cardiac hypertrophy, remodeling, and heart failure (71). Gαq mice have cardiac myocyte hypertrophy and increased hypertrophy-related gene expression (ANP, α-myosin heavy chain, α-skeletal actin). Mice with cardiac-directed expression of βAR signaling proteins have been crossbred with Gαq mice. For example, cardiac-directed expression of β_2AR (30-fold increased protein) in Gαq mice did not change βAR-stimulated AC activity, but improved basal left ventricular function and attenuated cardiac hypertrophy, and expression of ANP and α-skeletal actin mRNA (72). Even higher levels of cardiac β_2AR expression (140-fold increased) in Gαq mice increased βAR-stimulated AC activity, but worsened cardiac function and hypertrophy. Cardiac-directed expression of AC isoform V (1.5-fold increased) in Gαq mice increased βAR-stimulated AC activity and cardiac function. However, myocardial hypertrophy was unaffected and cardiac fibrosis was increased (73). In contrast, cardiac-directed expression of AC_{VI} (20-fold increased) in Gαq mice improved βAR-stimulated AC activity in cardiac myocytes, increased basal and βAR-stimulated left ventricular function, attenuated cardiac hypertrophy and ANP gene expression, did not exacerbate fibrosis, and significantly improved survival (51,52). Results of the studies described above in Gαq mice are summarized in Table 5.

It could be argued that cardiac-directed AC_{VI} expression in Gαq/AC_{VI} mice, by restoring βAR responsiveness and normalizing cardiac function, may have removed a physiological stimulus for cardiac hypertrophy. However, this result contrasts with the findings observed for cardiac-directed AC_V expression, which normal-

Table 5 Expression of Adrenergic Signaling Elements in Gαq Cardiomyopathy

	βAR-stimulated AC activity	Cardiac function	Hypertrophy	Fibrosis
Gαq/β₂AR 30-fold	↔	↑	↓	↔
Gαq/β₂AR 140-fold	↑	↓	↑	↑
Gαq/AC_V 1.5-fold	↑	↑	↔	↑
Gαq/AC_VI 20-fold	↑	↑	↓	↔

Abbreviations: AC, adenylyl cyclase; βAR, β-adrenergic receptor; β₂AR, β₂-adrenergic receptor.

ized cardiac function in Gαq mice but did not prevent pathological cardiac hypertrophy or ANP and α-MHC gene expression. Furthermore, cardiac-directed expression of a regulator of G-protein signaling, isoform 4 (RGS4), improved cardiac function in Gαq/RGS4 mice at four weeks of age and attenuated ANP gene expression, but by nine weeks of age cardiac function in Gαq/RGS4 mice was reduced to an extent similar to that of Gαq mice and expression of ANP was increased (74). Cardiac-directed expression of an εPKC translocation activator, ψεRACK, improved cardiac function in ψεRACK/Gαq mice and attenuated hypertrophy, but ANP and α-MHC gene expression were elevated similar to that of Gαq mice (75). Targeted deletion of MAPK/ERK kinase kinase 1 (MEKK1) in Gαq mice attenuated hypertrophy and hypertrophy-related genes, but did not improve basal left ventricular function (70). Overall, these results involving the use of different signaling components suggest that restoration of normal cardiac function does not fully explain our findings in Gαq/AC_VI mice, in which more extensive improvement and lack of disease progression occurs compared to other components tested. Thus, the attenuation of hypertrophy in Gαq/AC_VI mice may involve interaction of AC_VI with cell signaling pathways that prevent cardiac hypertrophy.

Gαq signaling in cardiac myocytes results in activation of phospholipase C (PLC), inositol 3-phosphate (IP₃), diacylglycerol, PKC, and MAPK pathways (69,70). MAPK signaling pathways involve a cascade of successively activated kinases resulting, ultimately, in phosphorylation and activation of three main terminal effector kinases: ERKs, c-Jun NH₂-terminal kinases (JNK), and p38 kinases (76). MAPKs phosphorylate multiple intracellular targets, including transcription factors, and reprogram cardiac gene expression. All three MAPK pathways are activated by Gαq-coupled receptors after pressure-overload stress is imposed on the left ventricle (77). Recently, Gαq- and ERK-mediated cardiac hypertrophy in adult rat cardiac myocytes was shown to be inhibited by a βAR, cAMP, and PKA-dependent mechanism, suggesting a role for AC-associated proteins in modulating MAPK signaling (78). Whether the mechanism by which cardiac-directed AC_VI expression has salutary effects in Gαq cardiomyopathy involved MAPK signaling is unknown, but is an attractive hypothesis.

Exogenous Gene Transfer of Adenylyl Cyclase Type VI

Data from studies of transgenic mice indicate that cardiac-directed expression of AC_VI may be beneficial to offset deleterious left ventricular remodeling. However, to apply these concepts in clinical settings would require that exogenous gene transfer could achieve an adequate amount of the transgene to evoke an effect, and that the strategy could be applied to models of heart failure other than genetic models in transgenic mice.

Can gene transfer methods provide enough exogenous AC_{VI} to the heart to increase global left ventricular function? We had previously shown that intracoronary delivery of adenovirus vectors provides an efficient means to obtain gene transfer to the heart (8). We then showed that intracoronary histamine increases the extent of gene transfer compared to the usual methods (11). Using this delivery protocol, we speculated whether we could alter left ventricular contractility and function in normal pigs (11). Animals were instrumented with high fidelity transducers in the left ventricle and flow probes to measure cardiac output. After recovery from surgery (10–12 days), animals underwent isoproterenol infusion and measurement of cardiac output and left ventricular dP/dt. Animals were studied again 12 days after gene transfer. Eighteen animals received intracoronary histamine (25 µg/min, 3-min) followed by intracoronary delivery of 1.4×10^{12} vp of an adenovirus encoding lacZ ($n = 8$) or AC_{VI} ($n = 10$). The vector was delivered by three separate infusions, one in each major coronary artery, with the proportions of the total virus delivered: 50% LAD, 20% LCx, and 30% RCA. Gene transfer of AC_{VI} without precedent histamine was ineffective. Gene transfer of lacZ had no influence on isoproterenol-stimulated left ventricular dP/dt or cardiac output. In contrast, animals that had received intracoronary adenovirus encoding AC_{VI} with precedent histamine showed marked increases in left ventricular contractile function (Fig. 3A) and cardiac output in response to isoproterenol infusion. In additional studies, we demonstrated that increased left ventricular function due to gene transfer of AC_{VI} was associated with a two-fold increase in left ventricular AC_{VI} protein (Fig. 3C). A corresponding two-fold increase in cAMP generation was also documented, an effect that persisted for the entire 18 weeks of the study (11) (Fig. 3D). Finally, gene transfer was found to be associated with increases in left ventricular function within six days, which increased by 29 days and were still substantial 57 days later (11) (Fig. 3B).

AC_{VI} Gene Transfer in Pacing-Induced CHF

To replicate the strategy envisioned for a future clinical study, it would be necessary to deliver the transgene in an adenovirus virus vector by intracoronary injection in subjects with severe heart failure. To achieve this, pigs underwent pacemaker and left ventricular pressure transducer implantation (12). Physiological and echocardiographic studies were performed thirteen days later, and pacing was initiated (220 bpm). Seven days later, isoproterenol-stimulated left ventricular dP/dt (a measure of contractile function) was reduced ($P < 0.0002$ vs. pre-pacing), documenting left ventricular dysfunction. Pigs then received intracoronary Ad.AC_{VI} (1.4×10^{12} vp; $n = 7$) or saline (PBS; $n = 9$) (randomized, blinded) preceded by an intracoronary infusion of nitroprusside (50 µg/min, 6.4 min) to increase gene transfer. Pacing was continued for 14 days and final studies were performed. The a priori key endpoint was a change in the left ventricular dP/dt during isoproterenol infusion (pre-Ad.AC_{VI} value minus value 14 days later). Pigs receiving Ad.AC_{VI} showed less fall in both peak positive left ventricular dP/dt ($P = 0.0014$; Fig. 4A) and peak negative left ventricular dP/dt ($P = 0.0008$). Serial echocardiography showed that Ad.AC_{VI} treatment was associated with increased left ventricular function and reduced left ventricular dilation (Table 6). AC-stimulated cAMP production was increased in left ventricular samples from Ad.AC_{VI}-treated pigs ($P = 0.006$; Fig. 4B) and gene transfer was confirmed by PCR. These data indicated that AC_{VI} gene transfer increases left ventricular function and attenuates deleterious left ventricular remodeling in pacing-induced heart failure (12).

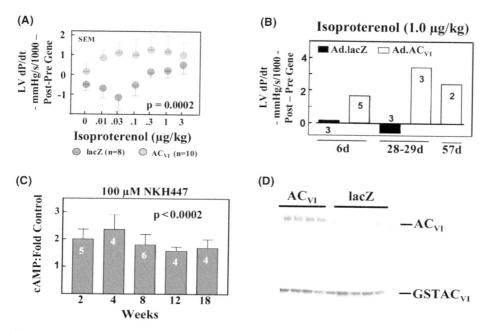

Figure 3 (**A**) Data from conscious pigs before and 12 days after intracoronary delivery of 1.4×10^{12} vp adenovirus encoding AC_{VI} (*red*) or lacZ (*blue*). The *y*-axis of the graph displays the post-gene values minus the pre-gene values. Peak positive LV dP/dt in response to doses of infused isoproterenol was measured. Studies were repeated two to three times both before and again 12 days after gene transfer (blinded analysis). Gene transfer was preceded by intracoronary histamine infusion. Data indicate that intracoronary gene transfer of AC_{VI} can increase left ventricular function. (**B**) The onset and duration of the physiological effect was studied before and 6, ~30, and 57 days after gene transfer of AC_{VI} or lacZ. The *y*-axis displays the difference in contractility between post- and pre-gene values. Peak positive LV dP/dt response to isoproterenol (1.0 µg/kg) was measured. Gene transfer of lacZ was not associated with changes in LV dP/dt 6 days or 28 days later. In contrast, AC_{VI} gene transfer was associated with increased LV function within six days, which increased by 29 days and was still substantial 57 days later. Numbers associated with bars denote animals studied at each time point. Blinded analysis. (**C**) Left ventricular samples from pigs receiving gene transfer 2 to 18 weeks before showed persistent increased cAMP generation compared to control animals, documenting a prolonged duration of effect of the transgene. NKH447 is a water soluble forskolin analog that directly stimulates adenylyl cyclase independently of β-adrenergic receptors and Gsα. Numbers associated with bars denote animals studied at each time point. (**D**) Ten to fourteen days after gene transfer, LV AC_{VI} protein content was increased two-fold ($P < 0.0007$) ($n = 4$ or 5 per group, as shown). Sucrose gradient centrifugation was used for left ventricular membrane preparation. GST-AC_{VI} was used to evaluate protein loading in each lane. *Abbreviations*: LV, left ventricle; AC, adenyl cyclase. *Source*: From Ref. 11.

Transcriptional Regulation and AC_{VI}

The translational physiological studies reviewed indicate that cardiac-directed expression of AC_{VI} increases stimulated cAMP production, improves heart function, and increases the period of survival in cardiomyopathy. In contrast, pharmacological agents that increase intracellular levels of cAMP have detrimental effects on cardiac function and survival. We wondered whether effects that are independent of cAMP might be responsible for these salutary outcomes associated with AC_{VI}

(A)

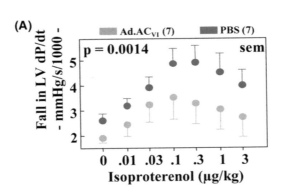

(B)

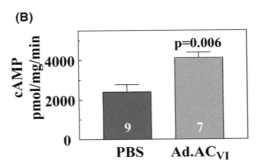

Figure 4 (**A**) Left ventricular peak +dP/dt. The y-axis represents mean values before pacing *minus* mean values obtained after 21 days of continuous pacing. PBS or Ad.AC$_{VI}$ was given after seven days of pacing, after documentation of heart failure. Pigs that received intracoronary Ad.AC$_{VI}$ (*light gray*) had less decrement in LV +dP/dt (less fall in contractile function) than pigs that received intracoronary PBS (*dark gray*). This was observed through a wide range of infused isoproterenol concentrations and was also evident when adenylyl cyclase was directly stimulated with NKH477 (*data not shown*). Peak negative LV dP/dt was also increased by Ad.AC$_{VI}$ (*data not shown*). (**B**) Left Ventricular cAMP Generation. Samples of LV from animals that had received Ad.AC$_{VI}$ showed increased cAMP generation in response to AC stimulation with NKH477 (10 μM), a water soluble forskolin analog that directly stimulates adenylyl cyclase independently of β-adrenergic receptors and Gsα. Bars represent mean values (net stimulation, i.e., basal subtracted), error bars denote SEM; numbers on bars denote number of animal per group. *Abbreviations*: AC, adenyl cyclase; phosphate buffered saline (PBS) *Source*: From Ref. 12.

expression. Studies on isolated cardiac myocytes were performed to elucidate the molecular mechanisms for these physiological changes.

Stimulation of cardiac myocytes with isoproterenol and forskolin (to increase cAMP generation) produced effects on gene transcription that were directionally opposite to that seen in the same cells after AC$_{VI}$ gene transfer whether unstimulated or stimulated with the same agents (Table 7). For example, while phospholamban, β$_1$AR, and ANF expression were increased by isoproterenol stimulation, they were decreased by AC$_{VI}$ gene transfer. Thus, AC$_{VI}$ expression appears to affect gene expression in a manner not recapitulated by cAMP generation alone.

In further studies (79), we found that gene transfer of AC$_{VI}$ downregulated mRNA and the protein expression of phospholamban, an inhibitor of SERCA2a, and that the CRE-like element in the phospholamban promoter was critical for

Table 6 LV Dimension, Function, and Response to Dobutamine

| | No dobutamine | | | | | | Dobutamine | | | | | | |
| | Ad.AC$_{VI}$ | | | PBS | | | Ad.AC$_{VI}$ | | | PBS | | | |
	Pre	21d	Abs Δ	Pre	21d	Abs Δ	Pre	21d	Abs Δ	Pre	21d	Abs Δ	P
EDD mm	48±1	61±3	13±2	45±1	62±2	18±2	45±1	58±4	13±4	42±1	62±2	19±2	0.043 0.073
ESD mm	28±1	50±3	22±3	25±1	53±2	28±2	21±1	43±4	22±4	18±1	48±2	30±2	0.009 0.71
Lat WTh %	68±4	32±4	36±8	71±4	20±7	51±6	71±5	51±7	20±11	83±7	36±9	47±12	0.048 0.34
IVS WTh %	57±5	49±4	8±5	57±2	46±6	11±6	67±5	52±5	14±9	65±4	54±5	11±8	0.94 0.66
FS %	40±1	18±2	23±3	43±1	14±2	29±2	53±2	26±4	27±5	58±2	22±3	36±3	0.03 0.13
Vcf circ/s	1.9±0.1	1.1±0.1	0.7±0.1	2.1±0.1	1.0±0.1	1.1±0.1	3.6±0.2	1.9±0.2	1.7±0.4	4.1±0.1	1.5±0.1	2.5±0.2	0.008 <0.0001
LV + dP/dt mmHg/s/1000	2.8±0.2	1.9±0.2	0.8±0.3	3.2±0.2	1.3±0.1	1.8±0.2	7.0±0.9	3.6±0.4	3.4±1.2	7.9±0.6	2.4±0.3	5.5±0.8	0.039 0.0002
LV − dP/dt mmHg/s/1000	−2.3±0.2	−2.2±0.2	0.1±0.4	−2.7±0.2	−1.6±0.1	1.1±0.2	−3.0±0.3	−2.8±0.4	0.2±0.5	−3.6±0.3	−2.1±0.1	1.5±0.4	0.005 0.462
Lat wall stress kdyn/cm²	50±3	211±31	161±29	52±2	298±37	246±38	31±3	143±34	112±31	24±2	197±23	173±24	0.031 0.069
IVS wall stress kdyn/cm²	52±3	78±5	27±6	50±4	148±9	98±10	32±3	96±14	65±12	24±2	110±9	86±9	<0.0001 0.18

Data entries (mean ± SEM) represent values before pacing (pre) and after 21 days of continuous pacing (21d) and absolute difference between these two values (Abs Δ). Ad.AC$_{VI}$ ($n = 6$); PBS ($n = 9$) except for the last four rows, where $n = 8$ for the PBS group. Data obtained with pacemakers inactivated. P, values, from two-way ANOVA, are for gene effect (top) and dobutamine effect (bottom).
Abbreviations: EDD, left ventricular end-diastolic diameter; ESD, left ventricular end-systolic diameter; FS, fractional shortening; Lat, WTh, systolic thickening of lateral wall; IVS WTh, systolic thickening of interventricular septum; Vcf, velocity of circumferential fiber shortening; LV +dP/dt, peak positive rate of left ventricular pressure development; LV −dP/dt, peak negative rate of left ventricular pressure development; kdyn, kilodynes.
Source: From Ref. 12.

Table 7 AC_{VI} Gene Transfer vs. Stimulation of cAMP Generation

	NKN or Iso stimulation	AC_{VI} gene transfer
PLB mRNA	↑	↓
β_1AR mRNA	↑	↓
ANF mRNA	↑	↓

Cultured neonatal cardiac myocytes underwent gene transfer with adenovirus encoding AC_{VI} or were stimulated with isoproterenol (Iso; 10 μM or NKH477 (a water soluble forskolin, analogue 10μM) 100 M). Gene expression of selected genes was evaluated using northern blotting.
Abbreviations: PLB, phopsholamban; β1AR, β1-adrenergic receptor; ANF, atrial natriuretic factor.

downregulation. Furthermore, AC_{VI} gene transfer was associated with increased expression of ATF3 protein, a suppressor of transcription, which binds to the phospholamban promoter and reduces its activity with subsequent reduced phospholamban expression (79) (Fig. 5). These findings indicate that AC_{VI} has effects on gene transcription that are not dependent on cAMP generation.

One of the elements of cardiac-directed AC expression that has been difficult to reconcile in the context of clinical heart failure trials is how such a strategy is protective when most other therapies that increase cAMP have been harmful. Studies of transcriptional regulation are likely to provide important insights. Favorable effects on heart function conferred by AC may not depend directly on cAMP generation, but may reflect instead alterations in transcription of genes that influence contractile function (53,79–81). Previous work has emphasized the importance of events distal to cAMP generation that are important determinants of left ventricular contractility. Reduced expression of phospholamban (79), increased expression of SERCA2a, increased myofilament sensitivity to calcium, and alterations in calcium handling (53) are examples of changes that may affect contractility, which do not directly or exclusively require changes in cAMP generation.

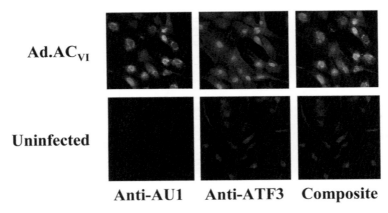

Figure 5 ATF3 protein localization after AC_{VI} gene transfer. Immunofluorescence staining with antiAU1 antibody for the expression of AC_{VI} (*green*) and anti ATF3 antibody (*red*) showed localization of AC_{VI} transgene in cell membrane and nuclear localization of ATF3 in cardiac myocytes overexpressing AC_{VI} (*top row*). Sparse amounts of endogenous AC_{VI} and ATF3 are seen (*bottom row*). *Source*: From Ref. 79. (*See color insert.*)

The precise molecular mechanisms by which AC_{VI} confers protection to the failing heart are unknown, but sufficient evidence indicates that the deleterious effects associated with other agents that increase intracellular stores of cAMP are circumvented by AC_{VI}. Several logical speculations can be rendered to provide possible explanations for these differences—effector rather than receptor-based gene transfer, reduced cAMP generation via inhibitory pathways that are simultaneously influenced by increased AC_{VI} expression, or effects of AC_{VI} on transcription of genes that alter contractile function. The answers to these important questions will require additional study in this evolving field.

SUMMARY AND FUTURE DIRECTIONS

Experimental studies suggest that gene transfer may be effective in attenuating deleterious left ventricular remodeling. Whether these effects will be ultimately applicable to clinical settings, and whether the net favorable effects will be more impressive than pharmacological agents, remains to be seen. However, the more specific mechanisms that could be exploited, the high local levels that could be achieved, and the potential to select a strategy that is tailored to a specific patient—all are potential advantages of gene transfer. The strategies by which one might reverse or attenuate left ventricular remodeling through gene transfer are better developed than are the vectors and delivery methods available to implement these strategies. But new vectors and delivery techniques will become available, and despite the somewhat disappointing initial course of clinical cardiovascular gene transfer—exacerbated by uncontrolled and unblinded studies that raised expectations too high—cardiovascular gene transfer is here to stay, and will likely be a major factor in cardiovascular therapeutics over the next twenty years.

For the attenuation of adverse left ventricular remodeling and—ultimately—its reversal, there are four gene transfer approaches that appear to be most promising: (i) angiogenic genes; (ii) genes that increase contractile function (phospholamban inhibition, SERCA2a, β-adrenergic receptor kinase inhibition and AC_{VI}); (iii) genes that reduce chronic apoptosis; and (iv) growth hormone and IGF-I.

ACKNOWLEDGMENTS

Drs. Hammond and Roth were supported by the NHLBI (1 P01 HL66941) and by Merit Awards from the Veteran's Administration.

REFERENCES

1. Hammond HK. Intracoronary gene transfer of fibroblast growth factor in experimental and clinical myocardial ischemia. Gene Therapy and Regulation 2002; 1:325–342.
2. French BA, Mazur W, Bolli R. Direct in vivo gene transfer into porcine myocardium using replication-deficient adenoviral vectors. Circulation 1994; 90:2414–2424.
3. Bonnetta L. Leukemia case triggers tighter gene–therapy controls. Nature Med 2002; 8:1189.
4. Fleury S, Simeoni E, Zuppinger C, Deglon N, von Segesser LK, Kappenberger L, Vassalli G. Multiply attenuated, self-inactivating lentiviral vectors efficiently deliver and express genes

for extended periods of time in adult rat cardiomyocytes in vivo. Circulation 2003; 107: 2375–2382.

5. Roth DM, Lai NC, Gao MH, Drumm MH, Jimenez J, Feramisco JR, Hammond HK. Indirect intracoronary delivery of adenovirus encoding adenylyl cyclase increases left ventricular contractile function in mice. Am J Physiol Heart Circ Physiol 2004; 287(1): H172–H177.

6. Grines CL, Watkins MW, Helmer G, Penny W, Brinker J, Marmur J, West A, Rade J, Marrott P, Hammond HK, Engler RE. Angiogenic gene therapy trial (AGENT): intra-coronary delivery of adenovirus encoding FGF-4 for patients with stable angina pectoris. Circulation 2003; 105:1291–1297.

7. Grines CL, Watkins MW, Mahmarian JJ, Iskandrian AE, Rade JJ, Marrott P, Pratt C, Kleiman N. A randomized, double–blind, placebo–controlled trial of Ad5FGF4 gene therapy and its effect on myocardial perfusion in patients with stable angina. J Amer Coll Cardiol 2003; 42:1339–1346.

8. Giordano F, Ping P, McKirnan MD, Nozaki S, Demaria AN, Dillmann W, Mathieu–Costello O, Hammond HK. Intracoronary gene transfer of fibroblast growth factor-5 increases blood flow and contractile function in an ischemic region of the heart. Nature Medicine 1996; 2:534–539.

9. Gao MH, Lai NC, McKirnan MD, Roth DA, Rubanyi GM, Dalton N, Roth DM, Hammond HK. Increased regional function and perfusion after intracoronary delivery of adenovirus encoding fibroblast growth factor 4: report of preclinical data. Hum Gene Ther 2004; 15:574–587.

10. McKirnan MD, Lai NC, Waldman L, Dalton N, Guo X, Roth DA, Hammond HK. Intracoronary gene transfer of fibroblast growth factor-4 increases regional contractile function and responsiveness to adrenergic stimulation in heart failure. Cardiac and Vas-cular Regeneration 2000; 1:11–21.

11. Lai NC, Roth DM, Gao MH, Fine S, Head BP, Zhu J, McKirnan MD, Kwong C, Dal-ton N, Urasawa K, et al. Intracoronary delivery of adenovirus encoding adenylyl cyclase VI increases left ventricular function and cAMP-generating capacity. Circulation 2000; 102:2396–2402.

12. Lai NC, Roth DM, Gao MH, Tang T, Dalton N, Lai YY, Spellman M, Clopton P, Hammond HK. Intracoronary adenovirus encoding AC VI increases left ventricular function in heart failure. Circulation 2004; 110:330–336.

13. Tanaka E, Hattan N, Ando K, Ueno H, Sugio Y, Mohammed MU, Voltchikhina SA, Mori H. Amelioration of microvascular myocardial ischemia by gene transfer of vascular endothelial growth factor in rabbits. J Thorac Cardio Surg 2000; 120:720–728.

14. Suzuki G, Lee T-C, Fallavollita JA, Canty JM. Adenoviral gene transfer of FGF-5 to hibernating myocardium improves function and stimulates myocytes to hypertrophy and reenter the cell cycle. Circ Res 2005; 96:767–775.

15. Muhlhauser J, Jones M, Yamada I, Cirielli C, Lemarchand P, Gloe TR, Bewig B, Signoretti S, Crystal RG, Capogrossi MC. Safety and efficacy of in vivo gene transfer into the porcine heart with replication-deficient, recombinant adenovirus vectors. Gene Ther 1996; 3:145–153.

16. Barr E, Carroll J, Kalynych AM, Tripathy SK, Kozarsky K, Wilson JM, Leiden JM. Gene transfer into the heart using replication-defective adenovirus. Gene Therapy 1994; 1:51–58.

17. Shah AS, Lilly RE, Kypson AP, Tai O, Hata JA, Pippen A, Silvestry SC, Lefkowitz RJ, Glower DD, Koch WJ. Intracoronary adenovirus-mediated delivery and overexpression of the beta(2)-adrenergic receptor in the heart: prospects for molecular ventricular assis-tance. Circulation 2000; 101:408–414.

18. Shah AS, White DC, Emani S, Kypson AP, Lilly RE, Wilson K, Glower DD, Lefkowitz RJ, Koch WJ. In vivo ventricular gene delivery of a beta-adrenergic receptor kinase inhibitor to the failing heart reverses cardiac dysfunction. Circulation 2001; 103: 1311–1316.

19. Tevaearai HT, Eckhart AD, Shotwell KF, Wilson K, Koch WJ. Ventricular dysfunction after cardioplegic arrest is improved after myocardial gene transfer of a beta-adrenergic receptor kinase inhibitor. Circulation 2001; 104:2069–2074.

20. Weig HJ, Laugwitz KL, Moretti A, Kronsbein K, Stadele C, Bruning S, Seyfarth M, Brill T, Schomig A, Ungerer M. Enhanced cardiac contractility after gene transfer of V2 vasopressin receptors in vivo by ultrasound-guided injection or transcoronary delivery. Circulation 2000; 101:1578–1585.

21. Donahue JK, Heldman AW, Fraser H, McDonald AD, Miller JM, Rade JJ, Eschenhagen T, Marban E. Focal modification of electrical conduction in the heart by viral gene transfer. Nature Med 2000; 6:1395–1398.

22. Iwanaga Y, Hoshijima M, Gu Y, Iwatate M, Dieterle T, Ikeda Y, Date MO, Chrast J, Matsuzaki M, Peterson KL, Chien KR, Ross J Jr. Chronic phospholamban inhibition prevents progressive cardiac dysfunction and pathological remodeling after infarction in rats. J Clin Invest 2004; 113:727–736.

23. Kaspar BK, Roth DM, Lai NC, Drumm JD, Erickson DA, McKirnan MD, Hammond HK. Myocardial gene transfer and long-term expression following intracoronary delivery of adeno-associated virus. J Gene Medicine 2005; 7:316–324.

24. Roth DM, Lai NC, Gao MH, Fine S, McKirnan MD, Roth DA, Hammond HK. Nitroprusside increases gene transfer associated with intracoronary delivery of adenovirus. Human Gene Ther 2004; 15:989–994.

25. Fazio S, Sabatini D, Capaldo B, Vigorito C, Giordano A, Guida R, Pardo F, Biondi B, Sacca L. A preliminary study of growth hormone in the treatment of dilated cardiomyopathy. N Engl J Med 1996; 334:809–814.

26. Duerr RL, McKirnan MD, Gim RD, Clark RG, Chien KR, Ross J Jr. Cardiovascular effects of insulin-like growth factor-1 and growth hormone in chronic left ventricular failure in the rat. Circulation 1996; 93:2188–2196.

27. Cittadini A, Isgaard J, Monti MG, Casaburi C, Di Gianni A, Serpico R, Iaccarino G, Sacca L. Growth hormone prolongs survival in experimental postinfarction heart failure. J Am Coll Cardiol 2003; 41:2154–2163.

28. Melmud S, Jackson I, Kleinberg D, Klibanski A. Current treatment guidelines for acromegaly. J Clin Endocrinol Metab 1998; 83:2646–2652.

29. Donath MY, Jenni R, Brunner HP, Anrig M, Kohli S, Glatz Y, Froesch ER. Cardiovascular and metabolic effects of insulin-like growth factor I at rest and during exercise in humans. J Clin Endocrinol Metab 1996; 81:4089–4094.

30. Chan JM, Stampfer MJ, Giovannucci E, Gann PH, Ma J, Wilkinson P, Hennekens CH, Pollak M. Plasma insulin-like growth factor-I and prostrate cancer risk: a prospective study. Science 1998; 279:563–566.

31. Jayasankar V, Pirolli TJ, Bish LT, Berry MF, Burdick J, Grand T, Woo YJ. Targeted overexpression of growth hormone by adenoviral gene transfer preserves myocardial function and ventricular geometry in ischemic cardiomyopathy. J Mol Cell Cardiol 2004; 36:531–538.

32. Chao W, Matsui T, Novikov MS, Tao J, Li L, Liu H, Ahn Y, Rosenzweig A. Strategic advantages of insulin-like growth factor-I expression for cardioprotection. J Gene Med 2003; 5:277–286.

33. Jayasankar V, Woo YJ, Bish LT, Pirolli TJ, Chatterjee S, Berry MF, Burdick J, Gardner TJ, Sweeney HL. Gene transfer of hepatocyte growth factor attenuates postinfarction heart failure. Circulation 2003; 108(suppl 1):II230–II236.

34. Helmer GA, McKirnan MD, Shabetai R, Boss GR, Ross J, Hammond HK. Regional deficits of myocardial blood flow and function left ventricular pacing-induced heart failure. Circulation 1996; 94:2260–2267.

35. Parodi O, De Maria R, Oltrona L, Testa R, Sambuceti G, Roghi A, Merli M, Belingheri L, Accinni R, Spinelli F. Myocardial blood flow distribution in patients with ischemic heart disease or dilated cardiomyopathy undergoing heart transplantation. Circulation 1993; 88:509–522.

36. Leotta E, Patejunas G, Murphy G, Szokol J, McGregor L, Carbray J, Hamawy A, Winchester D, Hackett N, Crystal R, et al. Gene therapy with adenovirus-mediated myocardial transfer of vascular endothelial growth factor 121 improves cardiac performance in a pacing model of congestive heart failure. J Thorac Cardiovasc Surg 2002; 123: 1101–1113.

37. Takahashi K, Ito Y, Morikawa M, Kobune M, Huang J, Tsukamoto M, Sasaki K, Nakamura K, Dehari H, Ikeda K, et al. Adenoviral-delivered angiopoietin-1 reduces the infarction and attenuates the progression of cardiac dysfunction in the rat model of acute myocardial infarction. Mol Ther 2003; 8:584–592.

38. Chatterjee S, Stewart AS, Bish LT, Jayasankar V, Kim EM, Pirolli T, Burdick J, Woo YJ, Gardner TJ, Sweeney HL. Viral gene transfer of the antiapoptotic factor Bcl-2 protects against chronic postischemic heart failure. Circulation 2002; 106(12 suppl 1):I212–I217.

39. Regula KM, Ens K, Kirshenbaum LA. Mitochondria-assisted cell suicide: a license to kill. J Mol Cell Cardiol 2003; 6:559–567.

40. Minamisawa S, Hoshijima M, Chu G, Ward CA, Frank K, Gu Y, Martone ME, Wang Y, Ross J Jr, Kranias EG, et al. Chronic phospholamban-sarcoplasmic reticulum calcium ATPase interaction is the critical calcium cycling defect in dilated cardiomyopathy. Cell 1999; 99:313–322.

41. del Monte F, Williams E, Lebeche D, Schmidt U, Rosenzweig A, Gwathmey JK, Lewandowski ED, Hajjar RJ. Improvement in survival and cardiac metabolism after gene transfer of sarcoplasmic reticulum $Ca^{(2+)}$-ATPase in a rat model of heart failure. Circulation 2001; 104:1424–1429.

42. Suarez J, Belke DD, Gloss B, Dieterle T, McDonough PM, Kim YK, Brunton LL, Dillmann WH. In vivo adenoviral transfer of sorcin reverses cardiac contractile abnormalities of diabetic cardiomyopathy. Am J Physiol Heart Circ Physiol 2004; 286(1):H68–H75.

43. Sunahara RK, Dessauer CW, Gilman AG. Complexity and diversity of mammalian adenylyl cyclases. Annu Rev Pharmacol Toxicol 1996; 36:461–480.

44. Defer N, Best-Belpomme M, Hanoune J. Tissue specificity and physiological relevance of various isoforms of adenylyl cyclase. Am J Physiol 2000; 279:F400–F416.

45. Ishikawa Y, Sorota S, Kiuchi K, Shannon RP, Komamura K, Katsushika S, Vatner DE, Vatner SF, Homcy CJ. Downregulation of adenylylcyclase types V and VI mRNA levels in pacing-induced heart failure in dogs. J Clin Invest 1994; 93:2224–2229.

46. Ping P, Anzai T, Gao M, Hammond HK. Adenylylcyclase and G protein receptor kinase expression during the development of heart failure. Am J Physiol 1997; 273:H707–H717.

47. Gao M, Ping P, Post S, Insel PA, Tang R, Hammond HK. Increased expression of adenylylcyclase type VI proportionately increases β-adrenergic receptor-stimulated cAMP in neonatal rat cardiac myocytes. Proc Natl Acad Sci (USA) 1998; 95:1038–1043.

48. Gao M, Lai NC, Roth DM, Zhou J, Anzai T, Dalton N, Hammond HK. Adenylylcyclase increases responsiveness to catecholamine stimulation in transgenic mice. Circulation 1999; 99:1618–1622.

49. Iwase M, Uechi M, Vatner DE, Asai K, Shannon RP, Kudej RK, Wagner TE, Wight DC, Patrick TA, Ishikawa Y, et al. Cardiomyopathy induced by cardiac Gsα overexpression. Am J Physiol 1997; 272:H585–H589.

50. Engelhardt S, Hein L, Wiesmann F, Lohse MJ. Progressive hypertrophy and heart failure in beta1–adrenergic receptor transgenic mice. Proc Natl Acad Sci 1999; 96: 7059–7064.

51. Roth DM, Gao MH, Lai NC, Drumm J, Dalton N, Zhou JY, Zhu J, Entrikin D, Hammond HK. Cardiac-directed adenylyl cyclase expression improves heart function in murine cardiomyopathy. Circulation 1999; 99:3099–3102.

52. Roth DM, Bayat H, Drumm JD, Gao MH, Swaney JS, Ander A, Hammond HK. Adenylyl cyclase increases survival in cardiomyopathy. Circulation 2002; 105:1989–1994.

53. Tang T, Gao MH, Roth DM, Guo T, Hammond HK. AC type VI corrects cardiac sarcoplasmic reticulum calcium uptake defects in cardiomyopathy. Am J Physiol 2004; 287:H1096–H1112.

54. Kawabe J, Ebina T, Toya Y, Oka N, Schwencke C, Duzic E, Ishikawa Y. Regulation of type V adenylyl cyclase by PMA-sensitive and-insensitve protein kinase C isoenzymes in intact cells. FEBS Lett 1996; 384:273–276.

55. Chern Y. Regulation of adenylyl cyclase in the central nervous system. Cellular Signaling 2000; 12:195–204.

56. Lai HL, Yang TH, Messing RO, Ching YH, Lin SC, Chern Y. Protein kinase C inhibits adenylyl cyclase type VI activity during desensitization of the A2a-adenosine receptor-mediated cAMP response. J Biol Chem 1997; 272:4970–4977.

57. Chen Z, Nield HS, Sun H, Barbier A, Patel TB. Expression of type V adenylyl cyclase is required for epidermal growth factor-mediated stimulation of cAMP accumulation. J Biol Chem 1995; 270:27525–27530.

58. Port JD, Bristow MR. Altered beta-adrenergic receptor gene regulation and signaling in chronic heart failure. J Mol Cell Cardiol 2001; 33:887–905.

59. Scheuer J. Catecholamines in cardiac hypertrophy. Am J Cardiol 1999; 83:70H–74H.

60. Morisco C, Zebrowski DC, Vatner DE, Vatner SF, Sadoshima J. β-adrenergic cardiac hypertrophy is mediated primarily by the β1-subtype in the rat. J Mol Cell Cardiol 2001; 33:561–573.

61. Schafer M, Frischkopf K, Taimor G, Piper HM, Schluter KD. Hypertrophic effect of selective β1–adrenoreceptor stimulation on ventricular cardiomyocytes from adult rat. Am J Physiol 2000; 279:C495–C503.

62. Engelhardt S, Hein L, Keller U, Klambt K, Lohse MJ. Inhibition of Na^+-H^+ exchange prevents hypertrophy, fibrosis and heart failure in β1-adrenergic receptor transgenic mice. Circ Res 2002; 90:814–819.

63. Bisognano JD, Weinberger HD, Bohlmeyer TJ, Pende A, Raynolds MV, Sastravaha A, Roden R, Asano K, Blaxall BC, Wu SC, et al. Myocardial-directed overexpression of the human $β_1$–adrenergic receptor in transgenic mice. J Mol Cell Cardiol 2000; 32:817–830.

64. Liggett SB, Tepe NM, Lorenz JN, Canning AM, Jantz TD, Mitarai S, Yatani A, Dorn GW II. Early and delayed consequences of $β_2$-adrenergic receptor overexpression in mouse hearts: critical role for expression level. Circulation 2000; 101:1707–1714.

65. Iwase M, Bishop SP, Uechi M, Vatner DE, Shannon RP, Kudej RK, Wight DC, Wagner TE, Ishikawa Y, Homcy CJ, Vatner SF. Adverse effects of chronic endogenous sympathetic drive induced by cardiac Gsα overexpression. Circ Res 1996; 78:517–524.

66. Steinberg SF. The molecular basis for distinct β-adrenergic receptor subtype actions in cardiomyocytes. Circ Res 1999; 85:1101–1111.

67. Condorelli G, Drusco A, Stassi G, Bellacosa A, Roncarati R, Iaccarino G, Russo MA, Gu Y, Dalton N, Chung C, et al. Akt induces enhanced myocardial contractility and cell size in vivo in transgenic mice. Proc Natl Acad Sci USA 2002; 99:12333–12338.

68. Morisco C, Zebrowski D, Condorelli G, Tsichlis P, Vatner SF, Sadoshima J. The Akt-glycogen synthase kinase 3β pathway regulates transcription of atrial natriuretic factor induced by β–adrenergic receptor stimulation in cardiac myocytes. J Biol Chem 2000; 275:14466–14475.

69. Molkentin JD, Dorn GW II. Cytoplasmic signaling pathways that regulate cardiac hypertrophy. Annu Rev Physiol 2001; 63:391–426.

70. Minamino T, Yujiri T, Terada N, Taffet GE, Michael LH, Johnson GL, Schneider MD. MEKK1 is essential for cardiac hypertrophy and dysfunction induced by Gq. Proc Natl Acad Sci USA 2002; 99:3866–3871.

71. D'Angelo DD, Sakata Y, Lorenz JN, Boivin GP, Walsh RA, Liggett SB, Dorn GW. Transgenic Gαq overexpression induces cardiac contractile failure in mice. Proc Natl Acad Sci USA 1997; 94:8121–8126.

72. Dorn GW, Tepe NM, Lorenz JN, Koch WJ, Liggett SB. Low-and high-level transgenic expression of β2-adrenergic receptors differentially affect cardiac hypertrophy and function in Gαq-overexpressing mice. Proc Natl Acad Sci USA 1999; 96:6400–6405.

73. Tepe NM, Liggett SB. Transgenic replacement of type V adenylyl cyclase identifies a critical mechanism of beta-adrenergic receptor dysfunction in the G alpha q overexpressing mouse. FEBS Lett 1999; 458:236–240.

74. Rogers JH, Tsirka A, Kovacs A, Blumer KJ, Dorn GW II, Muslin AJ. RGS4 reduces contractile dysfunction and hypertrophic gene induction in Gαq overexpressing mice. J Mol Cell Cardiol 2001; 33:209–218.

75. Wu G, Toyokawa T, Hahn H, Dorn GW II. ε Protein kinase C in pathological myocardial hypertrophy. J Biol Chem 2000; 275:29927–29930.

76. Widmann C, Gibson S, Jarpe MB, Johnson GL. Mitogen activated protein kinase: conservation of a three-kinase module from yeast to human. Physiol Rev 1999; 79:143–180.

77. Esposito G, Prasad SV, Rapacciuolo A, Mao L, Koch WJ, Rockman HA. Cardiac overexpression of a Gq inhibitor blocks induction of extracellular signal-regulated kinase and c-Jun NH_2-terminal kinase activity in in vivo pressure overload. Circulation 2001; 103: 1453–1458.

78. Xiao L, Shen M, Colucci W. Cross talk between β-adrenergic receptor (AR) and endothelin receptor (ET-R)-induced hypertrophic signaling in adult rat ventricular myocytes (ARVM). FASEB J 2003; 17:A210, 142.5.

79. Gao MH, Tang T, Guo T, Sun S, Feramisco JR, Hammond HK. AC type VI gene transfer reduces phospholamban expression in cardiac myocytes via activating transcription factor 3. J Biol Chem 2004; 279:38797–38802.

80. Yue P, Long CS, Austin R, Chang KC, Simpson PC, Massie BM. Postinfarction heart failure in the rat is associated with distinct alterations in cardiac myocyte molecular phenotype. J Mol Cell Cardiol 1998; 8:1615–1630.

81. Lowes BD, Gilbert EM, Abraham WT, Minobe WA, Larrabee P, Ferguson D, Wolfel EE, Lindenfeld J, Tsvetkova T, Robertson AD, et al. Myocardial gene expression in dilated cardiomyopathy treated with beta-blocking agents. N Engl J Med 2002; 346:1357–1365.

26

Stem Cell/Myoblast Transplantation for Myocardial Regeneration

Donald Orlic

Cardiovascular Branch, National Heart, Lung, and Blood Institute, National Institutes of Health, Bethesda, Maryland, U.S.A.

Marc S. Penn

Departments of Cardiovascular Medicine and Cell Biology, Cleveland Clinic Foundation, Cleveland, Ohio, U.S.A.

INTRODUCTION

Embryonic and adult stem cells are being investigated for their potential use in regenerative medicine. During the last five years animal studies suggest that adult stem cells undergo multiple patterns of development by a process referred to as plasticity or transdifferentiation. Data are extensive and reports on myocardial regeneration using adult bone marrow stem cells provide some of the most convincing evidence in support of stem cell plasticity. Studies in rodents suggest that adult bone marrow stem cells, directly transplanted into myocardium or cytokine-mobilized, can infiltrate acute infarctions and regenerate healthy tissue, leading to significant preservation of myocardial function. These data have suggested that there is great potential for adult stem cell therapy to prevent and treat congestive heart failure, a condition that affects over 10% of Americans over the age of 65 years.

Estimates of human cardiomyocyte renewal based on postmortem microscopic analysis of Y-chromosome positive cardiomyocytes in orthotopic female hearts transplanted into males range from 0.016% to 9%. This observation is believed to be the result of host stem cell differentiation into cardiomyocytes but host cell/cardiomyocyte fusion is considered as an alternate explanation. Although little is known regarding possible mechanisms that enable stem cells to differentiate into cardiomyocytes, clinical protocols are in progress at several medical centers throughout the world. Autologous bone marrow stem cells are being tested for their ability to regenerate myocardium following acute myocardial infarction. To date, the trials are of short term and do not involve adequate numbers of patients or controls; however, preliminary results suggest that transplantation of autologous bone marrow stem cells leads to improved cardiac output. Importantly, for this nascent therapy, the protocol appears to be safe with no procedural related deaths.

Additional clinical trials involve the transplantation of autologous skeletal myoblasts into the zone of infarction at the time of coronary artery bypass grafting. The myoblasts appear to engraft and although cardiac function improves, because of the manner in which the trials to date have been designed, it is difficult to distinguish benefits derived from the transplanted myoblasts versus benefits due to bypass surgery.

To improve upon these early outcomes, additional animal experiments are required to identify the environmental cues that induce stem cell trafficking and engraftment into injured myocardium and their differentiation into cardiomyocytes. Further understanding of the role of stem cells in myocardial regeneration requires the development of instrumentation and technologies to monitor myocardial repair over time in large animal models and patients. In this chapter, we review the current status of bone marrow stem cell and skeletal myoblast repair of infarcted myocardium. It is anticipated that these protocols may become a reliable clinical therapy for cardiac patients suffering from acute myocardial infarction in order to prevent remodeling and cardiac myocyte loss, as well as to develop clinical strategies for the treatment of congestive heart failure.

EMBRYONIC ORIGIN OF STEM CELLS

At the center of any regenerative process is the cell of origin. Thus, stem cells are a key element in tissue renewal. Although all cells originate from the fertilized oocyte, cellular characteristics that define tissue specificity emerge at an early stage of embryonic development. This leads to the formation of three embryonic germ layers, namely, the ectoderm, endoderm, and mesoderm. Developmental biologists early in the last century and particularly in the 1920s and 1930s (1) traced the migratory pathways of cells from these germ layers to their ultimate organ distribution. This led to the theory that similar tissue layers in multiple fetal and adult organs are derived from specific germ layers. Accordingly, stem cells of ectodermal origin exist in epithelial and neural tissues and stem cells of endodermal origin populate organs such as liver and gut. The stem cells of mesodermal origin give rise to cartilage, bone and muscle, and the hemangioblasts of bone marrow. It was recently demonstrated that hemangioblasts are derived in vitro from embryonic stem cells of the early blastocyst (2). As a consequence of their dual developmental potential, hemangioblasts are able to generate hematopoietic stem cells (HSC) and endothelial progenitor cells (EPC) (3–6).

The cornerstone of developmental biology that stem cells are tissue specific is now being re-analyzed based on recent data that support a process referred to as cellular plasticity or transdifferentiation. According to these findings, adult stem cells can be programmed to differentiate into cells of alternate tissues even across germ layer boundaries.

BIOLOGY OF ADULT STEM CELLS

In the adult, tissue renewal is a normal, continuous process accomplished through the activity of stem/progenitor cells. Stem or progenitor cells have been identified in numerous tissues and structures, including hair follicles of epidermis, mucosal glands of the gastrointestinal organs, skeletal muscle, and periosteum of compact bone, to name a few. All of these tissues are capable of repair following injury. In contrast, it is generally believed that the brain and heart are postmitotic organs that lack the capacity for self-renewal.

The existence of neural stem cells is now widely accepted (7,8) and evidence is accumulating in support of the existence of cardiac stem cells (9,10). Although the ability of these newly identified stem cell populations to regenerate neural and cardiac tissues has not been widely established, many researchers believe that we are on the verge of discoveries that will eventually lead to clinical reversal of debilitating neural and cardiac disorders.

MULTIPLE STEM CELL POPULATIONS EXIST IN BONE MARROW

Bone marrow stem cells are exceedingly rare. However, several well-established stem cell populations in bone marrow are being investigated, with the most clearly defined, to date, being the HSC that gives rise to the hierarchy of developing blood cells. This stem cell population can be enriched using antibodies that recognize specific surface markers. Mouse HSC are phenotypically classified as Lineage$^-$ c-kit$^+$ Sca-1$^+$, whereas the human HSC phenotype is Lineage$^-$ CD34$^+$ CD38$^-$. Based largely on theoretical considerations, it was proposed in the 1960s and 1970s that, throughout adult life, asymmetrical HSC divisions provide one cell for hematopoietic differentiation and one for maintenance of the HSC population (11).

Endothelial progenitor cells are the immediate ancestors of the endothelial cells that are involved in neovascularization and angiogenesis. They can be isolated from bone marrow and, in small numbers, from peripheral blood (12). Their phenotype includes the markers CD34, CD31, and von Willebrand factor.

HSC exist primarily in the bone marrow and enter the circulation at a low rate that can be significantly enhanced by cytokine treatment (13–17). In response to recombinant human granulocyte–colony stimulating factor (rhG-CSF), a cascade of enzymatic activity occurs within the bone marrow, resulting in the severance of docking proteins and the mobilization of HSC into the circulation (18–20) (Fig. 1). HSC have the capacity to interact with environmental cues through surface receptors. A number of chemokines and cytokines provide specific signals that induce HSC homing and differentiation.

Until recently, stromal cells of bone marrow were believed to serve a structural function and to secrete proteins for assembly into extracellular fibers. However, these cells, also designated as mesenchymal stem cells (MSC) (21–23), are now considered by many to be capable of multilineage differentiation. In vitro they can maintain an undifferentiated phenotype over many generations, but when cultured with the DNA demethylating agent 5-azacytidine, MSC can differentiate into cardiomyocytes. Because MSC may avoid detection by the host immune system, allogeneic bone marrow MSC may have potential clinical utility.

Another subset of multipotent bone marrow cells referred to as mesodermal adult cells or multipotent adult progenitor cells (MAPC) has been described (24). Multipotent adult progenitor cells co-purify with MSC. They proliferate extensively and differentiate into cells of all three germ layers in vitro. When injected into rodents they reconstitute bone marrow, liver, gut, lungs, and endothelium, and recently there is also evidence to suggest that they form new cells in the heart (25). Similar to MSC, MAPC may avoid detection by the host immune system; thus, allogeneic MAPC may have potential clinical utility as well.

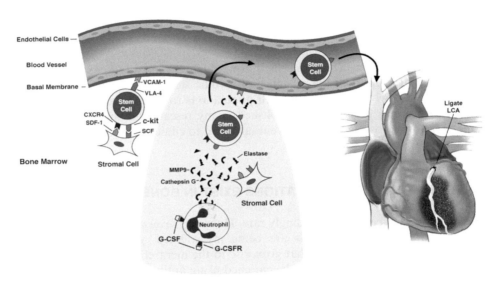

Figure 1 In the steady state, bone marrow stem cells are tethered to structural components in bone marrow, including stromal cells and vascular components. The ligand-receptor bindings that serve as docking proteins are severed during stem cell mobilization when cytokines, such as G-CSF, interact with the G-CSF receptor on neutrophils resulting in the release of enzyme containing cytoplasmic granules into the bone marrow environment. Unbound bone marrow stem cells enter the circulation and, as suggested in several studies, may traffic to sites of tissue injury in response to chemotactic signals such as SDF-1 and SCF. *Abbreviations*: SDF-1, stromal cell derived factor-1; SCF, stem cell factor.

CYTOKINES MOBILIZE STEM CELLS INTO PERIPHERAL BLOOD

Under steady-state conditions, approximately 0.01% of the total HSC population in mouse bone marrow enters the circulation at any given moment (15). Although this number is extremely low, circulating HSC may nevertheless have a role in homeostasis since their daily turnover rate is high with an estimated 10,000 cells passing through the bloodstream daily (16). The observation that the number of stem cells in the circulation can be significantly increased following multiple cytokine injections has been the basis for numerous laboratory and clinical trials. In mice, HSC mobilization in response to cytokine therapy has been quantified in a competitive repopulation assay. When recombinant rat stem cell factor (rrSCF) was injected into adult mice for seven days (26), the number of HSC increased 10-fold and 16-fold in blood and spleen, respectively, and decreased 2.7-fold in bone marrow. From this it appeared that the spleen functioned as a "sink" for mobilized HSC. In a subsequent study, mice that had undergone splenectomy two weeks prior to treatment with rrSCF and rhG-CSF had a 250-fold increase in HSC, from 29 to 7200, in total blood volume after five daily injections (15). In addition to HSC mobilization, white blood cell counts rose to 40,000 to 80,000 per microliter.

The ultimate distribution of the mobilized HSC in mice is not entirely clear but there are indications that, in addition to a return to bone marrow, some circulating HSC repopulate nonhematopoietic organs. Skeletal muscle is a known repository of HSC from bone marrow (27–29). Interestingly, even after a period of residence in skeletal muscle, HSC retain the capacity to reconstitute bone marrow and generate blood cells when transplanted into myeloablated mice. It also appears that bone marrow–derived HSC in skeletal muscle convert to satellite cells with myogenic properties (30,31).

Mobilization experiments were also done in a nonhuman primate model based on an assay for $Lin^- CD34^+$ stem/progenitor cells. Baboons were injected with rhG-CSF alone or in combination with rhSCF for five days and mobilized progenitor cells were harvested by apheresis. Both treatments resulted in an increase in progenitor colony forming cells (CFC) (32,33) but the combination of rhSCF and rhG-CSF resulted in mobilization of 14-fold more CFC than rhG-CSF alone. These striking increases in total stem cells in the blood of adult mice and baboons have created new research opportunities in the field of plasticity involving myocardial regeneration.

An alternate approach to cytokine-induced mobilization of bone marrow stem cells involves the reagent AMD-3100. This small molecular weight symmetrical bicyclam is an antagonist of the CXCR4 receptor (34). It interferes with the docking of $CXCR4^+$ stem cells to stromal cell derived factor-1 (SDF-1) in bone marrow and leads to their transit from bone marrow into the circulation. In addition, AMD-3100 does not induce high level mobilization of white blood cells and platelets as seen with cytokine treatment. A single dose of 80 mcg/kg of AMD-3100 results in a four-fold increase in circulating $CD34^+$ cells at six hours after injection (35). This rapid mobilization following a single dose of AMD-3100 may be superior to the rrSCF and rhG-CSF protocol, which requires five daily injections to achieve peak circulating stem cells since irreversible changes may occur in damaged tissues within five days of the onset of ischemia (36).

BONE MARROW STEM CELL REGENERATION OF TISSUES FOLLOWING ISCHEMIC INJURY

Bone marrow stem cells have long been considered to be multipotent based on their capacity to generate such diverse populations as erythrocytes, granulocytes, lymphocytes, and platelets. Recently, however, evidence suggests that HSC or an alternate population in bone marrow is involved in a transdifferentiation phenomenon. According to this hypothesis, environmental factors play a major role in determining the developmental fate of bone marrow stem cells. This is based, in part, on the response of stem cells to specific environmental factors in bone marrow and thymus, where stem cells give rise to erythrocytes or T-lymphocytes, respectively. We propose, without much evidence (Fig. 2), that bone marrow stem cells respond to environmental factors in heart, skeletal muscle, cartilage, and bone and generate, unexpectedly, non hematopoietic cell types.

Ischemia of even short duration can lead to severe tissue damage. The study of ischemic injury has been the focus of a series of studies that involve tissue regeneration. This hypoxic model is highly amenable to experimental analysis and mimics the clinical condition because the oxygen supply to a specific site can be interrupted by occlusion of an afferent blood vessel and, if desired, can be followed by reperfusion when the ligature is removed.

In several studies of tissue regeneration, adult mice were myeloablated and subsequently reconstituted with transplanted, genetically marked HSC. Reconstituted β-galactosidase$^+$ chimeric mice were subjected to tubule damage by occlusion of the renal artery. A transient 30-minute ischemic episode resulted in necrosis and apoptosis of the epithelium in the proximal tubules of the outer medulla. Presumed signals from the injured region induced bone marrow stem cells to home to the injured tubules,

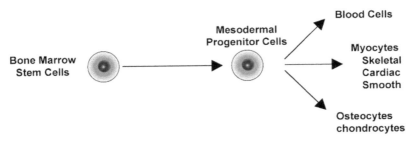

Figure 2 A number of recent studies propose that one or several stem cell populations with primitive developmental properties exist in adult bone marrow and peripheral blood. It is suggested that stromal/mesenchymal cells, hemangioblasts, or perhaps hematopoietic stem cells can be triggered to give rise to various progenitor cells that have the capacity to differentiate into multiple mesodermal derived cell types.

where they differentiated into β-galactosidase positive epithelial lining cells (37). The regenerated tubules reacquired the capacity to maintain urea nitrogen balance.

Local hypoxia has also been utilized to study the neovascularization of a retinal capillary network in chimeric adult mice previously reconstituted with bone marrow cells expressing enhanced green fluorescence protein (eGFP). Brief exposure to a highly focused argon laser beam resulted in photocoagulation in a limited region of retinal vessels. Regenerating eGFP[+] capillaries were observed in the region of injury within three weeks (5). In a follow-up study, human umbilical cord blood CD34[+] stem cells were used to generate xenogenic bone marrow in mice (6) and then tested for their ability to migrate to photocoagulated retinal vasculature. Regenerating vascular endothelium displayed human specific proteins, indicating the hemangioblast potential of the transplanted human stem cells.

A similar regenerative potential was reported when human cord blood CD34[+] stem cells were injected directly into the ischemic adductor muscles of cyclosporine-A immunosuppressed mice (38). The CD34[+] stem cells were recruited for angiogenesis and skeletal myogenesis. In a clinical trial, autologous bone marrow mononuclear cells injected into the gastrocnemius muscle of patients with chronic limb ischemia improved the ankle-brachial index and foot ulcers began to heal (39).

Taken together, these investigations provide some evidence for a potential activity of murine and human bone marrow, and peripheral blood stem cells to not only form hematopoietic cells and vascular endothelium, but also epithelium, myocytes, and other cell types. A critical condition that appears to enhance the regenerative capacity of bone marrow stem cells is the need for ischemia-induced tissue injury.

MYOCARDIAL INFARCTION

Myocardial infarction following occlusion of a coronary artery is a process that transitions from reversible injury at 15 to 30 minutes to subsequent irreversible injury (40,41). In the region distal to a coronary artery blockage, apoptotic cell death, necrosis, and scar formation occur in a progressive manner over a period of two weeks. Initially, a decrease in metabolic function can be observed in the subendocardium (42). If the perfusion defect is prolonged the zone of damaged myocardium will be extensive. After the initial event, impaired global function can lead to ventricular remodeling of the healthy myocardium, further reducing cardiac function (43). This can occur over a period of several weeks to several months after vessel occlusion in clinical populations (44).

In laboratory experiments, myocardial infarctions can be achieved by occlusion of a coronary vessel or through cryoinjury. Ligation of a major artery, usually the left anterior descending artery in mice, produces a visible infarction within a few hours. In mice, the blockage is often maintained for the duration of the experiment, but in larger animals, it is common to remove the ligature to obtain reperfusion of the injured tissue after one to two hours. This brief period of ischemia is sufficient to produce a discernable change in the contractile pattern in the zone of infarction. Alternatively, infarctions can be induced with a liquid nitrogen cryoprobe. The probe is placed on the epicardial surface of the left ventricle (LV) and the freezing temperature produces an area of dead myocardium. The immediate destruction of myocytes and coronary vessels in cryoinjury highlights a likely disadvantage to this approach. Any homing signals that would be released into the extracellular environment of a ligation-induced infarction by apoptotic myocytes and/or fibroblasts may not be released by instantly frozen cardiomyocytes. In support of this concept, there is an emerging body of evidence, which suggests that cytokine and chemokine signals may be generated during acute myocardial infarction (45) and that SDF-1 may be required for cardiac stem cell migration to the site of infarction (36).

MYOCARDIAL REGENERATION

Current therapies, including angioplasty and medication, improve the prognosis of patients following a myocardial infarction but do not lead to regeneration of dead myocardium. It is generally believed that heart muscle, unlike many other tissues, is terminally differentiated and that mitotic replacement of adult cardiomyocytes, if it occurs at all, can only occur at a low rate. Failure of injured myocardium to rapidly renew itself may form the basis in acute myocardial infarction for replacement of necrotic myocardium with scar tissue. Recent efforts to prevent scar formation through the regeneration of functional myocardium have focused on the transplantation of a wide variety of cell types (Fig. 3 and Table 1). The most promising reports on regeneration and improved cardiac function involve the use of autologous bone marrow cells and skeletal myoblasts. These cell types are currently being used in multiple laboratory and clinical trials. The emerging literature describes new instrumentation to deliver and monitor the activity of transplanted stem cells and myocytes in myocardial therapy.

INSTRUMENTATION TO IMAGE TRANSPLANTED CELLS IN MYOCARDIUM

Although improved function is the single most important aspect of cellular therapy for myocardial regeneration following an acute myocardial infarction, it is crucial, moving the field forward, to be able to understand the temporal sequence of cellular events involved in regeneration. This implies a need for high definition visualization of transplanted cell migration and development over time. Current experimental studies utilize histologic analysis of postmortem animal tissues taken at specific time

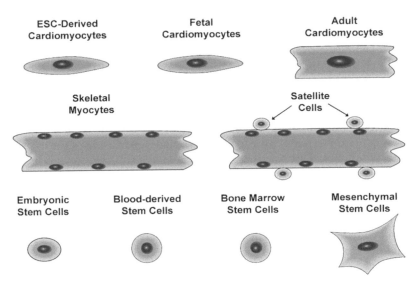

Figure 3 Numerous cell types have been utilized in an effort to regenerate injured myocardium. Some of the most convincing in vivo findings have been derived from experimental animal models and clinical trials in which skeletal myocytes or bone marrow stem cells have been transplanted. In vitro studies have demonstrated the ability of embryonic stem cells and mesenchymal stem cells to differentiate into cardiomyocytes but, to date, there have not been any attempts to treat patients with these cell types.

intervals after treatment. There is a need to develop technology for serial noninvasive imaging to analyze migration and for development of new cardiomyoctes from transplanted cells. It was recently demonstrated that a small bolus of gadolinium, delivered by endomyocardial injection, was visualized by real-time magnetic resonance imaging (MRI) (46). In a subsequent study, mesenchymal cells harvested from bone marrow of adult swine were labeled with iron particles and injected percutaneously into normal and acutely infarcted myocardium. Myocardial sites containing 10^5 labeled cells were seen by serial cardiac MRI at four to eight days after injection. The iron laden mesenchymal cells remained intact when viewed at four to eight days in formalin-fixed tissue samples (47). Efficient endosomal labeling of HSC with greater than one micron diameter iron oxide particles permits detection of single cells by MRI (48). It is hoped that serial MRI viewing will enable investigators to study survival, proliferation, and migration of transplanted cells and thus advance the field of bone marrow stem cell transplantation, transdifferentiation into cardiomyocytes, and functional recovery.

BONE MARROW STEM CELLS FOR THE REGENERATION OF INFARCTED MYOCARDIUM

Several recent papers establish the efficacy of using adult stem cells from bone marrow and peripheral blood to regenerate infarcted myocardium in mice and rats (49–53). In one study involving immune deficient rats, intravenous transplantation of human peripheral blood CD34[+] cells following an acute myocardial infarction resulted in neovascularization of the injured tissue (51). In another study, the bone

Table 1 Partial History of Attempts to Regenerate Myocardium in Animal Models of Heart Disease and Observed Outcomes

Donor cells	Species	Treatment	Outcome	Author	Year	References
Muscle satellite cells	Dog	LV cryoinjury	Generate cardiomyocytes	Chiu	1995	79
ES cell derived cardiomyocytes	Mouse	No injury	Graft survival	Klug	1996	80
Cultured fetal cardiomyocytes	Rat	LV cryoinjury	Improved systolic pressure	Li	1996	81
Skeletal myoblasts	Rabbit	LV cryoinjury	Improved cardiac function	Taylor	1998	58
Smooth muscle cells	Rat	LV cryoinjury	Improved cardiac function	Li	1999	82
Bone marrow cells	Rat	LV cryoinjury	Improved cardiac function	Tomita	1999	49
Fetal cardiomyocytes/skeletal myocytes	Rat	LAD occlusion	Improved cardiac function	Scorsin	2000	61
Lin⁻ c-kit⁺ bone marrow cells	Mouse	LCA occlusion	Improved cardiac function	Orlic	2001	52
Mobilized bone marrow cells	Mouse	LCA occlusion	Increased survival	Orlic	2001	53
Mobilized CD34⁺ bone marrow cells	Human[a]	LAD occlusion	Improved function	Kocher	2001	51
Side population bone marrow cells	Mouse	LAD occlusion	Regenerated myocardium	Jackson	2001	50
Mesenchymal stem cells	Human[a]	No injury	Graft survival	Toma	2002	83
Skeletal myoblasts	Sheep	CCA embolism	Improved cardiac function	Ghostine	2002	84
Fetal cardiomyocytes	Mouse	LV cryoinjury	Increased survival	Roell	2002	85
Endothelial progenitor cells	Mouse	No injury	Restored angiogenesis	Edelberg	2002	86
Muscle satellite cells	Rat	No injury	No cardiomyogenesis	Reinecke	2002	60
Akt1⁺ mesenchymal stem cells	Rat	LAD occlusion	Improved cardiac function	Mangi	2003	87

(Continued)

Table 1   Partial History of Attempts to Regenerate Myocardium in Animal Models of Heart Disease and Observed Outcomes (*Continued*)

Donor cells	Species	Treatment	Outcome	Author	Year	References
Mesenchymal stem cells	Rat	LAD occlusion	Low level trafficking	Barbash	2003	88
SDF-1[+] cardiac fibroblasts	Rat	LAD occlusion	C-kit+ stem cell homing	Askari	2003	36
Cardiac stem cells	Rat	LCA occlusion	Improved cardiac function	Beltrami	2003	9
Fetal cardiomyocytes	Mouse	No injury	Donor/host cell coupling	Rubart	2003	89
Myocardial progenitor cells	Mouse	LAD occlusion	Generate cardiomyocytes	Oh	2003	10
HGF[+] mesenchymal stem cells	Rat	LAD occlusion	Improved cardiac function	Duan	2003	90
Skeletal myoblasts/bone marrow cells	Rabbit	LV cryoinjury	Improved cardiac function	Thompson	2003	91
Bone marrow Sca-1[+] cells	Mouse	Doxorubicin	No cardiomyogenesis	Agbulut	2003	92
CD31[+] peripheral blood cells	Swine	LCA constrictor	Improved cardiac function	Kawamoto	2003	93
Mesenchymal progenitor cells	Rat	LAD occlusion	Improved cardiac function	Davani	2003	94
CD34[+] peripheral blood cells	Human[a]	LAD occlusion	Myocardial regeneration	Yeh	2003	95

[a]Cells injected into immune deficient rats or mice; HGF=hepatocyte growth factor.

Abbreviations: ES=embryonic stem; Lin⁻, lineage negative; LV, left ventricular; LAD, left anterior descending coronary artery; CCA, circumflex coronary artery; LCA, left coronary artery.

marrow of lethally irradiated adult mice was reconstituted with LacZ$^+$ CD34$^{low/-}$ c-kit$^+$ Sca-1$^+$ bone marrow stem cells. After permanent bone marrow engraftment was achieved, the left coronary artery was occluded for 60 minutes followed by reperfusion. Two to four weeks after infarction, 0.02% and 3% β-galactosidase$^+$ myocytes and endothelial cells, respectively, were present in the infarcted zone (50). This study suggested the existence of a naturally occurring but inadequate response by bone marrow stem cells to repair acutely infarcted myocardium. From these animal trials, it is clear that significant regeneration of myocardium requires some form of intervention.

Our group was interested in repair of injury to the myocardium of the LV of adult mice induced by ligation of the descending branch of the left coronary artery without reperfusion. This approach resulted in transmural infarctions that occupied up to 70% of the free wall of the LV and involved necrosis of myocytes and coronary vessels. Lineage$^-$ c-kit$^+$ bone marrow cells from male transgenic donor mice that carried the EGFP gene were injected into female recipients within three to five hours after coronary artery ligation. A band of regenerating myocardium at 9 to 11 days after surgery extended from the anterior to the posterior border of the infarction (Fig. 4). The developing myocardium consisted of Y-chromosome$^+$ EGFP$^+$ cardiomyocytes, endothelium and smooth muscle cells (52). The myocardial infarctions that were treated with Lin$^-$ c-kit$^-$ control cells, known to be devoid of primitive stem cell activity, did not regenerate myocardium. Transcription factors expressed during early cardiomyocyte development, cardiac specific filaments, and connexin 43 were observed in the immature myocytes. Hemodynamic functions improved 30% to 40% in hearts transplanted with Lin$^-$ c-kit$^+$ cells compared with control mice treated with Lin$^-$ c-kit$^-$ cells. Our data demonstrated the ability of primitive bone marrow cells to differentiate

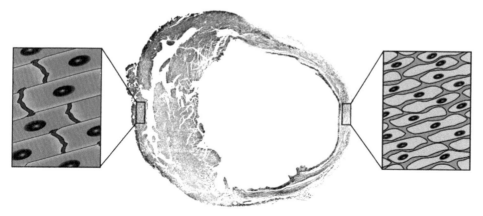

Figure 4 Occlusion of the left anterior descending branch of the left coronary artery in rodents resulted in necrosis of as much as 70% of the free wall of the left ventricle. When bone marrow stem cells were transplanted into the healthy myocardium adjacent to the infarction within hours after the onset of ischemia, or were mobilized by cytokine therapy, the dead tissue was replaced by regenerating myocardium. Histologic examination at four weeks after treatment revealed developing cardiomyocytes with many characteristics of fetal or neonatal cells. They were actively proliferating and were positive for cardiac specific filaments and early acting transcription factors. Although the new cardiomyocytes began to synthesize connexin43, they did not acquire the mature morphology and size seen in the cardiomyocytes of the uninjured right ventricle displayed in the rectangle on the left.

into cardiac myocytes, endothelial cells, and vascular smooth muscle cells following an induced infarction (52).

We also demonstrated myocardial regeneration in the mouse infarction model using cytokine-mobilized autologous stem cells (53). Recombinant rSCF and rhG-CSF were injected daily for eight days. After five daily injections, when circulating HSC reach a level 250-fold higher than the number detected without cytokine treatment (15), an infarction was induced by occlusion of the left coronary artery. It should be noted that there may be primitive stem cells other than HSC and EPC, which were mobilized by the combination cytokine therapy we used. As seen in our intramyocardial stem cell transplant study, a band of regenerating myocardium formed throughout most of the zone of infarction. The developing myocytes resembled fetal cardiac myocytes in morphology and gene expression. It is not clear why they failed to mature into adult-like myocytes during the four-week period following transplantation. One suggestion is that, although ischemic myocardium appears to generate the signals necessary for stem cell homing and early differentiation, the molecular regulatory molecules for late stage maturation of myocytes may not be present in the adult heart. This suggestion is supported by a recent report that at three months after infarction/cytokine therapy, the number of developing myocytes decreases via an apoptotic mechanism (54).

In contrast to the incomplete maturation of myocytes in this mouse model, the developing coronary vessels continued to enlarge over time following injury. At four weeks after treatment, many developing vessels were clearly arterioles, consisting of multilayers of smooth muscle cells. Some of the developing arterioles in the band of regenerating myocardium contained red blood cells in their lumen, suggesting possible anastomosis with spared coronary vessels. Cytokine therapy improved several cardiac functions, including ejection fraction (EF), LV end-diastolic pressure, LV end-systolic pressure, and resulted in improved survival.

These experiments demonstrate that some population(s) of mobilized adult bone marrow stem cells are able to regenerate myocardium in a mouse heart following an episode of ischemia. However, it is still unclear whether one or several stem cell populations are responsible for the regeneration of the different cell types in myocardium. It is also not clear whether some stem cells with regenerative capability originate in organs other than bone marrow, possibly even within the heart.

MYOCARDIAL INFARCTION: REGENERATION BY ADULT DIFFERENTIATED CELLS

While preclinical models have demonstrated the significant potential for myocardial regeneration through the use of stem cell transplantation and stem cell mobilization strategies, review of cell-based therapies for cardiac dysfunction would not be complete without a discussion of autologous, differentiated cell transplantation, which is currently under study in clinical populations.

Early studies with autologous, differentiated or committed cells were designed to assess the ability of adult myocardial tissue to engraft skeletal myocytes and fetal cardiomyocytes (55,56). These studies demonstrated that a cell line committed to differentiate into a skeletal muscle phenotype, specifically C2C12 cells, could engraft into normal myocardium. These studies further demonstrated that transplanted cells from an immortalized cardiomyocyte cell line (AT-1 cells) could engraft in normal myocardium, and, perhaps more importantly, could form intercalated disks connecting the engrafted cells to the host myocardium (57). These early observations offered the potential that transplanted cells could lead to formation of functional myocardium.

AUTOLOGOUS TRANSPLANTATION OF DIFFERENTIATED CELLS INTO ISCHEMIC MYOCARDIUM

The demonstration that autologous skeletal myoblast (SKMB) transplantation led to successful engraftment of cells within normal myocardium led a number of investigators to study the efficacy of transplantation of differentiated cells into ischemic or infarcted myocardium (58–61). Implanting SKMB into ischemic/infarcted myocardium has several theoretical advantages, including the fact that SKMB are significantly more resistant to ischemia than cardiac myocytes.

The mechanism by which LV functional improvement with SKMB transplantation is achieved is not clear and is still under investigation. Transplanted SKMB do not result in the development of new cardiac myocytes, and the transplanted cells do not integrate in the electrical syncitium of the myocardium (57). Increases in left ventricular ejection fraction (LVEF) with the introduction of SKMB into the infarct zone are modest (61). However, the combined improvement in systolic function along with the improvements in diastolic function (58) that have been demonstrated ultimately appear to lead to an overall improved cardiac performance. Despite the fact that the mechanism(s) for the improvement in cardiac function in response to SKMB transplantation is not known, the utility of autologous SKMB for the treatment of cardiac dysfunction is under investigation in clinical trials (see below).

AUTOLOGOUS CELL TRANSPLANTATION AS A PLATFORM FOR GENE TRANSFER

While clinical trials are ongoing to determine the efficacy of SKMB transplantation in patients with ischemic cardiomyopathy, preclinical studies are attempting to broaden the application of differentiated cell transplantation to include being a platform for gene transfer (36,62–65). Support for the concept of combining cell and gene therapy is rapidly evolving. Multiple studies have demonstrated that transplanting VEGF expressing SKMB results in increased neovascularization within the infarct zone (62–65). Furthermore, in a rodent model of ischemic cardiomyopathy, we have recently demonstrated that the transplantation of SKMB transfected with an adenoviral construct encoding VEGF-165 leads to significantly greater angiogenesis and recovery of LV function than the injection of the adenovirus or SKMB alone (63).

Recent studies have demonstrated that the molecular signals, which mediate stem cell–based regeneration of myocardial tissue, are expressed only transiently following an ischemic event (36). These results imply that in order to achieve optimal regeneration of myocardial tissue at a time remote from myocardial infarction, these molecular signals will need to be re-expressed. We have recently demonstrated that SDF-1 expression is sufficient to induce stem cell homing of CD34[+] and c-kit[+] bone marrow derived cells to injured myocardium. By transplanting SDF-1 expressing cardiac fibroblasts two months after myocardial infarction, we demonstrated the re-establishment of stem cell homing to myocardial tissue with a 20-fold increase in c-kit[+] cells, as well as an 80% increase in LV function (36). Combined data from these studies demonstrate the great potential of using autologous cells for gene trans-fer, and suggests that many adenoviral constructs being considered and/or tested in clinical trials should be reconsidered as potential adjuncts to cell therapy.

CLINICAL TRIALS IN CELL THERAPY

Differentiated Cell Transplantation

It has been recently reported that SKMB engraft into the infarct zone of patients with ischemic cardiomyopathy (66). Six patients undergoing left ventricular assist device (LVAD) placement received simultaneous autologous SKMB transplanta-tion. Of these patients, five had their LVAD explanted and immunohistochemistry was performed on the explanted heart to identify the presence of engrafted SKMB. Of the four patients who received 300 million cells, evidence for the presence of SKMB engraftment was found at five days and at three, four and six months after SKMB transplantation. In the one patient who received 2.2 million cells, perhaps not surprisingly, no evidence of SKMB engraftment was found.

Menasche and colleagues recently reported results for a European phase I trial of SKMB transplantation as an adjunct to coronary artery bypass grafting (CABG) (67). Patients enrolled in this trial had severe ischemic cardiomyopathy (EF ≤ 35%) and received, on average, 871 million cells (86% of which were SKMB). The inves-tigators reported that 63% of the injected scars showed improved systolic thickening (67). While encouraging, whether this improvement was independent of the fact that all patients underwent coronary revascularization is difficult to determine. Perhaps more importantly, in this phase I trial, 4 of the 10 patients developed sustained, monomorphic ventricular tachycardia between 11 and 22 days after surgery. We will have to await the results of future placebo-controlled trials to determine if these arrhythmias were in response to SKMB transplantation or to the severe underlying ischemic cardiomyopathy.

The first clinical trial of SKMB transplantation in the United States has demonstrated short-term safety and feasibility in a population of patients who are relatively young (mean age 55 years) and with significant LV dysfunction (mean EF = 23%) (68). This study was designed to assess the effect of escalating the dose of SKMB (10, 30, 100, and 300 million cells, three patients per group). In contrast to the European study, at least for a dose up to 300 million cells, SKMB transplantation does not appear to be arrhythmogenic. This finding opens up the possibility that the arrhythmia potential of cell transplantation could be dose dependent. As in the European study, this U.S. study was designed as a feasibility study, and we await larger randomized placebo-controlled studies to determine if there is any clinical benefit of SKMB transplantation in humans.

Stem Cell Transplantation

A number of modes of delivery, either via direct intramyocardial injection at the time of surgery, percutaneous direct myocardial injection, or percutaneous intracoronary infusion are being studied in clinical populations. A phase I trial in 10 patients within days of myocardial infarction suggested that intracoronary delivery of bone marrow derived mononuclear cells resulted in improved LV contractility and improved perfusion of the infarct zone (69). Expanding upon these data, it has been demonstrated that intracoronary infusion of either bone marrow derived mononuclear cells or peripheral blood derived endothelial progenitor cells within days of myocardial infarction resulted in improved regional LV function and viability within the infarct zone (70). Similar improvements in LV perfusion and function have also been seen with direct myocardial injection, either percutaneously or surgically, of bone marrow cells in patients with stable ischemic heart disease (71,72). Most importantly, feasibility and safety were demonstrated in these studies as no untoward effects were reported.

The benefits of these predominantly nonrandomized studies were recently supported in a randomized trial of adjunctive intracoronary bone marrow mononuclear cell (BMMNC) delivery reported at the 2003 Scientific Sessions of the American Heart Association in Orlando, Florida (73). Sixty patients were randomized to receive optimal care versus optimal care plus BMMNC infusions 3.5 days after successful percutaneous coronary intervention for acute myocardial infarction. Patients who underwent infusion of BMMNC demonstrated a greater improvement in LVEF as assessed by MRI six months later.

Recently, the first placebo-controlled trial for stem cell transplantation for congestive heart failure was reported (74). In this prospective, open-labeled, nonrandomized study, patients with end-stage ischemic heart disease received transendocardial injections of autologous bone marrow derived mononuclear cells guided by electromechanical mapping of the LV (74). The patients received, on average, 15 injections with a total of approximately 30 million cells (~2.4% were endothelial progenitor cells) delivered to each patient. In this study, virtually all clinical and diagnostic parameters were significantly improved by stem cell transplantation. There were significant improvements in NYHA, CCSAS, and treadmill times, as well as LV volumes and overall EF. Perhaps more importantly, there was no report of any adverse events in this relatively sick patient population. These data are exciting, and beg for validation in a large randomized, blinded clinical trial.

Stem Cell Mobilization

Stem cell mobilization has been used as a strategy to promote growth of collateral vessels in patients with chronic stable angina who were not eligible for CABG. In a randomized, double blind, placebo-controlled trial, the first dose of recombinant human granulocyte–macrophage (rhGM)-CSF was delivered by intracoronary infusion followed by daily injections of rhGM-CSF (10 mcg/kg) for two weeks (75). Following the two weeks of cytokine therapy, the investigators demonstrated a significant ~50% increase in collateral flow index in those patients who received rhGM-CSF compared to an insignificant ~25% decrease in the collateral flow index in those patients who received placebo. Furthermore, significantly fewer patients following treatment with rhGM-CSF exhibited objective signs of ischemia by electrocardiography during coronary artery balloon occlusion. No significant difference was seen in the group which received placebo (74).

To date, there are no clinical data on the use of rhG-CSF or rhSCF to regenerate or preserve cardiac function in patients with acute myocardial infarction. In contrast to the data in mice, work performed in nonhuman primate models of acute ischemia/reperfusion suggests an excess of mortality and little or no benefit to survivors treated with rhG-CSF combined with rhSCF (76,77). The negative effect of SCF could perhaps have been predicted since it is known to induce hypotension, which obviously could have profound negative clinical consequences in the setting of myocardial infarction.

At the Cleveland Clinic Foundation, we have initiated the RECOVER trial. In this randomized, double blind, placebo-controlled, dose-escalating trial, patients are randomized to rhG-CSF (5 mcg/kg/day or 10 mcg/kg/day) or placebo for five days following myocardial infarction. Eligible patients are those who present with a serum troponin elevation and myocardial infarction, and have an infarct related vessel rendered patent by mechanical or pharmacological intervention. Based on our data demonstrating the transient expression of stem cell homing signal(s) following myocardial infarction (36), drug therapy must be initiated within 48 hours of the onset of symptoms. We are quantifying LV function in these patients by echocardiography, and assessing regional LV contractility with myocardial strain and strain rate (36,78). To date, recruitment for the 5 mcg/kg/day dose is complete, and while many have expressed concern regarding the potential for adverse events associated with G-CSF administration so soon after myocardial infarction, at 30 days after randomization, we had no deaths in the treated group ($n = 6$) and one death in the placebo group ($n = 3$). Thirty days after myocardial infarction no difference in cardiac function was observed between those patients that received G-CSF and those that received saline. The effects of a higher dose of G-CSF (10 µg/kg/day for 5 days) is still under study.

CONCLUSION

In summary, this is an extraordinarily exciting time for the field of cardiovascular medicine and our patients. The dogma of the past, stating that myocardial tissue once lost can never be regenerated is all but gone. Our tremendous successes, in the last decade of the last century, in our understanding and ability to achieve myocardial reperfusion and improve survival in patients with acute myocardial infarction have left us with the ever increasing challenge of treating patients with congestive heart failure. This is a population that needs new and novel therapeutic strategies, which can restore functional capacity, so that these patients can return to enjoyable and productive lives. While the studies reviewed above offer great promise, cell, gene, and stem cell therapy are not yet ready for extensive clinical application. It is our hope that carefully designed mechanistic studies and the development of creative clinical strategies, combined with well designed clinical protocols, will ultimately deliver a revolution in the care of patients with heart disease. Achieving this goal will require the forging of new collaborations between the bench researcher and the clinician, as well as cross discipline collaborations between clinicians that have significantly different areas of expertise.

GLOSSARY

Embryonic stem cells: Pluripotent stem cells that are derived from the inner cell mass of the blastocyst. They give rise to cells of the three germ layers.

Adult stem cells: Multipotent stem cells that exist in self-renewing tissues. They proliferate for self-renewal and differentiate into multiple progenitor cell types.

Progenitor cells: Intermediate stem cells, derived from multipotent stem cells. They give rise to unipotential precursor cells committed to tissue specific functional cells.

Hemangioblasts: Bipotential stem cells that give rise to hematopoietic stem cells and endothelial progenitor cells.

Hematopoietic stem cells: Present in blood and bone marrow. They give rise to several distinct populations of blood forming progenitor cells.

Endothelial progenitor cells: Present in blood and bone marrow. They differentiate into endothelial cells during angiogenesis and neovasculogenesis.

Stromal cells: Structural cells within adult hematopoietic tissues that are currently being investigated for their myogenic potential.

Mesenchymal stem cells: Possibly analogous to stromal cells. They can be triggered in vitro to differente into cardiomyocytes.

Multipotent adult progenitor cells: Co-purify with marrow mesenchymal stem cells. Possess wide capacity for differentiation.

Plasticity/transdifferentiation: Describes the process by which adult stem cells that reside in one tissue differentiate into mature cells of an unrelated tissue.

REFERENCES

1. DeRobertis EM, Arechaga J. The Spemann Mangold Organizer: 75 Years On. Int J Dev Biol 2001; 45 (1) Special Issue.
2. Lacaud G, Robertson S, Palis J, Kennedy M, Keller G. Regulation of hemangioblast development. Annals NY Acad Sci 2001; 938:96–108.
3. Choi K, Kennedy M, Kazarov A, Papadimitriou JC, Keller G. A common precursor for hematopoietic and endothelial cells. Development 1998; 125:725–732.
4. Nishikawa SI, Nishikawa S, Hirashima M, Matsuyoshi N, Kodama H. Progressive lineage analysis by cell sorting and culture identifies FLK1+VE-cadherin+ cells at a diverging point of endothelial and hemopoietic lineages. Development 1998; 125:1747–1757.
5. Grant MB, May WS, Caballero S, Brown GA, Guthrie SM, Mames RN, Byrne BJ, Vaught T, Spoerri PE, Peck AB, Scott EW. Adult hematopoietic stem cells provide functional hemangioblast activity during retinal neovascularization. Nature Med 2002; 8:607–612.
6. Cogle CR, Wainman DA, Jorgensen ML, Guthrie SM, Mames RN, Scott EW. Adult human hematopoietic cells provide functional hemangioblast activity. Blood 2004; 103:133–135.
7. Johe KK, Hazel TG, Muller T, Dugich-Djordjevic MM, McKay RDG. Single factors direct the differentiation of stem cells from the fetal and adult central nervous system. Genes and Development 1996; 10:3129–3140.
8. Clarke DL. Generalized potential of adult neural stem cells. Science 2000; 288:1660–1663.
9. Beltrami AP, Barlucchi L, Torella D, Baker M, Limana F, Chimenti S, Kasahara H, Rota M, Musso E, Urbanek K, Leri A, Kajstura J, Nadal-Ginard B, Anversa P. Adult cardiac stem cells are multipotent and support myocardial regeneration. Cell 2003; 114:763–776.
10. Oh H, Bradfute SB, Gallardo TD, Nakamura T, Gaussin V, Mishina Y, Pocius J, Michael LH, Behringer RR, Garry DJ, Entman ML, Schneider MD. Cardiac progenitor cells from adult myocardium: homing, differentiation, and fusion after infarction. Proc Natl Acad Sci USA 2003; 100:12313–12318.
11. Lajtha LG, Pozzi LV, Schofield R, Fox M. Kinetic properties of haemopoietic stem cells. Cell Tissue Kinet 1969; 2:39–49.
12. Lin Y, Weisdorf DJ, Solovey A, Hebbel RP. Origins of circulating endothelial cells and endothelial outgrowth from blood. J Clin Invest 2000; 105:71–77.
13. Neben S, Marcus K, Mauch P. Mobilization of hematopoietic stem and progenitor cell subpopulations from the marrow to the blood of mice following cyclophosphamide and/or granulocyte colony-stimulating factor. Blood 1993; 81:1960–1967.

14. Morrison SJ, Wright DE, Weissman IL. Cyclophosphamide/granulocyte colony-stimulating factor induces hematopoietic stem cells to proliferate prior to mobilization. Proc Natl Acad Sci USA 1997; 94:1908–1913.

15. Bodine DM, Seidel NE, Gale MS, Nienhuis AW, Orlic D. Efficient retrovirus transduction of mouse pluripotent hematopoietic stem cells mobilized into the peripheral blood by treatment with granulocyte colony-stimulating factor and stem cell factor. Blood 1994; 84: 1482–1491.

16. Wright DE, Wagers AJ, Gulati AP, Johnson FL, Weissman IL. Physiological migration of hematopoietic stem and progenitor cells. Science 2001; 294:1933–1936.

17. Abkowitz JL, Robinson AE, Kale S, Long MW, Chen J. Mobilization of hematopoietic stem cells during homeostasis and after cytokine exposure. Blood 2003; 102:1249–1253.

18. Levesque JP, Takamatsu Y, Nilsson SK, Haylock DN, Simmons PJ. Vascular cell adhesion molecule-1 (CD106) is cleaved by neutrophil proteases in the bone marrow following hematopoietic progenitor cell mobilization by granulocyte colony-stimulating factor. Blood 2001; 98:1289–1297.

19. Levesque JP, Hendy J, Takamatsu Y, Simmons PJ, Bendall LJ. Disruption of the CXCR4/ CXCL12 chemotactic interaction during hematopoietic stem cell mobilization induced by GCSF or cyclophosphamide. J Clin Invest 2003; 111:187–196.

20. Heissig B, Hattori K, Dias S, Friedrich M, Ferris B, Hackett NR, Crystal RG, Besmer P, Lyden D, Moore MA, Werb Z, Rafii S. Recruitment of stem and progenitor cells from the bone marrow niche requires MMP-9 mediated release of kit-ligand. Cell 2002; 109:625–637.

21. Prockop DJ. Marrow stromal cells as stem cells for nonhematopoietic tissues. Science 1997; 276:71–74.

22. Prockop DJ. Adult stem cells gradually come of age. Nat Biotechnol 2002; 20:791–792.

23. Jiang Y, Jahagirdar BN, Reinhardt RL, Schwartz RE, Keene CD, Ortiz-Gonzalez XR, Reyes M, Lenvik T, Lund T, Blackstad M, et al. Pluripotency of mesenchymal stem cells derived from adult marrow. Nature 2002; 418:41–49.

24. Reyes M, Lund T, Lenvik T, Aguiar D, Koodie L, Verfaillie CM. Purification and ex vivo expansion of postnatal human marrow mesodermal progenitor cells. Blood 2001; 98:2615–2625.

25. Gojo S, Gojo N, Takeda Y, Mori T, Abe H, Kyo S, Hata J-I, Umezawa A. In vivo cardiovasculogenesis by direct injection of isolated adult mesenchymal stem cells. Exp Cell Res 2003; 288:51–59.

26. Bodine DM, Seidel NE, Zsebo KM, Orlic D. In vivo administration of stem cell factor to mice increases the absolute number of pluripotent hematopoietic stem cells. Blood 1993; 82:445–455.

27. Issarachai S, Priestley GV, Nakamoto B, Papayannopoulou T. Cells with hemopoietic potential residing in muscle are itinerant bone marrow-derived cells. Exp Hematol 2002; 30:366–373.

28. Kawada H, Ogawa M. Bone marrow origin of hematopoietic progenitors and stem cells in murine muscle. Blood 2001; 98:2008–2013.

29. McKinney-Freeman SL, Jackson KA, Camargo FD, Ferrari G, Mavilio F, Goodell MA. Muscle-derived hematopoietic stem cells are hematopoietic in origin. Proc Natl Acad Sci USA 2002; 99:1341–1346.

30. LaBarge MA, Blau HM. Biological progression from adult bone marrow to mononucleate muscle stem cell to multinucleate muscle fiber in response to injury. Cell 2002; 111:589–601.

31. Corbel SY, Lee A, Yi L, Duenas J, Brazelton TR, Blau HM, Rossi FMV. Contribution of hematopoietic stem cells to skeletal muscle. Nature Med 2003; 9:1528–1532.

32. Andrews RG, Briddell RA, Knitter GH, Opie T, Bronsden M, Myerson D, Appelbaum FR, McNiece IK. In vivo synergy between recombinant human stem cell factor and recombinant human granulocyte colony-stimulating factor in baboons enhanced circulation of progenitor cells. Blood 1994; 84:800–810.

33. Andrews RG, Briddell RA, Knitter GH, Rowley SD, Appelbaum FR, McNiece IK. Rapid engraftment by peripheral blood progenitor cells mobilized by recombinant

human stem cell factor and recombinant human granulocyte colony-stimulating factor in nonhuman primates. Blood 1995; 85:15–20.

34. Gerlach LO, Skerlj RT, Bridger GJ, Schwartz TW. Molecular interactions of cyclam and bicyclam non-peptide antagonists with the CXCR4 chemokine receptor. J Biol Chem 2001; 276:14153–14160.

35. Liles WC, Broxmeyer HE, Rodger E, Wood B, Hubel K, Cooper S, Hangoc G, Bridger GJ, Henson GW, Calandra G, et al. Mobilization of hematopoietic progenitor cells in healthy volunteers by AMD3100, a CXCR4 antagonist. Blood 2003; 102:2728–2730.

36. Askari AT, Unzek S, Popovic ZB, Goldman CK, Forudi F, Kiedrowski M, Rovner A, Ellis SG, Thomas JD, DiCorleto PE, et al. Effect of stromal-cell-derived factor 1 on stem-cell homing and tissue regeneration in ischaemic cardiomyopathy. Lancet 2003; 362:697–703.

37. Kale S, Karihaloo A, Clark PR, Kashgarian M, Krause DS, Cantley LG. Bone marrow stem cells contribute to repair of the ischemically injured renal tubule. J Clin Invest 2003; 112:42–49.

38. Pesce M, Orlandi A, Iachininoto MG, Straino S, Torella AR, Rizzuti V, Pompilio G, Bonanno G, Scambia G, Capogrossi MC. Myoendothelial differentiation of human umbilical cord blood-derived stem cells in ischemic limb tissues. Circulation Res 2003; 93:51–62.

39. Tateishi-Yuyama E, Matsubara H, Murohara T, Ikeda U, Shintani S, Masaki H, Amano K, Kishimoto Y, Yoshimoto K, Akashi H, et al. Therapeutic angiogenesis for patients with limb ischaemia by autologous transplantation of bone-marrow cells: a pilot study and a randomised controlled trial. Lancet 2002; 360:427–435.

40. Ho KK, Anderson KM, Kannel WB, Grossman W, Levy D. Survival after the onset of congestive heart failure in Framingham Heart Study subjects. Circulation 1993; 88:107–115.

41. Heyndrickx GR, Baig H, Nellens P, Leusen I, Fishbein MC, Vatner SF. Depression of regional blood flow and wall thickening after brief coronary occlusions. Am J Physiol 1978; 234:H653–H659.

42. Griggs DM Jr, Tchokoev VV, Chen CC. Transmural differences in ventricular tissue substrate levels due to coronary constriction. Am J Physiol 1972; 222:705–709.

43. Askari AT, Brennan ML, Zhou X, Drinko J, Morehead A, Thomas JD, Topol EJ, Hazen SL, Penn MS. Myeloperoxidase and plasminogen activator inhibitor 1 play a central role in ventricular remodeling after myocardial infarction. J Exp Med 2003; 197:615–624.

44. Pfeffer MA. Left ventricular remodeling after acute myocardial infarction. Annual Rev Med 1995; 46:455–466.

45. Deten A, Volz HC, Briest W, Zimmer HG. Cardiac cytokine expression is upregulated in the acute phase after myocardial infarction: experimental study in rats. Cardiovasc Res 2002; 55:329–340.

46. Lederman RJ, Guttman MA, Peters DC, Thompson RB, Sorger JM, Dick AJ, Raman VK, McVeigh ER. Catheter-based endomyocardial injection with real-time magnetic resonance imaging. Circulation 2002; 105:1282–1284.

47. Hill JM, Dick AJ, Raman VK, Thompson RB, Yu ZX, Hinds KA, Pessanha BS, Guttman MA, Varney TR, Martin BJ, et al. Serial cardiac magnetic resonance imaging of injected mesenchymal stem cells. Circulation 2003; 108:1009–1014.

48. Hinds KA, Hill JM, Shapiro EM, Laukkanen MO, Silva AC, Combs CA, Varney TR, Balaban RS, Koretsky AP, Dunbar CE. Highly efficient endosomal labeling of progenitor and stem cells with large magnetic particles allows magnetic resonance imaging of single cells. Blood 2003; 102:867–872.

49. Tomita S, Li R-K, Weisel RD, Mickle DAG, Kim E-J, Sakai T, Jia Z-Q. Autologous transplantation of bone marrow cells improves damaged heart function. Circulation 1999; 100(suppl II):II-247–II-256.

50. Jackson KA, Majka SM, Wang H, Pocius J, Hartley CJ, Majesky MW, Entman ML, Michael LH, Hirschi KK, Goodell MA. Regeneration of ischemic cardiac muscle and vascular endothelium by adult stem cells. J Clin Invest 2001; 107:1395–1402.

51. Kocher AA, Schuster MD, Szabolcs MJ, Takuma S, Burkhoff D, Wang J, Homma S, Edwards NM, Itescu S. Neovascularization of ischemic myocardium by human bone-marrow-derived angioblasts prevents cardiomyocyte apoptosis, reduces remodeling and improves cardiac function. Nat Med 2001; 7:430–436.

52. Orlic D, Kajstura J, Chimenti S, Jakoniuk I, Anderson SM, Li B, Pickel J, McKay R, Nadal-Ginard B, Bodine DM, et al. Bone marrow cells regenerate infarcted myocardium. Nature 2001; 410:701–705.

53. Orlic D, Kajstura J, Chimenti S, Limana F, Jakoniuk I, Quaini F, Nadal-Ginard B, Bodine DM, Leri A, Anversa P. Mobilized bone marrow cells repair the infarcted heart, improving function and survival. Proc Natl Acad Sci USA 2001; 98:10344–10349.

54. Kajstura J, Chimenti S, Bearzi C, Cascapera S, Limana F, Nadal-Ginard B, Leri A, Anversa P. Long-term cardiac repair after infarction by mobilization of bone marrow cells. Circulation (suppl) 2003; 108:IV156.

55. Koh GY, Klug MG, Soonpaa MH, Field LJ. Differentiation and long-term survival of C2C12 myoblast grafts in heart. J Clin Invest 1993; 92:1548–1554.

56. Koh GY, Soonpaa MH, Klug MG, Field LJ. Long-term survival of AT-1 cardiomyocyte grafts in syngeneic myocardium. Am J Physiol 1993; 264:H1727–H1733.

57. Soonpaa MH, Koh GY, Klug MG, Field LJ. Formation of nascent intercalated disks between grafted fetal cardiomyocytes and host myocardium. Science 1994; 264:98–101.

58. Taylor DA, Atkins BZ, Hungspreugs P, Jones TR, Reedy MC, Hutcheson KA, Glower DD, Kraus WE. Regenerating functional myocardium: improved performance after skeletal myoblast transplantation. Nat Med 1998; 4:929–933.

59. Reinecke H, MacDonald GH, Hauschka SD, Murry CE. Electromechanical coupling between skeletal and cardiac muscle. Implications for infarct repair. J Cell Biol 2000; 149:731–740.

60. Reinecke H, Poppa V, Murry CE. Skeletal muscle stem cells do not transdifferentiate into cardiomyocytes after cardiac grafting. J Mol Cell Cardiol 2002; 34:241–249.

61. Scorsin M, Hagege A, Vilquin JT, Fiszman M, Marotte F, Samuel JL, Rappaport L, Schwartz K, Menasche P. Comparison of the effects of fetal cardiomyocyte and skeletal myoblast transplantation on postinfarction left ventricular function. J Thorac Cardiovasc Surg 2000; 119:1169–1175.

62. Askari A, Goldman CK, Forudi F, Ellis SG, Thomas JD, Penn MS. VEGF-expressing skeletal myoblast transplantation induces angiogenesis and improves left ventricular function late after myocardial infarction. Mol Therapy 2002; 5:S162.

63. Askari A, Unzek S, Goldman CK, Ellis SG, Thomas JD, DiCorleto PE, Topol EJ, Penn MS. Cellular, but not direct, adenoviral delivery of vascular endothelial growth factor results in improved left ventricular function and neovascularization in dilated ischemic cardiomyopathy. J Am Coll Cardiol 2004; 43:1908–1914.

64. Suzuki K, Murtuza B, Smolenski RT, Sammut IA, Suzuki N, Kaneda Y, Yacoub MH. Cell transplantation for the treatment of acute myocardial infarction using vascular endothelial growth factor-expressing skeletal myoblasts. Circulation 2001; 104:I207–I212.

65. Yau TM, Fung K, Weisel RD, Fujii T, Mickle DA, Li RK. Enhanced myocardial angiogenesis by gene transfer with transplanted cells. Circulation 2001; 104:I218–I222.

66. Pagani FD, DerSimonian H, Zawadzka A, Wetzel K, Edge AS, Jacoby DB, Dinsmore JH, Wright S, Aretz TH, Eisen HJ, Aaronson KD. Autologous skeletal myoblasts transplanted to ischemia-damaged myocardium in humans. Histological analysis of cell survival and differentiation. J Am Coll Cardiol 2003; 41:879–888.

67. Menasche P, Hagege AA, Vilquin JT, Desnos M, Abergel E, Pouzet B, Bel A, Sarateanu S, Scorsin M, Schwartz K, Bruneval P, Benbunan M, Marolleau JP, Duboc D. Autologous skeletal myoblast transplantation for severe postinfarction left ventricular dysfunction. J Am Coll Cardiol 2003; 41:1078–1083.

68. Penn MS, McCarthy PM. Cell transplantation: first United States clinical experience. In: Kipshidze N, Serruys PW, eds. Handbook of Cardiovascular Cell Transplantation. London: Martin Dunitz, 2004; 21:293–300.

69. Strauer BE, Brehm M, Zeus T, Kostering M, Hernandez A, Sorg R V, Kogler G, Wernet P. Repair of infarcted myocardium by autologous intracoronary mononuclear bone marrow cell transplantation in humans. Circulation 2002; 106:1913–1918.

70. Assmus B, Schachinger V, Teupe C, Britten M, Lehmann R, Dobert N, Grunwald F, Aicher A, Urbich C, Martin H, et al. Transplantation of Progenitor Cells and Regeneration Enhancement in Acute Myocardial Infarction (TOPCARE-AMI). Circulation 2002; 106:3009–3017.

71. Stamm C, Westphal B, Kleine HD, Petzsch M, Kittner C, Klinge H, Schumichen C, Nienaber CA, Freund M, Steinhoff G. Autologous bone-marrow stem-cell transplantation for myocardial regeneration. Lancet 2003; 361:45–46.

72. Tse HF, Kwong YL, Chan JK, Lo G, Ho CL, Lau CP. Angiogenesis in ischaemic myocardium by intramyocardial autologous bone marrow mononuclear cell implantation. Lancet 2003; 361:47–49.

73. Wollert KC. Randomized-controlled clinical trial of intracoronary autologous bone marrow cell transplantation post myocardial infarction. Plenary Session III: Late-Breaking Clinical Trials. American Heart Association Annual Meeting, Orlando, FL, 2003; 9–12.

74. Perin EC, Dohmann HF, Borojevic R, Silva SA, Sousa AL, Mesquita CT, Rossi MI, Carvalho AC, Dutra HS, Dohmann H, et al. Transendocardial, autologous bone marrow cell transplantation for severe, chronic ischemic heart failure. Circulation 2003; 107: 2294–2302.

75. Seiler C, Pohl T, Wustmann K, Hutter D, Nicolet PA, Windecker S, Eberli FR, Meier B. Promotion of collateral growth by granulocyte-macrophage colony-stimulating factor in patients with coronary artery disease: a randomized, double-blind, placebo-controlled study. Circulation 2001; 104:2012–2017.

76. Orlic D, Arai AE, Sheikh F, Agyeman KO, McGehee J, Hoyt R, Sachdev V, Yu Z-X, San H, Metzger M, et al. Cytokine mobilized CD34+ cells do not benefit rhesus monkeys following induced myocardial infarction. Blood 2002; 100(28a):99.

77. Norol F, Merlet P, Isnard R, Sebillon P, Bonnet N, Cailliot C, Carrion C, Ribeiro M, Charlotte F, Pradeau P, et al. Influence of mobilized stem cells on myocardial infarct repair in a nonhuman primate model. Blood 2003; 102:4361–4368.

78. Greenberg NL, Firstenberg MS, Castro PL, Main M, Travaglini A, Odabashian JA, Drinko JK, Rodriguez LL, Thomas JD, Garcia MJ. Doppler-derived myocardial systolic strain rate is a strong index of left ventricular contractility. Circulation 2002; 105:99–105.

79. Chiu RC-J, Zibaitis A, Kao RL. Cellular cardiomyoplasty: Myocardial regeneration with satellite cell implantation. Ann Thorac Surg 1995; 60:12–18.

80. Klug MG, Soonpaa MH, Koh GY, Field LJ. Genetically selected cardiomyocytes from differentiating embryonic stem cells from stable intracardiac grafts. J Clin Invest 1996; 98:216–224.

81. Li R-K, Jia Z-Q, Weisel RD, Mickle DAG, Zhang J, Mohabeer MK, Rao V, Ivanov J. Cardiomyocyte transplantation improves heart function. Ann Thorac Surg 1996; 62:654–661.

82. Li R-K, Jia Z-Q, Weisel RD, Merante F, Mickle DAG. Smooth Muscle Cell Transplantation into Myocardial Scar Tissue Improves Heart Function. J Mol Cell Cardiol 1999; 31:513–522.

83. Toma C, Pittenger MF, Cahill KS, Byrne BJ, Kessler PD. Human mesenchymal stem cells differentiate to a cardiomyocyte phenotype in the adult murine heart. Circulation 2002; 105:93–98.

84. Ghostine S, Carrion C, Souza LC, Richard P, Bruneval P, Vilquin JT, Pouzet B, Schwartz K, Menasche P, Hagege AA. Long-term efficacy of myoblast transplantation on regional structure and function after myocardial infarction. Circulation 2002; 106(12 suppl 1):I131–I366.

85. Roell W, Lu ZJ, Bloch W, Siedner S, Tiemann K, Xia Y, Stoecker E, Fleischmann M, Bohlen H, Stehle R, et al. Cellular cardiomyoplasty improves survival after myocardial injury. Circulation 2002; 105:2435–2441.

86. Edelberg JM, Tang L, Hattori K, Lyden D, Rafii S. Young adult bone marrow-derived endothelial precursor cells restore aging-impaired cardiac angiogenic function. Circ Res 2002; 90:E89–E93.

87. Mangi AA, Noiseux N, Kong D, He H, Rezvani M, Ingwall JS, Dzau VJ. Mesenchymal stem cells modified with Akt prevent remodeling and restore performance of infarcted hearts. Nat Med 2003; 9:1195–1201.

88. Barbash IM, Chouraqui P, Baron J, Feinberg MS, Etzion S, Tessone A, Miller L, Guetta E, Zipori D, Kedes LH, et al. Systemic delivery of bone marrow-derived mesenchymal stem cells to the infarcted myocardium: feasibility, cell migration, and body distribution. Circulation 2003; 108:863–868.

89. Rubart M, Pasumarthi KB, Nakajima H, Soonpaa MH, Nakajima HO, Field LJ. Physiological coupling of donor and host cardiomyocytes after cellular transplantation. Circ Res 2003; 92:1217–1224.

90. Duan HF, Wu CT, Wu DL, Lu Y, Liu HJ, Ha XQ, Zhang QW, Wang H, Jia XX, Wang LS. Treatment of myocardial ischemia with bone marrow-derived mesenchymal stem cells overexpressing hepatocyte growth factor. Mol Therapy 2003; 8:467–474.

91. Thompson RB, Emani SM, Davis BH, van den Bos EJ, Morimoto Y, Craig D, Glower D, Taylor DA. Comparison of intracardiac cell transplantation: autologous skeletal myoblasts versus bone marrow cells. Circulation 2003; 108(suppl 1):II264–I271.

92. Agbulut O, Menot ML, Li Z, Marotte F, Paulin D, Hagege AA, Chomienne C, Samuel JL, Menasche P. Temporal patterns of bone marrow cell differentiation following transplantation in doxorubicin-induced cardiomyopathy. Cardiovasc Res 2003; 58:451–459.

93. Kawamoto A, Tkebuchava T, Yamaguchi J, Nishimura H, Yoon YS, Milliken C, Uchida S, Masuo O, Iwaguro H, Ma H, et al. Intramyocardial transplantation of autologous endothelial progenitor cells for therapeutic neovascularization of myocardial ischemia. Circulation 2003; 107:461–468.

94. Davani S, Marandin A, Mersin N, Royer B, Kantelip B, Herve P, Etievent JP, Kantelip JP. Mesenchymal progenitor cells differentiate into an endothelial phenotype, enhance vascular density, and improve heart function in a rat cellular cardiomyoplasty model. Circulation 2003; 108(suppl 1):II253–II258.

95. Yeh ET, Zhang S, Wu HD, Korbling M, Willerson JT, Estrov Z. Transdifferentiation of human peripheral blood CD34+-enriched cell population into cardiomyocytes, endothelial cells, and smooth muscle cells in vivo. Circulation 2003; 108:2070–2073.

27

Future Directions in Cardiac Remodeling

Barry Greenberg
Advanced Heart Failure Treatment Program, University of California,
San Diego, California, U.S.A.

What is known with great certainty about cardiac remodeling is that the process is as complex as it is important. The stimuli and triggers for activating remodeling are numerous and the signaling pathways through which structural changes develop are complicated. Although the past several years have brought important advances in our understanding of the causes, mechanisms, and consequences of cardiac remodeling, more work needs to be done in defining the underlying pathways that influence this processs. Moreover, additional treatments that can be combined with existing ones to prevent and/or reverse cardiac remodeling need to be developed.

The ultimate phenotypic changes that develop in the remodeling heart are now known to be the result of the interplay of genetic, hemodynamic, neurohormonal, and environmental factors. In some settings, remodeling may be a physiologic response that helps to compensate for increased load on the heart, while in others it leads to deterioration in cardiac function. Future directions should be aimed not only at unraveling the various triggers and pathways through which remodeling occurs, but also toward defining the aspects of this process that may be beneficial and need to be preserved as opposed to those that are deleterious and potentially targets for therapeutic interventions. In devising strategies aimed at altering the remodeling process, it is important to recognize that timing is likely to be a critical factor in determining when the activation of a particular pathway is beneficial and, alternatively, when it is likely to have adverse effects on cardiac function. In the final chapter of this comprehensive text some thoughts regarding the future directions for research and treatment of cardiac remodeling are presented.

NEUROHORMONAL BLOCKADE

Medical therapies that inhibit or reverse cardiac remodeling have been remarkably effective in improving the clinical course of patients with cardiac dysfunction. Neurohormonal antagonists that target the renin–angiotensin–aldosterone and sympathetic nervous systems accomplish both of these goals. As a consequence, angiotensin converting enzyme (ACE) inhibitors, angiotensin receptor blockers (ARBs), aldosterone antagonists, and beta blockers are now firmly entrenched as cornerstones of therapy for post–myocardial infarction (MI) and heart failure patients. However, in settings in which injury to the myocardium occurs, activation of neurohormonal systems is quite widespread. This has lead to the notion that inhibiting additional neurohormonal systems will further limit remodeling and improve clinical outcomes. To the surprise of many of us who were involved in the pivotal clinical trials, attempts at further neurohormonal blockade have met with limited (at best) success. Thus, clinical trials have failed to provide convincing evidence that drugs which combine inhibition of ACE and neutral endopeptidase (NEP), block endothelin receptors or inhibit tumor necrosis factor-alpha (TNF-α) favorably affect the remodeling process or improve the clinical course more effectively than current therapies. Although the reasons for this lack of success are uncertain, one explanation is that the efficacy of medical regimens that include ACE inhibitors, ARBs, aldosterone inhibitors, and beta blockers in various combinations imposes a rather narrow window of opportunity for additional improvement. This would be particularly important when newly developed neurohormonal blocking agents target systems, which activate common signaling pathways or lead to similar downstream effects as the renin–angiotensin–aldosterone and sympathetic nervous systems. In addition, the beneficial effects of some new investigational agents may not be of sufficient magnitude to counterbalance their adverse effects.

Evidence of cross-talk between neurohormonal systems and redundancy in their signaling pathways, however, suggests that novel approaches aimed at interrupting critical points in common downstream pathways might be an effective means of blocking the deleterious effects of multiple systems with a single drug. If such a common point in the signaling pathways of several systems could be identified (e.g., a critical phosphorylase or nuclear transcription factor), its inhibition could provide a very effective means of treating cardiac remodeling and improving the clinical course of patients with heart failure. The use of a single agent in place of several others would likely be more palatable to heart failure patients since many of them are elderly and the prescription of multiple drugs poses numerous difficulties. The use of a single agent would also improve compliance and reduce the incidence of side effects, both of which would lead to improved therapeutic efficacy.

An alternative approach to counteract the influence of neurohormonal stimulation on cardiac remodeling would be to enhance the effects of effector agents or signaling pathways that are "natural" inhibitors of cardiac cell activation and growth. Within the renin–angiotensin system there are several possible targets that might accomplish this. The angiotensin (Ang) II Type 2 (AT_2) receptor has been reported to activate antigrowth pathways (at least during certain phases of the remodeling process) and, development of selective agonists of this receptor could prove to have beneficial effects on remodeling. The AT_2 receptor also appears to be capable of limiting downstream signaling by the AT_1 receptor by promoting heterodimerization between the 2 receptor subtypes. Thus, factors that increase AT_2 receptor density could favorably affect cardiac remodeling in a manner similar to that of a dominant negative inhibitor of the pro-growth AT_1 receptor. Finally, a homolog of ACE known as ACE2 is present within the heart. This enzyme is resistant to the effects of ACE inhibitors and has substrate specificity for angiotensin peptides that differs markedly from ACE. In particular, ACE2 has a preference for Ang II. Increases in ACE2 activity would then be expected to favorably affect cardiac remodeling by reducing the influence of Ang II, a potent growth promoting peptide. An added benefit of increased ACE2 activity might also be related to the generation of Ang-(1-7), a peptide that has both vasodilatory and antigrowth effects on vascular cells.

PREVENTING LOSS AND REPLACING CARDIAC MYOCYTES

New medical therapies designed at targets that extend beyond the neurohormonal pathways noted above represent another approach for treating cardiac remodeling. Patients who demonstrate progressive left ventricular (LV) remodeling have evidence of ongoing myocyte loss due to apoptosis. As the pool of myocytes is diminished the remaining cells are exposed to increases in wall stress as well as neurohormonal stimulation, both of which stimulate further structural changes. Thus, strategies aimed at reducing myocyte loss due to apoptosis are likely to have favorable effects in preventing deleterious remodeling. Many of the steps in the signaling cascade that result in apoptotic myocyte death have been delineated and there are numerous potential targets for future therapies. An important drawback of this strategy is that there is considerable redundancy in the signaling cascades that eventually cause apoptosis to occur. Thus, interruption of an individual step within a single pathway might not be sufficient to inhibit the apoptotic process. Even more important, however, is the need for any potential therapy aimed at blocking apoptosis to be confined to cardiac myocytes, since programmed cell death remains an important mechanism for preventing unbridled growth of cells throughout the body and systemic blockade of apoptosis is likely to have numerous unfavorable effects, including an increased incidence of malignancies.

Since myocyte loss represents a potent stimulus for cardiac remodeling, therapies that are aimed at replacing myocytes represent an exciting potential therapeutic option. Stem cells provide a potential means of accomplishing this and, as a consequence, stem cell therapy to repair the damaged heart has emerged as an area of intense interest in the scientific community. There are, however, numerous stem cell populations ranging from embryonic stem cells to a variety of resident cells within the myocardium. Each of these stem cells has different inherent properties, characteristics, and levels of accessibility. In addition, each of these stem cells is likely to respond differently to various stimuli. Thus, identification of the cells that are most likely to differentiate into functioning myocytes and how to best direct them to the regions of myocardium where they are most needed is an issue of utmost importance. It also needs to be determined whether stem cells can transform into functioning contractile units that are integrated into the coordinated mechanical and electrical activity of the heart and under what conditions this can be best accomplished. Finally, information about the longevity of these cells once they become localized in the heart, and how they affect the structure and function of adjacent cells is also needed.

TARGETING THE EXTRACELLULAR MATRIX

Alterations in the composition and quantity of the extracellular matrix (ECM) occur during cardiac remodeling and it is now recognized that these changes have an important impact on the global structure and function of the heart. Moreover, ECM remodeling appears to be a far more dynamic process than was previously considered. Consequently, interventions that can alter or reverse changes in the ECM are now actively being investigated. One of the most intriguing of these interventions is the use of therapies designed to block the effects of the matrix metalloproteinases (MMPs) in post-MI patients. In response to the damage that results from the initial event, a cascade of MMPs are activated within the heart. This results in the breakdown of the collagen struts that maintain the integrity of the heart by tethering cardiac myocytes to the fibrous infrastructure. It leads to slippage between myocytes, chamber enlargement, and an increase in LV wall stress. These changes then serve as a trigger for further structural remodeling of the LV. Thus, the use of MMP inhibitors as a means of preventing remodeling is now actively being investigated in a number of settings. For this therapy to move forward, however, the identity of the MMPs that are critical for initiating and maintaining ECM remodeling needs to be defined, and the optimal timing and duration for their inhibition must be determined. The impact of MMP inhibition on cells and organs throughout the body also needs to be considered.

DEVICE THERAPY

The role of device therapy in treating heart failure patients has experienced considerable growth over the past few years. Not surprisingly, the most promising device therapies have been associated with favorable effects on cardiac remodeling. Cardiac resynchronization therapy (CRT) using a biventricular pacemaker device is the most advanced of these therapies and it is being employed with increasing frequency in clinical practice. Although CRT appears to have favorable effects in reversing the remodeling process, more work is needed to define the patients who are most likely to derive benefit from this form of therapy and also how to best optimize synchrony between and within the cardiac chambers. Long-term studies of CRT that include an adequate control population and an extended period of observation after the pacemaker is shut off will be needed to determine if the promising effects on remodeling are due to acute hemodynamic unloading versus more long-term effects on cardiac structure.

An additional approach to the unfavorable remodeling that occurs in patients with heart failure is the use of a mechanical device to restrain the growth of the LV. This concept, which is based on the now abandoned latissimus dorsi muscle wrap procedure is designed to provide support to the LV to relieve wall stress as well as to prevent further increases in chamber size. One such device, the ACORN, has been reported to improve the clinical course of patients with heart failure. The clinical improvements that were seen after placement of the ACORN device were associated with favorable effects on the remodeling process. A major drawback of this approach, however, is that it requires an open thoracotomy. An intriguing possibility that is now being tested in an ongoing clinical trial is that a device (i.e., the PARACOR device), which can be inserted and positioned over the heart using percutaneous techniques, can provide similar beneficial effects on cardiac structure and function as devices that require thoracotomy for placement.

Left ventricular assist devices (LVADs) were initially developed as a means of providing a 'bridge' to cardiac transplantation for patients with refractory end-stage heart failure. The results of the REMATCH trial indicate that LVADs can also be effective as destination therapy in some patients who are not considered to be candidates for transplantation. One of the interesting observations made in patients who have undergone LVAD placement is that the heart undergoes substantial reversal of the remodeling process. In some cases, recovery of cardiac function is of the magnitude such that the device can be removed and the patient taken off the transplant list. Whether these changes come about as a result of the mechanical unloading of the heart or whether they are due to changes in the neurohormonal milieu that accompanies improvement in the hemodynamic status brought about by the device is uncertain. Although the number of patients who achieve substantial recovery in cardiac function that is sustained over time appears to be small, further research directed toward defining the characteristics (both clinical and genetic) of these patients and other treatment factors, which might encourage recovery after LVAD, is warranted.

CARDIAC SURGERY

One of the most promising approaches in treating cardiac remodeling is through surgical reconstruction of the LV. Reconstructive surgery that recognizes the complex changes in shape that the heart undergoes throughout each phase of the cardiac cycle is now being performed. This approach helps to maintain favorable patterns of cardiac function while reducing overall LV wall stress and the severity of secondary mitral regurgitation. In addition, various approaches in repairing an incompetent mitral valve (either as an isolated procedure or combined with coronary bypass and LV reconstructive surgery) can favorably affect the remodeling process. As with other evolving approaches, better definition of the patients who are most likely to derive benefit from these surgical approaches will emerge, as the procedures are applied to increasing numbers of patients with LV dysfunction.

CONCLUDING THOUGHTS

Throughout this text the connection between remodeling and the development of heart failure has been emphasized from a variety of different perspectives. So intertwined are the process of remodeling and the natural history of heart failure that it is worth considering whether it might not be preferable to consider the former, the disease, while the latter represents its clinical manifestation. This approach is supported by the fact that over the past three decades virtually all important advances in the treatment of heart failure have been associated with the inhibition or reversal of the remodeling process.

Therapies known to improve the clinical course of heart failure patients have, by and large, been associated with favorable effects on the remodeling process. Thus, medical therapy with ACE inhibitors and beta blockers, revascularization strategies designed to restore blood flow to an acutely occluded coronary artery, and device therapy that improves the synchrony of cardiac contraction, have all been shown to benefit the clinical course of heart failure patients and to limit or even reverse cardiac remodeling. In contrast, therapeutic approaches, which have failed to demonstrate efficacy in treating remodeling, have been shown to offer little in the way of clinical benefit to heart failure patients. This raises the important issue of whether the effects of an intervention (either medical, surgical, or device based) on remodeling could be used as a reliable surrogate for efficacy in treating heart failure. At present, the information that is available would suggest that this is the case. Since reliable measures of remodeling can be easily obtained by noninvasive means, acceptance of remodeling as an end-point in clinical trials would greatly facilitate and speed the development of new therapeutic approaches for the treatment of heart failure. The impact of this approach on clinical trial design, which now heavily relies on hospitalization and mortality end-points, would be substantial and is likely to stimulate greater interest in a broader range of potentially effective therapies, since the time and expense factors would be substantially reduced. Although the use of remodeling end-points would be acceptable in establishing the efficacy of treatment, issues regarding safety and tolerability would still likely require exposure of a substantial number of patients to treatment, to ensure that untoward side effects are within an acceptable range. Perhaps, a hybrid approach in which provisional approval based on demonstrated efficacy in treating cardiac remodeling and limited safety data from randomized clinical trials could be combined with more extensive and rigorous analysis of the postapproval experience with the treatment. The latter could be derived from an ongoing registry database, the scope and purpose of which could be defined by regulatory agencies.

Index

7 transmembrane G-protein, 193

AAV. *See* Adenovirus and adenoassociated virus.
ACE, 110
ACE2, 554
  increasing levels of Ang-(1–7) by, 200
  regulation of Ang II levels by, 200
  regulation of blood pressure, 200
  role of, 199–200
ACE inhibition, 406
  augmentation of endogenous bradykinin by, 330
  degree of tissue, 329
  effects of, 197, 202
  reduction of increased collagen deposition by, 328
  reduction of myocardial hypertrophy, 328
ACE inhibitor enalapril, 203
  beneficial impact of, 333, 339
  contribution of endogenous bradykinin to, 338
  effects of, 329
ACORN device, 405, 556
Actin filaments, 76
Adaptive remodeling vs. Maladaptive remodeling, 5
Additional neurohormonal blockade clinical benefit from, 375
Adenovirus and adenoassociated virus, 505
Adjunctive therapeutic interventions, 453
Adrenergic nervous system, 349
  activation of, 349
Adrenergic receptor, 215–216
  in the pathophysiology of myocardial remodeling, 370
  role in heart failure, 218
  signal transduction, 224

[Adrenergic receptor]
  subtypes of, 224
Agonist, role of, 225
Aldosterone
  Aldosterone antagonist, 341
    in interstitial collagen, 332–333
    influencing ventricular remodeling, 342
  adverse effects of, 342
  CV effects of, 196
  in fibroblast responses, 332
  in the pathophysiology of ventricular remodeling, 327, 331
  in ventricular remodeling, 332
  role of, 196
  secretion of, 331–332
AMD-3100, 535
α_1-adrenergic blockade
  in preventing cardiac remodeling, 374
α_{1B}-adrenergic receptor
  overexpression of, 374
α_2- and βAR polymorphisms
  consequences in human heart failure, 228
α_{2A}AR Lys253, function of, 227
α_{2B}Del303–305, function of, 227
α_{2C}/β_1AR interactions, 230
α_{2C}AR polymorphisms, 230
α_{2C}Del324–327, effect of, 227
Ang II. *See* Angiotensin II.
Ang II receptors. *See also* angiotensin II, 325
  AT_1, 325, 327
  AT_2, 325–326
  AT_4, 325
Ang-(1–7), 199–200
  effects of, 201
  immunoreactivity, 201
Angiotensin (Ang) II Type 2 receptor, *See* AT_2 receptor
Angiotensin converting enzyme, 110
Angiotensin converting enzyme 2, *See* ACE2

Angiotensin converting enzyme inhibitors,
272
Angiotensin II, 419
as a potent vasoconstrictor, 327
effects of, 192, 327
functions, 191–192
hormonal mediator of the RAS,
325
mechanisms of, 192
production of, 192
role of, 216
stimulation of myocyte, fibroblast growth
responses by, 327
Angiotensin, 163
Annexin-V, 166, 170
Antiapoptotic Bcl-2, 166
Antioxidants, in human cardiomyopathy,
268, 270
ANZ carvedilol
ECHO substudy of, 308
Aortic regurgitation, 425. *See also*
volume loading.
Aortic stenosis, 11, 16, 122, 128,
418, 422
myocardial changes include, 418
Aortic valve replacement, 109, 421
Aortic valve stenosis, 108
Apoptosis, 6, 92–93, 139–140, 142, 144,
146, 218
apoptosis rate, 165, 168, 174
characteristic of, 105
AR *See* Aortic regurgitation.
Arg391 β_1AR, function of, 228
AS *See* Aortic stenosis.
Asymptomatic diastolic dysfunction, 288
Asynergia
extension of, 485
types of, 485
AT. *See* Angiotensin.
AT$_1$ receptor
AT$_1$ receptor blockade, 330
activation of, 325, 327
limiting increased interstitial collagen
content, 330
limiting LV hypertrophyAT$_1$, 330
AT1A promoter, 163
autophagic, 166
AT$_2$ receptor, 198, 554
cardioprotective role of, 331
gene expression, 199
pathways of, 199
Athlete's heart, 124
Athletic training, 124

Atrial natriuretic factor (ANF),
191, 419
Atrio-ventricular dyssynchrony. *See* AV
dyssynchrony.
Autophagic cell death, 95, 166
AV dyssynchrony, 462
AVR *See* Aortic valve replacement.

B2 kinin receptor, 329–330
Basement membrane, 27
components of, 27
Batista procedure, 405, 500
Bcl-2, role of, 220
Beat-to-beat variation
in reproducibility, 307
β_1AR polymorphisms, 229
β_2AR polymorphisms, 229
β-adrenergic receptor blockade, 352
β-adrenergic signal transduction
abnormalities of, 353
β-Adrenergic signaling, 223
β-adrenergic system suppression
anti-remodeling effects of, 367
βAR antagonists (β-blockers), 226
β-blockade
β-blockade's beneficial effects
on remodeling and mortality, 367
β-blockade's safety
in patients, 358
in improving indices of cardiac function,
355
pleiotropic effects of, 366
β-blockers, 272
acute administration of, 352
antiadrenergic potential of, 350
classes of, 350
Beta blockers effects, 407
long-term administration of, 352
orthostatic hypotension with vasodilating,
352
treatment with, 367, 373
BiDil, 6
Biomarkers, 454
Bisoprolol's effects
on mortality, 360
Blood volume, 175
BNP. *See* B-type natriuretic peptide.
Body surface area, 111
Bone marrow mononuclear cell. *see*
BMMNC.
Bradykinin, 198
B-type natriuretic peptide, 5

Bucindolol
neutral effects of, 364
pharmacological differences of, 364

CABG. *See* Coronary artery bypass grafting.
Cadherin promoter, 168
Calcineurin–NFAT signaling, 173
Calcium handling
in hypertrophy, 16
Candesartan, 340
Captopril
beneficial impact of, 333
Captopril therapy, 327
in reducing the extent of LV dilatation,
327–328
losartan compared with, 340
Cardiac ankyrin repeat protein, 83
Cardiac catheterization, 400
Cardiac genes
associated with the fetal phenotype, 140
Cardiac hypertrophy
Cardiac hypertrophy, 172, 174, 269–270
apoptosis as a mechanism of, 218
Gq signaling in, 216
PKC signaling in, 216
Cardiac magnetic resonance. *See* CMR.
Cardiac myocyte apoptosis. *See also*
cardiomyocyte apoptosis.
prevention of, 354
Cardiac myocyte hypertrophy, 242
role of IL-6 in, 243
role of TNF in, 243
Cardiac PKC, feature of, 221
Cardiac remodeling, 160, 553–558
changes within myocardium
during, 91
definition, 91
effect of metoprolol succinate on, 357
global cardiac remodeling, 146
infarction-induced remodeling,
143–144
ischemia-induced remodeling, 145
parameters, 141
pathophysiology, 141
reversal of, 355
triggers for remodeling, 139
Cardiac resynchronization therapy, 6
See also CRT
Cardiac Titin Isoforms, 77
Cardiocytes, 242
Cardiomyocyte, 10, 139–140, 142, 160, 163,
165, 168

Cardiomyocytes, 139–140, 142
Cardiomyocyte apoptosis
mechanisms for, 220
Cardiomyopathy, 28–29, 500
idiopathic dilated, 362
Cardiotrophin-1 (CT-1), 174
CARP. *See* Cardiac ankyrin repeat protein.
Carvedilol
beneficial clinical effects of, 370
beneficial effects of, 357
early functional improvement
from, 370
relative merits of, 370
Caspase 3, activation effect, 93
Caspase-cascade, 166, 168–170
Catalase, 271, 273
Catecholamines, function of, 227
Cause–effect relationship, 33–34
Cell death, 166, 168–170
cell death, contribution of different modes
to, 95
forms of, 104
types, 268–269
apoptosis, 268
necrosis, 268
pathophysiology of, 268
triggering factors, 268
Cell necrosis, 485
Cellular hypertrophy, 142, 146
Central sympathetic outflow
inhibition of, 374
Chamber dilatation, 292–293
Chimeric mice, 163
Chymase
a serine protease, 330
in cardiac Ang II formation, 330
CMR. *See* Magnetic resonance imaging.
Colchine, 75
Collagen 27–28
composed of, 28
endomysial collagen, 28
epimysial collagen, 28
perimysial collagen, 28
found in, 28
network, 28
secretion of, 28
COMPANION trial, 472
Compensatory hypertrophy, 10–11
Computed tomography
drawbacks of, 312
Concentric hypertrophy, 10–11, 14
Conductance-micromanometer, 175
Congestive heart failure, 1, 110–111

Connexins, 125, 130, 132
Contractile proteins, 16–17
Contrast angiography
    theoretical and practical limitations of, 304
Coronary artery bypass grafting, 170
Coronary artery disease, 3
Coronary artery occlusion, 167
Coronary revascularization, 488, 491
Costamers, 76
Coulter Channelyzer
    in validating volume measurements, 47
Coulter Histogram, 47
CRT, 556
    acute hemodynamic effects of, 466
    antiarrhythmic effects of, 470
    in heart failure monitoring, 474–475
    in myocardial oxygen consumption, 466
    indication of, 474
    nonresponders to, 476
Cryotherapy, 492–493
CT-1. *See* Cardiotrophin-1.
Cyotskeleton
    comprising, 73
    filaments of, 73
        actin, 76
        intermediate, 75
        microtubules, 73
Cytochrome C, 166–167
Cytokine
    interactions with renin angiotensin system,
        251
    overexpression of, 241
    proinflammatory cytokines, 241
Cytokine-induced stem cell transfer, 99

DCM. *See* Dilated cardiomyopathy.
Death-receptor pathway, 166
Decompensation, 215
Desmin filaments
    characteristics of, 76
    protein expression of, 76
Device therapy, 556
DHF. *See* Diastolic heart failure.
Diabetes mellitus, 287
Diastolic dysfunction, 288
Diastolic heart failure, 111
Differentiated cell transplantation, 544
Dilatation
    containment of, 500
Dilated cardiomyopathy, 29, 74, 109, 286,
        289
    causes of, 29
Disease

[Disease]
    localization of, 484
Diuretic therapy, 3, 6
Diuretics, 4
DNA fragmentation, 93–94
Dor procedure, 405
Dyspnea, 4
Dystrophin, 74, 76, 123

Early reperfusion, 395
Eccentric hypertrophy, 10–11
Echocardiography, 160, 176, 306, 484–485
    advantages of, 308
    for assessment of changes in LV function,
        308
    for assessment of LV remodeling, 308
    reproducibility of, 308
    transoesophageal, 488
ECM. *See* Extracellular matrix.
EDVI. *See* End diastolic volume index
EF *See* Ejection fraction.
Ejection fraction, 108, 160, 418, 485
    assessments of, 306
    measurement of, 307
Emergent CABG, 403
Enalapril, 338
End-diastolic volume
    examination of, 333, 338, 342
Endocardial border, 304
    digitization of the, 306
    techniques enhancing definition of, 313
        imaging and contrast echocardiography,
            307
        second harmonic imaging, 307
Endothelial progenitor cells, 533
Endothelial selectin (E-selectin), 419
Endoventricular circular patch plasty, 488
End systolic volume index, 487
End-systolic volume
    examination of, 331, 331, 340
Eplerenone
    effects of, 198
    in reduction of cardiomyocyte
        hypertrophy, 331
ERK. *See* Extracellular signaling related
        kinase pathway.
Erk2 heterozygous null mice, 175
ESVI. *See* End systolic volume index.
ET-1, 393
EVCPP. *See* Endoventricular circular patch
        plasty.
Exons, 78, 81
Extracellular matrix, 104, 122, 393

[Extracellular matrix]
action of, 191–190
Extracellular matrix proteins, 419
endothelial selectin. *See* E-selectin.
matrix metalloprotein *See* MMP.
Extracellular signaling related kinase
pathway (ERK), 17
Extracellular signal-related kinase (ERK),
518

F-actin. *See* Actin filaments.
Factor associated with neutral
sphingomyelinase (FAN), 244–245
FAK. *See* Focal adhesion kinase.
Familial hypertrophic cardiomyopathy, 77
FAN. *See* Factor associated with neutral
sphingomyelinase.
Fas ligand, 168
FFA metabolism, 354
FGF. *See* Fibroblast growth factor.
Fibrillar collagen, 165
Fibroblasts, 28, 163, 165
fibroblast growth factor, 165
fibroblast, functions of, 190
Fibronectin, 104, 123, 133
Fibrosis, 106, 163
Fluid retention, 487
Focal adhesion kinase (FAK), 220
Food and Drug Administration in the United
States, 362
Force–frequency relationship, 16
Frank–Starling mechanism, 9, 141
Free radical, 259
generation of, 260
toxicity due to, 260

Gadolinium late enhancement, 484
Gαq-mediated apoptosis, 220
Gene expression
correction of abnormalities of, 354
Gene transfer to increase calcium handling,
514
GLE. *See* Gadolinium late enhancement.
Gly49 β_1AR, 227
Gold standard
for measuring LV volumes, 316
G-protein coupled receptor signaling, 230
G-protein receptor kinases, role of, 230
Glutathione peroxidase, 261, 264, 271
GPx. *See* Glutathione peroxidase.
Gq signaling
in vivo inhibition effects of, 216–217

Gq/PLC/PKC signaling
in vivo analysis of, 216
G_q-mediated signaling, 176
GRK activity
modulation of, 231
GRK-phosphorylated receptors, 231

H&E sections, 46
Heart
assistance of failing, 501
dilation of, 501
transplantation, 501
Heart failure, 28, 140–141
changes in actin isoforms, role of, 64
impacts, 225–226
NYHA Class IV, 3
ACE-I, beta-blockers in the management
of, 303
changes in regulatory protein expression,
role of, 64
congestive, 1
development of, 28
Etiology, 3–4
myosin isoforms, role of, 59
remodeling related to progression of, 285
therapies
angiotensin converting enzyme (ACE)
inhibitors, 285
β-adrenergic blocking drugs, 285
tnt isoform shifts, role of, 64
Heart-to-body weight ratios, 172
Hematopoietic stem cells, 98. *See also* HSC
migration of, 99
Hemodynamics, 175
Hemodynamic stress, 171, 176
Hepatocyte growth factor, 512
Hereditary Titin diseases, 81
Heterodimerization, 199
HGF. *See* Hepatocyte growth factor.
Hibernating myocardium
in determining LVEF response, 365
Homozygotic Bax, 167
HSC, 533
Hydralazine, 272
Hypertensive cardiovascular disease, 287
Hypertrophic cardiomyopathy, 170, 286, 289
hypertrophy, 170
Hypertrophy, 170, 215
and demand, 10
and remodeling, 9
magnitude of, 11, 13
pressure–overload, 289
Hypertrophy signaling, 17, 215–216

IGF. *See* Insulin-like growth factor.
IGF-I. *See* Insulin-like Growth Factor-1
IGF-1, role of, 219
Imaging techniques, 160
  factors influencing, 303
In situ end labeling, 166
Index event, 285
  in the form of, 286–287
    hypertension, 287
    onset of diabetes mellitus, 287–288
    phenotypic expression of a mutation, 285
    the root cause of remodeling, 285
Infarct expansion
  caused by new myocardial necrosis, 291
  concept, 291
Infarct extension
Inflammatory mediators, 242
  effects, 242
    biology of cardiac fibroblasts, 246
    cardiac remodeling, 242
    extracellular matrix, 242
Insulin-like growth factor, 165
Insulin-like Growth Factor-1, 506
Interferon, 169
Interleukin-6, 194
Intermediate filaments, 75
Interstitial fibrosis, 190, 197, 288
Interventricular dyssynchrony, 463
Intraaortic balloon pump, 405
Intraventricular dyssynchrony, 462, 464, 466
Ischemia, 16
Ischemia-induced apoptosis, 160, 166
  ischemia-induced remodeling, 166
Ischemic cardiomyopathy, 288
  ischemic wall
  surgery of, 487
ISEL. *See* In situ end labeling.
Isotonic activity, 10

JAK. *See* Janus kinase.
Janus kinase, 163
Janus kinase/signal transducer and activator
  of transcription, 163
JNK. *See* JN kinase pathway.
JN kinase pathway, 17

lacZ gene, 163
LAD. *See* Left anterior descending
Langendorff perfusion systems, 175
Laplace equation, 10–11
  in detecting wall stress, 45
Laplace law, 485

Laplace's principle, 176
Laplacian stress, 22
Late LV dilatation, 390
Left anterior descending, 484
Left ventricle, 483
Left ventricular assist devices, 441. *See also*
    LVADs
  after implantation of, 441
    other consequences, 443
Left ventricular dilatation, 391
  studies of, 338
Left ventricular ejection fraction, 424.
    See also LVEF
Left ventricular end-diastolic diameter, 111
Left ventricular hypertrophy, 29
  patterns of, 29
    POH, 29
    VOH, 29
Left ventricular reconstruction, 488
Left ventricular remodeling, 325
Left ventricular reverse remodeling. *See* LV
    reverse remodeling
Life expectancy, 496
LV angiogram, 492
LV contractile function, in vivo, 449
LV dysfunction
  causes of progressive, 198
  following myocardial infarction, 367
LV end systolic volume index (LVESVI), 422
LV end-diastolic dimension (LVEDD), 422
LV end-systolic dimension (LVESD), 425
LV hypertrophy, 418, 287
LV Mass, 417
LV performance
  deterioration of, 303
LV remodeling, 289
  chronic, 286
  depending on nature of the index
    event, 290
  events of, 287
    adaptive response, 289, 293
    heart failure, 294
    index event, 285
    progressive dilatation, 294
    neurohormonal systems contributing,
      289
LV reverse remodeling, 467
  difficulties in evaluation of, 417
  related to, 419, 421
LV. *See* Left ventricle.
LVADs, 556 *See also* Left ventricular assist
    devices.
LVAD-associated improvement, 445
  in basal contractile function, 445

[LVAD-associated improvement]
mechanisms promoting, 453
Pathologic hypertrophy, 443
LVAD impacts in
chamber geometry, 445
extracellular matrix, 444
myocytes, 443
pathological structure, 443
LVAD support
cytokine pathways in, 450
gene expression in, 451
LVEDD. *See* Left ventricular end-diastolic diameter.
LVEDP. *See* Left ventricular end-diastolic pressure.
LVEDV
in mitral stenosis, 429
LVEF. *See* Left ventricular ejection fraction.
LVESD. *See* Left ventricular end-systolic diameter.
LVH. *See* LV hypertrophy.
LVR. S*ee* Left ventricular reconstruction

Magnetic resonance imaging, 110, 483–484
to examine the bisoprolol's effects, 361
Maladaptive remodeling. *See* Adaptive remodeling.
MAPC. *See* Multipotent adult progenitor cells.
MAPK. *See* Mitogen-activated protein kinases.
Matrix metalloproteinases. *See* MMPs.
Matrix metalloproteinases, 30, 107, 165,
classification of, 30
functions of, 30
profile of, 33
synthesis of, 30
Matrix metalloproteinases, in post-MI patients. *See* MMPs.
MCE. *See* Myocardial contrast echocardiography.
MCIP. *See* Myocyte-enriched calcineurin-interacting protein.
Medical therapy, 496
MEK 1/2 pathways, 17
MEK1–ERK1/2 signaling pathway, 174
MEK5–ERK5 pathway, 172
MEKK1. *See* Mitogen-activated protein kinase kinase kinase.
Mesenchymal stem cells. *See* MSC.
Metoprolol
in dilated cardiomyopathy, 355
Metoprolol tartrate

[Metoprolol tartrate]
in patients with heart failure, 356
Metroprolol CR/XL randomized interventional trial in congestive heart failure (MERIT-HF) study, 461
MHC synthesis, 19
MI. *See* Myocardial infarction.
Microtubular networks, 75
Microtubules, 73
role of, 73–74
Microvascular obstruction, *see* MO
Mitochondria, 262
Mitochondrial pathway mechanism, 220
Mitochondrial permeability transition, 166
Mitogen activated protein kinases (MAPK), 518
Mitogen-activated protein kinase kinase kinase, 173
Mitogen-activated protein kinases, 160
Mitral insufficiency, 491–492
Mitral regurgitation, 14–17. *See also* MR
Mitral repair, 488, 500
Mitral Stenosis, 429
MLP (muscle LIM protein), 231
M-mode echocardiography
in LV mass and dimensions, 308
MMPs, 392–393, 419, 555. *See also* Matrix metalloproteinases.
MMP activation, 31
measurement of, 33
MO, 399
Moxonidine
use of, 374
MPT. *See* Mitochondrial permeability transition.
MR, 419
MR severity
indicator of, 420
MRI
assessment of LV structure and function by, 315
gold standard for the noninvasive assessment of LV dimensions, 316
limitations in the routine utilization of, 316
MRI tomography
in vivo measurements of LV mass and volumes by, 305
MSC, 533
Mst1 transgene, 169
Multicenter insync randomised clinical evaluation (MIRACLE) trial, 468
Multipotent adult progenitor cells, 533
Myocardial calcium handling
improvement in, 354

Myocardial contraction
  enhancement of, 499
Myocardial contrast echocardiography
    MCE, 397, 400
Myocardial expansion, 291
  extreme forms of, 291
    LV aneurysm, 291
    myocardial rupture, 291
Myocardial extracellular matrix, 27
Myocardial function, 3
Myocardial hypertrophy, 28, 91, 97, 170
Myocardial infarction, 11, 143, 159, 160,
    389–390, 394, 536
Myocardial necrosis. *See* Infarct expansion.
Myocardial regeneration, 537
Myocardial remodeling, 28
Myocardium, 3, 5, 11–12
Myocardium, failing,
  stem cell migration in, 98
Myocyte, 46, 104
  additional considerations, 46
  assessment of shape alterations, 46
  changes during remodeling, 91
  characteristic changes identified by, 104
  considerations in, 46
  consists of, 104
  cross-sectional area of, 46, 48–49,
  degeneration of, 104
  description of, 104
  death
    by apoptosis, 92
    by necrosis, 92
  death factors, 92
  length:width ratio (L/W), 49
  lengthening of, 53
  shape alterations
    impacts of, 46
  spatial arrangement, 46
Myocyte hypertrophy, 287
Myocyte loss of, 191–190
Myocyte mitosis, 10
Myocyte necrosis, 393
Myocyte Replication, 97
Myocyte slippage, 52
Myocyte structural remodeling, 45–54
  a study on patients with hypertrophy, 46
  compensating mechanism, 49
  effects without slippage, 53
  factors of, 48
  illustrations, 52
  impacts in heart function, 52
  in female rats, 49
  in humans, 50–51
    gender differences, 50

[Myocyte structural remodeling]
  mean LV myocyte size, 51
Myocyte volume, 47
Myocyte-enriched calcineurin-interacting
  protein, 174
Myofibril, composition of, 61
Myosin
  composition of, 60
  failing and nonfailing
  failing and nonfailing molecular
    performance of, 59
  isoform, 58
    expression of, 58
    performance of, 58

N2B titin, 79–80
NADH/NADPH oxidase, 262, 265
Natriuretic peptide levels, 5
Natural antioxidant defense
  mechanisms, 270
Necropsy assessment
  with measurements of LV mass, 306
Necrosis, 106, 166
Necrotic cells, 92
Neurohormonal blockade, 553
Neurohormonal escape, 254
Neurohormones, 140–141, 486
Neurohumoral activation, 11
Neurohumoral control, 9
Neurohumoral systems, 9
New York Heart Association, 111
Nidogen, 27
Nitric oxide synthase (NOS), 262
Nitric oxide synthase-2 (NOS2), 165
Norepinephrine, in heart failure, 469
NOS2. *See* Nitric oxide synthase-2 (NOS2).
Nuclear changes, 113
  during hypertrophy, 113
NYHA Class IV. *See also* Heart failure.
NYHA. *See* New York Heart Association.

Obscurin, 82
Ohm's Law, 175
Oncosis
  characteristics of, 106
Overt heart failure, 113
Oxidative stress
  animal model for, 261
  consequences on cardiac remodeling, 268
  definition, 269
  human studies of, 261
  sources of, 261

p53-responsive genes, 219
PARACOR device, 556
Pathologic hypertrophy, 215
Pathological cardiac remodeling, 441
Pathological remodeling, 128
Patients, 108, 110, 112
    with moderately reduced function,
        108, 112
    with preserved cardiac function, 110
    with severely reduced function,
        108, 112
PCI. *See* Primary angioplasty/stenting.
Percutaneous transluminal coronary
        angioplasty, 170
Perivascular fibrosis, 288
    with type I collagen, 288
Pharmacological therapy
    race-specific responses to, 362
Pheochromocytoma, 289
Phosphatidylserine, 170
Phosphocreatine (PCr), 420
Phospholamban, 173
Phosphorylation of Contractile
        Proteins, 65
    protein kinase A, role of, 65
    protein kinase C, role of, 66
PHV, 420
Physiologic hypertrophy,
        124, 215
PKC
    role of, 221–222
    translocation and activation of, 221
PLB. *See* Phospholamban.
Postinfarction remodeling, 30
    phases of, 30
Postoperative therapy, 496
Pressure overload hypertrophy, 17
Pressure overload, 11, 16–17, 130
    cardiomyocyte response to, 132
    fibroblast response to, 130
Pressure–volume technique, 175
Primary angioplasty/stenting, 402
Probucol, 273
Pro-fibrotic effects of Ang II,
        193, 195
Prosthetic heart valve, *see* PHV
Protein synthesis, 17
PS. *See* Phosphatidylserine.
P-selection, 166
PTCA. *See* Percutaneous transluminal
        coronary angioplasty.
Pulmonary edema, 16
Pulse Doppler echocardiography, 110
Pulse pressure, 14

RAAS. *See* Renin-angiotensin-aldosterone
        system.
    alternative pathway, 199
    cardiac, 191
    circulatory, 190–191
    role of cardiac, 191, 289
Radionuclide ventriculography, 292
    on LV function, 308
Ramipril
    reduction of LV mass and fibrosis by, 329
RAS inhibition
    role of, 366
RAS. *See* Renin–angiotensin system.
Reactive fibrosis, 108
Receptor signaling, modulation in
        remodeling heart, 230
Reconstructive surgery post MI
    indications of, 497
Reconstructive surgery, 557
RECOVER trial, 546
Reduction of apoptosis, 513
Regeneration of myocardium, 541
    adult differentiated cells in, 542
    bone marrow stem cells in, 538
Regurgitant volume (RgV), 427
Relative wall thickness, 111
Remodeling, 9–10, 57, 103
    and prognosis, 5
    definition of, 103
    endothelial cells, in, 123
    gravity, effects on, 127
    left ventricular, 389
    leukocytes, in, 123
    pathologic irreversible, 390
    physiologic, 390
    postinfarction, 390
Reverse and slow, 5
    structural changes in, 2,
Renal sodium retention, 3–4
Renin-angiotensin axis, 163
Renin-angiotensin system (RAS), 163, 219
Renin–angiotensin–aldosterone system, 4
    role of, 189, 325
Renin-angiotensin axis, 487
    activation of, 487
Reparative fibrosis, 288
Reperfusion
    early, 395
    late, 398
Replacement fibrosis, 104
Reverse and slow remodeling, 5
Reverse remodeling
    factors affecting, 418
    mechanism of, 418

[Reverse remodeling]
  time-dependence of, 453
Roentgenographic evidence
  of cardiac enlargement, 294
ROS. *See* Reactive oxygen species.
ROS generation, 273
RVG assessment
  of serial changes in LV volumetrics, 311
RVG technique
  advantages of, 309
  disadvantages of, 309
  reproducibility of, 309
RWT. *See* Relative wall thickness.

Sample size determination
  for clinical trials, 305
Sarcomeres, 10
Sarcomere lengths, 75
Sarcoplasmic reticulum, 16, 104
SAVE trial
  ECHO substudy of, 308
Sc-35, 114–115
Secondary growth factors, 190, 194
Selective imidazoline receptor
  use of, 374
Sequential ligation, 145
Signal transducer and activator of
      transcription
  staining, 163
Signaling pathways, 17
Skeletal myoblast. *See* SKMB
SKMB, 543
SLs. *See* Sarcomere lengths.
SOD. *See* Superoxide dismutase.
SOLVD trial
  ECHO substudy of, 308
Spin-echo techniques
  anatomical assessment of the heart by, 312
Spontaneously hypertensive rat model, 79
SR. *See* Sarcoplasmic reticulum.
STAT. *See* Signal transducer and activator of
      transcription.
Statins, 272
Stem cell
  biology of, 532
  embryonic origin of, 546
  mobilization of, 545
  transplantation of, 544
Stem cell therapy, 555
Stroke volume, 9–10, 14
Stroke volume index, 429
Stromelysins, 31
Structure–function relationships, 115

[Structure–function relationships]
  in human myocardium, 115
  problem of reversibility, 116
Subendocardial ischemia, 111
Superoxide dismutase, 271
Surrogate end points
  as an index of clinical efficacy, 349
Sustained myocardial inflammation,
      247, 249
SVI. *See* Stroke volume index.
Sympatholysis, 362

Tamoxifen, 160, 173
Technetium sestamibi scintigraphy, 365
Terminal deoxynucleotidyl transferase-
      mediated dUTP nick end labeling, 166
TGFβ-1, 393
Thick filament composition of, 60
Thin filament, 61, 62
  activation of, 62
  composition of, 61
Thrombolytic therapy, 395
Tibialis muscular dystrophy, 81
TIMPs, 30, 33. *See also* Tissue inhibitors of
      matrix metalloproteinases.
Tissue inhibitors of matrix
      metalloproteinases, 30, 165
  profile of, 33–35
Tissue-based RAS, 190
Titin, 77
  adaptations during disease, 79
  cardiac titin isoforms, 77
  genomic structure of, 81
  passive force of, 77
    effects of, 76
  role in heart failure patients, 79
Titin-based protein complexes, 82
Titin-capping protein T-cap, 82
Titin–exon microarray, 80
TNF. *See* Tumor necrosis factor.
TNFα-1, 393
  TNF, negative inotropic effects of, 245
  role of, 420
  TNF-α receptors 1 and 2, 420
TNFR1–induced cardiac apoptosis, 243, 244
  intrinsic, 244
  neutral sphingomyelinase, 244
  pathways
Tomographic techniques
  in the diagnosis of cardiovascular disease,
      312
  MRI, 312
  ultrafast computed tomography, 312

Topographic changes, 291
  as a consequence of acute MI, 291
Transgenic mice, 168
Transmural infarcts
  leading to complex alterations, 290–291
Transmyocardial direct current shock, 145
Transthoracic echocardiography, *see* TTE
Treppe phenomenon, 354
TTE, 399
  three dimensional, 400
Tumor necrosis factor, 168, 420
Tumor necrosis factor-alpha, 193
TUNEL assay, 218
TUNEL technique, 94
TUNEL. *See* Terminal deoxynucleotidyl
  transferase-mediated dUTP nick end
  labeling.
Type 1 (AT_1) receptor, 190–191

Ubiquinone (co-enzyme Q_{10}), 274
Ubiquitin, 95
  protein cascade mechanism of, 95
Ubiquitination, 95, 97
Ubiquitin-related autophagic cell death, 104
  characteristics of, 104–106
Uninfarcted zone
  of the heart, 290, 292

Val-HEFT. *See* Valsartan-Heart Failure
  Trial.
Valsartan-Heart Failure Trial, 5
Vascular endothelial growth factor, 165
Vasodilator-Heart Failure Trial (V-HeFT), 6
VEGF. *See* Vascular endothelial growth
  factor.
Ventricular contractility, 13
Ventricular dimensions

[Ventricular dimensions]
  assessment of, 303
Ventricular dysfunction, 1
Ventricular function improvement
  time course of, 364
  wall motion scoring in assessing, 307
Ventricular remodeling, 505
  as therapeutic target in heart failure, 303
  definition of,
  role of tissue ACE in, 329
Ventricular size
  measurements of, 306
Ventricular volume
  measurements of, 304–305
V-HeFT. *See* Vasodilator-Heart Failure
  Trial (V-HeFT).
Vitamins, role in HF, 274
VOH. *See* Volume overload hypertrophy.
Volume loading, 122, 127
  cardiomyocytes, role in, 122, 127
  ECM remodeling in, 133
  fibroblast, role in, 122
Volume overload, 11, 14
Volume overload hypertrophy, 19, 29

Wall stress, 391
  effects of, 45
  in hypertrophy, 11
  increased, 390

Xanthine oxidase, 263
  inhibitors, 263
  role in heart failure, 263

Z-disks, 75

T - #0232 - 111024 - C0 - 254/178/29 - PB - 9780367454029 - Gloss Lamination